Newton, Robert R.

Ancient planetary observations and the validity of ephemeris time

# ANCIENT PLANETARY OBSERVATIONS

# ANCIENT PLANETARY OBSERVATIONS AND THE VALIDITY OF EPHEMERIS TIME

*Robert R. Newton*

THE JOHNS HOPKINS UNIVERSITY PRESS
Baltimore and London

Manufactured in the United States of America

The Johns Hopkins University Press, Baltimore, Maryland 21218
The Johns Hopkins University Press Ltd., London

Library of Congress Catalog Card Number 75–44392
ISBN 0-8018-1842-7

Library of Congress Cataloging in Publication data
will be found on the last printed page of this book.

# PREFACE

For a number of years, the independent variable used in the ephemerides of the sun and planets in the major national ephemeris publications has been called ephemeris time. In concept, ephemeris time is the quantity which, when used as the independent variable in the gravitational equations of motion of the earth and planets, causes the solutions of the equations to agree with observation. Put another way, it is assumed that gravitation is the only force field that needs to be considered in the motion of the earth (or sun) and planets, to the accuracy of observation. This means that the sun and planets have no secular accelerations with respect to ephemeris time.

Since the use of ephemeris time rests upon an assumption about the physical forces involved, the validity of the underlying concept must ultimately rest upon observation; it cannot be postulated. There have been several tests of the concept with respect to high frequency effects, by which I mean effects whose characteristic periods are a few years. It is generally considered that the concept has been justified by these tests, but a careful examination shows that the tests are far from conclusive. Further, there has been little concern about whether the concept is valid with regard to long term effects whose characteristic periods are measured in centuries or millennia.

The main concern of this work is to test the validity of the concept of ephemeris time with respect to long term effects. The test is based upon observations of the sun and planets, starting with the oldest observations available and continuing into medieval times. For brevity, I did not mention the sun in the title of this work, nor did I refer to the medieval period.

In order to keep the work to a reasonable length, it is necessary to stop about the year 1000. It happens that there is a gap of about 70 years between an observation made on 1019 September 18 and the next observation that I have found, and the date 1019 September 18 therefore makes a convenient terminus for the data used in the work. The "ancient" in the title can thus be considered to extend from Babylonian times to this date.

I imagine that only the words "Ancient Planetary Observations" will be used in citing the title of this work, for reasons of brevity. However, I have added the words "and the Validity of Ephemeris Time" to the title page in order that the reader may form an idea of the main purpose of this work from the title page.

References that were specifically used in the course of this work are listed in a separate section between the close of the appendices and the index. A work is cited in the text by giving the name of the author (or the names when there are several authors) followed by the date of publication. The name is underlined and the date is enclosed in square brackets. When it is necessary to give additional information such as a page or chapter number, this information follows the date within the brackets.

I thank my colleagues B.B. Holland, R.E. Jenkins, and A.D. Goldfinger for many useful discussions on general problems involved in this work. I also thank my colleagues G.B. Bush, A. Eisner, E.P. Gray, W.H. Guier, B.J. Hook, S.M. Krimigis, L.L. Pryor, I.H. Schroader, and M. Sturmanis, and my son F.P. Newton, for helping on various specific problems such as programming, or locating relevant literature, that are too numerous to mention.

I gratefully acknowledge the support of this work by the Department of the Navy through its contract with The Johns Hopkins University, and I thank the Director's Office of the Applied Physics Laboratory for encouraging me to carry out this study.

CONTENTS

CONTENTS

CONTENTS

CONTENTS

FIGURES

FIGURES

TABLES

TABLES

TABLES

TABLES

TABLES

# CHAPTER I

# BACKGROUND AND PURPOSES OF THIS WORK

## 1. The Measurement of Time

The concepts of time and of its measurement involve philosophical and psychological problems that are beyond the scope of this work. However, there are some considerations that are always present and usually tacit in astronomers' thinking about time, and a short reference to some of them will help the reader in understanding the purposes of this work.

We usually measure the flow of time by following the position of something that is moving. The motion may proceed always in one direction, or it may be repetitive. The flow of water in a water clock and the motion of the yet unburned surface of a standard candle are examples of motions in one direction that have been used to keep time. The swinging of a pendulum and the vibrating of a quartz crystal are examples of repetitive motions that have been used to keep time. It is arbitrary whether we consider some motions as one-directional or repetitive. For example, we say that it is noon when the sun is in the plane of the local meridian. In one sense, this is repetitive; in another sense, the observed phenomenon is produced by the continued rotation of the earth in the same direction.

A particular motion that is used in keeping time is said to furnish a time base or a measure of time.

We are always concerned with the question of whether a given base is uniform or not. It does not seem to be possible to say that a single time base is uniform. Apparently we need to have at least two time bases that we can call uniform before we can say that either one of them is. This need is inherent in the way that we decide upon uniformity of time flow.

Suppose that we have two independent time bases and that we compare the values of time that they furnish. If the times agree whenever we compare them, within the accuracy of observation, we tend to say that each one furnishes a uniform time base. We can then base our practical measure of time upon one of them that is chosen for convenience, or we may choose to use the average of the times furnished.

Suppose, however, that the two measures of time disagree. We are unable to say which, if either, is to be called uniform. In order to resolve the question, we turn to a third measure. If we are lucky, it agrees with one of the first two. If it does not, we keep adding new time bases until we find at least two that agree. The two that agree are now considered to furnish a uniform time base, while the other time measures are considered to be non-uniform.

## 2. Solar Time and Universal Time

Solar time is the time base furnished by the angular position of the rotating earth† with respect to the sun. Throughout history, solar time has probably been the principal measure of time used in ordinary life. Until this century, it was also the principal time base used in technical astronomy. Various types of clock have been used as practical time keepers, but people have tacitly regarded them as devices for keeping up with solar time during intervals when the sun could not be observed, or perhaps simply as devices that were easier to read than the position of the sun.

In quantitative work, it is necessary to distinguish between apparent solar time and mean solar time. Apparent solar time is the less regular (farther from uniform) of the two, and its greatest difference from mean solar time is about 16 minutes. We do not need to consider the reasons for the

---

†That is, solar time is furnished by the diurnal spin of the earth on its axis, as opposed to the yearly revolution of the earth in its orbit around the sun.

difference between the two time bases; the interested reader can find a discussion on pages 79-80 of the Explanatory Supplement [1961].

I shall assume that the reader is acquainted with the idea of local time, with standard time zones, and with the use of Greenwich mean solar time as the reference to which most parts of the world refer their local standard time. In the rest of this work, the term solar time, when it is used without modifiers, will usually indicate Greenwich mean solar time. It may be used with a different meaning occasionally when there seems to be little danger of confusion.

Sidereal time is the time base furnished by the angular position of the rotating earth with respect to the vernal equinox. Just as we have Greenwich mean solar time, we also have Greenwich mean sidereal time.

Universal time is an approximation to (Greenwich mean) solar time. It is calculated from the measurements of sidereal time, with the use of Newcomb's theory [Newcomb, 1895] of the sun. It is what solar time would be if the measurements of sidereal time were exact and if Newcomb's theory were exact, and many astronomers consider universal time and solar time to be synonymous.

However, universal time and solar time are conceptually different and it is desirable to distinguish them. Universal time is based upon observations of the equinox while solar time, as defined here, is based upon observations of the sun. The two measures agree exactly only if Newcomb's theory of the sun is exact. Since the data that will be used in this work use a time base derived from observations of the sun, I shall use the term solar time in preference to the term universal time, and I shall try to avoid references to universal time in the rest of this work.

## 3. The Accelerations of the Moon, the Sun, and the Planets

Acceleration denotes a change in velocity.

Mathematically, acceleration is the first derivative of velocity with respect to time. Since velocity in turn is the first derivative of position with respect to time, acceleration is the second time derivative of position. Acceleration can be either positive (increasing velocity) or negative (decreasing velocity).†

Each object in the solar system is continually changing its velocity (that is, it is continually being accelerated) because of the gravitational forces exerted on it by the other objects in the solar system. Accelerations that result from gravitational forces are not a subject of concern in this work. Here, when I refer to the acceleration of the moon, for example, I mean that part of its acceleration which results from non-gravitational effects.‡

Edmond Halley (1656-1742) was probably the first person to compare ancient astronomical observations with observations made in his own time and to conclude in consequence that there are non-gravitational accelerations in the solar system. On 1692 October 19 [MacPike, 1932, p. 229], Halley read a paper to the Royal Society in which he compared records of eclipses found in the work of Ptolemy [ca. 142] and al-Battani [ca. 920, pp. 56-57] with calculations that he had made from contemporaneous tables. He found systematic discrepancies between the recorded times and the calculated ones, and he concluded that the resistance of the "aether" to planetary motion does "in time become sensible." Starting from this, he stated an intention to prove that the "eternity of the World" is impossible.

According to the record, the Society was much

---

†I am not attempting to maintain the distinction between vector acceleration and scalar acceleration. It is necessary to maintain the distinction in the quantitative analysis of the data, but it is not necessary in this qualitative discussion.

‡It is to be understood that the term "gravitational" includes any corrections that are required by general relativity.

struck by this idea, and it instructed Halley "to publish a discourse" about it. Halley never did so, so far as I can discover. Several years later, in a different context, he wrote [Halley, 1695] that he could "pronounce in what Proportion the Moon's Motion does Accelerate" if he could establish accurate values for the longitudes of Alexandria and other cities where the old observations had been made. Many years passed before geodesists and astronomers knew the separation between Greenwich and, say, Alexandria with enough accuracy to permit confident comparison of modern calculations with ancient records.

We now know the longitudes of the places involved with high accuracy, and we have highly precise theories of the motions of the sun, moon, and planets. Thus we know what Halley could only suspect, namely that there is an acceleration of the moon that is not of gravitational origin.

It is doubtful that any astronomer nowadays believes in a "resistance of the aether" in the sense that Halley meant. Most astronomers probably believe that the acceleration of the moon arises mainly from friction in the lunar tide, although this belief presently rests on a somewhat shaky foundation. We can estimate the acceleration of the moon from astronomical data, and we can calculate the amount of tidal friction needed to produce the estimated acceleration. Various geophysicists [see Jeffreys, 1970, Chapter VIII, or Miller, 1966] have estimated the amount of tidal friction directly on the basis of detailed tidal studies. The largest direct estimate of tidal friction is only about half of what is needed by even the smallest estimate of the lunar acceleration, and it is only about a quarter of what is needed by most estimates.†

On the other hand, every other source of a lunar acceleration that has been thought of is too

†In these statements, I am using ephemeris time rather than solar time as the time base for calculating accelerations. See Sections I.4 and I.7 for further discussion of the importance of this point.

small by orders of magnitude. Thus belief in tidal friction as the main source of the lunar acceleration rests more upon the absence of an alternative source than upon our present ability to explain the acceleration by means of tidal friction.

Around 1900, it became clear that the sun and the planets, as well as the moon, have accelerated motions. It also became clear that the accelerations of all these bodies contain components of high frequency, with periods of the order of a few years, as well as secular components.

## 4. The Spin Fluctuation Hypothesis

It is natural to ask whether the apparent fluctuations or accelerations in the motions of the moon, sun, and planets can all be explained as the consequences of a single phenomenon. It may have been E. W. Brown†, in an address delivered in 1914 to the British Association for the Advancement of Science, who first suggested that the fluctuations may be highly correlated. Brown found that the fluctuations of Mercury and the moon are correlated and in the same direction. He found, on the other hand, that the fluctuations of the sun, though correlated with those of Mercury and the moon, are in the opposite direction. Fotheringham quotes him [1927, p. 143] as suggesting, apparently with diffidence, that there is "some kind of a surge spreading through the solar system and affecting planets and satellites the same way but to different degrees." Brown speculated that the "surge" might be magnetic in origin.

Later writers developed the hypothesis that the fluctuations or accelerations in the motions of the moon, sun, and planets (other than the secular acceleration of the moon) arise from non-uniformity in the solar time base and hence from accelerations in the rate of rotation of the earth. We may call this hypothesis the 'spin fluctuation hypothesis' [Newton, 1973a].

---

†See Fotheringham [1927]. I have not read Brown's address itself, but only Fotheringham's summary of it.

The spin fluctuation hypothesis may be tested by a method that is simple in concept but that has proved difficult to carry out because of the small size of the quantities involved. If we are to suppose that the solar time base is not uniform, we must suppose the existence of a better base; I shall turn to the question of an alternative base in Section I.7. Let $\delta t$ denote the departure of solar time from uniformity at some particular instant, and let $\omega_g$ denote the geocentric angular velocity of any object in the solar system at this instant. Then there will be an apparent fluctuation $\omega_g \delta t$ in the position of the object. That is, its position will depart from that given by a gravitational ephemeris (still excepting the secular behavior of the moon) by the amount $\omega_g \delta t$.

In order to test the spin fluctuation hypothesis, then, we determine the amount $\delta\lambda$ by which the geocentric position of any object deviates from a smooth ephemeris. We divide $\delta\lambda$ by the geocentric angular velocity $\omega_g$ to obtain an estimate of $\delta t$. If we obtain the same value of $\delta t$ from every object for the same value of solar time, the hypothesis is upheld and the resulting $\delta t$ is the deviation of solar time from uniformity.

There are two different aspects to the spin fluctuation hypothesis, the short term and the long term. The hypothesis may be valid for the short term variations, those with a time scale of a few years, but invalid for long term or secular variations. Conversely, it may be valid for the long term but not for the short term. Finally, it may be valid for both or valid for neither.

We should note that the existence of short term accelerations in the earth's spin has been well established. Since about 1955, it has been possible to monitor the rotation of the earth by means of several cesium clocks [Markowitz, 1970]. It is found that the cesium clocks agree well with each other but disagree with the rotation of the earth. Hence we conclude that the cesium clocks are more nearly uniform time keepers than the earth's spin is. Markowitz finds that the acceleration of the earth's spin undergoes large changes in intervals that are too short to resolve, but that are probably a few months. Once a value of the acceleration is established, it tends to hold for some time before another sudden change occurs. In Markowitz's data, the average interval between changes is about 4 years. The root-mean-square value of a sudden change is about 350 parts in $10^9$ per century.

However, the existence of these fluctuations does not necessarily mean that the spin fluctuation hypothesis is valid even in the short term, much less in the long term. We must still analyze the astronomical data in order to decide whether fluctuations in the positions of celestial objects can all be explained by the measured fluctuations in the rotation of the earth, or whether there is yet another effect operating in the solar system.

Many papers have been written in this century for the purpose of testing the spin fluctuation hypothesis. We may single out those by Fotheringham [1927], de Sitter [1927], and Spencer Jones [1939] for particular attention here.

Fotheringham studied fluctuations of the moon, the sun, Venus, and even the vernal equinox. He found, like Brown, that the fluctuations are highly correlated but he found, unlike Brown, that the fluctuations of the sun and moon are in the same direction. He then tested the spin fluctuation hypothesis, both with regard to the short term and the long term; this is the only earlier attempt to study the hypothesis in the long term that I have seen. It is only with the advantage of hindsight that we may smile at his conclusion (page 163): "We may therefore declare with confidence that while the Earth's rotation may be affected by a secular retardation, it is certainly not affected by a fluctuation." (By "fluctuation", Fotheringham meant only a short term variation.) Fotheringham agreed with Brown in attributing the fluctuations of the sun and planets to a surge of unspecified nature spreading through the solar system.

de Sitter denoted the lunar fluctuations by $B(\tau)$, say, and he wrote the fluctuations of other bodies as $B(\tau)$ multiplied by the appropriate ratio of angular velocities and by a quantity Q. He took Q to be a constant for a given body but he did not require it to be the same for all bodies. He considered that values of $B(\tau)$ for different times are uncorrelated if the values of $\tau$ differ by more than a few years. He found that Q has nearly the same value, about 1.25, for the sun, Mercury, and Venus. This finding is not compatible with the spin fluctuation hypothesis, which requires $Q = 1$.

Spencer Jones also worked in terms of the quantity Q that de Sitter had introduced. He used many of de Sitter's data and introduced additional data. Altogether, he used longitudes of the moon and of Venus, transits of Mercury, declinations of the sun, and right ascensions of the sun. He found

$$Q = 1.062 \pm 0.033. \qquad \text{(I.1)}$$

This is still in disagreement with the spin fluctuation hypothesis, but it is closer to agreement than any earlier result.

Thus all studies of the spin fluctuation hypothesis that have been published to date have resulted in disagreement between the data and the hypothesis, and almost every study has disagreed with almost every other study. We may not conclude from this, however, that the hypothesis is invalid. All the studies are subject to serious objections, and we can only say at the moment that the hypothesis has never been adequately tested, even with regard to the short term.

Different studies are individually subject to various objections, but it should be sufficient to discuss the paper of Spencer Jones, which is probably the most influential one of the series. It is necessary to outline Spencer Jones's method briefly.† Let $\tau$ be a value of solar time at which the mean longitude of the moon is observed to be $L_{MO}$. Let $L_{MG}$ denote the value of the mean longitude calculated from an ephemeris at the same epoch,

---

†Much of the following discussion has been published elsewhere [Newton, 1973a].

and let $\nu_M'$ denote the secular acceleration of the moon, taken with respect to solar time. Then the lunar fluctuation $B_M(\tau)$ is defined by

$$B_M(\tau) = L_{MO} - (L_{MG} + \tfrac{1}{2}\nu_M' \tau^2).\dagger \qquad (I.2)$$

The values assigned to the fluctuations $B_M(\tau)$ clearly depend in a sensitive way upon the value assigned to the acceleration $\nu_M'$.

Let us define analogous quantities to $B_M(\tau)$ for the sun and for a planet, and let us use the subscripts S and P in place of M to denote these quantities. It is not necessary to refer to a specific planet in this preliminary discussion. Further, let $n_M$, $n_S$, and $n_P$ denote the average angular velocities of the moon, the sun, and the planet. Spencer Jones formed the quantities

$$\begin{aligned} L_{SO} - L_{SG} &= \tfrac{1}{2}\nu_S' \tau^2 + Q(n_S/n_M)B_M(\tau) , \\ L_{PO} - L_{PG} &= \tfrac{1}{2}(n_P/n_S)\nu_S' \tau^2 + Q(n_P/n_M)B_M(\tau) . \end{aligned} \qquad (I.3)$$

He then found the values of the constants $\nu_S'$ and Q that best fit these equations in a statistical least-squares sense.‡ Two main points need to be made in discussing this treatment.

First, we note that the second of Equations I.3 is equivalent to

---

† $\tau$ is measured from 1900. See Section I.7 for a more precise statement about the origin of $\tau$.

‡ For simplicity in the discussion, I have omitted the terms in Equations I.2 and I.3 that are independent of $\tau$ and the terms that are linear in $\tau$. Spencer Jones adjusted the constant and linear terms simultaneously with the quadratic terms. It should be noted that $L_{PO}$ and $L_{PG}$ are heliocentric longitudes, although the earlier discussion was in terms of geocentric longitude. The second of Equations I.3 would take on a much more complicated form if we wrote it in terms of geocentric longitude.

$$L_{PO} - L_{PG} = (n_P/n_S)(L_{SO} - L_{SG}) . \qquad (I.4)$$

That is, Spencer Jones assumed that the spin fluctuation hypothesis is valid for the sun and planets, in both the long term and the short term senses, in spite of the fact that the purpose of his paper was to study the validity of the hypothesis.

The most serious objection, however, is to the treatment of the secular acceleration of the moon, which enters directly into the definition of the lunar fluctuations $B_M(\tau)$. Spencer Jones and the earlier investigators did not fit $\nu_M'$ to the data, although they could have done so. Instead, they assumed values of $\nu_M'$ _a priori_, and we need to look at the validity of the values that they assumed.

_Martin_ [1969] has analyzed more than 2000 lunar occultations that have been observed since 1627; _Newcomb_ [1875 and 1912] had already used many of the observations. Martin did not estimate $\nu_M'$ himself, but a straightforward calculation [_Newton_, 1972a] leads from his results to the estimate

$$\nu_M' = - 12.1 \pm 0.7 \quad ''/cy^2. \qquad (I.5)$$

The unit in which $\nu_M'$ is expressed in Equation I.5 is the second of arc per century per century.

A recent study [_Newton_, 1972, Section XVIII.4] has shown that $\nu_M'$ attained a value close to -12 about the year 1300. There have been no estimates made for years between 1300 and the period of Martin's data, but the agreement suggests that $\nu_M'$ has been roughly constant at about this value since 1300. The data show, however, that $\nu_M'$ was far from constant for years before 1300. It was approximately constant, but at a value around +10 rather than -12, from about -700 to about +600 or +700. It then began an abrupt decline that lasted to at least about 1300.

Spencer Jones and the others who wrote about

the spin fluctuation hypothesis did not know about the large change in $\nu_M'$ that has occurred within the historic period. Earlier studies of $\nu_M'$ had dealt mostly with the period around the year 0, so that Spencer Jones used a value derived from those studies, specifically the value +10.44. Thus Spencer Jones made an error of about 22 $''/\text{cy}^2$ in the value of $\nu_M'$. We see from Equation I.2 that his values of the fluctuations $B_M(\tau)$ should be changed by the amount $\frac{1}{2} \times 22\tau^2 = 11\tau^2$ seconds of arc. For the year 1700, say, his value of the lunar fluctuation should be changed by 44″. This should be compared with the largest value that he found for any fluctuation, which is about 17″.

The mean longitude of the moon contains terms that we may write as $A + B\tau$; A and B are clearly the mean longitude at the epoch and the mean motion at the epoch, respectively. We evaluate A and B by fitting the lunar theory to observation. Spencer Jones evaluated A and B by using the value 10.44 for the lunar acceleration $\nu_M'$ and by using lunar data from about 1680 to 1936. Since the value of $\nu_M'$ for the period covered by the data is about -12, Spencer Jones did not evaluate A and B correctly. He should not have allowed the value of $\nu_M'$ inferred from ancient data to affect the values of A and B. Instead, he should have estimated $\nu_M'$ from the data since 1680, and he should then have evaluated A and B with the use of this estimate of $\nu_M'$.

Similarly, when we deal with the accelerations of the sun and planets, we should not alter the constant and linear terms in the ephemerides in response to an acceleration estimated from ancient data, since we have strong reasons to believe that the accelerations have not been constant since ancient times. Instead, we should estimate the constant and linear terms entirely from recent data. We should also estimate the accelerations from recent data and compare with the values estimated from ancient data.†

†In a recent paper, Muller and Stephenson [1974] say that the lunar acceleration turns out to be constant within historic times, at about the value +12, if one adjusts A and B according to the ancient data. However, they have overlooked the recent lunar data that lead to -12.1 for the present value of $\nu_M'$.

The values assigned to the lunar fluctuations form the most important body of data in the studies of Spencer Jones and others. The large errors made in the values of the lunar fluctuations mean that the studies do not tell us anything about the validity of the spin fluctuation hypothesis.

The reader may object to my saying that the value in Equation I.1 is Spencer Jones's value of Q. Spencer Jones actually concluded that

$$Q = 1.025 \pm 0.033 \,, \tag{I.6}$$

rather than the value in Equation I.1. That is, he concluded that the spin fluctuation hypothesis was verified (Q = 1) within experimental error. However, the value in Equation I.1 is in fact the value that Spencer Jones found from the data that he used. In order to reach the value in Equation I.6, he rejected all measurements of the right ascension of the sun but retained all measurements of the declination, saying [Spencer Jones, 1939, p. 544] that "declination observations, in the case of the Sun, are far more likely to be free from systematic errors than are the observations of right ascension."

The analysis of the errors in measuring right ascensions and declinations is too complex to treat in detail here. Spencer Jones may well be right on this point, although I cannot find another writer who agrees with it, and it does not sound plausible. The basic reason is that the declination of the sun itself furnishes the origin from which right ascension is measured.

The raw observation that leads to a value of right ascension is an observation of the time when an object passes a crosshair or other fiducial mark in some type of transit instrument. Spencer Jones refers to differences between people and between different methods of marking the instant of time to be measured, and it is easy to see that different people may say "Now!" at different times if they are asked to say when the sun passes the meridian;

they may judge the sun's limbs differently, for example. It is not clear that any of these differences affect the measured right ascension of the sun.

Suppose, for simplicity in discussion, that an observer is always too late by an increment $\delta t$ in indicating the passage of the sun, but that he is exact about the passage of a star. This will affect the right ascension deduced for the star. However, the origin of right ascension is the position of the sun in right ascension when its measured declination is zero. If there are no other errors, the observer is wrong by $\delta t$ in judging the origin of right ascension, and he is wrong by the same amount at any other position. Since the right ascension of the sun at any other position is the difference of the two measurements, the error $\delta t$ cancels out. The right ascension is systematically in error only if the declination, which determined the origin, is also systematically in error.

Other writers make statements similar to those that I have just made. Schmeidler [1965, p. 77], for example, says that "observations of declinations are more affected by systematic error than observations of right ascension, and are therefore more difficult to make." Hemenway [1966] finds that errors in right ascension are about two-thirds of errors in declination, for objects at moderate declinations. It should be noted that both of these statements refer to the positions of stars. If they are accurate for stars, they are all the more accurate for the sun, which benefits from the cancellation of some errors that are present with star observations.

I am now almost ready to state the purposes of this work. Before I do so, however, it is convenient to mention a few geophysical matters briefly.

## 5. Some Geophysical Factors Connected with the Spin Fluctuation Hypothesis

In Section I.3 above, I remarked that friction in the tides induced by the moon is the only known source of an acceleration of the moon. Jeffreys [1970, Chapter VIII], for example, gives a general description of the detailed mechanism by which friction in the lunar tide accelerates the moon, and I shall not take the space to repeat the description here.

Friction in the lunar tide also accelerates the earth's spin, and therefore it affects the uniformity of solar time as a time base. This in turn means that the long term motion of the sun deviates from a gravitational ephemeris if solar time is used as the time base; in other words, friction in the lunar tide produces a secular acceleration of the sun with respect to solar time.

Although we cannot calculate accurately either the acceleration of the moon or of the sun that arises from friction in the lunar tide, we can calculate accurately the ratio of the accelerations. The ratio depends upon several physical parameters of the sun-earth-moon system, and its value is about 5. If the spin fluctuation hypothesis applied to the effects of tidal friction, the ratio would equal the ratio of the average angular velocities of the moon and sun, which is about 13. Hence the spin fluctuation hypothesis does not apply to the long term accelerations of the sun and moon.

Other phenomena affect the earth's spin but are not capable of affecting the orbital motion of the moon. For example, there is the tide induced in the solid earth and oceans by the gravitation of the sun, and there is the tide induced in the atmosphere, probably by solar heating rather than by gravitation. The periods of these tides are incommensurable with the orbital period of the moon, and hence their average effect on the moon is zero. Munk and MacDonald [1960, Chapter 11] and Newton [1972a] have discussed much of what is known or speculated about phenomena that affect the earth's spin but not the moon. The spin fluctuation hypothesis clearly applies to the accelerations that they produce.

Still other phenomena are potentially able to

affect the orbital motions of the earth and the other planets around the sun. Every such phenomenon that can be estimated is, however, many orders of magnitude too small to matter here. That is, to the accuracy with which we can hope to work at present, we know of no accelerations of the sun and planets except those that result from fluctuations in the earth's spin.

In sum, so far as long term effects are concerned, we expect the spin fluctuation hypothesis to apply to the accelerations of the sun and the planets but not to the acceleration of the moon.

The second part of this conclusion is well founded: The effects of the lunar tides upon the moon and the earth depend upon parameters that have nothing to do with the spin fluctuation hypothesis. The first part of the conclusion, by contrast, rests only upon the absence of any known mechanism. It is not safe to argue from the absence of knowledge, particularly in this field. After all, we have not had much success in explaining the observed accelerations of the moon and of the earth's spin.

In other words, with regard to the secular behavior of the sun and planets, we should look upon the spin fluctuation hypothesis as something to be tested by observation, not as something that can be established by theories of known mechanisms.

## 6. The Purposes and Plan of This Work

The main purposes of this work are as follows:

a. to find the secular acceleration of the sun with respect to solar time,

b. to find the secular accelerations of the planets with respect to solar time, and

c. to determine whether the secular accelerations are consistent with the spin fluctuation hypothesis.

There is a secondary purpose that will be described in Section IV.8.

There is a limitation that will be imposed upon the data used in meeting the first purpose. In describing this limitation, I continue to use $\nu_M'$ and $\nu_S'$ for the secular accelerations of the moon and sun, respectively.

Data that involve the position of the moon with respect to the star background yield an estimate of $\nu_M'$, as we mentioned in Section I.4 with regard to the work of Martin [1969]. Data that involve the position of the moon with respect to the sun yield relations that involve both $\nu_M'$ and $\nu_S'$. With a few trivial exceptions that do not have much value, data of this sort in fact yield the quantity

$$D'' = \nu_M' - \nu_S' . \tag{I.7}$$

$D''$ is the second derivative of the lunar elongation with respect to solar time. It is the quantity [Newton, 1972, Section XVIII.4] that we get from the analysis of eclipses, and most available data that yield $D''$ are eclipse data.

Since we can estimate both $\nu_M'$ and $D''$, we can estimate $\nu_S'$:

$$\nu_S' = \nu_M' - D'' . \tag{I.8}$$

The difficulty with using Equation I.8 is that $\nu_S'$ is considerably smaller than $\nu_M'$ and $D''$, and the inevitable errors in the estimates of $\nu_M'$ and $D''$ are magnified, on a relative basis, when we use it.

In order to reduce the error in $\nu_S'$ to an acceptable level if we attempt to estimate it from lunar and eclipse data, we must use a large amount of data. Using a large amount of data means using data over a long time span,† as I have pointed out elsewhere [Newton, 1973a]. On the other hand,

†Because we do not have large bodies of old data from short time spans.

since $\nu_M'$ and $D''$ are far from being constant in time, we must use only data from a limited time span in a single analysis. This basic contradiction means that we cannot derive $\nu_S'$ reliably from eclipse data or other data that involve the moon. Data from the historical period before about 600 may form an exception, but it is conservative to ignore this possibility.

Therefore I shall use only solar and planetary data in this work, except for carrying out secondary purposes.

To study both the long term and the short term aspects of the spin fluctuation hypothesis is beyond the scope of a work of reasonable length, and hence I shall not attempt to study the short term aspect here. With regard to the long term aspect, it was my original hope to cover the historical period since the beginning of quantitative astronomy. However, it soon became apparent that covering a period of this length in a single work is also impractical. Therefore I arbitrarily terminate this work with the observations of the Muslim astronomer al-Biruni. His last solar observation was made on 1019 September 18, so far as I know.

In sum, this work will be concerned with solar and planetary data from the Babylonian period through 1019 September 18. It is my intention to analyze later data in another work.

There is still another restriction on the data that will be used. The planets beyond Saturn were not known during the historical period that will be used, so I shall of course not use any data for them. In addition, Jupiter and Saturn move slowly and it is unlikely that they have accumulated deviations from a gravitational theory, within the time span adopted, that are large enough to study. Hence I shall not use any data concerning Jupiter and Saturn, except for conjunctions of them with other planets. However, I shall reduce any data concerning Jupiter and Saturn to tabular form, in the hope that they will be useful to somebody sometime.

In previous studies of ancient astronomical

data [Newton, 1970, 1972, 1972b], I have rigorously followed the practice of studying the texts, deciding upon their meanings, reducing the data to numerical form, and assigning weights, before doing any astronomical analysis of the data. I am more than ever convinced that this is a desirable approach. If a person allows himself any freedom of choice with regard to the data after he has analyzed them, he will have a built-in bias in spite of his best intentions. The reason is that a person looks skeptically at data whose analysis does not agree with what he expects to find, but it may not even occur to him to be skeptical about data that do agree. Thus a person is automatically biased toward data that agree with his hypotheses and against data that disagree.

Unfortunately, it is not possible to make all decisions about the texts without doing some astronomical calculations. In this work, the greatest but not the only need for using calculations in the textual analysis arises in dating problems. This work will depend heavily upon observations that are dated in the Babylonian and Muslim calendars. There is a fundamental ambiguity† of a day and occasionally of two days in relating either of these calendars to the Julian or Gregorian calendars. Even with calendars for which there is no ambiguity of this sort, there may still be problems about the conventions used, for example, in the hour at which the formal day begins, and hence there may still be a problem about the date.

In most of the cases that arise, the astronomical body being observed moves about 1° in a day. This is greater than expected errors in observation, and it is much greater than the effect of the unknown parts of the solar and planetary accelerations. Thus, if we can assign a date within a day or so on the basis of the text, we can often assign the exact date by means of an astronomical calculation. In a few cases, a planet will be at a point where its daily motion is comparable with errors in observation or with uncertain effects of the accelerations. If the date is uncertain in such a case, we cannot

---

†See Sections II.2 and II.5.

determine the exact date by calculation, and I must exert care in using the corresponding records.†

When it is necessary, then, I shall make astronomical calculations as an adjunct to the textual study for the purpose of resolving dates. It is possible to do this without learning much about the agreement, or lack of agreement, between the data and any hypotheses about the accelerations. Thus it should be possible to avoid the dangers of bias or of reasoning in a circle.

After I dispose of some preliminary matters in Chapters II and III, I shall study the ancient records from the textual standpoint in Chapters IV through VIII. I shall complete these chapters before I begin any quantitative analysis of the data, except for the analysis that is needed in finding the exact day of an observation. After I begin the quantitative analysis in Chapter IX, I shall make no changes in earlier chapters, except perhaps for minor editorial changes intended to clarify the exposition.

The weight assigned to an observation is based upon the level of error that is allowed by the method of observation. In many cases, we can estimate a standard error and hence assign a weight by an *a priori* study of the observing procedures. If we have a homogeneous body of observations, it is often better to base the weight upon the scatter that the observations show after they have been analyzed; that is, we can assign weights by an *a posteriori* study of the precision of the observations.

In dealing with ancient observations, a factor other than the accuracy of observation sometimes comes into the assignment of weight. This factor,

---

†Since I wrote this, I have discovered from the analysis of the Babylonian data that the standard deviation of a Babylonian measurement of position is between $0^\circ.5$ and $1^\circ$. This means that we cannot date a Babylonian planetary observation by analysis of the observation, except in unusual circumstances.

which I have called "reliability", is an estimate of the probability that the text, as it now stands, corresponds to the facts. In some cases, there is strong evidence of a hoax or of fraud, and these records will receive a reliability of zero. More commonly, we have to consider the possibility that there has been an accidental but significant error in the transmission of the text, or in its interpretation, or in some similar matter. In most cases, however, it is not even necessary to mention the reliability, and the reliability is taken to be as unity if it is not mentioned.

## 7. Atomic and Ephemeris Times; Some Notation

In writing dates in either the Julian or the Gregorian calendars, I shall write the year first, then the month, and finally the day of the month, without the use of commas to separate the various components of the date. In writing the names of the months in tables, I shall use only the first three letters of the English name of the month, without the use of a period following.

The number of the year will be written in astronomical rather than historical style. The two styles are the same for years of the common era. The year before the year 1 is the year 1 B.C.E. in historical style, but it is the year 0 in astronomical style. The year before that is 2 B.C.E. in historical style and -1 in astronomical style, and so on.

For examples, George Washington was first inaugurated as president on 1789 April 30, and Julius Caesar was assassinated on -43 March 15.

Except in a few specialized uses, the unit of time will be the century. Unless the contrary is explicitly stated, time is always taken with respect to the solar time base. $\tau$ will denote time in Julian centuries† from the epoch 1900 January

†A Julian century denotes 36525 days. The term century, when used as a unit of time, will mean a Julian century.

0.5.† A prime will denote differentiation with respect to solar time; I have already used this convention in Equation I.7, for example.

The symbol $\nu$ will denote an angular velocity with respect to solar time, and a subscript will identify the body involved. The subscripts M and S will refer to the moon and sun, respectively. For individual planets, I shall use the classical symbols ☿, ♀, and so on, as the identifying subscripts. I shall use the subscript P to denote a planet in general. An angular velocity will be in the units of seconds of arc per century (″/cy). For brevity, I shall usually omit a statement of the units.

It follows from these conventions that an acceleration with respect to solar time will be designated by writing the symbol for the corresponding angular velocity, with a prime after it. The units of acceleration are thus seconds of arc per century per century ($''/cy^2$), and the units will usually be omitted.

The important notation is summarized in Table I.1 at the end of this section. The table does not include symbols used only trivially or locally.

Angular velocities and accelerations are geocentric for the sun and moon, but heliocentric for the planets.

For examples, the symbol $\nu_M$ will denote the geocentric angular velocity of the moon with respect to solar time, in units of seconds of arc per century. Its numerical value is approximately 1 732 564 379. The symbol $\nu_{♂}'$ will denote the acceleration of the heliocentric angular velocity of Mars with respect to solar time. It will be expressed in units of seconds of arc per century per century. The estimation of its numerical value is one of the purposes of this work.

When we discuss physical sources of acceleration, as we did in Section I.3, we find that it is

†That is, from noon, Greenwich mean solar time, on the date 1900 January 0, which is the same as 1899 December 31.

convenient if not necessary to use a 'better' time base than solar time; I shall define 'good' and 'better' time bases in a moment after I develop a suitable notation. Let t denote a good time base. We need to find the relation between accelerations with respect to t and accelerations with respect to $\tau$.

The differential operators $d/d\tau$ and $d/dt$ satisfy the relation

$$(d/d\tau) = (dt/d\tau)(d/dt) \quad . \tag{I.9}$$

When we apply this to an angular coordinate, $\theta$ say, we have

$$(d\theta/d\tau) = (dt/d\tau)(d\theta/dt) \quad .$$

The left member of this is an angular velocity $\nu$, taken with respect to solar time. The factor $d\theta/dt$ in the right member is the corresponding angular velocity taken with respect to t. I shall use n as the symbol for such an angular velocity. Thus

$$\nu = (dt/d\tau)n \tag{I.10}$$

for any object.

When we apply Equation I.9 to Equation I.10, with the aid of the relation $dt/d\tau = (d\tau/dt)^{-1}$, we find

$$\nu' = (dt/d\tau)^2\dot{n} - n(dt/d\tau)^3(d^2\tau/dt^2) \quad .$$

If we are not interested in differentiating this further, but wish to use it only in numerical relations, we may now make approximations. In particular, $\tau$ is a good enough time base that we may take $dt/d\tau = 1$ to an accuracy of several significant figures, provided that we restrict ourselves to times within a few thousand years of the present. With these limitations, then,

$$\nu' = \dot{n} - n(d^2\tau/dt^2) \ . \tag{I.11}$$

The ratio of the second to the first term in the right member of Equation I.11 is a measure of the quality of a time base. The smaller this ratio, the better is the time base. A time base is 'good' in a particular discussion if the second member is negligible to the accuracy needed. A time base may be good in one context and poor in another.

In order to introduce solar time, let $R_S$ be the right ascension of the mean sun. In a secular sense, with the neglect of periodic terms, $d^2R_S/dt^2$ is the same as $\dot{n}_S$, the acceleration of the sun's longitude with respect to 'good' time. Let $\theta_e$ be the right ascension of the Greenwich meridian, and let $\omega_e$ be the corresponding angular velocity with respect to t:

$$\omega_e = 4.746\ 600 \times 10^{10} \tag{I.12}$$

in the units adopted here. Except for an additive constant, solar time $\tau$ is proportional to $\theta_e - R_S$:

$$\left.\begin{aligned} \tau &= (\theta_e - R_S)/K \ , \\ K &= 4.733\ 640 \times 10^{10} \ . \end{aligned}\right\} \tag{I.13}$$

The ratio $\omega_e/K$ is the ratio of an interval of sidereal time to the corresponding interval of solar time, and it is approximately equal to 1.002 738.

Now we can evaluate the derivative $d^2\tau/dt^2$. In a secular sense, that is, in a long term average,

$$d^2\tau/dt^2 = (\dot{\omega}_e - \dot{n}_S)/K \ .$$

Thus Equation I.11 becomes

$$\nu' = \dot{n} - (n/K)(\dot{\omega}_e - \dot{n}_S) \ . \qquad \text{(I.14)}$$

We may apply Equation I.14 to any object by attaching its subscript to $\nu$ and to n throughout, if there is not a subscript already attached. However, the acceleration $\dot{n}_S$ in the last parenthesis in the right member is always the solar acceleration. This situation arises because the sun obviously plays a unique role in defining the solar time base $\tau$.

The first term on the right of Equation I.14 is that which arises from the force system acting on the body in question; for example, in the case of the moon, it contains the contribution from tidal friction. The second term on the right arises from the lack of uniformity of the time base. We see that it is proportional to the angular velocity n, as we said that it should be when we were discussing the spin fluctuation hypothesis in Section I.4.

We saw in Section I.4 that atomic time is a better time base than solar time. I shall adopt atomic time as the 'good' time base in this study. van Flandern [1970], Oesterwinter and Cohen [1972], and Morrison [1973], in their studies of recent lunar occultations and other data, have made the same choice, either tacitly or explicitly. However, it is more common in the astronomical literature of the past few decades† to use ephemeris time. Ephemeris time means a time base in which

---

†Clemence [1971] attributes the idea of ephemeris time to André Danjon, in a paper published in 1929 that I have not consulted. Hers [1971], however, attributes the idea to R.T.A. Innes in 1925. Ephemeris time was defined in terms of the tropical year by means of two resolutions adopted in 1957 and 1958 [Explanatory Supplement, 1961, p. 70]. A quantity called ephemeris time has been used as the time base for the ephemerides since 1960. However, this time base uses the lunar month rather than the year and hence it does not agree conceptually with the earlier definition.

the term $\dot{n}$ (Equation I.14) is zero for the sun and for each planet. That is, when and if ephemeris time can be used, the sun and planets obey strictly a gravitational theory. This in turn means that the values of $\nu'$ are exactly proportional to the corresponding values of n; that is, it means that the spin fluctuation hypothesis is satisfied. Thus we can reword the third purpose listed in Section I.6 in the following manner:

> c. to determine whether the concept of ephemeris time is valid, with respect to the long term behavior of the sun and planets.

Until we have satisfied this purpose, then, we cannot use ephemeris time as the main time base in this work, because we do not know whether the ephemeris time base exists.

In the rest of this work, the symbol t for a 'good' time base will be used only for atomic time. In an operational sense, t has been available for use in astronomy only since about 1955. In spite of this, we do have some significant studies of accelerations in which t is the time base.

Occasionally it will be desirable to define a time base by the annual motion of the sun. When I use the term ephemeris time, as I did in an earlier footnote, I mean the time base defined this way, without implying the validity of the general concept. I shall develop the notation needed with this limited definition of ephemeris time where it occurs.

When we measure angles or time in minutes and seconds, we are using sexagesimal notation. There will be a number of occasions when it will be desirable to use sexagesimal notation without using the notation for minutes and seconds. The notation that will be used is best described by an example. The number that is written as 27;33,16,26 means

$$27 \times 60^{0} + 33 \times 60^{-1} + 16 \times 60^{-2} + 26 \times 60^{-3}.$$

This notation can be extended to any positive or

negative power of 60, but it is rarely used for a number that is as large as 60. Note that a semicolon is used to separate the coefficient of the zero power from the coefficients of the negative powers, but that a comma is otherwise used to separate the coefficients.

TABLE I.1

A SUMMARY OF THE PRINCIPAL SYMBOLS USED IN THIS WORK

Part 1. Calendrical Notation

When a date is written in a calendar other than the Julian or Gregorian calendars, the calendar is identified by a pair of letters placed in front of the date. The identifying letters are:

Aℓ.: The Egyptian calendar in which the year is 365¼ days. It is usually but not always referred to the era of Diocletian (284 August 29).

EH: The Muslim calendar referred to the era of the Hijra, which different writers take as 622 July 14, 15, or 16.

EN: The Egyptian calendar in which the year is 365 days, referred to the era of Nabonassar (-746 February 26).

EY: The Persian calendar referred to the era of Yazdigird (632 June 16).

SE: The Babylonian calendar referred to the era of Seleucus. This era is the first day of the Babylonian month that comes closest to -310 April 3.

Part 2. Astronomical Symbols

Astronomical symbols are used to denote the planets and the vernal equinox. They are

☿: Mercury

♀: Venus

♂: Mars

♃: Jupiter

♄: Saturn

♈: Aries (the vernal equinox)

Part 3. Other Non-alphabetic Symbols

A prime denotes a derivative with respect to solar time $\tau$.

A dot over a symbol denotes a derivative with respect to atomic time t.

Part 4. Latin Letters

a̲: the longitude of apogee

D: the mean lunar elongation

e: eccentricity. The eccentricity may have either its modern definition or the definition that was used in Hellenistic and medieval times, according to the context.

$e_1$, etc.: various small distances that appear in ancient models of planetary motion

L: a mean longitude. L is heliocentric for the planets and geocentric for the sun and moon.

M: as a subscript, M denotes the moon. Otherwise, it denotes a mean anomaly.

n: a mean motion with respect to time t

P: a subscript used to refer to a planet in general as opposed to a specific planet

r: the radius of an epicycle in an ancient or medieval model of motion

S: a subscript to identify the sun

t: atomic time

Part 5. Greek Letters

$\gamma$: the anomaly in the epicyclic model of planetary motion. It is not the quantity called anomaly in modern usage.

$\delta$: declination

$\varepsilon$: mean obliquity of the ecliptic

$\lambda$: a geocentric or a true heliocentric longitude, according to the context

$\nu$: a mean motion with respect to $\tau$

$\tau$: solar time

$\phi$: latitude of an observer

Note added in proof: This note pertains to the discussion in Section I.4 about the paper by Spencer Jones [1939]. I have made a new analysis of the data used by Spencer Jones, with $\nu_M{}'$ changed to the correct value. When this is done, there is no significant difference between parameters obtained from the measurements of the right ascension of the sun and those obtained from the measurements of declination. In fact, the measurements of right ascension show smaller residuals than the measurements of declination. The most important difference between the two sets of measurements seems to lie in their time span. The measurements of declination extend from 1761 to 1934 while the measurements of right ascension extend only from 1839 to 1935.

However, more does need to be said about the measurements of right ascension, as I have learned from discussions with R.L. Duncombe and T.C. van Flandern of the U.S. Naval Observatory. Tabulated values of the right ascension of the sun prior to about 1840 may be subject to errors made in the reduction of the data, and it would be necessary to go back to the raw data in order to use these measurements with assurance. Further, because measurements of right ascension of the sun are necessarily made with the sun shining on the instrument, there are effects due to heat and illumination that were poorly understood before about 1840. The errors made in the reduction of the data could be corrected, but the effects of heat and illumination probably could not be.

Thus there are likely to be large errors in measurements of the right ascension for years before about 1840 that cannot be corrected. These errors tend to vary from one observatory to another, and from one set of observers to another, and thus it may not be correct to call them systematic. However, they are so large that the data probably cannot contribute usefully to the study of the spin fluctuation hypothesis.

The sun also necessarily shines on an instrument used to measure its declination. Instruments used for declination measurements differ from those used for right ascension, and the errors introduced by the effects of heat and bright light are not the same. Measurements of the declination of the sun seem to have attained a high level of accuracy sooner than did those of the right ascension.

In summary, solar data earlier than those used by Spencer Jones, whether they are declinations or right ascensions, are probably not useful in studying the spin fluctuation hypothesis. Both right ascensions and declinations should be used in such a study. According to the new analysis, the data for both Mercury and Venus conflict significantly with the spin fluctuation hypothesis. However, an extensive new reduction of the Venus data (R. L. Duncombe, Astronomical Papers, XVI, Part I, 1972) gives results in agreement with the hypothesis, so that the difficulty, at least with Venus, seems to lie in an inaccurate original reduction of the data. It is my understanding the L. V. Morrison of the Royal Greenwich Observatory is preparing a new reduction of the Mercury data.

# CHAPTER II

# CALENDARS

## 1. Classification of Calendars, Conventions Used in Writing Dates

The calendars that will be used in this work can be put into four categories, as follows:

a. Lunar calendars. In these, the fundamental unit is the synodic month. Since the month is too short for many purposes, a longer unit of 12 months is also used. The long unit will be called a lunar year.

b. Solar calendars. In these, the fundamental unit is the tropical year. For convenience, the year is usually divided into 12 units called calendar months.

c. Mixed solar-lunar calendars. These will be called mixed calendars for brevity. In these, the fundamental unit is still the synodic month. Some years contain 12 synodic months and some contain 13. The number of months assigned to each year is chosen in a way that keeps the average length of the year equal to the tropical year.

d. Numerical calendars. By a numerical calendar, I mean one which is calculated on purely numerical principles. Numerical calendars, at least those that will be used here, are divided into months and years. However, no attempt need be made to keep either the month or the year in synchronism with the corresponding astronomical quantities.

The solar day is also a fundamental unit of time in all of the calendars mentioned above. The problem in calendars is that the day, the synodic month, and the tropical year have ratios that are not simply related. The makers of the first two types of calendar adopt various devices to try to keep two of the three quantities in the correct ratio. The makers of the third type try to keep all three in the correct ratio, on the average. The makers of the fourth type apparently quit trying and go to a calendar with simple numerical properties.

The distinction between the numerical calendars and the others needs a little more discussion. The Gregorian calendar, for example, is determined by a set of arithmetic rules and, in that sense, it might be called numerical. However, the rules that govern the Gregorian calendar were based upon astronomical relations and they were designed to be an accurate embodiment of those relations. Further, we may suspect that the calendar will be adjusted when it becomes necessary in order to keep it properly related to the seasons. The purely numerical calendars have elements that approximate the month and the year. However, the elements probably did not represent the state of astronomical knowledge at the time the calendars were devised. Further, the numerical calendars were not revised even when divergence from astronomical relations became obvious by the crudest observation.

I have already explained (Section I.7) the conventions that will be used in this work for writing dates in the Julian or Gregorian calendars. I shall use similar conventions for other calendars. That is, I shall write the year first, then the month, and finally the day of the month. In order not to burden the reader with the names of the months in other calendars, I shall use Roman numerals to designate the months. Thus, if a month is written with letters rather than Roman numerals, the reader will know that the date is in either the Julian or the Gregorian calendar.† I shall not use any notation to distinguish between the Julian and Gregorian calendars. The latest observation that will be used in this work was made on 1019 September 18 (Section I.6). Since this was before the in-

†I make a minor exception for the Greek month Skirophorion, which is the only month that we shall have to use in a certain Greek calendar. See Section II.4.

vention of the Gregorian calendar, that calendar will not be used in identifying data; it will only be used in minor background discussions if at all.

In using calendars other than the Julian or Gregorian, I shall indicate the calendar by means of a symbol placed in front of the year. I shall explain the symbol for each calendar in the section of this chapter that is devoted to that calendar. The calendrical symbols are also listed in Table I.1.

al-Biruni [1000] is a good general reference for the study of the calendars that will be used in this work.

## 2. The Babylonian Calendar

The Babylonian calendar was in use in the historical period around -2000, and it was probably well established by then. It still found some use as late as +75. Few calendars have been used for such a long span of time. Further, the Jewish calendar is a relative and probably a descendant, and so is the lunar calendar still used by the Christian church for determining the date of Easter.

The Babylonian calendar was a mixed calendar. It was regulated by direct astronomical observation rather than by computation, at least for most of its lifetime; in this respect it is unique† among the calendars that we shall use in this work. The official day began at sunset. The month was lunar, and the first day of a month began with the sunset at which the crescent moon was first seen in the western sky by officially recognized observers. If the moon was not seen for any reason, either because it was still too close to the sun or because the sky was obscured, the beginning of the month was deferred.

It is plausible that there was always an important qualification to this. It is quite possible for visibility near the horizon to be poor

†The Athenian calendar (Section II.4), which we shall use slightly, may be an exception. We do not know how this calendar was regulated.

several days in a row in Babylon, either because of clouds or dust storms. If the beginning of the month were deferred until clear weather, we would expect months with more than 30 days. Many existing documents give the lengths of various months and, so far as I know, no month had more than 30 days. Thus it is probable that the month never contained more than 30 days and thus it is probable that the rule went something like this: Watch for the crescent moon at the sunset that closes the 29th day of a month. If the moon is seen, the next month begins at sunset. If it is not seen, sunset begins the 30th day, and the next month begins at the following sunset. This was certainly the rule in the late Babylonian period.

An ordinary Babylonian year contained 12 lunar months, and thus had an average length of about 354 1/3 days. This is considerably shorter than the solar year. In order to keep the average year equal to the solar year, the Babylonians inserted an extra month, called an intercalary month, from time to time.

In early times, the need for an intercalary month was undoubtedly determined by observation of some sort. The intercalary month was sometimes called a repetition of the 12th month (Addaru) and sometimes a repetition of the 6th month (Ululu); these choices occur about equally in times around -2000 [Langdon and Fotheringham, 1928, p. 60]. Occasionally the intercalary month is a repetition of the first month in early times.

We do not know the exact basis of determining the need for an intercalary month. In the Jewish calendar, and in the Christian ecclesiastical lunar calendar, the intercalary month is chosen in a way that will not let the vernal equinox come after the middle of the first month. We might suppose that the ancestral Babylonian calendar had a similar basis, but two other possibilities have been suggested. Neugebauer [1946] has suggested that the purpose was to keep the sun in Libra on the first day of month VII; this would perhaps account for making month VI be the intercalary month. Sachs [1952] has suggested instead that the idea was to keep the heliacal rising of Sirius in month IV.

Whatever may have been the rule governing the insertion of an intercalary month, it would not be safe to assume that the rule was followed faithfully. In a parallel situation under the Roman republic, the need for an intercalary month was determined annually, presumably on the basis of some kind of observation. At the time when Julius Caesar instituted his calendar reform, a total error of about three months had been accumulated, and it would not be safe to assume that the Babylonians were any more conscientious.

In fact, Kugler [1909, pp. 283-284], in a statement that is meant to apply to middle Babylonian times, assigns mid-March to mid-June as the limits for the beginning of the Babylonian year. He based this upon a study of contracts relating to harvesting operations. We should not be surprised to find irregularity in the assignment of intercalary months. The decision to insert an intercalary month was a governmental act made by people who may have had strong political motivations, and we can think of many reasons why having a month added or omitted would confer political advantage.

Intercalation became standardized in late Babylonian times. At some time that is certainly before -365 [Parker and Dubberstein, 1956, p. 2], the Babylonians recognized that 235 (synodic) lunar months are nearly equal to 19 solar years. They introduced a regular scheme of intercalation based upon a 19-year cycle, in which 7 years out of each 19 are intercalary, and they maintained this scheme down to the end of the life of the calendar. They also standardized the choice of the intercalary month. In one year out of each cycle, month VI was intercalary; in the other intercalary years, month XII was.

However, introduction of a regular scheme of intercalation did not necessarily mean a regular scheme for the lengths of the months. We do not know whether calculation ever replaced observation in determining the beginning of the month. This fact means that exact synchronization of our calendar with the Babylonian calendar must proceed on a month-by-month basis.

Occasionally we have the date of some known event, as an eclipse, in the Babylonian calendar. Comparison of the dates then establishes the correspondence between the calendars for that month.

If we do not have a specific event, we can only proceed by trying to calculate when the new moon became visible at Babylon. Schoch [1928] has given tables for making the necessary calculations. Schoch bases his criterion of visibility upon the difference $\Delta A$ in the azimuths of the sun and moon at the instant of sunset. He then establishes a function $a_c(\Delta A)$. Finally he calculates the elevation angle $a$ of the moon at sunset, and the azimuth difference $\Delta A$, from ephemerides. He assumes that the crescent is visible if $a > a_c(\Delta A)$. Clearly there are several points of uncertainty in finding the beginning of the Babylonian month in this way:

1. As everyone recognizes, the beginning can be affected by meteorological conditions that impair visibility.

2. Schoch's criterion for visibility may not be correct. He says (page 95) that he bases his function $a_c(\Delta A)$ in part upon observations of the crescent that he has made. He does not say explicitly where he made the observations, but he says elsewhere with regard to other observations that he made them (the other ones, that is) at Heidelberg and Berlin. It is probable that he made the lunar observations either at these places or elsewhere in northern Europe. It is not clear that the same criterion applies to Babylon and to Germany. He also says that he used more than 400 Babylonian observations with dates from -2095 to 0, but I do not see how this is possible.† Further, the condition for visibility

†I believe that Schoch meant that he knew, or thought that he knew, the correspondences between our calendar and the Babylonian calendar for more than 400 months, and that he had determined the correspondences by analysis of the information in Babylonian astronomical texts. His statement seems highly optimistic to me. Apparently many of his

is probably a function of the observer.

3. The ephemerides may not have been accurate. Schoch says (page 94) that his times of new moon are accurate to ± 3 minutes for times back to -2000 and that the error can reach ± 5 minutes for times before -2000! † I suspect that these errors refer to the precision kept in his calculations, and not to his estimate of the true accuracy. However, his discussion on page 95 shows that he believed the true accuracy to be ± 8 minutes or better for an epoch near -300.

Schoch is neither the first nor the last writer to be optimistic about the accuracy with which he could calculate an ephemeris for ancient times. Now the time of a new moon depends upon the elongation‡ D of the moon from the sun. At present, when we take full advantage of all the data that have been analyzed in the more than 40 years since Schoch's work, the lunar elongation is uncertain by about the amount $1''.0\tau^2$, in which $\tau$ is solar time in centuries before 1900 [Newton, 1974a]. This makes an uncertainty of about 16 minutes in the time of a new moon near -300.

As we go back in time to -2000, the value of $\tau^2$ increases. So does the coefficient of $\tau^2$, be-

---

correspondences came from comparing his calculations with those found in Babylonian calculations of ephemerides; these correspondences are certainly not based upon observations. In other cases, it is known that he mistook predicted information for observed information; see Sachs [1948, p. 290], for example. Certainly he could not have had any known correspondences as far back as -2095 or close to it. Finally, knowing the first day of a Babylonian month by establishing its correspondence with our calendar does not often give useful information about lunar visibility. It does so only in the few cases, perhaps a few tens out of 400, when conditions were close to critical.

†The exclamation point is mine.

‡The angle between the moon and the sun, being taken as positive if the moon is east of the sun.

cause we are extrapolating far beyond the available data. To me, it does not seem possible to assert that we know the elongation near -2000 better than to about $4''.0\tau^2$, and it is speculation to assume that we know it this well. Thus the uncertainty in the time of a new moon near -2000 is at least 3.5 hours. With this much uncertainty, a calculation that a solar eclipse was visible at a particular place has a 50-50 chance of being wrong.

Schoch used $3''.5\tau^2$ for the quadratic term† in the lunar elongation. The best estimate that we can make at present is about $1''.5 \pm 1''.0$ for the coefficient of $\tau^2$ for an epoch near -300. Thus Schoch's values of the elongation D are probably in error by the amount $(2''.0 \pm 1''.0)\tau^2$. The expected error in the times of his new moons near -300 is thus about 32 minutes. This implies a probability of about 1 in 40 or 50 that his date of a new moon is wrong. That is, we may expect that dates calculated from his tables will be wrong for about 1 month in each three or four years, if there are no errors in his calculations other than those in D.

I do not mean for these remarks to be interpreted as adverse criticism of Schoch's work. His visibility tables are a valuable contribution to Babylonian chronology. The reliability of his tables is high, and we still have no better basis for converting the Babylonian calendar to ours for most months. His tables are the basis of the chronological tables of Parker and Dubberstein [1956]. These tables illustrate the month-to-month nature of converting the Babylonian calendar to ours. All that one can do is to give the date, in our calendar, of the first day of the Babylonian month, separately for each month for each year.

Documentary evidence may allow refining the tables of Parker and Dubberstein slightly. For example, their tables show successive months beginning on -273 October 7 and -273 November 6, separated by 30 days. However, there is a document

---

†He gave $9''.6$ (p. 95) for the coefficient of $\tau^2$, but he was including a contribution of $6''.1$ from gravitational effects.

[Epping and Strassmaier, 1892, p. 229] which says that the month in question had only 29 days, if the document has been transcribed and translated correctly. Until we have a larger amount of evidence of this sort, the slight improvements that may be possible are probably not worth the effort.†

I said above that we do not know whether the beginning of the month was ever determined by calculation. While we do not know, there are slight indications which suggest that calculation was used sometimes. The document just mentioned says that clouds prevented seeing the moon at sunset following the 29th day, but the new month began at this sunset in spite of this fact.

It remains to enlarge upon Section II.1 about the conventions that will be used in writing a date in the Babylonian calendar. The ordinary months will be designated by the Roman numerals I through XII. In all the years that concern us here, the intercalary month is a repetition of either month VI or month XII. If the intercalary month follows month VI, it will be designated by the symbol $VI_2$; if it follows month XII, it will be designated by $XII_2$. As usual, the day of the month, in Arabic numerals, will follow the designation of the month.

Alexander's general Seleucus became the undisputed monarch of Babylon sometime around -305, after a series of wars that settled the division of Alexander's empire. In a fashion that occurs often in such circumstances, he claimed an earlier date for the beginning of his reign, and the claimed date was subsequently adopted as an era. I shall designate a year of the Seleucid era by placing the

†It is not invidious to suggest that either the transcription or the translation might be wrong. Reading a cuneiform tablet is like reading a badly weathered inscription in stone; reading depends purely upon differences in relief of an eroded surface. The work of Epping and Strassmaier is now more than 80 years old, and in the meantime Babylonian scholars have learned much about the language that Epping and Strassmaier had no opportunity to know.

letters SE before the number of the year. Thus SE 1 I 1 will mean the first day of the first month of the first year of the Seleucid era. If Schoch's tables of lunar visibility are correct, we have the correspondence

SE 1 I 1 = -310 April 3 . (II.1)

This is certainly correct within a day or so.

The realm of Seleucus later collapsed, and the first king of the part of it known as Parthia was Arsaces I. Some cuneiform writers used a date connected with him as an era, without otherwise changing the calendar. The year 1 of his era equals SE 65. Thus we can readily change any Arsacid date to a Seleucid date, and I shall ignore the era of Arsaces.

We shall have to deal with a few Babylonian dates before the Seleucid era. In those times, years were designated as the year Y of a certain ruler. The year that was called a ruler's first year, at least in chronological usage, always began with the date I 1. I do not know whether his year 1 began with the date I 1 after he acceded or with the preceding date I 1, and indeed different rulers may have insisted upon different conventions. In any event, we now have a consistent set of 'regnal years' for chronological purposes, which are carefully set forth by Parker and Dubberstein [1956]. In effect, each ruler started a different era.

In order not to burden the reader with a multiplicity of eras, I shall convert all years before the Seleucid era into a negative (or 0) year of the Seleucid era, in the same way that we use zero or negative years with the common era. In doing so, I shall follow rigorously the enumeration of the regnal years given by Parker and Dubberstein. I hope that doing so will not deceive the reader and cause him to think that the Babylonians followed this practice. If the reader encounters a date such as SE -67 IX 17, he should realize that it was the 17th day of the 9th month of the year SE -67, which was originally written as the 26th year of the king Artaxerxes II.

The Babylonian astronomical observations that will be discussed in Chapter IV will allow us to establish the correspondence of the calendars for some Babylonian months. Until the analysis is complete, we shall not know the number of months for which we shall succeed, but the number is probably less than 100. It is certainly far below the number 400 that Schoch speaks of. The analysis may allow some further refining of the tables of Parker and Dubberstein, beyond the possibility that was mentioned above.

## 3. The Egyptian and Alexandrian Calendars

The Egyptians had at least one calendar [Parker, 1950] that was based upon the lunar month and that was used for fixing the dates of certain religious ceremonies, in a way that is analogous to the use of the Christian lunar calendar in fixing the date of Easter. From an early date, however, the Egyptians paralleled the religious lunar calendar with a purely numerical calendar, which was used for various administrative purposes. When I refer to the Egyptian calendar in this work, I shall mean only this numerical calendar.

Neugebauer [1957, p. 81] has called the Egyptian calendar "the only intelligent calendar which ever existed in human history." Every Egyptian year contained 12 months of 30 days each. After the end of the 12th month, 5 'epagomenal'† days, which did not belong to any month, were added to complete the year. Thus there were exactly 365 days in each calendar year, so that the calendar gradually precessed through all the seasons.

al-Biruni [1000] lists the names of the Egyptian months, as do several other writers. As I said in Section II.1 above, I shall designate the Egyptian months by numbers instead of by their names. The reader who needs the names of the months should have little difficulty in finding them.

The Egyptian calendar, like most calendars, can be used in conjunction with any convention for the

---

†Which means 'added on'.

hour at which the day changes. The Egyptian day of ordinary usage began [Parker, 1950, p. 10] at sunrise. Astronomers do not need to follow ordinary usage, and I could not decide upon the convention that Ptolemy [ca. 142] uses for beginning the day. When Ptolemy states a time between sunset and sunrise, he usually refers to the night between, for example, the 12th and the 13th of the month. However, he has other usages sometimes. For example, in Chapter IX.9, he refers to "the morning of the 24th" of the month, and in Chapter III.2 he refers to "the 3rd, at midnight before the 4th". These usages are consistent with beginning the day at either midnight or sunrise, but they do not seem consistent with beginning it at noon or at sunset.

In one place, Ptolemy [ca. 142, Chapter III.1] refers to the end of the sixth hour in a context where he clearly means noon; that is, he takes the day to begin at sunrise in this passage. Here, however, he is citing an observation made by Hipparchus. In doing so, he uses the convention for beginning the year that Hipparchus used, although he uses the incompatible Egyptian way of designating months and days. Similarly, he may be keeping Hipparchus's convention for beginning the day, while his own way may be different.

The Egyptians originally did not use an era for numbering the years, so far as we know. After Alexander conquered Egypt, the people in the resulting Hellenistic society sometimes used the era of Alexander, which is placed in the year -322, with a variety of conventions about the length of, and the beginning of, the year. Some Hellenistic writers also used the accession of Ptolemy II Philadelphus in -284 as an era.

Ptolemy† uses the 'era of Nabonassar' that I shall define in a moment, and he uses it in connection with the Egyptian calendar. Whenever we encounter a date referred to Alexander or to Ptolemy II Philadelphus in the work that concerns us here, we find it immediately translated into Ptolemy's

---

†The astronomer, that is, not one of the Hellenistic rulers of Egypt.

system. Therefore we need to consider only the latter. Let Y be the number of the year with respect to the era of Nabonassar, let M be the number of the month, and let D be the day of the month. Then the Julian day number JD at noon on any Egyptian date is [Newton, 1972b]

$$JD = 1\ 448\ 242 + 365Y + 30M + D. \qquad \text{(II.2)}$$

I shall denote a date written in the Egyptian calendar, if it is referred to the era of Nabonassar, by putting EN in front of the date. Thus the first day of the era is EN 1 I 1. For this date, Equation II.2 gives us

JD at noon on EN 1 I 1 = 1 448 368.

This is the date -746 February 26. I suspect that Ptolemy chose this as an era because he thought Nabonassar was a ruler whom he could date accurately and to whom he could assign a date earlier than any astronomical observation known to him. There is no evidence that the era of Nabonassar was ever used for any purposes other than astronomical ones.

When I write Egyptian dates, and when I use Equation II.2, I shall look upon the epagomenal days as forming a 13th month. Thus, for example, the 3rd epagomenal day of the year 880 will be written EN 880 XIII 3. The Julian day number at noon on this day is 1 769 835.

al-Biruni [1025] still uses the Egyptian calendar for some purposes. When he does so, he uses the convention that the day begins at noon, although he uses other conventions with some calendars.

The greatest value of the Egyptian calendar probably lies in the ease with which we can calculate the exact interval between two dates.† While

† So far as I can see, this property has little value except in the calculation of ephemerides. It is interesting that the Egyptians, who never calculated ephemerides so far as we know, devised the calendar that, of all known calendars, is the most useful for that purpose.

it is undesirable for some purposes, I think that we should acquit it of one charge that is often made against it. This is the charge that it was useless for telling the Egyptians when to expect the flooding of the Nile upon which their agriculture depended, and that they had to resort to observations of Sirius in order to know when to expect the floods. The time of maximum flooding, measured with respect to the tropical year, varies from its mean by more than 15 days, so that a definition of the time more closely than, say, 10 days is not significant. It takes 40 years for the Egyptian calendar to precess 10 days. Thus, if a farmer learned an average date when he was young, this date would serve him well all his working life.

We do not know when the 365-day calendar was introduced, but all estimates [Parker, 1950, Chapter IV] put its introduction before -2000. It is possible that the length of its year was chosen to agree with some phenomenon, such as the flooding of the Nile, that does not repeat accurately; this would keep the discrepancy between the calendar year and the astronomical year from being discovered for some time.† If this is so, the length of the calendar year was probably hallowed by tradition when the discrepancy was discovered, and an alteration intended to make the calendar agree with the astronomical year would have been resisted strenuously.

The first known attempt to change the average calendar year to $365\frac{1}{4}$ days was made by the "decree of Canopus" in -236 [Parker, 1950, p. 39], under the Hellenistic rulers of late Egypt. It failed completely. An attempt by Augustus was probably more successful. While it is clear that the old calendar of 365 days was still the "Egyptian calendar" to Censorinus in 238,‡ the reformed calendar of Augustus apparently had some contemporaneous adherents.

---

†Whatever the Egyptians of -2000 thought about the accuracy of their calendar year, they surely did not think that their calendar month represented the astronomical month accurately.

‡See Section V.8.

However, when Augustus's reformed calendar finally became the calendar of common use in Egypt, its relation to Augustus was usually ignored. In Egypt, the year of $365\frac{1}{4}$ days was almost always used with the era of Diocletian.

I shall use the term 'Alexandrian calendar' to denote the Egyptian calendar of $365\frac{1}{4}$ days when it is referred to the era of Diocletian.† In the Alexandrian calendar, one year in every four has 6 rather than 5 epagomenal days. I shall indicate a date in the Alexandrian calendar by putting the symbol "Aℓ." in front of the year. The day in the Julian calendar which is the first day in the year Y of the Alexandrian calendar [_Explanatory Supplement_, 1961, p. 430] is given by:

$$\text{A}\ell.\ \text{Y I 1} = \begin{cases} (283 + \text{Y})\ \text{Aug 30} & \text{if Y is divisible by 4,} \\ (283 + \text{Y})\ \text{Aug 29} & \text{otherwise.} \end{cases} \qquad \text{(II.3)}$$

Thus the year Y is a leap year of 366 days if it precedes a year that is divisible by 4.

The era of Diocletian is the date Aℓ. 1 I 1. From Equation II.3, we see that this corresponds to 284 August 29.

The formula for the Julian day number in the Alexandrian calendar is slightly more complicated than Equation II.2. In order to write it, we let $\mathcal{J}(x)$ be the greatest-integer function. That is, if x is real, $\mathcal{J}(x)$ is the integer that satisfies

---

†The terms 'reformed Egyptian' and 'Coptic' are also used for the calendar with $365\frac{1}{4}$ days in a year. It is not clear to me that these terms, or the term 'Alexandrian', for that matter, necessarily refer to the era of Diocletian. However, as we shall encounter it in this work, the calendar of $365\frac{1}{4}$ days is used only with the era of Diocletian, and we may safely connect the adjective 'Alexandrian' with that era.

$$x - 1 < \vartheta(x) \leq x . \quad (II.4)$$

Then

$$JD = 1\,824\,634 + 365Y + 30M + D + \vartheta(Y/4) . \quad (II.5)$$

Some astronomers adopted the reformed or Alexandrian calendar and some continued to use the old one, even after Diocletian. When we meet a date given in Egyptian terms in a period after the reign of Diocletian, we must start by determining whether the calendar is the old one or the Alexandrian one.

## 4. Greek Calendars

The Greeks in the times of classical antiquity had many calendars, of which only two will concern us.

The first is a calendar† that was used in Athens in the period around -430, and for several centuries thereafter. This calendar was a mixed calendar. The summer solstice was apparently more important to the Athenians than the vernal equinox, and they chose their intercalary month in a way that made the summer solstice come during the last month of the year. In other words, the new year began with the new moon that followed the summer solstice.

The Athenian day, like the Babylonian, began at sunset.

As well as I can make out from the literature on the Athenian calendar, these are about the only facts on which there is general agreement. Some authorities, including Pritchett and Neugebauer [1947], believe that the Athenian month, like the Babylonian month, began at the sunset when the crescent moon was first visible. Others believe that it began at the sunset next after the true conjunc-

†The Athenians used two calendars based upon entirely different principles, but only one of them concerns us.

tion of the moon and sun. Meritt [1961] is willing to accept either definition, if I understand his position.

There has been much controversy about whether the beginning of the month was determined by calculation or observation. In the writings that I have studied, those who uphold observation are willing to concede that it may have been supplemented by calculation on occasion, such as in bad weather. Those who uphold calculation are willing to concede that calculation was sometimes checked and corrected by observation. If this is so, it means that the beginning of the month was determined by a mixture of observation and calculation, from either viewpoint, and the only point at issue is the proportion of the mix.

Whatever may have been the rule that governed the months, the magistrates were free to alter it at will, within limits. Dates in the Athenian calendar are sometimes stated to be κατα θεον (according to [the] god)† and they are sometimes stated to be κατ᾽ αρχοντα (according to [the] archon). In some documents, the date is stated in both ways. Apparently the archons could decree, whenever they chose, that a certain date would be repeated once or even several times. It is as if our civil authorities could decree that several successive days should all be called December 25, for example.

There were some limits to this freedom, however. In all cases that are known, the basic arbitrary power was limited to repeating a day, and the archons had to come out right at the end of the year. Thus, if they repeated a day near the beginning of the year, they had to make up for it by omitting a day later on in the year. The only freedom they had in omitting a day, so far as is known, was to decide when the omission would come; the number of days omitted had to equal the number repeated.

---

†That is, dates that are κατα θεον are dates determined by the moon, whatever the rules of determination were.

Thus, in the last month of the year, named Skirophorion, which contained the summer solstice, there was little freedom to manipulate the calendar. Meritt [1961, p. 208] says that there is no evidence of a year in which days were repeated in Skirophorion, nor even of a year in which the mandatory omission of days was deferred to a point that late in the year. In other words, according to the available evidence, dates in Skirophorion are the same whether they are κατα θεον or κατ' αρχοντα .

Some writers object to the idea that the beginning of the month was determined by the true conjunction rather than by the visible crescent, on the basis that the Athenians could not find the time of true conjunction except on the rare occasions of a visible solar eclipse. Actually, it is possible to find and even to predict the time of true conjunction by means that were easily within the capability of Meton and his contemporaries. However, there is no direct evidence, so far as I know, that Meton actually used the method.

The method consists simply of observing the rising of the moon some morning near the end of a month and of measuring either the angle between the sun and moon or the difference between their rising times. From either the time or the angle, it is then a simple matter to predict when the sun and moon will coincide.

For example, suppose that the angle is measured, and that it is measured with an accuracy of 1°. Since the average length of the month is about 29.53 days, the Athenians would readily know, if they were using the angle, that it changes by about 12°.2 per day. Even if they made no attempt to use the true rather than the mean angular velocity of the moon, they could still predict the true conjunction within an error of a few hours if their error in the angle were 1°. They could get similar results by working directly with the difference between rising times.

However, it is not a purpose of this study to enter into the controversy about the nature of the

Athenian calendar. Fortunately, all that is needed for the purposes of this work is to find the earliest possible Julian dates that correspond to a few Athenian dates; it would be convenient but not necessary to know the exact dates. We can accomplish this limited objective without knowing whether conjunction or visibility determined the calendar, and without knowing whether dates are κατα θεον or κατ' αρχοντα.

I shall illustrate this with an example. In the summer of -431, what is the earliest possible time when the 13th day of Skirophorion† could have begun? With the aid of the ephemeris programs that are described in Sections IX.2 and IX.6,‡ we find that true conjunction occurred at about 09 hours, Athens mean time, on -431 June 16. If the months were determined by conjunctions, Skirophorion 1 thus began at sunset on -431 June 16. If the months were determined by visibility of the crescent, Skirophorion 1 probably began at sunset on -431 June 17.

Moving forward from day 1 to day 13 of the month, then, we find that sunset on -431 June 28 is the earliest time when Skirophorion 13 could have begun. This is the correct time if the month was fixed by conjunctions. The correct time is probably sunset on -431 June 29 if the month was fixed by first visibility of the crescent. All these statements are based upon the assumption that the date given is κατα θεον.

If it is correct that dates κατα θεον and κατ' αρχοντα are always the same in this, the last month of the year, the conclusions just stated are still valid. If the dates can differ in Skirophorion, the Greek date κατ' αρχντα is necessarily a later date in our calendar than the same Greek date κατα θεον,

---

†The reasons for choosing this example will appear at the end of this section.

‡The calculated phase of the moon depends upon $D''$ but not upon the individual accelerations $\nu_S'$ and $\nu_M'$, within reasonable limits. I took the value of $D''$ from Equation 17 of Newton [1972a]; this value is 3.38 $''/cy^2$. I then used $\nu_M' = 3.38\ ''/cy^2$ and $\nu_S' = 0$.

on the basis of all that is known about the Greek calendar. If the date Skirophorion 13 is κατ' αρχοντα, and if this can differ from the date κατα θεον, sunset on -431 June 28 is still the earliest possible time. It may not be the correct time, however.

Thus it seems that we may abandon the conventional calendrical correspondence, which says that Skirophorion 13 in this year equals -431 June 27. This conclusion will receive independent confirmation in Section V.3. There we shall find that -431 June 28 is the most probable equivalent. The odds against -431 June 27 as the equivalent are about 52 to 1, but the odds against -431 June 29 as the equivalent are only about 10 to 1.

According to numerous authorities, Meton introduced a 19-year cycle into the Athenian calendar beginning with the summer solstice in the year -431. We saw in Section II.2 that a 19-year cycle was introduced into the Babylonian calendar at roughly the same time, but probably several decades later. The two cycles are not quite the same, however. Both cycles involve the equation 19 years = 235 months. In the Babylonian calendar under this rule, the months continued to be determined by the moon and not by rule; the cycle merely required 7 intercalary years in each set of 19 years, and it specified the years in which intercalation was to occur.

Meton's cycle, on the other hand, regulated the allotment of days to the month as well as months to the year. At least, this seems to be the implication of the statement [Ptolemy, ca. 142, Chapter III.1, for example] that the year in the calendar of Meton and Euctemon had 365 + (1/4) + (1/76) days. If the allotment of days to the months were left free, it would not be possible to make such a statement.

About a century later, Callippus altered Meton's cycle by omitting one day in every fourth Metonic cycle. That is, Callippus shortened Meton's length of the year by 1 day in each 76 years, leaving him a year of $365\frac{1}{4}$ days [Ptolemy, Chapter III.1, for example]. Thus he anticipated the length of the Julian year by nearly three centuries.

It is clear from existing Athenian records [Meritt, 1961, Chapter I and Pritchett and Neugebauer, 1947, Chapter I] that the calendar actually used in Athens did not follow the rigid rules of Metonic and Callippic cycles. What probably happened was that the cycles were adopted in the astronomical literature in order to make dating unambiguous. This brings us to the second Greek calendar that we need to discuss. Ptolemy often gives the year of one of Hipparchus's observations as the "Yth year of the Pth Callippic cycle". Let JY denote the year, in the Julian calendar, in which year Y of cycle P began. JY = -329 when Y = P = 1, so that

$$JY = 76P + Y - 406 \; . \qquad (II.6)$$

However, Ptolemy gives the month and day by means of the Egyptian calendar. Since the beginning of an Egyptian year precesses through all the seasons, while the beginning of a Callippic year was always near the summer solstice, it is a tricky matter to convert a date given in this way. One method of conversion is the following:

1. From the year number and the cycle number, we calculate the year of the Julian calendar from Equation II.6. For example, the 32nd year of the 3rd Callippic cycle began in the Julian year -146.

2. The year began at the new moon following the day that the Athenians thought was the summer solstice. From this fact, we estimate the Julian day number at the beginning of the stated year. In the year -146, for example, the summer solstice came on about June 26 in the Julian calendar. Thus the year of the Callippic cycle began soon after Julian day number 1 667 908.

3. From the year in the Julian calendar, we guess a year in the Egyptian calendar, with respect to the era of Nabonassar (see Section II.3). Since the year 1 of this era began early in the year -746, a date in the year -146, Julian calendar, comes in about the year

601, era of Nabonassar.

4. The Egyptian year that we have just guessed is close to the correct year. We combine the guessed-at year with the stated Egyptian month and day, and we calculate the Julian day number for each possible combination, using Equation II.2. We then find which possible Julian day number came within one solar year after the day that we found in step 2 above. This gives the exact year.

For example, Ptolemy [ca. 142, Chapter III.1] gives the dates of the vernal equinox and of the autumnal equinox that Hipparchus measured in the 32nd year of the 3rd Callippic cycle. The autumnal equinox was at the midnight between the 3rd and 4th epagomenal day; let us say for simplicity that this is the date XIII 3 of some Egyptian year. The vernal equinox was at sunrise on the date VI 27 of the same Callippic year.

The second date is actually later, although a casual inspection suggests that it is earlier. As we have seen, the summer solstice in -146 came on about Julian day 1 667 908. The Julian day number of the day that began at noon on EN 601 XIII 3 is 1 668 000; this day began at noon on -146 September 26. The Julian day number of the day that began at noon on EN 601 VI 27 is 1 667 814.

Therefore EN 601 XIII 3 is correct for the autumnal equinox, since it comes about 92 days after the summer solstice, and the Julian date is -146 September 26. However, EN 601 VI 27 is too early, so the correct date must be EN 602 VI 27. This corresponds to -145 March 24.

This is a good place to revert to the solstice of Meton and Euctemon and to emphasize a point that will be justified in Sections V.3 and VIII.4. An ancient inscription called the Milesian parapegm† says that the solstice came on the Athenian date

†Quoted by Dinsmoor [1931, p. 311] and by Meritt [1961, p. 4], who cite the original sources where the inscription is translated and discussed.

Skirophorion 13 and on the Egyptian date that is equivalent to June 27. Since Ptolemy (Chapter III.1) gives the same Egyptian date, all students of the Athenian calendar have taken Skirophorion 13 to be equivalent to June 27 in the year -431, and they have encountered insuperable difficulties in doing so. The task of constructing a consistent theory of the Athenian calendar should be eased enormously by taking Skirophorion 13 to be equivalent to June 28 (or perhaps to June 29).

## 5. The Muslim Calendar†

The Arabic peoples have always used some type of lunar calendar, so far as we know. According to al-Biruni [1000, pp. 73-74], they used a purely lunar calendar until about the year that we call 400. At this time, they changed to a mixed solar-lunar calendar. al-Biruni says that they learned the necessary method of intercalation from the Jews, which probably means that they adopted the 19-year cycle (Section II.2). However, still according to al-Biruni, Muhammad‡ forbade the intercalation of months, and the Muslims returned to a purely lunar calendar, with 12 lunar months in each Muslim year.

The Muslim calendar is based upon the era of the Hijra. Let Y denote the number of the year with respect to this era, and let Y be written in the form

$$Y = 30N + R , \qquad (II.7)$$

in which N and R are integers. If R has one of the eleven values 2, 5, 7, 10, 13, 16, 18, 21, 24, 26, or 29, the year is a leap year; otherwise it is an

†Most of the information in this section appears in Section 4 of Newton [1972b].

‡There is a considerable variety in the way that Muslim names and terms, including the word 'Muslim' itself, are rendered in English. I use the form given by Kennedy [1956] whenever I can find the name or term in his survey. Otherwise, I use the form that appears in the immediate source that is cited.

ordinary year. The first month of each year has 30 days. After that, the months have alternately 29 and 30 days, until we reach the last month. The last (twelfth) month has 29 days in an ordinary year and 30 days in a leap year.

TABLE II.1

THE CALENDRICAL FUNCTION $F_Y(R)$

FOR THE MUSLIM CALENDAR

| R | $F_Y(R)$ | R | $F_Y(R)$ |
|---|---|---|---|
| 0 | -354 | 15 | 4 961 |
| 1 | 0 | 16 | 5 315 |
| 2 | 354 | 17 | 5 670 |
| 3 | 709 | 18 | 6 024 |
| 4 | 1 063 | 19 | 6 379 |
| 5 | 1 417 | 20 | 6 733 |
| 6 | 1 772 | 21 | 7 087 |
| 7 | 2 126 | 22 | 7 442 |
| 8 | 2 481 | 23 | 7 796 |
| 9 | 2 835 | 24 | 8 150 |
| 10 | 3 189 | 25 | 8 505 |
| 11 | 3 544 | 26 | 8 859 |
| 12 | 3 898 | 27 | 9 214 |
| 13 | 4 252 | 28 | 9 568 |
| 14 | 4 607 | 29 | 9 922 |

TABLE II.2

THE CALENDRICAL FUNCTION $F_M(M)$

FOR THE MUSLIM CALENDAR

| M | $F_M(M)$ | M | $F_M(M)$ |
|---|---|---|---|
| I | 0 | VII | 177 |
| II | 30 | VIII | 207 |
| III | 59 | IX | 236 |
| IV | 89 | X | 266 |
| V | 118 | XI | 295 |
| VI | 148 | XII | 325 |

Thus the extra day in leap years is not added for the purpose of keeping the calendar year close to the solar year on the average. Instead, it is added for the purpose of keeping the average calendar month close to the astronomical month. In a cycle of 30 years, there are 360 calendar months and there are 10 631 days, so that the average calendar month has $29^d.530\ 556$. This differs from the astronomical month by $0^d.000\ 043$,† so that the calendar will be in error by one day in about 23 000 months, about 1900 solar years.

We can readily calculate the Julian day number JD at noon on any Muslim date by means of the two

†The length of the month that we deduce from Ptolemy [ca. 142; see Section V.1 below] is in error by $0^d.000\ 005$ days. This does not necessarily mean that Ptolemy's accuracy was not available to the Muslims who devised the calendar. They would have had to go to a more complicated calendar with a cycle much longer than 30 years, in order to improve their accuracy appreciably.

functions $F_Y(R)$ and $F_M(M)$ that are given in Tables II.1 and II.2.† R is defined by Equation II.7 and M is, as usual, the number of the month. Then

$$JD = (JD)_{MO} + 10\ 631N + F_Y(R) + F_M(M) + D. \quad (II.8)$$

We can find the constant $(JD)_{MO}$ in Equation II.8 as soon as we know the era of the Hijra. I shall denote a date written in the Muslim calendar by placing EH in front of the date. If we use this convention, the era of the Hijra is the date EH 1 I 1 and $(JD)_{MO}$ is the Julian day number on date EH 1 I 0. Unfortunately, not even Muslim writers agree on the date EH 1 I 1, and three different choices appear in the literature. A few examples will be given.

Explanatory Supplement [1961, p. 433] says that the era is 622 July 16, but adds that "some Oriental chronologists" choose the date July 15. At the other extreme, in the astronomical tables of al-Khwarizmi [Neugebauer, 1962; Goldstein, 1967],‡ the epoch is local noon on the day EH 1 I 1 at longitude 76° approximately, and the mean elongation of the moon is 4°.325. This agrees almost exactly with calculation from modern tables if EH 1 I 1 = 622 July 14, and it disagrees by an impossibly large amount for any other choice of the epoch.

Finally, Adelard of Bath translated al-Khwarizmi's work into Latin in about the year 1135. One copy of this translation [Haskins, 1924] has an introduction that gives, among other information, the dates in the Julian calendar of the first day of several Muslim years from EH 520 to 529. These correspondences require the equation EH 1 I 1 = 622 July 15, and thus the introduction is incompatible with the very tables that it introduces.

The three possible values of the constant $(JD)_{MO}$ are

---

†These tables are copied from Newton [1972b], by permission of the Royal Astronomical Society.

‡Also see Section VI.2 below. The tables were prepared in about the year 820.

$$(JD)_{MO} = \begin{cases} 1\ 948\ 437 & \text{if} \quad EH\ 1\ I\ 1 = 622\ \text{July}\ 14, \\ 1\ 948\ 438 & \text{if} \quad EH\ 1\ I\ 1 = 622\ \text{July}\ 15, \\ 1\ 948\ 439 & \text{if} \quad EH\ 1\ I\ 1 = 622\ \text{July}\ 16. \end{cases} \tag{II.9}$$

Thus astronomical observations that are dated by means of the Muslim calendar cannot be used safely unless the observer's choice of era can be determined. We can often determine the era by analyzing the data, as we just did with al-Khwarizmi's tables. Further, many Islamic astronomers give dates in two different calendars, and we can test the choice of era in this way.

The ordinary day in the Muslim calendar is taken to begin at sunset, but the astronomical day is often taken to begin at noon. Confusion about this point, perhaps combined with the establishment of correspondences with calendars having still other definitions of the day, may account for some of the problems about the era of the Hijra.

The calendar that has been described is the Muslim civil calendar. In the Muslim religious calendar, the first day of the month is still determined [Explanatory Supplement, 1961, p. 433, or Kennedy, 1956, p. 144] by the visibility of the lunar crescent at sunset, separately for each locality. As Kennedy says, this may account for Islamic astronomers' interest in visibility problems, a matter that had been of great concern in Babylonian astronomy but that received little attention from Ptolemy.

## 6. The Persian Calendar

The last nominal ruler of the Sassanid empire in Persia was Yazdigird III, who came to the throne as a child on (presumably) 632 June 16. The Sassanid empire fell to the expanding forces of Islam before Yazdigird had any opportunity to rule, and he was killed in 651. The state religion of the Empire was Zoroastrianism, and Yazdigird, not surprisingly, was commemorated by the Zoroastrians as the last ruler to hold to their faith. The date of his accession is still used as their calendrical era by the surviving Zoroastrians.

The Persian calendar used under the Sassanid empire was apparently derived from the Egyptian calendar. According to al-Biruni [1000, pp. 54ff], the Sassanids met the intercalation problem by adding a month (30 days) in every 120 years rather than by adding a 6th intercalary day in every 4th year. After the fall of the empire, he says, the Persians (Zoroastrians) had to abandon intercalation because the intercalation required a ceremony of state and there was no one left to perform the ceremony. Thus the Persian calendar, by necessity, reverted to a calendar of exactly 365 days.

In this work, the only Persian calendar that we shall encounter is the calendar of 365 days referred to the era of Yazdigird. I shall denote a date in this calendar by placing EY in front of the date.

Because of its constant length, the Persian calendar played the same role in Islamic astronomy that the Egyptian calendar played in Hellenistic astronomy. Some Islamic astronomers, of whom al-Biruni is an example, used both the Egyptian and Persian calendars. al-Biruni [1025] used the Egyptian calendar even for observations that he made himself if he needed to combine them with the Hellenistic observations that were preserved by Ptolemy [ca. 142]. Otherwise, he commonly used the Persian calendar.

There are at least two variants of the Persian calendar, which differ in where they place the epagomenal days. In one variant, they are put after month IX [Carmody, 1960, p. 49]. al-Biruni [1000 and 1025] puts them after month VIII, although he notes [al-Biruni, 1000, pp. 55-56] that there are other practices. I have noticed only one example outside of the writing of al-Biruni in which month IX occurs,† and in this instance it is clear that

†This is the record numbered 6 in Section VII.6.

the epagomenal days come after month VIII. So far as this work is concerned, we can take the Persian calendar to mean that variant in which the five epagomenal days come after month VIII.

We can calculate the Julian day number JD at noon on a Persian date from the formula

$$JD = (JD)_{PO} + 365Y + 30M + D \; . \qquad (II.10a)$$

Since the epagomenal days do not come at the end of the year, the value of $(JD)_{PO}$ depends upon the value of month M:

$$(JD)_{PO} = \begin{cases} 1\ 951\ 667 & \text{if} \quad M \le VIII \; , \\ 1\ 951\ 672 & \text{if} \quad M \ge IX \; . \end{cases} \qquad (II.10b)$$

If the reader encounters a Persian date in some source that is not used in this work, he should remember that the epagomenal days do not necessarily come between months VIII and IX and hence that Equation II.10b may need an obvious modification.

Alternatively, we can write the formula for JD using the greatest-integer function defined in relation II.4:

$$JD = 1\ 951\ 667 + 365Y + 30M + D + 5\mathcal{J}(M/9) \; . \qquad (II.11)$$

This is the form that applies if the epagomenal days come after the month VIII.

When we use Equations II.10 or II.11, we count the epagomenal days as part of month VIII. Thus we assign 35 days to this month and 30 days to all other months.

If the reader needs the names of the months in the calendars that have appeared earlier in this chapter, he should have no trouble in finding them. The names of the Persian months are harder to find, and I list them in Table II.3. The spellings in Table II.3 are those used in the cited translation of al-Biruni [1000].

TABLE II.3

THE MONTHS IN THE PERSIAN CALENDAR

| | | | |
|---|---|---|---|
| I. | Farwardin Mah | VII. | Mihr Mah |
| II. | Ardibahisht Mah | VIII. | Aban Mah |
| III. | Khurdadh Mah | IX. | Adhar Mah |
| IV. | Tir Mah | X. | Dai Mah |
| V. | Murdadh Mah | XI. | Bahman Mah |
| VI. | Shahrewar Mah | XII. | Isfandarmadh Mah |

TABLE II.4

THE DAYS OF THE MONTH IN THE PERSIAN CALENDAR

| | | | | | |
|---|---|---|---|---|---|
| 1. | Hurmuz | 11. | Khur | 21. | Ram |
| 2. | Bahman | 12. | Mah | 22. | Badh |
| 3. | Ardibahisht | 13. | Tir | 23. | Dai-ba-din |
| 4. | Shahrewar | 14. | Gosh | 24. | Din |
| 5. | Isfandarmadh | 15. | Dai-ba-mihr | 25. | Ard |
| 6. | Khurdadh | 16. | Mihr | 26. | Ashtadh |
| 7. | Murdadh | 17. | Srosh | 27. | Asman |
| 8. | Dai-ba-adhar | 18. | Rashn | 28. | Zamyadh |
| 9. | Adhar | 19. | Farwardin | 29. | Marasfand |
| 10. | Aban | 20. | Bahram | 30. | Aniran |

The Zoroastrians had names for each day of the month, and for each epagomenal day, and astronomers often used the name rather than the number of the day. The names of the days are listed in Table II.4. The names are the same in each month. However, al-Biruni [1000, p.54] says that he never sees the same names for the epagomenal days in two different works.

## 7. The Days of the Week

The week is a numerical division that is not supposed to approximate any astronomical period. There are seven days of the week and there are seven classical planets. The present names of the days are derived from the planets, and Neugebauer [1957, p. 169] describes an astrological calculation that puts the planets into the sequence that they have in the week days. Thus it is possible that the week has an astrological origin.

Since the week does not need to approximate anything, there has never been any need to revise it, and the sequence of week days has never been interrupted since the week was introduced. Not even the transition between calendars has disturbed it. In the first introduction of the Gregorian calendar, for example, Thursday, 1582 October 4 was followed by Friday, 1582 October 15. When the Gregorian calendar was introduced into the English-speaking countries, Wednesday, 1752 September 2 was followed by Thursday, 1752 September 14.

Some peoples have numbered the days of the week and some have named them. The Jews usually numbered them. The Arabic peoples [al-Biruni, 1000, pp. 75-76] named them at one time but numbered them in the time of al-Biruni. Medieval Christians in Europe also numbered them. All these peoples had an exception to the numbering. The Christians named the first day of the week. The Jews and Arabs named the last day, and sometimes they named the sixth day as well.

In this work, I shall designate the days of the week entirely by means of numbers, and I shall indicate this practice by placing the medieval Latin word *feria* after the number. Thus, for example, "4th *feria*" will denote the day that is called Wednesday in English.

One advantage of numbering the days of the week is that the feria is simply related to the Julian day number. In fact,†

†The method of relating the feria to JD in Equation II.12 ensures that the feria will take on the values 1 to 7 rather than 0 to 6.

$$\text{feria} - 1 \equiv (\text{JD} + 1)(\text{mod } 7) \; . \qquad \text{(II.12)}$$

Neither a solar year nor a lunar year can contain an integral number of weeks. Thus the week day of a given calendar date is never the same for two successive years. In consequence, the week day, if given, is a valuable adjunct to the date, and it often allows us to detect and to remove errors that have occurred in the transmission of a text. Calendar reforms that interrupt the sequence of week days in order to make each year start on the same week day might be a boon to bankers, but they would be a disaster to chronologists. They would also put printers of calendars out of business.

## 8. Calculating the Julian Day Number by Integral Arithmetic

In Section II.5 above, I presented a method of calculating the Julian day number, for a date in the Muslim calendar, by means of tables or arrays. This follows the usage of *Explanatory Supplement* [1961, pp. 436-439], which uses tables in order to calculate the Julian day number for a date in either the Julian or the Gregorian calendar.

In Section II.5, however, I departed from the usage of the *Explanatory Supplement* in one important regard. *Explanatory Supplement* uses a double-entry table or array, with one argument for the month and the other argument for the year within a basic cycle; this cycle is 100 years for the Julian or Gregorian calendars.† Thus one table has 1200 entries. The same approach, if applied to the Muslim calendar in which the basic cycle is 30 years, would have required a table with 360 entries. Instead of using a double-entry table, I used the sum of two single-entry tables, one for the month and one for the year; this approach needs only 42 entries for the Muslim calendar.

---

†Except for a year that is a multiple of 400, in the Gregorian calendar. The cycle could be reduced to 4 years if one were interested only in the Julian calendar.

An alternative approach to calculating the Julian day number involves the use of integer arithmetic. That is, it makes use of the greatest-integer function $\mathcal{J}(x)$ defined in relation II.4. Equations II.5 and II.11, for calculating the Julian day number in the Alexandrian and Persian calendars, respectively, already use integer arithmetic.

With modern computing equipment, the amount of computer usage required with either approach is probably trivial and hence not a factor in choosing the approach. In some programming systems, one approach may be simpler, and this may afford a basis for choice. Usually, however, the choice probably reduces to a matter of taste.

With the Muslim calendar, it is necessary to use the tabular method to some extent, because of the rule governing leap years. Let $R_i$, with $i = 1$ to 11, denote the set of integers 2, 5, 7, 10, 13, 16, 18, 21, 24, 26, and 29. Then we can replace Equation II.8, Table II.1, and Table II.2 by

$$JD = (JD)_{MO} + 354Y + \sum_{i=1}^{11} \mathcal{J}[(Y + 29 - R_i)/30] + 29M + \mathcal{J}(M/2) + D - 383 . \qquad (II.13)$$

This is not a simple relation. The simplest procedure for calculating JD in the Muslim calendar is probably to take the yearly dependence from Table II.1 and the monthly dependence from Equation II.13.

Many ways have been devised to calculate the Julian day number for a date in the Julian or Gregorian calendars. The use of tables has already been mentioned. One way to proceed with the use of integral arithmetic is as follows. Let Y, M, D denote the year, month, and day respectively. Define U and $\mu$ by

$$U = Y - 1 + \mathcal{J}[(M + 9)/12] , \quad \mu = M + 10 - 12\mathcal{J}[(M + 9)/12]. \qquad (II.14)$$

Then U and $\mu$ are the year and month in the style for which the year begins with March 1 rather than with January 1.

For the Julian calendar,

$$JD = 1\ 721\ 087 + 365U + \mathcal{J}(U/4) + 30\mu + \mathcal{J}(\mu/2)$$

$$+ \mathcal{P}[\mu\mathcal{J}(\mu/7)] + \mathcal{J}(\mu/12) + D\ . \qquad (II.15)$$

In this, $\mathcal{P}(n)$ for integral n denotes the parity function; $\mathcal{P}(n) = 1$ if n is odd and $\mathcal{P}(n) = 0$ if n is even.

For the Gregorian calendar, we add

$$\mathcal{J}(U/400) - \mathcal{J}(U/100) + 2 \qquad (II.16)$$

to the right member of Equation II.15. Relations (II.15) and (II.16) should be valid in any interval.

Again, it is probably simplest to use a hybrid method, taking the yearly dependence from Equation II.15, modified if necessary for the Gregorian calendar, and using a table for the monthly dependence. We can form the monthly table by adding all the terms that contain $\mu$ in the right member of Equation II.15.

However, the interested reader should consult the formulas for the Julian day number that Oesterwinter and Cohen [1972] give.

# CHAPTER III

# A CLASSIFICATION OF PLANETARY OBSERVATIONS

## 1. Broad Classification of Planetary Observations; Synodic Periods

Most old planetary observations can be put into two broad classes. In one class, the observation concerns the position of the planet against the celestial sphere. An observation of this kind may locate the planet with respect to the fixed stars, or it may give the celestial coordinates. The other class concerns the position of the planet with respect to the sun.

The classes may be distinguished in many ways. One way is by means of the periods. Looked at geocentrically, the average period of an inner planet with respect to the stars is the same as the solar year. The average period of an outer planet is the same as its heliocentric period. The periods of motion with respect to the sun may be quite different.

I shall use the term 'synodic phenomenon' for a planet to mean an appearance in which the emphasis is upon the position of the planet relative to the sun, as seen from earth. Thus a synodic phenomenon involves the earth, the sun, and a planet.

A synodic phenomenon repeats, on the average, after an interval $P_S$ called the synodic period. If we use the notation of Section I.7, we have

$$P_S = 1\ 296\ 000/|\nu_S - \nu_P|\ \text{cy}\ , \qquad \text{(III.1)}$$

if the angular velocities are in the standard units. Because the angular velocities are not constant, synodic phenomena do not repeat at uniform intervals. The values given by Equation III.1 are merely the average intervals taken over many repetitions.

TABLE III.1

SYNODIC PERIODS OF THE PLANETS

| Planet | Synodic Period, days |
|---|---|
| Mercury | 115.877 |
| Venus | 583.921 |
| Mars | 779.938 |
| Jupiter | 398.885 |
| Saturn | 378.092 |

TABLE III.2

GEOCENTRIC ANGULAR VELOCITIES AT PERIGEE AND APOGEE

| Planet | Geocentric Angular Velocity, deg./day | |
|---|---|---|
| | at Apogee | at Perigee |
| Mercury | 1.8527 | -0.9766 |
| Venus | 1.2444 | -0.6263 |
| Mars | 0.7070 | -0.3574 |
| Jupiter | 0.2286 | -0.1317 |
| Saturn | 0.1238 | -0.0781 |

The periods calculated from Equation III.1 are listed in Table III.1, converted to days. In general discussion, as opposed to detailed analysis, I shall ignore the variability in the intervals of repetition.

I shall next take up several important types

of planetary phenomena, with discussions of the ways in which related observations appear in the literature. For descriptive purposes, I shall take the heliocentric orbits of the planets to be circles, with the periods related to the radii by Kepler's law. In actual computations, of course, I shall use accurate theories. For convenience in the description, I shall depart from the standard units of Section I.7, and shall measure angles in degrees and time in days.

## 2. Conditions at Apogee and Perigee

Apogee means the greatest distance from the earth. Apogee of a planet occurs when it is on the other side of the sun from us, that is, at conjunction for an outer planet or at superior conjunction for an inner planet.

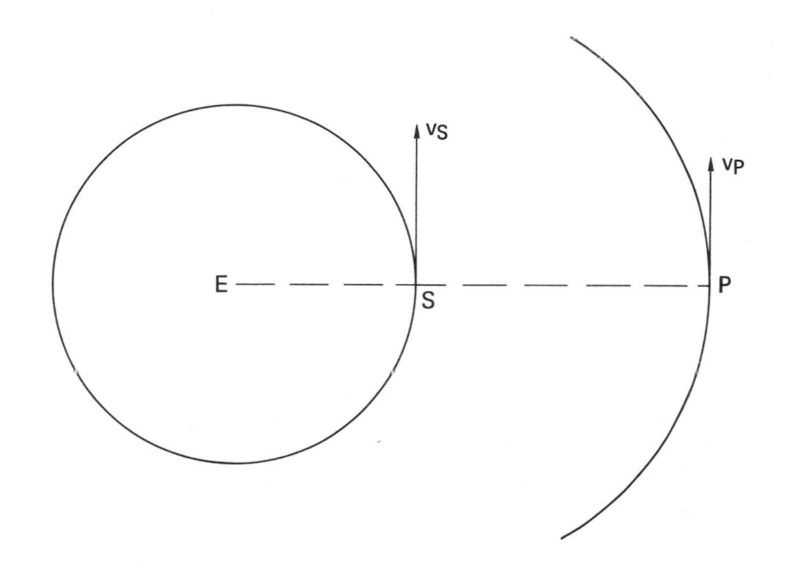


Figure III.1. Conditions at the apogee of a planet, drawn from the geocentric point of view. E is the earth. S is the sun and $v_S$ is its vector velocity with respect to the earth. P is the planet and $v_P$ is its vector velocity with respect to the sun.

At apogee (Figure III.1), the velocity vector $v_S$ of the sun with respect to the earth and the velocity vector $v_P$ of the planet with respect to the sun are in the same direction, and thus the velocity of the planet with respect to the earth is a maximum at this time. The geocentric angular velocity is the average of the angular velocities of the earth and the planet taken with respect to the orbital radii. The geocentric angular velocities at apogee are listed in the second column of Table III.2. The reader should notice that the angular velocity at apogee is greater than the angular velocity of the sun for the inner planets and less for the outer planets.

Observations at apogee are difficult and perhaps impossible without an instrument like a coronagraph. However, the Babylonians placed great importance upon observing the planets as close to apogee as possible, for reasons that had nothing to do with the circumstance that it is apogee, and they made many observations near apogee. I shall return to this matter in Section III.5 below.

Behavior of a planet at perigee is one of the matters that is perhaps understood more easily in a geocentric picture than in a heliocentric one. In a geocentric picture, the sun goes around the earth in a circle; this circle is the deferent. In turn, the planet goes around the sun in a circle called the epicycle. Thus, to the accuracy of coplanar and circular orbits, the epicycle-deferent picture gives an exact description. The radius of the deferent is the distance of the sun from the earth. The center of the epicycle moves around the deferent with the angular velocity of the sun. The radius of the epicycle is the heliocentric distance of the planet, and the angular velocity on the epicycle is the heliocentric angular velocity of the planet.

In this description, the appearances are quite different for inner and outer planets. For inner planets, the deferent circle is larger than the epicycle, and the angular velocity of motion around

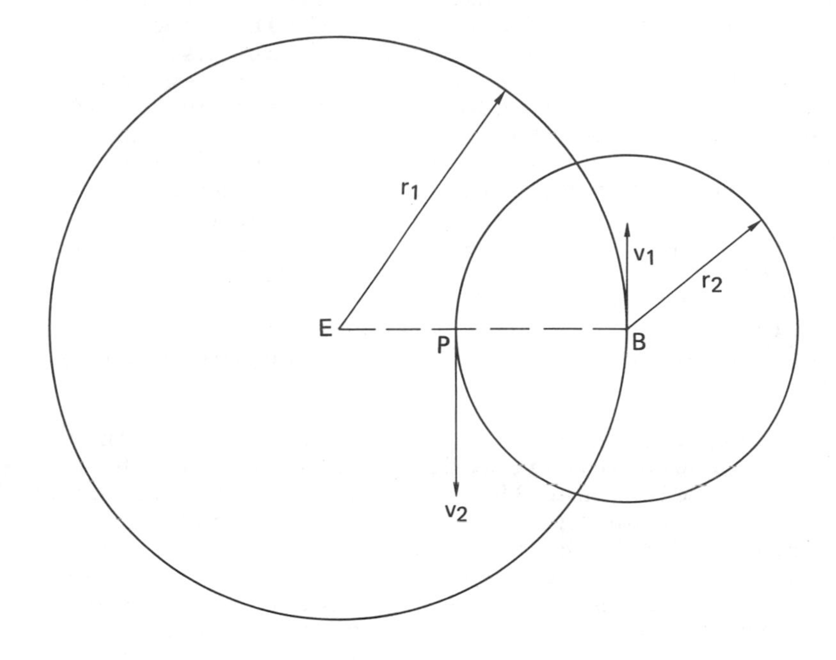


Figure III.2. Conditions at the perigee of a planet for the deferent-epicycle model of a planet. E is the earth. The circle of radius $r_1$ centered on E is called the deferent, and the point B moves around the deferent with the constant speed $v_1$, in the direction shown. The circle of radius $r_2$ centered on B is called the epicycle, and the planet P moves around the epicycle with the constant speed $v_2$, in the direction shown. B can be identified with the sun if P is one of the inner planets, but not if P is one of the outer planets. For any planet, we have $v_2 > v_1$.

the deferent is less than that around the epicycle.[†] For outer planets, these conditions are reversed.

However, there are two choices of the epicycle and deferent circles[‡] that give exactly the same geocentric motion. The choice described above, in which the sun moves on the deferent and is the center of the epicycle, is one choice. Ancient and medieval astronomers made the other choice for the outer planets. If we put their choice into modern terminology, we can say that the radius of the deferent circle, for an outer planet, was equal to the heliocentric distance of the planet, and that the angular velocity of the epicycle center around the deferent was equal to the heliocentric angular velocity of the planet. The epicyclic radius equalled the earth-sun distance, and the angular velocity around the epicycle equalled the mean motion of the sun.

Thus, for outer planets as well as for inner planets, the epicyclic radius was less than the deferent radius, and the epicyclic angular velocity was greater than the angular velocity around the deferent. For both inner and outer planets, this leads to the picture shown in Figure III.2 at perigee. $v_1$ is the speed of the center B of the epicycle in its motion along the deferent, and $v_2$ is the speed of the planet **P** relative to the center of the epicycle. Since $v_2 > v_1$, the planet is

---

[†] Most ancient and medieval astronomers did not put the center of the epicycle for Mercury and Venus at the distance of the sun. However, they did necessarily have the angular velocity of motion around the deferent equal to the mean motion of the sun, and they had the angular position (as opposed to the distance) of the epicycle center equal to the angular position of the sun, on the average. The statements made in this paragraph therefore apply to their description.

[‡] Neugebauer [1957, pp. 122ff] gives an excellent description of this matter, using geometrical relations and the principle of relativity of motion. In Appendix IV, I give a discussion based upon analytic geometry, for those who prefer this approach.

moving downward in the picture. That is, the geocentric angular velocity of a planet is negative at perigee, for both inner and outer planets.

The geocentric angular velocity at perigee is given for the classical planets in the third column of Table III.2. It is negative for all planets, as we have just said. On the other hand, the geocentric angular velocity is positive on the average, and it is positive at apogee, for all planets, as we can see from Table III.2.

Although we can choose the deferents and epicycles in a way that makes analogous formal relations for all the planets, perigee is a synodic phenomenon in which the configurations are quite different for inner and outer planets. For an inner planet, perigee occurs at the time of inferior conjunction, and we get the correct configuration in Figure III.2 if we identify B with the sun. For an outer planet, however, perigee occurs when the earth passes between the sun and the planet. That is, perigee for an outer planet occurs at opposition.

## 3. The Turning Points

The geocentric angular velocity of a planet is positive at apogee and negative at perigee, as we have seen. Therefore it must pass through zero between apogee and perigee, and it must pass through zero again between perigee and apogee. Thus, after a planet passes apogee, its geocentric longitude increases for awhile. That is, it moves eastward through the stars. Before it reaches perigee, however, its geocentric angular velocity goes to zero. At this point, it quits moving eastward and begins to move westward with respect to the stars. The point at which it does this is called the first turning point.

After it passes the first turning point, the planet proceeds on to perigee, where it attains its greatest negative angular velocity. It continues to move westward until its angular velocity again becomes zero. The point at which this occurs is called the second turning point. At the second turning point, the planet resumes its eastward motion among the stars, finally reaching its maximum geocentric angular velocity again at apogee.

TABLE III.3

APPROXIMATE CONDITIONS AT THE TURNING POINTS

| Planet | Elongation at a Turning Point, degrees | Time from Perigee, days |
|---|---|---|
| Mercury | 18.2 | 11.4 |
| Venus | 28.9 | 21.1 |
| Mars | 136.2 | 36.4 |
| Jupiter | 115.6 | 60.3 |
| Saturn | 108.8 | 68.8 |

TABLE III.4

CONDITIONS AT MAXIMUM ELONGATION

| Planet | Maximum Elongation, degrees | Time from Perigee, days |
|---|---|---|
| Mercury | 22.8 | 21.6 |
| Venus | 46.3 | 70.8 |

The turning points are synodic phenomena in that they occur for certain relations in the three-body system of the earth, sun, and planet. Approximate conditions at the turning points are summarized in Table III.3. The second column of the table gives the geocentric angle (elongation) between the sun and the planet at a turning point. The planet is east of the sun by the amount shown at one point, and it is west by the same amount at the other. The third column of the table shows the interval in days between perigee and either turning point.

The configurations of earth, sun, and planet at the turning points are quite different for inner and outer planets, as the numbers in Table III.3 indicate. The elongation is greater than 90° for the outer planets. Therefore the turning points of the outer planets are easily visible. Many early peoples looked upon the planets as deities, or at least as signs put in place by deities. Thus they regarded the manifest changes in course as mysterious and portentous matters, and we have many early records about the turning points of the outer planets.

The turning points of the inner planets, however, occur near inferior conjunction, when the planet is near the sun and when much of its dark side is turned to us. The turning points are thus hard to see, and we have few early records of them.

A measurement of the time when a planet is at a turning point is a measurement made when the angular velocity is zero. Thus a single measurement of time made near a turning point is almost useless for our purposes. We should need a systematic series of measurements made near a turning point, with an accuracy probably of a few minutes of arc, in order to extract useful information from the time measurements.

A turning point is also a sidereal phenomenon in the sense that it involves the motion of a planet with respect to the stars. We find that the sidereal position of a planet at a turning point is sometimes recorded, and we must ask whether the position gives useful information. The answer would depend upon our goal. If we had assumed that the spin fluctuation hypothesis is correct, and if we were only trying to find the acceleration of the earth's spin, the position of a turning point would not be useful. The reason is that the turning point is determined by the orbital motions of the sun and the planet, and these motions are not affected by the rotation of the earth.

However, if the accelerations of the sun and the planet with respect to universal time are not proportional to the respective angular velocities, the position of a turning point is affected. In fact, the position of a turning point with respect to the stars is a fairly sensitive function of the difference between the accelerations, and it may be that the positions will allow a useful test of the spin fluctuation hypothesis.

4. Maximum Elongation

Only the inner planets exhibit this phenomenon, which we can understand with the aid of Figure III.2. Since we are dealing with an inner planet, we can take the center B of the epicycle to be the actual sun. Let us draw lines from the earth E that are tangent to the epicycle. Clearly the planet can never be farther from the sun than the positions of these tangent lines, so the points of tangency are called the positions of maximum elongation.

The elongation of an inner planet thus oscillates between two limits. The elongation of an outer planet, on the other hand, has no limit.

The values of the maximum elongation, and the length of time between maximum elongation and perigee, are listed in Table III.4. The greatest elongation is necessarily greater than the elongation at a turning point, and so is the time between greatest elongation and perigee. We can verify these facts by comparing Tables III.3 and III.4.

The time of a maximum elongation, like the time of a turning point, does not help us in this study. The position of a planet at maximum elongation is independent of the acceleration of the planet, so the position will not help us test the spin fluctuation hypothesis. However, the position reflects one-for-one any variation in the longitude of the sun. Hence the positions of maximum elongation of Mercury and Venus may give us valuable information about the motion of the sun.

## 5. First and Last Visibility

When a planet is in conjunction with the sun, it is invisible to the naked eye, although it may remain visible with sophisticated instrumentation. Thus, before conjunction, there is a time of last visibility and, after conjunction, there is a time of first visibility. Either first or last visibility may occur in either the morning or the evening, so there are four combinations to consider.

Several ways of designating the combinations that are found in the literature are confusing and even self-contradictory. For example, we may find "western rising". Rising in the west is obviously impossible; a "western rising" of Venus or Mercury actually means the first observed setting. I shall use the terms 'first, morning', 'last, morning' 'first, evening', and 'last, evening'. It seems to me that these terms are self-explanatory and unambiguous.

Ptolemy [ca. 142, Chapters XIII.7 to XIII.9] devotes three chapters to the subject of first and last visibility. He assumes that the visibility of a planet near conjunction depends upon the distance of the sun below the horizon at the instant when the planet is on the horizon. He does not seem to have used a definite term for this distance. Some writers call it the "arc of vision" or the equivalent, but this does not indicate the direction along which the arc is to be measured. I shall use 'minimum vertical separation' and I may sometimes omit 'minimum'.

Ptolemy takes the minimum vertical separation to be 5° for Venus and 10° for Mercury. He does not give values of the vertical separation for the outer planets, so far as I noticed, although we can infer his values from other information that he does give.

The application of this simple assumption leads to considerable complications in the times of first and last visibility. To start with, the angle between the horizon and the ecliptic at sunrise or sunset varies with the time of year. During a given sidereal rotation of the earth, the horizon makes the smallest angle with the ecliptic when the pole of the ecliptic is nearest to the zenith. This condition occurs at noon at the winter solstice, at sunrise at the spring equinox, and so on. At sunset, the angle between the horizon and the ecliptic varies from a minimum of $90° - \phi - \varepsilon$ at the autumnal equinox to a maximum of $90° - \phi + \varepsilon$ at the vernal equinox. In these expressions, $\phi$ is the latitude of the observer and $\varepsilon$ is the obliquity of the ecliptic. In order for a planet to have the same vertical separation, its separation in longitude (and hence the time interval between conjunction and visibility) must be many times as great in the fall as in the spring as seen at Babylon, for example, other things being equal.

But other things are far from equal. The latitude of a planet, as well as its longitude, has an effect upon the vertical separation that is actually present. The effect of latitude is small when the ecliptic is inclined steeply to the horizon, but it can be important when the inclination is near its minimum.

Further, it seems unlikely to me that visibility can be governed by a single variable, quite apart from the possibility of clouds. It seems to me that the difference in azimuth should also come into the situation somewhat, as it seems to do with the visibility of the moon.† Also, an inner planet is in a crescent phase near inferior conjunction (perigee) and in a gibbous phase near superior conjunction, and the geocentric distances are different for the two kinds of conjunction. Thus it seems unlikely that the same value of vertical separation should govern first visibility near both types of conjunction, and Schoch [1928, p. 103] in fact uses different values for inferior and superior conjunctions.

---

†See Section II.2.

Schoch [1928] and Langdon and Fotheringham [1928]† have attempted to extend the work of Ptolemy. Schoch uses a different value of the vertical separation for European latitudes and for Babylon. If he is correct, this may reflect systematic differences in the state of the atmosphere. Both sources try to find a dependence upon the time of year. Langdon and Fotheringham [1928, p. 52] find that the vertical separation is smaller for last visibility than for first. They find (pp. 51-52) that the visibility is better in the fall and winter in Babylon but not necessarily in Europe. Schoch, however, finds that visibility is better in the winter and spring. As I have indicated in a footnote, these findings are often based upon a mis-reading of the records. In contrast to Schoch and to Langdon and Fotheringham, Ptolemy (Chapter XIII.7) finds that conditions at Babylonian latitudes are best in summer "when the sky is clearest".

Finally, as everyone recognizes, visibility can be spoiled completely by clouds.

At the summer solstice, the angle between the horizon and the ecliptic has its average value at either sunrise or sunset. Let us calculate the separation in geocentric longitude (elongation) between sun and planet needed to give the critical value of the vertical separation, for the summer solstice, assuming zero for the latitude of the

---

†The reader should be cautious about accepting the conclusions of Schoch, Langdon, and Fotheringham, particularly as they relate to Babylonian conditions. At the time that those scholars worked, a number of astronomical expressions in Babylonian writing had not been translated correctly. They often have wrong dates for the data, and they often use predicted quantities as if they had been observed. As Babylonian scholars learn more about Babylonian astronomy, they continually find more 'observations' that are actually predictions. It is possible that the present work uses some 'observations' that will ultimately turn out to be predictions. See Section IV.1 for more discussion of the studies of Schoch and of Langdon and Fotheringham.

TABLE III.5

APPROXIMATE CONDITIONS AT FIRST AND LAST VISIBILITY

| Planet | Elongation Used by Ptolemy, degrees | Time from Conjunction, days |
|---|---|---|
| Mercury | 11 1/2 | 13.3[a] |
| | | 5.9[b] |
| Venus | 5 2/3 | 21.9[a] |
| | | 3.5[b] |
| Mars | 14 1/2 | 52.0 |
| Jupiter | 12 3/4 | 16.8 |
| Saturn | 14 | 16.2 |

[a]Near superior conjunction

[b]Near inferior conjunction

planet. This should then give us something like the average elongation needed for first and last visibility. The values that Ptolemy uses for the average elongation are listed in Table III.5. The table also gives the approximate time in days needed to achieve the elongation listed.

In calculating the number of days, I combined the angular velocity of the sun with the angular velocities of the planets listed in Table III.2. Because the latter are extreme values, the number of days in Table III.5 is somewhat small, on the average. Since the table is intended only as an approximate guide and not as a source of exact information, this error should not matter much.

For the inner planets, the elongations at the

turning points are greater in magnitude than the elongations at first or last visibility, as we can see by comparing Tables III.3 and III.5. Therefore the turning points are visible in principle. However, the inner planets are near their vanishing points when they turn, so visibility is difficult and we do not expect many observations of the turning points of Mercury and Venus.

At apogee (superior conjunction), the inner planets are travelling faster than the sun and therefore they cross from the west to the east side. This means that the sequence of events is from last, morning visibility to conjunction to first, evening visibility. In all other cases, the planet crosses the sun from east to west. Thus we go from last, evening visibility to conjunction to first, morning visibility.

At critical visibility, a planet is travelling rapidly with respect to the sun in most cases; this is particularly so for an inner planet near inferior conjunction. Thus, if visibility were a precisely defined quantity that varied continuously, it would be valuable in a study of the accelerations. Unfortunately, critical visibility meets neither condition. If a planet is not seen one evening, for example, there is no chance to look for it until the next evening and, in the meantime, the planet has travelled a large distance. Further, the amount of disagreement about the conditions that govern visibility shows that it is not precisely defined.

In sum, it does not appear that we can use observations of first and last visibility for the purposes of this study, because we expect the observations to exhibit large errors. We cannot even expect the errors to have zero mean, because the conditions of first and last visibility may not be symmetrical with respect to the sun. I shall give observations of visibility mostly as a matter of record, in the hope that they will prove useful sometime. In some cases, they may help us judge or interpret a record of other phenomena.

TABLE III.6

A SUMMARY OF THE SYNODIC PHENOMENA OF THE PLANETS

| Phenomenon | Average Time from Apogee, days | | | | |
|---|---|---|---|---|---|
| | Mercury | Venus | Mars | Jupiter | Saturn |
| Apogee[a] | 0 | 0 | 0 | 0 | 0 |
| First visibility[c] | E,13.3 | E,21.9 | M,52.0 | M,16.8 | M,16.2 |
| Maximum elongation | 36.3 | 221.2 | --- | --- | --- |
| Turning point | 46.5 | 270.9 | 353.6 | 139.1 | 120.7 |
| Last visibility[c] | E,52.0 | E,288.5 | --- | --- | --- |
| Perigee[b] | 57.9 | 292.0 | 390.0 | 199.4 | 189.5 |
| First visibility[c] | M,63.8 | M,295.5 | --- | --- | --- |
| Turning point | 69.3 | 313.1 | 426.4 | 259.7 | 258.3 |
| Maximum elongation | 79.5 | 362.8 | --- | --- | --- |
| Last visibility[c] | M,102.6 | M,562.0 | E,727.9 | E,382.1 | E,361.9 |
| Apogee[a] | 115.9 | 583.9 | 779.9 | 398.9 | 378.1 |
| First visibility[c] | E,129.2 | E,605.8 | M,831.9 | M,415.7 | M,394.3 |

[a]Conjunction for an outer planet, superior conjunction for an inner one

[b]Opposition for an outer planet, inferior conjunction for an inner one

[c]An E indicates evening visibility, an M indicates morning visibility.

## 6. A Summary of Synodic Phenomena

For convenience, I summarize the synodic phenomena and their time relations in Table III.6, with the times rounded to the tenths of a day. I believe that the table is nearly self-explanatory. Maximum elongation, and vanishing visibility near perigee, do not apply to the outer planets, as I have indicated by lines. In the entries for first and last visibility, the letter E or M before the number of days indicates whether the visibility in question is in the evening or the morning, respectively.

## 7. Transits

A transit occurs when a planet passes between the earth and the sun; thus one occurs only for the inner planets. A transit is a synodic phenomenon, but I have not included it in the preceding discussion and tables because it need not occur during every synodic period. It occurs rarely, and only near inferior conjunction.

The disk of Mercury is so small that a Mercury transit probably cannot be observed without magnification. There are a few early reports of transits of Mercury, but modern writers usually assume that the object observed was a sunspot [Goldstein, 1969]. A transit of Venus can be observed without magnification, as Goldstein shows by quoting a report from a Honolulu newspaper of 1874 December 12. Thus pre-telescopic reports of transits of Venus cannot be ruled out within investigation. Goldstein investigates several medieval reports, and he finds only one that cannot be excluded with certainty. This is a report from probably the first part of the eleventh century. There is no particular reason to suppose that the event recorded was a transit; there is just not enough information to settle the matter one way or the other. Thus all known transit observations occurred after the invention of the telescope. Transits are not used in this work and I mention them only for completeness.

Transits are similar to eclipses in many ways and their occurrence is governed by the same general principles. In particular, they can occur only when the earth and the planet are both close to a node of the planetary orbit. Because it is unlikely that both earth and planet will be in this position at the same time, transits are rather rare.

On the average, there is one transit of Venus in 61 years. Only six have occurred since the birth of Galileo, the inventor of the telescope. The dates of these [Newcomb, 1902, p. 175] are: 1631 December 7, 1639 December 4, 1761 June 5, 1769 June 3, 1874 December 9, and 1882 December 6. There are no known observations of the first of these, so that only five transits of Venus have ever been observed, so far as we know. The next transit will be on 2004 June 8.

On the average, there are about thirteen transits of Mercury per century [Explanatory Supplement, 1961, p. 263]. In modern times, the earth is near a node of Mercury only near November 10 and May 8, so that transits are frequently designated as 'May transits' and 'November transits'. There are many telescopic observations of Mercury transits, and they have found considerable use in studies of the accelerations.

## 8. Double Conjunctions

By a double conjunction, I mean a conjunction of the moon with a planet or of two planets with each other. These are synodic phenomena in the sense that they take place without regard to the star background or to an inertial coordinate system. They are not subject to rules as simple as those for the synodic phenomena discussed earlier, since they involve the positions of four bodies.

Double conjunctions that involve the moon tell us mostly about the acceleration of the moon, because of the rapid motion of the moon. Thus we shall be concerned only with double conjunctions that involve two planets for the major purposes of this work.

In listing double conjunctions, I shall list them first in order of the planet that is nearer to the sun. Those that involve the same near planet will be listed chronologically.

## 9. Statements of Position

We sometimes have direct statements of a celestial coordinate of a planet at some epoch. The accuracy of such a statement depends upon both the accuracy of observing the planet and the accuracy of establishing the coordinate system used in giving the observation. Thus, as a general proposition, I prefer not to use direct statements of position, although there are some exceptions.

It is doubtful that any direct statement of a celestial coordinate before the Hellenistic period has enough precision to be useful in this study. From the Hellenistic period onward, however, we have some useful observations of this type.

We may also have statements of the coordinates of a planet in an earth-fixed system, such as azimuth and elevation. Most observations of this sort are solar observations, however, rather than planetary in the usual sense. I have already analyzed al-Biruni's collection of meridian elevations of the sun [Newton, 1972b] and have found them to be highly useful.

## 10. Planetary Conjunctions; Normal Stars

By a planetary conjunction, I mean a conjunction of a planet with a known star. Planetary conjunctions form what is probably the most useful class of pre-telescopic observations of the planets. They are useful because they let us infer the position of the planet without being dependent upon the accuracy with which the early observer established his coordinate system.

By about -400 and probably earlier, the Babylonians were systematically noting conjunctions of the moon and the planets with a set of stars that lie close to the ecliptic. Epping and Strassmaier [1892] use the term "normal stars" (*normalsterne*) for this set and they give (pp. 224-225) a list of the normal stars and their coordinates for the epoch

1800. Kugler [1924, p. 553] repeats the list, along with the coordinates for -300 and -100. Kugler also gives two stars that Epping and Strassmaier do not have in their list; presumably he found them in a source that Epping and Strassmaier did not have the opportunity to see.

There is no problem in identifying most of the normal stars. The Babylonian names for the stars are descriptive phrases. The name for the 27th star on the list, for example, can be translated as "the middle star on the brow of the Scorpion". In his star table, Ptolemy [ ca. 142, Books VII and VIII] lists a star with just this name and gives its coordinates, and this name continued in use into medieval times. There is no question, I think, that this star is δ Scorpii.

Since many conjunctions, involving different celestial bodies, are recorded with many normal stars, we can determine their relative coordinates with considerable accuracy. Since most of them can be identified by their names alone, we can thus find the coordinates of those, if any, that cannot be so identified. The coordinates then lead to the identifications. I have not followed in detail the way in which Epping and Strassmaier and Kugler reached their identifications, and I do not know how often they needed to use the coordinates, as opposed to the use of names alone.

A few problems remain. The Babylonian names for the stars in Cancer differ from those used by Ptolemy; since there are no outstanding stars in this part of the sky, the identifications may not be as secure as we like. Hopefully, the analysis† of the conjunctions will resolve any questions about these identifications.

The 'star' called the "middle of the Crab" in Babylonian records is apparently the 'star' that Ptolemy calls the nebulous mass in the center of the Crab. Kugler and Epping and Strassmaier identify this as ε Cancri, a term that is no longer

†I am writing this before analyzing any observations involving the normal stars.

used in standard catalogues. The source in question is the one that Becvar [1964, p. 285] identifies as the galactic cluster NGC 2632. It is also known as Praesepe and as the Beehive.†

The star that Kugler designated as $\eta$ Tauri is the star that Epping and Strassmaier call $\eta$ i. d. Pleiaden. The Pleiades cover a part of the sky that covers hardly more than 1° in either direction. It is not clear to me from the Babylonian name that they distinguished the individual members of the Pleiades, and Kugler [1924, p. 550] in fact translates the Babylonian name using the term "star cloud" (sterncumulus). However, $\eta$ Tauri is brighter than any other of the Pleiades, and it is approximately in their center. Hence there should be little error in taking the reference object to be $\eta$ Tauri, and I shall do so. There is further confusion in identifying the stars that Ptolemy calls the Pleiades. See Section V.6.

Another problem concerns the star that the Babylonians called the "horn of the Goat". Ptolemy lists three stars in the horn of the Goat, and they are clearly $\alpha^1$ Capricorni, $\alpha^2$ Capricorni, and $\beta$ Capricorni. Apparently the Babylonians counted all three as one star, even though they are separated by about 2°. However, they distinguished $\eta$ Geminorum and $\mu$ Geminorum, which are also separated by only 2°. The difference may be that the two stars in Gemini are separated mostly in longitude, so that their discrimination is useful in studying motion. In contrast, the three stars in Capricorn differ by only about 12′ in longitude. Thus their discrimination may not have been useful to the Babylonians, at the level of accuracy that they needed.

I shall use the term '$\alpha\beta$ Capricorni' for the object that the Babylonians designated as the "horn of the Goat". For its position and motion, I shall use the means of the three individual stars, weighting each star by the factor $2.512^{-m}$; m denotes the

---

†It is interesting that the Babylonians explicitly recognized the diffuse nature of this source in an observation of about -567 May 25. See Table IV.10.

visual magnitude.

There is a problem that probably results from a modern slip of the pen and not from trouble in understanding the Babylonian records. One star is called the "northern beam of the Scales". Epping and Strassmaier [1892] give this star the modern designation β Librae, but they give coordinates for a different star. Judging from the coordinates, this star is ν Librae. Kugler [1907, p. 29] gives the same name and coordinates as Epping and Strassmaier.

ν Librae is about ½ degree north of the ecliptic and is well-placed to be a normal star, but its magnitude is 5.28. β Librae is much brighter, with a magnitude of 2.74, but it is about 8¼ degrees north of the ecliptic.

Ptolemy gives a star with a name like the Babylonian name, along with its magnitude and position, and this star is clearly β Librae. Further, Kugler [1924, p. 553] calls this normal star β Librae, and gives it the coordinates of β Librae, in contrast to what he did in 1907. These facts make it likely that the star is β Librae, but there is even more definite proof. On several occasions, such as SE 83 IX 7 and 110 VI 4 (Table IV.8), the Babylonians recorded conjunctions of Venus with α Librae and a few days later they recorded conjunctions with the star in question. They also give the latitude differences on each occasion. It is clear from the records that the star we want is considerably farther north than α Librae, and thus it cannot be ν Librae.

In summary, the "northern beam of the Scales" must be β Librae, and it is probable that Epping and Strassmaier accidentally read off the coordinates of the wrong star when they prepared their table.

The normal stars in Kugler's list are given in Table III.7, along with some of their important properties. With three exceptions, the designations in the first column are those that appear in current editions of the American Ephemeris and Nautical

TABLE III.7

PROPERTIES OF THE NORMAL STARS

| Name | Visual Magnitude | Right Ascension 1950.0, degrees | Declination 1950.0, degrees | Proper Motions, degrees per century | |
|---|---|---|---|---|---|
| | | | | In Right Ascension | In Declination |
| η Psc | 3.72 | 22.200 821 | 15.088 720 | 0.00079 | -0.00006 |
| β Ari | 2.72 | 27.968 083 | 20.564 470 | 0.00288 | -0.00300 |
| α Ari | 2.23 | 31.087 154 | 23.226 964 | 0.00575 | -0.00400 |
| η Tau | 2.96 | 56.126 767 | 23.952 103 | 0.00058 | -0.00122 |
| α Tau | 1.06 | 68.262 067 | 16.410 420 | 0.00188 | -0.00525 |
| β Tau | 1.78 | 80.782 125 | 28.567 150 | 0.00079 | -0.00486 |
| ζ Tau | 3.00 | 83.663 596 | 21.113 889 | 0.00004 | -0.00061 |
| η Gem | 3.7[a] | 92.964 617 | 22.523 156 | -0.00200 | -0.00036 |
| μ Gem | 3.19 | 94.983 733 | 22.541 106 | 0.00167 | -0.00311 |
| γ Gem | 1.93 | 98.705 817 | 16.443 722 | 0.00125 | -0.00122 |
| α Gem | 1.58 | 112.852 725 | 31.999 742 | -0.00558 | -0.00283 |
| β Gem | 1.21 | 115.564 654 | 28.148 642 | -0.01975 | -0.00142 |
| θ Cnc | 5.57 | 127.186 679 | 18.264 803 | -0.00167 | -0.00167 |
| ε Cnc[b] | 3.7 | 129.375 | 19.867 | -0.002 | -0.001 |
| γ Cnc | 4.73 | 130.098 758 | 21.649 672 | -0.00308 | -0.00119 |
| δ Cnc | 4.17 | 130.461 446 | 18.339 397 | -0.00050 | -0.00647 |
| ε Leo | 3.12 | 145.754 150 | 24.005 450 | -0.00138 | -0.00042 |
| α Leo | 1.34 | 151.427 696 | 12.212 372 | -0.00708 | 0.00008 |
| ρ Leo | 3.85 | 157.544 842 | 9.564 506 | -0.00025 | -0.00017 |
| θ Leo | 3.41 | 167.904 442 | 15.703 161 | -0.00175 | -0.00231 |

[a]This is an approximate average.

[b]This is a cluster and has no precise properties. The mean motions given for it are approximate averages for nearby stars.

TABLE III.7 (Continued)

| Name | Visual Magnitude | Right Ascension 1950.0, degrees | Declination 1950.0, degrees | Proper Motions, degrees per century | |
|---|---|---|---|---|---|
| | | | | In Right Ascension | In Declination |
| β Leo | 2.23 | 176.627 521 | 14.851 611 | -0.01429 | -0.00331 |
| β Vir | 3.80 | 177.022 446 | 2.046 558 | 0.02058 | -0.00764 |
| γ Vir | 2.91 | 189.781 050 | - 1.175 439 | -0.01579 | 0.00022 |
| α Vir | 1.21 | 200.638 754 | -10.900 933 | -0.00121 | -0.00092 |
| $\alpha^2$ Lib | 2.90 | 222.026 862 | -15.835 164 | -0.00312 | -0.00197 |
| β Lib | 2.74 | 228.578 050 | - 9.199 703 | -0.00275 | -0.00064 |
| δ Sco | 2.54 | 239.342 992 | -22.480 936 | -0.00029 | -0.00069 |
| $\beta^1$ Sco | 2.90 | 240.631 279 | -19.670 125 | -0.00017 | -0.00058 |
| α Sco | 1.22 | 246.584 192 | -26.322 764 | -0.00021 | -0.00064 |
| θ Oph | 3.37 | 259.733 988 | -24.951 445 | -0.00008 | -0.00058 |
| αβ Cap[c] | | 304.184 338 | -13.859 950 | 0.00125 | 0.00008 |
| ψ Cap | 4.26 | 310.784 588 | -25.451 983 | -0.00150 | -0.00433 |
| ω Cap | 4.24 | 312.210 250 | -27.107 556 | -0.00017 | 0.00000 |
| γ Cap | 3.80 | 324.330 717 | -16.889 181 | 0.00546 | -0.00061 |
| δ Cap | 2.98 | 326.070 800 | -16.355 117 | 0.00758 | -0.00817 |

[c]This denotes a combination of three stars that the Babylonians did not usually distinguish.

Almanac. The exceptions are αβ Capricorni and ε Cancri, which have already been discussed, and θ Cancri, which is not in the American Ephemeris and Nautical Almanac. The three-letter designations of the constellations are those given on page 495 of the Explanatory Supplement [1961].

The designations $\alpha^2$ Librae and $\beta^1$ Scorpii are used in Table III.7 in order to identify the stars accurately. Since their counterparts with other superscripts are not involved in the observations that will be used here, the simpler designations $\alpha$ Librae and $\beta$ Scorpii will be used in the rest of this work.

The visual magnitudes appear in the second column. They are taken from Becvar, except for the variable star $\eta$ Geminorum. Its magnitude is taken from Smithsonian Astrophysical Observatory [1966]; the magnitude that appears in that reference is described as the arithmetic mean of the least and greatest magnitudes. For $\varepsilon$ Cancri, the magnitude given is the integrated visual magnitude listed by Becvar.

The remaining columns give the right ascensions, declinations, and proper motions in the units stated. Except for the galactic cluster that I am designating as $\varepsilon$ Cancri, these quantities are taken from Smithsonian Astrophysical Observatory [1966]. No information is available about the motion of the cluster. For it, I have used approximate averages of the proper motions of nearby stars, and I have indicated this by a footnote in Table III.7.

The coordinates in Table III.7 are based upon the mean equator and equinox of 1950.0.

## 11. Other Reference Stars

Because they are fairly bright and fairly close to the ecliptic, the normal stars have been useful reference points for the moon and planets throughout history and not just in Babylonian times. However, a few other stars have been used to some extent.

The earliest known Babylonian astronomical diary (see Section IV.4) uses the Hyades as a reference for the moon (see Table IV.4). The close pair of stars $\theta^1$ Tauri and $\theta^2$ Tauri seems to lie in about the middle range of longitude of the Hyades. I shall use the mean of their properties as the representation of the Hyades, and I shall use $\theta$ Tauri as the corresponding designation. This is not precise, but the Hyades are not used in a precise way in the records. It is possible that the position assigned to the Hyades is dominated by $\alpha$ Tauri, which is by far the brightest star in the Hyades. However, $\alpha$ Tauri is such an outstanding star that we would expect it to have a separate designation.

The same text, which dates from -567, uses a star whose name is translated as "the northerly star at the rear of the Lion". Ptolemy does not list a star with this name, and the editor and translator give no clue to its identity that I can find. A sketch of the stars with relation to the imaginary Leo suggests that the star may be $\delta$ Leonis. I shall test this identification in the analysis of the data, but I shall assign at most a low weight to the observations that depend upon it.

In an observation of 140 July 30, Ptolemy [ca. 142, Chapter X.1] refers to "the star at the center of the knee of the eastern twin." He does not list a star with exactly this description in his table, but he does give the position of the star in the text of the observation, and it must be $\zeta$ Geminorum.

In an observation of 136 December 25 [Chapter X.1], Ptolemy refers to "the most northerly of the stars in the quadrilateral after the one that is in line with the groin of the Waterbearer". Again it is difficult to identify any of these stars by means of the names in the star table. Again he gives, in the text of the observation, the position of the reference star,† and it must be $\phi$ Aquarii.

In the same passage, Ptolemy refers to the star in the hind leg of Aries. This is the last star listed under the constellation Aries in Ptolemy's star table. Manitius identifies this star

†That is, the most northerly in the quadrilateral. Manitius, in his translation, identifies the quadrilateral as $\phi$, $\chi$, $\psi^1$, and $\psi^3$ Aquarii. It is doubtful to me that these are the ones that Ptolemy meant, because $\phi$, $\chi$, and $\psi^3$ are so nearly in a straight line. I.H. Schroader of the Applied Physics Laboratory studied Aquarius carefully at my request. He feels that $\eta$, $\phi$, $\chi$, and $\psi^1$ are the ones that are most likely to strike the eye as a quadrilateral.

as 38 Arietis. However, 38 Arietis is so far from the position given by Ptolemy that I do not see how this identification can be correct. Halma identifies it as μ Arietis. This star is farther from the tabular place than 38 Arietis; it is even on the wrong side of the ecliptic. However, μ Ceti is barely on the Cetus side of the line between Aries and Cetus, and its position agrees closely with the position given by Ptolemy. I conclude that μ Ceti is the star that Halma meant and that is in Ptolemy's table. Apparently the designation of this star has changed since Halma's time, or else Halma read the notation μ from a star chart without noticing that it is actually in Cetus.

In describing observations made on -271 October 12 and 16, Ptolemy [Chapter X.4] refers to the star behind the one at the end of the southern or left wing of Virgo. This star is η Virginis.

TABLE III.8

PROPERTIES OF REFERENCE STARS OTHER THAN BABYLONIAN NORMAL STARS

| Name | Visual Magnitude | Right Ascension 1950.0, degrees | Declination 1950.0, degrees | Proper Motions, degrees per century | |
|---|---|---|---|---|---|
| | | | | In Right Ascension | In Declination |
| μ Cet | 4.36 | 40.558 758 | 9.904 172 | 0.00792 | -0.00083 |
| θ Tau[a] | 3.06 | 66.439 796 | 15.807 215 | 0.00296 | -0.00069 |
| ζ Gem | 3.9[b] | 105.285 946 | 20.645 403 | -0.00025 | -0.00008 |
| α CMi | 0.48 | 114.171 021 | 5.354 669 | -0.01971 | -0.02858 |
| δ Leo | 2.58 | 167.862 855 | 20.797 964 | 0.00425 | -0.00375 |
| η Vir | 4.00 | 184.336 504 | -0.389 069 | -0.00179 | -0.00061 |
| β Cap | 3.25 | 304.550 832 | -14.940 714 | 0.00112 | 0.00008 |
| φ Aqr | 4.40 | 347.933 233 | - 6.318 867 | 0.00104 | -0.00533 |

[a]This is a doublet that the observers did not separate into components.

[b]This is an approximate average.

Other reference stars that will appear in this work pose no difficulties in identification and their usage does not need discussion. The reference stars other than the normal stars are listed in Table III.8, which has the same form as Table III.7.

Note added in proof: There is another possibility for the three stars in the horn of the goat. The three stars may be $\alpha^1$ and $\alpha^2$ Capricorni seen as one star, plus $\nu$ Capricorni and $\beta$ Capricorni. This choice would not change the properties of the object that is called $\alpha\beta$ Capricorni, the Babylonian "horn of the Goat", by a significant amount.

# CHAPTER IV

# DATA FROM BABYLONIA AND ASSYRIA

## 1. Venus Observations under Ammizaduga

Ammizaduga was a king of what is called the first dynasty of Babylon, who reigned for about 21 years sometime around -1800. Many cuneiform inscriptions are known from the reign of this monarch. Some of them are contracts which allow us to find the years in which an extra month was intercalated. For example, if a contract is dated on a particular day of second Ululu (see Section II.2), in a certain year of the reign of Ammizaduga, we know that that year must have been intercalary, and we even know which month was intercalated, unless the scribe made an error. If there are no errors, we find in this way [Langdon and Fotheringham, 1928] that years 4, 5, 10, 11, 14, 19, and 20 of Ammizaduga's reign were intercalary. Since seven is about the right number of intercalary years for a reign of 21 years, it is possible that these are all of them.

If we knew the first year of the reign, we could calculate the beginning and end of each month in it, subject to the uncertainty of a day or so that results from basing the month upon direct observation. If we knew the principle that governed the beginning of the year, we could then assign the correct names to the months. However, the place of the months with respect to the solar year, as actually practiced, depends upon the conscientiousness of those who assigned the intercalations, and we must allow an uncertainty of about 90 days in placing specific Babylonian months within the context of a solar year, at this stage in history (Section II.2). Further, we do not know the first year of the reign.

Since the beginning of a Babylonian month was based upon the direct observation of the new moon at sunset, weather permitting, a statement of the day of the month is a fairly close statement of the phase of the moon on that day.

Several inscriptions give dates of some first and last visibilities of Venus (Section III.5) in the reign of Ammizaduga. Since the dates imply the phase of the moon, the inscriptions specify relations between the phases of Venus and the moon in their orbital motions. If the relations were sufficiently precise, it would be possible to find a unique time when they were satisfied, and this would give us the exact dates of the reign of Ammizaduga.

However, there is considerable uncertainty in calculating the phase of Venus at first or last visibility, as we saw in Section III.5. Thus several solutions of the phase relations are possible. Langdon and Fotheringham find the following five possibilities for the first year of Ammizaduga within what they consider to be the possible historical period: -1976, -1920, -1856, -1808, and -1800.

In order to settle among these possibilities, Langdon and Fotheringham turn to the contracts. A number of contracts specify that certain harvesting operations are to be performed between certain Babylonian dates. Langdon and Fotheringham assume that the dates when crops become ready for harvest are rather well fixed in terms of the solar calendar, so they then deduce the relation between the months of the lunar calendar and those of a solar calendar. Since the solutions listed above, based upon the phases of Venus and the moon, require different relations of the lunar months to the seasons, they settle upon -1920 as the only choice that fulfills both the Venus observations and the crop contracts.

It should be noticed that the calculations relating to the phases of Venus and the moon involve the accelerations of the earth and moon. Langdon and Fotheringham used the accelerations found by Fotheringham [1920], which apply for an epoch around the year 0. It is fairly sure [Newton, 1970, pp. 275-278] that Fotheringham's values are somewhat in error for the epoch 0. Further, the accelerations must be considered as time-dependent, and the accelerations that apply at the epoch -2000 need not be close to those that apply at the epoch 0. In other words, the uncertainties in the analysis are greater than Langdon and Fotheringham assumed, and hence we may expect that there are more possible solutions than they found.

I do not believe that we can draw any valid chronological conclusions from the Venus observations made under Ammizaduga, and for this reason I do not present the data. Even if we could assume the accelerations, I doubt that we can take the agricultural seasons to be sufficiently constant to let us draw unique conclusions. Even if we could assume the constancy of the seasons, we could still not use the data without reasoning in a circle. That is, we should have to assume the accelerations in order to determine the dates of the observations. We cannot then use the dates in order to infer the accelerations; if we were to do so, we should simply find the accelerations that we assumed in the first place.

I am not alone in doubting the conclusions of Langdon and Fotheringham. It is clear that Smith [1958] does not accept them. He says that the question is still open, but that to accept their conclusions would be to force the "abandonment of a great deal of Babylonian historical writing as worthless". Neugebauer is even more explicit. He says [Neugebauer, 1957, p. 139]: "The chronological conclusions of this work† have been disproved by subsequent archaeological evidence." However, he gives no details about the nature of the evidence.

It does not seem possible to use the Venus observations unless we have archaeological evidence that can furnish the dates closely without recourse to astronomical calculations. Such archaeological evidence has yet to be found.

I originally wrote this section in the early part of the year 1972. Because I doubt that the Venus tablets can be useful for the purposes of this work, I made no attempt to study the later history of the subject. Since the first writing, however, a summary of the subject has appeared

---

†That is, of Langdon and Fotheringham, [1928].

[Weir, 1972], which the interested reader may wish to consult. It is indeed clear that there are solutions beyond those considered by Langdon and Fotheringham. Besides the five solutions listed above, the following years have been advocated by various people for the first year of Ammizaduga: -2259, -1709, -1701, -1645, and -1581. I have no doubt that diligence would uncover other solutions.

I feel that the reader should be cautioned against an uncritical acceptance of Weir's conclusions. He decides (p. 47) that the solution for the year -1645 is the correct one, although he admits that the texts show many discrepancies with this solution. In reaching this conclusion, he concludes that the observations were made at a latitude of 34°.0, while the latitude of Babylon is 32°.55.† He also concludes (p. 62) that the movements of the moon required by the Venus tablets disagree with calculations based upon modern lunar theory, and he ascribes this to a change in the eccentricity of Venus's orbit.

I do not believe that the reign of Ammizaduga can be dated unless it can be done by evidence exterior to the Venus tablets. Using the Venus tablets for chronology is rather like playing the 'identification game' [Newton, 1970, Section III.2] with solar eclipses. Unless an unusual amount of detail is correctly given, we can find a possible eclipse date at average intervals of around 40 years. Thus, when the historical period itself is uncertain, we can find an eclipse to fit any preconception about the period involved. Here we see that nine solutions to the Venus tablets for years from -1976 to -1581 (neglecting the anomalously early solution of -2259) have been found. This is

---

†This means that the observation point would be at least 160 kilometers from Babylon. On page 47, Weir writes: "Now that the theory of continental drift has been accepted, . . 4000 years ago 34°N may have been the latitude of Babylon . ." The greatest rate of crustal motion is about 5 centimeters per year, and the greatest total motion in 4000 years is thus about 20,000 centimeters, which is 0.2 kilometers, not 160.

an average of one solution per 50 years. It is clear that we can find a solution to fit any desired historical period.

Of course, if it ever becomes possible to date the period within a narrow range by data other than the Venus tablets, the tablets may be able to furnish a precise dating within this range.

## 2. A Classification of Astronomical Texts from Seleucid Times

In Seleucid times (roughly -300 to 0), mathematical astronomy reached a fairly high level of sophistication in Babylonia, although, so far as we know, it did not reach the level attained somewhat later in the Hellenistic culture by Hipparchus and his successors. The existing knowledge about mathematical astronomy in Babylonia is presented extensively by Neugebauer [1955]. There are two interesting and important differences between the Babylonian and the Hellenistic mathematical astronomy.

First, the goal of Hellenistic astronomy was to construct a conceptual model of the solar system; the mathematical description then followed from the properties of the model. We still use this basic approach. The Babylonian approach, on the other hand, can be called purely phenomenological. The Babylonians developed mathematical functions that describe the phenomena which interested them but, so far as we know, they never tried to invent a model or picture to correspond to the functions.

Second, the main mathematical task of Hellenistic and of modern astronomy is to construct a coordinate system in the heavens and then to calculate coordinates of the heavenly bodies as functions of time. In contrast, the Babylonians were mainly interested in 'horizon phenomena' such as times of first and last visibility. For the planets, they were also interested in the other synodic phenomena described in Chapter III. They developed methods of calculating the times and positions of these phenomena without necessarily calculating position as a continuous function of time. When they wanted to calculate the position of a planet, for example, at some other time, they did so by interpolation in these results [Neugebauer, 1955, vol. II, pp. 279-287].

We do not know much about the history of how the Babylonian methods arose. They were apparently well developed in early Seleucid times, say by -200 and probably earlier. We believe that the Babylonians did not adopt a regular scheme of intercalary months (see Section II.2 above) until about -380. It is plausible that a sophisticated mathematical astronomy would soon lead to a regular scheme of intercalation, although conservatism could have delayed the adoption of such a scheme until long after the astronomers had developed it. Thus we may tentatively assign the -4th century as the period of the most rapid development of Babylonian mathematical astronomy. It is likely that crude methods, of which we have no direct knowledge, existed at an earlier time.

Since the Babylonians had a well-developed mathematical astronomy, a Babylonian statement about an astronomical event could originate either from computation or from observation. Therefore we must proceed carefully before we accept a Babylonian 'datum' as an observation, as Neugebauer and Sachs have often emphasized. For example, Neugebauer [1948] has worked out the Babylonian methods of calculating equinox and solstice times and he concludes: " . . . *no*† solstitial or equinoctial date which is found *in* 'observation' texts can be evaluated as an observation and utilized for comparison with modern calculation." Sachs [1948] has established a classification of Babylonian astronomical texts from Seleucid times and finds: "A somewhat startling result is that all classes of Seleucid astronomical texts contain at least some predictions."

In his statement about equinoxes and solstices, Neugebauer puts the word "observation" in quotation marks. This refers to the fact that many Babylonian texts were labelled "observation texts" by enthusiastic early Assyriologists. This misleading

†The emphasis is mine.

terminology is still found in some places, but its usage fortunately seems to be declining.

There are four main classes of astronomical text from the Seleucid period that are entirely computational or predictive in nature.

1. Astronomical tables. These are calculations dealing with a particular body, such as the moon or a planet, or with a particular phenomenon. They usually cover a period of many years. They are analogous to, for example, the tables of Jupiter and Saturn given in Appendix III at the end of this work. They are extensively edited by Neugebauer [1955].

2. Procedure texts. These are texts that describe the methods by which the tables were calculated. They are also edited by Neugebauer [1955].

3. Almanacs. These are tables that give a variety of items or tables of astronomical information, but only for a single Babylonian year. They are analogous to the American Ephemeris and Nautical Almanac, for example. Sachs [1948] describes the contents of 14 almanacs, which are all the ones that he had been able to see at the time of writing. Most of them have been published, but a few remain unpublished.

4. NS almanacs. These are a particular type of almanac. In addition to the information found in ordinary almanacs, they give the times of conjunctions of the planets with the normal stars (Section III.10), but they give the time only to the nearest morning or evening. They also give the distance by which the planet misses the star in its passage. Some of the NS almanacs differ from the ordinary almanacs in the amount of lunar information that they give. Sachs [1948] describes 13 NS almanacs, some of which are still unpublished.

Two classes of text contain at least some observations.

1. Diaries. Sachs [1948] lists five texts in this class, aside from a few fragments that are too

small to be dated. Of these, one has not been published at all, two have been published only in the form of copies of the cuneiform inscriptions, and two have been published both in the form of copies and in German translation. I am able to use only the last two. Each diary gives a large amount of information, but usually for a period not longer than half a year. A separate section is devoted to each month. The amount and kind of information are not entirely standardized, but the following are typically given: weather conditions, particularly around the first and last of the month, conjunctions of the moon and planets with the normal stars, first and last visibilities, the height of the river, and the prices of barley, sesame, and wool.

The diaries also list some noteworthy political events. Thus they form a sort of annals that I did not know about when I made an earlier study of annals [Newton, 1972, Chapter III]. Their existence strengthens the finding that annals are not a Christian invention.

The meteorological, economic, and political statements in the diaries are probably the result of observation rather than prediction. It seems likely that many of the astronomical statements also come from observation, but it is certain that some do not. For example, a diary for the last half of the year SE 79 says approximately the following for month VIII [Epping and Strassmaier, 1892, p. 241]: "Evening of the 28th, first appearance of Venus; it happened for 17 degrees on the 26th." If the first appearance of Venus as the evening star was on the 28th, it cannot have been seen at all on the 26th. The average interval of invisibility before the first evening appearance (Table III.6) is more than a month, and no interval can be as short as 2 days.

A plausible interpretation of the passage is that the appearance was predicted for the 28th but that Venus was actually seen on the 26th. This interpretation will be discussed at more length in the next section.

Epping and Strassmaier [1892] translate some

passages to read that, because of clouds, the first day of some months in the diary was determined by calculation rather than observation; see Section II.2 above. If the translation is correct, these are additional items of calculation in the diaries. I am not qualified to judge the accuracy of the translation.

2. Goal-year texts. Sachs [1948] has proposed the name 'goal-year' for the class of texts that earlier German writers† have called *planetarische Hilfstafeln*. Sachs's term seems the more descriptive one to me, and I shall follow other recent writers in using it. In order to understand the goal-year texts, we must look at the subject of resonant periods of the planets.

TABLE IV.1

RESONANT PERIODS OF THE PLANETS

| Planet | Synodic Period | | N | N Synodic Periods | | ΔD |
|---|---|---|---|---|---|---|
| | in years | in months | | in years | in months | days |
| Mercury | 0.31726 | 3.9241 | 145 | 46.003 | 569.00 | 0.0 |
| Venus | 1.59872 | 19.7742 | 5 | 7.994 | 98.87 | - 3.8 |
| Mars | 2.13539 | 26.4121 | 22 | 46.979 | 581.07 | + 2.1 |
| | | | 37 | 79.009 | 977.25 | + 7.4 |
| Jupiter | 1.09211 | 13.5080 | 65 | 70.987 | 878.02 | + 0.6 |
| | | | 76 | 83.000 | 1026.61 | -11.5 |
| Saturn | 1.03518 | 12.8030 | 57 | 59.005 | 729.82 | - 5.3 |

Table IV.1 gives the necessary information about synodic and resonant periods. The first column in the tables gives the names of the planets that were known in antiquity. The second column gives the synodic period measured in years and the third column gives the same period in months.

---

†Schaumberger [1933a, p. 101] uses the term "goal-year" (*Zieljahr*) for the year but not for the text.

The fourth column gives an integer N; for Mars and Jupiter, it gives two integers apiece. The next two columns give the length of N synodic periods, both in years and in months. These are the resonant periods. The meaning of the last column will be explained in a moment.

We see that each value of N synodic periods in Table IV.1 is close to a time interval of $N_Y$ years, in which $N_Y$ is an integer. Then, after $N_Y$ years, the planet returns nearly to its original position with respect to the sun.† Further, since $N_Y$ is an integer, the sun returns, except for small effects, to the same position among the stars. Still further, if the position of the sun repeats with respect to the stars, and if the position of the planet repeats with respect to the sun, the position of the planet must repeat with respect to the stars.

In other words, $N_Y$ years is a period of near-resonance between the synodic planetary phenomena and the geocentric position of the planet. The planetary phenomena and conjunctions that occur in any year will repeat closely in the solar calendar $N_Y$ years later.

However, the Babylonian calendar was based upon lunar months regulated by direct observation of the moon. Hence, we must be concerned with N synodic periods measured in months as well as years. Let us consider Venus, for example, for which N = 5. Five synodic periods equal 98.87 months. The difference between this and an integral number of months is 0.13 months = 3.8 days. That is, when a phenomenon involving Venus recurs, it recurs 3.8 days earlier in the lunar month, on the average.‡

†The planet returns to exactly the same position except for small effects, after 1 synodic period, and it therefore does so after N synodic periods. Since N periods is nearly equal to $N_Y$ years, the planet almost repeats its synodic position in $N_Y$ years.

‡Since some Babylonian years contain 12 months and some contain 13, the number of months in $N_Y$ years is not constant. Thus, although the recurrence is on nearly the same day of the month, the name of the month itself may change.

This is the meaning of the value -3.8 in the column labelled ΔD in Table IV.1.

Now suppose that a person wishes to make a systematic set of observations of the planets, and that he wants a guide to supply approximate times and positions of the phenomena chosen for observation. He can prepare such a guide by extracting the desired information from the astronomical tables. Alternately, he can copy logs of observations from earlier years. Thus he can copy a set of observations of Mercury from 46 years before, a set for Venus from 8 years before, and so on through a set of observations of Saturn from 59 years before. If he brings these sets of observations together into one document, he has a handy guide to assist him in planning his observations.

The texts under discussion have just this character. That is, for each text there is a year Y with the property that the Mercury data are given for year Y - 46, that the Venus data are given for year Y - 8, and so on through the Saturn data, which are given for the year Y - 59. The number of years before Y for each planet is the number in the fifth column of Table IV.1.

In other words, the information given is just what a person would need for help in planning observations in the year Y. Hence Sachs [1948] calls Y the goal year and he uses the name "goal-year texts" for the set of texts under discussion. He describes seven goal-year texts from the Seleucid period.

The reader may have noticed that I have not named Mars and Jupiter explicitly in this discussion, and he may remember that there are two lines apiece for Mars and Jupiter in Table IV.1. The pairs of values have an interesting property that I do not know how to interpret. 37 periods for Mars and 76 periods for Jupiter are close to being integral numbers of years but not to integral numbers of months. The other values, namely 22 periods for Mars and 65 for Jupiter, have just the opposite property.

Neugebauer and Sachs [1967, p. 207] suggest an explanation. They say that the intervals of 37 periods for Mars and of 65 periods for Jupiter are used only for synodic phenomena, and that the intervals of 22 periods and 76 periods, respectively, are used only for conjunctions with the normal stars. The reader should note that the interval stated for the synodic phenomena, for example, is the one that is nearly an integral number of years for one planet and nearly an integral number of months for the other.

It seems to me that there are a few counter-examples to the statement of Neugebauer and Sachs. The goal-year text for the year 118 of the Seleucid era has conjunctions of Mars with the normal stars from the time that was 22 periods (47 years) earlier, in accordance with the statement. On the other hand, the goal-year text for SE 236 has conjunctions of Mars with the normal stars for the year SE 157, which is 37 periods earlier. In the same text, we find both phenomena and conjunctions of Jupiter given for a time 76 periods (83 years) earlier, and we find conjunctions given for both 65 periods (71 years) and for 76 periods earlier.

## 3. Calculation, Prediction, and Observation in Seleucid Times

We must now turn to the task of trying to decide which statements in the Babylonian astronomical texts are observations and which are not. In doing so, we must distinguish carefully between prediction and calculation. It is possible to make astronomical predictions without being able to calculate astronomical ephemerides. In fact, prediction based directly upon observation is quite simple. It is first necessary to study a set of observations in order to find a fundamental period. Once the period is found, we can predict a phenomenon simply by saying that it will occur P days, say, after the last time.

For example, once they started watching for it, it would be easy for a group of people to recognize the first evening visibility of Mercury and to note that successive appearances are separated by about 116 days, on the average. Then, if they saw it on day D, they had only to count forward 116 days and predict that it would appear again on day D + 116. In fact, people could learn to do this before they learned that Mercury as an evening star and as a morning star is the same body.

Let us return to the question of the purpose of the goal-year texts.

We start our inquiry by noting a puzzling feature of the goal-year texts. If they were prepared as guides in planning observations, and if they had been prepared with a writing medium like ink on paper, we should expect to find that the observer sometimes noted an observation directly in the guide. However, the goal-year texts were written in clay. Once a clay tablet is dry, as I understand the matter, it is impossible to add a remark. Thus it is surprising to find that the texts contain what seem to be current observations, that is, observations made in the goal year Y.

One remark that seems to be a current observation is found under month VIII of the year SE 110, in the goal-year text for year SE 118 [Epping and Strassmaier, 1891a, p. 93]: "Arahsamnu† 17, Venus in Libra in first morning visibility; for 12 degrees‡ it happened on the 14th." The reader should notice that this remark has the same form as the remark quoted in the preceding section from the diary for the year SE 79. Other remarks found in goal-year texts say that the phenomenon was not looked for; sometimes clouds are given as the reason and sometimes no reason is given.

Sachs [1948] says that remarks of this sort were completely misunderstood in the literature until Kugler [1924, pp. 539ff] correctly determined the meanings of many astronomical and meteorological terms.

A plausible interpretation of the passage about Venus is that the user of the text had a statement, from some source, that a first, morning visibility of Venus occurred on the date SE 110 VIII 17 and

†This is the name of month VIII.

‡These are degrees of time, so that 12 degrees equals 48 minutes. The editors note that the numeral 12 cannot be read clearly.

that he used this statement to alert him to begin watching a few days earlier in SE 118. He did so and succeeded in seeing Venus for 48 minutes on SE 118 VIII 14. The reader should remember that synodic phenomena of Venus recur about 4 days earlier in the Babylonian month after a resonant period, and in fact we find that the second day of the month, in remarks about Venus, is usually a few days earlier than the day given first. On the other hand, we see from Table IV.1 that Mercury phenomena should recur on the same day of the month; in remarks about Mercury, we find that the two days are often the same. These facts strengthen the suggested interpretation.

The reader should also remember that the goal-year texts contain statements about both synodic phenomena and conjunctions with the normal stars. Remarks that seem to be current observations never occur in connection with the conjunctions. They occur only with the synodic phenomena, and then only with Mercury, Venus, and Saturn. I do not know whether the point is significant or not, but the two planets for which current remarks never occur are the ones that have two intervals apiece in Table IV.1, namely Mars and Jupiter.

Thus we have a puzzling circumstance. We seem forced to conclude that the goal-year texts, in the form that we now have them, were prepared after the fact and not as guides to assist in observation. In this case, why did the writer copy the large body of information relating to the earlier years? We might think that he wanted to confront prediction with observation, which would certainly be a reasonable motive. If so, why did he do so for only planetary phenomena and for only three of the five planets? Why did he not do so for the conjunctions? I do not know of a plausible explanation for the existence of the goal-year texts in the form that we have them.

Now we must turn to the source of the predicted information in the goal-year texts, that is, the information from one resonant period before the goal year. There are two obvious possible sources. The information could have come from the almanacs or other types of table. Alternatively, it could have come from diaries or other collections of observations. If the latter is the case, the data in the goal-year texts are genuine observations even though they are being used as predictions. I tend to think that the data are observations, for two reasons.

First, the information about conjunctions of planets with normal stars includes the angular distance of the planet above or below the star. This is equivalent to a statement about the latitude of the planet. However, the astronomical tables and procedure texts edited by Neugebauer [1955] make no reference to latitudes, and hence the information could not have been deduced from them.

Unfortunately, we cannot state categorically that the Seleucid Babylonians were unable to calculate latitudes of the planets. Neugebauer and Sachs [1967] have described a fragmentary text designated as B.M. 37266, which cannot be dated accurately. They say this about the text: "This small fragment is of unique interest, as it contains the only text known to us which proves the existence of a theory of planetary latitudes, however primitive in structure." In modern terms, the theory gives latitude as a simple periodic function of longitude. The latitude is constant at 1° 20′ between longitudes 165° and 225°, and it is constant at -1° 20′ between 345° and 45°. From 45° to 165°, and from 225° to 345°, the latitude varies linearly with longitude.

This function fits the facts fairly well for Jupiter and poorly for Saturn, although the text refers specifically to Saturn. It is moderately accurate to make latitude a function only of geocentric longitude for planets as distant as Jupiter and Saturn, but this simple scheme is poor for the nearby planets. Thus it may be that the Babylonians had a theory of latitudes for Jupiter and Saturn but not for the more difficult nearby planets; in fact we do not have a satisfactory theory of latitude for the nearby planets until we come to Kepler. It may also be that the Babylonians had a theory of latitudes for the nearby planets and that we have not yet discovered the texts containing it.

If the Babylonians did not have a theory of latitudes for the near planets, then the data in the goal-year texts for at least the near planets probably came from observation.

Second, if the maker of a goal-year text used calculated tables for the planetary information, why did he use tables for an earlier year instead of for the goal year? Why should he go to the trouble of using different but carefully chosen years for each planet, and in fact of using two years apiece for two of the planets? I cannot think of any plausible reason why he should follow this course. Thus it sounds plausible that the maker took the information from diaries or other observational sources, choosing a year for each planet in which the data would be nearly the same as for the goal year.

In sum, there are two reasons for thinking that the goal-year texts are based upon observation, but both reasons are somewhat weak. The first reason is based upon the absence rather than the presence of knowledge, and an argument with this basis is always weak. The second reason is based upon trying to imagine how or why a person with a greatly different background and culture would go about doing something. It is always hard to say how even a person with a similar background would do something, and we can put little confidence in such an argument when the background differs widely. This is especially so when we cannot think of a good reason for the activity in the first place, as is the case with the goal-year texts.

In sum, I shall accept the planetary information in the goal-year texts as being observational, but with low weight. I shall not try to assign weights or reliability here, however. I shall first see what weight will result from the size of the residuals when I have completed the quantitative analysis of the information. It may turn out that the weight on this basis alone will be so low that we do not need to worry further about the matter.

Although I have directed most of the discussion toward the goal-year texts, the information in both the goal-year texts and the diaries is fundamentally the same. In both types of text, we have a large body of planetary conjunctions and we have a large body of synodic phenomena for the planets. For Mercury, Venus, and Saturn, but apparently never for Mars or Jupiter, we also seem to have comparisons of the synodic phenomena in one year with those in a suitably chosen earlier year. The main difference is that most of the events in a diary refer to a single Babylonian year while those in a goal-year text refer to a variety of years which are, however, related by a simple scheme.

There does not seem to be a great risk in accepting the planetary information, at least tentatively. If the information came from calculated ephemerides, there are three possibilities. If the parameters used in the ephemerides were adjusted frequently, then they are essentially smoothed observations and it is legitimate to use them. If they were not adjusted frequently, the errors are probably so large that they will show up readily when we come to the detailed analysis. The risk is that the information came from a time when the errors had grown to the point of affecting our conclusions seriously but not to the point that they can be detected. This does not seem likely.

If the predictive information came from observation, the main risk is that of mistaking the year. An error in the year will almost surely show up in the detailed analysis.

There are clearly many problems connected with the Babylonian texts. Many of them are probably the result of having such a small sample of material for study. Sachs [1948] gave a complete inventory of all the astronomical texts from Seleucid times, other than ephemerides, known to him, whether published or unpublished. His inventory contained only 41 items. Luckily, the situation should improve in the reasonably near future. A large body of material was discovered shortly after 1948, and seven years later Sachs [1955, p. vi] was able to write: "The complete bibliography of the non-mathematical astronomical texts of the Hellenistic period† could recently be presented on a page

†That is, of the Hellenistic or Seleucid period in Babylonia-Assyria.

and a half. The present volume† contains more than 900 copies of the same type of texts from the identical period."

It is my understanding that Sachs and his colleagues are editing this new material and preparing it for publication. The material should be of great value in studies of the accelerations and in many other studies when it becomes available.

## 4. A Summary of the Data from Late Babylonian Times

Twenty-seven of the items in Sachs's inventory are almanacs or NS almanacs, leaving only fourteen items with some observational material. Seven of these are goal-year texts that have been published.

TABLE IV.2

A SUMMARY OF LATE BABYLONIAN SOURCES

| Reference | NW 1915 | K 1907 | K 1907 | ES 1892 | ES 1892 | Sch 1933b | Sch 1933b | Sch 1933b | ES 1891a | K 1907 | ES 1891b | ES 1890 |
|---|---|---|---|---|---|---|---|---|---|---|---|---|
| Diary or Goal Year, Seleucid Era | -256 | -211 | -67 | 38 | 79 | 81 | 90 | 91 | 118 | 140 | 225 | 236 |
| Moon | -256 | -211 | -67 | 38 | 79 | 63[b] | 72 | 73 | 100 | | 207 | |
| Mercury | -256 | -211 | -67 | | 33, 79 | 35 | | 45, 91 | 72, 118 | 94, 140 | 179, 225 | 190, 236 |
| Venus | -256 | -211 | | 30, 38 | 71, 79 | 73 | | 83 | 110, 118 | 132, 140 | 217 | 228, 236 |
| Mars Phenomena | | -211[a] | -67 | | | | | 12, 44? | | 61 | | |
| Mars Conjunctions | -256 | | -67 | 38 | | | | 44 | 71 | 93 | | 157 |
| Jupiter Phenomena | -256 | -211[a] | -67 | | 79 | 10 | 19 | | | 69, 140 | 154 | 153 |
| Jupiter Conjunctions | -256 | | | 38 | | | | | | 57 | 142 | 153, 165 |
| Saturn | | -211[a] | | 38 | 79 | | | 32 | 59 | 81, 140 | | 177, 236 |
| Double Conjunctions | | -211 | -67 | 38 | 79 | 35[c] | | | | | 179[d] | |

[a] Also contains entries well into the next year.

[b] There are eclipses and visibility conditions in this and later years, but no conjunctions.

[c] Mercury is the inner body, Jupiter is the outer.

[d] Mercury is the inner body, Venus is the outer.

† That is, the volume from which this quotation is taken.

The remaining seven are diaries. Five of the diaries are either unpublished or too fragmentary to be useful. In this section, I shall summarize the data from the two diaries that are useful and available, as well as from the seven goal-year texts.

There are also some texts from late but pre-Seleucid times. Kugler [1907, pp. 80-84] gives a text for the years 18 to 21 of Artaxerxes II (SE -75 to -72). This text relates only to Jupiter, and it does not contain any remarks about the weather or any other remarks that suggest current observation. Thus it is not safe to use this text. There are also three texts that resemble diaries but, unlike the diaries from Seleucid times, they do not seem to have any predictive features. I shall refer to these texts as diaries, although this may be technically incorrect.

The information contained in the twelve texts is summarized in Table IV.2. The heading of each column gives the reference in which the text is published. The abbreviations in these headings are:

ES = Epping and Strassmaier,

K = Kugler,

NW = Neugebauer and Weidner,

Sch = Schaumberger.

One point should be noted about the publication of Schaumberger. The transcriptions of the texts are in the reference Schaumberger [1933b]. Parts of each text are translated in the same reference, but for the translations of the other parts we must go to Schaumberger [1933a].

All the references except Kugler [1907] are journal articles and need no page identification beyond that in the list of references at the end of this work. In Kugler's work, the relevant pages are as follows: The diary for SE -211 is on pages 70-71, the diary for SE -67 is on pages 76-79, and the goal-year text for SE 140 is on pages 84-87.

The first row of the table gives the diary year or the goal year, as the case may be, with respect to the Seleucid era (see Section II.2). The texts with year SE 79 and earlier in this row are all diaries and the other texts are all goal-year texts. This chronological separation of the types of text is an accident, however. There are known diaries for years even later than SE 200, but they have not been published [Sachs, 1948].

The remaining rows give the years of the Seleucid era for which data of the indicated kinds are found in the various texts. If there is no entry in a particular position, this means that the text in question has no usable data of the type in question. Thus, for example, the goal-year text for the year SE 90 has data for the moon from year SE 72, and it has data for Jupiter from the year SE 19, but it has no other usable data. At one time it had more data, and fragments of Jupiter observations from the year SE 7 can be made out, but not enough of the text has been preserved for them to be useful.

For each of Mars and Jupiter, I have two rows in Table IV.2, one for synodic phenomena and one for conjunctions, in accordance with the suggestion of Neugebauer and Sachs [1969]. As I have mentioned, their suggestion does not seem to be rigorously followed.

In several places in the table, two years occur. These are the instances in which two items of information are given about the same phenomenon, in accordance with the discussion in the preceding section. Under Mars phenomena in the goal-year text for year SE 91, I have the entry "44?" because the text contains a conjunction with the Pleiades under the date SE 44 VI 9. Part of this entry is broken away, but what remains seems to contain a reference to a turning point. Hence I have put an entry under phenomena for the year 44, but with a question mark since the entry cannot be read with certainty.

In many places, we have observations from dates lying a few days outside the year indicated. I have not tried to indicate these places in Table IV.2.

In the diary for the year SE -211, however, the data for Jupiter and Saturn continue well into the year SE -210, and the data for Mars continue into the year SE -209.

I have listed double conjunctions separately from conjunctions with the normal stars in Table IV.2. It is not surprising to see double conjunctions in the diaries. However, in two instances we have double conjunctions in goal-year texts. This is surprising, because the repetition intervals for double conjunctions are not the simple ones listed in Table IV.1.

## 5. Double Conjunctions

Most of the records of double conjunctions are accompanied by remarks that are usually transcribed as num du and su du. Epping and Strassmaier [1892, pp. 222-223] translate these as "earlier" and "later", respectively. That is, they suggest, the observer could not always be observing when the conjunction occurred. Therefore he observed at some time near the time of conjunction, and noted whether the conjunction would occur later, or whether it had occurred earlier. These meanings may be correct, but I suspect that the matter is more complicated, for two reasons.

First, the need to use "earlier" and "later" is present just as much for conjunctions of the moon and planets with the normal stars as for their conjunctions with each other. However, so far as I noticed, num du and su du are used only with double conjunctions and never for conjunctions with the normal stars.

Second, if the interpretation of Epping and Strassmaier is correct, num du and su du are redundant. For example, there is a record dated SE 38 VII 26 which says that the moon was a stated distance east of Jupiter at the morning, and this statement is accompanied by num du. If the moon is east of Jupiter, the conjunction has already occurred, and it is not necessary to add "earlier".

When a person must make voluminous records, all of a similar nature, I think that he tends to develop short, concise, patterns, and indeed the Babylonian entries have this nature. Economy of style is even more important with a writing medium like clay than with pen and paper. _num du_ and _su du_ are fairly difficult characters to make. Each requires something like ten strokes of the wedge-forming instrument, and it seems unlikely to me that the scribe deliberately chose to exert this effort when he knew that it was redundant.

Neither of these arguments is conclusive. However, they create some doubt in my mind that the records have been read correctly and, for safety, I shall not use the double conjunctions. In order to preserve them, I have tabulated them and put the table in Appendix V.

If the reader will consult Table A.V-1, which lists the double conjunctions, he will see that there are seven double conjunctions from the year SE -211. The comments _num du_ and _su du_ are absent from the conjunctions for this year, but they are present in both surviving observations from the year SE -67. This suggests that use of the comments was introduced sometime between these two years, which are approximately -522 and -378 of the common era.

The comments are occasionally absent in later years. In some cases, this may be because the text is broken or because the scribe forgot to include them. There is also a tendency for the difference in longitude of the two bodies to be small when the comments are missing. This fact is consistent with the interpretations of Epping and Strassmaier. If the longitude difference was small, the scribe might have felt justified in saying that the conjunction was "now" rather than earlier or later, and he could imply this by omitting a comment.

## 6. Conversion of Dates; Lunar Observations

As we saw in Section II.2, the conversion of dates from the Babylonian calendar to our calendar must usually proceed on a month-to-month basis. Further, because the beginning of the month was often and perhaps always determined by direct observation of the lunar crescent, the conversion may be uncertain by a day, for either of two reasons.

First, clouds or other weather conditions may have prevented seeing the moon on what would otherwise have been the first evening of the new month. Second, we must usually establish the evening of first visibility by calculation, and there will be months when the condition was near a critical case. That is, there is sometimes an uncertainty of a day in calculating when the moon would have been visible even in ideal weather.

The combination of the two reasons means that sometimes the conversion of dates may be wrong by two days.

In order to convert a Babylonian date with certainty, then, we must have some means, other than calculation of first lunar visibility, for establishing a correspondence of calendars. Unfortunately, a correspondence established in this way does not extend outside of the Babylonian month for which it was done, in the absence of other information.

In some instances, we could perhaps establish a correspondence by means of some historical event. Such instances are probably rare, and I have made no attempt to search for them. For this work we must establish correspondences by astronomical means.

The moon moves about 13° per day, with respect either to the stars or to the sun. A relatively crude measurement of lunar position or of elongation should be accurate to about a degree. On the basis of lunar observations that have already been analyzed [Newton, 1970], I think that we can calculate the position of the moon with an uncertainty that does not exceed 30′ for late Babylonian times.† Thus there should be an uncertainty of at most 1°.5 in comparing a modern calculation with an ancient observation, and we should have no difficulty in establishing the exact day of a lunar observation, in the absence of textual errors. We can then

---

†This is greater than an estimate that I made in Section II.2. There I made what I believe to be a realistic estimate. Here I have deliberately made what I believe to be a pessimistic estimate.

extend the results of this calculation to an entire Babylonian month.

Lunar observations thus help us in establishing the dates of solar and planetary observations from Babylonian astronomy. Hence lunar data are valuable for the purposes of this work, even though its purposes do not include a study of the lunar motion. There are two additional reasons for analyzing the lunar data that I shall take up in the next section. Because of their usefulness, I have listed lunar observations from late Babylonian times in Tables IV.3 and IV.4 in this chapter, instead of in an appendix. Because the tables are rather lengthy, I have placed them at the end of this chapter in order not to break up the main text. I shall undertake a detailed discussion of the tables in a moment, but there is another item that it is useful to take up first.

We see by reference to Table IV.2 that we do not often have lunar observations in the same years as planetary observations, much less in the same months. We shall have to determine the dates of most planetary observations without the aid of lunar data, and we must do so by analysis of the planetary observations.

Mercury and Venus move about 1° per day, on the average. If the observations have errors much less than 30′, something that we have no right to expect for such early observations, we can establish the day unambiguously for Mercury and Venus data. If we have several observations within the same month, we can correlate them and probably succeed when we could not have succeeded with individual observations. In some instances, we can correlate observations in different months, when the records tell us the numbers of days in each month.

Mars moves only about 30′ per day, on the average. Thus it may be that we can use Mars observations only from months in which we also have other data.

The reader may object that we are reasoning in a circle if we date the planetary observations by planetary calculations, and if we then use the dates in order to study the accelerations. This objection is well taken, but we can probably avoid the need for circular reasoning. Before I analyze the Babylonian planetary data, I shall first analyze the data from later times, which can be dated without ambiguity. I shall then analyze the planetary data that can be dated with the aid of lunar data. If these analyses agree, we can then calculate a position of a planet in late Babylonian times with an uncertainty that does not affect the dating of an observation of Mercury or Venus. Then, if the data have enough precision, we can improve our knowledge of the accelerations further by using the observations that can be dated only by internal evidence.

We can now turn to a detailed discussion of Table IV.3, which gives the synodic lunar phenomena, and of Table IV.4, which gives conjunctions of the moon with the normal stars.

The first column in both tables gives the Babylonian date, referred to the Seleucid era in the manner described in Section II.2. By using the years listed for the moon in Table IV.2, the reader can tell the published source from which each observation is taken. The second column in each table is called the tentative Julian date, that is, the date in the Julian calendar that tentatively corresponds to the Babylonian date. When the editor of a given source gave the Julian date of an observation, I have used the date that he gave. There are a few exceptions that I have noted, in which the editor made an accidental error in converting the date. When the editor did not give a Julian date, or when he made an obvious slip, I have supplied the Julian date with the aid of Parker and Dubberstein [1956]. Two further points should be noted with regard to the dates.

First, most of the editors converted the dates with the use of the old astronomical day that begins at noon. I have converted their dates to the present usage of beginning the day at midnight.

Second, even after this conversion is made, the date given by the editors often disagreesby a day with the date inferred from Parker and Dubberstein. I do not know whether the editors found their dates by analysis of the data, in the manner that I have just described, or whether they used older tables of visibility than the ones that Parker and Dubberstein used. I have not noted these places of disagreement.

The third column in Table IV.3 gives a brief description of the synodic observations. Some are eclipses, and some are the times of moonrise and moonset.

The first lunar eclipse is that of -522 July 16, which I used earlier [Newton, 1970, Chapters V and X]. It is interesting that this is the only Babylonian eclipse used by Ptolemy [ca. 142, Section V.14] of which the cuneiform record has been found, and the apparent disagreement of the two sources has caused a lively controversy. The records usually do not say what phase of an eclipse corresponds to the time given, although I have specifically noted this fact in the table only for the eclipse of SE -211 X 14 (= -521 January 10). It is tolerably certain that the time given for the eclipse of -522 July 16 is for the beginning, but we cannot take this as a general rule.

Some of the eclipses were undoubtedly calculated, as we know because some of the records say that the eclipse did not happen. Some of the eclipses are listed by Oppolzer [1887] and some are not, such as the one tabulated for SE 63 VII 15. In this connection, it is interesting to observe from Table IV.2 that the lunar part of a goal-year text deals only with synodic phenomena, and that it is for a year 18 years before the goal-year. This clearly implies that the Babylonians in Seleucid times had some knowledge of the cycle of 223 months, the one that is usually miscalled the "saros", as well as the cycle of 235 months that appears in the calendar. The interested reader should consult Neugebauer [1957] for further discussion.

Some of the times are given by means of the double-hour and some are given by means of the degree of time, which equals four minutes. I have converted some times to modern units and left others as they were.

In several of the sources, we find tables of the time of moonrise and moonset for days near the beginning and end of the month. I suspect that these tables were calculated, and thus I have not included the information from them in Table IV.3.

In other sources, we find isolated statements about the visibility of the moon. For example, on SE -67 VIII 1, we find that the moon was visible for 14° 30′ (that is, for 58 minutes) in the evening. This could mean that the observer first saw the moon at some time, and that it set 58 minutes later. More plausibly, it means that the time of moonset was 58 minutes after sunset. If the first interpretation is the correct one, the time interval depends upon when the observer happened to catch sight of the moon. If we calculate the time intervals between moonset and sunset and compare them with the stated intervals, we expect much more scatter if the first interpretation is correct than if the second one is. Thus we should be able to decide upon the correct interpretation.

If the second interpretation is correct, the time intervals could be either calculated or observed. We must ultimately decide between calculation and observation by textual analysis of the records. Here, our main interest in the intervals is that of using them for dating purposes. Even if they are calculated, they are valid for this purpose unless they are based upon grossly wrong ephemerides.

The comments in the last column of Table IV.3 are self-explanatory, I believe, except for the comment "cloudy". In some instances, we have "cloudy" accompanying a quantitative statement about the moon. This could imply that the statement was calculated; this is an ever-present danger, as I have said more than once before. It could also imply that there were clouds but that they did not interfere with the observation.

The texts use several different terms to indicate unfavorable conditions for visibility. I have not tried to distinguish these terms from each other, and I have used the comment "cloudy" to indicate any of them.

In Table IV.4, the third column, for the time of day, uses only three symbols. These correspond to the fact that the original texts used only three terms for the time of day. The translators usually translated them as "evening", which I have denoted by E, as "morning", which I have denoted by M, and as "beginning of the night", which I have denoted by BN. I shall take up the astronomical significance of these terms in the next section.

In the column for the position of the moon relative to the body that it had conjoined, I believe that the directions east, west, over, and under are self-explanatory.† Occasionally, however, we find the terms that I have rendered as "back" and "forward"; it is interesting that these seem to occur only when there is a simultaneous statement of over or under. These meanings are ambiguous. If the recorder focussed his attention upon the motion of the moon relative to the stars, then "back", for example, means that the moon had not reached the reference object. Hence it was west of it. If the recorder focussed his attention upon the apparent diurnal rotation, however, "back" means that the moon was east of the reference object. Luckily, back and forward occur only when the east-west separation was small, and the ambiguity should not interfere seriously with using the observation.

Two units are used in specifying the positions in Table IV.4. Translators have used a variety of terms in translating each of them, and therefore I have not tried to translate either one. I have merely denoted them by the abbreviations "am." and "u.". The "u.", which is the abbreviation of "uban", is the unit that Neugebauer [1955, v. 1, p. 39] calls the "finger". It equals one-twelfth of a degree. A unit meaning one-twelfth, and called "finger" or "digit" or some such term, has been used astronomically in many languages, and it is

†The east-west direction is to be taken as parallel to the ecliptic.

still used in some modern literature for measuring the magnitude of an eclipse.

The abbreviation "am." stands for "ammat".† It can mean either 2° or $2\frac{1}{2}$° in the Babylonian literature [Neugebauer, 1955, v. 1, p. 39], and unfortunately both meanings were used contemporaneously. We shall have to try to determine, separately for each text, what meaning a particular scribe attached to the term. The latitudes given for the planetary conjunctions, which will be discussed in the next section, give us what are probably our best data for doing this.

Certain items of information were missing from the original text, either because they have been lost or because the scribe failed to supply them. In many cases, the translator of a text supplied the missing information by calculation. I have tried to note these cases, but I have probably missed some. The moon moves so fast that there should be little uncertainty in supplying the information, but it is desirable to note when it is present in the text and when it has been restored.

Occasionally the text does not give the time of day. I have suggested the time of day in these cases, on the basis of the day of the month.

## 7. The Planetary Data

The observations of Mercury, Venus, and Mars from late Babylonian times are summarized in Tables IV.5 through IV.10. There are two tables for each planet. The first gives the synodic phenomena and the second gives the conjunctions with the normal stars. Since I do not plan to use observations of Jupiter and Saturn in this work, I have relegated them to Appendix V. I have put Tables IV.5 through IV.10 at the end of the chapter.

Much of the discussion of the lunar tables in

---

†I do not find standard forms in English for the terms that I have called "uban" and "ammat". I shall arbitrarily use these forms in the singular, with ubans and ammats in the plural.

the preceding section applies to the planetary tables also.

When I have suggested morning or evening in the planetary tables, I have done so on the basis of the implied longitude. For example, in the first observation in Table IV.6, Mercury was in conjunction with $\alpha$ Leonis and the time was near the summer solstice. Thus it is almost certain that Mercury was an evening star.† When the reader uses the stars in this way, he should remember that precession has carried the equinoxes and solstices a considerable distance from where they were in late Babylonian times.

The translators have supplied some items of information in the planetary observations by calculation, just as they did in the lunar observations. The accuracy of doing this for the planets is questionable. I have given the calculated items in the tables in some cases, with a note that they were calculated. In other cases, I have omitted the information entirely. There should be no risk in restoring the name of a normal star by calculation.

As I mentioned in the preceding section, the ammat has two possible values. Most of the planetary conjunctions give the distance in ammats by which the planet passed over or under the normal star. Our ignorance of the accelerations hardly affects the calculated value of a planetary latitude. Therefore we should be able to calculate the interval between planet and star with high accuracy and therefore determine the size of the ammat used in a particular text. We can then use this size when the ammat is used to give longitude, on the assumption that at least it has the same size within a given text.

In a few places, the translators have noticed that there seems to be a scribal error. For example, the scribe may have written "over" when the translators think that he meant to write "under". I have noted some of these places when the matter

---

†Because $\alpha$ Leonis was east of the summer solstice.

seemed obvious. Otherwise, I have tried to give only the original statement of the scribe, on the assumption that the translators were being guided by calculation and that it would be useful to check their conclusions.

For reasons that I discussed in Section III.5, simple statements of the dates of first and last visibility of the planets are probably not useful for this study, although I shall calculate the positions of the sun and planets on those dates as a matter of interest. However, we also have statements about the settings of the planets, such as the one on SE 91 IV 7 in Table IV.5. This statement, which resembles a number of statements about the moon (Section IV.6), says that Mercury was visible for 15° in the evening. This probably means that Mercury set 15° (60 minutes) after the sun, but it should be possible to settle the meaning unambiguously by an analysis of the lunar data. If the lunar data show that these statements mean what we expect, they should be useful in a study of the accelerations.

In analyzing these data, I shall assume that the Babylonians used the upper limb of the sun in defining sunrise and sunset. The exact definition is not important for the lunar data, if they were consistent in dealing with the sun and moon, but there is a difference of roughly 16′ in position for a planet, and this is potentially important. However, we can keep the assumption from introducing a bias into the results if we give equal weight to risings and settings.

Only three statements are ever made about the time of a lunar conjunction with a normal star, or of a double conjunction.† It was in the morning, which I have indicated by M in the tables, or it was in the evening, indicated by E in the tables, or it was at the "beginning of the night", which is indicated by BN. There is never an "end of the night", which we might expect as the symmetrical point to BN.

---

†See Appendix V.

BN probably means some kind of twilight. Three kinds of twilight are defined in Chapter 13 of the Explanatory Supplement [1961]. Astronomical twilight ends at night (or begins in the morning) when the center of the sun is 18° below the horizon. At this condition, there is more illumination from the stars than from the sun, and photographic astronomy is possible. This condition does not concern us.

Civil twilight ends when the center of the sun is 6° below the horizon. At this condition [Explanatory Supplement, p. 400], "the brightest stars are visible and the ...horizon is clearly defined." At the end of nautical twilight, which is defined to occur when the sun is 12° below the horizon, the horizon "is in general not visible and it is too dark for the observation of altitudes with reference to the horizon."

Most of the normal stars do not come into the category of "the brightest stars", as we can see by inspecting their magnitudes in Table III.7. Thus we do not expect to see them at the end of civil twilight. However, a normal star used to specify a lunar conjunction is close to the bright moon, and the moon gives the eye something to focus upon. Thus a person might see one sooner than he would if he were merely looking around the sky for visible stars.†

It is plausible that BN means the condition almost midway between civil and nautical twilight, that is, that it means about the time when the sun is 9° below the horizon. At the latitude of Babylon, the interval between civil and nautical twilight is about 30 minutes, as we can see from the tables of sunrise, sunset, and twilight in the American Ephemeris and Nautical Almanac. Hopefully, then, defining BN as the midpoint is not in error by more than 15 minutes.

However, BN does not concern us explicitly in this work, since it is used only with lunar conjunctions or double conjunctions, so far as I

†I thank I.H. Schroader of the Applied Physics Laboratory for pointing this out to me.

noticed. With all of the planetary conjunctions, and with most of the lunar conjunctions, the only time designations used are "morning" and "evening". Morning and evening do not sound as definite as "beginning of the night", although it may be that "beginning of the night" is merely a synonym for evening.

With observations of the moon or the outer planets, evening could possibly mean any time between sunset and midnight. With observations of Mercury and Venus, the possibilities are more limited. Roughly speaking, Mercury cannot be more than an hour and a half from the sun; this may account for the fact that there are few conjunctions for Mercury but many synodic phenomena. Venus can be about 3 hours from the sun, and we find that there are more conjunctions of Venus than there are synodic phenomena.†

Tentatively, I shall assume that morning and evening mean the same twilight condition as the beginning of the night, namely the time when the sun is 9° below the horizon. For simplicity, I shall take this as the time that is 45 minutes from sunrise or sunset, whichever is appropriate. The exact time varies but slightly with the seasons at the latitude of Babylon, and the variation is probably small compared with other errors.

With observations of Mercury, it does not seem possible that the assumption is in error by a serious amount, because of the small range of time each night when Mercury is visible. With observations of Venus, the error in the assumption could conceivably reach somewhat more than two hours. However, the situation is not as bad as this fact first suggests. When Venus is far enough from the sun to make such errors possible, it is near maximum elongation. This automatically means that the time of the observations will have low weight, because the position hardly depends upon the acceleration of

---

†Because the synodic period of Mercury is much less than that of Venus, there are more synodic phenomena of Mercury to be recorded.

Venus.† Further, for both Mercury and Venus, the errors should be opposite for morning and evening observations, and thus they will tend to cancel.

There is a way to test the assumed meanings of morning and evening. We can probably relate a calculated position of the moon to a Babylonian observation of the moon with a bias of no more than about 30′ and a scatter that is probably about 1°. These correspond to about 1 hour and 2 hours of time, respectively. Thus, if we calculate the times of all the lunar conjunctions, we should be able to find whether the times tend to cluster or whether they occur all through the night. When we have done this, we can be guided in our interpretation of the times in the planetary data.

I believe that Tables IV.5 through IV.10 can be readily understood with the aid of the discussions of the other tables.

## 8. A Synopsis of the Situation

We have discovered that there are questions about the dates, the times of day, and the size of a metric unit in the Babylonian observations. There is also uncertainty about whether some of the statements of position are predictions or observations. The reader may feel that so many uncertainties will destroy the reliability of any conclusions that we may draw from the Babylonian records.

If a reader has developed this feeling of malaise from the earlier sections of this chapter, I can sympathize with him. However, I do not feel that the situation is at all hopeless. The thing that should make possible the unscrambling of the situation is the fact that we have data for the moon and the planets, and the fact that their angular velocities range over more than an order of magnitude. Further, we have observations that give both latitude and longitude. Thus we can use the lunar data to set useful limits to the uncertainties in time, because of the rapid motion of the moon. We

---

†However, observations near maximum elongation depend usefully upon the acceleration of the sun.

can use the latitudes to set useful limits to the uncertainties in the metric unit. In this way we can use the longitudes of the planets with relatively small uncertainties of interpretation, with one large exception.

The exception arises from the circumstance that we do not have lunar data to go with the planetary data to give us the date in most instances. If the errors in the planetary positions are considerably smaller than the daily motions, we can find the dates from the planetary positions alone, particularly if there are several planetary observations in the same Babylonian month. However, there will always be some question about finding the date from the observation for isolated observations, and we shall have to use the standard tables for the Babylonian months in many cases.

The proposed program of analysis rests upon the tacit assumption that the Babylonians were consistent in their usage of terms, and that our problem is to discover what that usage was. It is possible that their usage does not meet standards of consistency that will be useful to us. If this is the situation, the proposed analysis will uncover this fact. Then we shall not be able to use the Babylonian data, but we shall have been saved from the possibility of misusing it.

It was not one of the original goals of this work to study the lunar acceleration. However, we shall have to analyze much lunar data in order to use the planetary data, and from that analysis it will be a simple step to infer a lunar acceleration. Therefore I shall adopt a secondary goal of inferring the lunar acceleration from the Babylonian observations, provided that the observations turn out to be usable.

## 9. Some Alleged Observations from the Year -424

Kugler [1914, pp. 233-242] published and discussed a text that he dates to the months IV through IX of the year 40 of Artaxerxes I (the year SE -113). The text makes statements about the number of days in each month, and about the dates of full moons, of first and last visibilities of the planets and of Sirius, of the summer solstice and the autumnal equinox, and of one lunar and one solar eclipse. It has been claimed that the statement about the lunar eclipse is a highly accurate and reliable observation of the lunar eclipse of -424 October 9, and much weight has been put upon this alleged observation. However, I do not see any reason to assume that the text contains an observation of a lunar eclipse at all.

I did not originate this conclusion. Kugler reached it in the cited reference sixty years ago, and he emphasized in at least two places that the text contains no observations, but only calculations. I have simply verified his arguments, with the aid of more accurate eclipse calculations than he had available.

There are several reasons for assuming that the text is a collection of calculated results, perhaps prepared as a guide for planning observations. One reason is that the text contains no remarks about weather or about other interferences with observation. The most cogent reason is probably one that is connected with the two eclipses. The text mentions a lunar eclipse beginning at 10° (40 minutes) after sunset on day VI 15, twelve days after the autumnal equinox. In the next line, the text mentions a solar eclipse on day VI 28, but with no time of day given. If we assume that one eclipse was observed, we must assume that the other was observed also, because the text furnishes no basis for discriminating between the statements in this regard.

If the lunar eclipse is that of -424 October 9, the solar eclipse is the penumbral eclipse of -424 October 23.† However, it is not possible for both statements to represent observations, whether made at Babylon or anywhere else. This conclusion is independent of any assumption about the accelerations, as Kugler pointed out. It depends only upon the time interval between the eclipses and not

---

†The statements of date are consistent. The Babylonian day began at sunset, whereas I begin the day at midnight when I use the Julian calendar.

upon the actual time calculated for either.

For example, suppose we adjust the accelerations so that the lunar eclipse began at sunset in Babylon rather than 40 minutes later. This would mean that the observation, if it has any accuracy at all, was made at some place well to the east of Babylon. Under this assumption, the solar eclipse was not visible at any place east of Athens.†

Therefore at least one of the eclipse statements was calculated. If one was calculated, the only safe assumption is that the other one was calculated also. The reader may assume, if he wishes, that the lunar eclipse was observed. If he does so, he must recognize that this is a matter of opinion with no supporting evidence, and hence he is not entitled to give high weight to his assumption.

While Kugler concluded from his calculations that the solar eclipse was invisible if the lunar eclipse began at the time stated, his calculations were in fact not accurate enough to support his conclusion. He made two errors in his calculations. First, he thought that the time stated for the lunar eclipse is within 4 minutes of the time calculated from Oppolzer's Canon [Oppolzer, 1887]. However, he made a mistake in copying from the Canon, and the actual discrepancy is 24 minutes.

Second, he did not recognize the errors that are present in Oppolzer's times. Here I am not referring to errors that may result from not knowing the secular accelerations. I am referring instead to errors that arise because Oppolzer had to simplify the theory of the moon considerably in order to carry out his mammoth program of eclipse computations. According to Oppolzer, syzygy for the solar eclipse occurred 3 minutes earlier, mean time, than did syzygy for the lunar eclipse. According to my calculations, it should be 55 minutes earlier.

---

†I do not know the extreme of position at which the eclipse became visible. I did not pursue the calculations for points farther west than Athens.

From the combination of these two errors, Kugler thought that the syzygy for the solar eclipse was so far after sunset that he did not need to calculate local circumstances. In fact, syzygy was $1^h\ 12^m$ closer to sunset than Kugler thought, and it is not obvious from the time above that the eclipse was invisible in Babylon.

It is interesting that Kugler used the word "alleged" in referring to these records, just as I have done, but for a different reason. He referred, in the caption of his relevant section, to "alleged records from the middle of the second millenium before Christ". The text in its present form has no indication of year, and students before Kugler had assigned it to a year near -1500. Kugler assigned the year SE -113 from an analysis of the astronomical information in the text. Now it is likely that the errors in the Babylonian ephemerides, the ones upon which the information is based, are greater than Kugler thought. It is certain that the errors in the modern ephemerides that he used are greater than he thought. I have not attempted to discover whether the errors are enough to bring the year into question. Because of this uncertainty, the conclusions of this section must be summarized with caution.

In summary, if the year is SE -113, it is certain that at least one of the eclipse records in the text is calculated, and there are various reasons to conclude that all records in the text are calculated. If the year is not SE -113, it is mathematically possible that both eclipse records represent observations. In either case, however, there is no basis for the assumption that the text records an observation of the lunar eclipse of -424 October 9. If the year is right, we are dealing with calculations and not observations. If the year is wrong, we are not dealing with the date -424 October 9 at all.

Therefore I shall not use the subject text in this work.

## TABLE IV.3
## SYNODIC LUNAR PHENOMENA FROM LATE BABYLONIAN SOURCES

| Babylonian Date (Seleucid Era) | Tentative Julian Date | Event | Comments |
|---|---|---|---|
| -256 I 1 | -567 Apr 22 | Visible for 14° | The number is hard to read. |
| -256 I 14 | -567 May 6 | Set 4° after sunrise | |
| -256 I 26 | -567 May 18 | Still visible for 23° | |
| -256 III 1 | -567 Jun 20 | Visible for 20° | |
| -256 III 15 | -567 Jul 5 | Set 7½° after sunrise | There is the start of a remark about an eclipse, of which the rest is lost. |
| -256 XI 1 | -566 Feb 12 | Visible for 14½° | |
| -256 XI ? | -566 Feb | Set 17° after sunrise | The day of the month was at the end of a line and was broken off. |
| -256 XII 1 | -566 Mar 14 | Visible for 25° | |
| -256 XII 12 | -566 Mar 26 | Set 1½° after sunrise | |
| -211 IV 14 | -522 Jul 16 | Lunar eclipse, 3 1/3 hr. after sunset, northern half eclipsed | Used in earlier work |
| -211 X 14 | -521 Jan 10 | Total lunar eclipse, 5 hr. before sunrise | Phase at this time is unknown. |
| -211 | -522/-521 | Times moon was visible near first and middle of each month, for months I through $XII_2$ | Chronology is suspicious. |
| - 67 VIII 1 | -378 Oct 27 | Visible 14° 30′ in evening | |
| - 67 VIII 14 | -378 Nov 9 | Visible 9° 30′ before sunset | |
| - 67 VIII 14 | -378 Nov 10 | Set 4° 30′ after sunrise | |
| 38 VII 1 | -273 Oct 6 | Visible 15° in evening | |
| 38 VII 14 | -273 Oct 19 | Rose 11° before sunset | Cloudy |
| 38 VII 15 | -273 Oct 21 | Visible 5° after sunrise | |
| 38 VII 27 | -273 Nov 2 | Visible 24° in morning | Cloudy |
| 38 VIII 16 | -273 Nov 19 | Rose 5° 30′ after sunset | Cloudy |
| 38 XII 27 | -272 Mar 29 | Visible 15° in morning | |
| 63 I 13 | -248 Apr 18 | Predicted lunar eclipse, did not happen | Statement of time ambiguous |
| 63 I 28 | -248 May 4 | Solar eclipse 6 hr. after sunrise | Eclipse was visible, but this may be predicted. |
| 63 VII[a] 15 | -248 Oct 13 | Lunar eclipse predicted for 2 hr. after sunset; did not happen | Not listed by <u>Oppolzer</u> |
| 63 VII 29 | -248 Oct 27 | Solar eclipse predicted for 4 hr. 28 min. after sunset | |
| 72 VIII 14 | -239 Nov 3 | Lunar eclipse began 12 min. before sunrise | |

[a]The text has month IV; this is clearly an error.

TABLE IV.3 (Continued)

| Babylonian Date (Seleucid Era) | Tentative Julian Date | Event | Comments |
|---|---|---|---|
| 73 | -238/-237 | Times moon was visible near first and middle of each month, for months VI through $XII_2$ | May be calculated |
| 79 VII 1 | -232 Oct 3 | Visible 15° 30′ in evening | |
| 79 VII 13 | -232 Oct 15 | Rose 5° 40′ before sunset | |
| 79 VII 14 | -232 Oct 16 | Visible 3° 50′ after sunset | Cloudy |
| 79 VIII 14 | -232 Nov 14 | Rose 2° 30′ before sunset | |
| 79 VIII 14 | -232 Nov 15 | Set 6° 40′ after sunrise | Cloudy |
| 79 VIII 28 | -232 Nov 29 | Visible 11° in morning | Cloudy |
| 79 VIII 29 | -232 Nov 30 | Solar eclipse predicted for 44° after sunrise; did not happen. | Not last day of month |
| 79 IX 13 | -232 Dec 13 | Rose 12° 30′ before sunset | Cloudy |
| 79 IX 13 | -232 Dec 14 | Set 1° 40′ before sunrise; lunar eclipse predicted for 4 hr. 56 min. after sunrise | Cloudy |
| 79 IX 14 | -232 Dec 15 | Set 14° 30′ after sunrise | Cloudy |
| 100 | -211/-210 | Times moon was visible near first and middle of certain months | May be calculated |
| 100 I 13 | -211 Apr 30 | Lunar eclipse about 20° before sunrise, eclipsed on south side | |
| 100 I 28 | -211 May 15 | Solar eclipse predicted for 35° before sunset; did not happen. | |
| 100 VII 15 | -211 Oct 24 | Lunar eclipse; time is hard to read. | Not clear whether observed or not |
| 207 V 13 | -104 Aug 13 | Lunar eclipse 14° before sunrise | Not clear whether observed or predicted |
| 207 X 14 | -103 Jan 8 | Lunar eclipse predicted for 25° before sunset | Not listed by Oppolzer |

## TABLE IV.4

## LUNAR CONJUNCTIONS FROM LATE BABYLONIAN SOURCES

| Babylonian Date (Seleucid Era) | Tentative Julian Date | Time of Day | Body Conjoined | Position Relative to Body Conjoined | Comments |
|---|---|---|---|---|---|
| -256 I 1 | -567 Apr 22 | E | Hyades | East? | |
| -256 I 8 | -567 Apr 29 | BN | β Vir | West, 1 am. | |
| -256 II 1 | -567 May 22 | E | β Gem | Under, 4 am. | |
| -256 III 5 | -567 Jun 24 | BN | δ Leo? | East, 1 am. | Star is not certain. |
| -256 III 8 | -567 Jun 27 | E | β Lib | Under, 2½ am. | |
| -256 III 10 | -567 Jun 29 | E | α Sco | Over, 3½ am. | |
| -256 XI ? | -566 Feb/Mar | | α Leo | Over, 1 am. | Day is missing but it can be calculated. α Leo was inside a halo. |
| -256 XII 2 | -566 Mar 15 | E | Pleiades | Under, 4 am. | Assume η Tau. |
| - 67 VIII 1 | -378 Oct 27 | E | β Ari | Under, 2 am., 10 u. | See Table IV.10. |
| - 67 VIII 10 | -378 Nov 5 | BN | η Psc | West | Distance missing |
| - 67 VIII 14 | -378 Nov 9 | E | α Tau | East, 2/3 am. | |
| - 67 VIII 17 | -378 Nov 13 | M | β Gem | Under, 1 am. | |
| - 67 VIII 19[a] | -378 Nov 15 | M | ε Leo | West, 2/3 am. | Cloudy |
| - 67 VIII 20 | -378 Nov 16 | M | α Leo | East, 1½ am. | Cloudy |
| - 67 VIII 24 | -378 Nov 20 | M | α Vir | East, 1 2/3 am. | |
| 38 VII 3[a] | -273 Oct 8 | E | θ Oph | East, 1 am. | |
| 38 VII 8[a] | -273 Oct 13 | | δ Cap | East, 1½ am. | Probably evening |
| 38 VII 13 | -273 Oct 18 | BN | η Psc | Under 2½ am., back ½ am. | Cloudy |
| 38 VII 15 | -273 Oct 21 | M | η Tau | Under, 3 am. | |
| 38 VII 16 | -273 Oct 22 | M | α Tau | Over, 2 u. | Cloudy |
| 38 VII 18 | -273 Oct 24 | M | γ Gem | West, 1 am. | Cloudy |
| 38 VII 19 | -273 Oct 25 | M | α Gem | Under 6 am., forward ½ am. | |
| 38 VII 22 | -273 Oct 28 | M | α Leo | Under 2½ am., back 8 u. | |
| 38 VII 24 | -273 Oct 30 | M | β Vir | East, 1 am., 8 u. | Cloudy |
| 38 VII 25 | -273 Oct 31 | M | γ Vir | East, 1½ am. | Cloudy |
| 38 VII 26 | -273 Nov 1 | M | α Vir | East, 2 am. | See Table A.V-3. |
| 38 VII 27 | -273 Nov 2 | M | α Lib | West, 1 am. | |
| 38 VIII 6 | -273 Nov 9 | BN | δ Cap | Over, 2 am. | Cloudy |
| 38 VIII 17 | -273 Nov 21 | M | α Gem | West, 2 am. | |
| 79 VII 6 | -232 Oct 8 | BN | αβ Cap | West, 1½ am. | |
| 79 VII 12 | -232 Oct 14 | BN | η Psc | West, 1½ am. | |
| 79 VII 14 | -232 Oct 17 | M | η Tau | East, 1½ am. | |
| 79 VII 15 | -232 Oct 18 | M | α Tau | East, 2 2/3 am. | |
| 79 VII 16 | -232 Oct 19 | M | η Gem | West, 1 am., 8 u. | |

[a]The translator supplied the day of the month by calculation.

TABLE IV.4 (Continued)

| Babylonian Date (Seleucid Era) | Tentative Julian Date | Time of Day | Body Conjoined | Position Relative to Body Conjoined | Comments |
|---|---|---|---|---|---|
| 79 VII 19 | -232 Oct 22 | M | δ Cnc | East, 1 am., 4 u. | |
| 79 VII 22 | -232 Oct 25 | M | θ Leo | East, 2 am. | |
| 79 VII 23 | -232 Oct 26 | M | β Vir | East, 2 2/3 am. | |
| 79 VIII 4 | -232 Nov 4 | BN | αβ Cap | West, 3½ am. | |
| 79 VIII 14 | -232 Nov 14 | BN | α Tau | East 2½ am. | |
| 79 VIII 15 | -232 Nov 16 | M | γ Gem | Over, 2 am., 8 u. | Cloudy |
| 79 VIII 23 | -232 Nov 24 | M | α Vir | East, 1 am. | |
| 79 VIII 26 | -232 Nov 27 | M | β Sco | West, 1½ am. | Cloudy |
| 79 VIII 27 | -232 Nov 28 | M | α Sco | East, 1 am. | |
| 79 IX 4 | -232 Dec 4 | BN | δ Cap | East, 3½ am. | |
| 79 IX 8[a] | -232 Dec 8 | | η Psc | Under 4 am., forward ½ am. | Probably evening |
| 79 IX 9 | -232 Dec 9 | BN | α Ari | East, 1½ am. | |
| 79 IX 10 | -232 Dec 10 | BN | η Tau | Under 3 am., back ½ am. | |
| 79 IX 11 | -232 Dec 11 | BN | α Tau | East, 1 am. | Cloudy |
| 79 IX 13 | -232 Dec 13 | BN | γ Gem | Over 3 am., 8 u., back 2/3 am. | Cloudy |
| 79 IX 14[a] | -232 Dec 15 | M | β Gem | East, 2½ am. | |
| 79 IX 16 | -232 Dec 17 | M | α Leo | Over 2/3 am., back ½ am. | See Table A.V-5. |
| 79 IX 18 | -232 Dec 19 | M | β Vir | Over, 1½ am. | |
| 79 IX 19 | -232 Dec 20 | M | γ Vir | West, 1 am. | |
| 79 IX 20 | -232 Dec 21 | M | α Vir | West, 1 am. | |
| 79 IX 22 | -232 Dec 23 | M | α Lib | Over, 1½ am. | |
| 79 IX 23 | -232 Dec 24 | M | β Lib | East, 2 1/3 am. | Cloudy |
| 79 IX 24 | -232 Dec 25 | M | β Sco | East, 2 am. | Cloudy |
| 79 IX 25 | -232 Dec 26 | M | θ Oph | West, 1 am. | |
| 79 X 5[a] | -231 Jan 4 | BN | η Psc | West, 1 2/3 am. | |
| 79 X 8 | -231 Jan 7 | BN | α Tau | West, 1 am., 4 u. | |
| 79 X 9 | -231 Jan 8 | BN | ζ Tau | West, 2 am., 8 u. | |
| 79 X 13 | -231 Jan 13 | M | α Leo | West, 3 2/3 am. | Cloudy |
| 79 XI 6 | -231 Feb 3 | BN | α Tau | West, 3 am. | |
| 79 XI 9 | -231 Feb 6 | BN | α Gem | East, 5 u. | |
| 79 XI 17 | -231 Feb 15 | M | α Vir | Over, 1½ am. | |
| 79 XII 2 | -231 Mar 1 | | α Ari | East, 3 am. | Presumably in the evening |
| 79 XII 4 | -231 Mar 3 | | α Tau | East, 1½ am. | Presumably in the evening |
| 79 XII 6 | -231 Mar 5 | | γ Gem | East, 1 am. 8 u. | Presumably in the evening |
| 79 XII 16 | -231 Mar 16 | M | β Sco | West, 3½ am. | |

[a]The translator supplied the day of the month by calculation.

TABLE IV.5

SYNODIC PHENOMENA INVOLVING MERCURY,
FROM LATE BABYLONIAN SOURCES

| Babylonian Date (Seleucid Era) | Tentative Julian Date | Event | Comments |
|---|---|---|---|
| -256 II 10 | -567 May 31 | Probably first evening visibility | |
| -211 IV 1 | -522 Jul 3 | Moon visible 3 am. west of Mercury | Not clear if this is first visibility of Mercury or not; see Table A.V-1. |
| - 67 VIII 22 | -378 Nov 18 | First, morning, in Sagittarius | |
| 33 VIII 9 | -278 Nov 9 | First, evening, in Sagittarius | Cloudy in 79 |
| 33 VIII 14 | -278 Nov 14 | Last, evening, in Sagittarius | Not seen from first to last in 79 |
| 33 X 8 | -277 Jan 7 | Last, morning, in Capricorn | Cloudy in 79 |
| 33 XI 24 | -277 Feb 20/21 | Last, in Pisces, time of day not clear, visible $\frac{1}{2}$° | Cloudy in 79; this entire passage is unclear. |
| 35 I 22 | -276 May 6 | Evening, at beginning of Gemini | Text broken, event is missing. |
| 45 I 18 | -266 Apr 12 | Last, evening, in Taurus | Probably not visible in 91 |
| 45 IV ? | -266 | First, evening, in Cancer | Day of month missing; see 91 IV 7. |
| 45 V 20 | -266 Aug 10 | Last, evening, at beginning of Virgo | Not watched for in 91 |
| 45 VII[a] 5 | -266 Sep 24 | Last, morning, in Virgo | |
| 45 VIII 29 | -266 Nov 16 | First, evening, in Sagittarius | |
| 45 IX 14 | -266 Dec 1 | Last, evening, in Sagittarius | |
| 45 $XII_2^a$ 10 | -265 Mar 25 | Last, evening, in Aries | |
| 72 I 2 | -239 Mar 28 | Last, evening, in Aries | Not watched for in 118 |
| 72 II 14 | -239 May 10 | First, morning, in Taurus | See 118 II 14. |
| 72 II 27 | -239 May 23 | Last, morning, in Taurus | See 118 II 27. |
| 72 III 28 | -239 Jun 21 | First, evening, in Cancer | See 118 III 25. |
| 72 V 3 | -239 Jul 25 | Last, evening, in Leo | |
| 72 VI 1 | -239 Aug 23 | First, morning, in Leo | See 118 V 29. |
| 72 VI 24 | -239 Sep 15 | Last, morning, in Virgo | Not watched for in 118 |
| 72 VIII 20 | -239 Nov 8 | First, evening, in Sagittarius | Not watched for in 118 |

[a]The translator supplied the month by calculation.

TABLE IV.5 (Continued)

| Babylonian Date (Seleucid Era) | Tentative Julian Date | Event | Comments |
|---|---|---|---|
| 72 VIII 29 | -239 Nov 17 | Last, evening, in Sagittarius | Not watched for in 118 |
| 72 IX 17 | -239 Dec 5 | First, morning, in Sagittarius | Not watched for in 118 |
| 72 X 27 | -238 Jan 14 | Last, morning, in Capricorn | Not watched for in 118 |
| 72 XII 1 | -238 Feb 16 | First, evening, in Pisces | Not watched for in 118 |
| 72 $XII_2$ 1 | -238 Mar 17 | Last, evening, in Aries | |
| 72 $XII_2$ 23 | -238 Apr 9 | Last, morning, in Taurus[b] | See 118 $XII_2$ 23. |
| 91 IV 7 | -220 Jun 30 | Visible for 15° in evening | See 45 IV. |
| 94 I 16 | -217 Apr 9 | Last, morning, in Aries | Not watched for in 140 |
| 94 II 19 | -217 May 11 | Event missing, probably first evening visibility | See 140 II 17. |
| 94 IV 26 | -217 Jul 18 | First, morning, in Cancer | See 140 IV 25. |
| 94 $VI_2$ 9 | -217 Sep 26 | First, evening, in Taurus | Mercury cannot be in Taurus in September. |
| 94 VII 12 | -217 Oct 30 | First, morning, in Libra, 1½ am. east of α Librae | See 140 VIII 11, also Table IV.6. |
| 94 VIII 21 | -217 Dec 7 | Last, morning, in Sagittarius | Not seen in 140 |
| 94 IX 27 | -216 Jan 11 | First, evening, in Aquarius | Probably not seen in 140 |
| 94 X 18 | -216 Jan 31 | Visible for 15° in evening | Year may be 140. See 140 XI 18. |
| 94 XI 7 | -216 Feb 19 | First, morning, visible 2 2/3 am. | See 140 XII 7. |
| 94 XII 10 | -216 Mar 23 | Last, morning, in Pisces | |
| 118 II 14 | -193 May 12 | First, morning, in Taurus | See 72 II 14. |
| 118 II 27 | -193 May 25 | Last, morning, in Taurus | See 72 II 27. |
| 118 III 25 | -193 Jun 21 | First, evening, visible for 15° | See 72 III 28. |
| 118 V 29 | -193 Aug 23 | First, morning, visible for 17° 40′ | See 72 VI 1. |
| 118 $XII_2$ 23 | -192 Apr 10 | Last, morning, in Taurus[b] | See 72 $XII_2$ 23. |
| 140 II 17 | -171 May 10 | Visible for 16° in the evening | See 94 II 19. |
| 140 IV 25 | -171 Jul 16 | Visible for 15° in the morning | See 94 IV 26. |
| 140 VIII 11 | -171 Oct 30 | Visible for 17° in the morning | See 94 VII 12. |
| 140 XI 18 | -170 Feb 1 | Last, evening, in Aquarius | See 94 X 18. |
| 140 XII 7 | -170 Feb 21 | First, morning, visible for 14° 30′ | See 94 XI 7. |

[b]There is probably a serious error in the text.

TABLE IV.5 (Continued)

| Babylonian Date (Seleucid Era) | Tentative Julian Date | Event | Comments |
|---|---|---|---|
| 179 II 8 | -132 May 20 | First, evening, in Gemini | See 225 II 6. |
| 190 II 6 | -121 May 18 | First, morning, in Taurus[b] | See 236 II 6. |
| 190 II 20 | -121 Jun 1 | Last, morning, in Taurus | See 236 II 20. |
| 190 III 20 | -121 Jun 29 | First, evening, in Cancer | See 236 III 17. |
| 190 IV 23[c] | -121 Aug 1 | Last, evening, in Leo | Not seen in 236 |
| 190 V 24 | -121 Sep 2 | First, morning, in Leo | See 236 V 21. |
| 190 VIII 8 | -121 Nov 12 | First, evening, in Sagittarius | Cloudy, not seen in 236 |
| 190 VIII 23 | -121 Nov 27 | Last, evening, in Sagittarius | Cloudy, not seen in 236 |
| 190 IX 7 | -121 Dec 12 | First, morning, in Sagittarius | Cloudy, not seen in 236 |
| 190 X 13 | -120 Jan 16 | First, morning, in Capricorn[d] | Not seen in 236 |
| 190 XI 24 | -120 Feb 24 | First, evening, in Pisces | See 236 XI 23. |
| 190 XII 18 | -120 Mar 19 | Last, evening, in Aries | Cloudy, not seen in 236 |
| 225 II 6 | - 86 May 20 | Visible for 15° 30′ in the evening | See 179 II 8. |
| 236 II 6 | - 75 May 18 | First, morning, in Taurus[b] | See 190 II 6. |
| 236 II 20 | - 75 Jun 1 | Last, morning, in Taurus | See 190 II 20. |
| 236 III 17 | - 75 Jun 27 | Visible for 15° in the evening | See 190 III 20. |
| 236 V 21 | - 75 Aug 31 | Visible for 16° 40′ in the morning | See 190 V 24. |
| 236 XI 23 | - 74 Feb 24 | Visible for 15° in the evening | See 190 XI 24. |

[b]There is probably a serious error in the text.

[c]The published source gives the date as 190 IV 23 in the transcription of the text but as 190 IV 1 in the translation. Both dates are followed by question marks.

[d]This is probably an error for last, morning.

TABLE IV.6

CONJUNCTIONS OF MERCURY FROM LATE BABYLONIAN SOURCES

| Babylonian Date (Seleucid Era) | Tentative Julian Date | Time of Day | Body Conjoined | Position Relative to Body Conjoined | Comments |
|---|---|---|---|---|---|
| -256 III 1 | -567 Jun 20 | | α Leo | Mercury and Mars were both 4 am. west of α Leo; Mercury was south of Mars. | Probably evening |
| 45 XII[a] 23 | -265 Mar 8 | E | β Ari[b] | Under | Distance missing |
| 72 IV 9 | -239 Jul 2 | E | α Leo | Over, 6 u. | |
| 79 IX 9 | -232 Dec 10 | M | θ Oph | Over | Distance missing |
| 94 III 1 | -217 May 23 | E | α Gem | Under, 3½ am. | |
| 94 III 4 | -217 May 26 | | β Gem | Under, 2½ am. | Probably evening |
| 94 III 17 | -217 Jun 8 | E | δ Cnc | Over, 2 u. | |
| 94 VII 12 | -217 Oct 30 | M | α Lib | East, 1½ am. | See Table IV.5. |
| 94 VIII 1 | -217 Nov 17 | M | β Sco | Over | Distance missing |
| 179 II 14 | -132 May 26 | E | α Gem | Under, 3 am. | |
| 179 II 17 | -132 May 29 | E | β Gem | Under, 2 am. | |
| 190 III 26 | -121 Jul 5 | E | α Leo | Over, 8 u. | |
| 190 XII 6 | -120 Mar 7 | E | β Ari | Under, 2½ am. | |
| 190 XII 10 | -120 Mar 11 | E | α Ari | Under, 3½ am. | |

[a]The translator supplied the month by calculation.

[b]The translator supplied the star by calculation.

TABLE IV.7

SYNODIC PHENOMENA INVOLVING VENUS, FROM LATE BABYLONIAN SOURCES

| Babylonian Date (Seleucid Era) | Tentative Julian Date | Event | Comments |
|---|---|---|---|
| -256 II 1 | -567 May 22/23 | Maximum elongation | |
| -211 III 10 | -522 Jun 12 | Last, evening, at the beginning of Leo | |
| -211 III 27 | -522 Jun 30 | First, morning, at the beginning of Cancer | |
| -211 VI 24 | -522 Sep 23/24 | Greatest elongation | |
| -211 XII 7 | -521 Mar 4 | Last, morning, at the beginning of Pisces | |
| -210 I[a] 23 | -521 Apr 7 | First, evening, at the beginning of Taurus | |
| 30 VIII 1 | -281 Nov 4 | Last, evening, at the end of Scorpio | Cloudy, not watched for in 38 |
| 30 VIII 13 | -281 Nov 17 | First, morning, in Scorpio | See 38 VIII 11. |
| 38 VIII 11 | -273 Nov 15 | Visible for 10° in the morning | See 30 VIII 13. |
| 71 VIII 28 | -240 Nov 27 | First, evening | See 79 VIII 26. |
| 73 I 11 | -238 Apr 26 | First, evening, in Taurus | |
| 73 X 10 | -237 Jan 15 | Last, evening, in Aquarius | |
| 79 VIII 26 | -232 Nov 26 | Visible for 17° in the evening | See 71 VIII 28. |
| 83 V 11 | -228 Aug 1 | Visible for 8° | Time of day is missing; year may be 91. |
| 83 V 13 | -228 Aug 3 | Event missing, probably a last visibility | |
| 110 VII 15 | -201 Oct 5 | Visible for 9° in the evening | Year may be 118. |
| 110 VII 19 | -201 Oct 9 | Last, evening, in Libra | Year may be 118. |
| 110 VIII 17 | -201 Nov 6 | First, morning, in Libra | See 118 VIII 14. |
| 118 VIII 14 | -193 Nov 5 | Visible for 12° (?) in the morning | See 110 VIII 17. |
| 132 I 15 | -179 Apr 9 | Last, morning, in Aries | Not seen in 140 |
| 132 V 1 | -179 Jul 20 | First, evening | |
| 132 XI 23 | -178 Mar 6 | Last, evening, at the end of Pisces | Text broken, data may be missing. |
| 132 XI 30 | -178 Mar 14 | First, morning | See 140 XII 29. |
| 140 XII 29 | -170 Mar 15 | Visible for 8° in the morning | See 132 XI 30. |

[a]The month is probably wrong.

## TABLE IV.7 (Continued)

| Babylonian Date (Seleucid Era) | Tentative Julian Date | Event | Comments |
|---|---|---|---|
| 217 VIII 23 | - 94 Nov 28 | Last, evening | Cloudy, not watched for in 225 |
| 217 VIII 27 | - 94 Dec 3 | First, morning, in Sagittarius | Cloudy, not watched for in 225 |
| 228 II 11 | - 83 May 21 | First, evening, in Gemini | Not seen in 236 |
| 228 XI 10 | - 82 Feb 10 | Last, evening, at the end of Aquarius | |
| 228 XI 15 | - 82 Feb 16 | First, morning, in Aquarius | See 236 XI 12. |
| 236 XI 12 | - 74 Feb 14 | Visible for 12° in the morning | See 228 XI 15. |

## TABLE IV.8
## CONJUNCTIONS OF VENUS FROM LATE BABYLONIAN SOURCES

| Babylonian Date (Seleucid Era) | Tentative Julian Date | Time of Day | Body Conjoined | Position Relative to Body Conjoined | Comments |
|---|---|---|---|---|---|
| -256 II 18 | -567 Jun 8 | | α Leo | Over, 1 am., 4 u. | Probably evening |
| -256 X 19 | -566 Feb 2 | | αβ Cap | Under, 2½ am. | Probably morning |
| -256 XI 4 | -566 Feb 15 | E | γδ Cap | Over, ½ am. | |
| 38 VII 3 | -273 Oct 8 | E | θ Oph | East, 1 am. | Also a lunar conjunction, day supplied by calculation |
| 73 III 24 | -238 Jul 6 | E | ρ Leo | Over | Distance missing |
| 73 IV[a] 9 | -238 Jul 21 | E | β Vir | Under, 4 u. | |
| 73 VI 24 | -238 Oct 3 | E | θ Oph | Under | Distance missing |
| 73 VIII[a] 22 | -238 Nov 28 | E | γ Cap | Over | Distance missing |
| 79 IX 2 | -232 Dec 2 | E | αβ Cap | East, 4 u., under about 2 am. | Also a lunar conjunction, day supplied by calculation |
| 79 IX 15 | -232 Dec 15 | E | γ Cap | Over, 6 u. | Cloudy |
| 79 IX 17 | -232 Dec 17 | E | δ Cap | Over, 6 u. | |
| 79 XI 20 | -231 Feb 17 | E | α Ari | Under, 3½ am. | |
| 83 II[a] 26 | -228 May 19 | E | δ Cnc | Over | Distance missing |
| 83 III[a] 23 | -228 Jun 15 | | ρ Leo | Over, ½ am.; a longitude separation may be missing. | Probably evening |
| 83 VII 19 | -228 Oct 8 | M | β Vir | Over, ½ am., slightly east | |
| 83 IX 7 | -228 Nov 24 | | α Lib | Over, ½ am. | Probably morning |
| 83 IX 12 | -228 Nov 29 | | β Lib | Under, 2; unit missing | Probably morning |
| 83 XI 15 | -227 Jan 30 | M | αβ Cap | Under, 1 2/3 am. | |
| 110 I 14 | -201 Apr 10 | E | α Tau | Over, 2½ am. | |
| 110 I 23 | -201 Apr 19 | E | β Tau | Under, 2 | Unit missing |
| 110 II 3 | -201 Apr 29 | E | η Gem | Over, 1 am. 4 u. | |

[a]The translator supplied the month by calculation.

## TABLE IV.8 (Continued)

| Babylonian Date (Seleucid Era) | Tentative Julian Date | Time of Day | Body Conjoined | Position Relative to Body Conjoined | Comments |
|---|---|---|---|---|---|
| 110 II 5 | -201 May 1 | | μ Gem | Over, 1 am. 4 u. | Probably evening |
| 110 II 8 | -201 May 4 | E | γ Gem | Over, 4 am. | |
| 110 II 17 | -201 May 13 | E | α Gem | Under, 3½ am. | |
| 110 II 21 | -201 May 17 | E | β Gem | Under, 2½ am. | |
| 110 III 3 | -201 May 28 | E | δ Cnc | Over, 2/3 am. | |
| 110 III 13 | -201 Jun 7 | E | ε Leo | Under, 3½ am. | |
| 110 III 21 | -201 Jun 15 | E | α Leo | Over, 2/3 am. | |
| 110 III 28 | -201 Jun 22 | E | ρ Leo | Over, 20 u. | |
| 110 IV 7 | -201 Jul 1 | E | β Leo | Under, 4½ am. | |
| 110 IV 15 | -201 Jul 9 | E | β Vir | Over, 2 u. | |
| 110 IV 28 | -201 Jul 22 | E | γ Vir | Under, 1½ am. | |
| 110 V 11 | -201 Aug 3 | E | α Vir | Over, 10 u. | Date may be 110 V 12. |
| 110 VI 4 | -201 Aug 26 | E | α Lib | Under, 2 am. | May be 2½ am. |
| 110 VI 10 | -201 Sep 1 | E | β Lib | Under, 5 am. | |
| 110 IX 4 | -201 Nov 23 | M | β Lib | Under, 2 am. | |
| 110 IX 25 | -201 Dec 14 | M | β Sco | Over, ½ am. | |
| 110 X 2 | -201 Dec 20 | M | α Sco | Over, 2 am.[b] | |
| 110 X 14 | -200 Jan 1 | M | θ Oph | Over, 2 am. | |
| 110 XI 22 | -200 Feb 8 | M | αβ Cap | Under, 1½ am. | |
| 110 XII 9 | -200 Feb 25 | M | γ Cap | Over, 1 am. 8 u. | |
| 110 XII 11 | -200 Feb 27 | M | δ Cap | Over, 1 am. 8 u. | |
| 132 VI 7 | -179 Aug 25 | E | α Vir | Over, 1 am. | |
| 132 VI[a] 29 | -179 Sep 16 | E | β Lib | Under, 3½ am. | |
| 132 $VI_2$ 10 | -179 Sep 26 | E | δ Sco | Over; east 4 u. | |
| 132 $VI_2$ 16 | -179 Oct 2 | E | α Sco | Over, 2 am. | |
| 132 $VI_2$ 25 | -179 Oct 11 | E | θ Oph | Over | Distance missing |
| 132 VIII 1 | -179 Nov 16 | E | αβ Cap | Under, 2½ am. | |
| 132 VIII 15 | -179 Nov 30 | E | γ Cap | Over | Distance missing |
| 132 VIII 18 | -179 Dec 3 | E | δ Cap | 1 u., east 6 u. | First direction missing |
| 217 I 30 | - 94 May 12 | E | γ Gem | Over, 3 am. | May be 3½ am. |
| 217 II 9 | - 94 May 21 | E | α Gem | Under, 4 am. | |
| 217 II 13 | - 94 May 25 | E | β Gem | Under, 3 am. | |
| 217 II 25 | - 94 Jun 6 | E | δ Cnc | Over 20 u.[c] | |
| 217 III 11 or 12 | - 94 Jun 22 or 23 | E | α Leo | Over, 1 am. | The date is hard to read. |
| 217 III 18 | - 94 Jun 29 | E | ρ Leo | Over, 1 am. | |
| 217 IV 6 | - 94 Jul 16 | E | β Vir | Over, 5 u. | The date may be 217 IV 7. |

[a]The translator supplied the month by calculation.

[b]The translation says 3 am. but the transcription says 2 am.

[c]The record also says that δ Cnc was invisible.

TABLE IV.8 (Continued)

| Babylonian Date (Seleucid Era) | Tentative Julian Date | Time of Day | Body Conjoined | Position Relative to Body Conjoined | Comments |
|---|---|---|---|---|---|
| 217 IV 18 | - 94 Jul 28 | E | γ Vir | Under, 1 am. | |
| 217 IV 29 | - 94 Aug 8 | E | α Vir | Over, 1 am. | |
| 217 V 18 | - 94 Aug 27 | E | α Lib | Under, 1 am. | |
| 217 V 22 | - 94 Aug 31 | E | β Lib | Under, 4 am. | |
| 217 VI 11 | - 94 Sep 19 | E | α Sco | Over, 10 u. | Distance is hard to read. |
| 217 VI 23 | - 94 Oct 1 | E | θ Oph | Under, 1 am. | |
| 217 XI 10 | - 93 Feb 13 | M | αβ Cap | Under, 1 am. | |
| 217 XI 27 | - 93 Mar 2 | M | γ Cap | Over, 2 am. | |
| 217 XI 29 | - 93 Mar 4 | M | δ Cap | Over, 2 am. | |
| 228 V 2 | - 83 Aug 9 | E | γ Vir | Under, 1 am., 14 u. | |
| 228 V 11 | - 83 Aug 18 | E | α Vir | Over, 1 am. | |
| 228 VI 1 | - 83 Sep 6 | E | α Lib | Under, 8 u. | |
| 228 VI 5 | - 83 Sep 10 | E | β Lib | Under, 3½ am. | |
| 228 VI 16 | - 83 Sep 21 | E | δ Sco | Under, 2 u. | |
| 228 VI 22 | - 83 Sep 27 | E | α Sco | Over, 2 am. | |
| 228 VII 1 | - 83 Oct 6 | E | θ Oph | Under, 2 u. | |
| 228 VIII 10 | - 83 Nov 14 | E | αβ Cap | Over, 3 am. | |
| 228 VIII 25 | - 83 Nov 29 | E | γ Cap | Over, 1 u. | |
| 228 VIII 27 | - 83 Dec 1 | E | δ Cap | Over, 1 u. | |

TABLE IV.9

SYNODIC PHENOMENA INVOLVING MARS, FROM LATE BABYLONIAN SOURCES

| Babylonian Date (Seleucid Era) | Tentative Julian Date | Event | Comments |
|---|---|---|---|
| -211 II 28 | -522 Jun 1 | Last, evening, west of Gemini | |
| -211 VI 13 | -522 Sep 13 | First, morning, in the feet of Leo | |
| -210 V 12 | -521 Aug 2/3 | Turning point | |
| -209 II 9 | -520 May 20 | Last, evening, east of Leo | |
| - 67 VIII 22 | -378 Nov 17/18 | Turning point | |
| 12 ? ? | | Turning point, 1½ am., west of the Pleiades | Date missing |
| 12 VII 15 | -299 Oct 8/9 | Opposition | |
| 12 VIII 22 | -299 Nov 13/14 | Turning point, 4 am., under β Arietis, ½ am. back | |
| 44 VI 9 | -267 Sep 8/9 | ————————, 1 am. from Pleiades | Event and direction missing |
| 61 VII 13 | -250 Oct 4/5 | Turning point, 2/3 am. over η Geminorum | |
| 61 VIII 26 | -250 Nov 16/17 | Opposition | |
| 61 IX 27 | -250 Dec 17/18 | Turning point, in Taurus | |

TABLE IV.10

CONJUNCTIONS OF MARS FROM LATE BABYLONIAN SOURCES

| Babylonian Date (Seleucid Era) | Tentative Julian Date | Time of Day | Body Conjoined | Position Relative to Body Conjoined | Comments |
|---|---|---|---|---|---|
| -256 II 3 | -567 May 24 | | ε Cnc | Entered ε Cnc | Probably evening |
| -256 II 5 | -567 May 26 | | ε Cnc | Came out of ε Cnc | Probably evening |
| -256 III 1 | -567 Jun 20 | | α Leo | Mars and Mercury were both 4 am. west of α Leo; Mercury was south of Mars. | Probably evening |
| - 67 VIII 1 | -378 Oct 27 | E | β Ari | Under, 2 am. 10 u. | See Table IV.4. |
| 38 VII 22 | -273 Oct 27 | E | αβ Cap[a] | , 1 2/3 am. | Direction missing |
| 44 III 11[b] | -267 Jun 14 | M | η Psc | Under, 2 am. | |
| 44 III 21 | -267 Jun 24 | M | β Ari | Under, 4 am. | |
| 44 XII[c] 4[b] | -266 Feb 28 | E | β Tau | Under | Distance missing |
| 44 XII[c] 24 | -266 Mar 20 | E | η Gem | Over | Distance missing |
| 45 I 3 | -266 Mar 28 | E | γ Gem | Over, 2 am. | |
| 71 VI 9 | -240 Sep 10 | M | β Leo | Under, 4½ am. | |
| 93 II 17 | -218 May 21 | M | η Psc | Under, 3½ am. | |
| 157 I 28 | -154 May 14 | M | η Psc | Under, 3 am. | |
| 157 II 8 | -154 May 24 | M | β Ari | Under | Distance missing |
| 157 II 13 | -154 May 29 | M | α Ari | Under, 3½ am. | |
| 157 III 9 | -154 Jun 24 | M | η Tau | Under, 2 am. | |
| 157 IV 18 | -154 Aug 1 | M | β Tau | Under, 2½ am. | |
| 157 IV 23 | -154 Aug 6 | M | ζ Tau | Under | Distance missing |

[a]The translator supplied the star by calculation.

[b]The translator supplied the day by calculation.

[c]The translator supplied the month by calculation.

## CHAPTER V

## DATA FROM HELLENISTIC ASTRONOMY

### 1. Did Ptolemy Observe Anything?

The earliest astronomical observation that we know of from Hellenistic astronomy is a measurement of the summer solstice by Meton and Euctemon in Athens, which is usually dated -431 June 27.† The last one that we know of is a conjunction of Venus and Jupiter observed in Alexandria by Ammonius and Heliodorus on 510 August 21.‡ All the Hellenistic observations that are useful in this study and that have dates before the year 142 come to us only through the writing of Ptolemy [ca. 142] (the Syntaxis), and many of them are observations that Ptolemy claims to have made himself. Thus we must start the study of Hellenistic astronomy by considering a question that Delambre [1817, volume 2, p. xxv] has raised: "Did Ptolemy do any observing? Are not the observations that he claims to have made merely computations from his tables, and examples to help in explaining his theories?"‡

In this work, we are interested only in solar and planetary observations, and the solar observations have already been analyzed [Newton, 1970, Chapter II]. Thus it would be possible to proceed by analyzing the planetary observations, deducing planetary accelerations from them, and deciding whether the deduced accelerations are plausible. Since this procedure is dangerously close to reasoning in a circle, I decided instead to try a test

†There are strong grounds for believing that this date comes from a late calculation and not an observation. See Sections V.2 and VIII.4.

‡The record in its existing form does not give any of the date unambiguously except the year. The rest of the date is a tentative restoration and may be in error. See Section V.9. It is ironic that we cannot date rigorously either the earliest or the latest Hellenistic observation.

‡The translation is mine.

case. In the test case, which is published elsewhere [Newton, 1973], I consider the observations that Ptolemy claims to have made himself and that are involved in his estimates of the solar and lunar parallaxes, as well as the observations of equinoxes and solstices that Ptolemy claims to have made.

Altogether, the observations that are included in the test include:

a. measurements of the times of equinoxes and solstices,

b. a measurement of the lunar perturbation known as the evection,

c. several measurements of the obliquity of the ecliptic,

d. a measurement of the latitude of the site where Ptolemy says that he made his observations,

e. several measurements of the lunar zenith distance made for the purpose of finding the inclination of the lunar orbit,

f. a measurement of the meridian altitude of the moon, made for the purpose of finding the maximum lunar parallax, and

g. several measurements of the apparent diameter of the solar disc.

In the circumstances where I have used the phrase "several measurements" in the above, Ptolemy says that he made a number of observations but gives only the average.

The results of the test are quite unambiguous. On one hand, many errors in the alleged measurements are enormous. The errors in time are as large as 30 hours or more, when a plausible error would be 2 or 3 hours. Errors in position are correspondingly large. The observation in item f above, for example, is in error by more than 40′ in circumstances where a plausible error is 5′ or less.

On the other hand, to an accuracy that often amounts to exactness, the alleged observations agree with the hypothesis that they were fudged in order to make Ptolemy's erroneous theories of the sun and moon compatible with known observations of pre-Ptolemaic astronomers. The probability that the agreement happened by chance is of the order of $10^{-200}$. For simplicity in discourse, then, we are justified in neglecting this small probability and in saying that Ptolemy's alleged observations are clearly fraudulent.†

In describing Ptolemy's theories as erroneous, I am not referring to a matter such as the distinction between a heliocentric and a geocentric picture of the solar system. This distinction may be important in philosophy, but it is not basically a scientific matter and it is unimportant for the purposes of this work. What I mean by erroneous is that Ptolemy's theories systematically disagree with observation, and the disagreement is so large that it would have been found by any competent observations made in Ptolemy's own time.

However, the test investigation does not answer all questions about Ptolemy's work. Ptolemy gives data concerning several lunar eclipses that he claims to have observed himself. We can analyze the data from these eclipses and compare them with ancient eclipse data that are entirely independent of Ptolemy, and we find [Newton, 1970, Chapters X and XIII] that the agreement is good. Furthermore, the data agree well with Ptolemy's theories. How is

---

†Delambre's gentlemanly suggestion, that the pseudo-observations were devised as aids to understanding, is unfortunately not tenable. It is clear that they were fabricated for the fraudulent purpose of proving Ptolemy's erroneous theories. From the parallax data alone, we cannot tell whether Ptolemy was the perpetrator of the fraud or the victim of a trusted associate. From other data, however [Newton, 1974b], we can tell with high assurance that Ptolemy himself was the perpetrator.

this possible if Ptolemy's theories give seriously wrong positions for the sun and moon?

This point is closely related to one that Newcomb [1875] raises. Newcomb did not use the lunar occultations preserved by Ptolemy, even though none of them were 'observed' by Ptolemy himself, because he feared that they formed a biased sample that Ptolemy had selected to support his erroneous value of the precession of the equinoxes. On the other hand, Newcomb did use Ptolemy's records of lunar eclipses.

TABLE V.1

ERRORS IN PTOLEMY'S VALUES OF THE MEAN MOTIONS

| Body | Mean Motion with Respect to the Equinox, degrees per Egyptian year | | |
|---|---|---|---|
| | Ptolemy | Modern[d] | Ptolemy-Modern |
| Sun | 359.756 876 6 | 359.761 277 4[a] | -0.004 400 8 |
| Moon | 4809.379 508 5 | 4809.384 732 3[a] | -0.005 223 8 |
| Mars | 191.281 794 6 | 191.285 974 8[a] | -0.004 180 2 |
| Jupiter | 30.339 689 2 | 30.342 246 6[b] | -0.002 557 4 |
| Saturn | 12.223 317 2 | 12.226 753 9[c] | -0.003 436 7 |
| Stars | 0.010 000 0 | 0.013 839 9[a] | -0.003 839 9 |

[a] From Explanatory Supplement [1961, Chapter 4].

[b] From Hill [1895].

[c] From Hill [1895a].

[d] The value for the stars is calculated for the epoch 100. The other values are calculated for the epoch 1900.

It is interesting to note that Ptolemy is wrong about the angular velocity of every astronomical body that he considers, as we see from Table V.1. Here I give Ptolemy's value of the average angular velocity, with respect to the equinox, of the sun, the moon, the outer planets, and the fixed stars. I do not give the angular velocities of Mercury and Venus, because their geocentric mean motion is necessarily the same as the sun's. The mean motion of the sun is found in Chapter III.1, that of the moon is found in Chapter IV.3, those of the planets are found in Chapter IX.3, and that of the fixed stars, which is of course the value of the precession, is found in Chapter VII.2 of the *Syntaxis*.

Ptolemy's values are compared with modern values in Table V.1. Except for the precession, the values are those that apply at the epoch 1900. For the precession, however, the change since Ptolemy's time is appreciable, and I give the value for the epoch 100 as calculated from the modern source cited.

We see that Ptolemy's value is too small in every instance, that the error is approximately the same size for every body, and that the errors for the sun, moon, and stars are particularly close to each other. We now have to ask: Are the errors in Table V.1 highly correlated, so that they can be explained and hence removed as the consequence of a simple error? Or are they the consequence of many errors that all happened to be about the same? Or, as Newcomb feared, are they the consequence of using biased samples chosen to substantiate erroneous hypotheses? When we follow through Ptolemy's work, we see that the first question suggests what is almost surely the correct explanation, namely that all the errors in mean motion flow from the fraudulent solar data.

First, Ptolemy (Chapter III.1) uses the length of the year that Hipparchus found and that he 'confirmed', and from it he derives the mean motion of the sun with respect to the equinox. Next, he uses a study that Hipparchus had made of lunar eclipses (Chapter IV.2). Hipparchus had used eclipses separated in time by 126 007 days plus 1 hour, an interval that had been chosen to correspond to nearly equal values of the lunar anomaly. This interval is 4267 months, and thus we find that there are 29.530 593 days per month; this is in error by only about 0.000 005 days.†

Thus we see that Hipparchus, and Ptolemy after him, put both the mean sun and moon in the wrong place with respect to the equinox but in highly accurate positions with respect to each other.

Next in point of development is the precession of the equinoxes. The astronomer Timocharis, observing in Alexandria around the year -290, measured the longitudes of various stars by measuring their separation from the moon during total lunar eclipses. Hipparchus did the same about 160 years later. Ptolemy, about 260 years later yet, claims that he also measured the stars with respect to the moon, but that he did so using the astrolabe under ordinary circumstances rather than during eclipses.

Thus the moon was the secondary standard of longitude for all these observations and the sun was the primary standard. The equinox, and hence the longitude, was introduced only through the medium of the length of the tropical year.

Finally, Ptolemy (Chapter IX.3) estimated the mean motions of the outer planets by using the resonance periods‡ listed in Table IV.1, combined

---

†As many writers [Aaboe, 1974, for example] have pointed out, division of the time interval by 4267 months leads to 29;31,50,8,9 days per month, in sexagesimal notation. However, Ptolemy says that the division leads to 29;31,50,8,20 days per month. The latter value comes from Babylonian ephemerides [Neugebauer, 1955, vol. I, p. 70, for example]. Thus it seems that this value, and some planetary parameters as well, come from Babylonian sources although Ptolemy says that they come from Hellenistic observations.

‡For Mars and Jupiter, Ptolemy used only the resonances of 79 and 71 years, respectively. See Table IV.1.

with estimates of the number of degrees by which the resonance fails to be exact. As I read him, Ptolemy implies that the data used are essentially those of Hipparchus, but that he has improved them by his own observations. Thus the mean motions of the planets are also based directly upon the mean motion of the sun.

In sum, Hipparchus and Ptolemy after him, had rather accurate representations of the average motions of the sun and moon with respect to each other and to the stars, and they had slightly less accurate representations of the average motions of the outer planets with respect to the other bodies. The estimate that Hipparchus and Ptolemy used for the precession of the equinoxes is directly related to their estimate of the difference between the sidereal and tropical years, with little contribution from other sources of error.

Thus Ptolemy did not need to use a biased sample of lunar observations in order to support his erroneous motion of the equinox. Paradoxically, accurate observations would have given better support to his theory than a biased sample would have done.

We can also see why my first studies [Newton, 1970] did not reveal a discrepancy between other observations and the observations of lunar eclipses that Ptolemy claims to have made. From Table V.1, we see that the error in the elongation of the moon from the sun, when calculated by means of Ptolemy's theories, grows at the rate of about 0°.0008 per year. If the observations were fudged, the calculations involved in the fudging probably started from genuine observations made in the time of Hipparchus about 265 years before. The error in the fudged values of the elongation would thus have grown to about 0°.2, and fudged times of the eclipses would have been in error by about $25^m$. The first analysis would not have been able to detect an error this small.

However, in a later analysis [Newton, 1974b], I tested the observations of lunar eclipses that Ptolemy claims to have made by a more delicate method, in which I compare the observations with calculations from Ptolemy's tables of the sun and moon. In the same paper, I tested the measurements of 18 stellar declinations that Ptolemy claims to have made himself. It turns out that the eclipse observations are definitely fudged. The stellar declinations give an unexpected result. Of the 18 declinations given, Ptolemy uses 6 in order to estimate the precession of the equinoxes. These 6 declinations are unquestionably fudged, but the 12 declinations that Ptolemy does not use are almost certainly genuine. They lead to an accurate value (52.8 ± 2.0 ″/year) for the precession, rather than to the value 36 ″/year that Ptolemy derives from the declinations that he uses.

In sum, four classes of Ptolemy's alleged observations have been tested by rigorous procedures. It turns out that all of these observations that Ptolemy actually uses are fraudulent; that is, they have been fudged for the purpose of establishing erroneous theories. This result constitutes sufficient grounds for assigning a weight of zero to the planetary observations that Ptolemy claims to have made. Therefore I shall not use any of Ptolemy's claimed observations in this study. I shall analyze them, however, in order to study their authenticity.

Besides the observations that he claims to have made himself, Ptolemy has transmitted the results of many observations that he attributes to earlier astronomers. We must now turn our attention to these records.

## 2. Did Ptolemy Preserve Anything?

We must now decide whether the observations that Ptolemy attributes to other astronomers are valid or whether they too have been fudged. If we decide that they are valid, we must still answer Newcomb's basic question that was posed in the preceding section. Are they an unbiased sample or have they been selected to support an erroenous theory?

In the present work, we are directly interested in Newcomb's question only with regard to the planetary data; other data are relevant only to the extent that they supply clues about the planetary data. On the basis of Table V.1, we can give the same answer about the planetary data that we gave about the lunar data in Section V.1: In the selection of older data to support his erroneous theories, Ptolemy would have been better served by an unbiased sample than by a biased one.

This by itself is not sufficient grounds for using the older observations. However, there are several reasons that lead me to believe in the validity of the older data in general.

First, Ptolemy's theories had to agree to some extent with observation, at least with regard to properties that were easily observable.† Hence they had to be based upon some valid observations, and Ptolemy quotes a number of older observations that he presumably used in the establishment of his theories. If he only pretended to use them, they must have been fudged, because they agree exactly with his theories. In other words, if the old observations are not valid, Ptolemy must have started by using some other old observations, which would have been valid, as the basis of his theories. He then concealed the existence of these observations while going to the trouble of fudging some ostensible old observations.

It is conceivable that Ptolemy did do this to some extent. It is possible that he based his theories upon certain old observations and that he fudged some additional 'old observations' in order to increase the apparent base for his work.‡ Even if he did this, we shall probably not err seriously if we take most of his old observations as valid. The results deduced from the valid observations will themselves of course be valid. The fudged observations would have been calculated from valid

†Otherwise his errors would have been obvious even to laymen.

‡In particular, it is possible that Ptolemy fudged the solstice of -431 June '27', perhaps in order to invoke the prestige of Meton and Euctemon. See Section VIII.4 below.

observations made at epochs fairly close to those of the ostensible observations, and the results inferred from them should not be seriously in error. On the average, then, the results should be acceptable.

Second, a number of observations that Ptolemy attributes to other astronomers have been used in studies of the accelerations. These observations include equinoxes, lunar occultations, and lunar eclipses from Greek sources, plus some lunar eclipses from Babylonian sources. There are also some pre-Ptolemaic observations that come to us through independent testimony, and the results of all these observations are reasonably consistent. Thus, if the cited observations in the Syntaxis are not valid, they are at least fairly accurate fudges.

Third, Ptolemy preserved several equinox observations attributed to Hipparchus, as well as a few other observations, that do not agree with his theories and models. They do agree within reasonable amounts, however, with what we calculate from modern theory. Thus there is a presumption that these observations are valid.

Fourth, we have what is probably the strongest argument in favor of viewing the old observations as valid. Many Greek works on astronomy that are now lost still existed in Ptolemy's time. Further, they were well known, widely distributed, and readily available to many of Ptolemy's contemporaries, and they would have constituted a strong constraint on Ptolemy's liberty of action.

For example, Hipparchus's work on the length of the year and the precession of the equinoxes was probably in Ptolemy's library and in the libraries of many of his contemporaries; since Ptolemy had not yet written the Syntaxis, there was no reason to discard the work of Hipparchus and other pre-Ptolemaic astronomers. If Hipparchus wrote that he measured the time of an autumnal equinox as sunset on the day that we call -142 September 26, many people would have known this or could easily have verified it. If Ptolemy had wanted to alter this to the following midnight in order to make it agree with other observations, he would have known that he ran a large risk of being detected. For this reason, if for no other, Ptolemy probably did not alter the datum in question.

This is not to say that all the older observations are necessarily valid. In a few cases of particular importance to him, Ptolemy could have taken the chance of altering an old datum; he could always have pleaded an accidental error if he were detected and questioned. However, it does not seem likely that he could have systematically altered the record of history.

The reader may object that Ptolemy would have been equally inhibited in fudging new observations, because his contemporaries could have detected the fraud by independent observations. This would be a strong argument if applied to many stages in history. In the time of Ptolemy, however, observational astronomy had almost vanished; after him, the only Hellenistic observations that we have are the "eclipse of Theon" [Newton, 1970, Section V.5] on 364 June 16 and the few observations around the year 500 that will be presented in Section V.9 below. If a person in the time of Ptolemy realized that observational astronomy had almost vanished, he might have felt fairly safe in fudging the data.

Further, checking data by means of independent observation can be done only by people with the necessary instrumentation. Checking the reliable transmission of old data, in contrast, can be done by anyone who has a copy of the literature.

In sum, I shall assume that Ptolemy transmits correctly the observations that he attributes to other astronomers, unless there is a particular reason to be suspicious, and it is not likely that failures of the assumption will lead to serious error. There is reason to be suspicious of the solstice attributed to Meton and Euctemon. In addition, there are the records of the eclipses allegedly observed in Babylon on -382 December 23, -381 June 18, and -381 December 12. These records show such large internal inconsistencies that I did not use them in the earlier study of Babylonian eclipses [Newton, 1970, Section V.2], and they have aroused a lively controversy in the literature. It is conceivable that these observations were also fudged.

TABLE V.2

THE EQUINOX OBSERVATIONS OF HIPPARCHUS

| Date | Hour | Observed JD -1 600 000 | Tabular JD -1 600 000 | $\Delta t$ hours |
|---|---|---|---|---|
| -161 Sep 27 | 18 | 62 522.167 | 62 521.837 | -7.9 |
| -158 Sep 27 | 06 | 63 617.667 | 63 617.564 | -2.5 |
| -157 Sep 27 | 12 | 63 982.917 | 63 982.807 | -2.6 |
| -146 Sep 27 | 00 | 68 000.417 | 68 000.472 | 1.3 |
| -145 Sep 27 | 06 | 68 365.667 | 68 365.714 | 1.1 |
| -142 Sep 26 | 18 | 69 461.167 | 69 461.441 | 6.6 |
| -145 Mar 24 | 06 | 68 178.677 | 68 178.990 | 7.5 |
| -144 Mar 23 | 12 | 68 543.927 | 68 544.232 | 7.3 |
| -143 Mar 23 | 18 | 68 909.177 | 68 909.474 | 7.1 |
| -142 Mar 24 | 00 | 69 274.427 | 69 274.716 | 6.9 |
| -141 Mar 24 | 06 | 69 639.677 | 69 639.959 | 6.8 |
| -140 Mar 23 | 12 | 70 004.927 | 70 005.201 | 6.6 |
| -134 Mar 24 | 00 | 72 196.427 | 72 196.655 | 5.5 |
| -133 Mar 24 | 06 | 72 561.677 | 72 561.897 | 5.3 |
| -132 Mar 23 | 12 | 72 926.927 | 72 927.140 | 5.1 |
| -131 Mar 23 | 18 | 73 292.177 | 73 292.382 | 4.9 |
| -130 Mar 24 | 00 | 73 657.427 | 73 657.624 | 4.7 |
| -129 Mar 24 | 06 | 74 022.677 | 74 022.867 | 4.6 |
| -128 Mar 23 | 12 | 74 387.927 | 74 388.109 | 4.4 |
| -127 Mar 23 | 18 | 74 753.177 | 74 753.351 | 4.2 |

## 3. Equinox and Solstice Observations

In an earlier study [Newton, 1970, Section II.1], I analyzed the equinox observations made by Hipparchus and preserved by Ptolemy [ca. 142, Chapter III.1], and I did so under the tacit assumption that the spin fluctuation hypothesis is valid. It is desirable to repeat the study without making this assumption. The changes needed are changes in wording only; in particular we want to put the results in terms of the acceleration of the sun rather than in terms of the difference between solar and ephemeris times.

The results of the study are summarized in Table V.2. The first column gives the date of the observation, and the second column gives the hour† of the observation in apparent solar time at the place of observation. The place is presumably Rhodes,‡ although it is possible that the three earliest observations were made in Alexandria. The difference in local time at Rhodes and Alexandria is negligible compared with the errors in the data.

The third column in Table V.2 gives the time of the observation by means of the Julian day number and fractions, with 1 600 000 subtracted for convenience. This column is now in terms of Greenwich mean solar time.

Fotheringham [1918] analyzed a number of star declinations that Hipparchus had measured and that have been preserved in places such as Chapter VII.3 of Ptolemy [ca. 142]. He concluded that there was an error in the setting of Hipparchus's equatorial plane, of amount

---

†Hipparchus recorded the times as midnight, sunrise, midday, or sunset. I have replaced these terms by conventional numbers for the hours, for two reasons. One is simply for brevity. The other is to remove the ambiguity in the date when the time is midnight.

‡I shall use the coordinates of the city of Rhodes. We do not know where on the island Hipparchus made his observations.

$$\delta_H = -0^\circ.073 \pm 0^\circ.018 \ . \qquad (V.1)$$

The sign of $\delta_H$ means that the declination of an object was actually negative when Hipparchus thought that it was zero. The value in Equation V.1 includes an estimate for the effect of refraction.

The fourth column in Table V.2 is called the "tabular JD", again with 1 600 000 subtracted for convenience. The tabular JD gives the value of the time to use in Newcomb's tables in order to obtain $-0^\circ.073$ for the declination of the sun. If it turns out that the spin fluctuation hypothesis is valid, this is the same thing as ephemeris time. The fifth column is the difference between the tabular time and the observed time, converted to hours.

Since Hipparchus rounded his times to the nearest quarter of a day, and since it is clear that the true precision of his observations is better than this, the errors in his equinox measurements are highly correlated. For this reason, I did not use a statistical method in analyzing the data. Instead, I set limits to the value of $\Delta t$ in Hipparchus's lifetime from the fact that the long series of vernal equinoxes shows no discrepancy with the length of the Julian (or Callippic) year, while the short series of autumnal equinoxes shows a discrepancy between the years -145 and -142.† Thus I estimated $\Delta t = 4.2 \pm 1.0$ hours at the mean epoch of the observations in Table V.2.

During 4.2 hours, the mean sun moves 621″. That is, the acceleration of the sun with respect to solar time has carried it through 621″ in about 20.4 centuries, so that

$$\nu_S' = 2.98 \pm 0.71 \ ''/\text{cy}^2 \ . \qquad (V.2)$$

†The autumnal equinoxes before -146 clearly belong to a different population from the later ones. Hipparchus was probably adjusting his equatorial plane before -146, and he may have been in Alexandria rather than Rhodes.

Fotheringham [1918] analyzed Hipparchus's equinox data by a statistical method and found $\nu_S{}' = 3.9 \pm 0.5$. I believe that the estimate in Equation V.2 is to be preferred, since the problem is essentially not a statistical one.

Ptolemy [ca. 142, Chapter III.1] says that Hipparchus measured the time of the summer solstice of the year -134, and that he found the interval from the vernal equinox to the summer solstice to be $94\frac{1}{2}$ days. Ptolemy does not explicitly state the time that Hipparchus found, but we can readily reconstruct it. From Table V.2, we find that Hipparchus measured the vernal equinox of that year at the midnight beginning March 24. Hence he must have found that the time of the solstice was noon on -134 June 26.

In the same place, Ptolemy says that Hipparchus derived his length of the year by comparing this solstice with one measured by Aristarchus 145 years before. He found that the interval was $\frac{1}{2}$ day less than what it would have been if the year were exactly equal to $365\frac{1}{4}$ days. In other words, the year is $365\frac{1}{4}$ days less 1/290 day, which Hipparchus, and Ptolemy after him, rounded to 1/300. Again, Ptolemy does not give the time of Aristarchus's solstice, but we readily find that it must have been at $\frac{1}{4}$ day after noon on -279 June 26.

I shall return to the question of these solstices in Section VIII.4. There it will appear that Ptolemy cites altogether four solstices, of which three were measured by astronomers earlier than himself. They are the two that have just been mentioned, plus one from the summer of -431 that is attributed to Meton and Euctemon, and plus one from +140 that Ptolemy claims to have measured himself. It will also appear, with high probability, that two of the four solstices are fudged. It is virtually certain that the fudged solstices are those of -431 and +140 and that the other two are genuine. Thus I shall use the solstices of Aristarchus and Hipparchus in the inference of the solar acceleration.

Tannery [1893, p. 156] gives a table in which he lists dates of both solstices and both equinoxes as observed by Euctemon, Eudoxus, Callippus, and Hipparchus in the years -431, -380, -329, and -144, respectively. Tannery cites Ptolemy as the authority for the dates in -144, which he calls "supposed, according to Ptolemy"; Ptolemy, however, gives no dates in -144 except that of the vernal equinox listed in Table V.2. I have not tried to find the data attributed to Eudoxus, which do not seem to have the character of genuine observations. They look like conventional calendrical dates, rather like the traditional Roman "8th calends" for the beginning of each season.†

The dates attributed to Euctemon and to Callippus give the appearance of real data in Tannery's tabulation, but it is clear from his discussion that they are not. Tannery gives two sets of dates for each of these astronomers, one "according to Leptinus" and one "according to Geminus". I cannot identify Leptinus with any person listed by Pauly-Wissowa [1894], but Geminus has an entry and his known work [Geminus, ca. -100]‡ has been published.

Geminus gives a variety of astronomical information. For example, in Chapter I he gives the lengths of the seasons as 94½, 92½, 88 1/8, and 90 1/8 days, respectively, beginning with the vernal equinox. It is clear that these are Hipparchus's values, but they imply that Hipparchus made observations of the winter solstice; this is a matter that I do not find mentioned by Ptolemy. For another example, in Chapter VIII Geminus gives the length of the synodic month that Ptolemy [Chapter IV.2] attributes to Hipparchus.

The information that we want here is given in the parapegm with which Geminus closes his work. This is a conventionalized calendar based upon the solar year and beginning at the summer solstice. It

---

†See the extensive discussion by Jones [1943], for example, or the shorter discussion by Newton, [1972, Section II.3].

‡I use the date -100 for purposes of citation only. Pauly-Wissowa gives Geminus's dates only as lying within the approximate range -100 to +200. Taking +200 as one limit seems conservative.

gives much important information, such as the weather for each day, and it gives the day upon which the sun enters each zodiacal sign. For Libra, Capricorn, and Aries, which mark the beginnings of fall, winter, and spring, Geminus tells us that these are the days according to both Euctemon and Callippus. In other words, Callippus used the same lengths of the seasons as had Euctemon and presumably Meton a century before him.†

It is now clear what Tannery has done. He started by taking the date -431 June 27 for the summer solstice, which is the date given by Ptolemy [Chapter III.1] for the solstice observed by Meton and Euctemon. Tannery has then assigned the same date, June 27, for the summer solstice as measured by Callippus, and he has then filled in the table by adding the lengths of the seasons to the date June 27 in the years -431 and -329.

It is common to say that Greek astronomers had no clear idea of the lengths of the seasons until the time of Hipparchus. Dinsmoor [1931, p. 318] has an interesting variant of this statement. On the authority of Ptolemy, he assumes that Meton and Euctemon found the time of the solstice to be sunrise on -431 June 27. Since the solstice was actually at about 10 hours on -431 June 28, Athens time, Dinsmoor assumes that Meton and Euctemon did not know how to measure solstices in -431, and that they merely took the solstice to be the midpoint between the equinoxes. However, in order to preserve his theory of the Athenian calendar, Dinsmoor has to assume that Meton and Euctemon learned how to measure the solstice accurately within a few years after -431.

This seems rather unlikely to me. It seems unlikely that Meton and Euctemon would establish

†Tannery gives different times for the vernal equinox according to Euctemon and according to Callippus. He says that Geminus does not give a vernal equinox according to Euctemon and that he has filled in the date from another source. Actually, it is the summer solstice that Geminus does not name specifically. There is no reason to, because it is the first day of the year, by definition.

a calendar based upon the solstice when they did not know how to measure it. If they did so, it is unlikely that their enlarged abilities would come soon after -431 rather than at some other time. Further, it is the solstices and not the equinoxes that can be measured by primitive methods.† I do not see how it is possible to find the equinoxes until the solstices have been firmly established. Once the methods have been developed, equinoxes can be found with more precision, but not necessarily more accuracy, than solstices. Solstices can be measured without bias using primitive methods, while equinoxes cannot.

If Geminus's information on the matter is reliable, Euctemon (and presumably Meton) could measure both equinoxes and solstices with considerable accuracy. Beginning with the summer solstice, the intervals attributed to Euctemon are 92, 89, 89, and 95 days, while the corresponding intervals as calculated from modern theory are 92.3, 88.7, 90.2, and 94.1 days. The largest discrepancy is about 1 day, but the matter is more interesting than this figure indicates. Euctemon's intervals between solstices are 181 days (summer to winter) and 184 days (winter to summer), while the modern values are 181.0 and 184.3, respectively.‡ Here, the biggest error is 0.3 days. The intervals between equinoxes, however, are 187 (spring to fall) and 178 (fall to spring) days, according to Euctemon, while the modern numbers are 186.4 and 178.9. In other words, the errors, even in -431, are larger for the equinoxes and smaller for the solstices.‡

†In this context, a primitive method is one that is based upon first principles and that does not require pre-existing standards.

‡Because of rounding, the values do not add up to 365.24 days.

‡Dinsmoor [1931, p. 317] quotes a papyrus document called "Ars Eudoxi" that apparently dates from around -200. According to Dinsmoor's quotation of it, the intervals determined by Euctemon were 90, 90, 92, and 93 days, beginning with the summer solstice. This gives 180 days from the summer to the winter solstice and 185 days from the winter to the summer. These intervals are not as accurate as those given by Geminus, but they are still not

We should note that Geminus uses at least two independent sources. The intervals that he attributes to Euctemon and Callippus are not rounded off from the intervals attributed to Hipparchus. We should also note that Hipparchus's intervals show little improvement over those attributed to Euctemon and Callippus.

Thus, on the basis of the documentary information that we have, the situation is the opposite of what was earlier stated. In the -5th century, Athenian astronomers were already able to measure the solstices, and perhaps the equinoxes, with errors that are less than a day. The accuracy of the solstices was better than that of the equinoxes.

This is a useful place to estimate the standard deviation in an ancient Greek measurement of a solstice time. Hipparchus's errors in both intervals between solstices are 0.36 days. Euctemon's errors are somewhat smaller, but this is probably an accident due partly to the fact that his total year is rounded to 365 days in the source that gives his intervals. If we take 0.36 days as the standard error in the interval between solstices, and if the interval was obtained by subtracting two measured times, the standard deviation of a measured time was $0.36/\sqrt{2} \approx 0.25$ days = 6 hours.

If we calculate the times of the solstices attributed to Aristarchus and Hipparchus, using $\nu_S' = 3.0$ ″/cy$^2$, we find that the time of Aristarchus is too early by about 7 hours and that the time of Hipparchus is too late by about 8 hours.

---

consistent with errors of more than a day in individual measurements. Further, the interval from the winter to the summer solstice is too long by about 17 hours. This interval is consistent with a measurement of the summer solstice that is too late by some hours, but it is almost impossible to reconcile the interval with a measurement that is too early by 28 hours. It would require a winter solstice that is too early by 45 hours. Still further, the intervals certainly do not result from taking the solstices to be the midpoints between the equinoxes.

Thus we are justified in using 7 hours as the standard deviation in the measured time of an ancient Greek solstice, even in the time of Meton and Euctemon, and we find that there was little or no improvement in the rest of Hellenistic astronomy.

The seasons given by Geminus would supply useful data for this study if we had a reliable starting epoch for them. Unfortunately, a reliable starting epoch is just what we lack.

The most extensive information about a starting epoch is furnished by the inscription called the Milesian parapegm.† This inscription commemorates the completion of the 17th Metonic cycle.‡ It says that the first cycle began with the summer solstice when Apseudes was archon at Athens; we know from independent evidence that the solstice must therefore be the one in -431. The inscription further says that the solstice was on Skirophorion 13, and that this was the date VII 21 in the Egyptian calendar. Diodorus‡ also gives the Greek date, while Ptolemy [ca. 142, Chapter III.1] also gives the Egyptian date. So far as ancient written sources are concerned, then, the Greek date and the Egyptian date have equal amounts of support.

The parapegm also gives the solstice in -108, which is 323 (= 17 × 19) years later. The solstice came, it says, on Skirophorion 14, which was the date X 11 in the Egyptian calendar. The parapegm must have been prepared near in time to -108, which is about 80 or 90 years before the time of Diodorus. Thus the parapegm may be the only independent source of the solstice data. It is important to remember that the parapegm is after Hipparchus but before Ptolemy.

---

†Both Dinsmoor [1931, p. 311] and Meritt [1961, p. 4] quote this inscription.

‡This fact does not prove that Meton's cycle was adopted, either in -431 or at any other time. It merely proves that there was an ancient tradition to that effect.

‡Dinsmoor [1931, pp. 309 and 311] and Meritt [1961, p. 4] give the quotation from Diodorus.

TABLE V.3

THE SOLSTICE DATA GIVEN BY THE MILESIAN PARAPEGM

| Time Calculated From Modern Theory | Greek Date[a] | | Egyptian Date[b] |
|---|---|---|---|
| | If Determined by Conjunction | If Determined by Visibility | |
| -431 Jun 28.4 | -431 Jun 28 | -431 Jun 29 | -431 Jun 27 |
| -108 Jun 25.5 | -108 Jun 27 | -108 Jun 28 | -108 Jun 26 |

[a]The Greek day given by the parapegm began at sunset on the date given; the date depends upon the assumption made about the basis of the Greek calendar (see Section II.4).

[b]The Egyptian day given by the parapegm began at sunrise on the date given.

The information given by the parapegm is summarized in Table V.3. The first column gives the time of the solstice as calculated from modern theory using the tentative value $\nu_S = 3$ ″/cy$^2$; it is doubtful that the error in the calculation is greater than the rounding error used in the table. The second column gives the equivalent of the Greek date if the Greek calendar is based upon conjunctions, while the following column gives the date as based upon first visibility. The last column gives the Julian equivalent of the Egyptian date.

So far as I know, every previous student of the subject has accepted the Egyptian dates as the ones actually given by the record. In order to reconcile the Egyptian date with the calculated date, they have assumed that the astronomers could not measure solstices accurately. We might conceivably accept this explanation for -431 in spite of the evidence of Geminus, but it clearly does not work for -108. Here, we are about 30 years after Hipparchus and nearly two centuries after Aristarchus, who were certainly able to measure solstices with errors of hours rather than days.†

We should also notice that the Greek dates, on whichever basis they rested, have the same relation to the Egyptian dates in both years.

The parapegm may commemorate Meton's cycle, but it clearly does not use it. If it did, the Greek date, being lunar, would be the same in both years. More importantly, Meton's year is 365.26316 days.‡ If his year were being used, the date would move forward with respect to the Julian calendar by four days in 323 years. If the Callippic year were being used instead of Meton's year, the date would not move at all with respect to the Julian calendar.

Instead, the year is the Hipparchan year, which falls behind the Callippic and Julian years by 1 day in 300 years. In fact, the Egyptian dates listed in Table V.3 are just those that a person would calculate from Hipparchus's observations and his length of the year.‡‡

It is unlikely that this agreement happened by

---

†We must remember that the Egyptian day given in the record did not begin until -108 June 26¼, approximately. Hence the Egyptian date is necessarily wrong.

‡That is, 365 + (1/4) + (1/76) days. See Section II.4.

‡‡There is no question about the date that would be calculated for -431, but there is some question about the one for -108. The day that would be stated depends upon the convention used for the beginning of the day and upon whether fractions of a day were rounded or truncated. The Egyptian day conventionally began at sunrise. In terms of days and fractions, the Egyptian date calculated for -108 would be X 10 3/4 of the appropriate year. Apparently the calculator rounded this to X 11. It is also possible that an observation of reasonable accuracy would give a time greater than X 10½, which could then have been rounded to X 11. Every way that I have thought of to get the date X 11, whether by observation or calculation, involves rounding upward.

chance. Calculation of the Egyptian dates from Hipparchus's theory of the sun could have happened quite naturally and with no intention to mislead, as we can show by reference to a fictitious example. Suppose that some British agency around the year 1910 had decided to erect a monument commemorating the signing of the Magna Carta in 1215, and to refer in the inscription to the summer solstice of 1215† and to the summer solstice of 1915, the 700th anniversary of the traditional date, when the monument would be dedicated. The time of the then-future solstice of 1915 would have to be calculated. The agency might search medieval records in an attempt to find an observation of the solstice in 1215. It is more likely, however, that the agency would write to the Astronomer Royal who would supply both solstices by calculation.

Similarly, we can imagine that the agency responsible for the Milesian parapegm turned to an astronomer to supply the dates. Around -108, the best theory of the sun was that of Hipparchus, and it would be natural for an Hellenistic astronomer to calculate the needed dates from Hipparchus's theory.

I do not advocate any particular explanation of the Greek dates listed in Table V.3. The reason for this is not difficulty in thinking of an explanation for them. On the contrary, it is easy to think of many explanations; the difficulty is in finding a sound basis of choice among competing explanations. The possibilities include the following:

a. The astronomer calculated the Greek dates from the Egyptian ones. If so, he made a systematic error. Perhaps, not unnaturally since he was dealing with Egyptian dates, he calculated the lunar dates by using the

---

†According to standard histories, a preliminary document but not the Great Charter itself was signed at Runnimede on 1215 June 15. Calculation from Newcomb's theory gives 09 hours on 1215 June 15, Runnimede time, as the time of the solstice.

principles of the Egyptian lunar calendar rather than the Athenian one.†

b. The Greek date in -431 may have been either an observed one or one that had been hallowed by tradition. If it were based upon observation, it probably corresponded to -431 June 28, because that is the date of the solstice and the measurement should have been fairly accurate. If so, the Greek calendar may have been based upon conjunctions, or as Meritt [1961, Chapter II] urges, the Athenians may not have been highly systematic in choosing the starting date for the month. If Ars Eudoxi rather than Geminus is right about Euctemon's observations, the date may have been -431 June 29; this agrees with a calendar based upon lunar visibility.

c. The Greek date Skirophorion 14 in -108 could have been calculated directly and easily from the date Skirophorion 13 in -431, without specifically using the Egyptian dates and without knowing anything about the Athenian calendar except that it was lunar. When rounded to whole days, 323 Hipparchan years equal 117 975 days. There are 3995 months in this interval, and 3995 mean synodic months take 117 974.72 days in the Hipparchan theory. The next step is to compare the mean anomaly of the moon at the beginning and end of the interval, and the comparison shows that this particular set of 3995 months was short and contained almost exactly 117 974 days. Thus an astronomer would know, without paying attention to the intervening months and years, that the date in -108 was 1 day later in a lunar calendar than the day in -431.‡

†The Egyptian lunar month was based upon the last visibility of the old moon rather than upon conjunction or the first visibility of the new moon. See Section II.3.

‡Because it is not easy to reconcile the Greek and Egyptian dates given in the parapegm if the later year is -108, van der Waerden [1960] concludes that the year is -105, when the Greek and Egyptian dates do apparently coincide. However, -108 is

In each explanation that I have thought of, the date Skirophorion 14 in -108 turns out to be calculated and not observed.

At this point, we can estimate, with reasonable assurance, the probability that the solstice of Meton and Euctemon was actually observed on any particular day. The best estimate that we can make of the actual time of the solstice is 10 hours on -431 June 28, Athens time. Sunset in the latitude of Athens at the time of the solstice is about $19\frac{1}{2}$ hours, local time.

If the solstice measurement came at a time that would cause us to equate the Greek date to -431 June 29, it must have been after sunset on -431 June 28, since sunset is presumably the time when the Athenians began their day. This would make the error in the time equal to $+9\frac{1}{2}$ hours, in the sense observed minus actual, and we have seen that the standard deviation of the measurement is close to 7 hours. The odds against the error being positive and greater than $9\frac{1}{2}/7 \approx 1.36$ standard deviations are about 10 to 1.

If the solstice measurement came at a time that would cause us to equate the Greek date to -431 June 27, it must have been before sunset on -431 June 27. This means an error that is negative and greater in magnitude than $14\frac{1}{2}$ hours. The odds against such an error are about 52 to 1.

In sum, the observation was made on the Greek date that we equate to -431 June 28, with high probability. The odds against using -431 June 29

---

the concluding year of a Metonic cycle and therefore -108 rather than -105 is far more likely for the year of an inscription commemorating that cycle. Further, since there is no difficulty in finding the Greek date by calculation if the year is -108, and since there is no need to assume that the Greek and Egyptian dates are equivalent, we are justified in accepting -108 as the year intended.

are 10 to 1,[†] the odds against using -431 June 27 are 52 to 1, and the odds against any other date are overwhelming.

Finally, then, we conclude the following from the Milesian parapegm:

a. The Egyptian dates that lead to -431 June 27 and -108 June 26 for summer solstices were calculated from Hipparchus's theory of the sun. There is little reason to assume that -431 June 27 is an observed date. There is even less reason to assume that this date equals Skirophorion 13 in the Athenian calendar.

b. The only date that may be the result of observation is Skirophorion 13 in the year -431. If it came from observation, -431 June 28 is its most likely equivalent, so far as the solstice observation is concerned. June 29 is suggested by lunar theory, however.

c. No time of day is given for the solstice in the parapegm. Even if we could unambiguously assign a date to the observation, we could not use it in estimating the solar acceleration for lack of the hour.

Thus the dates in the Milesian parapegm are not useful for the purposes of this study. It is doubtful that they are useful for the study of chronology either; in fact they have probably damaged such study.

The only Hellenistic times of equinoxes and solstices that have not yet been considered in this section are those that Ptolemy claims to have measured himself. Since these times were clearly

---

[†]The reader should remember, however, that sunset on -431 June 28 is the earliest possible time when the day Skirophorion 13 could have begun. This suggests that -431 June 29 is the correct equivalent of Skirophorion 13, in spite of the odds.

fudged, it follows that the only useful times are those in Table V.2, which are due to Hipparchus, and the times of the solstices attributed to Aristarchus and Hipparchus.

## 4. Direct Statements of Solar Position

In various chapters of the *Syntaxis*, Ptolemy preserves five values of the position of the sun that were allegedly read on the longitude circle of an astrolabe. He attributes three of the readings to Hipparchus and two to himself. I shall summarize the five records in the order that they occur in the text; the chapter number is given as the first item in each summary. Hipparchus's observations were made on Rhodes and Ptolemy's observations were allegedly made in Alexandria.

1. Chapter V.3. Ptolemy observed the sun and moon at $5\frac{1}{4}$ equal hours before noon on EN 886 VII 25 (= 139 February 9). The sun was seen at 18 5/6 degrees† of Aquarius and the moon was seen at 9 2/3 degrees of Scorpio. The mean place of the moon was 17°20′ of Scorpio. The mean place of the sun was 16° 27′ of Aquarius and its exact place was 18°50′ of Aquarius.

2. Chapter V.3. Hipparchus observed the sun and moon on the Egyptian day XI 16 of a certain year; I shall return to the question of the year in a moment. The observation was made when 2/3 of the first unequal hour had passed, and Ptolemy equates the time to 6 1/6 equal hours, apparent time, before noon, or $6\frac{1}{4}$ hours, mean time, before noon. The sun was seen at 8 7/12 degrees of Leo and the moon at 12 1/3 degrees of Taurus. The mean place of the sun was 10°27′ of Leo and its exact place was 8°20′.

There is extensive confusion about the year of

---

†I write fractions of a degree in the form that Ptolemy used, except that I use a single fraction in the places where he used a sum of fractions. Ptolemy rarely if ever used a fraction with a numerator other than unity, except for thirds and fifths. Thus, for example, he might write $\frac{1}{2} + \frac{1}{3}$, which I would change to 5/6.

this observation. Heiberg's text and Manitius's translation† both give it as the 50th year of the 3rd Callippic cycle.‡ This makes the date equal to -128 August 5. Halma, in both his text and translation, gives the year as the 52nd year of the 3rd cycle; this makes the date equal to -126 August 5. Finally, in a statement that reads the same way in all versions, Ptolemy says that the observation was made 619 years, 314 days, and 17 3/4 hours mean time after the epoch of Nabonnassar. This puts the observation on -127 August 5 and hence in the 51st year of the cycle.

Since the moon is in a quite different part of the sky at the same time in successive years, we can find which year is correct by calculating the position of the moon. The position of the mean moon is in fact sufficient for this purpose. On August 5, the mean moon was in Sagittarius in -128, in Taurus in -127, and in Virgo in -126. Thus the correct year is the 51st year of the 3rd Callippic cycle and the date is -127 August 5.

3. Chapter V.5. Hipparchus observed the sun and moon at the beginning of the 2nd unequal hour on the Egyptian date VIII 11 in the 197th year after the death of Alexander.‡ Ptolemy takes this time to be 5 2/3 equal hours, apparent time, before noon. The sun was seen at 7 3/4 degrees of Taurus and the moon was seen at 21 2/3 degrees of Pisces. The mean place of the sun was 6°41′ of Taurus and its exact place was 7°45′.

4. Chapter V.5. Hipparchus observed the sun and moon at 9 1/3 unequal hours on day X 17 of the same year, so that the date is -126 July 7. Ptolemy takes the time to be 4 equal hours after noon,

---

†See the entry under <u>Ptolemy</u> [ca. 142] in the list of references.

‡See Section II.4 and especially Equation II.6.

‡Ptolemy counts the era of Alexander from the date EN 425 I 1, as we learn from his chronological table that precedes the <u>Syntaxis</u>. Thus the date is EN 621 VIII 11 = -126 May 2.

apparent time. The sun was seen at 10 9/10† degrees of Cancer and the moon at 29 degrees of Leo. The mean sun was at 12°5′ of Cancer and the exact sun was at 10°40′.

5. Chapter VII.2. Ptolemy observed the sun and moon at sunset, which was 5½ equal hours after apparent noon, on EN 886 VIII 9 = 139 February 23. The sun was seen at 3° of Pisces and the moon was 92 1/8 degrees from the sun.‡ Half an hour later, the star $\alpha$ Leonis was seen to be 57 1/6 degrees east of the moon. The exact sun was at 3 1/20 degrees of Pisces, but the mean sun is not given.

In order to verify my interpretation of the dates of these observations, which do not always agree with those given by Manitius in the notes to his translation, I calculated the mean positions of the sun and moon from Ptolemy's theory. In modern notation, the mean longitudes of the sun and moon according to Ptolemy's theory are

$$\begin{aligned} L_S &= 330^\circ.75 + 0^\circ.985\ 635\ 278\ 4\ D_P\ , \\ L_M &= 41^\circ.37 + 13^\circ.176\ 382\ 215\ 2\ D_P. \end{aligned} \qquad (V.3)$$

In these relations, $D_P$ is the number of days, in solar time, from the epoch of Nabonassar. The mean motions are based upon Table V.1. Ptolemy's values of $L_S$ and $L_M$ at the epoch of Nabonassar, that is, when $D_P = 0$, are given in Chapters III.7 and IV.8, respectively, of the _Syntaxis_.

The values of $L_S$ and $L_M$ calculated from Equations V.3, using the dates that I have stated,

†It is unusual to see a decimal fraction in Hellenistic astronomy, but all cited texts agree on the reading. The position is actually written as 11 degrees minus 1/10.

‡Since the observation was allegedly made at sunset, the moon was east of the sun by 92 1/8 degrees. A fraction with the base 8 is even more unusual in Hellenistic astronomy than one with the base 10.

agree closely with the values given by Ptolemy. Since the calculated values serve no purpose beyond confirming the dates, it does not seem necessary to give details.

We must now decide what Ptolemy meant by the phrase that I have translated as the "exact sun", and we must decide whether the observations attributed to Hipparchus are genuine or fudged. We start by calculating the position of the sun at the time of each observation, using Ptolemy's solar tables for the purpose.

The calculation involves several parts. First we need the mean longitude of the sun. In order to determine better what Ptolemy did, I use the values that he stated rather than those calculated from Equation V.3. Next we subtract the longitude of the solar apogee, which Ptolemy takes as fixed at 65° 30′ (Chapter III.4).† The difference is the

TABLE V.4

STATEMENTS OF SOLAR POSITION PRESERVED BY PTOLEMY

| Date | Apparent Time after the Epoch of Nabonassar, days | Longitude of the Sun | | | | | |
|---|---|---|---|---|---|---|---|
| | | As Read | | "Exact" | | Calculated | |
| | | ° | ′ | ° | ′ | ° | ′ |
| +139 Feb 9 | 323 228.781 | 318 | 50 | 318 | 50 | 318 | 44 |
| -127 Aug 5 | 226 249.743 | 128 | 35 | 128 | 20 | 128 | 20 |
| -126 May 2 | 226 519.764 | 37 | 45 | 37 | 45 | 37 | 45 |
| -126 Jul 7 | 226 586.167 | 100 | 54 | 100 | 40 | 100 | 43 |
| +139 Feb 23 | 323 243.229 | 333 | 00 | 333 | 03 | 333 | 04 |

†That is, Ptolemy thought that apogee precesses with the equinox, always remaining at the same position with respect to it. This is a consequence of the fact that he 'verified' Hipparchus's values for the lengths of the individual seasons as well as his value for the length of the year.

independent variable in Ptolemy's table (Chapter III.6) of the προσθαφαιρεσις,† which is the quantity that Ptolemy adds to the mean longitude in order to obtain the longitude of the true sun.

The results of the calculations are summarized in Table V.4. The first column of the table gives the date in the Julian calendar. The second column gives the apparent time of the observation, measured in days after the epoch of Nabonassar.‡ The third column gives the longitude that was read on the astrolabe, the fourth column gives what I have translated as the exact longitude of the sun, and the fifth column gives the longitude calculated in the manner just described.

We see that the "exact" longitude agrees closely with the calculated longitude for all observations except that of 139 February 9. The discrepancies that exist for the other dates can easily be explained as the result of differences between my calculations and those of Ptolemy with regard to ways of interpolating in the table of the προσθαφαιρεσις, or with regard to rounding.# For the observation of 139 February 9, either there has been an error in transmitting the text, or Ptolemy made a simple error, or he rounded to the nearest multiple of 10′, rather than of 5′, at each of two points in the calculation.

---

†Manitius and Halma merely transliterate this term in slightly different ways. I shall not attempt to translate it either. The προσθαφαιρεσις is the function that is called the "equation of the center" in more recent writing, but this term [Neugebauer, 1957, p. 192] "appears first in Latin translations of Arabic treatises."

‡This is not quite the same as $D_P$ in Equation V.3, which is in mean time.

#Ptolemy often, but by no means always, shows indifference to quantities less than 5′. He may have regarded an angle of about this size as the practical limit to the accuracy of observations. Half of the discrepancy for 139 February 9 may be the result of Ptolemy's value of the solar parallax, which was 3′.

With regard to the readings that Ptolemy claims to have made himself, the "read" position agrees with the "exact" position in one instance and it disagrees by a small amount in the other. Thus it is likely that the "read" position was derived from the calculated "exact" position. The small discrepancy in one instance may be the result of rounding, or of scribal error. Alternatively, it may have been introduced intentionally in order to lend verisimilitude to the record.

With regard to the readings that Hipparchus made, we must first decide whether Hipparchus would have calculated the same position as Ptolemy. Since Ptolemy adopted the values of Hipparchus for the mean motion of the sun, for the lengths of the seasons and hence for the eccentricity and apogee, and for the times at the equinoxes and solstices, calculations made by Hipparchus and Ptolemy should agree except perhaps for small differences caused by differences in the mathematical techniques used.

We see that the "exact" position agrees with that read by Hipparchus in one instance and that it disagrees by about 15′ in two instances. It seems unlikely that differences in the method of computation, made with identical physical parameters, would yield this much difference. Thus it is likely that the "read" values of Hipparchus were not obtained by calculation.

We must now consider the structure and use of the astrolabe that Ptolemy describes, which in essentials is probably the same as the one that Hipparchus used. Ptolemy gives us the necessary information in Chapter V.1 of the Syntaxis.

First, there is a circle marked with positions for two pairs of poles. One pair represents the poles of the equator. The circle is mounted so that it can rotate about an axis through these poles, and the support of the astrolabe is oriented so that these poles point to the celestial poles in the heavens. The other pair of poles represents the poles of the ecliptic, which are placed at an appropriate angle from the poles of the equator.

A second circle is mounted rigidly with respect to the first circle in a position corresponding to the plane of the ecliptic. That is, the line joining the poles of the ecliptic passes through the center of the second circle and is normal to its plane. For brevity, I shall call this second circle the ecliptic circle. The ecliptic circle carries pointers and sights that allow measuring the latitude and longitude of any object, once the ecliptic circle is correctly oriented.

As the astrolabe is rotated about the poles of the equator, there is one and only one position in which the ecliptic circle is parallel to the plane of the ecliptic. Thus the first task in using the astrolabe is to orient the ecliptic correctly. Then the circle can be used to read celestial coordinates.

If the user has an accurate clock, he can set the astrolabe from a knowledge of the time. However, for the observations in Table V.4, Ptolemy describes a different method. The user first calculates an approximate longitude of the sun and turns the astrolabe until the sun reads approximately this value on the longitude scale. He then sets the astrolabe accurately by rotating until the sun is in the plane of the ecliptic circle. He judges this by watching for the position in which the interior of the circle is completely unlighted by the sun. Then, with the astrolabe held in this position, he reads the position of the sun and the moon. There is a variant of this procedure for use at night.

In order to simplify the discussion, let us suppose that the observation was made exactly at the vernal equinox. Then, when the user makes the ecliptic circle pass through the sun,† he necessarily finds the sun at the vernal point on the

†Since the method of aligning the ecliptic circle with the sun is a null method (no light on the interior of the circle), it is capable of high precision. I have made some tests with a simplified model of an astrolabe that George B. Bush of the Applied Physics Laboratory made for me, and I believe the circle can easily be set to a precision of a minute or two of arc in the declination of the sun.

circle. Since the longitude of the vernal point is 0° by definition, he therefore finds the correct longitude, within measurement error, independent of any error in his tables of the sun.

When we generalize the argument to any position of the sun, we find that the user should measure the correct longitude at any time of the year, within measurement error. In fact, the process basically is equivalent to measuring the declination $\delta$ of the sun and finding the longitude $\lambda$ by solving the equation

$$\sin \delta = \sin \varepsilon \sin \lambda \ .$$

In this, $\varepsilon$ is the obliquity of the ecliptic.

For the two observations that Ptolemy claims to have made himself on 139 February 9 and 139 February 23, the dates are within 40 days of the vernal equinox, and the longitude is well determined by the declination. Aside from an error in aligning the ecliptic plane, there can be systematic errors in aligning the axis of the instrument with the polar axis of the earth and in the value used for the obliquity $\varepsilon$. The error in finding $\lambda$ might be as large as 0°.5 under these circumstances. However, the error in Ptolemy's solar tables is more than 1° at the time of these observations.

Hence the "calculated" and "read" values in Table V.4 should differ by more than 0°.5 degrees if the observations are genuine. The fact that the values agree closely shows that the observations are fudged. This analysis did not form part of the test described in Section V.1, and it furnishes additional evidence that the observations claimed by Ptolemy are not genuine.

In the time of Hipparchus's observations, Ptolemy's tables are fairly accurate because they are in fact just Hipparchus's own tables. Hence the "calculated" and "read" values should agree within ordinary error, and we see from Table V.4 that they do so agree. Thus the observations attributed to Hipparchus are probably genuine. If they were well distributed throughout the seasons, I would use them in this work. However, they are bunched near the summer solstice, so that there is serious danger of a systematic error that cannot be removed. For this reason, and not because of doubts about their validity, I shall not use the observations attributed to Hipparchus in Table V.4.

## 5. Observations of Mercury

Ptolemy establishes the parameters of his model of Mercury in Chapters IX.7 through IX.11 of the Syntaxis, and he uses seventeen observations in order to do so. He claims to have made eight of the observations himself. One observation was made by Ptolemy's near contemporary, Theon. Six were made by an Alexandrian astronomer of the -3rd century named Dionysios.† Two observations with dates fairly close together are unattributed. From Ptolemy's use of them, it is clear that they were made in Alexandria, and they were made about twenty years after the observations that are specifically attributed to Dionysios. Thus it is possible that Dionysios made them; if he did, the dates of his observations are peculiarly distributed.

When he gives the times of the observations of Mercury, Ptolemy states them only as morning or evening, with one exception. This is almost exactly the practice that the Babylonians (Section IV.7) followed. I shall assume that the times meant are 15 minutes from sunrise or sunset, whichever is appropriate, just as I did with the Babylonian data. Luckily, there are almost exactly as many observations in the morning as in the evening, so that the assumption should not introduce a bias into the estimates of the accelerations.

†We do not know anything about Dionysios [Pauly-Wissowa, 1894, v. 5, p. 991] except what we can learn from the observations quoted by Ptolemy. There was a Theon in Alexandria at the right time [Pauly-Wissowa, 1894, v. 5, Second Series p. 2067] whose name is known from other sources. It is uncertain whether this Theon is the one whom Ptolemy quotes. There was also an astronomer named Theon in Alexandria in the +4th century, who was clearly not Ptolemy's contemporary.

We can compare this assumption with the assumption that Ptolemy made about the times. Ptolemy calculated the position of the mean sun for each observation. We can calculate the mean sun from the first of Equations V.3 for the time assumed for each observation and compare with the value that Ptolemy gave. On the average, the value from Equations V.3 is 0°.043 smaller than Ptolemy's value for the morning observations and 0°.035 larger for the evening ones. Thus it is likely that Ptolemy took the times to be sunrise and sunset. This does not mean that he really thought that sunrise and sunset were the correct times. Instead, he probably thought that the precision involved in using the correct times was not needed in a planetary theory.

I shall now summarize the facts of the observations as Ptolemy gives them, in the order that they occur in the text. All the observations were made at Alexandria. Here I use only the modern designations of the stars that are involved. The bases for the identifications, and the coordinates and proper motions of the stars, are given in Sections III.10 and III.11. The chapter designation given as the first part of each summary is the number assigned by Heiberg and Manitius.

'Maximum elongation' in these records means a maximum with respect to the mean sun, not the true sun.†

1. Chapter IX.7. Ptolemy observed in the evening between the 16th and 17th of month VII in the 16th year of Hadrian. Mercury was at maximum elongation. Its position was found by comparison with $\alpha$ Tauri, and its longitude was found to be 331° 0′.‡ The date of the observation is EN 879 VII 16 = 132 February 2.

---

†I shall give a more precise definition of 'mean elongation' in Section XIII.5.

‡Ptolemy almost always gives a longitude by stating a position within a zodiacal sign. I preserved this usage in the preceding section. For simplicity, I shall give the longitude in the contemporary manner in the remaining records.

In Ptolemy's star table, which has the epoch EN 885 I 1,† α Tauri has the longitude 42 2/3 degrees. Ptolemy's rate of precession is 1° per century. Since the observation is about 5½ years before the epoch of the table, the longitude that Ptolemy would have assigned at the time of the observation is about 42°37′. Thus Mercury was 71° 37′ west of α Tauri, according to the record. In accordance with the decision made in Section V.1, this observation receives a reliability of zero.

2. Chapter IX.7. Ptolemy observed in the morning of day XI 19 in the 18th year of Hadrian. Mercury was again at maximum elongation, but in the opposite direction. Compared with α Tauri, it was at longitude 48 3/4 degrees. The date of the observation is EN 881 XI 18 = 134 June 4. Ptolemy's position for α Tauri was therefore about 42°38′, so that Mercury was 6°7′ east of α Tauri. This observation receives a reliability of zero.

3. Chapter IX.7. Ptolemy observed in the evening between the 20th and 21st of month XI in the first year of Antoninus Pius. Mercury was again at maximum elongation. Compared with α Leonis, it was at longitude 97°0′. The date is EN 885 XI 20 = 138 June 4. In Ptolemy's table, α Leonis is at 122½ degrees, and we may neglect the precession since we are so near to the epoch of the table. Hence Mercury was 25°30′ west of α Leonis. The reliability assigned to this is zero.

4. Chapter IX.7. Ptolemy observed in the morning between the 18th and 19th of month VII in the 4th year of Antoninus Pius. Mercury was at maximum elongation. Compared with α Scorpii, it was at longitude 283½ degrees. The date is EN 888 VII 19 = 141 February 2. The longitude of the star in the table is 222 2/3 degrees,‡ which would have increased to about 222°42′ by the time of the observation. Hence Mercury was about 60°48′ east of α Scorpii, if we accept the reading of Heiberg and

---

†The epoch commemorates the year of the accession of Antoninus Pius.

‡Halma has 222 1/3, but Heiberg and Manitius have 222 2/3.

Manitius. The reliability assigned to the observation is zero.

5. Chapter IX.7. Dionysios observed Mercury in the morning between EN 486 IV 17 and 18.† Mercury was in conjunction‡ with δ Capricorni and 3 moons to the north. Ptolemy says that it was hence at longitude 292°20′; he does not give the latitude. We should of course use the longitude of δ Capricorni as calculated from modern data. The observation receives a reliability of unity and its date was -261 February 12. Mercury was at maximum elongation.

The moon was used fairly often as a unit of angular measure in Greek writing. It is possible that it was used carefully, which would require that we calculate its apparent diameter carefully. However, I doubt that it was used with enough accuracy to warrant such care; it is probably adequate to use 1 moon = 30′.

6. Chapter IX.7. Dionysios again observed Mercury at maximum elongation in the evening of EN 486 VI 30 = -261 April 25.‡‡ It was 3 moons more advanced in longitude than the straight line that joins β Tauri and ζ Tauri. In latitude, it was south of β Tauri by more than 3 moons. This observation receives a reliability of 1.

†Dionysios stated his dates in terms of a late Greek calendar that I did not mention in Section II.4. Since Ptolemy converts each date into its Egyptian equivalent, I shall give only the latter. The interested reader can find what is known about Dionysios's calendar in volume 2, pages 406-408, of Manitius's translation of the Syntaxis. There are so few dates known in this calendar that we cannot determine the structure of its months unambiguously.

‡Ptolemy says explicitly that Mercury and δ Capricorni had the same longitude.

‡‡Ptolemy had to find the date by converting from Dionysios's calendar. He wrote the Egyptian date using the name of month VII. As Manitius points out, he made a simple slip. The data clearly show that month VI is correct.

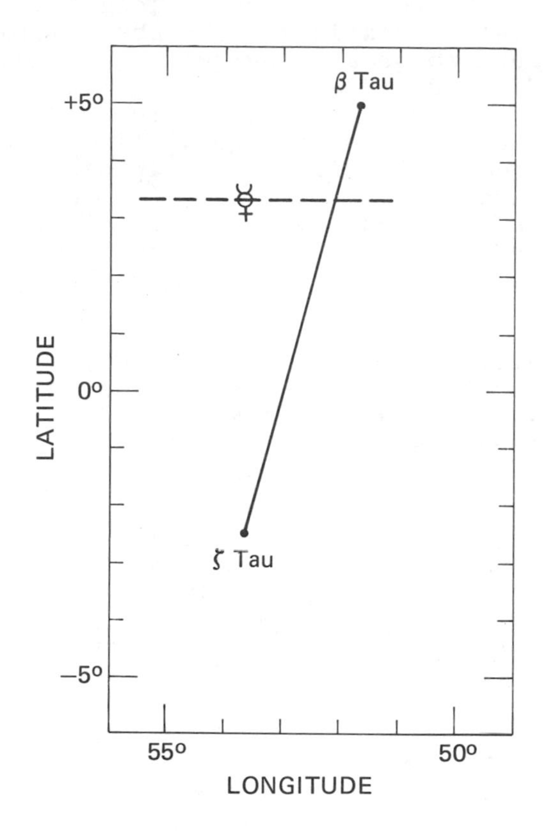


Figure V.1. The configuration of Mercury, β Tauri, and ζ Tauri, as observed by Dionysios in the evening of -261 April 25. The stellar coordinates are those used by Ptolemy rather than values calculated from modern tables.

The configuration is shown in Figure V.1. The epoch of the observation is almost exactly four centuries before the epoch of Ptolemy's star table, so we may subtract 4° from his tabular longitudes in analyzing his interpretation of the observation. He thus took β Tauri to be at longitude 51 2/3 degrees and latitude +5 degrees. He took ζ Tauri to be at longitude 53 2/3 degrees and latitude $-2\frac{1}{2}$ degrees. Mercury was on the horizontal line that passes 1°40′ south of β Tauri. It was $1\frac{1}{2}$ degrees east of the intersection of this line with the line joining β Tauri and ζ Tauri. Thus its longitude was almost exactly 53 2/3 degrees, which is the value that Ptolemy used. This is just the longitude of ζ Tauri.

We should refer the position of Mercury directly to a star instead of using Ptolemy's longitude. Thus we may say that Mercury was 5 5/6 degrees north and 0 degrees east of ζ Tauri.

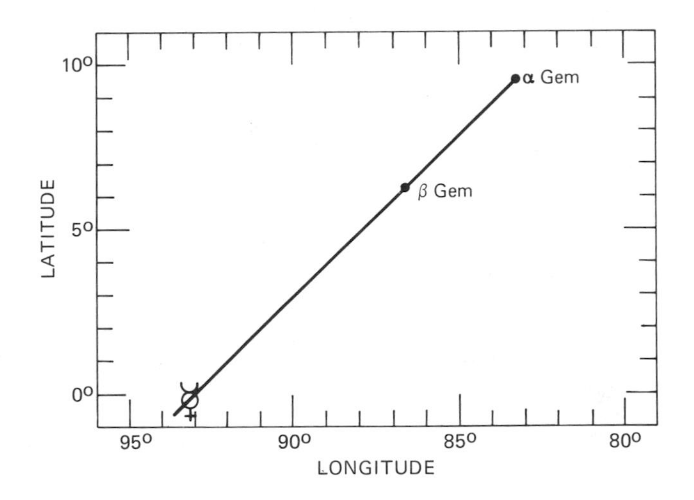


Figure V.2. The configuration of Mercury, α Geminorum, and β Geminorum, as observed by Dionysios in the evening of -256 May 28. The stellar coordinates are those used by Ptolemy rather than values calculated from modern tables.

7. Chapter IX.7. Dionysios again observed Mercury at maximum elongation in the evening of EN 491 VIII 5 = -256 May 28. It was on the extension of the straight line joining the heads of the twins ($\alpha$ and $\beta$ Geminorum). Its distance from the southern one, measured along the extension of the line, was equal to twice the distance between them, less 1/3 of the width of the moon. It was thus at longitude 89 1/3 degrees, according to Ptolemy.

The configuration is shown in Figure V.2. The abscissa in the figure is the longitude given in Ptolemy's star table, before correcting for precession. Twice the distance from $\alpha$ to $\beta$ Geminorum, measured along the extension from $\beta$ Geminorum, gives the point of intersection marked E in the figure. Moving back toward $\beta$ Geminorum by 10′, which is about 1/3 of the width of the moon, brings us to the point marked with ☿. This point has the longitude 93°12′. In order to compare it with the longitude Ptolemy gave, we should subtract his value of the precession, which is 3°56′. The result is 89°16′, whereas he gives 89 1/3 degrees. I suspect that he did not bother to move back from point E by "1/3 of the width of the moon".

It should be sufficiently accurate to say that Mercury was 6½ degrees east and 6½ degrees south of $\beta$ Geminorum. This observation receives a reliability of 1.

8. Chapter IX.7. This is the last Mercury observation that Ptolemy attributes to Dionysios. He observed Mercury at maximum elongation in the evening of EN 486 X 30 = -261 August 23. It preceded Spica by a little more than 3°, so that it was at longitude 169½ degrees by Ptolemy's reckoning. In Ptolemy's table, Spica ($\alpha$ Virginis) has the longitude 176 2/3 degrees,† so that Ptolemy would have assigned it the value 172 2/3 at the epoch of the observation. Thus Ptolemy takes the observation to mean that Mercury was 3°10′ west of $\alpha$ Virginis, and we may as well do the same. The observation receives a reliability of 1.

---

†Here and in several other places, Halma's Greek text agrees with the other cited sources in reading 2/3, while his translation reads 1/3.

9. Chapter IX.7. Ptolemy does not name the observer, although he specifically names Dionysios for each of the four preceding observations. Further, the calendar used in the original statement of the date, before Ptolemy translates it to the Egyptian calendar, is different from the one used in the observations attributed to Dionysios. Thus, while it is physically possible that Dionysios made this observation, it seems unlikely to me that he did so. According to the observation, Mercury was ½ 'ammat'† above $\alpha$ Librae in the morning of EN 512 I 10 = -236 October 30. Ptolemy says that this put Mercury at longitude 194 1/6 degrees. In his table, $\alpha$ Librae has longitude 198°. Thus he has taken the precession to be 3 5/6 degrees, whereas I calculate 3°44′. Mercury was again at maximum elongation.

In sum, the observation says that Mercury was 0° east and ½ ammat north of $\alpha$ Librae. The observation receives a reliability of 1.

10. Chapter IX.7. It is probable that the person who made this observation made the preceding one, since he uses the same calendar and the same unit of measurement. This is also a measurement of Mercury at maximum elongation. In the morning between EN 504 I 27 and 28, and thus on -244 November 19,‡ Mercury was ½ ammat above $\beta$ Scorpii. According to Ptolemy, the longitude of Mercury was therefore 212 1/3 degrees.

---

†I gather from the context that the unit used here is the same as the Babylonian unit found in the preceding chapter. Use of this unit rather than the 'moon' for measuring angles seems to decrease further the likelihood that Dionysios made this observation. The reader may remember from Section IV.6 that the unit may equal either 2 or 2½ degrees in Babylonian usage. Manitius, in the translation cited, takes the unit to be 2°.

‡Manitius always gives the year in historical style rather than astronomical style. In a few places, such as this one, his year is wrong by 1. In each such place that I noticed, it appears that he calculated the year in astronomical style and forgot to change the number before stating it in historical style.

In Ptolemy's table, the longitude of β Scorpii is 216 1/3 degrees. Therefore Ptolemy rounded off the precession in 381 years to 4°.

In sum, the observation says that Mercury was 0° east and ½ ammat north of β Scorpii. It receives a reliability of unity.

11. Chapter IX.8. Ptolemy observed in the morning between III 14 and 15 in the 19th year of Hadrian (EN 882). Mercury was again at maximum elongation. When it was compared with α Leonis, Mercury was found to be at 170 1/5 degrees in longitude. In Ptolemy's table, α Leonis is at 122½ degrees, and Ptolemy probably neglected the precession for an observation this close to his epoch. Hence the record says that Mercury was 47°42′ east of α Leonis. The date is 134 October 3. This observation receives a reliability of zero.

12. Chapter IX.8. Ptolemy observed in the evening of IX 19 of the same Egyptian year (135 April 5), when Mercury was again at maximum elongation. When it was compared with α Tauri, Mercury was found to have longitude 34 1/3 degrees. In the star table, α Tauri has longitude 42 2/3 degrees. Thus the record says that Mercury was 8 1/3 degrees west of α Tauri. The observation receives a reliability of zero.

13. Chapter IX.9. Theon observed Mercury at maximum elongation in the evening of XII 18 in the 14th year of Hadrian (EN 877), so that the date was 130 July 4. Mercury was more advanced in longitude than α Leonis by 3 5/6 degrees, so that it was at longitude 126 1/3 with respect to Ptolemy's equinox. Thus Ptolemy neglected the precession for this observation, which was made about 7 years from the epoch of his tables. The observation receives a reliability of unity.

14. Chapter IX.9. Ptolemy says that he observed Mercury at maximum elongation in the morning of EN 886 XII 24 = 139 July 8. When compared with α Tauri, Mercury was found at longitude 80 1/12 degrees. Thus Mercury was 37 5/12 degrees east of α Tauri. The observation receives a reliability of zero.

There is a problem about the date. Ptolemy says that the mean longitude of the sun was 100 1/3 degrees, but I get 103°.200 when I calculate the mean longitude from Equations V.3 (in Section V.4). When we look at Ptolemy's use of this observation in his Chapter IX.9, it is clear that he uses 100 1/3 degrees for the mean longitude of the sun, and not a value near 103°. The date of the observation should therefore be 139 July 5 = EN 886 XII 21. The explanation of the error is simple. The correct day (21) of the Egyptian month is written as $\overline{\text{KA}}$ in Greek numerals. It would be easy for an early copyist, or even for Ptolemy himself in writing his book from his notes, to confuse this with $\overline{\text{K}\Delta}$, which is the numeral for 24.

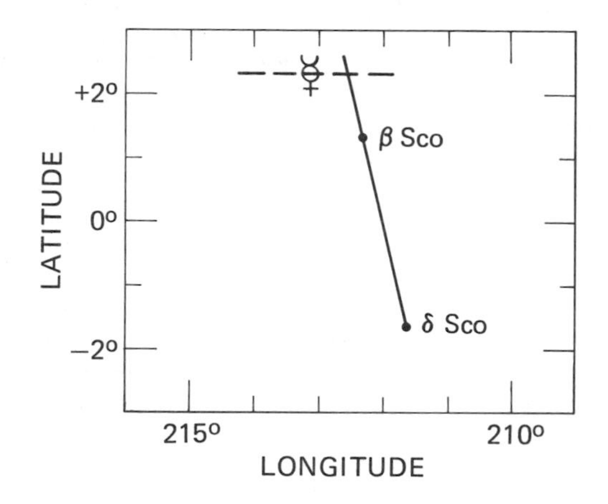


Figure V.3. The configuration of Mercury, β Scorpii, and δ Scorpii, as observed by Dionysios in the morning of -264 November 15. The stellar coordinates are those used by Ptolemy rather than values calculated from modern tables.

15. Chapter IX.10. Ptolemy observed Mercury during the night between EN 886 XI 2 and 3, at 4½ equal hours before midnight. Thus the date is EN 886 XI 2 = 139 May 17. Ptolemy implies that he found the time by means of the astrolabe. This is the first observation of Mercury when it was not at maximum elongation. Ptolemy compared Mercury with $\alpha$ Leonis and found that it was at longitude 77½ degrees. He also says that Mercury was 1 1/6 degrees behind the center of the moon. Since $\alpha$ Leonis has longitude 122½ degrees in Ptolemy's table, the observation says that Mercury was 45 degrees west of $\alpha$ Leonis. This observation has a reliability of zero.

16. Chapter IX.10. Dionysios observed Mercury in the morning between EN 484 I 18 and 19, so that the date of the observation is -264 November 15. Mercury was not at maximum elongation. It was 1 moon to the east of the line joining $\beta$ Scorpii and $\delta$ Scorpii and it was 2 moons north of $\beta$ Scorpii. Thus, says Ptolemy, Mercury was at longitude 213 1/3 degrees.

The configuration is shown in Figure V.3. Ptolemy specifically takes the precession between the observation and the epoch of his table to be 4°. We see from the figure that Mercury was at longitude 213°06′ according to the observation. Either Ptolemy did not calculate the position carefully or there has been an error in the transmission of the text.

According to the observation, then, Mercury was 1° north and 46′ east of $\beta$ Scorpii. The observation receives a reliability of unity.

17. Chapter IX.10. Dionysios reported that Mercury moved one-half of the width of the moon to the east during the four days after the preceding observation. Thus, in the morning of -264 November 19, Mercury was 61′ east of $\beta$ Scorpii, while presumably remaining 1° north. This observation receives a reliability of unity.

We have now reduced each observation to a form in which it states the position of Mercury relative to a known star. We can also estimate the solar time of each observation. In forming this estimate, I take 'morning' and 'evening' to mean 45$^m$ from sunrise and sunset, respectively. In order to estimate the local (Alexandria) mean times of sunrise and sunset, I first find the date of each observation in the Gregorian calendar. I then

TABLE V.5

OBSERVATIONS OF MERCURY PRESERVED BY PTOLEMY

| Date | Hour | JSD − 1 000 000 | Position[a] | Reliability |
|---|---|---|---|---|
| 132 Feb 2 | E | 769 303.182 | 71°37′ west of α Tau | 0 |
| 134 Jun 4 | M | 770 155.593 | 6°7′ east of α Tau | 0 |
| 138 Jun 4 | E | 771 617.238 | 25°30′ west of α Leo | 0 |
| 141 Feb 2 | M | 772 590.671 | 60°48′ east of α Sco | 0 |
| -261 Feb 12 | M | 625 769.668 | 0° east, 1½° north of δ Cap | 1 |
| -261 Apr 25 | E | 625 842.220 | 0° east, 5 5/6° north of ζ Tau | 1 |
| -256 May 28 | E | 627 702.234 | 6½° east, 6½° south of β Gem | 1 |
| -261 Aug 23 | E | 625 962.225 | 3°10′ west of α Vir | 1 |
| -236 Oct 30 | M | 635 161.641 | 0° east, ½ am. north of α Lib | 1 |
| -244 Nov 19 | M | 632 259.652 | 0° east, ½ am. north of β Sco | 1 |
| 134 Oct 3 | M | 770 276.631 | 47°42′ east of α Leo | 0 |
| 135 Apr 5 | E | 770 461.212 | 8⅓° west of α Tau | 0 |
| 130 Jul 4 | E | 768 725.243 | 3 5/6° east of α Leo | 1 |
| 139 Jul 5[b] | M | 772 015.597 | 37°25′ east of α Tau | 0 |
| 139 May 17 | 19.5 | 771 964.227 | 45° west of α Leo, 1 1/6° east of the moon | 0 |
| -264 Nov 15 | M | 624 950.650 | 0°46′ east, 1° north of β Sco | 1 |
| -264 Nov 19 | M | 624 954.652 | 1°1′ east, 1° north of β Sco | 1 |

[a]East and west are measured parallel to the ecliptic, not the equator.

[b]Ptolemy incorrectly gives the date as the equivalent of 139 July 8.

use the tables of sunrise and sunset in the American Ephemeris and Nautical Almanac, assuming that the times are independent of the year if the Gregorian dates are used. Finally, I transfer the time to Greenwich. In calculating the times, I take the coordinates of Alexandria to be 31°.2N, 30°.0E.

The observations are summarized in Table V.5. The first column in the table gives the date in the Julian calendar. The second column gives the time of the observation at Alexandria. For every observation except one, the time is given as merely morning (M) or evening (E). For the observation of 139 May 17, however, Ptolemy gives the time as 4½ equal hours before midnight; I take this to be apparent time at Alexandria. The third column gives the time of the observation by means of the Julian day number and fraction, in terms of Greenwich mean solar time. Finally, the last two columns give the position of Mercury relative to a reference star, and the reliability that the observation will receive.

The first fourteen observations of Mercury in Table V.5 are at the position that Ptolemy calls maximum elongation. If the maximum were measured from the true sun, these fourteen observations would give us information about the acceleration of the sun only. Since the elongation is not necessarily a maximum with respect to the true sun, they may also give us some information about the acceleration of Mercury, but with low sensitivity. Only the last two measurements are likely to give us useful information about the acceleration of Mercury.

## 6. Observations of Venus

Ptolemy uses eleven observations in order to establish the parameters in his model of the motion of Venus. Six of the observations were allegedly made by him. Three were made by Theon, the contemporary who also made some of the observations of Mercury. Two were made by Timocharis. As Pauly-Wissowa [1894, v. 6, Second Series, p. 1936] says, all we know about Timocharis is what we can deduce from his observations: He observed in Alexandria around the year -270.† I shall now summarize the facts as they appear in the records.

1. Chapter X.1. Theon observed Venus in the evening between EN 879 VIII 21 and 22. Venus was at its maximum elongation. It was west of the center of the Pleiades by a distance equal to the width of the Pleiades, and it was slightly south of them. The most easterly and westerly of the Pleiades are 28 Tauri and 17 Tauri, respectively, and the point midway between them is 7′ west of $\eta$ Tauri. The width of the Pleiades is almost exactly 1°. Hence we may say that Venus was 1°7′ west of $\eta$ Tauri, and slightly to the south.‡ The date of the observation is 132 March 8.‡ The observation would receive a reliability of unity except for the confusion about the Pleiades. In view of the confusion, I shall lower the reliability to $\frac{1}{4}$.

2. Chapter X.1. Ptolemy observed Venus at maximum elongation in the morning between I 11 and 12 in the 14th year of Antoninus Pius. Venus was $\frac{1}{2}$ moon north and east of $\zeta$ Geminorum. There are two problems in understanding the record.

Let us dispose of the simpler problem first. The text leaves it unclear whether Venus was $\frac{1}{2}$ moon from $\zeta$ Geminorum in a northeast direction, or whether it was $\frac{1}{2}$ moon north and also $\frac{1}{2}$ moon east. Ptolemy's calculations based upon this observation

---

†Timocharis observed four lunar conjunctions [Ptolemy, ca. 142, Chapter VII.3] with dates ranging from -294 to -282.

‡I am unable to determine exactly which stars Ptolemy considered as the Pleiades. The identifications that Manitius and Halma make are contradictory. Ptolemy lists only four Pleiades, whereas standard mythology [Graves, 1955, v. 1, pp. 152-154, for example] says that there were six, a seventh having allegedly vanished about the time of the Trojan War. Ptolemy assigns a spread of $1\frac{1}{2}$ degrees to his four Pleiades, whereas the stars that I take to be the Pleiades cover only about 1°.08. See also Section III.10.

‡Manitius gives 132 March 7.

make it clear that he means the latter.

The other problem concerns the year. Halma's and Heiberg's texts agree that the year was the 14th year of Antoninus Pius, but this is many years later than any other observation recorded by Ptolemy. Thus there is reason to suspect that the year is an error. Since a single error is more likely than a multiple error, the year is probably either the 4th year of Antoninus Pius or the 14th year of Hadrian, if it is not the 14th year of Antoninus Pius. We can test the matter by using the mean longitude of the sun that Ptolemy calculates, plus the fact that Venus was at maximum elongation. The only one of the three possibilities that fits is the 4th year of Antoninus Pius, as Manitius concludes in the notes to his translation.

Hence the date of the observation is probably EN 888 I 12 = 140 July 30. This record receives a reliability of zero.

3. Chapter X.1. Theon observed Venus at maximum elongation in the morning between EN 875 III 21 and 22. It was east of β Virginis, and its distance from β Virginis was equal to the length of the Pleiades less the diameter of Venus.† Ptolemy takes this to mean that Venus was 1°25′ east of β Virginis. I shall take the width of the Pleiades to be 1°, as I did in the first observation above, and hence I shall take Venus to be 55′ east of β Virginis.

Because of the uncertainty about what the Pleiades mean, I shall give this observation a reliability of only $\frac{1}{4}$. This is a conservative action: Since Venus was west of the reference point in the first observation and east of it in this one, the uncertainty in the width of the Pleiades cancels when the observations are combined. The detailed

---

†This does not imply a knowledge of the physical size of Venus. Ancient astronomers attributed a diameter to celestial objects that was a function of the apparent brightness. It is clear from Ptolemy's calculations that he used 5′ for the 'diameter' of Venus.

analysis of the observations may tell us what the width of the Pleiades meant to the Hellenistic astronomers.

The date of this observation is 127 October 12.

4. Chapter X.1. Ptolemy observed Venus at maximum elongation during the evening between EN 884 VI 9 and 10; hence the date of the observation is 136 December 25. Venus was west of $\phi$ Aquarii by δυο μερη εγγιστα σεληνης διχομηνου (about two-thirds of the width of the full moon). Ptolemy goes on to say that Venus was hence 2/5 of a degree west of $\phi$ Aquarii.

Halma has two slips in his translation of this passage. δυο μερη can mean two parts, as Halma translates it, but it also has the specific meaning of two-thirds, and Halma did not notice that Ptolemy uses it with this meaning in his calculations. Halma also takes σεληνης διχομηνου to be half of the width of the moon. However, διχομηνος means the half-month, not the half-moon, and the entire phrase means the moon at this time, that is, the full moon.

Manitius's translation leaves the impression that Ptolemy took 36′ for the diameter of the moon. Actually, Ptolemy (Chapter VI.5) takes the diameter of the moon to vary between 31 1/3 and 35 1/3 minutes. I think that Ptolemy took 2/3 of some value between these limits, perhaps the average, and then rounded to the nearest convenient fraction of a degree.

After he specifies the position of Venus with respect to $\phi$ Aquarii, Ptolemy goes on in the same sentence to say: ". . and she seemed to efface this star by her brightness." This remark gives me pause. Someone could know as an academic matter that Venus is so bright that it dims a nearby star like $\phi$ Aquarii and could have inserted the remark in order to lend verisimilitude to a fudged record. If he did this, he had imagination and it would be a shame if he wasted his talents on the forging of scientific data. This kind of remark is usually taken as the sign of an eye witness.

All of Ptolemy's alleged observations (that he actually uses) that have been tested so far prove to be fraudulent. In spite of this, it is of course possible that he did make some genuine observations. Of all the observations that he claims to have made, this seems like the one with the best claim to be genuine, as judged from the text alone before doing the astronomical analysis. However, I shall keep with the decision of Section V.1 and give this a reliability of zero.

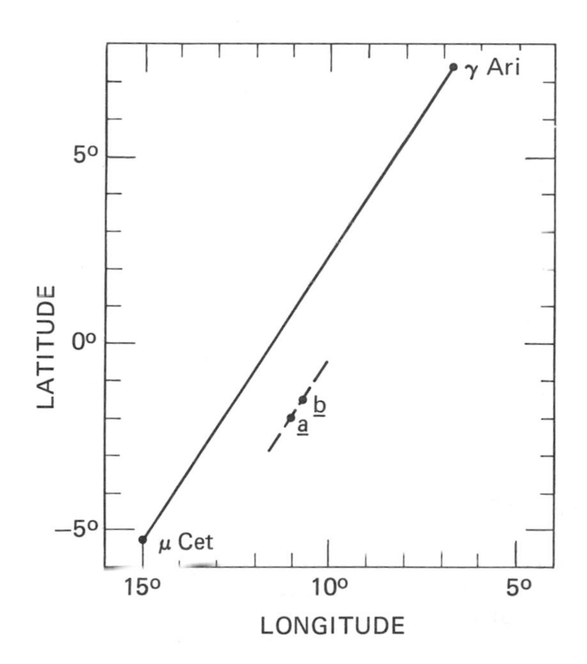


Figure V.4. The configuration of Venus, γ Arietis, and μ Ceti as observed by Theon in the morning of 129 May 20. The stellar coordinates are those used by Ptolemy rather than values calculated from modern tables. Point a is the point that corresponds to the described position of Venus, but point b is the position that Ptolemy uses.

5. Chapter X.2. Theon observed Venus at maximum elongation in the morning between EN 876 XI 2 and 3. Hence the date of the observation is 129 May 20. The reference for the observation was the line joining γ Arietis and μ Ceti.† Venus was 1 2/5 degrees west of this line, and it was twice as far from γ Arietis as from μ Ceti.

The configuration is shown in Figure V.4. The coordinates used for the stars are taken from Ptolemy's table, with no allowance for precession. There is a question about whether the distance of 1 2/5 degrees is to be measured in a direction parallel to the ecliptic or perpendicular to the line joining the two stars. The broken line in Figure V.4 is 1 2/5 degrees from the solid line along the perpendicular. Point *a* is the point on this line that is twice as far from γ Arietis as from μ Ceti, while point *b* is the position that Ptolemy assigned to Venus. Thus this line is the one that Ptolemy meant, but it seems that he did not find the position with great care.

Hence the record says that Venus was 3°.9 west of μ Ceti and 3°.35 north of it. The observation receives a reliability of unity.

6. Chapter X.2. Ptolemy observed Venus at maximum elongation in the evening between EN 884 V 2 and 3, so that the date of the observation is 136 November 18. When Venus was compared with the horns of Capricorn, it was found to be at longitude 282 5/6 degrees. Ptolemy lists four stars that are connected with the horns, and he leaves us free to refer the position of Venus to anyone of them. Thus we may say that Venus was $5\frac{1}{2}$ degrees east of β Capricorni. The observation receives a reliability of 0.

7. Chapter X.3. Ptolemy observed Venus at maximum elongation in the morning between EN 881 VIII 2 and 3, so that the observation was on 134 February 18. When Venus was compared with α Scorpii, it was found to be at longitude 281 11/12 degrees. Ptolemy assigns the longitude 222 2/3 to α Scorpii,

---

†See Section III.11.

and he probably neglected the precession for this observation, so that Venus was 59¼ degrees east of α Scorpii. This observation receives a reliability of 0.

8. Chapter X.3. Ptolemy observed Venus at maximum elongation in the evening between EN 887 VIII 4 and 5, so that the observation was on 140 February 18. When Venus was compared with α Tauri, it was found to be at longitude 13 5/6 degrees. Ptolemy puts α Tauri at longitude 42 2/3 degrees, and he probably neglected precession, so that Venus was 28 5/6 degrees west of α Tauri. This observation receives a reliability of 0.

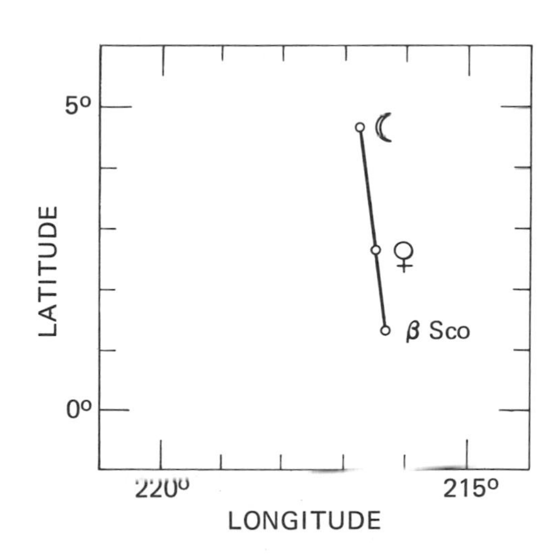


Figure V.5. The configuration of Venus, β Scorpii, and the moon that Ptolemy claims to have observed in the morning of 138 December 16. The stellar coordinates are those used by Ptolemy rather than values calculated from modern tables.

9. Chapter X.4. Ptolemy observed Venus past maximum elongation at 4 3/4 equal hours in the morning between EN 886 V 29 and 30, so that the date is 138 December 16. Venus was on the line joining β Scorpii and the center of the moon, and it was 1½ times as far from the moon as it was from β Scorpii.

The configuration is shown in Figure V.5, which is based upon Ptolemy's values for the coordinates of β Scorpii and the moon. Since his tables give the position of the moon fairly well relative to the stars, this is a reasonable procedure. The error in the position of Venus that results from using a poor position of the moon is only 4/10 of the error in the lunar position. In calculating the position of the moon, Ptolemy includes the effect of parallax. The position that is 4/10 of the way from β Scorpii to the moon in Figure V.5 agrees exactly with the position that Ptolemy derived for Venus.

Thus the observation says that Venus was 10′ east and 1°20′ north of β Scorpii. The observation will receive a reliability of zero.

10. Chapter X.4. Timocharis observed Venus past maximum elongation in the 12th hour (in unequal hours) in the night between EN 476 XII 17 and 18, so that the observation was on -271 October 12. Venus was touching η Virginis. This observation receives a reliability of unity.

I have been taking the time of a morning or evening observation to be 45 minutes from sunrise or sunset, whichever is appropriate. 45 minutes before sunrise occurs during the 12th hour of the night, so that I shall continue with the usual assumption for this observation.

11. Chapter X.4. Timocharis observed Venus 4 days after the preceding observation and found that it had moved 4°40′ to the east. That is, Venus was 4°40′ east of η Virginis. Ptolemy does not preserve any details about how Timocharis obtained this result. For this reason, I shall use a reliability of ½ rather than 1, because there is no way to check the reduction of the observation.

TABLE V.6

OBSERVATIONS OF VENUS PRESERVED BY PTOLEMY

| Date | Hour | JSD − 1 000 000 | Position[a] | Reliability |
|---|---|---|---|---|
| 132 Mar 8 | E | 769 338.200 | 1°7′ west of η Tau | 1/4 |
| 140 Jul 30 | M | 772 403.605 | 15′ east, 15′ north of ζ Gem | 0 |
| 127 Oct 12 | M | 767 728.635 | 55′ east of β Vir | 1/4 |
| 136 Dec 25 | E | 771 091.161 | 24′ west of φ Aqr | 0 |
| 129 May 20 | M | 768 314.597 | 3°.9 west, 3°.35 north of μ Cet | 1 |
| 136 Nov 18 | E | 771 054.159 | 5½° east of β Cap | 0 |
| 134 Feb 18 | M | 770 049.664 | 59¼° east of α Sco | 0 |
| 140 Feb 18 | E | 772 241.191 | 28 5/6° west of α Tau | 0 |
| 138 Dec 16 | 4.75 | 771 811.611 | 10′ east, 1°20′ north of β Sco; ¼° west of moon | 0 |
| -271 Oct 12 | M | 622 359.633 | Touching η Vir | 1 |
| -271 Oct 16 | M | 622 363.635 | 4°40′ east of η Vir | 1/2 |

[a]East and west are measured parallel to the ecliptic, not the equator.

The observations of Venus are summarized in Table V.6, which has the same form as Table V.5. All but three of the observations were at maximum elongation.

## 7. Observations of Mars

Ptolemy uses the same pattern of observations in establishing the parameters of his model for all three outer planets known to him. First he uses three observations of oppositions. Next he uses one observation when the planet was not in opposition, but when it was in conjunction with the moon or nearly so. He claims to have made these four observations himself. Finally he uses one observation that had been made in Alexandria about four centuries before his own time.

When Ptolemy refers to the maximum elongation of Mercury or Venus, he means a maximum with respect to the mean sun. Similarly, he takes opposition to be with respect to the mean rather than to the true sun.† I shall call this mean opposition. Ptolemy does not outline the procedure by which he found mean opposition. Since the mean sun is not observable, the procedure must have involved calculated positions to some extent.

I do not know why he chooses a position when the planet was near the moon for the fourth observation. I doubt that he does so in order to enhance the accuracy, for two reasons. First, it does not help to have this observation be more accurate than the three observations of mean opposition. Second, he should be able to locate the planet more accurately by means of a point star than by means of the extended disc of the moon.

I shall now give the facts of the observations, in the order that they occur in the text.

1. Chapter X.7. The first mean opposition was observed at 1 equal hour after the midnight between EN 878 V 26 and 27; hence the observation was on 130 December 15. Mars was at longitude 81°. This observation receives a reliability of zero.

2. Chapter X.7. The second mean opposition was at 3 hours before the midnight between EN 882 VIII 6 and 7, so that the date was 135 February 21. Mars was at longitude 148°50′. This observation receives a reliability of zero. Ptolemy does not say in the record whether the hours were equal or unequal, but his calculations show that he means equal ones.

---

†These statements are not quite correct. The precise definition of elongation, as Ptolemy uses the term, is given in Section XIII.5. Mean opposition occurs when the elongation is 180°; see Section XIII.6 for more discussion.

3. Chapter X.7. The third mean opposition was at 2 hours before the midnight between EN 886 XI 12 and 13, so that the date was 139 May 27. Mars was at 242°34′. This observation receives a reliability of zero. Again Ptolemy's calculations show that equal hours are meant.

Ptolemy does not give any indication of the methods by which the longitudes in these records were measured. This omission increases the case of fudging the data, and we should be particularly suspicious of these records. He writes the longitudes using minutes, just as I write them above. The appearance of 34′ in the last record is suspicious, since Ptolemy rarely uses such precision except with calculated values. However, all cited forms of the text agree in this reading.

4. Chapter X.8. Ptolemy made this observation about 3 days after the preceding one, namely at 3 equal hours before midnight on 139 May 30. When he adjusted his astrolabe, he saw Spica ($\alpha$ Virginis) at its proper longitude† on the ecliptic circle. He then saw Mars at longitude 241 3/5 degrees, and 1 3/5 degrees east of the center of the moon. Hence Mars was 64°56′ east of $\alpha$ Virginis and the moon was 63°20′ east of $\alpha$ Virginis. Although this record has considerable detail, the detail is of a kind that can easily be fudged. This record receives a reliability of zero.

5. Chapter X.9. According to Ptolemy, this observation was made in the morning between EN 476 III 20 and 21, and hence on -271 January 18. Mars was described as touching $\beta$ Scorpii. I shall use the conventional time of 'morning' in analyzing this record, which receives a reliability of unity. Ptolemy does not name the observer. The date suggests Timocharis but the style of the record suggests Dionysios.

The only date near this time when Mars could

---

†In Ptolemy's table, the longitude of $\alpha$ Virginis is 176 2/3 degrees. Halma has the correct value in his text but he has 176 1/3 degrees in his translation.

be described as touching β Scorpii is -271 January 16, and the latter is the date that I shall use except in discussing Ptolemy's own computations. More details will be given in Section XII.4.

TABLE V.7

OBSERVATIONS OF MARS PRESERVED BY PTOLEMY

| Date | Hour | JSD — 1 000 000 | Position[a] | Reliability |
|---|---|---|---|---|
| 130 Dec 15 | 01 | 768 888.455 | Mean opposition at 81° | 0 |
| 135 Feb 21 | 21 | 770 418.302 | Mean opposition at 148°50′ | 0 |
| 139 May 27 | 22 | 771 974.332 | Mean opposition at 242°34′ | 0 |
| 139 May 30 | 21 | 771 977.290 | 64°56′ east of α Vir, 1 3/5 degrees east of the moon | 0 |
| -271 Jan 16[b] | M | 622 092.675 | Touching β Sco | 1 |

[a]East is measured parallel to the ecliptic, not the equator.

[b]Ptolemy gives the equivalent of -271 January 18 for the date, and he uses -271 January 18 in his calculations in his Chapter X.9. Calculation of the position of Mars shows that the correct date is nonetheless almost surely -271 January 16. See Section XII.4 below.

The observations are summarized in Table V.7, which has the same form as Tables V.5 and V.6.

This completes the observations of Mercury, Venus, and Mars that are found in the Syntaxis. The observations of Jupiter and Saturn are given in Appendix VI.

## 8. An Heliacal Rising of Sirius

I have not paid much attention to heliacal risings, whether of planets or of stars, because of a suspicion that the circumstances governing them are not known with enough precision to make them useful for our purposes. There is one record concerning the first visibility of Sirius that I wish to discuss, however, because it has played an important part in some theories of Egyptian chronology and because it has been widely interpreted as an observation.

Censorinus [238] wrote a short work whose title can be translated as A Book About Birthdays. He tells us the year in which he wrote it in several different ways. For example, he says that he is writing in the 265th year after the year in which Gaius Octavius received the title of Augustus on the 16th calends of February (= January 17), and we know that this event occurred on -26 January 17.

He points out (Chapter XVIII) that many peoples have had a "great year" in addition to their ordinary year, and that the great year† has been determined in many different ways. For example, the Egyptian year had 365 days exactly (see Section II.3), and the first day of their year is called Thoth 1. Because 365 days does not equal the true length of the year, the date Thoth 1 gradually works its way through all the seasons and through all the days of the Julian calendar.

We know that the first visibility (heliacal rising) of Sirius in the morning was watched for with interest by the ancient Egyptians, presumably because it came at about the time that the annual flooding of the Nile began. Thus, says Censorinus, the Egyptians begin their great year with a year in which the heliacal rising of Sirius comes on Thoth 1. In the year in which he is writing, he says (Chapter XXI), Thoth 1 came on the 7th calends of July (= June 25). One hundred years ago, he says, Thoth 1 came on the 12th calends of August (= July 21), which is "the time when Sirius usually rises in Egypt."‡ This means that the current year (238)

---

†"Great year" is sometimes used to mean a year containing an intercalary day or month. Censorinus uses it in a different sense, to denote a cycle containing many years.

‡". . . quo tempore solet Canicula in Aegypto facere exortum."

is the 100th year of the Egyptian great year.

This has been interpreted as a record of an observation that Sirius rose heliacally on 139 July 21. Aside from the question about the year,† it seems to me that this is not the record of an observation at all. Censorinus does not say that Sirius rose heliacally on 139 July 21 (assuming that 139 is the year intended). He says that July 21 is the day when Sirius usually rises. This sounds as if he is referring to a conventional date for the rising, rather like the conventional choice of March 21 as the date of the vernal equinox in our calendar, or the conventional dates of the 8th calends of January, April, July, and October‡ for the equinoxes and solstices in the Roman calendar. Conventional dates of this sort can be seriously in error. For example, March 21 was never the date of the vernal equinox in the Julian calendar at any time when it was used as such.

Further, Censorinus uses the present tense in referring to the rising of Sirius. In the paraphrases and quotations given above, I have kept the tenses that Censorinus used. Thus, 139 (still assuming that this was the year) July 21 <u>was</u> on Thoth 1, but July 21 is the day when Sirius usually <u>rises</u>.‡ If July 21 is to be interpreted as an observed date at all, this construction suggests that the observation was made on 238 July 21 rather than 139 (or 138) July 21. However, we should not put much stress on the matter of the tenses; the tense used in a particular passage may easily be a matter of accident rather than of intent, as every writer knows.

---

†At one point, Censorinus writes "100 years" and at another point he writes "the 100th year". The first usage implies 138 for the year and the second implies 139.

‡See Section II.3 of <u>Newton</u> [1972], for example.

‡We should notice that Censorinus's contemporary Solinus [<u>Parker</u>, 1950, p. 47] gives the range July 20-22 for the heliacal rising of Sirius, presumably in reference to his own time.

Finally, if either 138 or 139 July 21 is intended to represent the date of an observation, it is wrong. When we convert the date 238 June 25 to the Egyptian calendar, we find (Section II.3 above) that it equals EN 986 I 1, in agreement with Censorinus's statement about that year.† If he was in the 100th year of the great year, the great year began on EN 887 I 1, which equals 139 July 20, not 139 July 21. In fact, Thoth 1 came on July 20 in each of the years 136 through 139. It came on July 21 only in the years 132 through 135 at this stage in history.

Since 138 and 139 July 21 both came on Thoth 2, neither could have been the beginning of a "great year" or "Sothic cycle" of the Egyptians.

It would take too long to examine all the possible explanations of Censorinus's error. The following explanation is perhaps the most plausible: The year 238 was the 100th year of the great year, and Censorinus inadvertently referred to "100 years ago", but whether "100" or "100th" is correct is not important here. He was correct that Thoth 1 came on 238 June 25, and he was correct that July 21 was the average date on which Sirius rose heliacally in his own time. Hence he concluded that July 21 was the day on which the first year of the great year began. He forgot, or perhaps did not know, that the sidereal year‡ is longer than the Julian year and that the date of heliacal rising moves steadily later in the Julian calendar by about 1 day in $1\frac{1}{2}$ centuries. Thus the date of 139 July 20 for an heliacal rising is consistent with the date 238 July 21 for an heliacal rising, and it is also a date that coincided with the Egyptian date Thoth 1. Thus 139 July 20 could readily have been the beginning of a great year.

In sum, if Censorinus's remarks about the Egyptian great year contain any observation of the

---

†Since Thoth is the name of month I in the Egyptian calendar.

‡The sidereal year of 365.25636 days, rather than the tropical year of 365.24220 days, governs the heliacal rising of a star.

heliacal rising of Sirius, they probably contain two. One was on 238 July 21 and the other was on 139 (or perhaps 138) July 20. Either 138 or 139 July 21 is highly unlikely for an heliacal rising. For safety, I shall not use any of these possibilities.

## 9. The Close of Hellenistic Astronomy

After the emperor Constantine accepted Christianity in 313, Hellenistic religion and Hellenistic philosophy did not immediately vanish. Particularly at Athens and Alexandria, the school of philosophy called Neoplatonism continued to be active for more than two centuries. The last figure of note in this school was Proclus (ca. 410-484 or 485), who held the chair of philosophy at Athens from about 450 to his death. The school survived him by less than 50 years. When Justinian ordered the closing of the school in 529, Damascius, who was then head of the school, went to Persia [Pauly-Wissowa, 1894, v. 2, p. 847] with six of his colleagues, in the hope of finding more favorable conditions. The scholars were disappointed in their hopes, but their situation aroused the sympathy of the Persian king Chosroes or Khosrau I, who made them the subject of diplomatic negotiations with Justinian. By stipulation in a treaty between the two monarchs, the seven neoplatonists were allowed to return to the Roman Empire, probably to Alexandria. They ended their lives ". . in peace and obscurity; and, as they left no disciples, they terminate the long list of Grecian philosophers,.." [Bury, 1908, v. 4, p. 266].

Two brothers Ammonius and Heliodorus were among the students of Proclus at Athens. They subsequently [Pauly-Wissowa, 1894, v. 1, pp. 415 and 1863, v. 8, p. 18] taught at Alexandria where they numbered Damascius among their students before he went to Athens. Pauly-Wissowa describes Heliodorus as "more undistinguished" than his brother. In spite of his non-distinction, however, it is to Heliodorus that we owe the last known observations of Hellenistic astronomy.

We deduce from surviving manuscripts that Heliodorus owned a copy of the *Syntaxis*. In his copy, he wrote down the record of an astronomical observation that Proclus made in Athens in 475, probably while Ammonius and Heliodorus were among his students. He also recorded six observations made in Alexandria from 498 to 510. From the language, we judge that Heliodorus made five of the observations himself and that he made one jointly with Ammonius.† Heliodorus's copy of the Syntaxis is lost, but it apparently served as the source from which additional copies were made. The copier fortunately preserved both the Syntaxis and the records that Heliodorus added.

Several existing manuscript copies of the Syntaxis contain the observations preserved by Heliodorus. Halma discusses them on page L of the introduction to his edition of the Syntaxis, and Heiberg [1907] also discusses them. Halma writes that he has published the observations, but I have not seen a copy of Halma's publication. Heiberg [1907, pp. XXXV-XXXVII] has published them, and Delambre [1817, v. 1, pp. 318ff] gives a translation of them into French.

The observations are dated by means of the Egyptian calendar referred to the era of Diocletian. This creates a presumption that the dates are in the Alexandrian calendar (see Section II.3) rather than in the older Egyptian calendar with the 365-day year. Since the two calendars differ by about 130 days in the time of Heliodorus, the observations themselves tell us unambiguously that Heliodorus used the Alexandrian calendar.

In Heiberg's text, the observations are in chronological order except for the observation made by Proclus, which appears between the second and third observations made by Heliodorus. I shall now state the facts that appear in the records, with the observation by Proclus put first.

†I noticed only one direct clue to the identity of the writer. In one place, he wrote "Heliodorus saw", but he wrote this in the first person. He does not name Ammonius, but he refers to "the most beloved brother" in one record. He does name Proclus in the record from 475, using the familiar term ο Θειος.

1. The moon occulted Venus on Aℓ.192 III 21 = 475 November 18. As seen from Athens, the record says, the apparent longitude of the moon was 283°, and it was 48° from the sun.

The record does not give the time, and it does not note that an occultation by the moon may last for about an hour. Since the longitude of the sun on November 18 is about 236°, it is clear that the observation was made in the evening. For simplicity, I shall assume that the center of the occultation was 45 minutes after sunset. This may be considerably in error.

If the moon was 48° from the sun when it was occulting Venus, Venus was also at an elongation of 48°. Thus this is implicitly an observation of Venus as well as of the moon. However, the elongation is likely to have been calculated, and I shall give a reliability of zero to the record when it is considered as an observation of Venus. If it has value, it is as an observation of the moon. Note that Venus must have been near maximum elongation.

2. Heliodorus names himself explicitly in this record, which receives a reliability of 1. He observed Mars and Jupiter in the early part of the night between Aℓ.214 IX 6 and 7, and hence on 498 May 1. In the 2nd hour of the night, he saw Mars and Jupiter so close together that there was no space visible between them.

Since the observation was made in the 2nd hour, I shall assume that the time was $1\frac{1}{2}$ unequal hours, or $1^h20^m$ of equal time, after sunset. Sunset at Alexandria on the date given was about $18^h40^m$ local mean time, so the observation will be assigned to $20^h00^m$ local mean time. Delambre assigns it to $21^h$. I do not see how this can be correct, because this is more than $2\frac{1}{2}$ unequal hours after sunset.

3. Heliodorus refers to himself and to his "most beloved brother" in this record. At the 1st hour of the night between Aℓ.219 VI 27 and 28, and hence on 503 February 21, the moon was west of Saturn. The brothers measured the time with the astrolabe after the moon had passed Saturn, and found it to be 5 3/4 unequal hours. They concluded that the conjunction was at 5 1/8 hours, presumably unequal ones. Saturn seemed

to bisect the illuminated curve of the moon. From this, I conclude that Saturn and the moon were in conjunction, in both latitude and longitude, at $23^h18^m$ local mean time, according to the record.

Delambre seems to suggest that Ammonius and Heliodorus did not understand what they were doing very well. The astrolabe should give sidereal time, he says, from which one finds the true time, and it is useless to go on to find the time in unequal hours. I do not understand his argument. Local sidereal time is the right ascension of a point on the local meridian. However, the astrolabe, at least as Ptolemy used it, gave the longitude rather than the right ascension of the point of the zodiac that was on the meridian.† Thus the astrolabe did not directly give the time in equal hours, and Heliodorus might well have deduced the unequal hours first.

The record numbered 1 in Section VII.6 furnishes an example, this one from Muslim astronomers of unquestioned competence, in which the observers measured the position of a star and thence calculated the time in unequal hours.

Since this is primarily an observation of the moon, I shall not use this record in this study. If I did use it, I should give it a reliability of unity.

4. On Aℓ.225 I 30 (= 508 September 27), Jupiter passed 3 digits‡ to the north of Regulus ($\alpha$

---

†The name astrolabe has been used at various times in history to denote various types of instrument, and it is likely that some type used in some historical period did give sidereal time. However, when Ptolemy gives the datum that he used to specify the time, it is specifically the longitude of the point where the meridian intersects the ecliptic. It is likely that Heliodorus and Ammonius used the astrolabe in the same way. Unfortunately, Heliodorus does not state the measurement from which he found the time.

‡The digit usually means 1/12 of a degree in this context.

Leonis). Since this is an observation of Jupiter, I should not use it even if it gave the time of the conjunction.

5. In the evening between Aℓ.225 VII 15 and 16,† and hence on 509 March 11, α Tauri was west of the bright arc of the moon by at most 6 digits. I shall still assume that the time is $45^m$ after sunset, although we are dealing with the moon and with a star of the first magnitude, and I shall assume that α Tauri was 4 digits west of the edge of the moon. This record would receive a reliability of 1 if I used it.

6. After sunset on Aℓ.225 X 19 (= 509 June 13), Mars was 1 digit west and 2 digits south of Jupiter. According to the tables, Heliodorus says, the conjunction should not have been until the 23rd of the month, but, at that time, Mars and Jupiter were far apart. I shall assume, as usual, that the time was $45^m$ after sunset and give this record a reliability of unity.

7. This record, which seems to involve two separate observations, is carelessly written. It does not give the month, and it does not give the time of day of either observation. Further, the text as given by Heiberg and the translation as given by Delambre differ on an important point, and neither of their forms of the record makes physical sense. I shall first summarize the facts as they appear in Heiberg's text:

> In the year 226 of Diocletian, Venus was west of Jupiter by 8 digits and on the 28th it was east by 10 digits. There seemed to be no difference in the latitudes. According to the ephemerides, the conjunction should have been on the 30th, but by then the planets were far apart.

Delambre's translation differs by saying that

---

†Heiberg's text has the letter sigma in place of the Greek numeral for 6. Since it is easy to make this error, either in typesetting or in copying a manuscript, it seems safe to read 16.

Venus was west of Jupiter by 20 digits in the year 226. By combining the two versions,† we can make a tentative restoration of the text: On the 20th of the month, whatever it was, Venus was west of Jupiter by 8 digits and on the 28th it was east by 10 digits. One trouble with this restoration is that it makes Venus move only 18 digits (= $1^\circ.5$) with respect to Jupiter in 8 days. This is possible if Venus is close enough to a turning point, and we can find out by calculation if this is so.

Most of Heliodorus's observations were made soon after sunset. Even the observation of 503 February 21 began soon after sunset, but Heliodorus and Ammonius waited for several hours in order to observe the end of the occultation. Thus it is plausible that Heliodorus made both observations in this record by following his usual habits, and I shall assume that the observations were at $45^m$ after sunset.

Lack of the month is the least serious problem that confronts us with this record, because the month can be restored unambiguously by calculation. Delambre, citing the authority of Bullialdus,‡ says that the month is August, which would put the observations in month XII of the Alexandrian calendar. It is unlikely that this is in error, and I shall accept it unless detailed calculation forces a different conclusion.

Thus I shall tentatively assume that the record contains two independent observations. The first was made at $45^m$ after sunset on A$\ell$.226 XII 20 (= 510 August 13), when Venus was west of Jupiter by 8 digits in longitude. The second was made $45^m$ after sunset on A$\ell$.226 XII 28 (= 510 August 21), when Venus was east of Jupiter by 10 digits in longitude. I shall assign a reliability of 0.5 to each observation, unless detailed calculation forces me to abandon either or both interpretations. A

---

†It is unlikely that the numeral for 8 was confused with the numeral for 20. Thus it is possible that there has been an accidental compression of the text, but it is odd that Heiberg and Delambre give readings so nearly alike if this is so.

‡Presumably Ismael Bouillaud, 1605-1694.

change in the month is unimportant from the standpoint of reliability. Otherwise, if it is necessary to change the interpretation of either observation, I shall change the corresponding reliability to zero.

TABLE V.8

THE LAST HELLENISTIC OBSERVATIONS

| Date | Hour, Local Mean Time | JSD − 1 000 000 | Event[a] | Reliability |
|---|---|---|---|---|
| 475 Nov 18 | 17.92 | 894 873.181 | Moon occulted Venus | 1 |
| | | | Venus 48° from sun | 0 |
| 498 May 1 | 20.00 | 903 073.250 | Conjunction of Mars and Jupiter | 1 |
| 503 Feb 21 | 23.30 | 904 830.388 | Moon occulted Saturn | 1 |
| 508 Sep 27 | —— | 906 875.417 | Jupiter was 3 digits north of α Leonis | 0 |
| 509 Mar 11 | 18.85 | 907 040.202 | α Tauri was 4 digits west of the edge of the moon | 1 |
| 509 Jun 13 | 19.83 | 907 134.243 | Mars was 1 digit west, 2 digits south of Jupiter | 1 |
| 510 Aug 13[b] | 19.48 | 907 560.228 | Venus was 8 digits west of Jupiter | 0.5 |
| 510 Aug 21[b] | 19.33 | 907 568.222 | Venus was 10 digits east of Jupiter | 0.5 |

[a]East and west are measured parallel to the ecliptic, not to the equator.

[b]These dates are tentative. See the accompanying discussion.

The observations are summarized in Table V.8, which should be mostly self-explanatory. Some of the observations are not relevant to this study. Nonetheless, I have included them here instead of relegating them to an appendix because there are not enough observations to warrant the use of separate tables. The reliability listed in Table V.8 is that which I should assign if I were using the observation; thus I have assigned a reliability different from zero to some observations that will not be used here.

# CHAPTER VI

# FROM HELIODORUS TO AL-BIRUNI

## 1. General Remarks about the Period

As we saw in Section V.9, Proclus observed an occultation of Venus by the moon in the evening of 475 November 18; the observation was made in Athens. This may be the last quantitative astronomical observation made in Europe for more than six centuries.

Charlemagne and his court were interested in astronomy, and there is a record [Loiselianos, ca. 814]† of several eclipses and of an occultation of Jupiter by the moon that occurred in the years 806 and 807. The record gives dates and/or the positions of the sun and moon. Since the dates and positions are those that we find in the ecclesiastical tables of the sun and moon, it is unlikely that they were measured ones.

Herimannus Contractus, or Herman of Reichenau (1013-1054), is credited [Bernoldus, 1100] with making instruments for measuring time and for making astronomical observations, and he is reputed to have made measurements of the equinoxes and solstices. For a long time, Europeans thought that he was the inventor of the astrolabe.‡ If Herimannus did make such observations, and if any of his observations survive, I have not found them.

Spain, though geographically part of Europe, formed part of the Islamic culture during much of the period of interest in this section. The earliest astronomical writing in Spain dates [Kennedy, 1956, p. 129] from about 1010. Writing does not necessarily imply observation, however. The earliest observations in Spain may have been made

†See also Newton [1972, pp. 392ff].

‡By this time, 'astrolabe' may denote an instrument for measuring elevation angles above the horizon, rather than an instrument for measuring celestial latitude and longitude. See Section V.4.

[Kennedy, 1956, p. 128] in Toledo by "a group of Arab and Jewish scientists under the leadership of one Qadi Sa'id" around 1060. The making of these observations is inferred from one extant document; if the observations were indeed made, the records of them have vanished.

Walcher of Malvern [Haskins, 1924, pp. 114-115] observed the total lunar eclipse of 1092 October 18. He used an astrolabe to measure the time, and he determined that totality began at $11^h03^m$; the language that he uses suggests that this is local apparent solar time in unequal hours. This is the first instrumented European† observation known to me after 475 November 18. Otherwise, in the intervening six centuries, we have only notices of eclipses, occasionally with an hour given, and we have a few qualitative records of other phenomena, but we find no indications of systematic measurements.

Walcher used his observation of 1092 October 18, and perhaps some other ones that have not survived, as the basis for calculating tables of the moon.

The six centuries between Heliodorus, the student of Proclus, and Walcher were not barren ones for astronomy. There was considerable astronomical activity in many places. Unfortunately, little of the work of this period has been published at all, and still less has been published in a language that makes it available to western scholars.

The Chinese, for example, were making systematic observations during this period. Their records of eclipses have been catalogued [Wylie, 1897 and Hoang, 1925], but the records from this time period have apparently not been studied with the detail needed to make them useful in a study of the accelerations. Gaubil [1732] has published some measurements of the winter solstice made during these centuries, and I have analyzed them elsewhere [Newton, 1970, Section II.4]. It is probable

†It is probably legitimate to call this European, since I am writing after 1973 January 1.

that there are, or at least once were, many more such measurements.

The analysis of the published Chinese solstice observations shows that their standard error is around 7 or 8 hours. Thus they are comparable in accuracy with the solstice measurements made by Aristarchus and Hipparchus (Section V.3). However, because they are much later in time, they are not as useful in the inference of the solar acceleration. The estimate of $\nu_S'$ formed from them is [Newton, 1970, p. 31]

$$\nu_S' = 2.21 \pm 15.49 \ .$$

Because of the large standard deviation assigned to this estimate, I shall neglect it. It is unfortunate that there are no more Chinese solar observations available for analysis in this work.

India was a center of much mathematical and astronomical activity in these centuries. Part of the importance of Indian astronomy during this perod lies in its role of preservation and transmission. The popularity of the Syntaxis in the Hellenistic culture caused the loss of almost all pre-Ptolemaic astronomy from the Greek literature. However [Kennedy, 1956, p. 170], some of the earlier Hellenistic astronomy "found its way to India and has been identified ..." It may ultimately be through the Indian literature that we shall recover an important part of Hellenistic astronomy before Ptolemy.

Indian astronomy and mathematics are also important in their own right. Apparently the number system that we call Arabic was developed in India around the 7th century, and the same time and place may have seen the development of algebra into the discipline that we recognize by that name. It is from India rather than from Hellenistic Egypt that Islamic astronomers probably derived their first inspiration; Persia may have served as the route of transmission. At least, the oldest Islamic zijes [Kennedy, 1956, p. 168] seem to derive from Indian sources. Indian astronomy during the centuries under discussion is distinguished by its use of the sidereal rather than the tropical year.

Astronomical activity did not necessarily stop in the areas farther to the west. In 7th-century Armenia, for example, Ananias of Sirak [Hewsen, 1968] wrote calendrical, chronological, and astronomical works and he made some independent observations. His work is not available in a European language.

The astronomy of the period between Heliodorus and Walcher that is most familiar to western scholars, however, is undoubtedly Islamic. Islamic astronomy survives in an enormous number of sources. Kennedy [1956] catalogues over a hundred zijes. Zijes form only one class of Islamic astronomical work, although it is the class that Kennedy considers to be the most significant. As of the date when Kennedy wrote, only two of these works had been published. One has been published in the original language and in a Latin translation. The second, which no longer exists in the original language, has been published in a Latin translation and in an English translation of the Latin translation. At least one has been published since Kennedy wrote, in the original language but not in translation. Fragments of a few others have been published, some in the original language and some as translations into European languages.

The earliest Islamic observations that we know of were made by Ahmad bin Muhammad an-Nihawandi at the town of Jundishapur† in about the year 790. Ibn Yunis [1008, p. 140] says that he knows of no observations of the solar mean motion between Ptolemy and the establishment of the observatory at Baghdad except those of an-Nihawandi. an-Nihawandi was a contemporary of Charlemagne, and it is interesting that the study of astronomy revived in Europe and in western Asia at about the same time.

The Asian revival, unlike the European one, was maintained and, in fact, the 9th century produced

---

†Near the present Ahwaz, in Iran.

more known Islamic works on astronomy than any other.[†] To a great extent, this is probably due to the establishment [Nasr, 1968, p. 69] by the caliph al-Ma'mun, around the year 815, of a state-supported center of learning in Baghdad that included a library and an observatory. The oldest surviving Islamic observations were made at this center; those of an-Nihawandi have been lost.

It is often stated that the Crusades were responsible for the discovery of Islamic learning by Europeans. Actually, the Crusades had little to do with the matter. The main route of transmission from Islam to Europe was through Spain,[‡] with a secondary route through Sicily. Both Spain and Sicily are areas where Islamic and Christian peoples lived side-by-side for long periods, often in a spirit of tolerance and cooperation. Further, the transmission of astronomy, at least, began well before the Crusades. Gerbert of Auvergne (940?-1003), who became Pope Sylvester II (pope 999-1003), wrote extensively on mathematics, astronomy, and other sciences, and he used technical Arabic terms in doing so.[‡‡] Since it is unlikely that he did much of this writing after he became pope, he must have been acquainted with at least some Islamic science before 1000, a century before the Crusades. A few decades later, Herimannus Contractus, who was mentioned above, also shows some acquaintance with Islamic learning. Finally, the documented observation of Walcher was made in 1092, several years before the Crusades began.

Islamic astronomy flourished for many centuries after al-Ma'mun founded the observatory at Baghdad. This work would be far too long if I attempted to cover the entire period of extensive Islamic activity. I shall arbitrarily stop with the work of al-Biruni, who will be discussed in the next section.

---

[†]See the chart of Kennedy [1956, p. 168], which gives the distribution in time and place of the zijes that he catalogues.

[‡]The reader is referred to the studies of Haskins [1924] and of Thorndike [1923], for example.

[‡‡]See Haskins [1924, p. 115], for example.

His last observation that I know of was made in 1019. Since I have found no instrumented observations between 1019 and the observation made by Walcher in 1092, al-Biruni provides a convenient terminus for the present work. The sequel can then begin after the revival of astronomical activity in Europe.

In the next section, I shall describe the main Islamic astronomers whose work figures in this study.

## 2. Some Important Islamic Astronomers Whose Work Will Be Used Here

Around 815, as I have already mentioned, the caliph al-Ma'mun established an observatory. It was located in the Shamasiya quarter† of Baghdad, and the first observations there were made by a group of astronomers headed by Yahya ibn abi Mansur. The first usable Islamic observation that I have found was made by this group; it is an observation of the autumnal equinox on 829 September 19. The table that resulted from the work of the Yahya group became known in medieval Europe under the name Tabula Probata. Part of this work [Kennedy, p. 132] still exists in the form of one manuscript, where it is mixed indiscriminately with other writing.

There is a near-contemporary of Yahya known as Habash [Kennedy, p. 126], ".. about whose person and accomplishments the fog of history has closed thickly." Ibn Yunis [1008] quotes Habash's writing in several places. In one place (pp. 42ff), ibn Yunis quotes Habash as saying that the authors of the Tabula Probata observed only the sun and the moon, and that it was only he (Habash, that is) who observed the other planets some time later. Since Habash's work, though extant, has never been published, the only ones of the early Islamic observa-

†I follow the usage of Kennedy [1956] in rendering Muslim terms into English, for all terms that I find there. Otherwise I follow the immediate source being cited. I omit accent marks. Most of the information in this section comes from Kennedy or from Nasr [1968].

tions that can be used here are those made under Yahya and his successor al-Marvarudi. al-Marvarudi transferred his activities to Damascus, but others continued to observe at Baghdad.

We may pass over intervening figures and go to Abdallah Muhammad al-Harrani al-Battani. His zij has been published in the original language and in a Latin translation [al-Battani, ca. 920].† His writing introduced Europeans to the use of the sine function in preference to the chord function.‡ al-Battani was probably not responsible for this innovation, but he was responsible for other innovations in trigonometry, and it may have been his treatment of trigonometry that accounted for his popularity in medieval Europe, where his name was spelled Albatenius or Albategnius.

al-Battani says that he made many diligent observations of the obliquity, the times of the equinoxes and solstices, of eclipses, and of the positions of the planets. However, his zij preserves very few observations. We have the time of the autumnal equinox of 882 September 18 observed at Raqqa. We have the times and magnitudes of the lunar eclipse of 883 July 23, the solar eclipse of 891 August 8, the solar eclipse of 901 January 23, and the lunar eclipse of 901 August 3. For the two

---

†The other zij that had been published when Kennedy wrote is that of al-Khwarizmi, whose name has given rise to the technical term 'algorism'. al-Khwarizmi was a near-contemporary of Yahya, but his astronomy is largely Indian and non-Ptolemaic in origin [Kennedy, pp. 128 and 170ff]. His zij no longer exists in the original language, but there is a Latin translation made in the 12th century. This translation has been published, as well as translations of it into yet other European languages. Since Kennedy's survey, the zij of al-Biruni (Al-Qanunu'l-Mas'udi, finished in 1030) has been published in the original language by the Osmania Oriental Publications Bureau, Hyderabad, India, in 3 volumes, 1954-1956.

‡As the name implies, the chord of an angle $\alpha$ is the chord that subtends a central angle $\alpha$ in a circle of unit radius. Thus, chord $\alpha = 2 \sin \frac{1}{2}\alpha$.

eclipses in 901, al-Battani uses observations made in both Antioch and Raqqa. Most of his observations seem to be of good quality. However, he reports the total eclipse of 901 August 3 as partial, and his reported times for this eclipse are inconsistent with other contemporary data [Newton, 1970, pp. 220 and 235].

ibn Yunis [1008] gives the values of the mean motion of the sun that were found by many different Islamic astronomers. The values cluster about two points, depending upon whether the Islamic astronomer combined contemporaneous data with the equinox observations of Ptolemy or of Hipparchus. al-Battani, for example, used Ptolemy's data. ibn Yunis [p. 142], working less than two centuries after Yahya, concluded that Ptolemy's data were not compatible with the Islamic observations, and he used Hipparchus's data instead. Other astronomers even through Copernicus [1543], however, tried to invent theories that would reconcile Ptolemy's equinoxes, as well as his obliquity, with other data. These attempts gave rise to the medieval theories of the "trepidation of the equinoxes".†

The idea of trepidation seems to have arisen with Thabit bin Qurra, a contemporary of al-Battani.‡ Nasr [1968, pp. 44-45] gives the following interesting biographical sketch of Thabit: "Thabit hailed from the community of Sabaens in Harran, where there was a religious cult centered around the symbolism of the planets. This cult, much interested in the Pythagorean mathematical and mystical tradition, survived well into the Islamic period. Like many members of this community Thabit was well versed in mathematics and astronomy. Due to religious differences with his community he set out for Baghdad and was fortunate in meeting on the way the influential mathematician, Muhammad ibn Musa ibn Shakir,♯ who,

---

†There are minor perturbations in the equinoxes and in the obliquity that are called the nutations. These are quite different from the so-called trepidations.

‡Also see Kennedy [1956, p. 136].

♯See Section VII.2 for another reference to ibn Shakir.

recognizing his ability, took him under his patronage. Thabit soon gained fame in Baghdad and became the court astronomer."

Passing over several astronomers of the 10th century, we come to two writers of the first rank who wrote in the early part of the 11th century. Although they were contemporaries, they do not seem to have known of each other, perhaps because one worked in Cairo while the other worked in Iran and areas to the north and east of there.

As Kennedy says, the zij of ibn Yunis "is a famous work which is extant only in fragments." The totality of the fragments probably contains about half of the original work. The entry identified as <u>ibn Yunis</u> [1008] in the list of references is apparently the largest portion of the work that has been published in one place. Kennedy [p. 126] lists places where other portions can be found. He mentions a translation of a considerable part of the work that was made early in the 19th century and that has vanished without being published. I have seen only the part of ibn Yunis's zij that is cited in the list of references.

Kennedy writes of ibn Yunis: "What is of almost unique interest is the author's wide knowledge of the work of his predecessors, his acute critical faculty, and an attitude toward observational errors and computational precision which is modern in tone and which would be completely foreign, say, to Greek astronomy." ibn Yunis documents thoroughly the discrepancies that exist between observations of the sun, of the planets, and of eclipses and the tables of motion found in the work of many of his predecessors. In doing so, he has preserved much information about Islamic astronomical writings that would otherwise have been lost, and this aspect of his work is of first importance to historians of science. He has also preserved a large body of observations that would otherwise have been lost, and this aspect of his work is of first importance in the present study.

The other Islamic writer whose work will be used here is Abu al-Raihan Muhammad bin Ahmad al-Biruni. al-Biruni was born in 973 in Khwarizm, a district to the south of the Aral Sea. He spent most of his twenties and thirties in north-central Iran, and then returned to Khwarizm for several years. Finally he moved to Afghanistan in 1017, where he probably died some time after 1025.

Kennedy [1956, p. 133] lists a zij by al-Biruni with the translated title The Masudic Canon, and he later discusses it extensively [pp. 157-159]. al-Biruni also wrote a work on chronology [al-Biruni, 1000] and a work on geodetic surveying [al-Biruni, 1025]. The work on chronology has been of considerable assistance in establishing the dates of various astronomical observations used here. The work on surveying contains a large number of observations of the sun that al-Biruni used as geodetic controls, as well as a few other observations, and I have analyzed these data elsewhere [Newton, 1972b]. It seems likely, on the basis of what Kennedy says, that the data found in the Masudic Canon are also found in others of al-Biruni's works. As I noted above, the Masudic Canon has been published in the original language but not in a European language.

The interested reader may consult Kennedy [1956] and Nasr [1968] for information about many more figures in Islamic mathematics, astronomy, and other fields of learning.

## 3. The Probable Hoax of al-Kuhi

The purpose of this section is to discuss an Islamic writer whose work I shall not use and to give the reason for this action. The writer is Abu Sahl al-Kuhi, who worked in Baghdad around the year 990. Nasr [1968, p. 149] describes al-Kuhi as an outstanding Muslim algebraist "who made a thorough study of trinomial equations", and he also mentions al-Kuhi's astronomical observations. al-Biruni [1025, pp. 69-70] quotes some of these observations.

al-Biruni first quotes al-Kuhi's measurement of a summer solstice. al-Kuhi says that the summer solstice of 988 June 16 occurred at 1 hour after sunset, Baghdad time, and that the solstitial point reached a maximum elevation angle of 80°10′ above the horizon at Baghdad. al-Biruni then further quotes al-Kuhi as saying that he (al-Kuhi) had made a thorough investigation of the summer and winter solstices. From his investigation, he had found the latitude of Baghdad to be 33°41′20″, and he had found the obliquity of the ecliptic to be 23° 51′20″.

al-Biruni distrusts these statements deeply. He points out that the value of the obliquity disagrees violently with other recent measurements, which cluster around 23°35′, but that it agrees to the second of arc with Ptolemy's value. al-Biruni believes that al-Kuhi did not make the claimed investigation at all, except perhaps for making the single observation of the solstice of 988 June 16, and he thinks that al-Kuhi fudged the latitude of Baghdad by using Ptolemy's value of the obliquity.

An elementary analysis made with the aid of modern data shows that al-Biruni was justified in his suspicion. The latitude of Baghdad is about 33°20′ rather than 33°41′20″. The obliquity calculated from Newcomb [1895]† for the epoch 990 is about 23°34′, rather than 23°51′20″. The error in the latitude is about 21′ and the error in the angle between the solstices (twice the obliquity) is about 34′. Both these errors are incredible. In fact, if al-Kuhi found 80°10′ for the elevation angle of the summer solstice, he would have had to find 32°27′20″ for the elevation angle of the winter solstice. This is in error by about 40′, an amount that exceeds the apparent diameter of the sun.

It is plausible that al-Kuhi did actually measure the summer solstice, as al-Biruni speculates. If the latitude is 33°20′ and if the obliquity is 23°34′, the elevation angle of the solstitial point is 80°14′, except for a correction of about 10″ for parallax and refraction. An error of 4′ in a single reading of an elevation angle is credible, although it is not creditable; al-Kuhi's contemporaries routinely achieved an accuracy of 1′ in the altitude of the sun [Newton, 1972b]. Then, if al-Kuhi

†I have estimated the obliquity at the epoch 880 from the analysis of a large body of Islamic data, [Newton, 1972b]. The discrepancy with Newcomb's theory is only about 10″ ± 20″, and is not statistically significant.

read 80°10′ for the elevation angle, and if he took 23°51′20″ for the obliquity, he would calculate 33° 41′20″ for the latitude, and this is exactly his value.

There is independent evidence that tends to confirm this speculation. A manuscript that is in the Escorial, or at least that was there around 1800, contains two observations attributed to al-Kuhi. A student of Caussin's named Reiche copied them and sent them to Caussin around 1800. Caussin quotes them on pages 222-223 of his edition of ibn Yunis [1008], and Delambre [1819, p. 94] also quotes them.

One of the observations is the summer solstice that we have just been discussing. The other is an observation of the next autumnal equinox, which is reported at 4 hours after sunrise on the date EY 357 VII 4 (= 988 September 18). This is too late by around 14 hours.

Suppose that al-Kuhi set up a meridian circle or similar instrument which was aligned with the local vertical within a few minutes of arc. If he had the correct elevation angle of the summer solstice, and if he thought that the obliquity is about 17′ larger than it really was, he thought that the equatorial plane was about 17′ farther south than it really is. If he measured the time when the sun passed through this wrong equatorial plane going south, he would find a time about 17 hours later than the correct equinoctial time. This is rather close to what he did find.

In sum, it seems to me that al-Kuhi's claimed measurements of the obliquity of the ecliptic and of the latitude of Baghdad are hoaxes, beyond a reasonable doubt. It is possible that he did make two genuine observations, one of the summer solstice and one of the time when the declination of the sun was about -17′; the latter is the condition that he thought marked the equinox. However, it is not safe to trust one who has committed a hoax, and it is possible that al-Kuhi fudged these observations also. For safety, I shall ignore the observations attributed to al-Kuhi.

Thus Ptolemy and al-Kuhi must both stand convicted of hoaxes. It is ironic that al-Kuhi's hoax was committed for the purpose of 'proving' one of the fraudulent results of the earlier one.

# CHAPTER VII

# ISLAMIC DATA THROUGH AL-BIRUNI

## 1. Equinox and Solstice Data

The Islamic equinox and solstice data that will be presented here have been analyzed in other works [Newton, 1970, Section II.3 and 1972b]. As a result of new information, it is possible to improve the earlier analyses, particularly with regard to decreasing the biases connected with the observation of an equinox. The data that will be analyzed are found in four Islamic works.

al-Battani [ca. 920] gives an observation of the autumnal equinox that he made himself.

al-Battani's near contemporary Thabit [ca. 880, p. 68] lists five Islamic observations of equinoxes or solstices and three observations of the sun at other times. The latter three observations will be taken up in Section VII.3. Apparently all the observations were made by the Yahya group at Baghdad.

ibn Yunis [1008, pp. 130-132] gives a collection of ten Islamic measurements of equinoxes and solstices made by various people. On page 223 he also gives the summer solstice and the autumnal equinox that were reported by al-Kuhi. The latter two are the reports that I shall not use because they are associated with al-Kuhi's probable hoax (see Section VI.3).

al-Biruni [1025, pp. 267-270] closes his work with a list of 22 observations of the autumnal equinox. These include the 6 observations that Ptolemy [ca. 142, Chapter III.1] attributes to Hipparchus as well as the two fudged observations (see Section V.1) that he claims to have made himself. The Hellenistic records are followed by 14 Islamic records. Two of the recorded observations were made by al-Biruni himself, and he gives us the raw data associated with one of them in another place [al-Biruni, 1025, p. 87]. I shall use the raw data in the next section and therefore I shall not use the corresponding equinox. This leaves 13 Islamic records from al-Biruni for use in this section.

TABLE VII.1

EARLY ISLAMIC EQUINOX AND SOLSTICE DATA, IN LOCAL APPARENT SOLAR TIME

| Date | Observer | Place | Julian Day Number -2 000 000 | Hour | Reli-ability |
|---|---|---|---|---|---|
| EN 1577 VIII 25 | Yahya | Baghdad | 24 112 | 6 4/5[a] | ½ |
| EY 199 V 25 | Yahya | Damascus? | 24 477 | $12^h\ 8^m$ | 0 |
| EY 199 V 25 | | Baghdad | 24 477 | 13[b] | 1 |
| EY 199 XI 19 | | Baghdad | 24 656 | 02 | 1 |
| EY 200 V 25 | Sanat al-Shams | Baghdad | 24 842 | 19 | 1 |
| EY 200 XI 19 | | Baghdad | 25 021 | 08 | 1 |
| EY 201 II 23 | | Baghdad | 25 115 | 00 | 1 |
| EY 201 V 25 | al-Marvarudi | Damascus | 25 207 | 23.3[c] | 1 |
| EY 213 V 28 | | Baghdad | 29 590 | 21 11/30 | 1 |
| EY 220 V 30 | al-Makki | Neyshabur | 32 147 | 12 | 1 |
| EN 1607 IX 2 | Bani Musa | Samarra | 35 069 | 12 | 1 |
| EN 1630 IX 8 | al-Battani | Raqqa | 43 470 | 1 1/4 | 1 |
| EN 1636 IX 9 | bin 'Ismat | Balkh | 45 661 | 13 3/5 | 1 |
| EN 1718 IX 29 | as-Sufi | Shiraz | 75 611 | 11 | 1 |
| EN 1719 IX 29 | as-Sufi | Shiraz | 75 976 | 18 | 1 |
| EN 1722 IX 30 | al-Wafa | Baghdad | 77 072 | 09 | 1 |
| EN 1767 X 11 | al-Biruni | Ghazni | 93 508 | 07 | 1 |

[a] The record has 12 4/5 hours; 6 4/5 hours is inferred from other evidence.

[b] One record has 11 hours.

[c] One record has 23 1/5 hours.

Unfortunately there is considerable duplication in the four sources, and there are only 17 independent observations. These are summarized in Table VII.1. In preparing Table VII.1, I have used ibn Yunis's form of the record for the records that he gives. This form is identical to Thabit's for the records that Thabit gives. I have preferred ibn Yunis's records to al-Biruni's, because the latter has transformed the dates to the old Egyptian calendar used by Ptolemy, and it is unlikely that this is the calendar actually used in making the observations.

al-Biruni gives the feria for each equinox that he lists. However, he has a wrong date in one instance and the feria listed corresponds to the wrong rather than to the right date. This feria was therefore calculated and was not part of the original record. Since ibn Yunis and Thabit rarely give the feria of an equinox or solstice, it is likely that the feria was not in most of the original records, and hence it is likely that al-Biruni calculated most of the feria. The feria is useful in spite of this, because the redundancy it gives allows us to detect typographical errors.

The dates of the observations appear in the first column of Table VII.1. The corresponding Julian day number appears in the fourth column. The conventions used in these two columns are as follows: (a) The Julian day number is connected with the calendar date by the formal relations given in Chapter II. (b) The Julian day number given is the integer value at noon, local time, on the day of the observation.†

ibn Yunis gives the records that are dated by means of the Persian calendar, as well as the one that is dated EN 1630 IX 8. The remaining records are given by al-Biruni but not by the other sources listed.

The second column gives the observer when he is known, and the third column gives the place of the observation. The fifth column gives the hour of the observation, measured from midnight, and the last column gives the reliability, as defined in Section I.6.

---

†If the time of the observation is midnight, the day given is the following one, not the preceding one.

The five observations at Baghdad from EY 199 V 25 through EY 201 II 23 appear in _Thabit_ as well as in _ibn Yunis_; the two autumnal equinoxes in this series (EY 199 V 25 and EY 200 V 25) also appear in _al-Biruni_. No source connects an observer with this series. ibn Yunis says that the records are given by Thabit while al-Biruni says that the record of EY 200 V 25 is found in the book called _Sanat al-Shams_. Since the translator of al-Biruni translates the title of this book as "The Solar Year", this may be the work by _Thabit_ [ca. 880], whose Latin title may be translated as "About the Solar Year". It is plausible that these five observations are part of the observations made under Yahya that led to the _Tabula Probata_ (see Section V.2).

al-Biruni names Yahya as the observer for the date EN 1577 VIII 25. ibn Yunis does not give the result of this observation explicitly, but he does discuss it (pages 38-40), and he says that it was made at Baghdad by the astronomers of al-Ma'mun. al-Biruni names al-Marvarudi as the observer at Damascus on EY 201 V 25. ibn Yunis includes this observation in his list, and he also refers to it on pages 38-40, where he ascribes it to other astronomers† at Damascus.

ibn Yunis gives the observation of EY 199 V 25 in his list on pages 130-132, and he attributes it to Yahya but does not say where it was made. Caussin, the translator of ibn Yunis, deduces that the observation was made at Damascus, "on the basis of the known history of Yahya and of the difference between this observation" and the one made at Baghdad on the same date.

Caussin is the only writer I have seen who associates Yahya with Damascus. The usual association is that of Yahya with Baghdad and al-Marvarudi with Damascus. Of course, it is possible that observations at both places were under the sponsorship of al-Ma'mun and under the general authority of Yahya before he was succeeded by al-Marvarudi. However, it seems possible to me that the Damascus

---

†That is, it was not made by the ones who made the observation of EN 1577 VIII 25 at Baghdad.

record of EY 199 V 25 was calculated from the Baghdad observation of the same date by using 13° for the difference in longitude.† For safety, I shall give this record a reliability of zero.

ibn Yunis says that the observation of EY 213 V 28 is from the second series of observations made at Baghdad, after the death of al-Ma'mun, but he does not name an observer. al-Biruni calls this an anonymous observation made at Baghdad. He has the month and day correct, but he has the year wrong, and the feria that he gives goes with the wrong year. In other words, he probably calculated the week day.

al-Biruni says that the time of the observation on EN 1577 VIII 25 was 4/5 of an hour after midday. ibn Yunis says that the astronomers of al-Ma'mun used this observation to find that the mean motion of the sun is 359;45,44,14,24 in a Persian year of 365 days. We can tell from the size of the mean motion that the astronomers used Ptolemy's data rather than Hipparchus's. Now Ptolemy [ca. 142, Chapter III.1] reports that the autumnal equinox came at 1 hour after sunrise on EN 887 III 9. If we calculate the time of the equinox 690 Egyptian or Persian years later, using the mean motion just given and using 2 hours‡ for the time difference between Alexandria and Baghdad, we get exactly 4/5 hours after sunset.

This disagrees with the time that al-Biruni gives. Further, both times disagree seriously with the other Islamic observations. I shall return to the question of this observation in Section VIII.5, where I shall study the observation with the aid of tables that were constructed from it. From this

---

†The actual difference in longitude is about 8°7′, but the early astronomers could have used a value as large as 13°. Two centuries later, al-Biruni used a value as large as 10° (see Newton [1972b, Table I]).

‡The correct difference is about an hour. al-Biruni's calculations show that he used about $1\frac{1}{5}$ hours, and it is possible that earlier Islamic astronomers took the difference to be 2 hours.

study, I shall conclude that the observed time was actually 4/5 hours after sunrise, and this is the time that is entered in Table VII.1. Because of the questions that are connected with this conclusion, I give it a reliability of only ½.

ibn Yunis lists the time of the observation on EY 199 V 25 at Damascus as 0;0,20 days after noon. Both this observation and the other one at Damascus on EY 201 V 25 are stated to a precision of 0;0,5 days, and this tends to suggest that they are by the same person and hence that the first one is valid. I shall give the first one a reliability of zero in spite of this. 0;0,20 days is equal to 8 minutes of time.

ibn Yunis lists the time of the observation on EY 201 V 25 as 0;28,15 days after noon. This is 11.3 hours after noon. al-Biruni gives the time as 11 1/5 hours after noon. Because ibn Yunis's form of statement is the same for both observations, I have used his time in Table VII.1. The difference is trivial.

Both Thabit and ibn Yunis give the time of the observation at Baghdad on EY 199 V 25 as 7 hours of the day, that is, 7 hours after sunrise. Since this is at an equinox, the time is 13 hours after midnight. al-Biruni gives the time as 1 hour before noon. al-Biruni converted all the times to a time stated with respect to noon. It appears that he accidentally wrote "before noon" when he meant to write "after noon".

The only other time that needs comment is the time on EY 213 V 28 at Baghdad. ibn Yunis gives the time as 0;23,25 days after noon, which is the same as $9^h22^m$. I have written $22^m$ as 11/30 hours in Table VII.1, while al-Biruni wrote it as (1/5) + (1/6) hours. There are minor problems about two other times listed in Table VII.1 which I have discussed elsewhere [Newton, 1972b], and which need not detain us here.

In addition to assigning reliability to the various observations, we must assign *a priori* estimates of the standard error of a measurement. The equinox times are based upon measuring the elevation angle of the sun, and the analysis of a large body of solar elevations [Newton, 1972b] shows that the standard error was about 1′ for most Islamic observers. This corresponds to about 1 hour in the time of an equinox, and this is the standard error that I shall assign to the data in Table VII.1, except for the observations of as-Sufi. A sizeable group of observations made by him will be presented in the next section. It is clear from their self-consistency [Newton, 1972b] that the standard error in as-Sufi's measurements is about 2′, corresponding to about 2 hours in the time of an equinox. I shall use this error estimate for as-Sufi's equinoxes.

TABLE VII.2

ISLAMIC OBSERVATIONS OF THE MERIDIAN ELEVATION OF THE SUN

| Date | Observer | Place | Julian Day Number 2 000 000 | Elevation, degrees | Reliability |
|---|---|---|---|---|---|
| EY 201 I 5 | al-Marvarudi? | Damascus | 25 067 | 72;7,50 | 0 |
| EY 201 I 5 | Abu-al-Hasan | Baghdad | 25 067 | 72;14 | 0 |
| EY 201 II 21 | al-Marvarudi | Damascus | 25 113 | 80;4,10 | 1 |
| EY 201 II 22 | al-Marvarudi | Damascus | 25 114 | 80;4,30 | 1 |
| EY 201 II 23 | al-Marvarudi | Damascus | 25 115 | 80;4,28 | 1 |
| EY 201 IV 9 | al-Marvarudi? | Damascus | 25 161 | 73;2,4 | 1 |
| EY 201 IV 9 | Abu al-Hasan | Baghdad | 25 161 | 73;7 | 1 |
| EY 201 VIII 24 | al-Marvarudi | Damascus | 25 296 | 32;55,0 | 1 |
| EY 201 VIII 25 | al-Marvarudi | Damascus | 25 297 | 32;54,58 | 1 |
| EY 201 VIII 26 | al-Marvarudi | Damascus | 25 298 | 32;55,28 | 1 |
| EH 243 II 19 | Bani Musa | Samarra | 34 245 | 79;22 | 1 |
| EY 226 VIII 30 | Bani Musa | Samarra | 34 427 | 32;13 | 1 |
| EY 228 VIII 31 | Bani Musa | Samarra | 35 158 | 32;13 | 1 |
| EY 237 VIII 33 | Bani Musa | Baghdad | 38 445 | 33;5[a] | 0.1 |
| EY 238 III 1 | Bani Musa | Baghdad | 38 628 | 80;15[a] | 0.1 |
| EY 257 IX 1 | bin 'Ismat | Balkh | 45 748 | 29;46 | 1 |
| EY 258 III 6 | bin 'Ismat | Balkh | 45 933 | 76;54 | 1 |
| EY 328 III 28 | al-Hirawi | Rayy | 71 505 | 78;3 | 0 |
| EY 328 III 29 | al-Hirawi | Rayy | 71 506 | <78;5[b] | 0 |
| EY 328 III 30 | al-Hirawi | Rayy | 71 507 | 78;6 | 0 |

[a]This may be the elevation of the solstitial point rather than of the sun.
[b]The text says almost 78;5.

TABLE VII.2 (Continued)

| Date | | | Observer | Place | Julian Day Number 2 000 000 | Elevation, degrees | Reli-ability |
|---|---|---|---|---|---|---|---|
| EY 328 | IV | 1 | al-Hirawi | Rayy | 71 508 | <78;6[c] | 0 |
| EY 328 | IV | 2 | al-Hirawi | Rayy | 71 509 | 76;5[d] | 0 |
| EH 349 | X | 21 | al-Hirawi | Rayy | 72 046 | 30;47 | 0 |
| EH 349 | X | 23 | al-Hirawi | Rayy | 72 048 | >30;46[e] | 0 |
| EH 349 | VIII | 11 | | Isfahan? | 71 977 | 50;0 | 0.1 |
| EY 338 | IX | 22 | as-Sufi | Shiraz | 75 334 | 36;50 | 1 |
| EY 338 | IX | 23 | as-Sufi | Shiraz | 75 335 | 36;49 | 1 |
| EY 338 | IX | 24 | as-Sufi | Shiraz | 75 336 | 36;50 | 1 |
| EY 339 | III | 25 | as-Sufi | Shiraz | 75 517 | <83;59[f] | 1 |
| EY 339 | III | 26 | as-Sufi | Shiraz | 75 518 | 83;59 | 1 |
| EY 339 | III | 27 | as-Sufi | Shiraz | 75 519 | <83;59[f] | 1 |
| EY 339 | IX | 23 | as-Sufi | Shiraz | 75 700 | 36;49 | 1 |
| EY 363 | IV | 1 | al-Khujandi | Rayy | 84 283 | 77;57,40 | 1 |
| EY 363 | IV | 2 | al-Khujandi | Rayy | 84 284 | 77;57,40 | 1 |
| EY 363 | IX | 27 | al-Khujandi | Rayy | 84 464 | 30;53,35 | 1 |
| EY 363 | IX | 30 | al-Khujandi | Rayy | 84 467 | 30;53,32 | 1 |
| EY 385 | V | 2 | al-Biruni | Jurjaniya | 92 344 | 69;11 | 1 |
| EY 385 | VI | 2 | al-Biruni | Jurjaniya | 92 374 | 61;43 | 1 |
| EH 407 | III | 25 | al-Biruni | Jurjaniya | 92 396 | 53;58 | 1 |
| EH 407 | III | 26 | al-Biruni | Jurjaniya | 92 397 | 53;35 | 1 |
| EY 385 | VII | 2 | al-Biruni | Jurjaniya | 92 404 | 50;55 | 1 |
| EY 385 | VII | 10 | al-Biruni | Jurjaniya | 92 412 | 47;44 | 1 |
| EY 385 | VII | 25 | al-Biruni | Jurjaniya | 92 427 | 41;53 | 1 |
| EY 385 | VII | 26 | al-Biruni | Jurjaniya | 92 428 | 41;30 | 1 |
| EY 385 | IX | 26 | al-Biruni | Jurjaniya | 92 493 | 24;28 | 1 |
| EH 409 | VI | 1 | al-Biruni | Jayfur | 93 169 | 45;0 | 1 |

[c]The text says almost 78;6.

[d]This is probably an error for 78;5.

[e]The text says more than 30;46.

[f]The text says almost 83;59.

The preceding discussion applies to the random error of an observation. We must also be concerned with the possibility of a bias. A bias in finding the time of an equinox could result from an observer having an incorrect value of his latitude, or from having non-uniform graduations on his measuring circle, and probably from other sources. Refraction is probably not serious, because we can probably calculate it with enough accuracy. I shall return to the question of biases in Chapter VIII. It will turn out that the biases can be effectively eliminated for some of the observations in Table VII.1 and that they cannot be for others. This fact will also affect the final weighting of the observations.

## 2. Meridian Elevations of the Sun

al-Biruni [1025] has preserved a large number of measurements of the elevation angle of the sun when it was in the meridian plane. I have already analyzed these observations [Newton, 1972b] for the purpose of estimating both the acceleration of the sun and the rate of change of the obliquity. Since there is nothing to add to the earlier analysis, I shall merely summarize the data here for the sake of completeness. The summary appears in Table VII.2. The information used in preparing this table is taken from Table V of the earlier work.

al-Biruni always gave the feria of an observation, and he nearly always gave both the date in the Muslim calendar and the date in the Persian calendar. Even though the feria was probably calculated in some instances, his giving it helps detect accidental errors of writing. The feria and the Persian date, when the latter is given, always agree. This fact allows us to infer the date that al-Biruni took for the era of the Hijra.† We find that he used 622 July 15 about half the time and that he used 622 July 16 about half of the time. I did not notice an instance in which he used 622 July 14. In a few instances the Muslim date does not agree with either the Persian date or the feria for any of the three possibilities. In these instances, there has obviously been a large error in the transmission of the records.

Table VII.2 has the Muslim date of an observation only in the instances when al-Biruni failed to

†The dates 622 July 14, 622 July 15, and 622 July 16 have all been used for the era of the Hijra. See Section II.5.

give the Persian date. In these cases, we must rely upon the feria alone in choosing among the three possibilities for the era.

There still remain a few problems in determining the date of an observation. In most cases, the correct date can be found from other evidence, and I refer the reader to the earlier work for the solutions. Here I shall mention only the cases where there is still some question, because it is necessary to lower the reliability assigned to the corresponding records.

al-Biruni gives observations at both Damascus and Baghdad on EY 201 I 5. The difference between the elevation angles is about right for the difference in latitude between the two places, so it is likely that the observations were made on the same date. However, the elevation angles are wrong by about 45 minutes of arc, an incredibly large error. Hence the dates must be wrong, and we do not have enough information to restore the correct date. These observations are given a weight of zero.

The Bani Musa, who are listed as observers for the observations of EH 243 II 19 through EY 238 III 1, were the three sons of Musa ibn Shakir.† They observed first in Samarra and later from what al-Biruni [1025, p. 64] calls "their house on the bridge" in Baghdad. ibn Yunis [1008, p. 132] says that two of them prepared a zij but that the other, named Ahmad, made his own zij. According to al-Biruni, only two of the brothers made the observations listed in Table VII.2, and they were Muhammad and Ahmad. Thus, although the brothers seem to have divided up more than once, they divided in different ways for different purposes.

The language used for the Bani Musa's observations on EY 237 VIII 33 and EY 238 III 1 leaves it unclear whether the elevation is that of the sun or of the solstitial point. I think that the elevation of the sun is meant but, because of the ambiguity of the record, I give the records a reliability of only 0.1.

†Presumably the ibn Shakir who is credited with sponsoring Thabit. See Section VI.2.

al-Hirawi made seven observations in Rayy that are listed in Table VII.2 with dates from EY 328 III 28 to EH 349 X 23. There is much confusion in the records about the dates, however, and the listed dates are likely to be incorrect. Since there does not seem to be enough information to let us establish the correct dates, I give these records a reliability of zero.

There are several questions about the anonymous observation on EH 349 VIII 11. The most important one concerns the place where the observation was made. The record says that it was made in Kashan, but the error in the observation is essentially impossible if this is the correct place. The observation must have been made at a place about 1°20′ farther south than Kashan. I have suggested Isfahan because it is the only large city near Kashan with about the right latitude, and it is plausible that al-Biruni could have written Kashan inadvertently when he meant Isfahan. Because of the questions about the record, I give it a reliability of 0.1.

It is interesting that al-Biruni also wrote the wrong place name for the observations that he made himself on EY 385 V 2, EY 385 VI 2, and EY 385 VII 2, which I shall discuss below.

al-Biruni lists an observation attributed to al-Kuhi on EY 357 III 30. I have omitted this observation from Table VII.2 for the reasons given in Section VI.3.

al-Khujandi, who made the observations at Rayy on EY 363 IV 1 through EY 363 IX 30, built [al-Biruni, 1025, pp. 70-71] a meridian circle with a radius of 40 cubits (about 20 meters) that was graduated at intervals of 10 seconds of arc. He also built a device that allowed him to observe both edges of the shadow cast in the circle and to bisect the distance between them. al-Biruni (page 75) says that this instrument surpassed all others in grandeur and that it should have given the most accurate solar measurements known. However, as al-Khujandi had told al-Biruni personally, the structure carrying the aperture sagged and he was not able to keep the aperture at the center of the circle.

Over a short range of angle, having an off-center aperture is equivalent to having an error in the direction of the vertical, but the error is not the same at high and low elevation angles. Thus we may take the errors to be correlated for the two winter readings, and we may take them to be correlated for the two summer readings, but we must take the summer and winter errors to be uncorrelated.

al-Biruni made the observations of EY 385 V 2, EY 385 VI 2, and EY 385 VII 2 as members of a group separated by 30 days. He said that all three readings were made in Khwarizm. However, we know that he was in Jurjaniya at the time. Further, he made a number of calculations with the resulting data, and he always took the latitude of the observing site to be 42°17′. This is exactly the value that he always used for the latitude of Jurjaniya, whereas he used a quite different value (41°35′40″) for the latitude of Khwarizm. Hence there is no appreciable doubt that the place is Jurjaniya, and we do not need to lower the reliability.

The observation of EY 385 VII 10 is the observation from which al-Biruni inferred the time of an autumnal equinox. He calculated, from the latitude and the meridian elevation, that the declination of the sun was +1′ at noon and hence that the equinox was 1 hour later.

Except for the observations of as-Sufi at Shiraz, between EY 338 IX 22 and EY 339 IX 23, we can assume that the random error of the observations has a standard deviation of 1′. as-Sufi's observations have more scatter, and for them we should use 2′. In addition, we must assume that there are errors in the local vertical. Except for the observations of al-Khujandi at Rayy, we can assume the same error for all of a cohesive set of observations made by the same person at the same place. Thus we can take the errors in the local vertical to be parameters that can be estimated from the data at the same time that the acceleration of the sun is estimated. For al-Khujandi, we must assume separate errors for his winter and for his summer observations.

TABLE VII.3

MISCELLANEOUS OBSERVATIONS OF THE SUN
FROM ISLAMIC SOURCES

| Date | Place | Julian Day Number -2 000 000 | Hour | Quantity[a] Measured | Value, degrees |
|---|---|---|---|---|---|
| EY 200 X 4 | Baghdad | 24 976 | 2 1/5 | L | 315;0 |
| EY 201 I 5 | Baghdad | 25 067 | 21 | L | 45;0 |
| EY 201 IV 10 | Baghdad | 25 162 | 0 5/6 | L | 135;0 |
| EH 349 VIII 11 | Isfahan? | 71 977 | 12 | L | 198;37 |
| EY 385 IV 6 | Jurjaniya | 92 318 | 16.09[b] | E | <36;30[c] |
| EH 407 III 26 | Jurjaniya | 92 397 | 12 | L | 165;11 |
| EY 385 VII 25 | Jurjaniya | 92 427 | 12 | L | 194;51 |
| EY 385 IX 26 | Jurjaniya | 92 493 | 13.52[b] | E | 21;10 |
| EY 385 IX 26 | Jurjaniya | 92 493 | 14.66[b] | E | 14;50 |

[a]L denotes a longitude of the sun and E denotes an elevation.

[b]Calculated from an observed azimuth of the sun.

[c]The text says slightly less than 36;30.

## 3. Other Solar Data

There are a few other Islamic observations of the sun, which are summarized in Table VII.3. Some of them are measurements of the longitude of the sun, while others are measurements of the elevation angle of the sun when it was at an azimuth other than due south.

Thabit [ca. 880, p. 72] has preserved measurements of the times when the sun was at the middle of the zodiacal signs Aquarius, Taurus, and Leo. The measurements were made on EY 200 X 4, EY 201 I 5, and EY 201 IV 10. They are in the middle of the equinox and solstice measurements made at Baghdad under Yahya (Table VII.1), and it is plausible to assume that they are part of the same set of systematic observations.

Thabit says that the interval between the first and third measurements is 185 days plus 22 2/3 hours.

The interval between the first and third times listed in Table VII.3 is 185 days plus $22^h38^m$, and it is plausible that Thabit rounded $38^m$ to 2/3 hours. This agreement confirms both times involved in the calculation.

Thabit also says that the interval between the first and second measurements is 91 days plus 18 1/6 hours. However, the interval between the first and second times in Table VII.3 is 91 days plus $18^h48^m$. It is unlikely that the first time is in error, as we just saw, so that there must be an error either in the second tabulated time or in the time interval. If the time interval is right, the time must be $3^h38^m$ before midnight, whereas it is written as $3^h$ before midnight. It is possible that the fraction of an hour was accidentally omitted, but I think it more likely that Thabit either made an accidental error in subtraction or he wrote down 18 1/6 hours when he meant to write 18 5/6 hours.

I have already discussed the record [al-Biruni, 1025, p. 86] of a meridian observation of the sun made on EH 349 VIII 11, presumably at Isfahan. The same record says that the longitude of the sun at that epoch was 198;37 degrees. It is possible that the longitude was obtained from a zij. If so, it was probably not obtained from Habash's zij, since the longitude is inconsistent with that zij, as we shall see in a moment. Likewise, it was probably not obtained from any other zij derived from the Yahya measurements. It is also inconsistent with al-Battani's tables. I shall assume that the stated longitude was found by measurement, but I shall assign a reliability of only 0.1 to this interpretation; it is dangerously close to circular reasoning.

al-Biruni (page 96) also gives the longitude of the sun at the times of his own meridian observations on EH 407 III 26 and EY 385 VII 25 in Table VII.2, and he does so in a useful and interesting manner. He first gives the longitude as calculated from Habash's zij. He then subtracts 11′ from the calculated longitudes because this ". . is the difference between what I have found by observation and the computation by this zij, .." The longitudes listed for these dates in Table VII.3 are the values that are found after the subtraction of 11′. They are not literally observations that al-Biruni made on the dates stated, but they represent a body of observations that he had made at times close to those listed, and we may use them as if they were observations for the purpose of inferring the acceleration of the sun.

I shall discuss Habash's zij at more length in Section VIII.5. It will appear that the error in his mean longitude of the sun was no more than about 1′ in his own time, say in EY 200. Two centuries later, in al-Biruni's time, the error had grown to 11′. At the time of the Isfahan observation just mentioned, the error would have been about 7′. However, when we compare the stated longitude with that computed from modern tables, we find a discrepancy of only 3′, and the discrepancy is still less when we make an allowance for the acceleration of the sun. Thus it is not likely that the stated longitude was found from Habash's zij.†

Finally, al-Biruni gives three values of the elevation angle of the sun measured at azimuths other than due south. One of these measurements was made on EY 385 IV 6, at the time when the azimuth was 0°; al-Biruni normally measured azimuth from due west and we may assume that he did so here. He made two other measurements on EY 385 IX 26, at azimuths of 67;30 degrees and of 52;30 degrees from the west. In Table VII.3, the times listed for these observations are calculated from the azimuths. In the analysis, we may take the times as exact and assign all of the error to the measurement of elevation.

Except for the observation that was presumably made at Isfahan, I assign a reliability of 1 to the observations listed in Table VII.3.

---

†al-Battani's zij was accurate in his own time but it accrued error rapidly because he combined his data with Ptolemy's. By the time of the Isfahan observation, the error in his solar tables was about 10′.

It is not safe to assume that the measurements in Table VII.3 are as accurate as those in Table VII.2. The determination of a longitude, with the methods available to the Muslim astronomers, requires combining two independent sightings and hence it is not as accurate as the measurement of an elevation angle. Further, when an elevation angle is measured out of the meridian plane, the angle is affected by both the north-south and the east-west errors in the local vertical, whereas a meridian elevation is affected by only one error component. It is plausible to assign 2′, rather than 1′, as the *a priori* estimate of the standard deviation for the measurements in Table VII.3.

## 4. Observations Involving Mercury

There are a few planetary conjunctions attributed to al-Kuhi† which Caussin, the editor and translator of *ibn Yunis* [1008], lists on pages 222-223 of the cited edition of *ibn Yunis*. *Delambre* [1819, p. 94] also quotes them. They accompany the solstice and equinox observations that were discussed in Section VI.3. We saw in that section that at least some of the observations that al-Kuhi claims to have made are in fact hoaxes. Since it is not safe to use observations made by a person who is known to have committed a hoax, I shall ignore al-Kuhi's planetary conjunctions.

Except for the observations attributed to al-Kuhi, the only planetary observations that I have found in Islamic sources are those preserved by ibn Yunis. In this section, I shall discuss the observations that involve Mercury, in the order that they appear in the text. The first item that appears in each discussion is the page in *ibn Yunis* on which the record begins.

1. Page 96. This observation was made at Baghdad by Abdallah al-Mahani, an astronomer and mathematician who is mentioned by *Kennedy* [1956, p. 136] and by *Nasr* [1968, p. 149]. It was made

---

†Caussin, in the cited edition of *ibn Yunis* [1008], and *Delambre* [1819, p. 94] refer to al-Kuhi as Abousahel.

on the 5th feria, EH 244 VI 10. al-Mahani also gave the Persian date in his original record. The year was EY 227, and the day of the month was the 5th, but the month itself is missing from the present form of the record. If we use 622 July 15† for the beginning of the Muslim era, we find that EH 244 VI 10 = EY 227 VI 5, which is the 5th feria. Hence we can reconstruct the Persian month without ambiguity.

In the morning of the 5th feria, Mercury was slightly more than 1° from Venus. The record does not give the direction, but calculation shows that Mercury was east of Venus. As I did with the Babylonian observations in Chapter IV, I shall take "morning" in a record like this to mean 45 minutes before sunrise. I shall take "slightly more than 1°" to be 1°.1. The record says that the distance had been the same the morning before.

2. Page 108. Abu al-Qasim Amajur, his son Abu al-Hasan Ali Amajur, and their freed slave Muflih made a long series of astronomical observations and wrote five works on astronomy that are listed by Kennedy [1956, p. 125]. All of their works have been lost, but ibn Yunis has preserved many of their observations, with dates ranging from 885 to 933. The observations through 901 (EH 288) were made at Shiraz and those in 923 (EH 311) and later were made at Baghdad.

The Amajurs made three closely related observations, one each of Mercury, Venus, and Mars, for which ibn Yunis [p. 108] writes the year as EH 356. Caussin says that this is a scribal error for EH 306. This is a plausible error because, in the notation used at the time, zero was written as a dot and five was written as a small circle. Calculation shows that the planets were nowhere close to the observed positions on the stated date in EH 356 but that they were close to the recorded positions on the same date in EH 306. Further, EH 356 is considerably too late for an Amajur observation. Thus we may confidently take the year to be EH 306.

---

† See Section II.5.

Although the record gives the feria as well as the Muslim date, we need even more information before we can assign the dates in our calendar, because we do not know the convention that the Amajurs used for the beginning of the day.

The observations of Mercury and Venus were made at the same time, in the morning of EH 306 VII 18, at $5^h50^m$, equal hours, before noon of the 6th feria. The observation of Mars was made at 2 unequal hours after the beginning of the night of the 7th feria, EH 306 VII 26. In the record of EH 306 VII 18, the phrase "before noon of the 6th feria" sounds as if the following noon is the beginning of the 6th feria and that the observations themselves were made while it was still the 5th feria. However, it is possible that "$5^h50^m$ before noon" and "6th feria" are both intended to apply to the time of the observation. In order to resolve the question, we turn to the calculated behavior of Mercury and Venus.

If we calculate the Julian day number from Equation II.8 in Section II.5, tentatively using $(JD)_{MO}$ = 1 948 438, we get 2 056 715, which is the day number at noon on 918 December 24; this day was the 5th feria. When we calculate the positions of Mercury and Venus at $6^h10^m$, local apparent time, on this day and on the days before and after, we find that both recorded positions agree well with the calculated positions on 918 December 24 and that they disagree by impossibly large amounts for the adjacent days. Hence there is no doubt that 918 December 24 is the correct day, and the meaning of the passage seems to be that noon on that day was the beginning of the 6th feria, and hence of EH 306 VII 19, in the conventions used by the Amajurs.

At the time of the observation, the elevation angle of $\alpha$ Scorpii was 24° above the eastern horizon. I shall take the time of the observation from this fact rather than using the stated time. The observers measured the separation of Mercury from $\alpha$ Scorpii. According to their tables, the longitude of $\alpha$ Scorpii was 234°31′, and thus they found the longitude of Mercury to be 254°20′. In other words, they found Mercury to be 19°49′ east of $\alpha$ Scorpii in longitude.

We do not know whether this observation was made in Shiraz or Baghdad and hence we do not know the time of the observation. The difference in time between the two places is about 32 minutes, so, if we use the average of the times for the two places, the error that we shall make is only 16 minutes. An error of this size is acceptable, although we should certainly prefer to avoid it if we could.

3. Page 160. This Amajur observation poses so many problems of interpretation that it is unusable. First, the date is given as the 3rd feria, EH 289 VI 8, EY 271 II 3. However, the two calendar dates are a week apart and both are the 4th feria.† Caussin, in a note, says that "3" in the Persian date is an error for "8". However, EY 271 II 8 is not compatible with the feria, and it is not compatible with EH 289 VI 8 for any choice of the era.

We cannot resolve the date by means of the observation. The record says that Mercury and Venus were in conjunction, but they were not in conjunction on any of the possible dates nor at any nearby time. The most that we can say is that the separation in longitude was near a minimum on EY 271 II 8.

The record does say that there was about a <u>shebr</u> between the planets, and Caussin says that a <u>shebr</u> is about 1°. A statement of separation accompanying a conjunction usually refers to the latitude, and 1° is about the separation in latitude of the planets at the time in question. Since it is also about the minimum separation in longitude, it is possible that the longitude is meant here. However, even if we should make this assumption, and there is no real basis for making it, the statement of separation is not precise enough to make the observation useful.

Finally, we do not know whether the observation was made in Shiraz or Baghdad.

---

†For the Muslim date, I have used 622 July 15 as the era.

4. Page 180. The remaining observations of Mercury were made by ibn Yunis himself in Cairo. The first one of these consists of two parts. On the 2nd feria, EH 375 II 1, EY 354 IV 5, Mercury was about 1° south of Venus in latitude, and had only a short distance to go before overtaking Venus in longitude.† On the 3rd feria, Mercury had passed Venus in longitude. ibn Yunis estimates that the conjunction occurred in the morning, presumably about sunrise, of the 2nd feria, whose day number is 2 081 002 according to the dates given.

On this day, however, Mercury was less than 20′ south of Venus in latitude. It was already more than 1° east of Venus and was "pulling away", and it is clear that the conjunction had happened some time before. Thus the date of the conjunction is seriously in error. It is possible that the date is exactly 1 week too late, but I have not tested the matter by calculation. We must regretfully conclude that this observation is not usable.

5. Page 190. After sunset, when the altitude of Venus was 10°, on the 4th feria, EH 384 XI 28, EY 363 X 16, Mercury was about 1 shebr behind Venus. ibn Yunis thinks that Mercury would have actually eclipsed Venus at the time of the conjunction, and he is indeed correct that the separation in latitude was quite small at the time of the conjunction. Unfortunately, he does not tell us when the conjunction occurred, and the estimate of 1 shebr is not sufficiently precise to make this a usable observation.

6. Page 200. ibn Yunis observed a conjunction of Mercury and Venus after sunset in the night before the 2nd feria, EH 390 VI 13. The exact time of the conjunction, he estimates, was 8 equal hours after the noon of the 1st feria, EY 369 III 5. Mercury was about 1/3 degree to the north of Venus in latitude. This observation seems to be accurate in all respects, and the time is carefully stated.

---

†The time of the observation is not given, but it must have been shortly after sunset, since Mercury and Venus were only about 24° east of the sun.

7. Page 202. ibn Yunis observed Mercury and Venus about 1 hour after sunset. From this observation, he concluded that they were in conjunction at the midnight before the 2nd feria, EH 391 VII 7, EY 370 III 9. All of the items of calendrical information conflict with each other and with the astronomical positions.

The Julian day number that corresponds to EH 391 VII 7 is 2 086 825, if we take 622 July 15 as the era, and this is feria 1. The day number that corresponds to EY 370 III 9 is 2 086 816, and this is feria 6. It is true that EH 391 VII 7 can be made to come on feria 2, in agreement with the record, if we use 622 July 16 as the era, but this conflicts violently with the astronomical requirements. Although the record says that the conjunction was at the midnight before feria 2, calculation shows that it occurred around sunrise or later on EY 370 III 14, which was the 4th feria.

I cannot think of any plausible explanation of the errors in this record, and it must therefore be taken as unusable.

Delambre,† who made an extensive study of ibn Yunis's work, concluded that he was excellent in

TABLE VII.4

OBSERVATIONS OF MERCURY PRESERVED BY IBN YUNIS

| Date | Place | Julian Day Number -2 000 000 | Hour, local mean time | Position of Mercury |
|---|---|---|---|---|
| EY 227 VI 5 | Baghdad | 34 707 | 5.08 | 1°.1 east of Venus |
| EH 306 VII 19 | Baghdad or Shiraz[a] | 56 716 | 6.24 | 19°49′ east of $\alpha$ Sco |
| EY 369 III 5 | Cairo | 86 447 | 19.91 | In conjunction with Venus |

[a]The midpoint will be used in the analysis.

†Quoted by Kennedy [1956, p. 126].

the computational aspects of astronomy but that he was an inferior observer. The Mercury observations confirm this assessment. Of the four observations of Mercury that he made and transmitted to us, three are recorded so poorly that they are unusable. It is possible, of course, that the errors in the records were made by later copyists and not by ibn Yunis in his original writing, and we should be cautious about criticizing ibn Yunis because of these errors.

Altogether, then, we have records of 7 observations of Mercury made by Islamic astronomers, 3 of which are usable. The usable records are summarized in Table VII.4, whose format should be obvious. Somewhat arbitrarily, I shall use 0°.15 (9′) as the *a priori* estimate of the standard deviation in the position of Mercury.

## 5. Observations Involving Venus

The observations from *ibn Yunis* that involve both Venus and Mercury were discussed in the preceding section. In this section, we shall discuss only the observations of Venus that do not involve Mercury. The first item in each discussion is the page in *ibn Yunis* on which the record begins.

1. Page 94. This observation was made by al-Mahani at Baghdad, toward dawn on the 1st feria, EH 244 V 15, EY 227 V 10. The day number at noon on that day was 2 034 682. Venus had 2/5 degrees to go before reaching Saturn. I shall take the time of this observation to be 45 minutes before sunrise.

al-Mahani also says that Saturn was 2/3 degrees from $\alpha$ Leonis. By combining these statements, we can find the distance from Venus to $\alpha$ Leonis. The wording suggests, however, that the position of Venus relative to Saturn was the quantity directly measured. Hence it is safer not to use the position relative to the star.

2. Page 96. This observation was also made by al-Mahani at Baghdad. Venus and Mars were in conjunction and seemed to touch each other at the

beginning of the night before the 2nd feria, EH 250 I 2, EY 232 X 26. The day that is identified has number 2 036 678. In terms of our calendar, the observation was made the day before, on feria 1, EY 232 X 25. I shall take the time to be 45 minutes after sunset on that day, the day that is numbered 2 036 677.

3. Page 108. This is an Amajur observation that goes with an observation of Mercury discussed in the preceding section. It was made at $5^h50^m$ before the noon of the 6th feria, EH 306 VII 19, the day whose number is 2 056 716. Venus was 4° 29′ east of $\alpha$ Scorpii. Since we do not know whether the observation was made at Shiraz or Baghdad, I shall use the average of the times for the two places. This is equivalent to assuming that the observation was made at the midpoint between them.

4. Page 156. This observation was made by Habash at Baghdad. In the evening of EY 199 X 2, Venus was at 22°42′ of Sagittarius. That is, Venus was at longitude 262°42′.

Something is seriously wrong with this record. Caussin says that the date is 831 January 25. Actually, the day stated is 831 January 29. Aside from the fact that Caussin uses a different convention for the beginning of the day, he has forgotten to include the intercalary days in his conversion of the date.

At the time stated, the longitude of Venus was about 271° rather than about 263°, and the longitude of the sun was about 313°. Thus Venus was not even an evening star. We must now ask whether any simple error can explain the record in the form that we now have.

The error cannot be in the zodiacal sign. Venus was at about 1° of Capricorn rather than about 23° of Sagittarius. Further, the error cannot be in the number of the year. At any year around this time, the sun was at about longitude 313° on the Persian date given. If Venus were at 262°42′, as stated, it would necessarily be a morning star rather than an evening star.

It would have been fairly easy for Habash to write "evening" when he meant "morning", but this does not explain the record either. There are still errors both in the zodiacal sign and in the number of degrees used to write the longitude.

The error cannot be in the day of the month. Venus could not have been an evening star on any day during the month specified.

That leaves the month itself as the only possible source of a single error. If we try the date EY 199 VII 2 rather than EY 199 X 2, we find that Venus was an evening star, that it was near its maximum elongation, and that its position agrees well with that found in the record. Thus a simple error in the month, alone of all the possibilities, explains the record. The day of the observation is thus the day numbered 2 024 514. I shall take the time to be 45 minutes after sunset at Baghdad.

5. Page 156. This observation was also made by Habash at Baghdad, although it is more than 30 years after the preceding observation. The time is dawn on the 1st feria, EH 250 IX 6, EY 233 VII 7. The Persian date stated is equivalent to 864 October 22, which was the 1st feria. Since this date also agrees excellently with the astronomical observation, there is no question that this is the correct date.

If the era is taken as 622 July 15, the Muslim date is equivalent to the 3rd feria, 864 October 10, and this is the date that Caussin assigns to the observation. At the time I wrote my earlier discussion of this observation [Newton, 1970, p. 165], I did not know how to use the Persian calendar and hence I also assigned 864 October 10, without noticing the discrepancy with the feria. In the analysis of the observation, however [page 205], I changed the date to 864 October 22. I did this on the basis of the calculated position of Venus without realizing that this is in fact the date that Habash originally recorded.

The error in the Muslim date tells us, I believe, how Habash actually recorded dates. He must have used the Persian calendar in making his original record, and then converted the Persian date to a Muslim one. In this case, he must have accidentally used the calendar for EH 251, in spite of the fact that he correctly wrote EH 250 for the year.

At the time of the observation, Habash writes, Venus and Mars were so close together that they made, "so to speak, a single planet." I shall follow the convention of taking the time to be 45 minutes before sunrise.

6. Page 158. This is an Amajur observation made at Shiraz.† The time is 1 hour before sunrise on the 6th feria, EH 272 III 26, EY 254 V 30. The latter day has the number 2 044 557, which is the 6th feria, and it is thus the correct date. Caussin assigns the day before, which he derived from the Muslim date by using 622 July 15 for the era. There was a "perfect occultation" of $\alpha$ Leonis by Venus at this time, according to the record. The record also says that Venus was more than 1° from $\alpha$ Leonis on both the morning before and the morning after, and this statement agrees with the calculated velocity of Venus. However, I shall use only the observation of the occultation and not the other two observations.

7. Page 158. This is another Amajur observation, which I shall assign to Shiraz. It has two errors in the statement of the date, but I believe that we can determine the correct date without ambiguity in spite of this fact. The observation is

†The place is not specifically stated. Amajur observations on and after 923 June 1 are specifically placed at Baghdad. A conjunction of Venus and Jupiter in October of 901 is specifically placed at Shiraz. In earlier work [Newton, 1970, Section VI.3], I took both Shiraz and Baghdad as possible places for the observations that are not specifically placed. If we do not assign Shiraz for all the observation in and before 901, however, we are tacitly assuming that the Amajurs may have moved several times, along with all of their observing apparatus. On reflection, this seems unlikely, so I shall use Shiraz for the earlier observations.

dated on the 2nd feria, EH 383 IX 13. The year EH 383 is a century too late for an Amajur observation, so we conclude that the year should be EH 283. It would be easy to make this mistake with any system of notation, but it is particularly easy in the Arabic notation of the time.

Caussin assigned the date 896 October 23 to this observation which equals EH 283 IX 13 if 622 July 15 is the era. However, 896 October 23 is the 7th feria rather than the 2nd. October 25, when the day number was 2 048 620, was the 2nd feria, and this day also agrees accurately with the circumstances of the observation. Hence it seems safe to conclude that the observation was made on 896 October 25.

According to the record, Venus and Mars were in conjunction, and Venus was to the north in latitude by less than 4 digits (20′). Calculation shows that Venus was to the north by about 17′, which agrees excellently with the record.

The record then goes on to give longitudes of Venus, Mars, and the sun, stated to the minute of arc. I believe that these are values calculated from some set of tables, especially since the stated longitudes of Venus and Mars differ by 4′.

The record does not give an hour. It is clear from the longitudes that the observation could only have been made in the morning, so I shall use the conventional time of 45 minutes before sunrise.

8. Page 158. This is another Amajur obsevation, and the record explicitly states that it was made at Shiraz. According to the record, the observation was made at dawn on the 3rd feria, EH 288 X 19. If the era is 622 July 15, this day was the 2nd feria. We can make EH 288 X 19 come on the 3rd feria if we use 622 July 16 as the era, but this date does not agree with the circumstances. This 3rd feria would be day number 2 050 427; on that day Jupiter was about 8° east of Venus in longitude, instead of being in conjunction as the record says. The conjunction actually occurred on day 2 050 434, which was the 3rd feria. Thus the Amajurs made a

mistake of exactly a week, in terms of the calendar that they were using. They should have written the date as EH 288 X 26, which is the date of the conjunction if 622 July 16 is taken as the era. I shall use EH 288 X 27 as the date in order to use 622 July 15 for the era of the Hijra.

On the basis of an observation made some time before sunrise, the Amajurs estimated that the conjunction would occur at sunrise. Venus was about 1 <u>fetr</u> north of Jupiter; calculation shows a separation of about 26′.

9. Page 180. This and the remaining observations of Venus were made by ibn Yunis at Cairo. The time is 8 equal hours after the noon of the 7th feria, EH 377 II 7, EY 356 IV 1. The Persian date is the day numbered 2 081 728, which is the 7th feria. The Muslim date contains a serious error. Caussin thinks that ibn Yunis wrote the day of the month as "7" when he meant to write "17", and this is a plausible error.

EH 377 II 17 is equivalent to 987 June 17 if we take 622 July 15 as the era. Caussin assigned this as the date of the observation, overlooking the point that this is the 6th rather than the 7th feria. In my earlier work [Newton, 1970, p. 167], when I did not know how to use the Persian calendar, I followed Caussin in using 987 June 17 as the date, and I also overlooked the matter of the feria. If the observation were simply dated on the 7th feria, we could assume that ibn Yunis was using the day that begins at sunset, so that the day would be 987 June 17 in our calendar. However, this interpretation does not seem possible. The record says explicitly that we are to start at the noon of the 7th feria, which must be the noon of 987 June 18, and then we are to go to the time that is 8 hours later.

The event observed is a conjunction of Venus and $\alpha$ Leonis.

10. Page 182. Venus and Saturn were in conjunction about $\frac{1}{2}$ hour before sunrise on the 6th feria, EH 377 IX 28, EY 356 XI 2. This day has the

number 2 081 944. Venus was north of Saturn by about 1 digit (5′) in latitude, according to the record, and this agrees well with the results of calculation. Jupiter preceded† Venus and Saturn by about 1°.

Calculation would show us without ambiguity whether Jupiter was east or west of Venus. Thus the record locates Venus with respect to Jupiter as well as to Saturn. Since the wording suggests that ibn Yunis was not trying to give precisely the position with respect to Jupiter, I shall use only the position with respect to Saturn.

11. Page 184. This observation was made about 1 hour after sunset on the 2nd feria, EH 380 III 26, EY 359 IV 7. The Persian date has Julian day number 2 082 829, which is the 2nd feria; the Muslim date is the day before, if we use 622 July 15 for the era. However, the Muslim date given in the preceding record is the same as the Persian date if we use 622 July 15 for the era. Thus ibn Yunis is not consistent in his use of the Muslim calendar.

If ibn Yunis wrote in terms of a day that begins at sunset in this record, the day changed just before the observation, and the day we want is the day number 2 082 828. However, calculation shows clearly that the correct day is the one numbered 2 082 829, which corresponds to 990 June 23. At 1 hour after sunset, according to the record, Venus and $\alpha$ Leonis were in conjunction in longitude and Venus was about 1 shebr (about 1°) north in latitude.

ibn Yunis says that he observed Venus for several days in succession and that he is certain about the accuracy of the observation.

12. Page 186. This observation was made on the 4th feria, EH 381 X 13, at 6 equal hours after

†This usually means to be to the west of, but the term is ambiguous when it is applied to the planets. The meaning depends upon whether precedence is taken with respect to the planetary motion or to the diurnal rotation of the heavens.

the noon of the 3rd feria. Thus ibn Yunis is clearly beginning the day at sunset, in contrast with what he did in the preceding record. The date agrees with the feria if we use 622 July 16 as the era, and the date given has day number 2 083 377. In terms of our convention for beginning the day, however, the observation was made on the day with number 2 083 376.

The observation concerns a conjunction of Saturn and Venus. Unfortunately, the record is worded in such a way that we cannot tell whether they were actually in conjunction at the stated time, or whether they were separated by "1 *shebr* and 2 knots", whatever that may mean. Thus we cannot safely use the record.

13. Page 186. This observation was made at 1 equal hour before sunrise on the 7th feria, EH 382 VII 17, EY 361 VII 4. If we use 622 July 15 for the Muslim era, all statements of the date lead to the day numbered 2 083 646; ibn Yunis seems almost to alternate between 622 July 15 and 622 July 16. At the time of the observation, Venus was about ½ degree south of α Leonis in latitude, and it had passed the star by about 1/3 degree in longitude. Since the motion of Venus was direct at the time, this means that Venus was to the east.

14. Page 190. Venus and Jupiter were in conjunction at 7 equal hours after the noon of the 3rd feria, EH 385 V 10, EY 364 III 26. Venus was north in latitude by about 1 *fetr*, which Caussin takes as about 40′. If we use 622 July 15 for the era of the Hijra, all statements lead us to the day numbered 2 084 643 for the observation. Venus was actually to the north by about 33′.

15. Page 194. Venus and α Leonis were in conjunction at $7^h40^m$, in equal hours, after the noon of the 3rd feria, EH 385 V 19, EY 364 IV 3. By comparison with the preceding record, we see that the day of the month in the Muslim date should have been "17" rather than "19". Since the numerals for "7" and "9" have little resemblance in the notation of the time, the mistake was probably simple inadvertence. ibn Yunis also says that Venus was north of α Leonis in latitude, by about

2/3 or 3/4 degrees, or about 1 fetr.

16. Page 194. For some reason, ibn Yunis has given us another record of the observation in item 14 above, and this record conflicts with the earlier one. According to this record, the conjunction occurred on the night before the 3rd feria, EH 385 V 10, EY 364 III 26; that is, the conjunction was on the 2nd feria according to the record. The record goes on to say that Venus had passed Jupiter on the 4th feria. Since the record skips from the 2nd to the 4th feria, it is likely that "before" in "before the 3rd feria" was written in mistake for "after". This record can be ignored.

17. Page 196. This record has several errors in the dating, but we can find the date without ambiguity in spite of them. ibn Yunis says that he watched Venus and Jupiter carefully for many days, and that they were in conjunction about ½ hour after sunset on the night before the 1st feria, EH 386 VII 22, EY 365 IV 2. Actually, the conjunction was around the time of sunset on the night before the 1st feria, EH 386 VII 29,† EY 365 VI 3. In giving the Muslim date, ibn Yunis gave the day that was exactly 1 week early. In giving the Persian date, he wrote the wrong month to start with.‡ In addition, he apparently gave the day of the observation itself, although he says he is giving the day after. Caussin assigns the date 996 August 8, which was the 7th feria. That is, Caussin gives the day before the one stated in the record, because the record date is stated to be the day after the observation; Caussin did not notice the error of a week. In terms of our calendar, the conjunction was actually on 996 August 15, which was the 7th feria and had day number 2 085 074.

ibn Yunis also says that Venus and Jupiter seemed to touch, and that Venus was north of Jupiter by about 5′. The value of 5′ may have come from a

†Assuming 622 July 15 for the era.

‡Since ibn Yunis writes out the Persian months by name, the mistake was not the simple one of interchanging I and V in a Roman numeral. Besides, ibn Yunis would not have used Roman numerals.

calculation. However, ancient and medieval astronomers frequently assigned a finite semi-diameter to the planets, presumably as the result of an optical illusion, and it may be that ibn Yunis thought that 5′ was the separation of the centers when the planets seemed to touch at the edges. The actual separation in latitude was about 4′.9.

18. Page 196. The date of this observation is given as the 2nd feria, EH 387 VI 14, EY 366 III 9. Although ibn Yunis writes out the names of the months, it is easy to confuse the names of month V and month VI in the Muslim calendar, and this is just what ibn Yunis does, as Caussin pointed out. Thus the Julian day number of the observation is 2 085 356, which was the 2nd feria. According to the record, Venus eclipsed Saturn "in a manner that cannot be doubted" at 2/3 equal hours before sunrise. Calculation shows that Saturn was about 3′ north of Venus in latitude.

19. Page 198. This observation was made at 1 hour after sunset on the night before the 6th feria, EH 388 VI 26, EY 367 IV 10. The 6th feria was therefore the day numbered 2 085 752, and the observation was therefore made on the day with number 2 085 751. The record also says that the time was 8 equal hours after the noon of that day. Venus was in conjunction with $\alpha$ Leonis, and the planet was about 1° north of the star in latitude, according to the record.

20. Page 198. Venus and Mars were in conjunction near the beginning of the night before the 3rd feria, with Venus being about 1 digit (5′) north of Mars in latitude. More specifically, Venus had already passed Mars in longitude by about $\frac{1}{4}$ degrees† at 8 equal hours after the noon of the 2nd feria, EH 388 VII 7, EY 367 IV 20. Hence the day number of the observation is 2 085 762.

21. Page 200. There are some ambiguities in this record, but luckily there is also enough redundancy that we can remove them. ibn Yunis

†Venus and Mars were moving eastward at the time, but Venus was moving more rapidly. Hence Venus was east of Mars.

observed Venus and Mars at the end of the night before the 2nd feria, at $16^h30^m$, in equal hours, after the noon of the 1st feria, EH 389 IV 20. It is not clear whether the Muslim date is intended to refer to the 1st feria or to the 2nd feria. Later, however, ibn Yunis says that the 1st feria was EY 368 I 24. Hence the observation was on the next day, on the day numbered 2 086 042. Venus and Mars were separated by about 1°, but it is not clear which planet was to the east. Since the difference between the two possibilities amounts to 2° in the relative position, we can resolve this uncertainty by calculation. Venus was to the east of Mars.

22. Page 202. ibn Yunis left this record incomplete, and we are not able to use it. At 7 equal hours before noon of the 2nd feria, EY 369 VII 5, Venus still had a small distance to go before being in conjunction with $\alpha$ Leonis. The day specified has Julian number 2 086 567. ibn Yunis probably intended to compare the positions of Venus and $\alpha$ Leonis on consecutive days in order to estimate the time of the conjunction. It is not safe to estimate the separation of the two bodies involved.

23. Page 204. Again we have an incomplete record. Venus had not yet reached $\alpha$ Leonis at about 1 hour after sunset that ended the 7th feria, EY 370 IV 22.† Thus the observation was on the day numbered 2 086 859. ibn Yunis says that the conjunction occurred sometime during the night, but he does not estimate the hour. He also gives the Muslim date, but it clearly contains an error.

24. Page 206. ibn Yunis observed Venus and Jupiter at $1^h30^m$, in equal hours, before sunrise on the 7th feria, EH 392 VI 2, EY 371 II 3. If we use 622 July 15 for the era, the two calendar dates are equivalent, but the date is the 6th rather than the 7th feria. Calculation of the positions of Venus and Jupiter shows clearly that the feria is correct. Hence the observation was on EY 371 II 4,

†ibn Yunis actually dates the observation by means of the day that began at sunset, namely the 1st feria, EY 370 IV 23.

which is equivalent to the day with Julian number 2 087 146. At the time stated, Venus was heading directly for Jupiter and had about 1/5 degree to go before reaching it. ibn Yunis determined, he says, that the conjunction took place in both latitude and longitude at 2 equal hours before noon. Calculation shows that the latitudes differed by about 2′.

If ibn Yunis had implied that he found the time by interpolation between two or more observations, we should use his time estimate on the basis that it represents more than one observation. Since he does not imply this, we should assume that the position given before sunrise represents the 'raw' form of the observation.

25. Page 208. ibn Yunis observed Venus and Saturn on two successive mornings. On this basis, he determined that they were in conjunction at 8 equal hours after the noon of the 3rd feria, EH 392 IX 2, EY 371 IX 1. He slipped in writing the Persian date; it should have been EY 371 V 1.† The conjunction was on the day numbered 2 087 233.

He also gives the details of one of the observations. Venus was about 1/3 degree east of Saturn at a time that was slightly more than 1 hour before the following sunrise. In my earlier work [Newton, 1970, Section VIII.3], I used the observation itself. Since ibn Yunis clearly implies that he estimated the time of the conjunction by interpolating between two observations, further reflection leads me to use his estimated time.

26. Page 208. Here ibn Yunis estimates the time of a conjunction between Venus and Mars without giving the observations themselves. He estimates the conjunction at 12 equal hours after the

†The month was specified by giving its name. As the reader can see by consulting Table II.3, the names of the Persian months V and IX have several letters in common, and the confusion may have arisen for this reason. On the other hand, ibn Yunis had just finished writing the name of month IX in the Muslim calendar, and this may have caused him to think of the corresponding month in the Persian calendar.

noon of the 5th feria, EH 393 III 1, EY 371 X 23. The day specified has Julian day number 2 087 410.

27. Page 208. Here we actually have two observations of Venus and Jupiter. Venus was about 1/3 degree west of Jupiter at 1/3 hour after sunset on the 5th feria and it was about 2/3 degrees east of Jupiter at the same time on the next day. From these observations, ibn Yunis concludes that the planets were in conjunction at 14 equal hours after the noon of the 5th feria, EH 393 IV 13, EY 371 XII 5. ibn Yunis remarks that Venus would not have eclipsed Jupiter at the conjunction, and calculation shows that Venus was in fact about 25′ north of Jupiter in latitude.

The main question about this record is whether to use the individual observations or ibn Yunis's estimate of the conjunction. I made the latter choice in earlier work [Newton, 1970, Section VIII.3]. Since the estimate is based upon two observations, I should have given it double weight, but I did not do so. Upon reflection, I have decided to use the individual observations here. The observations were made on the days with Julian numbers 2 087 452 and 2 087 453.

28. Page 210. ibn Yunis found that Venus and $\alpha$ Leonis were in conjunction at 14 equal hours after the noon of the 6th feria, EH 393 VIII 15, which was June 18 of year 1314 of Alexander. Venus was about $\frac{1}{4}$ degree north of $\alpha$ Leonis.

The method of giving the date needs comment. ibn Yunis uses the Syrian calendar in giving this date. The Syrian calendar, at this stage of history, differed in three minor ways from the Julian (Roman) calendar:

a. The Syrians used the Roman months but gave them different names [al-Biruni, 1000]. Six of the names are taken from the Jewish calendar; I do not know the origin of the other names. I have written "June" in place of its Syrian equivalent in the record above.

b. The Syrians began the year with the equivalent of October 1.

c. Although ibn Yunis refers to Alexander, the era used was actually that of Seleucus (Section II.2). Since the years do not begin at the same time, the beginning of year 1 of Seleucus in the Syrian calendar is not identical with its beginning in the Babylonian calendar; it began on the preceding October 1 in the Syrian calendar. Specifically, the Syrian year 1 of Seleucus began on -311 October 1. Thus the Syrian date in the record above is equivalent to 1003 June 18, which was the 6th feria and had day number 2 087 572.†

Except for one or two oversights, I have already analyzed these observations of Venus [Newton, 1970, Sections VI.3 and VIII.3], and I have repeated the texts of the observations here for the primary purpose of making this work be self-contained. The astronomical analysis to be used here will differ in several respects from the earlier analysis:

a. In several instances in the earlier work, I had to assign dates only upon the basis of astronomical calculations. Here, I am able to confirm most of the dates by a more extensive study of the calendars involved. In one instance, the date finally assigned here differs from that used earlier.

b. The earlier analysis assumed the validity of the spin fluctuation hypothesis (Section I.4). In contrast, one of the main purposes of this work is to investigate the validity of the hypothesis.

---

†ibn Yunis gives the date of this observation in still a third way. I have not tried to identify this third calendar, but it may be a variant of the Coptic calendar (Section II.3).

## TABLE VII.5

## OBSERVATIONS OF VENUS PRESERVED BY IBN YUNIS

| Date | Place | Julian Day Number -2 000 000 | Hour, local mean time | Position of Venus |
|---|---|---|---|---|
| EY 199 VII 2 | Baghdad | 24 514 | 17.95 | At longitude 262°42′ |
| EY 227 V 10 | Baghdad | 34 682 | 4.85 | 0°.4 west of Saturn |
| EY 232 X 25 | Baghdad | 36 677 | 18.52 | Touching Mars |
| EY 233 VII 7 | Baghdad | 36 929 | 5.45 | Touching Mars |
| EY 254 V 30 | Shiraz | 44 557 | 4.72 | Occulting $\alpha$ Leo |
| EH 283 IX 15 | Shiraz | 48 620 | 5.48 | Less than 20′ north of Mars |
| EH 288 X 27 | Shiraz | 50 434 | 6.03 | In conjunction with Jupiter |
| EH 306 VII 19 | Shiraz or Baghdad[a] | 56 716 | 6.25 | 4°29′ east of $\alpha$ Sco |
| EY 356 IV 1 | Cairo | 81 728 | 19.99 | In conjunction with $\alpha$ Leo |
| EY 356 XI 2 | Cairo | 81 944 | 6.40 | About 5′ north of Saturn |
| EY 359 IV 7 | Cairo | 82 829 | 20.08 | About 1° north of $\alpha$ Leo |
| EY 361 VII 4 | Cairo | 83 646 | 4.80 | $\frac{1}{3}$ degree east and $\frac{1}{2}$ degree south of $\alpha$ Leo |
| EY 364 III 26 | Cairo | 84 643 | 18.97 | About 40′ north of Jupiter |
| EY 364 IV 3 | Cairo | 84 650 | 19.66 | About 2/3 degree north of $\alpha$ Leo |
| EY 365 VI 2 | Cairo | 85 074 | 19.12 | About 5′ north of Jupiter |
| EY 366 III 9 | Cairo | 85 356 | 4.33 | Occulting Saturn |
| EY 367 IV 9 | Cairo | 85 751 | 20.08 | About 1° north of $\alpha$ Leo |
| EY 367 IV 20 | Cairo | 85 762 | 20.05 | $\frac{1}{4}$ degree east and about 5′ north of Mars |
| EY 368 I 25 | Cairo | 86 042 | 4.49 | 1° east of Mars |
| EY 371 II 4 | Cairo | 87 146 | 3.92 | 1/5 degree west of Jupiter |
| EY 371 V 1[b] | Cairo | 87 233 | 20.07 | In conjunction with Saturn |
| EY 371 V 2[b] | Cairo | 87 234 | 4.10 | About 1/3 degree east of Saturn |
| EY 371 X 23 | Cairo | 87 410 | 24.18 | In conjunction with Mars, about $\frac{1}{2}$ degree south |
| EY 371 XII 5 | Cairo | 87 452 | 18.25 | About 1/3 degree west of Jupiter |
| EY 371 XII 6 | Cairo | 87 453 | 18.27 | About 2/3 degree east of Jupiter |
| 1003 Jun 18 | Cairo | 87 572 | 25.99 | About $\frac{1}{4}$ degree north of $\alpha$ Leo |

[a]The midpoint will be used in the analysis.

[b]These entries do not represent independent observations, and only the first one will be used in the analysis.

c. The present analysis will be based upon the accurate ephemeris programs described in Chapter IX below, whereas the ephemerides used in the earlier work were based upon the mean Kepler elements. The algebraic average error introduced by the use of mean elements is zero. However, the error in a finite set of observations could conceivably differ from zero. Since the results obtained from the Venus observations are important in the study of the spin fluctuation hypothesis, I am removing this possible source of error in the current work.

The usable observations of Venus are summarized in Table VII.5, which has the same format as Table VII.4. The observations are listed in chronological order in the table, rather than in the order in which they occur in the text. I shall use 0°.15 as the standard deviation of an observed position, as I did for Mercury.

Since the Syrian calendar differs from the Julian calendar only in the notation used for the year and the month, I give the direct Julian equivalent when ibn Yunis gives the Syrian date. Since the Muslim date is ambiguous, because different writers use different choices of the era, I give the Muslim date only when the date is not given in any other form. The Muslim dates that are listed in Tables VII.4 and VII.5 are based upon 622 July 15 as the era of the Hijra.

Some of the records give the time of the observation with respect to noon. I have assumed that the time stated in this way is local apparent time, which I have converted to local mean time by using the equation of time. I calculate the equation of time separately for each observation by means of the ephemeris program of the sun.

The other records give the time only with respect to sunrise or sunset. In these cases, I begin as usual by finding the Gregorian date of the observation. I then find the time of sunrise or sunset by using the table of sunrise and sunset that is found in the <u>American Ephemeris and Nautical</u>

Almanac, assuming that the times are independent of the year if the dates are Gregorian. This table gives sunrise and sunset in terms of local mean time. When the record does not give a specific interval from sunrise or sunset, I have used 45 minutes, for the reasons that are explained in Section IV.7. This differs from my earlier practice, in which I took the interval to be zero in the absence of a specific statement.

## 6. Observations Involving Mars

In this section, I discuss the Islamic observations involving Mars that do not also involve an inner planet. The observations that do involve either Mercury or Venus have been given in the preceding sections. The observations that involve either Jupiter or Saturn without involving a closer planet are presented in Appendix VII.

1. Page 110. The Amajurs made a set of closely related observations of Mercury, Venus, and Mars; the first two have been discussed in the preceding sections. The year appears in the records as EH 356, but we have already shown that the correct year is EH 306.

The observation of Mars was made on a 7th feria, 8 days after a 6th feria that is related to the observations of Mercury and Venus. We concluded in Section VII.4 that the 6th feria in question began at noon on the day numbered 2 056 716. Hence the 7th feria on which Mars was observed began at noon on the day numbered 2 056 724. The Amajurs used EH 306† VII 26 as the designation of this day, but this is not consistent with their other usage. If we use Equation II.8 for the day number, and if we keep 622 July 15 as the era, we must use EH 306 VII 27 as the Muslim date.

At the time of the observation, Procyon ($\alpha$ Canis Minoris) was at an altitude of 28° above the eastern horizon. The record says that the time was therefore 2 unequal hours after the beginning of the night. The latter time is close to $19^h.48$,

†This is the corrected year rather than the year that appears in the record.

local mean time, and this agrees within a few minutes with the time calculated from the altitude of $\alpha$ Canis Minoris.

The Amajurs observed Mars as well as $\alpha$ Canis Minoris. According to their star tables, $\alpha$ Canis Minoris was at longitude 101°01′, and they inferred from this that Mars was at longitude 65°12′. This means that they found Mars to be 35°49′ west of $\alpha$ Canis Minoris in longitude, and this is the datum that I shall use for the observation. We do not know whether the observation was made in Shiraz or Baghdad, and I shall therefore use the midpoint as the place of observation.

2. Page 162. The Amajurs made this observation in the morning of the 2nd feria, EH 290 III 26, EY 271 XI 20. The feria and the Persian date agree, while the Muslim date is 2 days early if 622 July 15 is the era. We may accept the Persian date, which gives the day numbered 2 050 937.

We do not know with certainty whether the observation was made in Baghdad or Shiraz. However, it is made only a short time after the observation numbered 8 in the preceding section, and the record explicitly puts that observation at Shiraz. Hence I shall assume that this observation was made at Shiraz also.

The time is given only as the morning of the day stated; I shall take this to be 45 minutes before sunrise, as usual. At this time, Mars and Saturn were in conjunction, with Mars being south of Saturn by half of the lunar diameter. The calculated separation in latitude is about 18′.

3. Page 162. This is the last of the Amajur observations of Mars that we have, and it unfortunately seems to be garbled. The date of the observation is given as the 4th feria, EH 297 I 1, and the observation is a conjunction of Mars with $\alpha$ Leonis. If we use 622 July 15 as the era, the Muslim date yields 2 053 332 for the day number, and this was the 3rd feria rather than the 4th feria. There was indeed a conjunction of Mars with $\alpha$ Leonis on day 2 053 332, and finding the date is not the main problem; the day was either

909 September 19 or 20.

The record says that the altitudes of Mars, $\alpha$ Leonis, and $\alpha$ Orionis† were all equal to 60° at the time of the observation. Since $\alpha$ Orionis is about 60° west of the common position of Mars and $\alpha$ Leonis, this is a possible configuration. However, the time when it occurred was more than an hour after sunrise. Aside from the question of whether the stars could even have been seen, the record distinctly says that the observation was made during the night of the 4th feria. Hence we must conclude that this record is not usable.

The remaining observations were made by ibn Yunis at Cairo.

4. Page 178. Jupiter and Mars were in conjunction in the night before the 6th feria, EY 352 II 22. The day specified has number 2 080 229, and the observation was actually made on day 2 080 228 in our conventions. ibn Yunis says that he determined the time of the conjunction as the beginning of the night, which suggests that he observed at some other time or times and found this time by calculation or extrapolation. Caussin says that the Arabic phrase used means the end of twilight, and I shall continue to take this as 45 minutes after sunset. The record says that Mars was north of Jupiter by about a shebr, and calculation puts Mars to the north by about 45′.

5. Page 180. ibn Yunis observed Jupiter and Mars on the 2nd feria, EH 377 VI 14, EY 356 VII 25, October 10 of the year 1299 of Alexander. All forms of the date yield day number 2 081 842. He goes on to say that the conjunction was at 7 equal hours after the noon of the first feria, which would be after the noon of day 2 081 841. According to calculation, Mars still had more than 0°.5 to go to reach Jupiter at this time, and it does not seem possible that ibn Yunis would make an error of this size. Thus I believe that he meant to write "7 equal hours after the noon of the 2nd feria."

---

†Called the shoulder of Orion in the text.

6. Page 182. ibn Yunis observed Jupiter and Mars at about 1 hour after sunset on a certain night, when Mars was going straight toward Jupiter. From this observation, ibn Yunis estimates that Mars eclipsed Jupiter at noon of the 1st feria, EY 358 IX 27; this is the day numbered 2 082 639. ibn Yunis gives a statement of the separation at the time of the observation, but the statement reads as if it was not intended to be accurate, and it is probably better to use the estimated time of conjunction. Although ibn Yunis thought that this was an eclipsing conjunction, calculation puts Mars about 7′ south of Jupiter.

7. Page 184. ibn Yunis says that he observed Mars and $\alpha$ Leonis in the morning of the 3rd feria and that Mars had already passed the star. He estimates the conjunction at midnight of the night before the 3rd feria, EH 378 VI 4, EY 357 VII 4. The Julian day number at noon on the day specified in this way is 2 082 186, and the record clearly says that the conjunction was near the preceding midnight. Unfortunately the record stands in contradiction to calculation. At the time specified, Mars was still about $0^\circ.5$ from $\alpha$ Leonis, and the conjunction occurred near the following midnight rather than the preceding one.

Thus it seems that ibn Yunis was writing in terms of beginning the day at noon, and that he wrote "before the 3rd feria" when he meant "after the 3rd feria". However, it is disturbing to conclude that he made the same mistake twice in records so near each other.

In sum, I shall take the time of the conjunction as the midnight that ended the day 2 082 186 in terms of Cairo time; this is about 2 hours before midnight in terms of Greenwich time.

8. Page 184. ibn Yunis says that he observed Mars and $\alpha$ Leonis for several consecutive days. They were close to conjunction before sunrise on the 6th feria. He observed the conjunction at $1^h20^m$ before sunrise on the 7th feria, EH 380 VI 6, EY 359 IV 15. He also gives the date in terms of the Alexandrian calendar referred to the era of Diocletian (Section II.3), but does not give the Syrian date. He

accidentally writes the name of month IV in the Persian calendar where he should have month VI. Thus the observation was made before sunrise on the day numbered 2 082 897, which was the 7th feria.

ibn Yunis does not say whether the interval of $1^h20^m$ was in equal hours or unequal ones. ibn Yunis uses unequal hours in a record of the lunar eclipse of 1001 September 5 (see Newton [1970, p. 147]). Otherwise, in every place that I have noticed, he uses only equal hours, even when he is measuring time with respect to sunrise or sunset. Thus it is fairly safe to assume that he is using equal hours here.

ibn Yunis says that Mars was south of $\alpha$ Leonis by about 1 fetr. He equates this to "4 open digits", which is a term that I do not remember seeing anywhere else. Actually, Mars was to the north by about 46′.

9. Page 186. ibn Yunis says that he watched Mars and Saturn several days before the 1st feria; I presume that he watched them for several consecutive days and not just once on a day that was several days earlier. The conjunction was from 12 equal hours to 18 equal hours after the noon of the 1st feria, EH 381 VIII 21, EY 360 VIII 13. Here we have a calendrical oddity. In order to make the Persian date be correct, we must put the epagomenal days before month VIII, but to do so conflicts with ibn Yunis's usage in the very next record. This suggests that ibn Yunis did not use a physical form of the Persian calendar with the days spread out on a page or in a diary, such as we commonly use today. It suggests instead that he calculated the Persian date and that he accidentally put the epagomenal days in the wrong place in making the calculation. In the form of the Persian calendar that we are using, the correct date of the 1st feria in question is EY 360 VIII 18. This is the day that is numbered 2 083 325 at noon.

Rather unusually, ibn Yunis assigns an interval rather than an instant to the conjunction. We should use the midpoint of the interval, which is 15 hours after the preceding noon. In our terms, then, we take the time of the conjunction to be $03^h$, in apparent time at Cairo, on the 2nd feria, EY 360 VIII 19, day number 2 083 326.

10. Page 186. ibn Yunis saw Mars and Saturn in conjunction at the beginning of the night before the 6th feria, EH 383 IX 2, 7 equal hours after the noon of the 5th feria, EY 362 VIII 6. Here, for at least the Muslim calendar, ibn Yunis is using a day that begins at sunset. The Persian date agrees with the calendrical data if the epagomenal days have not yet occurred, unlike the situation that we found in the preceding record. The day number of the conjunction is 2 084 043. Mars was south of Saturn by about 4 open digits, according to the record. The calculated separation was about 18′, which is close to 4 ordinary digits.

11. Page 188. ibn Yunis observed Mars and Jupiter after sunset of the 5th feria, EH 384 IV 18, and saw that Mars had not yet reached Jupiter. After sunset on the 6th feria, he saw that Mars had passed Jupiter by 1/6 degree. He observed them on both evenings at about 40 minutes, in equal hours, after sunset. Hence he decided that the conjunction occurred at noon on the 5th feria. Presumably he meant noon on the 6th feria, EH 384 IV 19, which was noon on the day numbered 2 084 268. Since ibn Yunis estimated this time on the basis of two observations, of which he has given us only one, I shall use his estimate of the time.

12. Page 190. This is an independent record of the same conjunction, but this time with a little more detail. ibn Yunis also gave us two records of a conjunction of Venus and Jupiter, which are items 14 and 16 in the preceding section. Apparently ibn Yunis has preserved some observations made by another person at Cairo, in spite of the fact that they occur in a section that he marks "Observations I have made". One of the Venus-Jupiter observations was garbled and could not be used. Both of these observations are usable and they will receive equal weight.

In this record, observations were made on both the 5th and the 6th feria, and the conjunction is estimated as being at the time of the observation on the 6th feria. The separation is not given for either observation. The time is not given for the 6th feria, but the time on the 5th feria is given as $\frac{1}{2}$ equal hour after sunset. I believe that the implication is that this is the time on the 6th feria also.

This record says that the planets were separated by about 1 shebr in latitude, but does not give the sense of the separation. Calculation puts Mars about 50′ to the north.

13. Page 192. This record confirms the speculation made in the preceding item. ibn Yunis says that he has this observation "from a person in whom I have great confidence, and I have no doubt of its accuracy." The unknown observer says that he watched Mars and Saturn in order to determine the time of their conjunction. It occurred after the midnight before the day of the 3rd feria, EH 385 V 10, EY 364 III 26. This is the day numbered 2 084 643. The planets rose at 7 hours of the night, and their height was 6° at the time of the conjunction. The record says that they were separated by about 1 digit in latitude, but it does not give the direction. According to calculation, Mars was about 12′ to the south.

When the altitude of Mars was 6° (to the east), the hour angle of the true sun was -11.06. Hence the time was $0^h.91$ after midnight on the day that is numbered 2 084 643, in mean time at Cairo.

14. Page 196. ibn Yunis says that he observed Mars and $\alpha$ Leonis on the 3rd feria, EH 388 VI 16, EY 367 III 30, and that their conjunction had not yet occurred. The next night, an hour after sunset, Mars had passed $\alpha$ Leonis. ibn Yunis does not give the relative position on either night, and he does not estimate the time of the conjunction. We could take the time to be the average of the two times, but it seems safer to ignore this record.

15. Page 204. ibn Yunis says that he watched Mars and Saturn attentively for many days in order to determine the time of their conjunction. He timed it for noon of the 7th feria, EH 391 VIII 25, EY 370 V 6. This is on the day numbered 2 086 873.

Mars was about 1° north of Saturn, he says, but the calculated separation is about 1°26′.

The quality of this observation seems to be poor in both latitude and longitude. A preliminary calculation indicates that Mars had already passed Saturn by the midnight preceding the noon that is stated. However, the day and time are clearly stated, and I do not see any a priori justification for rejecting the record. It is possible that ibn Yunis simply made a blunder of 1 day in his estimation of the time.

16. Page 204. ibn Yunis observed that Mars was 2° west† of $\alpha$ Leonis at 8 equal hours after the noon of the 7th feria, EH 392 IV 27, EY 370 XII 29; this is the day with number 2 087 111. Actually, Mars was about 2° east of $\alpha$ Leonis. The discrepancy may be a simple slip, or it may conceivably be an ambiguity in the intended meaning of "precede".

We may wonder why ibn Yunis chooses to record this configuration instead of waiting for the conjunction. Calculation shows that Mars was near a turning point, and it may be that ibn Yunis was recording what he took to be the position of this point. However, this record is closely related to the next record, which suggests a slightly different interpretation.

We should note that ibn Yunis labels this record as a "conjunction of Mars and $\alpha$ Leonis", although this label is clearly not to be taken literally.

17. Page 206. This a garbled record that ibn Yunis also labels as a conjunction of Mars and $\alpha$ Leonis. He saw them, he says, at 7 equal hours after the noon of the 7th feria, EH 392 V 5, EY 371 I 6. The calendrical information clearly indicates the day numbered 2 087 118, just a week after the preceding record. Calculation shows that Mars was more than $1\frac{1}{2}$ degrees east of the star, and it is not credible that ibn Yunis could have mistaken this configuration for a conjunction. However,

†More specifically, he says that Mars preceded $\alpha$ Leonis.

while the motion of Mars was retrograde, as ibn Yunis notes, it was much closer to being stationary than it had been a week before. Since ibn Yunis sometimes mentions "conjunction" when he does not mean the term literally, we may speculate that he did not mean it here, and that instead he was following the motion closely, looking for the exact stationary point, taking advantage of the close proximity of the point to $\alpha$ Leonis. Continuing the speculation, he observed the relation of Mars to $\alpha$ Leonis on this day, in order to compare with a week before, but failed to include the

TABLE VII.6

OBSERVATIONS OF MARS PRESERVED BY IBN YUNIS

| Date | Place | Julian Day Number -2 000 000 | Hour, local mean time | Position of Mars |
|---|---|---|---|---|
| EY 271 XI 20 | Shiraz[a] | 50 937 | 5.62 | In conjunction with Saturn |
| EH 306 VII 27 | Shiraz or Baghdad[b] | 56 724 | 19.48 | 35°49′ west of $\alpha$ C Mi |
| EY 352 II 21 | Cairo | 80 228 | 19.52 | In conjunction with Jupiter |
| 987 Oct 10 | Cairo | 81 842 | 18.77 | In conjunction with Jupiter |
| EY 357 VII 4 | Cairo | 82 186 | 23.87 | In conjunction with $\alpha$ Leo |
| EY 358 IX 27 | Cairo | 82 639 | 12.00 | In conjunction with Jupiter |
| EY 359 VI 15 | Cairo | 82 897 | 4.27 | In conjunction with $\alpha$ Leo |
| EY 360 VIII 19 | Cairo | 83 326 | 2.75 | In conjunction with Saturn |
| EY 362 VIII 6 | Cairo | 84 043 | 18.76 | In conjunction with Saturn |
| EH 384 IV 19 | Cairo | 84 268 | 11.94 | In conjunction with Jupiter |
| EH 384 IV 19 | Cairo | 84 268 | 19.48 | In conjunction with Jupiter |
| EY 364 III 26 | Cairo | 84 643 | 0.91 | In conjunction with Saturn |
| EY 370 V 6 | Cairo | 86 873 | 12.07 | In conjunction with Saturn |
| EY 370 XII 29 | Cairo | 87 111 | 20.13 | 2° east of $\alpha$ Leo |

[a]The text does not give the place explicitly, but this is a strong inference.

[b]The midpoint will be used in the analysis.

information in the record through some accident.

Whether this speculation be correct or not, it is clear that the record does not have enough information to be useful.

This concludes the observations of Mars that are preserved by ibn Yunis. The quality of his own Mars records is not as bad as that of his Mercury records, but is certainly not good enough to let us praise his qualities as an observer. We continue, however, to be thankful for his acumen as a theorist and calculator.

The usable records relating to Mars are given in Table VII.6, which has the same format as Tables VII.4 and VII.5. The remarks made at the end of the preceding section also apply here.

# CHAPTER VIII

# THE USE OF SOLAR TABLES IN THE STUDY OF THE ACCELERATIONS

## 1. Preliminary Remarks

If an astronomer wishes to construct a set of tables, say of the sun, he may begin by developing a theory or model of its motion. This model contains a number of parameters that can be found, at least in our present state of knowledge, only by using observations of the sun. A modern astronomer uses more observations than there are parameters, and he uses statistical procedures to find the set of parameters that gives the 'best fit' of the model to the entire set of observations.

Now suppose that an astronomer in the year 3000 wishes to estimate the average acceleration of the sun, say, between 1850 and 3000. Suppose further that all records of solar observations near the year 1850 have been lost, but that Newcomb's parameters and his table of perturbation coefficients [Newcomb, 1895] have been preserved. Our future astronomer can calculate a position of the sun from Newcomb's theory for, say, the epoch 1850 January 1.0. This calculated position is not a direct observation, but the astronomer can use it for many purposes as if it were one. It represents the best estimate that a near-contemporary could make of the position of the sun at that epoch on the basis of many observations. In that sense, it may even be superior for some purposes to any single observation. Worded another way, it is the smoothed representation of a large number of observations.†

†This is not a recommendation to use smoothed observations or theoretical values when original observations are available. It is probably true that some useful information is always lost whenever data are smoothed. Thus, before we use smoothed data, we must consider whether the lost information is likely to be relevant to the problem at hand, and we must consider whether we have an alternative.

Similarly, under highly specific conditions, we may use ancient and medieval tables and models in the study of the accelerations. From the standpoint of this study, there are two main differences between ancient tables and those of Newcomb. First, the older tables are much less accurate. Second, early astronomers did not have statistical methods at their disposal; in consequence, they often but by no means always used exactly as many observations as they had parameters to find, instead of using a surplus of observations.

We can set forth two main conditions that must be satisfied if we are to use tables or models in the inference of the accelerations:

1. We must understand the tables or models, and their associated terminology, well enough to use them with confidence.

2. We must know the approximate epochs of the underlying observations, and we must use the tables only for an epoch that is close to the observations.

The stringency of the second condition is related to the accuracy of a table. If a table is accurate in most of its parameters, and in particular if it is accurate in the mean motions, we get almost the same error in position over a wide range of epoch. Thus the discrepancy, $\Delta L$ say, between a value from the old table and a value from modern tables is almost independent of the epoch. To estimate the acceleration from $\Delta L$, we multiply by $2/\tau^2$, in which $\tau$ is the time before 1900. Since $\tau$ changes slowly as we change the ancient epoch, the estimate of the acceleration is almost independent of the epoch for which $\Delta L$ is calculated.

There are two main reasons why I shall not attempt to use Babylonian tables in this study. One reason is that we have many records which seem to incorporate valid observations, so that there is no obvious gain from the use of tables. The second reason is connected with the problem of using the tables at the correct epoch. Even as late as

Hipparchus, as we shall see in Section VIII.3, the errors in basic parameters are so great that we must use an epoch that is within about a decade of the underlying observations. With the Babylonian tables, the epochs of the underlying observations are apparently not known even to the century. Thus, until we know more about the origin of the tables, we cannot use them in a study of the calculations.

In this chapter I shall consider only solar tables. I shall take up planetary tables in Chapter XIII, after the analysis of the Hellenistic data.

## 2. The Solar Equation

To the accuracy that is needed in dealing with pre-telescopic observations, we may take the latitude of the sun as being identically zero. That is, we may assume that the sun moves in the plane of the ecliptic, and we may locate it by giving its geocentric longitude and radius vector in that plane. In this section, we are interested only in its longitude.

To an accuracy of about 2°, the longitude $\lambda_S$ of the sun is the same as the longitude $L_S$ of the mean sun. The difference between $\lambda_S$ and $L_S$ is the quantity that Ptolemy called the προσθαφαιρεσις. In more recent writing (see Section V.4), it is called "the equation of the center for the sun", or, more simply, "the solar equation". If we use $e_c(\tau)$ to denote the solar equation as a function of time, we can write

$$\lambda_S = L_S + e_c(\tau) \ . \qquad \text{(VIII.1)}$$

In Newcomb's theory, $L_S$ is a quadratic function of $\tau$; the quadratic term arises from planetary perturbations. The acceleration of the sun, in the sense that we use the word in this work, contributes another quadratic term that is about the same size as the planetary contribution. The nature of $e_c(\tau)$ over geological time is not known, but it can be taken as a periodic function over a time as short as historical time.

For an elliptic orbit of eccentricity e, we can expand $e_c(\tau)$ (in degrees) as a power series in e having the form

$$e_c = (180/\pi)[2e \sin M + (5/4)e^2 \sin 2M + \ldots]. \quad \text{(VIII.2)}$$

In this,

$$M = L_S - \Gamma , \quad \text{(VIII.3)}$$

with $\Gamma$ being the longitude of perigee. For the sun, e is about 0.0167, and the second term within brackets in Equation VIII.2 contributes about 0°.02, slightly more than 1 minute of arc. If we neglect it, we have

$$e_c = (360e/\pi) \sin M \text{ degrees} . \quad \text{(VIII.4)}$$

At the present time, the coefficient of sin M is about 1°.919.

According to Salam and Kennedy [1967], medieval astronomers used three different forms for the equation of the center.

(a) The method of sines. In this,

$$e_c = E \sin M . \quad \text{(VIII.5)}$$

E is a constant that we may call the maximum equation. This form is identical with Equation VIII.4, although it presumably resulted from different considerations.† It apparently originated in India.

(b) The declination method. In this, we start

---

†Medieval astronomers customarily measured the mean anomaly M from apogee rather than from perigee. In what immediately follows, I shall refer their results to perigee in order to maintain parallelism with the modern forms.

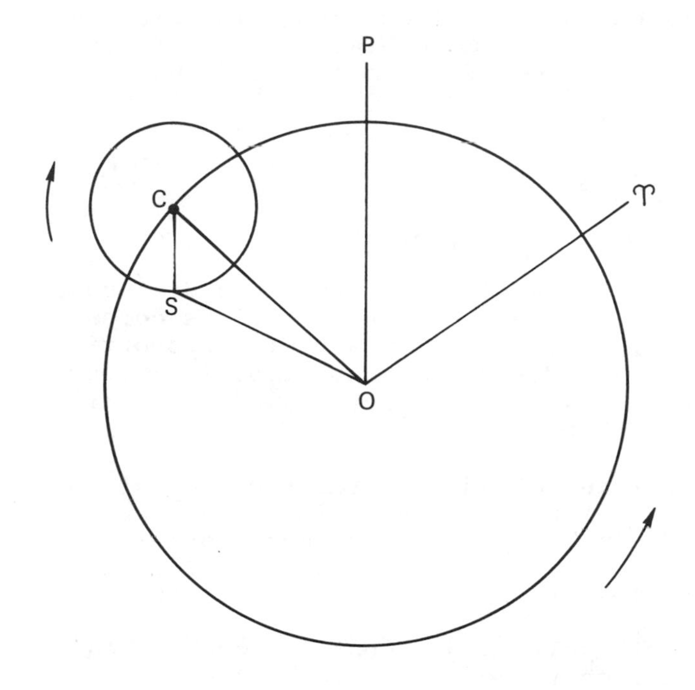


Figure VIII.1. The solar epicycle model. The point C moves around the large circle (deferent) at a uniform rate, with respect to ♈, in the indicated direction, with a period of 1 tropical year. The sun S moves around the small circle (epicycle) at a uniform rate, with respect to the line OC, in the indicated direction, with a period of 1 anomalistic year. ♈ denotes the vernal equinox and P denotes the position of perigee. The line CS is always parallel to the line OP.

from the fact that zijes often had a table giving the declination $\delta$ of the sun as a function of its longitude $\lambda_S$. In the method of declinations, the astronomer started by calculating the mean anomaly M in the usual manner. He then used this value in place of $\lambda_S$ in the declination table and read off the corresponding value of $\delta$; call this value $\delta(\lambda_S \rightarrow M)$, say. Then

$$e_c = (E/\varepsilon)\delta(\lambda_S \rightarrow M) \quad , \qquad \text{(VIII.6)}$$

in which $\varepsilon$ is the obliquity of the ecliptic.

Since $\sin \delta = \sin \varepsilon \sin \lambda_S$, the main difference between this method and the method of sines comes from the fact that $\sin \delta$ is not strictly proportional to $\delta$. The origin of the method of declinations is unknown, but Salam and Kennedy suspect that it is connected with Iran.

I mention this method only for the sake of completeness. So far as we know, it was not used by any of the Islamic astronomers whose work will be used here. al-Khwarizmi (see Section VI.2) is perhaps the best-known astronomer who used it.

(c) The epicyclic and eccentric models. The epicycle model used to represent the motion of the sun is shown in Figure VIII.1. In this figure, ♈ is the vernal equinox and **P** is perigee; thus ∠ **PO**♈ = Γ, the longitude of perigee. C, the mean sun, rotates uniformly around the large circle (deferent), and it takes 1 tropical year for C to move from ♈ back to ♈ again. S, the true sun, rotates uniformly around the small circle (epicycle) in the opposite sense to the motion of C. When C is on the line **OP**, S is also, and the sun is at perigee. S makes one full rotation with respect to the line **OC** in 1 anomalistic year. Thus, whether **P** precesses with respect to ♈ or not, the line CS is always parallel to the line **OP**.

We have the following interpretations of angles in Figure VIII.1:

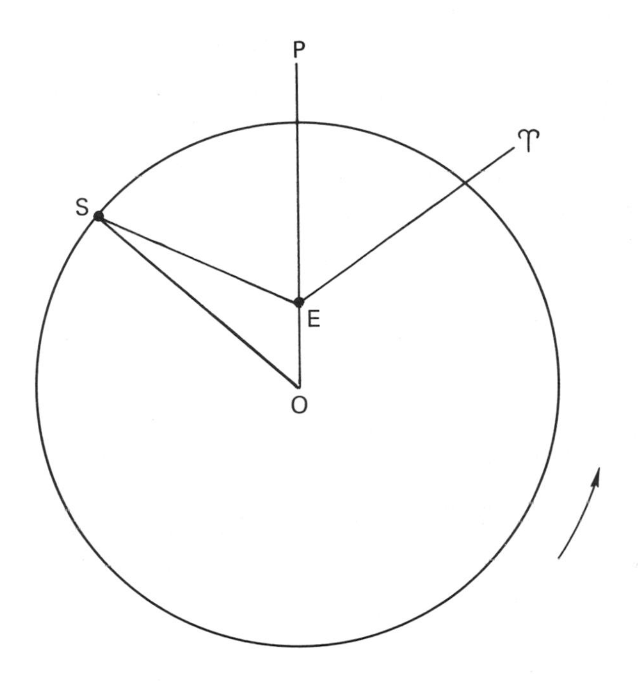


Figure VIII.2. The solar eccentric model. The sun S moves uniformly around the circle. The center of the circle is at O but the earth from which it is viewed is at E.

$\angle$ CO♈ = $L_S$, the longitude of the mean sun;

$\angle$ SO♈ = $\lambda_S$, the longitude of the true sun;

$\angle$ COP = $\angle$ SCO = M, the mean anomaly of the sun;

$\angle$ SOC = $e_c$, the equation of the center of the sun.

In accordance with customary Hellenistic and Islamic usage, let us take 60 for the radius OC of the deferent, and let us use r for the radius CS of the epicycle. From the law of sines, we find

$$r/\sin e_c = 60/\sin (e_c + M) .$$

After some straightforward manipulation, we change this to

$$\tan e_c = r \sin M/(60 - r \cos M). \qquad \text{(VIII.7)}$$

We also see that the maximum value of $e_c$, namely E, occurs when the angle CSO is a right angle. Let $M_E$ denote the value of M at this condition. Then

$$\sin E = \cos M_E = r/60. \qquad \text{(VIII.8)}$$

Ptolemy and many later astronomers seemed to prefer the eccentric model to the epicycle model for the sun. The eccentric model is shown in Figure VIII.2. The sun S moves uniformly around the circle whose center is O, but the earth E is not at the center of the circle. Instead, E is a distance r away from O in the direction of the perigee P. We have these angles:

$\angle$ SE♈ = $\lambda_S$, the longitude of the sun;

$\angle$ SOP = M, the mean anomaly of the sun;

$\angle$ OSE = $e_c$, the equation of the center.

$L_S$, the mean longitude, is equal to $\angle$ PE♈ + M, but it does not appear directly as a physical angle in Figure VIII.2. When we take the radius SO to be 60, we find that $e_c$ is given by Equation VIII.7. In other words, the triangle OCS in Figure VIII.1

is congruent to triangle SOE in Figure VIII.2, and hence the epicycle and eccentric models give exactly the same results, as Ptolemy [ca. 142] shows in his Chapter III.3.†

The main difference between Equation VIII.5 for the method of sines and Equation VIII.7 for the eccentric (or epicycle) model is that the right member of Equation VIII.7 is not symmetrical about the value M = 90°. Instead, it is symmetrical about the value 90° - E. Thus, if we have an astronomer's table of $e_c$ as a function of M, we can tell the model that he used even if we have lost his description of the model. We shall see an example of this in Section VIII.5.

In most of the examples that will concern us here, we have available neither an explicit description of the model nor a table of $e_c$. What we usually have are the values of E and the position of perigee (or apogee). We may feel sure from the known historical circumstances that we are dealing with either the sine method or the eccentric method (Equations VIII.5 and VIII.7). We are concerned with values of E near 2°, and, for such values, the difference between the two models can reach about 3′. The difference is greatest when M is an odd multiple of 45°, and it is just large enough to need attention in some circumstances.

Luckily we shall need to use the corresponding equations of the center only at the equinoxes and solstices, and only at times in history when perigee was between 260° and 270°. Under these circumstances, the values of M are near multiples of 90°, where the difference between Equations VIII.5 and VIII.7 is near zero.‡

---

†Knowledge of this goes back at least three centuries before Ptolemy. See Neugebauer [1957, p. 155].

‡It is rather sure from the known historical circumstances that the model, in all the examples that will concern us, was the eccentric model. However, because of the circumstances mentioned, we are not dependent upon assuming this.

## 3. An Example from the Work of Hipparchus

This section will actually deal with the solar tables prepared by Ptolemy, but, since we know that Ptolemy based his tables upon fudged data that were derived from the work of Hipparchus, I shall identify the tables with the name of Hipparchus. We know a great deal about the observations from which the tables were derived and do not need to use the tables themselves in a study of the accelerations. However, an example based upon a well-known case will give us confidence that we are handling the tables correctly.

First, we must choose the epoch at which we use the tables. The care that we must exert in choosing the epoch is related to the accuracy of the tabular mean motion of the sun. We know that Hipparchus used $365\frac{1}{4}$ days minus 1/300 day for the length of the tropical year. This is in error by about 6.4 minutes, and the corresponding mean motion is in error by about 1600″ per century. In order to estimate the acceleration of the sun with useful accuracy, we should have its position correct within about 160″. Thus we have a tolerance of only about a decade in choosing the epoch; that is, the epoch that we use must be within a decade of the mean epoch of the underlying observations.

We know in fact that one of the basic observations was that of the autumnal equinox in the year -145, which Hipparchus put at 6 hours, local time, on September 27. The Julian day number of this observation, in terms of Greenwich mean time, is 1 668 365.667. The example that I shall work out in this section is finding the mean longitude of the sun at this epoch by using the Hipparchan tables in two different ways. From the mean longitude, we can then estimate the solar acceleration.

The mean longitude that Ptolemy derives from Hipparchus's observations is given by Equations V.3 in Section V.4 and is repeated here for convenience:

$$L_S = 330°.75 + 0°.985\ 635\ 278\ 4\ D_P. \qquad \text{(VIII.9)}$$

In this, $D_P$ is the time in days from the epoch of Nabonassar, which has the Julian day number 1 448 637.917 in terms of Greenwich mean time. At the epoch of Hipparchus's equinox, Equation VIII.9 gives

$$L_S = 182°.172 .$$

We can also estimate $L_S$ by using the parameters of the eccentric model. Ptolemy, and presumably Hipparchus before him, puts perigee at 245°.5,† and he takes the eccentric radius to be 2.5 for a deferent radius of 60. This makes E (Equation VIII.8) equal to 2°.388. For comparison, Newcomb's theory, evaluated at Hipparchus's epoch, gives 245°.98 for the position of perigee and 2°.012 for the maximum equation E. We see that Hipparchus did a good job of placing perigee. He had a bias in the setting of his equatorial circle (Equation V.1) that caused him to put the autumnal equinoxes about 4 hours too late and the vernal equinoxes about 4 hours too early. This in turn caused him to over-estimate the maximum equation of the center.

According to Ptolemy, Hipparchus found E and perigee by using an autumnal equinox, a vernal equinox, and a summer solstice. At an autumnal equinox, $\lambda_S = 180°$. We calculate $e_c$ from Equation VIII.7‡ and find that $L_S = 182°.173$ at an autumnal equinox. Similarly, we find $L_S = 357°.827$ at a vernal equinox and $L_S = 90°.990$ at a summer solstice.

The change between $L_S$ at an autumnal equinox and the succeeding vernal equinox is 175°.654, and

---

†Ptolemy assumed that perigee has a constant position with respect to the equinoxes. That is, his model requires perigee to share in the precessional motion of the equinox with respect to the stars.

‡Equation VIII.7 requires the use of $M = L_S - \Gamma$. In using this, we must start with $\lambda_S$ rather than $L_S$. This requires that we evaluate $e_c$ by successive approximations, or by an equivalent method such as a series.

the increase to the next summer solstice is 93°.163. Since the mean motion of the sun is about 0°.98565 per day, the corresponding time intervals are $178^d.21$ and $94^d.52$.† For comparison, Hipparchus measured intervals of $178\frac{1}{4}$ and $94\frac{1}{2}$ days, respectively. The close agreement gives us confidence that we are interpreting and using the model correctly.

For the epoch of Hipparchus's equinox, we get 182°.172 for $L_S$ by using Equation VIII.9 and we get 182°.173 by using the tabular position of perigee and the value of E; the difference is trivial. Since there are fewer mathematical operations between the observations and the latter value, I shall use the latter in an estimate of the solar acceleration.

The mean longitude in Newcomb's theory is given by

$$L_S = 279°.6967 + 36\ 000°.768\ 925\tau + 0°.000\ 302\ 5\tau^2. \qquad \text{(VIII.10)}$$

$\tau$ is -20.442 281 54 at the epoch of Hipparchus's equinox, and $L_S$ = 181°.969 at this epoch. Newcomb's formula refers to the geometric position while the value from Hipparchus's work necessarily refers to the apparent position. Thus we subtract the aberration, about 0°.006, from Newcomb's result in order to put the values on a comparable basis. The difference between the two values of $L_S$ is then 0°.210 = 756″. Hence the acceleration of the sun is

$$\nu_S' = 2 \times 756/(20.442)^2 = 3.62\ ''/\text{cy}^2\ . \qquad \text{(VIII.11)}$$

Although this value seems at first sight to be derived from the equinox of -145 September 27, it differs greatly from the value that we find for that equinox in Table V.2. According to that table, the difference between the recorded time and the time

† It should be understood that I am referring here to values in the Hipparchus-Ptolemy model, not to values given by modern theory.

calculated from Newcomb's theory is $1^h.1$. This is equivalent to 162″ in the position of the sun, and the corresponding acceleration is 0.78 ″/cy$^2$ rather than 3.62. The reason for the discrepancy is that the value 3.62 from Equation VIII.11, whichever way it is derived, actually depends upon an autumnal equinox, a vernal equinox, and a summer solstice. The precision of observing a summer solstice is not as high as the precision of observing an equinox, but this is balanced by the fact that the value of $L_S$ at the time of the equinox is not sensitive to the measured time of the equinox, because of the relation of perigee and apogee to the solstice points.

The most valuable feature of using $L_S$ is that the value found is independent, to very high accuracy, of the bias in the setting of the equatorial plane. In other words, an observation of an autumnal equinox and an observation of a vernal equinox are much more valuable than two observations of the same equinox. I shall recognize this fact by giving a weight of 4 to a determination of the mean longitude, when compared with a single observation of an equinox. The weight of 4 is arbitrary, but I do not see a way of establishing a weight by a rigorous procedure.

The value of $\nu_S'$ from Equation VIII.11 agrees fairly well with the value 3.0 ± 0.7 from Equation V.2. The latter value may be preferred because it is based upon all the observations in Table V.2. Use of the value of $L_S$ can eliminate the effect of a bias in the equator, but it cannot eliminate the effects of other errors in the measured times. It can only average them in a statistical fashion.

## 4. Was the Solstice of -431 June 27 Fudged?

I shall digress for a section in order to consider an interesting question that is raised by the example of Hipparchus's solar tables.

According to Ptolemy [ca. 142, Chapter III.1], Hipparchus measured the time of the summer solstice in the year -134. He found that the interval from the vernal equinox to the summer solstice is $94\frac{1}{2}$ days. Since the vernal equinox (Table V.2) was at the midnight that began the day -134 March 24, the time found for the solstice must have been noon on -134 June 26 (see Section V.3).

As we noted in the preceding section, Hipparchus also found 178¼ days for the interval from an autumnal equinox to a vernal one. There are three pairs of equinoxes in Table V.2 from which the interval may be derived. The pair -146 September 26 to -145 March 24 gives 178¼ for the interval, the next pair also gives 178¼, but the last pair gives 178½. Thus it seems that both Hipparchus and Ptolemy used the results from the first two pairs.

Ptolemy presumably fudged his equinoxes by starting from Hipparchus's -146/-145 pair, so his equinoxes necessarily show the interval of 178¼ days. However, I have been assuming that he fudged his summer solstice by starting with the observation commonly attributed to Meton and Euctemon on -431 June 27, at 06 hours, Athens time. If so, the time of Ptolemy's summer solstice has nothing to do with Hipparchus's observations. How does it happen, then, that Ptolemy found the same interval as Hipparchus, namely 94½ days? This section, and the discussion of the Milesian parapegm in Section V.3, are written in response to this question.

If Hipparchus measured the summer solstice as Ptolemy says he did, he found it at noon on -134 June 26, at day number 1 672 291.0 in terms of local time. Ptolemy could have used this as the basis for fudging his own solstice. If we add 274 years to this time, using 365.246 667 days per year, we get $2^h.08$ after midnight on 140 June 25, and Ptolemy reported $2^h$. Thus we can get Ptolemy's claimed equinox by starting either from the solstice of -134 attributed to Hipparchus or from that of -431 attributed to Meton and Euctemon.

This is just what we should expect if Hipparchus had found the length of the year by using the solstices of -431 and -134. However, according to Ptolemy [Chapter III.1], Hipparchus derived his estimate of the year by using his solstice of -134 and a measurement of the summer solstice by Aristarchus of Samos in -279. He did not base the year upon measurements of the equinox, perhaps because he did not have any old equinox data available to him, or perhaps because he thought that solstices were more accurate.

Ptolemy does not tell us the time that Aristarchus measured, but we reconstructed it in Section V.3. The time was $\frac{1}{4}$ day after noon on -279 June 26.

TABLE VIII.1

FOUR ALLEGED OBSERVATIONS OF THE SUMMER SOLSTICE

| Reported Epoch | | Number | Comparison Epoch | |
|---|---|---|---|---|
| Date | Hour | of Years | Date | Hour |
| -431 Jun 27 | 06 | 0 | -431 Jun 27 | 06.0 |
| -279 Jun 26 | 18[a] | 152 | -279 Jun 26 | 17.8 |
| -134 Jun 26 | 12[a] | 297 | -134 Jun 26 | 12.2 |
| +140 Jun 25 | 02 | 571 | 140 Jun 25 | 02.3 |

[a]These hours are not explicitly stated in the reports but are determined from the results of calculations that Ptolemy based upon them.

Thus we have four alleged observations of the summer solstice, which are summarized in Table VIII.1. The reported epochs of the solstices are given in the first column. Ptolemy explicitly states the times of the first and last records, those of -431 and +140. For the other two, he explicitly states only the years. However, he gives us the results of calculations based upon the other two times, and we can reconstruct them readily and without ambiguity.

When we take the number of years given in the

second column, and add this many multiples of 365.246 667 days to the first time, which we treat as exact, we get the "comparison epochs" listed in the last column. Thus we see that all four solstices are consistent with a year of 365.246 667† days with a maximum discrepancy of less than half an hour.

We may assume with confidence that Ptolemy's alleged measurement was fudged, so there is no significance to the fact that it agrees with Hipparchus's length of the year. The length that Hipparchus assigned to the year could have been found from any two of the remaining observations. The exact agreement of a third measurement is, however, rather unlikely. Thus we need to ask whether one of the earlier records of the solstice is fudged as well as the one that Ptolemy attributes to himself.

If one of the old records is fudged, there are several indications that it is the one attributed to Meton and Euctemon:

a. The error in the observation attributed to Hipparchus, in the sense of recorded minus correct time, is about +8 hours, and the error in the one attributed to Aristarchus is about -7 hours. These errors are reasonable. The error in the observation attributed to Meton and Euctemon, however, is about -28 hours.

In an earlier work [Newton, 1970, Section II.4], when I was accepting what Ptolemy said about earlier astronomers without question, I thought that an error of 28 hours was acceptable, since this is the earliest observation of a solstice that is known. However, I have shown in Section V.3 that the standard deviation in a Hellenistic measurement of a solstice time is about 7 hours, even in the time of Meton and Euctemon. In fact, we would be justified in using an even smaller standard deviation for this particular measurement. If Meton began his cycle in -431, the year reported for the measurement, he must have had a large body of earlier data upon which he

---

†This is $365\frac{1}{4}$ days less 1/300 day.

based his cycle. Thus the solstice time reported for -431 is not the result of single measurement. It must have come from smoothing a number of earlier measurements.

We can now assess the probability that the agreement shown in Table VIII.1 happened by chance. The solstice of -431 agrees with the other two within half an hour, but with a total error of about 28 hours. Thus the error lies in a range whose width is about 1/7 of a standard deviation and that is centered at 4 standard deviations. The probability that this happened by chance is negligibly small.

b. According to Ptolemy, Hipparchus estimated the length of the year by using the solstices of -134 and -279. If the measurement attributed to Meton and Euctemon existed in his time, why did he not use it? If he knew of it, he might have suspected its accuracy and he might have preferred to base his length of the year upon the measurement by Aristarchus in -279. However, he would surely have been struck by the exact agreement of Aristarchus's measurement with the measurement of -431. I find it virtually impossible to believe that Hipparchus would have mentioned only the observations of -134 and -279 if he knew of the observation of -431 in the form that it takes in Table VIII.1, particularly since he was concerned about the possibility that the year might vary in length.

It is likely that a tradition in Hipparchus's time attributed an observation of the solstice to Meton and Euctemon. The traditional observation may not have had a time of day associated with it, so that Hipparchus would have considered it crude. It may have had the Greek date Skirophorion 13 associated with it and scholars in Hipparchus's time may have had as much difficulty in using it as do scholars of today. Thus Hipparchus might not have known either the day or the hour of the observation.

c. The observation attributed to Meton and Euctemon is probably the only one that could have been fudged. Aristarchus and Hipparchus worked in times when there were well-established calendars, and they apparently attached the times of day to the observations that they made. The observations attributed to Meton and Euctemon, although they were probably good in their own time, could well have been recorded in a form that made them unusable by later scholars.

We saw in Section V.3 that the date -431 June 27 is given in a stone inscription that was presumably carved around -108, since it commemorates an event of that year. The same inscription gives -108 June 26 as the date of the solstice in that year. We concluded in Section V.3 that someone calculated both dates from Hipparchus's theory, but not with the intention of deceiving astronomers. He merely calculated them because he needed them for the inscription and he did not know any other way to find them.

Whether this conclusion is correct or not, the date -431 June 27 was established by Ptolemy's time, and the only possible freedom left was in the hour. Since the existing inscription does not give the hour, it may be that Ptolemy fudged the hour for the purpose of 'confirming' Hipparchus's solar theory.

However, this is basically an argument *ex silentio*, and the conclusion of such an argument may be accepted only with grave reservations. Most of the astronomical literature from the period between Hipparchus and Ptolemy has been lost, and it is dangerous to infer that no part of that literature contained the hour, since the hour could have been calculated at the same time as the date.

In summary, it is almost certain that the solstice of -431 June 27 at 06 hours was fudged in the sense that the date and time were calculated from Hipparchus's solar theory. The date, namely -431 June 27, was already calculated by about the year -108. The reason for the calculation was an innocent one, but the reason was soon lost and later astronomers down to the present have accepted the date as the result of observation. It is possible that the hour was calculated at the same time, or at some time between -108 and the time of Ptolemy, with the result being transmitted in a work that is now lost. It is also possible that the hour was fudged specifically for the purpose of being included in the Syntaxis, with the intent of deceiving contemporaneous scholars.

## 5. The Solar Tables of Yahya and Habash

There is a manuscript that Kennedy [1956, p. 145] calls "the purported" zij of Yahya, but its authenticity is doubtful. Kennedy says that we should assume that the material in it is "not† a part of the original zij of Yahya unless there is reason for thinking otherwise." However, ibn Yunis [1008] discusses Yahya's zij in several places, and we can reconstruct a moderate amount of information about it even in the absence of an authentic copy.

We are better off with regard to Habash's zij. ibn Yunis discusses it, and in addition there are two manuscript copies, one in Berlin and one in Istanbul. Much of the material in the Berlin copy is clearly of later date than Habash and is thus not authentic. Much of the material in the Istanbul copy, however, probably is the authentic work of Habash; at least it agrees with the description that ibn Yunis gives of Habash's zij. In addition, al-Biruni [1025] gives a result that he calculated from Habash's zij.

Since Habash and Yahya presumably used the same solar observations,‡ we expect their solar tables to be basically the same, with at most minor differences resulting from slight differences in the analysis of the data. This expectation is borne out, in spite of the fact that there is much confusion about the content of both zijes.

All sources that I have seen say that Yahya and Habash both used 1;59 degrees for E, the maximum solar equation, but here the unanimity ends. On pages 38-40, ibn Yunis says that Yahya found 83;39 for the solar apogee, but on pages 214-216 he says that the value is 82;39. The difference is not consequential for our purposes, and I shall use 83;39 in calculations. We know that Habash

---

†The emphasis is Kennedy's.

‡See Section VI.2.

used the epicyclic or eccentric model of the sun.† Thus, for an autumnal equinox occurring in Habash's own time, we find from the relations in Section VIII.2 that

$$L_S = 181°.971 \quad \text{(VIII.12)}$$

al-Biruni [1025, p. 96] says that Habash's zij gives 165;22 degrees for the longitude of the sun at noon on the day EH 407 III 26 (see Section VII.3). This time has Julian day number 2 092 397.0 in terms of Baghdad time. This is about two centuries after the observations upon which Habash based his zij, and the longitude of apogee is about 3°.5 greater‡ than 83;39. Thus, at the time that al-Biruni noted,

$$L_S = 167°.309 \quad \text{(VIII.13)}$$

In Section VII.1 above, I quoted ibn Yunis as saying that the astronomers of al-Ma'mun found

---

†We can tell this because Habash's table of the solar equation has a maximum at 88° from perigee, instead of being symmetrical about 90°. See Section VIII.2, and see Salam and Kennedy [1967].

‡Habash and Yahya put the solar apogee at a considerably greater longitude than Ptolemy did, but it is usually assumed that they did not explicitly recognize the motion of apogee. al-Battani is usually credited in modern writing with being the first person to recognize the motion explicitly, although ibn Yunis [1008, pp. 34ff] implies that the Bani Musa did so about 30 years earlier than al-Battani; see Section VIII.7 below. Since al-Biruni was certainly acquainted with the motion, since it is simple to introduce the motion into a pre-existing zij, and since many people took liberties with Yahya's and Habash's zijes, I have assumed that al-Biruni used a motion of apogee when he used Habash's zij. It is not important for the considerations of this section whether the assumption is correct or not.

TABLE VIII.2

A RECONSTRUCTION OF THE EQUINOX OBSERVATIONS USED AS THE BASIS OF HABASH'S ZIJ

| Year | Julian Day Number[a] - 2 000 000 | |
|---|---|---|
| | As Reconstructed | As Recorded |
| EN 1577 | 24 111.777 | 24 112.033 |
| EY 199 | 24 477.018 | 24 477.042 |
| EY 200 | 24 842.259 | 24 842.292 |
| EY 201 | 25 207.500 | 25 207.495[b] |

[a] In terms of Baghdad time.

[b] This observation was made in Damascus. I have transferred it to Baghdad time by using 10° for the difference in longitude; this is the value used by al-Biruni [1025].

359;45,44,14,24 for the mean motion† of the sun. Let us use this mean motion, in combination with the values in Equations VIII.12 and VIII.13, to calculate back to the solar observations that Habash might have used in establishing his tables. Presumably he based his value for the mean longitude at the epoch upon one of the first four autumnal equinoxes listed in Table VII.1. These are in the years EN 1577, EY 199, EY 200, and EY 201, which correspond to our years 829 through 832. The results of the calculation are summarized in Table VIII.2.

The first column in Table VIII.2 lists the years. The second column gives the times calculated in the manner just described, and the third column gives the recorded times; both columns are given in terms of the Julian day number for Baghdad mean time. The agreement between the reconstructed and recorded times is good except for the first year, which is

† I shall always give the mean motion in degrees per Persian year in this discussion.

the one that caused trouble in Section VII.1. For this year, the recorded equinox is seriously in error.

Since the reconstructed equinox times agree well with the accurate observations of the equinox, it appears that Habash based his zij upon an accurate observation, which is what we should expect if his zij was still useful two centuries later. So far as we can learn from the reconstruction, Habash could have used any of the equinoxes listed in Table VIII.2. However, ibn Yunis [1008, pp. 38-40] explicitly says that the first equinox in Table VIII.2 was the one that was used. This is not possible if the recorded value listed in Table VIII.2 is the value that was actually observed. Hence we must conclude that the reconstructed time is closer to the real observation than the recorded time is. The recorded time is given as 4/5 hours after noon, while the reconstructed time is almost exactly 4/5 hours after sunrise. Thus it is plausible that the latter is the time that was actually observed, and that there was a simple error in the recorded time.†

Whether the reconstruction of the original form of the first observation is correct or not, we may still conclude that Habash and/or Yahya based their zijes upon fairly accurate observations. However, there is a problem. For brevity, let me use the term 'correct Islamic' to denote any accurate observation made around the period of Table VIII.2, and let me use the term 'wrong Islamic' to denote the first recorded time in Table VIII.2; this time is too large by about 6 hours. The problem is that we cannot get the mean motion attributed to Habash by any accurate use of either a 'correct Islamic' or a 'wrong Islamic' observation of the equinox. ibn Yunis [pp. 38-40] says that the mean motion was found by combining the Islamic observation with the observations of Ptolemy. However, if we calculate the mean motion by using one

---

†Perhaps the original had 4/5 hours after the beginning of the day. Later astronomers tended to begin their day at noon, and they might not have realized that the original observers used a different convention.

of Ptolemy's equinoxes, we get a mean motion that is considerably larger than 359;45,44,14,24 whether we use a 'correct Islamic' or a 'wrong Islamic' observation. If we assume that ibn Yunis made a slip of the pen and that Hipparchus's observations were the ones actually used, we get a mean motion that is much smaller, again regardless of whether the Islamic observation was 'correct' or 'wrong'.

The situation is thus a rather complex one, but I believe that all the facts can be accounted for by the following explanation, which has four parts:

1. The first Islamic observation in Table VII.1 was made at 4/5 hours after sunrise on EN 1577 VIII 25 (829 September 19).

2. By an accidental error, the time of this observation was soon altered to 4/5 hours after noon.

3. Yahya and/or Habash misinterpreted the time of Ptolemy's observations, and took them to be 12 hours early.

4. Other Islamic observers soon realized that the mean motion used in the original tables was not compatible with Ptolemy's data or with the now-erroneous record of 829 September 19, and they made what they thought were appropriate changes.

The first part of the explanation is required only because of ibn Yunis's statement about the Islamic observation that was used. If there is an accidental error in ibn Yunis's statement of the year of the observation, the first part of the explanation is not needed. However, it seems unlikely that the Islamic astronomers had an error of 6′ in the setting of their equatorial plane in 829, which the existing form of the record would require, and that they corrected it by 830.

We know that the existing record has the time

given in the second part of the explanation. Whether the error in the time is an accidental error in writing or a real error in observation is unimportant for the rest of the explanation.

TABLE VIII.3

VARIOUS POSSIBILITIES FOR THE MEAN MOTION OF THE SUN

| Form of Ptolemy's Record Used | Form of the Islamic Record Used | Mean Motion, degrees per Persian year |
|---|---|---|
| Correct | Correct | 359;45,46,45 |
| Correct | Wrong | 359;45,45,14 |
| Wrong | Correct | 359;45,44,14 |
| Wrong | Wrong | 359;45,42,30 |

For definiteness in discussing the last two points, let me assume that the Islamic astronomers used Ptolemy's record of an autumnal equinox at 07 hours (1 hour after sunrise) on EN 887 III 9 (139 September 26); it would make only a trivial difference if we assumed that they used Ptolemy's other autumnal equinox. Let me call this the 'correct Ptolemy' equinox.† Let me use the term 'wrong Ptolemy' equinox to denote a time that is 12 hours earlier. We can get four different values of the mean motion by combining both 'correct' and 'wrong' Ptolemy equinoxes with both 'correct' and 'wrong' Islamic equinoxes. The values that we get, rounded to the thirds position, are given in Table VIII.3, which should be self-explanatory.

The value of the mean motion found in any

†I do not mean to imply that this is a valid measurement, but merely that it is the time reported by Ptolemy.

particular instance can differ slightly from the corresponding value in Table VIII.3, since the value depends upon which Islamic observation is used, upon the assumed time difference between Alexandria and Baghdad, and perhaps upon other minor matters. The last combination in Table VIII.3 does not seem to occur, but values close to the other three combinations are found.

For example, the mean motion that we have used in the calculations so far is the one that ibn Yunis gives on pages 38-40, immediately adjacent to his description of how it was found. We see that it is based upon 'wrong Ptolemy' and 'right Islamic' values. On pages 214-216, on the other hand, where ibn Yunis gives a detailed comparison of his solar, lunar, and planetary parameters with those attributed to Yahya, he says that Yahya's solar motion is 359;45,45,14. I thought at first that one of the forms was a misprint. I suspect now, however, that ibn Yunis used two different forms of Yahya's zij in writing the different parts of his work; the value on pages 214-216 corresponds to 'correct Ptolemy' and 'wrong Islamic' values.

Another example is furnished by the Berlin manuscript of Habash's zij, which Kennedy [1956] and Salam and Kennedy [1967] discuss. This copy shows clear signs that it has been extensively altered by astronomers later than Habash. Kennedy [p. 152] quotes the length of the year that is used in preparing the solar tables in this copy; the corresponding mean motion is almost exactly 359;45, 45,10,5.

Kennedy [pp. 152-153] also quotes a marginal note in the same manuscript that gives three values for the length of the year; I shall change them to the equivalent mean motions. The value 359;45,45, 32,8 comes, the note says, from the observation at Damascus, the value 359;45,45,10,5 comes from the Shamasi observations in Baghdad, and the value 359;45,46,33,14 is the value of Yahya. The first statement in the note seems to be wrong. The mean motion clearly comes from using 'correct Ptolemy' and 'wrong Islamic', but the only Damascus observations that we know (Table VII.1) are 'correct Islamic' ones. The other two may come from the observations cited, but that is not the main point. Both are based upon the 'correct Ptolemy', but the second value still uses a 'wrong Islamic' observation while the third value clearly uses a 'correct Islamic' one.

We can only speculate about the origin of the 'wrong Ptolemy' observation, if indeed it existed. The time that Ptolemy reports is 1 hour after sunrise. We know that Ptolemy sometimes followed the Egyptian convention of beginning the day at sunrise, and Egyptian scribes who copied him may have unconsciously written some of his times in the terms that they were used to. That is, it is possible that the Islamic astronomers had a copy of Ptolemy in which the time was given as 1 hour after the beginning of the day. Since Muslims begin their day at sunset, the time could easily have been misunderstood as 1 hour after sunset, just 12 hours from the time that Ptolemy meant.

With regard to the tables of Yahya and Habash, it remains only to discuss the relation between their parameters and the lengths that they measured for the seasons. There are two pairs of values in Table VII.1 from which we can find the length of fall plus winter; both pairs give the same value, $178^d.541$. There is one pair from which we can find the length of spring, and it gives $93^d.667$. Since Yahya and Habash had $E = 1;59 = 1°.983$, and perigee at 263°.65, we calculate that the mean longitude of the sun is 181°.971 at an autumnal equinox, 358°.029 at a vernal equinox, and 90°.219 at the summer solstice. Their value of the mean motion is almost exactly equal to 0°.985 650 per day. Hence their parameters lead to $178^d.621$ for fall plus winter, and to $93^d.532$ for spring. The comparison with the observed values is good, but not as good as it was for Hipparchus's parameters. Perhaps Yahya and Habash used data that are not in Table VII.1, or perhaps they did not solve the equations (which cannot be solved in closed form) in the same way.

Since we have both autumnal and vernal equinoxes in Table VII.1, we can avoid the effects of a bias in the setting of the equatorial plane. Hence I shall use the observed values rather than values derived from the tables of Yahya and Habash. The main purposes of this section have been to give us confidence in the use of the tables and the underlying models, to let us reconstruct the original form of the observation of 829 September 19, and to explain the variety of values of the mean motion that are attributed to Yahya and Habash.

## 6. The Solar Table of al-Marvarudi

al-Marvarudi made the equinox observation on EY 201 V 25 that is listed in Table VII.1, as well as a number of other observations listed in Table VII.2. He may also have made the observation in Damascus on EY 199 V 25, but I shall not assume that he did so.

ibn Yunis [1008, pp. 38-40] says that astronomers in Damascus, presided over by al-Marvarudi, observed the sun in the year EH 217 and that they found its mean motion to be 359;45,46,33,50,43. He says further that they found the obliquity to be 23;33,52, the maximum equation to be 1;59,51, and the position of apogee to be 82;1,37.

We see by reference to Table VIII.3 that the mean motion comes from combining a correct reading of Ptolemy's records with an accurate Islamic measurement. The equinox observation of EY 201 V 25 is for the autumnal equinox in the year that ibn Yunis mentions, and we know that Islamic astronomers preferred to base the mean motion upon the autumnal equinox. Thus it is highly likely that al-Marvarudi's parameters are based in part upon the observation of EY 201 V 25. I now wish to calculate the value of $L_S$ from al-Marvarudi's parameters at the epoch of this observation.

We do not have an explicit statement that tells us which method al-Marvarudi used in dealing with the solar equation. Since his contemporaries Yahya and Habash had gone to the Ptolemaic eccentric method, it is likely that al-Marvarudi had done the same. The value of $L_S$ at an autumnal equinox that results from the eccentric method is

$$L_S = 181°.978 \quad \text{(VIII.14)}$$

Since the value that results from using the method of sines is 181°.968, the difference in method is unimportant for our purposes.

The values of $L_S$ at a vernal equinox and at the summer solstice are 358°.022 and 90°.277, respectively. These correspond to intervals of $178^d.607$ and $93^d.598$.

## 7. The Solar Table of the Bani Musa

The Bani Musa, or at least two of them, observed the autumnal equinox on the date EN 1607 IX 2 (= 855 September 19), as we see from Table VII.1, and they also made some observations near the solstices, as we see from Table VII.2. ibn Yunis [1008, pp. 134ff] says that the brothers combined their data with those of Hipparchus and found that the solar mean motion is 359;45,39,58,2. This is equal to 0°.985 646 855 per day, while Newcomb's value for the epoch 855 is 0°.985 647 162.

Hipparchus's autumnal equinoxes are listed in Table V.2. We may be sure that the Bani Musa did not use any of the first three, which are clearly defective. It makes only a trivial difference whether they used the equinox of -146 or that of -145, but it does make a difference whether they used one of these or the equinox of -142. The equinox of -145 has day number 1 668 365.75 in Alexandria time. The length of the year that corresponds to the Bani Musa's mean motion is $365^d.242$ 377. If we add 1004 of these years to Hipparchus's time, we find 2 035 069.096. If we take the equinox of -142, which has day number 1 669 461.25, and add 1001 years, we find 2 035 068.869. The day number of the Bani Musa's equinox is 2 035 069.0 in Samarra time, which differs by slightly less than an hour from Alexandria time. It looks as if the Bani Musa used the average of Hipparchus's equinoxes of -145 and -142, which is probably the most accurate course that was open to them. Their measurement of the mean motion seems to be the most accurate that had yet been made or that would be made for some time to come. It is considerably better than al-Battani's measurement, for example, in part because al-Battani used Ptolemy's times rather than Hipparchus's.

ibn Yunis also says that the Bani Musa found 2;0,50 for the maximum equation E, that they found 1° in 66 years for the motion of the solar apogee, and that they found the apogee to be at 80;44,19 for the epoch EY 1 I 1. This implies that they specifically recognized the motion of apogee, although al-Battani is usually credited with this discovery. It is possible that the Bani Musa did not themselves deal with the motion, and that later astronomers read a rate of motion into their data.

We may assume that the Bani Musa found E and apogee from the lengths of the seasons. It is not likely that they took the times in Table VII.2 as solstice times, since the tabulated times are all at noon. It is more likely that they used other data which al-Biruni [1025], who is our source for the Bani Musa data, did not transmit to us, but it is desirable to test the matter. From the Bani Musa observations in Tables VII.1 and VII.2 combined, we can infer 89 days for the interval from the autumnal equinox to the winter solstice, and 182 days from then to the summer solstice. From the maximum equation and the apogee, updated from EY 1 I 1 to the epoch of the data, we calculate 182°.003 for the mean longitude at the autumnal equinox, 269°.794 at the winter solstice, and 90°.206 at the summer solstice. The corresponding intervals are $89^d.069$ and $183^d.039$. Thus it seems that the Bani Musa did not use the tabulated times as the solstice times, but that they used more accurate times.

The only number that we need is the value of the mean longitude at one of the epochs. I shall use

$$L_S = 182°.003 \qquad \text{(VIII.15)}$$

as the mean longitude at the epoch of the equinox listed in Table VII.1.

## 8. The Solar Table of al-Battani

al-Battani [ca. 920, Chapter XXVII] observed the autumnal equinox on EN 1630 IX 8 (= 882 September 18) at Raqqa, combined this observation with Ptolemy's equinoxes, and found approximately

359;45,46 for the mean motion of the sun in a Persian or Egyptian year of 365 days. His equinox is listed in Table VII.1. The error in the mean motion is large because al-Battani used Ptolemy's equinoxes, not because of serious error in his own measurement.

In his Chapter XXVIII, al-Battani says that he has made many diligent observations for the purpose of finding the lengths of the seasons. He finds 178 days, 14½ hours, for the interval from an autumnal to a vernal equinox, he finds 186 days, 14 3/4 hours for the interval from there back to the autumnal equinox, and he finds 93 days, 14 hours from the vernal equinox to the summer solstice. He finds 1;59,10 for the maximum equation E of the sun, and he finds 82;15 for the position† of apogee.

As many writers have noted, al-Battani's seasons are inconsistent. They add up to 365 days, 5¼ hours, which is almost exactly ½ hour less than al-Battani's value for the length of the year. I shall return to this point in a moment.

Since al-Battani does not give the data from which he found the lengths of the seasons, there could be a risk that he used data from other sources. However, his seasons and his solar parameters are different from those attributed to any other source, and thus we may assume that his observations are, at least in the main, independent of others.

His values of E and of apogee lead to mean longitudes of 181°.968 at an autumnal equinox, 358°.032 at a vernal equinox, and 90°.268 at a summer solstice. When we combine these with his mean motion, we find 93 days, 13.9 hours, for the duration of spring. This agrees excellently with the value that he states. We also find 178 days, 15.05 hours, for the interval from an autumnal to a vernal equinox, and we find 186 days, 14.73 hours for the interval from a vernal to an autumnal

---

†There is some confusion about whether al-Battani's apogee was 82;15 or 82;17. The difference is trivial for our purposes, and I shall use 82;15.

equinox. The latter agrees almost exactly with the value that he gives. Thus it seems that his interval of 186 days, 14 3/4 hours is correctly stated and that the other interval, from the autumnal to the vernal equinox, is incorrectly stated. The matter is still somewhat puzzling, however. It seems that he should have written 15 hours where he wrote 14½, but this does not look like a common error in writing.

The value that we shall use from al-Battani's tables is the mean longitude at the epoch of his autumnal equinox. The epoch is given in Table VII.1, and the mean longitude is

$$L_S = 181°.968 . \qquad \text{(VIII.16)}$$

## 9. The Solar Table of al-Biruni

Table VII.1 contains an observation of the autumnal equinox made by al-Biruni at Ghazni on EN 1767 X 11 (= 1019 September 18). Table VII.2 contains a number of observations that he made at various times of the year at Jurjaniya. These include an observation on EY 385 VII 10 from which he also found the time of an autumnal equinox. We can also readily find the parameters in his solar model, and Kennedy [1956, p. 158] gives most of them.

Since we have many observations at Jurjaniya, al-Biruni's parameters would be useful for this work only if they involved the observation at Ghazni. However, according to al-Biruni [1025, p. 07], he found the mean motion from the observations at Jurjaniya. I have not seen a statement about where he made the observations from which he derived his other parameters, but the observations at Jurjaniya are well suited for the purpose. Thus it is not safe to assume that the maximum equation and the position of apogee involve the Ghazni observation, and I shall not use al-Biruni's solar parameters in this study.

## 10. The Tables of ibn Yunis

In my first study of the accelerations, I used [Newton, 1970, Section II.4] the values of the mean longitudes of the sun and moon that ibn Yunis [1008] gives on page 222 of the cited edition. Later I discovered the extensive tabulation of equinox data that al-Biruni [1025, Section XXVI] gives, and I noted that many of them are also given by ibn Yunis [pp. 130-132]. This led me to suspect [Newton, 1972b, p. 127] that there was a well-known stock of Islamic equinox data which ibn Yunis used in deriving his Hakémite tables. This would mean, if so, that ibn Yunis's value of the mean longitude of the sun was dependent upon data that had been analyzed separately and hence that the value should not be used as an independent observation.

In reaching this tentative conclusion, I overlooked ibn Yunis's own statements about his parameters. On page 142, he says that he combined Yahya's observation of the autumnal equinox in EY 199† with one of Hipparchus's equinoxes and that he calculated 359;45,39,54 for the solar mean motion. He says on the same page that he made his own observation of the autumnal equinox, that he combined this with Hipparchus's equinox, and that he found 359;45,40, 3,44 for the mean motion. From the consistency of the values, as compared with the inconsistency that results from using Ptolemy's data, he concludes that Hipparchus's data are better than Ptolemy's. He does not give the year in which he made his own observation.

On pages 214-216, ibn Yunis gives a detailed comparison of his solar parameters with those of Yahya. He repeats the value of the mean motion that has been stated, and he says that his value‡ of the maximum equation is 2;0,30 and that his position of apogee is 86;10 for the year EY 372. This is approximately the year 1003, whereas the epoch for his mean longitudes is 1000 November 30. This does not necessarily mean that he made his observations in either 1000 or 1003, but the independence of his parameters from any others known means that he based the values of E and of apogee upon independent observations,

†This is the observation at Baghdad on EY 199 V 25 in Table VII.1.

‡That is, the value used in the Hakémite tables.

and we have his explicit statement about an independent equinox observation.

It follows, then, that his mean longitude of the sun is based upon a number of observations that are independent of any found in Tables VII.1, VII.2, and VII.3. Thus we may use his value of the mean longitude as an independent datum, in the absence of the raw data. This value is

$$L_S = 254;45,57,6 = 254°.7659 \qquad \text{(VIII.17)}$$

at the epoch 1000 November 30, Cairo mean noon.

ibn Yunis gives a large number of eclipse observations that he made himself. It is possible that he found his lunar parameters by combining these observations with his solar parameters. Thus it may be that we should not use the mean longitude of the moon as an independent quantity.

TABLE VIII.4

VALUES OF THE MEAN LONGITUDE OF THE SUN DERIVED FROM ISLAMIC TABLES

| Astronomer | Place | Local Apparent Time | | Mean Longitude, |
|---|---|---|---|---|
| | | Day Number | Hour | degrees |
| al-Marvarudi | Damascus | 2 025 207 | 23.3[a] | 181.978 |
| Bani Musa | Samarra | 2 035 069 | 12 | 182.003 |
| al-Battani | Raqqa | 2 043 470 | $1\frac{1}{4}$ | 181.968 |
| ibn Yunis | Cairo | 2 086 642 | 12[b] | 254.766 |

[a]Actually stated as 0;28,15 days after noon.

[b]Mean rather than apparent time.

## 11. A Summary of the Mean Longitudes of the Sun

In the preceding sections, I have found four values of the mean longitude of the sun that are useful in studying the acceleration of the sun. The values are given in Equations VIII.14 through VIII.17. They have been obtained from the tables, or from abstracts of the tables, of four Islamic astronomers, and they are summarized in Table VIII.4.

The first column of the table gives the astronomer and the second column gives the place for which the tables apply. The third column gives the Julian day number, in terms of local time, and the fourth column gives the local hour at which the mean longitude applies. Finally, the last column gives the mean longitude.

The times of the first three entries are the times of autumnal equinoxes listed in Table VII.1. The dates of the observations can be found from that table. The date of the fourth entry is EH 391 I 1 (= 1000 November 30); a Muslim date is uncertain unless we know the convention that the writer used for the era of the Hijra, but the position of the moon given in conjunction with the solar position resolves the date unambiguously.

The source says explicitly that the hour for the last entry is mean time. It is the only entry in Table VIII.4 that is taken from a formal set of ephemeris tables. The other entries are obtained by starting from an observed time when the observer measured the longitude of the true sun to be 180°, and applying the observer's form of the solar equation to the true longitude. Thus the other times are presumably in apparent solar time.

It should be emphasized that the longitudes in Table VIII.4 are apparent rather than geometric. That is, they correspond to the direction in which the mean sun would be seen at the stated time, and thus they need to be corrected for aberration before they are compared with values from Newcomb's tables.

In the inference of the solar acceleration, each value in Table VIII.4 will be given the weight of four individual observations of an equinox or solstice, for reasons that were explained in Section VIII.3.

# CHAPTER IX

## EPHEMERIS PROGRAMS USED IN THE ANALYSIS

### 1. The Time Base

In all the ephemeris programs that are now used in the national ephemeris publications, the time base is considered to be ephemeris time. However, ephemeris time cannot be used in this work, because a major purpose of this work is to investigate whether the concept of an ephemeris time base rests upon a valid foundation. Instead, we must use solar time as the base for ephemeris calculations.

In spite of this, we can take over the standard ephemeris programs with only one minor modification for each body. When the accepted ephemeris theories were first devised, they were in fact devised on the basis of solar time. With a trivial exception that will be taken up in a moment, the adoption of ephemeris time meant that the theories remained intact, with the only change being to adopt a new name for the independent (time) variable in the theories.

For the purposes of this work, then, we use solar time as the independent variable in the ephemeris theories. We calculate the position of a celestial body for any solar time $\tau$, entering the standard theories with this value of the time, and we then assume that the calculated longitude of the body is in error by the amount $\frac{1}{2}\nu'\tau^2$. We do not assume any relation between $\nu'$ for one body and $\nu'$ for a different body, and we estimate $\nu'$ for each body by comparing observed and calculated positions of that body.

The exception concerns the moon. When Brown [1919] developed his lunar theory, he found discrepancies between observation and his tables that he was unable to remove by theoretical considerations. In consequence, he added his "great empirical term" to his theoretical expressions. When Eckert, Jones, and Clark [1954] adapted Brown's theory for use with ephemeris time, they removed the great empirical term, made a few corresponding adjustments of the parameters, and added the term $-11''.22T^2$ in accordance† with Spencer Jones's study of the spin fluctuation hypothesis [Spencer Jones, 1939].

When I use the lunar theory as revised by Eckert, Jones, and Clark, I shall add $11''.22$ multiplied by the square of the time, in order to remove Spencer Jones time from its role in the theory and to allow the use of solar time. I then assume that the longitude calculated from the theory is in error by the amount $\frac{1}{2}\nu_M{}'\tau^2$.

## 2. The Sun and the Inner Planets

The programs used in the analysis of the solar and planetary data fall into four classes from the standpoint of the theories involved. The first class is that used for the sun and the inner planets.

For the sun, I use the theory of Newcomb [1895], which, so far as the orbital motion is concerned, is still used in the American and British ephemeris publications [Explanatory Supplement, 1961, p. 98]. For simplicity, I keep only periodic perturbations that amount to $0''.03$ of arc or more in the longitude; for the terms that are this large in the longitude, I also keep the corresponding perturbations in the latitude and in the radius vector, regardless of their size. I keep all secular and long-period terms.

Newcomb's theory gives directly the mean ecliptic coordinates of date. These are transformed, when necessary, into the mean equatorial coordinates of date by the use of his expression for the mean obliquity as a cubic function of time.

It may be necessary to refer the coordinates either to the mean or true systems of date. In

---

†T means ephemeris time, as it is currently defined in the lunar theory. In order to avoid using the term 'ephemeris time', I shall call this "Spencer Jones time".

calculating the nutations needed to go from the mean to the true systems, I keep all terms in the nutations on pages 44-45 of Explanatory Supplement that are as large as 0″.01.

In previous work, I have tested this ephemeris program of the sun only in an implicit fashion, by using it in a program to calculate the paths of solar eclipses, and comparing the paths with those calculated by the staff of the Naval Observatory (see Newton [1970, pp. 186-187]). For the present work, I have made a direct comparison in two parts. First, I compared coordinates with those given in the tables of the sun in the American Ephemeris and Nautical Almanac for 1958, at monthly intervals. Second, I extracted coordinates of the sun from the eclipse calculations made by the Naval Observatory, for twelve solar eclipses ranging in date from -1062 July 31 to 590 October 4, taking care to see that each calendar month was represented. Between them, the comparisons should test both the long-term and the short-term accuracy with which the computing program represents Newcomb's theory. I found no discrepancy as large as 1″ of arc.

The comparisons involve the acceleration of the moon in a slight way, because there are lunar perturbations in the theory of the sun. Since the calculations used as a standard were based upon Spencer Jones time, I took $\nu_M' = -22.44$ in making the comparisons. This is legitimate, even though -22.44 is certainly not the acceleration of the moon with respect to solar time. It is legitimate because time enters into the calculations only as a mathematical quantity, and the physical interpretation of the variables does not affect the mathematics.

In the analysis of the solar (and planetary) data that will be used in this work, I shall use $\nu_M' = +10$. This is the value that almost all students of the subject have found to be appropriate for classical antiquity.† It is far from being correct for the past 1000 years, as we saw in Section I.4, but, for times this recent, the error is

†That is, for times around the year 0.

inconsequential for our purposes. Suppose that the error in $\nu_M'$ has been 20 $''/cy^2$ over a span of ten centuries. This makes an error of 1000″, less than a third of a degree, in the longitude of the moon. Since the perturbation on the sun due to the moon has an amplitude of less than 7″, an error of 1000″ in the position of the moon makes an error of less than 0″.05 in the position of the sun.†

In some cases it is easier to calculate aberration within the ephemeris program and in some cases it is easier to find it by an external calculation. In either case, I calculate the aberration in solar longitude by the formula

$$\text{Aberration in solar longitude} = 0°.00569/r_S. \tag{IX.1}$$

That is, we subtract this quantity from the geometric longitude in order to obtain the apparent longitude. $r_S$ denotes the radius vector of the sun. Formula IX.1 includes the effects of the eccentricity upon both the radius vector and the orbital angular velocity of the sun. I neglect the effect of aberration upon latitude and solar distance.

Older observations nearly all use apparent rather than mean solar time as the time base. In order to work with these observations, we need the equation of time. I calculate this by directly comparing the right ascension of the sun with the right ascension of the fictitious mean sun [Explanatory Supplement, 1961, p. 73]. I include the aberration in this calculation, whether I include it in the ephemeris position or not.

For Mercury and Venus I also use the theories of Newcomb [1895a and 1895b], which have the same

†It is clear that we cannot make the approximation $\nu_M' = 10$ when we are analyzing observations of the moon. We can make it only when we want an approximate position of the moon in order to calculate a small perturbation term in the position of some other body.

basic structure as his theory of the sun. In these theories, Newcomb derives expressions for the mean Kepler elements that are quadratic or cubic functions of time, and he also derives expressions for periodic perturbations in heliocentric longitude, latitude, and radius vector. In order to find a position, a person first calculates the mean elements for the epoch in question. He then calculates what the position would be if the mean elements described the orbit exactly. Finally, he evaluates the perturbations and adds them to the position calculated from the mean elements.

Newcomb's theories give the heliocentric positions of Mercury and Venus, whereas we shall need the geocentric positions in order to compare with observation. I calculate the geocentric positions by combining the ephemeris programs of the planets with the program of the sun that was just described.

For Mercury, I keep all perturbations whose amplitudes are greater than 0″.05 in longitude or latitude, or greater than $5 \times 10^{-8}$ in the common logarithm of the radius vector. I have calculated positions at 1-month intervals for the calendar year 1958 and compared them with the values tabulated in the American Ephemeris and Nautical Almanac. I found one discrepancy equal to 1″ of arc; the average discrepancy was 0″.54.

For Venus, I first used the same criteria as for Mercury and did the same comparison. I found errors in the geocentric position slightly in excess of 2″.† In an attempt to find the origin of the errors, I enlarged the program to include all of the perturbations given by Newcomb. However, I still find occasional errors as large as 2″, with an average error of about 0″.76. I presume that errors this large result either from an undetected error that I have made in transcribing the perturbation coefficients into the ephemeris program or from an unlisted erratum. They do not result from the solar program used in changing from heliocentric to geocentric coordinates.

---

†This is after I used the errata listed in v. 11 of the Astronomical Papers.

While it would be pleasing not to have this error, it is not important for our purposes. No individual observation that will be used has a precision as good as 2″. Thus the error in the ephemeris program should contribute little to the final residuals. Further, the error is definitely oscillatory and should average rapidly to zero for observations extending over an appreciable time interval.

It would be possible to have an error in the ephemeris program of either Mercury or Venus that is negligible for the year 1958 but that grows to a large value for an ancient or medieval time. An error of this sort would be important for the purposes of this work, but it would not be detected by the tests described above. I have no accurately calculated ancient positions of Mercury or Venus to use for testing my programs, as I had for the sun.

In order to test the Mercury and Venus ephemerides for ancient times, I resort to positions calculated from the mean orbital elements as listed on page 113 of the *Explanatory Supplement* [1961]. These are derived from Newcomb's theories, just as my programs for Mercury and Venus are. However, by taking the numerical values of the elements directly from the *Explanatory Supplement* and using them in calculations done on a hand calculator, I am able to test whether I have made a significant error in programming the theories.

When I compared the heliocentric longitudes of Mercury and Venus that are calculated from the mean elements with those from the ephemeris programs used in this work, I found no discrepancy greater than 40″. Since this is about the greatest possible value that the perturbations can have, the discrepancy probably results from the omission of the perturbations from the test calculations. Even if it were real, however, it would be negligible. I did the calculations for an epoch near -200. An error of 40″ at that epoch would affect the inferred accelerations by less than 0.2 $''/\mathrm{cy}^2$, and the average discrepancy was only 15″.

## 3. Mars

It was my original intention to use the theory of Mars developed by Newcomb [1898] and corrected by Ross [1917], since this is the theory used in the American and British ephemeris publications. However, I was dissuaded from this by reservations that have been expressed about the long-term accuracy of the Newcomb-Ross theory. Therefore I turned to the theory of Mars that has been developed by Clemence [1949 and 1961].

Clemence [1961, pp. 265-266] says that his goal "was to attain a precision of 0″.01 in the geocentric position of the planet for five centuries or more before and after 1850." It is possible that he did not achieve his goal. He compared his theory with the result of a careful numerical integration carried out by Herget, in a work that is apparently unpublished, and found oscillatory discrepancies in longitude that average about 0″.02 and that attain a peak of 0″.042 near 1920. He finds no satisfactory explanation of the discrepancies, but he feels that the least unsatisfactory explanation is the possibility of error in the theories of Jupiter and Saturn. Because Jupiter is massive and fairly close to Mars, errors in its ephemeris could conceivably give noticeable errors for Mars.

In any case, we may concur with Clemence that the theory of Mars is "considerably more precise than any other general planetary theory yet constructed."

The general method used by Clemence is Hansen's method, as modified by Hill for his theories of Jupiter and Saturn [Hill, 1895 and 1895a]. This method starts with Kepler elements chosen for some epoch; Clemence, following Hill, chose the epoch that we now designate as 1850 January 0.5 (Julian day number 2 396 758.0). The plane defined by the node and inclination at this epoch can be called the fundamental plane.

First we calculate the longitude in the fundamental plane by using the Kepler elements as if they were absolutely constant. However, in using the elements in order to calculate the longitude at some time $t$, we do not use the time $t$ itself† but rather a "perturbed time" $t + \delta z$. The first task in the Hansen-Hill method is to develop perturbation formulae for $\delta z$ as functions of $t$. With the same time $t + \delta z$ that we use to calculate the longitude, we also calculate a radius vector $\overline{r}$. We then calculate a quantity $\nu$ defined so that the actual radius vector r is given by

$$r = \overline{r}(1 + \nu) \quad . \qquad \text{(IX.2)}$$

The second task is to develop perturbation formulae for $\nu$.

The position given by r and the longitude lies in the fundamental plane. The third coordinate is analogous to latitude, but measured from the fundamental plane instead of from the ecliptic. It is found by calculating a third perturbation u.

u and $\nu$ have the dimensions of angle. In order to put the time perturbation on the same basis, Clemence multiplies $\delta z$ by the mean motion to give a perturbation that he writes as $n\delta z$, which is also an angle. Clemence then gives tables of perturbation coefficients for $n\delta z$, $\nu$, and u in units of 0″.0001.

I did not count the number of perturbation terms that Clemence determined, but it must be well into the thousands. With only a few small exceptions, I keep all terms that can apparently contribute as much as 0″.01 to any one of the three quantities $n\delta z$, $\nu$, or u over a period of 10 centuries. If I keep a term with a certain set of arguments because it contributes 0″.01 or more to any of these quantities, I keep its tabulated contributions to all three quantities. Altogether I keep 191 terms in each of the three quantities.

Clemence gives two sets of Kepler elements for the epoch 1850 January 0.5. The first set [Clemence, 1949, pp. 231-232] was the one used in evaluating

†In describing Clemence's theory, I am using his notation rather than the standard notation adopted in Section I.7.

the perturbation coefficients, and it was probably derived on the basis of earlier theories. After he completed the theory, Clemence [1961, pp. 328-333] fitted the Kepler elements to a number of observations with dates ranging from 1802 to 1950, using the assumption that the perturbation terms would not need to be altered by the fitting. From this fitting, he derived the set of Kepler elements that he gives on page 333. Finally, he recalculated all the first-order perturbation formulae and evaluated them at the times of all the observations. He found that the largest difference between "the new perturbations and the old ones is 0″.004, and the probable value of a difference is 0″.001."

He concludes that another iteration in this procedure is not necessary.

He also points out that Newcomb's theories of the obliquity and equinox are used at several points in the fitting process. Hence the set of Kepler elements must be derived again if Newcomb's theories are altered.

I use Clemence's final set of elements in my ephemeris program for Mars. Since the paper with the first-order perturbations [Clemence, 1949] was published before the fitting was done, I presume that the perturbation coefficients in the program are the original ones and not the ones that were derived after the fitting.

Clemence's theory gives directly the position referred to the (fixed) plane defined by the basic Kepler elements, with the x-axis at the position of the basic perihelion. In using his results for the purposes of this work, it is necessary to change to either ecliptic or equatorial coordinates of date. Since Clemence [1961, p. 318] gives the rotation matrix for transforming to the mean equatorial system of 1950.0, one's first thought is to use this matrix and then to use the program (Section IX.5 below) for transforming stellar coordinates. However, the rotation matrix is worked out for the provisional set of Kepler elements rather than the final one. It is thus necessary to alter either the rotation matrix or the program for stellar coordinates, which is set up to start with 1950.0 coordinates.

I decided to do the latter. Thus, in the Mars ephemeris program, I transform Clemence's coordinates first to the mean ecliptic system of 1850.0 and then to the system of date, using an obvious modification of the program described in Section IX.5.

Clemence [1961, pp. 318-321] lists the coordinates from Herget's numerical integration at 80-day intervals between Julian ephemeris dates 2 422 080.5 and 2 434 960.5. I took the first and last of these, and some intermediate values, calculated the corresponding right ascensions and declinations, and transformed to ecliptic coordinates of date using the program of Section IX.5. This gives me a check on the Mars ephemeris program. The greatest discrepancy is about $0''.5$ of arc.

Finally, it is necessary to change to geocentric coordinates. I do this using the ephemeris program of the sun, as with Mercury and Venus. At the closest approach of Mars to earth, the errors in the geocentric coordinates may be somewhat larger than those in the heliocentric ones.

The check described does not test the long-term accuracy of the ephemeris program, and I do not know of a satisfactory way to test it. We cannot use Newcomb's mean elements because the perturbations of Mars are too large to allow a meaningful comparison between an accurately calculated position and one calculated from mean elements. Further, the accuracy of the mean elements is questionable. I can only exert care in checking those parts of the program that affect the long-term behavior of Mars.

## 4. Jupiter and Saturn

It is not part of the purpose of this work to estimate accelerations of Jupiter and Saturn. However, Jupiter and Saturn are involved in some of the observations that will be used in estimating the accelerations of Mars and the inner planets. Hence it is necessary to have ephemeris programs for Jupiter and Saturn.

Originally I intended to program the theories of Hill [1895 and 1895a] for Jupiter and Saturn, and I maintained this intention through the completion of the theory of Jupiter. However, I decided not to use Hill's theory of Saturn when I read the following remark of Eckert, Brouwer, and Clemence [1951, p. v]: "Particularly in the case of Saturn, Hill's tables are now good to only five significant figures and will not long suffice for the most ordinary applications." If the tables of Saturn had degraded, little more than half a century after their preparation, to the point that they do not suffice for modern applications, they may not be accurate enough for comparison with observations ten centuries old, even though the observations did not have modern accuracy.

It should be noted that the two planets whose theories have caused the most trouble are the ones closest to Jupiter, the largest planet.

Because of the defects in Hill's theory of Saturn, I decided to adopt the method of Eckert, Brouwer, and Clemence [1951], which will be abbreviated EBC in the rest of this section. This method is a numerical integration of the equations of motion of the five outer planets simultaneously, with the masses of the four inner planets being lumped in with the mass of the sun. With this method, it is necessary to obtain ephemerides of all five planets in order to obtain the ephemeris of any one. Hence I decided to use this method for the ephemeris of Jupiter also.

EBC do their numerical integration in rectangular coordinates, with a time step of 40 days. They believe that the errors accrued in integrating backward for three centuries, from the middle of the 20th century to 1653 July 14.0† (Julian day number 2 325 000.5), may amount to a few parts in the 9th decimal place in rectangular coordinates. For Jupiter, with a mean distance of 5.2, this is of the order of $10^{-4}$ seconds of arc. For the present work, we need to extend the integration for about fifteen more centuries, for a total of about 18 centuries. Even if the error is growing as the

---

†EBC use the old astronomical day that began at noon, and they write this as 1653 July 13.5.

cube of the time, which is unlikely, the error in the 2nd century would be only of the order of 0″.02, which is negligible compared with the errors of observation.

The preceding paragraph applies only to the errors introduced by the strictly computational processes. There are also errors of observation, which cause errors in the initial conditions used in the integration. We may assume that EBC designed their integration method so that the computational errors would be negligible in comparison with the observational ones. It is a primary postulate in this work, and in others connected with the accelerations, that the errors in calculating ancient positions from modern data are negligible in comparison with the ancient errors of observation, as well as being negligible in comparison with the accumulated effects of the accelerations.

In carrying the integration of EBC back in time from 1653, I decided that the safest procedure is to try to preserve complete consistency with them, even though the precision kept in the calculations is perhaps illusory and certainly unnecessary. In order to explain how I did this, it is necessary to give a little detail about their integration.

The method that they use is a modification of Cowell's method. It is purely a predictor method, as opposed to a predictor-corrector method, and it is based upon the assumption that 10th differences of the integrand can be neglected in any single step of the integration. Let $X(t)$ denote any coordinate, let $\ddot{X}(t)$ denote its second derivative (the integrand), and let us adjust the time scale so that the time step is unity.

Suppose that we want to evaluate $X(t_o + 1)$. In order to do this, we must have the values of $\ddot{X}(t_o - k)$, and hence of $X(t_o - k)$, for k from 0 to 9. In addition, we must have the value of a function that EBC denote by $''X(t)$, for $t = t_o$ and $t = t_o-1$. The two values $''X(t_o)$ and $''X(t_o-1)$ supply the necessary arbitrary constants. Thus we must have altogether twelve quantities in order to start the integration. We then find the value of

TABLE IX.1

WEIGHTING FACTORS USED IN THE NUMERICAL INTEGRATION OF THE OUTER PLANETS

| k | 159 667 200$A_k$ |
|---|---|
| 0 | + 113 257 214 |
| 1 | - 470 982 463 |
| 2 | +1 207 940 748 |
| 3 | -2 069 052 724 |
| 4 | +2 449 952 000 |
| 5 | -2 023 224 114 |
| 6 | +1 148 601 188 |
| 7 | - 428 607 748 |
| 8 | + 94 880 274 |
| 9 | - 9 458 775 |

$X(t_o + 1)$ from the pair of equations:

$$''X(t_o + 1) = 2''X(t_o) - ''X(t_o - 1) + \ddot{X}(t_o),$$
$$X(t_o + 1) = ''X(t_o + 1) + \sum_{k=0}^{9} A_k \ddot{X}(t_o - k). \qquad (IX.3)$$

The weighting factors $A_k$ have the common denominator 159 667 200. The values of 159 667 200$A_k$ are listed in Table IX.1. The values of the $A_k$ may be checked by the relation

$$\sum_{k=0}^{9} A_k = 1/12 . \qquad (IX.4)$$

In order to calculate the accelerations, let i denote the ith planet, i = 5 to 9. Let $X_i$ denote one of its rectangular coordinates, let $\ddot{X}_i$ denote

TABLE IX.2

MASSES USED IN THE EPHEMERIDES OF THE OUTER PLANETS

| $i$ | $m_i$ |
|---|---|
| 0 | 0.000 005 976 820 000 00 |
| 5 | 0.000 954 786 104 043 04 |
| 6 | 0.000 285 583 733 150 56 |
| 7 | 0.000 043 727 316 454 59 |
| 8 | 0.000 051 775 913 844 88 |
| 9 | 0.000 002 777 777 777 78 |

the corresponding component of acceleration, and let $r_i$ denote the distance of the planet from the sun. Let $m_i$ denote the mass of the ith planet, $i = 5$ to 9, in units of the sun's mass, and let $m_o$ denote the total mass of the four inner planets. EBC lump $m_o$ with the sun's mass in the integration. The values of the masses used, including $m_o$, are listed in Table IX.2. The value used for the Gaussian constant k is

$$k = 0.017\ 202\ 098\ 95\ . \qquad \text{(IX.5)}$$

In the numerical integration, the values listed for k and for the masses are treated as if they were exact.

Let $\Delta_{ij}$, i and j = 5 to 9, be a square array. When $i \neq j$, I define $\Delta_{ij}$ to be the distance between the ith and jth planets. When $i = j$, I define $\Delta_{ii} = 1$. The use of $\Delta_{ij}$ allows us to write the accelerations in a highly symmetrical fashion. Finally, let the time step used in the integration be w days. Then

$$\ddot{X}_i = -k^2w^2(1 + m_o)(X_i/r_i^3) - k^2w^2\sum_{j=5}^{9} m_j(X_j/r_j^3)$$

$$- k^2w^2 \sum_{j=5}^{9} m_j((X_j - X_i)/\Delta_{ij}^3) \quad . \tag{IX.6}$$

The reader should notice that he can start the integration if he has values of the coordinates for 11 successive time points. From the coordinates, he first calculates the accelerations for the same times. From the coordinates and the accelerations, he then calculates the necessary pair of values of "X from the second of Equations IX.3. The reader might think that this would be the natural way for me to carry the numerical integration backward in time from 1653. However, EBC give the coordinates only to 9 places of decimals. From these, I can get the values of "X only to 9 places, whereas EBC used 14. This accuracy would probably be sufficient, but it would not give me the consistency with EBC that I wish to maintain.

On pages 322-326, EBC give the basic data that they used for the integration. These consist of 21 successive values of each acceleration, for times at intervals of 40 days between Julian days 2 429 600.5 and 2 430 400.5, tabulated to 14 decimal places. In addition, they give values of the "X to 14 places, but the values are those needed for integrating forward in time.

In order to find the values of the "X needed for integrating backward, I first integrated forward a short time, tabulating the coordinates to 14 decimal places. When I combined these coordinates with the accelerations, I was able to calculate the values of "X needed for backward integration, also to 14 decimal places. The values of the accelerations that I used in starting the backward integration, along with the values of the "X, are listed in Tables A.I-1 through A.I-5 in Appendix I. Specifically, these are the values to use with the choices $t_o$ = 2 430 000.5 and w = -40 days.

With the starting values given in Appendix I, I carried the integration backward in time to Julian day 1 720 960.5.† This integration furnished a striking illustration of the progress that was made in both computers and programming between 1951 and 1972, when I did the integration. EBC do not say how much expenditure of effort and computing time went into their work, but it was obviously considerable. In contrast, the preparation of the programs used here was the part-time effort of one person for a few weeks, and the computer time was approximately 190 seconds.

In case anyone should wish to carry the integration still farther back in time, I give the 'continuing conditions' in Appendix II. These are the necessary values of the $\ddot{X}$ and the $''X$ for the choices $t_o$ = 1 720 960.5 and w = -40 days. That is, if the reader uses the values in Appendix II, along with Equations IX.3, the first values that he will generate will be for Julian day 1 720 920.5.

Some of the values generated in the numerical integration are listed in the tables in Appendix III. Since EBC have published their tables forward in time from Julian day 2 325 000.5, there is no point in duplicating their publication. Therefore the tables in the appendix, which begin with date 1 720 960.5, are carried only to a short time after 2 325 000.5; it is convenient to the reader to provide some overlap with the tables of EBC.

Further, the planets Uranus, Neptune, and Pluto had not been discovered by date 2 325 000.5, and it is unlikely that there were any 'pre-discovery' observations earlier than this, since the telescope had been invented only a short time before. Therefore there does not seem to be any point in giving tables for the three outermost planets; they are carried in the integration only to include their perturbations on Jupiter and Saturn.

Finally, the tables would be bulky if they used the interval of 40 days. Therefore they use the

---

†This time is the midnight that begins -101 September 26.

largest intervals that are consistent with convenient interpolation. The tables will be used to give angular positions, not Cartesian ones. Thus in using them, we start by converting to right ascension, declination, and radius vector, and we are concerned with interpolation for these coordinates, not rectangular ones.

Therefore I give positions at intervals of 160 days for Jupiter and 320 days for Saturn. These allow interpolation to an accuracy of about 1″ of arc, if one uses an interpolation method that neglects sixth differences. It should be noted that linear interpolation at intervals of 40 days does not give this accuracy, so that higher order interpolation is necessary with any interval that is convenient in tabulation. The integral parts of the tabular Julian dates are multiples of 160 and 320, respectively.

The tables in Appendices I, II, and III have been reproduced directly from the computer output.

In order to use the tables in Appendix III for the purposes of this work, I first convert a set of six consecutive entries to heliocentric right ascension, declination, and radius vector, and interpolate in order to find these quantities at a desired time. Since these coordinates are referred to the mean equatorial system of 1950.0, I then use the program of Section IX.5 to convert to the coordinate system of date. Finally, I convert to geocentric coordinates in the familiar way.

I did not achieve the goal of full consistency with EBC. When I compare my integration with theirs back to day 2 325 000.5, I find that the differences are trivial for Jupiter; they could easily have come from different ways of rounding fractions, for example. For Saturn, however, there is a discrepancy that seems to grow linearly with time as we go back from day 2 430 000.5, and that amounts to about 12 in the ninth decimal place around day 2 325 000.5. I have not been able to find the origin of this discrepancy.

The discrepancy is annoying, but it is unimportant for our purposes. If it grows linearly, as it seems to be doing, it amounts to about 6 in the eighth place at the earliest dates in Appendix III. This corresponds to about 0″.001.

I do not know of any independent values of position for Jupiter and Saturn that have accuracies comparable with what we expect from the numerical integration, and hence I do not know of an independent way to test the tables in Appendix III. We do not even have mean Kepler elements of high accuracy for Jupiter and Saturn. All that I can do is to try to insure against an accidental computer error in the numerical integration. In order to do this, I waited a few days after the integration had been done until I knew that maintenance operations had been performed on the computer. I then did the integration again, taking care that different segments of computer storage were used in the two runs. The results of the two runs were identical.

There is a convenient way to test the long-term accuracy of Hill's theories. As examples of the use of his tables, he works out the positions of Jupiter and Saturn at $3^h.672$ after Greenwich mean noon on 1007 October 31. I find discrepancies of about 3″ for Jupiter and about 18″ for Saturn. Since 18″ is well below the accuracy of observations made with the naked eye, it would probably have been acceptable to use Hill's tables in this work. However, it is more satisfactory to avoid this error by using the tables in Appendix III.

I have described this as a test of Hill's theories. We may also regard it as a test which says that the numerical integration is accurate enough for the purposes of this work, at least back to the time of the Islamic data.

## 5. The Stars

A number of the observed positions to be used here are given with reference to the positions of stars. The most accurate star catalogues available give coordinates with respect to the mean equatorial coordinate system of 1950.0. It is therefore necessary to have a program that converts stellar coordinates in this system to coordinates in either the ecliptic or equatorial system of the date of the observation.

I have given the method of doing this elsewhere [Newton, 1970, pp. 189-190]. Briefly, I use the proper motions to give the position at an earlier epoch but still referred to the system of 1950.0. Next, I change to mean ecliptic coordinates, still based upon the system of 1950.0. Then I convert to ecliptic coordinates for the former epoch, using the transformations listed on page 38 of Explanatory Supplement [1961].† Finally I convert back to equatorial coordinates, using the mean obliquity of date, when it is necessary.

I described a test of the accuracy of this method in the earlier work. It is doubtful that the error exceeds about 3″ of arc over an interval of 22 centuries, provided that the latitude does not exceed about 5°; the error can be greater for objects at higher latitudes. Thus the method should be adequate for the planets and the stars that are likely to be used as reference points for them.

## 6. The Moon

The reader may remember that it is not a major purpose of this work to study the acceleration of the moon. However, it is necessary to analyze many Babylonian observations of the moon in order to determine the correspondences of their calendar with the Julian calendar. Since I must analyze the Babylonian lunar data for chronological purposes, I decided in Section IV.8 to make the slight additional effort needed to estimate the acceleration of the moon from the same data.

In order to calculate lunar positions, therefore, I programmed the lunar theory of Brown [1919], modified in the way described in Section IX.1 above, and with the neglect of most of the terms in Brown's theory. This program is the one that I have used in earlier studies [Newton, 1970 and 1972] of eclipses; high accuracy is not needed either in that work or in the present work. I tried to keep

†It is desirable to caution the reader that the definitions of the quantities called c and c′ are accidentally interchanged in the reference.

all secular terms in the theory, but I kept only 329 periodic terms out of the 1500 or so that Brown originally derived. The error in the program is typically around 10″, but it sometimes peaks at about 35″. I do not know what the maximum possible error is, but I feel sure that it is less than the error in an observation made with the naked eye.

TABLE IX.3

COORDINATES OF PLACES USED IN THE ANALYSIS

| Place | Latitude, degrees | Longitude, degrees |
|---|---|---|
| Alexandria | 31.18 | 29.84 |
| Athens | 38.00 | 23.73 |
| Babylon | 32.55 | 44.42 |
| Baghdad | 33.33 | 44.43 |
| Balkh[a] | 36.77 | 66.83 |
| Cairo | 30.05 | 31.25 |
| Damascus | 33.50 | 36.32 |
| Ghazni | 33.55 | 68.47 |
| Isfahan | 32.58 | 51.68 |
| Jurjaniya[b] | 42.33 | 59.88 |
| Jayfur[c] | 34.50 | 69.17 |
| Neyshabur | 36.22 | 58.82 |
| Raqqa | 35.95 | 39.05 |
| Rayy | 35.58 | 51.45 |
| Rhodes | 36.43 | 28.23 |
| Samarra | 34.22 | 43.87 |
| Samos[d] | 37.68 | 26.83 |
| Shiraz | 29.63 | 52.57 |

[a]The modern Wazirabad.

[b]These are the coordinates of the modern Kungrad. This is close to the place that al-Biruni [1025] describes; we cannot identify Jurjaniya with certainty.

[c]Jayfur is described in modern sources as being near the modern Kabul; these are the coordinates of Kabul.

[d]These are the coordinates of Pagondhas, which is close to the ancient town of Samos.

## 7. Coordinates of Places Where Observations Were Made

Table IX.3 gives the latitudes and longitudes of the places where the observations were made that will be used in the analysis. The coordinates are taken from the *Times Atlas* [1955]. The source gives the coordinates in degrees and minutes, which I have converted to degrees and decimal fractions.

# CHAPTER X

# ANALYSIS OF THE LUNAR DATA FROM LATE BABYLONIAN SOURCES

## 1. Calendrical Correspondences That Are Established by the Lunar Data

For reasons that were discussed in Section II.2, the analysis of Babylonian astronomical data is a 'boot strap' process: In order to analyze an observation, we must assign it a date, but, in order to assign it a date, we must first analyze it. In this section, I shall presuppose the results of this process for the lunar data in order that I may proceed to summarize the calendrical results.

If we assume that there are no scribal errors in the recording of the data, each valid lunar observation determines the date in the Julian calendar that corresponds to the recorded date in the Babylonian calendar. This correspondence in turn establishes the calendrical correspondence for the entire Babylonian month in which the observation occurred. Further, in a few instances, we have two adjacent months whose lengths are both recorded, with one containing observations while the other does not. In these instances, the observations occurring in one month establish the correspondence for both months.

Correspondences for a total of 26 months can be established by means of the lunar data given in Tables IV.3 and IV.4. The information concerning these 26 months is summarized in Table X.1. The first column gives the Babylonian date, referred to the Seleucid era. In giving the Babylonian date, I have referred to day 0 of the month rather than to day 1. Reference to day 0 is simpler than reference to day 1. In order to find the calendrical correspondence for day D of any month, we have only to add D days to the dates given in the table. We do not have to remember, with a consequent risk of forgetting, to subtract 1 day after adding D days.

## TABLE X.1

## BABYLONIAN MONTHS WHOSE CORRESPONDENCES ARE ESTABLISHED BY LUNAR DATA

| Babylonian Date (Seleucid Era) | | | Date in Julian Calendar | Julian Day Number | Number of Observations Used |
|---|---|---|---|---|---|
| -257 | $XII_2$ | 0 | -567 Mar 24 | 1 514 044 | |
| -256 | I | 0 | -567 Apr 22 | 1 514 073 | 4 |
| -256 | II | 0 | -567 May 22 | 1 514 103 | 1 |
| -256 | III | 0 | -567 Jun 20 | 1 514 132 | 3 |
| -256 | X | 0 | -566 Jan 14 | 1 514 340 | |
| -256 | XI | 0 | -566 Feb 12 | 1 514 369 | 1 |
| -256 | XII | 0 | -566 Mar 14 | 1 514 399 | 3 |
| -255 | I | 0 | -566 Apr 12 | 1 514 428 | |
| -211 | IV | 0 | -522 Jul 3 | 1 530 581 | 1 |
| -211 | X | 0 | -522 Dec 27 | 1 530 758 | 1 |
| - 67 | VII | 0 | -378 Sep 28 | 1 583 264 | |
| - 67 | VIII | 0 | -378 Oct 27 | 1 583 293 | 3 |
| 38 | VI | 0 | -273 Sep 7 | 1 621 594 | |
| 38 | VII | 0 | -273 Oct 6 | 1 621 623 | 15 |
| 38 | VIII | 0 | -273 Nov 4[a] | 1 621 652 | 3 |
| 38 | XII | 0 | -272 Mar 2[a] | 1 621 771 | 1 |
| 72 | VIII | 0 | -239 Oct 20 | 1 634 056 | 1 |
| 79 | VI | 0 | -232 Sep 3 | 1 636 566 | |
| 79 | VII | 0 | -232 Oct 3 | 1 636 596 | 10 |
| 79 | VIII | 0 | -232 Nov 1[a] | 1 636 625 | 9 |
| 79 | IX | 0 | -232 Dec 1 | 1 636 655 | 17 |
| 79 | X | 0 | -231 Jan 0 | 1 636 685 | 4 |
| 79 | XI | 0 | -231 Jan 29 | 1 636 714 | 2 |
| 79 | XII | 0 | -231 Mar 0 | 1 636 744 | 4 |
| 100 | I | 0 | -211 Apr 17 | 1 644 097 | 1 |
| 207 | V | 0 | -104 Aug 0 | 1 683 284 | 1 |

[a]This is one day earlier than the date inferred from Parker and Dubberstein [1956].

The day listed in the first column of Table X.1 begins at a sunset. The day in the Julian calendar that is listed in the second column of the table is the day that begins at the following midnight. Thus, for example, the Babylonian day that is being called† -257 $XII_2$ 0 began at sunset on -567 March 23, while -567 March 24 is the day listed in the table.

The Julian day number listed in the third column of Table X.1 is the number of the day that begins at noon on the date listed in the second column. Thus, for example, the day number was 1 514 044.0 at noon on -567 March 24. This means that the day identified by the day number begins about 18 hours after the day in the Babylonian calendar. If only the day number and the Babylonian date were involved, a different correspondence would probably be adopted. However, if we wish to use correspondences between the Babylonian and Julian calendars, as well as a day number, the method of correspondence adopted is perhaps the most useful one.

The last column in Table X.1 gives the number of observations that were used in establishing the correspondence. When no number is given, it means that the correspondence is based upon the adjoining month, plus a statement about the number of days in one or both months. For the months that are established in this way, as well as for the months with only one observation, an error in the statement of a single number could affect the calendrical correspondence. The months with more than one observation are probably secure.

I mentioned in Section IV.6 that the editors of the Babylonian texts often give a date in the Julian calendar to correspond with the Babylonian date of an observation, that these dates sometimes disagree with the tables of Parker and Dubberstein [1956], and that the editor's dates, rather than those of Parker and Dubberstein, are used in preparing Tables IV.3 and IV.4. The dates given in

†I word the matter this way in order to remind the reader that the notation used in Table X.1 is not a notation that the Babylonians ever used.

Tables IV.3 and IV.4 prove to be correct in every instance.

The dates established by the astronomical analysis agree with the tables of Parker and Dubberstein in all but three instances. In these instances, the date given in Table X.1 is one day earlier than the date inferred from Parker and Dubberstein, and in no instance is the confirmed date later. Thus none of the discrepancies is the result of weather; weather can delay the beginning of the month but can never advance it. There seem to be only two possible explanations of the discrepancies: Either there is an error in the record,† or there is an error in Schoch's criterion for the visibility of the lunar crescent, the criterion upon which Parker and Dubberstein base their tables.‡ If there are no errors in the records, this means that the moon can be seen more easily than Schoch assumed.

If the weather did affect the visibility of the lunar crescent at the beginning of any of the months in Table X.1, its effect is cancelled by an error in the record, because none of the dates of observation is late. Since an accidental cancellation of two errors is unlikely, we tentatively conclude that the weather did not interfere with the visibility as much as 1 time in 26.

If the months were concentrated in the summer, this conclusion might be plausible. However, the months are distributed rather uniformly throughout the year. In particular, 9 of the 26 dates are in the months of November through February. Both the Babylonian records and modern weather data indicate that cloudiness is common during these months, and for the skies to be clear near the horizon on 9 occasions out of 9 during these months is quite unlikely. This suggests that the Babylonians were

†Including the possibility that an over-zealous observer thought that he saw the moon when he didn't, and including the possibility that the beginning of the month was calculated rather than observed.

‡See the discussion in Section II.2.

willing to establish the beginning of the month by calculation, at least during most of the period covered by Table X.1, if cloud cover interfered with visibility.†

There are several runs of consecutive months in which the Babylonian records give the length of each month. These runs are grouped together in Table X.1. In several runs, we can establish the beginning of two or more months independently by analysis of the lunar data. In all such cases, the lengths of the months stated in the records are consistent with the dates established by the lunar data.

## 2. Analysis of the Babylonian Eclipse Data

Table IV.3 contains a number of references to eclipses, both solar and lunar. We can be sure that many of the references are derived from calculation and not observation. The reference to a lunar eclipse on SE 63 I 13, for example, contains a statement that the eclipse did not happen. In the reference to a solar eclipse on SE 63 VII 29, to give another example, the time of the eclipse is after sunset, and this must represent a calculation rather than an observation. There are five references to eclipses, all lunar, which we may plausibly accept as records of observations.

All the records of eclipses are accompanied by the statement of an hour. For the eclipse of SE 72 VIII 14 (-239 November 3), we are told explicitly that the recorded time is the beginning. We have no a priori knowledge of the phase at the other recorded times.

Partly because of our ignorance of the phase, and partly because of the small number of eclipses involved, I shall not use these eclipses in an

---

†If the Babylonians determined the month by observation when possible, and resorted to calculation only when the weather interfered, a simple scheme of calculation would have been sufficient. The alternation of 29- and 30-day months would provide accurate interpolation between observations.

estimate of the accelerations. I shall use them only to establish the calendar for the months involved.

TABLE X.2

ECLIPSES FROM LATE BABYLONIAN SOURCES

| Oppolzer's Canon Number | Date, Julian Calendar | Time Inferred From the Record | | Calculated Time, Hours | |
|---|---|---|---|---|---|
| | | If hours are equal | If hours are unequal | Beginning | Middle |
| 1056 | -522 Jul 16 | 19.54 | 18.82 | 19.14 | 20.41 |
| 1057 | -521 Jan 10 | -0.91 | -1.79 | -0.08 | 1.79 |
| 1487 | -239 Nov 3 | 3.02 | 3.02 | 3.19 | 5.01 |
| 1528 | -211 Apr 30 | 1.01 | 1.15 | -0.01 | 1.37 |
| 1698 | -164 Aug 13 | 1.44 | 1.54 | 0.22 | 1.19 |

The five eclipses are summarized in Table X.2. The first column gives the canon number assigned by Oppolzer [1887], and the second column gives the date in the Julian calendar. The reader can find the Babylonian dates by referring to Table IV.3.

The next two columns give values of the Greenwich mean time[†] in hours that we infer from the record, on the basis of two different assumptions about the method of stating the time. The time is always given in the records by means of the time interval from sunset or sunrise, whichever is nearer. The third column in the table gives the time on the assumption that the time interval is in equal hours. The next column gives the time on the assumption that the interval is in unequal hours.

[†]In some instances, I have used a negative hour in Table X.2 in order to keep the date that Oppolzer assigns.

The final two columns give the calculated times, in Greenwich mean time, of the beginning and middle of the eclipses. In finding these times, I did not calculate the full circumstances of the eclipses. I first calculated the time of syzygy, and the middle time listed is just this time. I then took the half-duration of the eclipse from Oppolzer and subtracted it from the time of syzygy in order to find the time of the beginning.

The calculated time of syzygy depends upon the difference between the acceleration $\nu_M'$ of the moon and the acceleration $\nu_S'$ of the sun, and I use the symbol D″ to denote this difference. In an earlier work [Newton, 1972a], I have derived an empirical formula for the variation of D″ with time from about -700 to the present. In calculating the syzygies for Table X.2, I calculated the value of D″ separately for each eclipse from this formula.

In the eclipse of -521 January 10, both possibilities for the recorded time are much earlier than the beginning of the eclipse, so there is probably an error in the record. In the eclipses of -239 November 3 and -522 July 16, it is clear that the recorded time is that of the beginning, and indeed the beginning is explicitly stated in the record for -239 November 3. For the other two eclipses, it seems safe to conclude that the recorded time is that of the middle.

Table X.2 does not allow us to choose between equal and unequal hours. The choice of unequal hours agrees with the records slightly better than the choice of equal hours, but not by a significant amount.

The formula used for D″ seems to be rather accurate. If we omit the eclipse of -521 January 10, for which the record is almost surely in error, the average discrepancy between the recorded and calculated times, whether the hours are equal or unequal, is 5 minutes or less.

## 3. Analysis of Moonrise and Moonset Data from Late Babylonian Sources

Aside from the eclipses, the synodic lunar data listed in Table IV.3 consist of times of moonrise and moonset at times near new moon and full moon. I noted the occurrence of several systematic tables of moonrise and moonset in Table IV.3, and I have not used these for fear that they are calculated. When we set these aside, we have left 27 observations† from astronomical diaries, in which the lunar phenomena are interspersed with other astronomical phenomena.

In the analysis of these observations, let us use the term 'lunar event' to denote either moonrise or moonset, and let us use 'solar event' to denote either sunrise or sunset. Let us further use the notation

$h_{S1}$ = hour angle of the sun at the time of the lunar event,

$h_1(\delta_M)$ = hour angle of the moon at the time of the lunar event,

$h_0(\delta_S)$ = hour angle of the sun at the time of the solar event,

$\alpha_{M1}$ = right ascension of the moon at the time of the lunar event,

$\alpha_{S1}$ = right ascension of the sun at the time of the lunar event.

Even after we establish the correspondence of dates that was discussed in Section X.1, it is still necessary to proceed by successive approximations. In order to get started, I converted the Julian calendar dates in Table IV.3 to the Gregorian calendar. I then found approximate times of the solar events from the tables of sunrise and sunset in the American Ephemeris and Nautical Almanac, on the assumption that the times are independent of the year if the dates are in the Gregorian calendar. The times found this way are in local mean time, which I converted to Greenwich mean time by using

†The day is missing from an observation in the month SE-256 XI. The testing of a few possibilities quickly shows that the day must have been SE -256 XI 15, and the observation must have been in the morning of -566 February 27.

44°25′ for the longitude of Babylon.

I then took the recorded time intervals between the solar and lunar events from Table IV.3 and applied them to the times just found in order to get a starting approximation to the times of the lunar events. Using these times, I then calculated the topocentric right ascension and declination of the sun and moon as seen from Babylon.

This calculation gives a good estimate of $\alpha_{M1}$, $\alpha_{S1}$, and the declinations of the sun and moon at the time of the lunar event. From the declination $\delta_M$ of the moon, and from the fact that the moon is always on the horizon at the time of the lunar event, we can calculate $h_1(\delta_M)$ by using

$$\cos h = -\tan \delta \tan \phi \,, \tag{X.1}$$

in which $\phi$ is the latitude. $h_1$ is negative at moonrise and positive at moonset.

Similarly, the sun is on the horizon at the time of the solar event, and we could calculate $h_0(\delta_S)$ using Equation X.1 if we had $\delta_S$, the declination of the sun at this time. I neglected the change in declination of the sun between the times of the lunar and solar events, and took $\delta_S$ to be the declination of the sun that was calculated in the preceding paragraph.

We now have all the needed quantities at the time of the lunar event except $h_{S1}$, the hour angle of the sun. To calculate it, we use the fact that the sum of the right ascension and the hour angle is the same for all bodies at a given instant. Using this fact at the time of the lunar event, we find that

$$h_{S1} = h_1(\delta_M) + \alpha_{M1} - \alpha_{S1} \,. \tag{X.2}$$

Let $\Delta H$ denote the time of the lunar event minus the time of the solar event; the time intervals listed in Table IV.3 are presumably measured values of $\Delta H$. Since the times are in solar time, we have, by definition of solar time, that

$$\Delta H = h_{S1} - h_0(\delta_S) \ .$$

When we use Equation X.2 in this, we find

$$\Delta H = h_1(\delta_M) + \alpha_{M1} - \alpha_{S1} - h_0(\delta_S) . \qquad (X.3)$$

$h_0(\delta_S)$ is now an improved estimate of the local mean time of the solar event, and $h_{S1}$ is an improved estimate of the time of the lunar event. We could use these times and repeat the calculations just outlined, thus obtaining a new estimate of ΔH. The original estimates are so close, however, that I did not bother to repeat the calculations.

In Section IV.6, I pointed out that a recorded time interval could conceivably be the interval between the time when the observer happened to catch sight of the moon and the time when it set, but that it is more likely to be a measurement of ΔH, that is, of the interval between sunset and moonset. When we compare the calculated values of ΔH with the intervals given in the records, we find a maximum discrepancy that is slightly greater than 3° (12 minutes in time). This shows that the recorded intervals are almost surely measurements of ΔH.

This also shows that the dates are correct, because ΔH changes by about 12° per day. With the dates confirmed, we may turn to the use of the data in a study of the accelerations.

The time intervals listed in Table IV.3 are, as I have said, measured values of ΔH. The notation $(\Delta H)_e$ will henceforth be used to mean a measured value, while ΔH, without a subscript, will mean a value calculated from Equation X.3. ΔH clearly depends upon the accelerations of the sun and moon used in the ephemeris calculations, and it is necessary to look at the nature of the dependence.

We can be sure that ΔH depends upon the difference D″ more than it depends upon $\nu_M'$ and $\nu_S'$ individually. However, Equation X.3 shows us that ΔH does depend upon the individual accelerations somewhat, for two different reasons. First, Equation

X.3 involves the difference of right ascensions rather than the difference of longitudes. Since the sun and moon are not in the same place, the transformation between longitude and right ascension is not the same for the two bodies, and their right ascensions are hence affected differently by equal accelerations. Second, an acceleration changes the declination and hence the hour angle in Equation X.3, and the change in hour angle is not the same for the two bodies.

TABLE X.3

ANALYSIS OF SYNODIC LUNAR PHENOMENA FROM LATE BABYLONIAN SOURCES

| Date | Local Mean Time, hours | Lunar Event | Observed $\Delta H$, degrees | $\Delta H_0$ degrees | $10^2 \times \frac{\partial(\Delta H)}{\partial D''}$ |
|---|---|---|---|---|---|
| -567 Apr 22 | 19.45 | S | 14 | 15.39 | 11.70 |
| -567 May 6 | 5.50 | S | 4 | 1.57 | 5.17 |
| -567 May 18 | 3.52 | R | -23 | -23.35 | 6.10 |
| -567 Jun 20 | 20.48 | S | 20 | 22.33 | 10.21 |
| -567 Jul 5 | 5.42 | S | 7 1/2 | 5.68 | 8.83 |
| -566 Feb 12 | 18.58 | S | 14 1/2 | 16.63 | 10.16 |
| -566 Feb 27[a] | 7.65 | S | 17 | 20.22 | 5.70 |
| -566 Mar 14 | 19.72 | S | 25 | 25.36 | 10.47 |
| -566 Mar 26 | 6.17 | S | 1 1/2 | - 1.62 | 5.52 |
| -378 Oct 27 | 18.28 | S | 14 1/2 | 14.24 | 6.14 |
| -378 Nov 9 | 16.48 | R | - 9 1/2 | - 4.43 | 4.30 |
| -378 Nov 10 | 6.65 | S | 4 1/2 | 0.33 | 8.20 |
| -273 Oct 6 | 18.73 | S | 15 | 13.75 | 5.39 |
| -273 Oct 19 | 16.72 | R | -11 | - 6.39 | 4.52 |
| -273 Oct 21 | 6.10 | S | 5 | 1.44 | 7.29 |
| -273 Nov 2 | 4.65 | R | -24 | -28.89 | 8.28 |
| -273 Nov 19 | 17.37 | R | 5 1/2 | 6.52 | 5.66 |
| -272 Mar 29 | 4.95 | R | -15 | -14.55 | 4.50 |
| -232 Oct 3 | 18.82 | S | 15 1/2 | 17.36 | 3.71 |
| -232 Oct 15 | 17.15 | R | - 5 2/3 | - 2.32 | 4.66 |
| -232 Oct 16 | 17.77 | R | 3 5/6 | 6.96 | 4.73 |
| -232 Nov 14 | 16.88 | R | - 2 1/2 | 0.37 | 5.13 |
| -232 Nov 15 | 6.88 | S | 6 2/3 | 2.18 | 8.75 |
| -232 Nov 29 | 5.77 | R | -11 | -15.19 | 7.11 |
| -232 Dec 13 | 16.08 | R | -12 1/2 | - 9.14 | 5.91 |
| -232 Dec 14 | 6.75 | S | - 1 2/3 | - 5.27 | 8.40 |
| -232 Dec 15 | 7.83 | S | 14 1/2 | 11.31 | 7.89 |

[a] Day supplied by calculation.

Thus we want to analyze the data in Table IV.3 using two independent parameters, and we clearly want one of the parameters to be the difference parameter $D''$. It is immaterial whether the second parameter is $\nu_M'$ or $\nu_S'$. I arbitrarily chose to use $\nu_S'$, and I inferred estimates of $D''$ and of $\nu_S'$, and of their standard deviations, from the data. The standard deviation in the estimate of $\nu_S'$ came out to be 95 $''/\text{cy}^2$, which is so large that the estimate is essentially useless. In other words, the dependence upon the individual accelerations is large enough to detect in the analysis but it is not large enough to give us useful estimates of the individual accelerations. Hence we need to consider only the dependence upon the difference parameter $D''$.

The results of the analysis are summarized in Table X.3. The first column gives the date of the observation in the Julian calendar, in terms of beginning the day at midnight. The second column shows the time of the lunar event, in local mean time. This column also shows us the nature of the solar event. The solar event is sunrise if the hour is a morning hour and it is sunset if the hour is an evening hour. The third column shows the nature of the lunar event, with an R denoting rise and an S denoting set. The fourth column shows the measured value of $\Delta H$, in degrees rather than time units. The fifth column, called $\Delta H_o$, is the value calculated for $\Delta H$ when $D''$ (and $\nu_S'$) are set equal to zero, and the sixth column is the value of the partial derivative $\partial(\Delta H)/\partial D''$, multiplied by 100 for convenience, with the differentiation being done numerically.

In the analysis, we take

$$\Delta H = \Delta H_o + D''[\partial(\Delta H)/\partial D''] , \qquad (X.4)$$

and we find the value of $D''$ that makes $\Delta H$ fit the observed valuesbest in the least-squares sense.

The result† is

$$D'' = 5.6 \pm 8.0 \ ''/cy^2 \ . \qquad (X.5)$$

The root-mean-square value of the residuals in $\Delta H$ is 3.0°. Equation X.5 applies at the mean epoch -370, approximately.

I shall defer discussion of the result until after the analysis of the lunar conjunctions; only one further point needs to be noted here. In the analysis, I have tacitly assumed that the measured time intervals are in equal hours. If this assumption is wrong, the resulting effect on $D''$ should be small, for two reasons. First, the time intervals are fairly small, so that the error does not have a chance to become large. Second, the sign of the error is different for morning and evening observations. Since morning and evening observations occur in almost exactly equal numbers, the error caused by using the wrong kind of hour should almost cancel.

## 4. The Value of the Ammat As Inferred from Lunar Conjunctions

The first problem in analyzing the lunar conjunctions from the late Babylonian sources was establishing the correspondence of the Babylonian calendar with ours; the solution of this problem is discussed in Section X.1 above. The second problem is finding the value of the Babylonian unit of angle called the ammat. According to the discussion of the ammat in Sections IV.6 and IV.7, the unit was equal to 2 degrees in some texts and to 2½ degrees in others.

In a considerable fraction of the records of conjunctions, whether lunar or planetary, the observer has recorded the difference in latitude

†The value in Equation X.5 is actually the value found by fitting $D''$ and $\nu_S'$ simultaneously. The value found from a single-parameter fit using Table X.3 might differ slightly, but it should not differ by an amount that is statistically significant.

between the moon or the planet and the reference star. Since the latitude of the moon or of a planet is almost independent of the acceleration, the latitude data should allow us to infer the value of the angular unit. We expect the planetary data to be more precise and hence more useful for this purpose, because the planets appear as points rather than as an extended disk, but the lunar data are surprisingly good.

In some instances when the latitude difference is recorded, it is recorded by using a combination of ammats and ubans. Since we know that 1 uban = 5′ (1/12 degree or 1 digit; see Section IV.6), we could use these in inferring the value of the ammat. For simplicity, however, I have used only the instances in which the angle is given by means of the ammat without the admixture of the uban.

TABLE X.4

DATA FOR INFERRING THE VALUE OF THE AMMAT

| Date | Greenwich Mean Time, hours | Reference Star | Latitude of Moon Minus Latitude of Star: Recorded, ammats | Latitude of Moon Minus Latitude of Star: Calculated, degrees |
|---|---|---|---|---|
| -567 May 22 | 16.67 | β Gem | - 4 | - 7.57 |
| -567 Jun 27 | 16.96 | β Lib | - 2½ | - 4.28 |
| -567 Jun 29 | 16.96 | α Sco | + 3½ | + 7.64 |
| -566 Mar 15 | 15.86 | η Tau | - 4 | - 7.64 |
| -378 Nov 13 | 2.69 | β Gem | - 1 | - 2.30 |
| -273 Oct 18 | 15.26 | η Psc | - 2½ | - 7.79 |
| -273 Oct 25 | 2.42 | α Gem | - 6 | -15.08 |
| -273 Oct 28 | 2.46 | α Leo | - 2½ | - 3.47 |
| -273 Nov 9 | 14.89 | δ Cap | + 2 | + 4.87 |
| -232 Dec 8 | 14.71 | η Psc | - 4 | -11.16 |
| -232 Dec 17 | 3.17 | α Leo | + 2/3 | + 2.66 |
| -232 Dec 19 | 3.19 | β Vir | + 1½ | + 3.82 |
| -232 Dec 23 | 3.22 | α Lib | + 1½ | + 3.80 |
| -231 Feb 15 | 3.09 | α Vir | + 1½ | + 6.01 |

There are 16 observations of lunar conjunctions in which the latitude is given by means of the ammat only. Two of these, which involve the star $\eta$ Tauri, are omitted for reasons that will be explained in the next section. Their omission or inclusion does not affect the value inferred for the ammat. The data from the remaining 14 observations are summarized in Table X.4. The first two columns of the table give the date and hour of the observation, in terms of Greenwich mean time. For reasons that were explained in Section IV.7, I take the time to be 45 minutes after sunset or 45 minutes before sunrise, whichever is appropriate. The third column identifies the reference star, the next column gives the recorded difference in latitude in ammats, and the last column gives the calculated difference in degrees.

The first question that concerns us is whether the observer did a good job of referring the angles to the center of the lunar disk, or whether he was unconsciously biased. For example, did he tend to be biased toward the near edge of the disk, thus tending to record an angle that is too small in absolute value?

In order to answer this question, let $A_i$ be a recorded angle in ammats and let $d_i$ be a calculated angle in degrees. We assume that the relation between $A_i$ and $d_i$ is linear, thus,

$$d_i = a + bA_i \ .$$

If we find a and b by a least-squares process, using the data in Table X.4, we expect to find a small value of a whether there is a bias or not, because positive and negative values of $d_i$ and $A_i$ tend to balance. In order to find a bias, we infer values of a and b using the relation above, but replacing $d_i$ and $A_i$ by their absolute values.

When we apply this process to the last ten observations in Table X.4,† we find $a = 0°.43 \pm 0°.78$. Thus there is no bias in the measurement

†The reason for this restriction will be explained later in this section.

that is statistically significant. Since we expect _a_ to be zero on _a priori_ grounds if the observers were careful, it is simpler to set _a_ = 0 and to find only _b_ (= $d_i/A_i$) by a statistical analysis.

The reader can readily convince himself that the value of the ammat changed between -567/-566 and the later observations in Table X.4. One way is to plot the values of the $d_i$ and $A_i$. The points from the older data seem to the eye to fall on a different line from the rest of the points. This impression is confirmed by a formal statistical analysis. For the four oldest observations, we find

$$d_i/A_i = 1.95 \pm 0.07 \ . \qquad (X.6)$$

For the remaining observations, we find

$$d_i/A_i = 2.58 \pm 0.16 \ . \qquad (X.7)$$

The difference between the values is significant even though there are only four observations in the first sample.

One thing we notice instantly is that the quality of the observations seems to have deteriorated between -567/-566 and the later years. The root-mean-square residual of the $d_i$ is only about ½ degree for the observations in the earlier year, but it is 1°.4 for the later observations. The increased scatter is not a consequence of having a mixture of units in the later data, with 2° per ammat being used in some records and 2°.5 in others, as we can see in two ways. One way is to calculate individual values of the ratio $d_i/A_i$ for the later observations and to note that there is no value close to 2° per ammat. The other way is to note that the central value in Equation X.7 should be less than 2.5 if it resulted from mixing values of 2 and of 2.5. However, the value is actually a little above 2.5.

Since the values of $d_i/A_i$ in Equations X.6 and X.7 do not differ significantly from 2 and 2.5 respectively, and since we have _a priori_ grounds for believing that these are the correct values, we conclude that

$$1 \text{ ammat} = \begin{cases} 2 \text{ degrees in the year } -567/-566, \\ 2\frac{1}{2} \text{ degrees in the other years.} \end{cases} \qquad (X.8)$$

More specifically, the first year is the year SE -256. The other years are SE -67, SE 38, and SE 79.† If the value of the ammat never became standardized, the conclusion embodied in Equation X.8 may not carry over to years that are not in Table X.4.

In testing for a bias earlier in this section, I used only the last ten observations in Table X.4. We now see that the reason for this is the change in the unit between the earlier observations and the last ten. Mixing observations from the different samples could have confused the question about a bias.

## 5. Analysis of Lunar Conjunctions from Late Babylonian Sources

Now that we have established the Babylonian dates and the size of the angular unit, we can turn to the analysis of the lunar conjunctions listed in Table IV.4. The analysis is relatively straightforward.

The hours of the observations must first be found. In order to assign an hour, I converted the dates to the Gregorian calendar and found times of sunrise and sunset in the manner described in Section X.3. I then assigned a time to an observation that is either 45 minutes earlier or later, whichever is appropriate. The reasons for using 45

†The year SE -67 (= -378/-377) is ambiguous. There is only one value of $d_i/A_i$ for that year, and that value is 2.30. Since this is closer to $2\frac{1}{2}$ than to 2, and since the year is closer to SE 38 than to SE -256, I assume that the correct value for SE -67 is $2\frac{1}{2}$. Because there are only a few observations from this year, it would not be important for our purposes if the assumption should prove to be wrong.

minutes are explained in Section IV.7. I then transferred to Greenwich mean time.

Now let $\lambda_M$ denote the longitude of the moon, let $\lambda_{st}$ denote the longitude of a star, and let $\Delta\lambda = \lambda_M - \lambda_{st}$. The observations in Table IV.4 give observed values, $\Delta\lambda_e$ say, of $\Delta\lambda$ at the times that have been assigned, and we wish to compare the values of $\Delta\lambda_e$ with theoretical values.

In order to find theoretical values of $\Delta\lambda$, I calculated $\lambda_{st}$ and $\lambda_M$ with the use of the ephemeris programs described in Sections IX.5 and IX.6, and I found $\Delta\lambda$ from $\lambda_M$ and $\lambda_{st}$. Since $\Delta\lambda$ clearly depends upon $\nu_M'$, the acceleration of the moon, we take

$$\Delta\lambda = \Delta\lambda_0 + \nu_M'[\partial(\Delta\lambda)/\partial\nu_M'] . \qquad (X.9)$$

In this, $\Delta\lambda_0$ is the value of $\Delta\lambda$ calculated with $\nu_M' = 0$. In order to find the partial derivative $\partial(\Delta\lambda)/\partial\nu_M'$, I calculated $\Delta\lambda$ with two values of $\nu_M'$ and found the derivative numerically.

TABLE X.5

ANALYSIS OF LUNAR CONJUNCTIONS FROM LATE BABYLONIAN SOURCES

| Date | Greenwich Mean Time, hours | Reference Star | Observed $\Delta\lambda_e$, degrees | $\Delta\lambda_0$, degrees | $10^2 \times \frac{\partial(\Delta\lambda)}{\partial\nu_M'}$ |
|---|---|---|---|---|---|
| -567 Apr 29 | 16.39 | β Vir | -2.00 | - 2.58 | 8.298 |
| -567 May 22 | 16.67 | β Gem | 0 | - 0.12 | 9.594 |
| -567 Jun 27 | 16.96 | β Lib | 0 | - 0.24 | 7.812 |
| -567 Jun 29 | 16.96 | α Sco | 0 | 3.44 | 7.660 |
| -566 Mar 15 | 15.86 | η Tau | 0 | - 1.12 | 9.037 |
| -378 Nov 9 | 14.91 | α Tau | 1.67 | 1.30 | 6.546 |
| -378 Nov 13 | 2.69 | β Gem | 0 | - 1.11 | 6.894 |
| -378 Nov 15 | 2.71 | ε Leo | -1.67 | - 2.40 | 7.161 |
| -378 Nov 16 | 2.72 | α Leo | 3.75 | 1.61 | 7.317 |
| -378 Nov 20 | 2.77 | α Vir | 4.17 | 3.88 | 7.981 |
| -273 Oct 8 | 15.49 | θ Oph | 2.50 | 2.45 | 6.818 |
| -273 Oct 13 | 15.36 | δ Cap | 3.75 | 3.87 | 5.985 |
| -273 Oct 18 | 15.26 | η Psc | 1.25 | 0.32 | 6.074 |
| -273 Oct 21 | 2.37 | η Tau | 0 | - 3.60 | —— |
| -273 Oct 22 | 2.39 | α Tau | 0 | - 0.49 | 6.453 |

TABLE X.5 (Continued)

| Date | Greenwich Mean Time, hours | Reference Star | Observed $\Delta\lambda_e$, degrees | $\Delta\lambda_o$, degrees | $10^2 \times \frac{\partial(\Delta\lambda)}{\partial\nu_M}$, |
|---|---|---|---|---|---|
| -273 Oct 24 | 2.41 | γ Gem | -2.50 | - 3.28 | 6.698 |
| -273 Oct 25 | 2.42 | α Gem | -1.25 | - 0.90 | 6.800 |
| -273 Oct 28 | 2.46 | α Leo | 0.67 | 1.43 | 7.054 |
| -273 Oct 30 | 2.49 | β Vir | 3.17 | 3.47 | 7.179 |
| -273 Oct 31 | 2.51 | γ Vir | 3.75 | 4.17 | 7.210 |
| -273 Nov 1 | 2.52 | α Vir | 5.00 | 5.28 | 7.214 |
| -273 Nov 2 | 2.54 | α Lib | -2.50 | - 1.52 | 7.174 |
| -273 Nov 9 | 14.89 | δ Cap | 0 | - 0.79 | 6.106 |
| -273 Nov 21 | 2.81 | α Gem | -5.00 | - 4.81 | 6.912 |
| -232 Oct 8 | 15.49 | αβ Cap | -3.75 | - 4.65 | 6.066 |
| -232 Oct 14 | 15.34 | η Psc | -3.75 | - 3.53 | 7.340 |
| -232 Oct 17 | 2.34 | η Tau | 3.75 | - 0.49 | —— |
| -232 Oct 18 | 2.36 | α Tau | 6.67 | 5.00 | 7.286 |
| -232 Oct 19 | 2.37 | η Gem | -3.17 | - 3.76 | 7.112 |
| -232 Oct 22 | 2.41 | δ Cnc | 2.83 | 3.22 | 6.422 |
| -232 Oct 25 | 2.44 | θ Leo | 5.00 | 6.96 | 5.918 |
| -232 Oct 26 | 2.46 | β Vir | 6.67 | 5.94 | 5.815 |
| -232 Nov 4 | 14.96 | αβ Cap | -8.75 | - 8.15 | 5.976 |
| -232 Nov 14 | 14.84 | α Tau | 6.25 | 7.19 | 7.251 |
| -232 Nov 16 | 2.74 | γ Gem | 0 | - 1.49 | 7.094 |
| -232 Nov 24 | 2.86 | α Vir | 2.50 | - 0.65 | 5.766 |
| -232 Nov 27 | 2.89 | β Sco | -3.75 | - 4.08 | 5.722 |
| -232 Nov 28 | 2.91 | α Sco | 2.50 | 1.34 | 5.758 |
| -232 Dec 4 | 14.71 | δ Cap | 8.75 | 7.73 | 6.346 |
| -232 Dec 8 | 14.71 | η Psc | -1.25 | - 0.21 | 6.872 |
| -232 Dec 9 | 14.69 | α Ari | 3.75 | 3.49 | 6.979 |
| -232 Dec 10 | 14.69 | η Tau | 1.25 | - 4.21 | —— |
| -232 Dec 11 | 14.69 | α Tau | 2.50 | 0.80 | 7.036 |
| -232 Dec 13 | 14.71 | γ Gem | 1.07 | 0.87 | 6.983 |
| -232 Dec 15 | 3.16 | β Gem | 6.25 | 6.47 | 6.809 |
| -232 Dec 17 | 3.17 | α Leo | 1.25 | - 2.24 | 6.497 |
| -232 Dec 19 | 3.19 | β Vir | 0 | - 2.66 | 6.146 |
| -232 Dec 20 | 3.21 | γ Vir | -2.50 | - 3.80 | 5.998 |
| -232 Dec 21 | 3.21 | α Vir | -2.50 | - 4.86 | 5.878 |
| -232 Dec 23 | 3.22 | α Lib | 0 | - 1.94 | 5.749 |
| -232 Dec 24 | 3.24 | β Lib | 5.83 | 5.77 | 5.740 |
| -232 Dec 25 | 3.24 | β Sco | 5.00 | 3.97 | 5.762 |
| -232 Dec 26 | 3.26 | θ Oph | -2.50 | - 2.14 | 5.824 |
| -231 Jan 4 | 14.91 | η Psc | -4.17 | - 3.73 | 6.800 |
| -231 Jan 7 | 14.89 | α Tau | -2.83 | - 3.60 | 6.876 |
| -231 Jan 8 | 14.91 | ζ Tau | -5.67 | - 4.24 | 6.863 |
| -231 Jan 13 | 3.34 | α Leo | -9.17 | - 7.73 | 6.488 |
| -231 Feb 3 | 15.31 | α Tau | -7.50 | - 7.01 | 6.876 |
| -231 Feb 6 | 15.33 | α Gem | 0.42 | - 5.08 | 6.644 |

TABLE X.5 (Continued)

| Date | Greenwich Mean Time, hours | Reference Star | Observed $\Delta\lambda_e$, degrees | $\Delta\lambda_0$, degrees | $10^2 \times \frac{\partial(\Delta\lambda)}{\partial\nu_M'}$ |
|---|---|---|---|---|---|
| -231 Feb 15 | 3.09 | α Vir | 0 | 10.42 | 5.833 |
| -231 Mar 1 | 15.67 | α Ari | 7.50 | 6.10 | 7.148 |
| -231 Mar 3 | 15.69 | α Tau | 3.75 | 3.51 | 6.979 |
| -231 Mar 5 | 15.74 | γ Gem | 3.17 | 2.81 | 6.702 |
| -231 Mar 16 | 2.54 | ß Sco | -8.75 | - 8.89 | 5.713 |

The results of the calculations are summarized in Table X.5. The first column gives the date of the observation in the Julian calendar; the date in the Babylonian calendar can be found by reference to Table IV.4. The second column gives the hour assigned to the observation, in Greenwich mean time, and the third column gives the reference star used in the observation. The fourth column gives the observed value of $\Delta\lambda_e$, the fifth column gives the value of $\Delta\lambda_0$, and the sixth column gives the value of the partial derivative of $\Delta\lambda$ with respect to $\nu_M'$, multiplied by 100 for convenience in tabulating. Several specific comments need to be made about the table.

In two of the observations in Table IV.4, namely those of -567 April 22 (= SE -256 I 1) and -378 November 5 (= SE -67 VIII 10), the direction of the moon from the reference star is given but not the distance. These observations are omitted from Table X.5. Calculation shows that the directions given in Table IV.4 are correct.

In the observation of -567 June 24 (= SE -256 III 5), the name of the reference star was not one of the names commonly used for the Babylonian normal stars. The star γ Virginis fits the observation better than any other star that I can find, but, for safety, I have omitted this observation from Table X.5. The descriptive name given to the star in the record does not seem to fit γ Virginis; see Section III.11.

We noted that there is a recorded conjunction of Mars with ß Arietis on -378 October 27 (= SE -67 VIII 1). The text also refers to the moon, but it is not clear whether there was also a lunar conjunction with β Arietis and hence with Mars or not. The calculations make it clear that the moon was in a quite different part of the sky, and hence I have omitted this record from Table X.5.

Table IV.4 contains a conjunction of the moon with α Leonis in the month SE -256 XI, but the day of the month is missing. The correct day is probably SE -256 XI 11 (= -566 February 22/23), but we do not know from the record whether the observation was in the morning or evening. By inadvertence, I calculated the position of the moon for the evening and morning of SE -256 XI 14. Instead of taking the trouble to repeat the calculations, I have omitted this observation from Table X.5.

In the observations of SE 38 VII 3, SE 79 IX 8, and SE 79 XII 2, 4, and 6, the time of day is missing. In preparing Table IV.4 I guessed, on the basis of the day of the lunar month, that all the times were in the evening, and the calculations show that the guess was correct in each instance. The corresponding dates in the Julian calendar are -273 October 13, -232 December 8, and -231 March 1, 3, and 5. I have included these observations in Table X.5 without comment.

In several observations, all made when the moon and the reference star were close together in longitude, the text uses the terms "back" and "forward" rather than "east" and "west". In preparing Table X.5, I have taken back as equivalent to east and forward as equivalent to west. These correspond to the old usage of "precede" and "follow" in which "precede", for example, means to precede in the apparent diurnal motion and hence to be west of. As it happens, the occurrences of back and forward are approximately balanced, and the interchange of their definitions would not make a significant change in the results of the analysis.

If the reader will prepare a histogram of the differences between $\Delta\lambda_o$ and the observed $\Delta\lambda$, he will see that they follow a normal distribution as well as can be expected out to about 2°, but that there are considerably more cases than we expect in which the absolute value exceeds 2°. This suggests that there are two populations represented in the table. In one population, the errors are those of observation, with a standard deviation of about 1° or less. In the other population, the errors seem to be dominantly recording errors. There are about 10 such cases, with the exact number depending upon where we begin to make the distinction between the two populations.

In the observation of -231 February 15 (= SE 79 XI 17), the difference is more than 10°. This is so large that we expect an error in the date. The error cannot be in the correspondence of the Babylonian calendar, however, because we have a number of other observations from the same Babylonian month, and their dates are confirmed by the analysis. Thus there may have been an error in writing down the Babylonian date,† or there may have been an error in recording the star involved. The other errors are too small to allow for an error in either the date or the star; thus they must have been in the recording of the separation.

There does not seem to be any systematic error in recording, except for the records that involve η Tauri, on -273 October 21 (= SE 38 VII 15), -232 October 17 (= SE 79 VII 15), and -232 December 10 (= SE 79 IX 10).‡ In two of these records, namely those on -273 October 21 and -232 December 10, the difference in latitude as well as the difference in longitude is given. The differences in latitude agree well with the assignment of η Tauri as the star, but not the differences in longitude. All three observations seem to require a star about 4° west of η Tauri.

---

†That is, the correct date may have been SE 79 XI 16.

‡η Tauri also appears in Table X.5 in the observation of -566 March 15 (= SE -256 XII 2). In the original of this record, however, the star was not identified. The record refers to what we call the Pleiades, which I took as equivalent to η Tauri in longitude. In the three observations listed above, a specific star is named.

I can offer no explanation of this situation. It is unlikely that there were recording errors that were almost equal in all three cases, and this suggests that our modern identification of the star is wrong. Now the moon was full in two of the cases and it was well beyond the first quarter in the other case, so that a star used as a reference would have had to be rather bright. However, there are no stars as bright as the fifth magnitude in the right part of the sky. The observations cover a span of more than 40 years, so the reference star is not likely to have been a nova. Further, if it had been a nova, there is a good chance that its position would be marked by a strong radio source, and there are no significant radio sources in the right place.†

Since there definitely seems to be something wrong with the observations involving $\eta$ Tauri, and since I do not know how to correct the situation, I shall not use these observations in the inference of the lunar acceleration. I have included the observations in Table X.5, but I have omitted the values of $\partial(\Delta\lambda)/\partial\nu_M'$ in order to alert the reader to these observations. The reader may use these observations if he wishes. In order to do so, he may supply values of $\partial(\Delta\lambda)/\partial\nu_M'$ by interpolation between neighboring values. I omitted the observations involving $\eta$ Tauri entirely from Table X.4, which we used in finding the value of the ammat.

We now wish to find the value of $\nu_M'$ that gives the best fit of $\Delta\lambda$ from Equation X.9 to the observed values of $\Delta\lambda$ in Table X.5. If we omit all observations for which the residual is greater than 2°, on the assumption that these residuals come from recording errors rather than from random errors of observation, we find

$$\nu_M' = 3.4 \pm 1.7 \quad ''/\text{cy}^2 \, . \qquad \text{(X.10)}$$

The root-mean-square residual of these observations is 0°.85. This estimate applies at the mean epoch -280.

---

†I thank Mr. B.L. Gotwols of this Laboratory for making a search of the catalogues of radio stars and for supplying this information.

If we use all of the observations except those involving $\eta$ Tauri, which show strong evidence of systematic error, we find

$$\nu_M' = 4.3 \pm 3.6 \ \ ''/cy^2 . \qquad (X.11)$$

Thus, if we include the records that seem to contain recording errors, except for those that involve $\eta$ Tauri,† we do not change the central estimate of $\nu_M'$ by an important amount, but we more than double the estimate of the standard error.

I think that the evidence favors the omission of the records containing what I have called recording errors. Accordingly, I believe that Equation X.10 gives the best estimate of $\nu_M'$ that we can form from the Babylonian lunar conjunctions. The fact that the recorded positions agree with the calculated positions within less than 1° for most of the observations indicates that my interpretation of the times of observation is reasonably accurate. Therefore we are justified in using the same interpretation with the planetary records that will be analyzed in Chapter XIV.

## 6. Discussion

The residuals in both the moonrise/moonset times, in the lunar latitudes, and in the conjunctions are surprisingly large. In earlier studies of observations made with the naked eye and without clocks, we have found errors in position to be typically around 15′ or perhaps less. Here, however, the root-mean-square residual is 3° (12 minutes) in the moonrise/moonset data, and it is 0°.85 in the lunar conjunctions if we exclude the large errors, those which seem to come from errors in recording rather than from observation.

In the case of the moonrise/moonset data, we are dealing with an 'horizon' phenomenon which can be affected by clouds, dust, or other meteorological

---

†If we include these also, we find $\nu_M' = 7.2 \pm 3.8$. The difference between this value and that in Equation X.11 is nearly 1 standard deviation. This is not highly significant.

conditions. This, rather than inaccurate time keepers, may be the explanation of the size of the errors.

In the case of the lunar conjunctions, we do not have explicit statements of time. In the absence of stated times, I adopted the convention of assigning a time that is 45 minutes from sunrise or sunset, and the first impulse is to say that the residuals in the conjunction data come from errors in the conventional assignment of time. However, as we saw in Section X.4, the residuals in the measurements of latitude are also around a degree, and they are unaffected by errors in the time. Hence we must conclude that the standard error in a Babylonian measurement of angle, at the period of the data in question, is a substantial fraction (certainly more than half) of a degree.

This level of error sounds inconceivable if the Babylonians used any kind of instrument with a graduated circle. Tentatively, then, it seems as if the Babylonians merely looked at the sky and estimated angular distances, perhaps with the aid of a standard held, say, at arm's length.

On the basis of earlier analyses, I have suggested [Newton, 1972a] the following empirical formula for the variation of $D''$ between -700 and +650:

$$D'' = 9\frac{1}{8} + \frac{1}{4}\tau , \qquad (X.12)$$

in which $\tau$ is, as usual, time in centuries from 1900. The mean epoch of the moonrise/moonset times in Table X.3 is about -370, so that $\tau$ is about -22.7 for these data. This value of $\tau$ when used in Equation X.12 gives $D'' = 3.45$. This should be compared with the value $5.6 \pm 8.0$ (Equation X.5) that is found by analysis of the moonrise/moonset data.

The value from Equation X.12 rests mainly upon two classes of data. One class consists of early Chinese records of solar eclipses [Newton, 1970, Sections IV.3 and XIII.2]. This class suffers from the disadvantage that the places of observation are not known; it is generally assumed that the observations were made at many places and I used an approximate centroid of China in their analysis. The other class consists of observations made by astronomers before Ptolemy but that come to us only through the writing of Ptolemy. Any data that come to us through Ptolemy are somewhat suspect. It is gratifying to have a confirmation of Equation X.12 even though the standard deviation of the new estimate is rather large.

The mean epoch of the lunar conjunctions in Table X.5 is about -280. For this epoch, Equation X.12 gives $D'' = 3.7$. In order to compare with the analysis of the conjunctions, we need $\nu_M'$ rather than $D''$. If we use the tentative values $\nu_S' = 3.0$ from Equation V.2, we get $\nu_M' = 6.7$. For comparison, the best estimate that I adopted from the analysis of the lunar conjunctions (Equation X.10) is $3.4 \pm 1.7$. The disagreement is nearly two standard deviations. However, if the uncertainty in Equation X.12 is about 1.5, the disagreement is about equal to the sum of the estimated standard errors, and is not highly significant.

Since we have estimates of both $\nu_M'$ and $D'' = \nu_M' - \nu_S'$, we can in principle combine them in order to estimate $\nu_S'$. If we ignore the effect of the slight difference in epoch between the two estimates, we find

$$\nu_S' = -2.2 \pm 8.2 .$$

The large value of the standard deviation comes from the fact that we had to subtract two estimated values in order to find $\nu_S'$ in this way. This illustrates the central difficulty of trying to find $\nu_S'$ by analysis of lunar data.†

I shall not use the estimate given above. Instead, I believe that the best use to make of the estimates of $\nu_M'$ and $D''$ is to combine them into a single estimate of $\nu_M'$, using an independent estimate of $\nu_S'$ in the process. Tentatively, until we have

---

†See the discussion of this point in Section I.6.

analyzed the solar and planetary data, we can accept the estimate obtained from Hipparchus's equinox data in Section V.3. This estimate, from Equation V.2, is 3.0 ± 0.7. When combined with Equations X.5 and X.10, it yields

$$\left.\begin{aligned} \nu_M{}' &= 3.6 \pm 1.7 \ , \\ \nu_S{}' &= 3.0 \pm 0.7 \ . \end{aligned}\right\} \qquad \text{(X.13)}$$

We see that the lunar data do not affect the estimate of $\nu_S{}'$, and thus what they give us is the estimate 3.6 ± 1.7 for the lunar acceleration $\nu_M{}'$. I propose this as the best estimate of $\nu_M{}'$ that we can form from the lunar data. It applies at the approximate epoch -300.

For this epoch, the best estimate of D″ from other data [Newton, 1972a] is slightly less than 4, and the corresponding estimate of $\nu_M{}'$ is therefore slightly less than 7. The estimate of $\nu_M{}'$ from the Babylonian lunar data is thus considerably smaller than the estimate formed from other data.

# CHAPTER XI

# ANALYSIS OF THE ISLAMIC DATA

## 1. Tabular Values of Solar Position

The studies in Chapter X allow us to assign dates to some, but by no means all, of the Babylonian observations of the planets, and we must date the remaining observations by some other means. It was my original hope that the errors in the Babylonian planetary observations would be substantially less than the daily motions, at least for Mercury and Venus, and that this would let us date the planetary observations by comparing the recorded and calculated positions. Since calculating the positions requires an approximate knowledge of the planetary accelerations, I decided in Section IV.6 to analyze the solar and planetary data backward in time, in order to have estimates of the accelerations to use in studying the Babylonian data.

Unfortunately, the errors in the lunar observations found in the preceding chapter are more than 0°.5. If the errors in the planetary observations are as large, we shall not be able, in most cases, to date the records by means of the recorded positions; we shall then have to rely upon the theoretical dates given by Parker and Dubberstein [1956]. However, there may be a few cases in which we can find the dates from the recorded positions, and I shall therefore stay with the decision to analyze the data in reverse chronological order.

TABLE XI.1

TABULAR VALUES OF THE MEAN LONGITUDE OF THE SUN

| Date | Hour, local mean time | Place | Apparent $L_S$, degrees: from Newcomb's theory | Apparent $L_S$, degrees: Tabular value | $\nu_S'$, "/cy$^2$ |
|---|---|---|---|---|---|
| 832 Sep 18 | 23.17 | Damascus | 181.940 | 181.978 | 2.40 |
| 859 Sep 19 | 11.87 | Samarra | 181.908 | 182.003 | 6.32 |
| 882 Sep 18 | 1.12 | Raqqa | 181.901 | 181.968 | 4.66 |
| 1000 Nov 30 | 12.00 | Cairo | 254.729 | 254.766 | 3.30 |

Table VIII.4 has four values of the mean longitude of the sun that were obtained from the solar tables of four Islamic astronomers. The analysis of these values is summarized in Table XI.1, which should be almost self-explanatory. Three of the hours in Table VIII.4 were in apparent local time; they have been converted to local mean time by means of the equation of time. The coordinates of the places listed are given in Table IX.3. In order to find the values of $L_S$ "from Newcomb's theory", I calculated the mean longitude from his expression for the geometric mean longitude and subtracted 0°.0057 for the aberration.

The individual values of $\nu_S'$ in Table XI.1 lead to the estimate

$$\nu_S' = 4.17 \pm 0.86 \ ''/cy^2 . \qquad \text{(XI.1)}$$

The corresponding epoch can be taken as 890.

It is interesting to calculate the precision of the tabular values of $L_S$ that appear in the table. If we add $2''.1\tau^2$ to Newcomb's values of $L_S$ and subtract the results from the Islamic tabular values, we have the best estimates we can form of the errors in the Islamic values. The standard deviation of the errors estimated in this way is 0°.026 = 1′.6. This is the distance that the sun moves in about 38 minutes.

Thus the tabular values of $L_S$ are more precise than measurements of the equinoxes, as we hoped they would be. In earlier work [Newton, 1970, Table II.3, p. 24], I used nine Islamic observations of equinoxes and one observation of a solstice. The standard deviation of the nine observed equinox times is 1.43 hours = 86 minutes. This is slightly more than twice the standard deviation of the $L_S$, and thus we are justified in giving four times the weight to a value of $L_S$ that we give to a measurement of an equinox.

2. Equinoxes and Solstices

Table VII.1 contains 17 observations of equinoxes

of solstices made by Islamic astronomers. The three listed under the dates EY 201 V 25, EN 1607 IX 2, and EN 1630 IX 8 are involved in preparing Tables VIII.4 and XI.1, which deal with values of the mean longitude $L_S$, and they should not be used again. The observation listed under EY 199 V 25, with the place given as "Damascus?", is given a reliability of 0 because there is some question about whether it is an observed result or whether it was calculated from the observation at Baghdad on the same date. There is no point in continuing with this entry, and we are left with 13 observations to analyze. The observations fall naturally into four main groups that require different methods of handling.

The first group contains the first five observations made at Baghdad, with dates EN 1577 VIII 25, EY 199 V 25, EY 199 XI 19, EY 200 V 25, and EY 200 XI 19. These are presumably part of the observations made under Yahya, sometimes called the Mumtahan observations, that were used in preparing the zijes of Yahya and Habash (see Section VIII.5). There are three autumnal equinoxes, of which the first has a weight of 0.5 rather than 1, and there are two fully weighted vernal equinoxes. Since the vernal and autumnal equinoxes are almost balanced in number, the simplest procedure is to start by analyzing the vernal and autumnal ones separately, and then to average the individual results, giving equal weights to each even though the numbers of observations differ slightly.

If we do this, we do not need to worry about biases in the data. We can take the observed solar longitude to be 0° at a vernal equinox and 180° at an autumnal one.

The second group contains the single observation of a summer solstice. There is no obvious way to bias the time of a solstice, so we can take the observed longitude to be 90° at the solstice.

The remaining observations are all of the autumnal equinox, and thus they form a severely biased sample. Obvious sources of bias include refraction, parallax, and an error in the orientation of the astronomer's equatorial plane, which is the same as an error in the direction of the local vertical. In order to estimate the effects of these biases, we must have a model of the observing methods used.

al-Biruni [1025] outlines a number of methods for finding the latitude, which is equivalent to establishing an equatorial plane, or at least to establishing the intersection of the meridian plane with the equator. They include measuring the maximum and minimum zenith distances of stars near the pole, measuring the zenith distance of the sun as a function of its azimuth on a given day, and measuring the zenith distances of the summer and winter solstices.†

In four cases, we know from explicit statements that the astronomers involved used the last method. We can readily analyze the bias in this method. Let $Z_a$ denote a measured zenith distance and let $Z_c$ denote the correct distance.‡ We assume that there is a source of error $\eta$ in the measurement, and that $\eta$ is a known function $\eta(Z_a)$ of the measured zenith distance. We adopt the sign convention for $\eta(Z_a)$ by taking

$$Z_c = Z_a + \eta(Z_a) . \qquad \text{(XI.2)}$$

Now let $Z_a(S)$, $Z_a(W)$, and $Z_a(E)$ denote the measured zenith distances of the summer solstice, the winter solstice, and of either equinox, respectively. The observer took his measured latitude $\phi_a$ to be

---

†This was usually done by measuring the zenith distance of the sun at local apparent noon on several successive days near each solstice. If the time found for the solstice was not at one of the noons, the astronomers could readily make the needed corrections to find the position of the solstice from the position of the sun.

‡The following discussion is generalized slightly from the discussion in an earlier work [Newton, 1972b].

$$\phi_a = \tfrac{1}{2}[Z_a(S) + Z_a(W)] . \qquad (XI.3)$$

When we correct for the error $\eta$, the corrected latitude $\phi_c$ is

$$\phi_c = \phi_a + \tfrac{1}{2}[\eta(Z_a(W) + \eta(Z_a(S))] . \qquad (XI.4)$$

The observer thought that the equinox occurred when $Z_a = \phi_a$; that is, he thought that the declination of the sun was zero at this time. However, the correct declination $\delta_c(E)$ at this time is $\phi_c - Z_c$, which is

$$\delta_c(E) = \tfrac{1}{2}[\eta(Z_a(W)) + \eta(Z_a(S))] - \eta(Z_a(E)). \qquad (XI.5)$$

We see from Equation XI.5 that an error in aligning the observing instrument with the vertical does not cause an error in the equinox, if the error is constant. However, there could be an error in the equinox if the alignment can change with the seasons, because of temperature variations, for example.

The solar parallax [Newton, 1972b, p. 118] causes $\delta_c(E)$ to be about $0^\circ.0001$. This is too small to concern us much, but we may as well include it. We do have to be concerned about parallax, but in a different way. In al-Biruni's time, astronomers still thought that the solar parallax was about $0^\circ.05$ ($3'$) rather than about $0^\circ.0024$ (about $8''.8$). If the astronomers had 'corrected' their observations according to a parallax of $0^\circ.05$, they would have introduced a serious error which we should have to undo. Fortunately, al-Biruni [1025, p. 80] says specifically that all of the data in Table VII.1 are the original observations, with no emendation for parallax.

The remaining obvious systematic error is refraction. Earlier, I calculated the refraction separately for summer solstice, winter solstice, and equinox observations at each observer's location, using standard tables for the variation of temperature and barometric pressure at each location in order to do so. I then applied Equation XI.5 in order to calculate a value of $\delta_c(E)$ for each equinox. Since all the Islamic observations, with one exception, were made at places with similar climates, it turns out that we can use $\delta_c = 0°.0040$ for all of the observations except that at Ghazni. Because of its high altitude, the value of $\delta_c$ there is only 0°.0028.

For the resultant of all the biasing errors, then, we can use $\delta_c = 0°.0029$ for Ghazni and $\delta_c = 0°.0041$ for all other sites involved. This means that the longitude of the sun should be taken as 179°.9928 for the autumnal equinox observed at Ghazni and as 179°.9898 for all the other autumnal equinoxes.†

In the earlier work, I applied this method to all of the equinoxes listed in Table VII.1. Further reflection leads me to doubt that this is the best procedure. There are many ways to establish the latitude, and it is unsafe to assume that the method of averaging the solstices was used unless this is specifically stated. If some other method were used, we would have to know the method and estimate the refraction correction for that method. Let $\delta_c(O)$ denote the correction to the solar declination that is needed because of the method used to find the latitude. Then the correct declination of the sun at the measured time of an equinox would be $\delta_c(O) - \eta(Z_a(E))$. $\eta(Z_a(E))$ is about 0°.0100, except at Ghazni, and thus we should subtract this quantity instead of adding 0°.0040, as Equation XI.5 requires for refraction. Thus using Equation XI.5 under the wrong circumstances introduces an error of about 0°.0140 into the declination of the sun, and this changes the time of a solstice by nearly an hour.

Of the last seven equinoxes in Table VII.1, we know from direct statements that the method of averaging the solstices was used for bin 'Ismat's observation on EN 1636 IX 9 at Balkh, for as-Sufi's

---

†That is, the correction in longitude is $\delta_c/\sin \varepsilon$.

observations on EN 1718 IX 29 and EN 1719 IX 29 at Shiraz, and al-Biruni's observation on EN 1767 X 11 at Ghazni. I shall apply Equation XI.5 to these observations, and they will constitute the third group. Observations in this group will receive full weight.

The fourth group in Table VII.1 consists of the anonymous observation on EY 213 V 28 at Baghdad, al-Makki's observation on EY 220 V 30 at Neyshabur, and al-Wafa's observation on EN 1722 IX 30 at Baghdad. We do not know how the equatorial plane was established for these observations, and I shall give them a weight of zero. As a matter of interest, however, I shall analyze them by using $\delta_C$ = -0°.0100. This is equivalent to taking 180°.0217 for the longitude of the sun.

TABLE XI.2

ISLAMIC EQUINOX AND SOLSTICE OBSERVATIONS

| Date | Hour, mean time at Greenwich | Julian Day Number -2 000 000 | Longitude of Sun, degrees from Newcomb's theory | observed value[a] | $\nu_S'$ "/cy$^2$ |
|---|---|---|---|---|---|
| 829 Sep 19 | 3.707 | 24 112 | 180.0015 | 180 | - 0.09[b] |
| 830 Sep 19 | 9.904 | 24 477 | 180.0166 | 180 | - 1.05 |
| 831 Mar 17 | - 0.832 | 24 656 | 359.9146 | 0 | + 5.38 |
| 831 Sep 19 | 15.904 | 24 842 | 180.0274 | 180 | - 1.73 |
| 832 Mar 16 | 5.165 | 25 021 | 359.9263 | 0 | + 4.65 |
| 832 Jun 18 | - 2.979 | 25 115 | 90.0362 | 90 | - 2.28 |
| 844 Sep 18 | 18.271 | 29 590 | 179.9737 | 180.0217 | + 3.11[c] |
| 851 Sep 19 | 7.945 | 32 147 | 179.8531 | 180.0217 | +11.05[c] |
| 888 Sep 18 | 9.011 | 45 661 | 179.9344 | 179.9898 | + 3.90 |
| 970 Sep 18 | 7.363 | 75 611 | 180.0051 | 179.9898 | - 1.27 |
| 971 Sep 18 | 14.361 | 75 976 | 180.0501 | 179.9898 | - 5.04 |
| 974 Sep 18 | 5.906 | 77 072 | 179.9651 | 180.0217 | + 4.76[c] |
| 1019 Sep 18 | 2.303 | 93 508 | 179.9169 | 179.9928 | + 7.05 |

[a]The basis for these values is described in the accompanying text.

[b]This value receives a weight of 0.5.

[c]These values receive weights of 0.

The analysis of the equinox and solstice observations is summarized in Table XI.2. The third column in the table gives the Julian day number at noon on the dates listed in Table VII.1; the day number serves to tie the tables together. The first column gives the date in the Julian calendar and the second column gives the time of the observation in Greenwich mean time. In order to find the values of Greenwich mean time, I assumed that the times listed in Table VII.1 are in apparent local solar time. I first applied the equation of time, as calculated from the ephemeris of the sun, in order to get local mean time, which I then converted to the meridian of Greenwich by using the longitudes listed in Table IX.3 for the places of observation.

The fourth column gives the longitude of the sun as calculated from Newcomb's theory, less the amount of aberration. The fifth column gives the longitude of the sun that is implied by the observation, in accordance with the preceding discussion, and the last column gives the value of $\nu_S'$ that results from comparing the observed and calculated longitudes.

In the ultimate use of the results from Table XI.2, I shall use each observation separately. However, it is interesting to examine the results obtained from the various groups of observations that were described above.

The first group contains the first five observations in Table XI.2. The values of $\nu_S'$ show that there is a bias in the setting of the equator. All observations of the autumnal equinox in this group give negative values of $\nu_S'$, while both observations of the vernal equinox give positive values. The mean value from the autumnal observations is -1.13 and the mean from the vernal observations is +5.01. If we treat these equally, the result is

$$\nu_S' = 1.94 \quad ''/\text{cy}^2 . \tag{XI.6}$$

There is no convenient way to attach an error estimate to this value. Since the bias in the equator is effectively removed by the method of analysis, this value is entitled to as much consideration as one of the values in Table XI.1. We see that the value in Equation XI.6 is slightly outside the range of values in Table XI.1, but not by a serious amount.

The second group contains the single observation of a solstice in Table XI.2. The value of $\nu_S'$ for this observation is -2.28. The change in time that is needed in order to bring this value into line with other values is about 1 hour. Since this is a plausible error for a single observation of a solstice, the deviation of this value from the others is not statistically significant.

The third group consists of the observations of the autumnal equinox on 888 September 18, 970 and 971 September 18, and 1019 September 18. This group yields two positive values of $\nu_S'$ and two negative ones. The value of $\nu_S'$ from this group is

$$\nu_S' = 1.16 \pm 2.33 \quad ''/\text{cy}^2 . \qquad \text{(XI.7)}$$

This does not differ significantly from the other values.

The fourth group consists of the observations on 844 September 18, 851 September 19, and 974 September 18. This group will receive a weight of zero. The estimate of $\nu_S'$ from this group alone is

$$\nu_S' = 6.31 \pm 2.42 \quad ''/\text{cy}^2 . \qquad \text{(XI.8)}$$

The difference between this value and the other ones is probably not significant in view of the small sample size, which suggests that the method of treating refraction may be valid. Nonetheless, it is safer to ignore the estimate in Equation XI.8, and the corresponding observations in Table XI.2, for the reasons that have been outlined.

One interesting point about the first group should be noticed. As we noted, there is a bias between the autumnal and vernal equinoxes when we ignore refraction. If we correct for refraction by using Equation XI.5, we increase the bias between the two kinds of equinox, but we decrease it if we take $\delta_C = -0^\circ.0100$. This suggests that the observers at Baghdad established their equator by using observations of the stars, or by using some other method that does not involve averaging solstice observations.

The standard deviation of the values of $\nu_S'$ in Table XI.2 about their mean is 4.80. Since the observations were made about 10 centuries before 1900, the standard error in the corresponding values of the solar longitude is about 240″, or 4′. This in turn corresponds to an error of about 1.6 hours in the measurement of the times. This agrees well with the estimate stated at the end of Section XI.1.

In order to combine the values of $\nu_S'$ in Table XI.2 with values from other observations, it is necessary to assign weights, and I shall assign them through the medium of assigned standard deviations. Since the assignment of weights or of standard deviations does not affect the final results in a critical way, I can be rather arbitrary in this matter. The standard deviation of a value in Table XI.1 about the mean is 1.71, and I shall use this as a starting point.

I shall treat the first group in Table XI.2 as the equivalent of a single tabular value from Table XI.1. Since this group contains four observations with a reliability of 1 (see Table VII.1) and one observation with a reliability of $\frac{1}{2}$, the standard deviation of a single observation should be $1.71\sqrt{4.5} = 3.63$ when the reliability is 1 and it should be $3.63\sqrt{2}$ when the reliability is $\frac{1}{2}$.

For the remaining observations in Table XI.2 that will be used, I shall assume that the standard deviation is twice that of a tabular value from Table XI.1, namely 3.42.

## 3. Meridian Elevations of the Sun

Table VII.2 contains 46 observations of the meridian elevation of the sun. Nine of the observations were assigned a reliability of zero because of problems connected with the texts in which they are found. This leaves 37 observations to be analyzed.

We know from explicit statements that many of the observations were made by means of a quadrant oriented to lie in the plane of the meridian. This tells us, among other things, that the time of each observation is local apparent noon. It seems unlikely that the error in aligning the quadrant with the meridian is as large as 1°. Even if we assume that it is this large, the corresponding error in the time is only 4 minutes. The effects of such a small time error are negligible compared with the effects of other errors, and we are justified in neglecting the error in the time. It is plausible that all of the measurements were made with a quadrant, which had been a well-established instrument since the time of Ptolemy and perhaps since the time of Hipparchus.

The observations in Table VII.2 are subject to the set of systematic errors that was discussed in the preceding section. This set includes the north-south component of error in the direction of the local vertical, solar parallax, and refraction.

There are many instances in Table VII.2 in which several measurements were made by the same observer in the same place within a short interval of time. It is plausible to assume that measurements having this property were made on the same instrument and that they thus have the same error in the direction of the vertical. In an earlier study [Newton, 1972b], I identified 13 groups of observations, which are distinguished by an index $j$. For the jth group, suppose that the vertical reference established by the observing instrument points south of the true vertical by an angle $v_j$. Then the correct elevation angle is equal to the observed angle plus the error $v_j$.

In the earlier study, I divided the observations made by as-Sufi at Shiraz into two groups, one for observations at high elevation angles and one for observations at low angles, because of a suspicion that his high elevations were not consistent with his low ones. He might have done a poor job of graduating his quadrant, for example. Later in that study, I also tested the effect of putting all of as-Sufi's observations into one group and found no significant change. Here I shall put all of his observations into one group, leaving 12 groups.

I similarly divided al-Khujandi's observations at Rayy into two groups. Here, however, we have the explicit statement by al-Khujandi that his instrument had different errors at high and low elevations; see Section VII.2. Therefore I shall continue to divide al-Khujandi's observations into two groups.

I also assigned one group to the observations that al-Kuhi claimed† to have made at Baghdad on EY 357 III 30 (= 988 June 16), but I used these observations with low weights because of the suspicious circumstances. Further reflection leads me to give these observations a weight of zero, because they almost surely constitute a hoax. Deletion of this group leaves 11 groups from Table VII.2 to be analyzed.

The values of the $v_j$ will be found as a by-product of the data analysis. The values found in this way do not necessarily represent errors made by the observers; they may only represent errors in the values of latitude that I use, which are taken from Table IX.3. These are nominal values that approximately represent the centers of the places in question. The observer could readily have been at a slightly different latitude.

Solar parallax is only a few seconds of arc and is thus hardly large enough to matter in the analysis of naked-eye observations. However, it can be included by a simple calculation, and I have therefore corrected the observed angles for parallax. Fortunately, we have explicit statements that none of the observers involved in Table VII.2 had 'corrected' his measurements by using the

---

†See Section VI.3.

seriously wrong value of 3′ for the horizontal equatorial parallax, and we need to consider only the correct effect of parallax.

In correcting for refraction, I used the tables given by Allen [1962], which include the dependence of refraction upon surface temperature and pressure. I corrected each observation separately [Newton, 1972b, Section 7], using tables of average temperature and pressure for each place involved for each month of the year; details can be found in the reference.

In the analysis of the observations in Table VII.2, I start by correcting the stated elevation angle for refraction and parallax. Use of this corrected angle, in combination with the latitude of the observing sites listed in Table IX.3, gives an 'observed value' of the solar declination. Let $\delta_{ei}$ denote the declination found in this way for the *i*th observation, and suppose that the *i*th observation is part of the *j*th group. Then the 'correct declination' is given by

$$\text{'correct declination'} = \delta_{ei} - v_j . \qquad \text{(XI.9)}$$

A value of solar time must now be associated with each observation. The date of each observation is given in Table VII.2, and the time of day of each observation is local apparent noon. The time of day can be converted into Greenwich mean time (solar time) by using the equation of time and the longitudes listed in Table IX.3. It does not seem necessary to give any details. The symbol $(ST)_i$ will be used to denote the solar time of the *i*th observation.

It is simpler to analyze the data in terms of ephemeris time than in terms of solar time. An analysis based upon ephemeris time is legitimate since the data involve the sun but not the planets. We want to find the ephemeris time $(ET)_i$ at which the sun had the correct declination given by Equation XI.9, and we define $(\Delta t)_i$ for the *i*th observation by

$$(\Delta t)_i = (ET)_i - (ST)_i . \qquad \text{(XI.10)}$$

Let $\delta_{0i}$ and $\dot{\delta}_{0i}$ be the values of the declination and of its time derivative that we calculate from Newcomb's theory† on the assumption that $(\Delta t)_i = 0$. This is equivalent to assuming that $(ET)_i = (ST)_i$. The value $\delta_{0i}$ is not the value to compare with the correct declination from Equation XI.9, however, because it has been calculated with a wrong value of ephemeris time. It has been calculated with the value $(ST)_i$, whereas the value of ephemeris time that we should use is $(ST)_i + (\Delta t)_i$, as we see from Equation XI.10. The value to be compared with the one from Equation XI.9 is $\delta_{0i} + \dot{\delta}_{0i}(\Delta t)_i$.

The symbol $\tau$ is used in this work to denote solar time, measured in Julian centuries, from Newcomb's fundamental epoch.‡ Since we expect $(\Delta t)_i$ to be approximately proportional to $\tau^2$, it is convenient to introduce a parameter $\alpha$ defined by

$$(\Delta t)_i = \alpha \, \tau_i^{\,2} \, , \qquad \text{(XI.11)}$$

in which $\tau_i$ is the value of $\tau$ for the ith observation.

Now the declination $\delta_{0i} + \dot{\delta}_{0i} \, \alpha \, \tau_i^{\,2}$ can be compared with the 'correct declination' from Equation XI.9, and the difference between the two is the ith residual $r_i$. Thus we have

$$(\dot{\delta}_{0i} \tau_i^{\,2})\alpha + v_j = (\delta_{ei} - \delta_{0i}) + r_i \, . \qquad \text{(XI.12)}$$

By the standard least-squares procedure, we find the values of $\alpha$ and of the $v_j$ that minimize the sum of the $r_i^2$ for all the observations.#

The information needed to apply Equation XI.12

---

†The calculation must include the effect of solar aberration.

‡See Section I.7.

#The method of analysis just outlined is the method that I used in the earlier study of the meridian elevations [Newton, 1972b].

## TABLE XI.3

## ISLAMIC OBSERVATIONS OF THE MERIDIAN ELEVATION OF THE SUN

| Date | Julian Day Number -2 000 000 | Solar Declination degrees observed | Solar Declination degrees Newcomb | $\dot{\delta}_O T^2$ deg cy$^2$ per day | Weight |
|---|---|---|---|---|---|
| Group 1 at Daghdad | | | | | |
| 832 Aug 3 | 25 160.8792 | 16.4463 | 16.5686 | -32.210 | 1 |
| Group 2 at Damascus | | | | | |
| 832 Jun 16 | 25 112.8982 | 23.5674 | 23.5853 | + 1.134 | 1 |
| 832 Jun 17 | 25 113.8983 | 23.5730 | 23.5917 | + 0.346 | 1 |
| 832 Jun 18 | 25 114.8985 | 23.5724 | 23.5913 | - 0.441 | 1 |
| 832 Aug 3 | 25 160.9017 | 16.5309 | 16.5622 | -32.210 | 1 |
| 832 Dec 16 | 25 295.8995 | -23.6048 | -23.5913 | - 0.509 | 1 |
| 832 Dec 17 | 25 296.8999 | -23.6054 | -23.5918 | + 0.394 | 1 |
| 832 Dec 18 | 25 297.9002 | -23.5971 | -23.5844 | + 1.297 | 1 |
| Group 3 at Samarra | | | | | |
| 857 Jun 17 | 34 244.8774 | 23.5811 | 23.5861 | + 0.375 | 1 |
| 857 Dec 16 | 34 426.8785 | -23.5903 | -23.5846 | - 0.562 | 1 |
| 859 Dec 17 | 35 157.8786 | -23.5903 | -23.5847 | - 0.122 | 1 |
| Group 4 at Baghdad | | | | | |
| 868 Dec 16 | 38 444.8771 | -23.6064 | -23.5871 | - 0.279 | 0.1 |
| 869 Jun 17 | 38 627.8759 | 23.5813 | 23.5873 | + 0.294 | 0.1 |
| Group 5 at Balkh | | | | | |
| 888 Dec 14 | 45 747.8142 | -23.4917 | -23.5655 | - 1.819 | 1 |
| 889 Jun 17 | 45 932.8137 | 23.6641 | 23.5850 | + 0.212 | 1 |
| Group 6 at Isfahan | | | | | |
| 960 Oct 6 | 71 976.8474 | - 7.3257 | - 7.2344 | -33.749 | 0.1 |
| Group 7 at Shiraz | | | | | |
| 969 Dec 15 | 75 333.8541 | -23.5498 | -23.5685 | - 0.599 | 0.5 |
| 969 Dec 16 | 75 334.8545 | -23.5664 | -23.5714 | + 0.087 | 0.5 |
| 969 Dec 17 | 75 335.8548 | -23.5498 | -23.5665 | + 0.773 | 0.5 |
| 970 Jun 16 | 75 516.8534 | 23.6115 | 23.5688 | + 0.492 | 0.5 |
| 970 Jun 17 | 75 517.8536 | 23.6156 | 23.5710 | - 0.106 | 0.5 |
| 970 Jun 18 | 75 518.8537 | 23.6115 | 23.5664 | - 0.703 | 0.5 |
| 970 Dec 16 | 75 699.8544 | -23.5664 | -23.5708 | + 0.082 | 0.5 |
| Group 8 at Rayy | | | | | |
| 994 Jun 16 | 84 282.8566 | 23.5422 | 23.5664 | + 0.355 | 1 |
| 994 Jun 17 | 84 283.8567 | 23.5422 | 23.5673 | - 0.212 | 1 |
| Group 9 at Rayy | | | | | |
| 994 Dec 14 | 84 463.8568 | -23.5452 | -23.5530 | - 1.267 | 1 |
| 994 Dec 17 | 84 466.8579 | -23.5460 | -23.5638 | + 0.681 | 1 |

TABLE XI.3 (Continued)

| Date | Julian Day Number -2 000 000 | Solar Declination degrees observed | Solar Declination degrees Newcomb | $\dot{\delta}_O \tau^2$ deg cy$^2$ per day | Weight |
|---|---|---|---|---|---|
| Group 10 at Jurjaniya | | | | | |
| 1016 Jul 11 | 92 343.8364 | 21.5119 | 21.4439 | -13.010 | 1 |
| 1016 Aug 10 | 92 373.8357 | 14.0431 | 14.0119 | -24.791 | 1 |
| 1016 Sep 1 | 92 395.8318 | 6.2901 | 6.2751 | -29.578 | 1 |
| 1016 Sep 2 | 92 396.8316 | 5.9065 | 5.8953 | -29.710 | 1 |
| 1016 Sep 9 | 92 403.8301 | 3.2390 | 3.1949 | -30.458 | 1 |
| 1016 Sep 17 | 92 411.8283 | 0.0540 | 0.0506 | -30.846 | 1 |
| 1016 Oct 2 | 92 426.8252 | - 5.7994 | - 5.8459 | -30.235 | 1 |
| 1016 Oct 3 | 92 427.8251 | - 6.1829 | - 6.2329 | -30.126 | 1 |
| 1016 Dec 7 | 92 492.8312 | -23.2346 | -23.2949 | - 5.110 | 1 |
| Group 11 at Jayfur | | | | | |
| 1018 Oct 14 | 93 168.7979 | -10.5109 | -10.1872 | -28.373 | 1 |

is summarized in Table XI.3. The first column gives the date of the observation in the Julian calendar rather than in the Persian or Muslim calendars. The second column gives the Greenwich mean time of the observation by means of the Julian day number and fraction thereof. In order to connect Table XI.3 with Table VII.2, round the day number in Table XI.3 upward to the next integer. The result is the day number that identifies an observation in Table VII.2. The third, fourth, and fifth columns give the values of $\delta_{ei}$, $\delta_{Oi}$, and $\dot{\delta}_{Oi}\tau_i^2$, respectively, while the last column gives the weight attached to each observation.

The important quantity that comes out of the analysis is the parameter $\alpha$, which turns out to be 0.000 951 ± 0.000 301. By means of a tedious but straightforward computation, we find that $\alpha$ is related to $\nu_S'$ by†

†The only tricky part of the analysis is deciding whether to use an angular velocity of the earth with respect to the sun or to the stars at certain points. In the earlier study of these data, [Newton, 1972b], I used a relation equivalent to $\nu_S' = 7116\alpha$, which differs from the value above by the factor 1.0027. I believe that the value given here is correct, but the difference is not important at the present stage of accuracy.

$$\nu_S' = 7097\alpha \ . \qquad \text{(XI.13)}$$

Hence we find

$$\nu_S' = 6.75 \pm 2.13 \ \ ''/\text{cy}^2 \ . \qquad \text{(XI.14)}$$

This is considerably larger than most estimates of $\nu_S'$ made in the earlier sections of this chapter, but the difference is not highly significant in view of the large standard deviation given in Equation XI.14.

In addition to the value of $\alpha$, the analysis of Table XI.3 yields the values of thirteen v's. If we attempted to combine individual observations from Table XI.3 with other data in the concluding analysis of all the data, it would be necessary to carry along all the v's. Since this would complicate the analysis considerably, I shall use Equation XI.14 rather than the individual observations; this contrasts with the practice to be followed with the observations in the preceding sections.

The values of the v's, and some other information that can be obtained from Table XI.3, are not connected with the main purposes of this work. However, they have some interest, and I shall digress for a section in order to consider them. The reader should note that three of the groups in Table XI.3 contain only one observation each. These groups do not contribute to the value of $\alpha$. $\alpha$ is determined only by the groups that contain more than one observation. An isolated observation gives only the corresponding value of v.

The residual for an isolated observation is necessarily zero, since the value of v is free to adjust exactly to the observed declination. The root-mean-square residual for the other observations is 0°.0121, slightly less than 1′. This may be misleading, however. Most of the observations in Table XI.3 were made near a solstice. All those that were not made near a solstice were made between the summer solstice and the winter solstice, and hence they were made when the declination was decreasing. If the sample had been balanced between autumn and spring observations, we would have a more reliable value of $\nu_S'$, but we would probably find larger residuals.

TABLE XI.4

NORTH-SOUTH COMPONENT OF THE ERRORS IN THE LOCAL VERTICAL

| Group | Place | v, degrees |
|---|---|---|
| 1 | Baghdad | -0.0917 |
| 2 | Damascus | -0.0140 |
| 3 | Samarra | -0.0053 |
| 4 | Baghdad | -0.0127 |
| 5 | Balkh | +0.0772 |
| 6 | Isfahan | -0.0592 |
| 7 | Shiraz | +0.0253 |
| 8 | Rayy[a] | -0.0247 |
| 9 | Rayy[b] | +0.0131 |
| 10 | Jurjaniya | +0.0603 |
| 11 | Jayfur | -0.2967 |

[a]Summer observations.
[b]Winter observations.

## 4. Errors in the Vertical; Secular Change of the Obliquity

The values of the $v_j$, which are the north-south components of the error in the local vertical (including any error in the latitude assigned in Table IX.3), are listed in Table XI.4. The values listed here differ slightly from those found in the earlier study [Newton, 1972b] because of slight changes in the method of analysis.

Most of the values are a few minutes of arc, which is the size that we should expect. This fact confirms that the corresponding latitudes from Table IX.3 are nearly correct, except for Jayfur. I was not able to identify Jayfur, which al-Biruni describes as a "village adjacent to Kabul", and so I used the coordinates of Kabul in the table.

The value of v found for Jayfur suggests that it was at least 30 kilometers away from Kabul.

The values of v for summer and winter readings at Rayy differ by only 0°.0378, slightly more than 2′. It seems that the deformation in the meridian instrument was not as great as the observer al-Khujandi feared.

The difference between meridian elevations of the sun at the summer and winter solstices is twice the obliquity $\varepsilon$ of the ecliptic, provided that we have corrected the measured elevation angles for parallax and refraction. Thus we can use the data in Table XI.3 for studying the secular change in the obliquity, if we impose two limitations upon the data. First, we must use only data from the groups in Table XI.3 that contain observations near both solstices. Second, we should not use observations made more than a few days away from a solstice. If we use observations made when the declination is changing rapidly with solar longitude, the uncertainty in the calculated declination caused by the uncertainty in $\nu_S'$ will cause a large uncertainty in $\varepsilon$. An uncertainty in $\nu_S'$ and hence in the longitude has a negligible effect on the declination near the solstices.

When we impose these restrictions on the data in Table XI.3, we find that we can use 6 observations from Group 2 and all the observations from Groups 3, 4, 5, and 7, but we cannot use any other data from the table. I have described the analysis of the surviving data in Section 12 of the earlier study [_Newton_, 1972b], and it does not seem necessary to give the details here. The result is

$$\dot{\varepsilon} = -47''.9 \pm 2''.0 \text{ per century} . \qquad \text{(XI.15)}$$

Newcomb's value [_Explanatory Supplement_, 1961, p. 98] is

$$-46''.845 \text{ per century} , \qquad \text{(XI.16)}$$

for the linear term only; the higher order terms are negligible in this discussion. Thus the data from Table XI.3 confirm Newcomb's theory of the obliquity to rather high accuracy.

The mean epoch of the data in Table XI.3 is about 880, and the value of $\varepsilon$ inferred from the table for this epoch is

$$\varepsilon = 23^\circ\ 35'\ 14'' \pm 21'' \ . \tag{XI.17}$$

The value from Newcomb's theory is $23^\circ 35' 3''.54$. So far as I know, the value in Equation XI.17 is the earliest accurate value of the obliquity.

## 5. Miscellaneous Solar Observations

Table VII.3 contains nine observations of the sun from Islamic sources. Six of the observations are solar longitudes and the other three are elevation angles measured at some azimuth other than due south.

In treating the elevation angles, I first corrected them for refraction and parallax in the manner described in Section XI.3. I then assumed that the north-south component of error in the direction of the vertical is the value given in Table XI.4 and that the east-west component is zero. The latter is not a particularly good assumption, but there are not enough data to let us deduce an east-west component. By making these assumptions, I can calculate the elevation angle with respect to the correct horizon, and thence I can calculate the declination of the sun.

The analysis of the data in Table VII.3 now follows the same lines as the analysis in Section XI.3. We have a solar position which will now be denoted by $\theta$; it may be either a longitude or a declination. $\theta_{ei}$ will denote the value of $\theta$ deduced from the observation, $\theta_{Oi}$ will denote the value of $\theta$ calculated from Newcomb's theory, and $\dot{\theta}_{Oi}$ will denote the change of $\theta$ in one day. I then use Equation XI.12, with $\delta$ changed to $\theta$ and with $v = 0$. The contribution of v has been absorbed into $\theta_{ei}$ in the manner described above for the declinations, and v makes no contribution to the longitudes.

By the use of Equation XI.12, we thus get a value of $\alpha$ from each observation, which is converted into $\nu_S{}'$ by means of Equation XI.13.

In Section VII.3, I decided to use 2′ as the a priori estimate of the standard deviation for the miscellaneous observations. In addition, I decided to assign a reliability of 0.1 to the observation on EH 349 VIII 11 because of difficulties connected with the text. The combination of this reliability with a standard deviation of 2′ is equivalent to using a reliability of 1 and a standard deviation of 6″.32.

TABLE XI.5

MISCELLANEOUS OBSERVATIONS OF THE SUN

| Date | Julian Day Number -2 000 000 | Solar Position degrees | | $\dot{\theta}_O T^2$ deg cy$^2$ per day | $\nu_S{}'$ ″/cy$^2$ |
|---|---|---|---|---|---|
| | | observed | Newcomb | | |
| 832 Jan 31 | 24 975.4794 | 315.0000 | 314.9191 | 114.84 | + 5.00 ± 3.09 |
| 832 May 1 | 25 067.2480 | 45.0000 | 45.0524 | 109.39 | - 3.40 ± 3.24 |
| 832 Aug 4 | 25 161.4138 | 135.0000 | 135.0786 | 110.03 | - 5.07 ± 3.23 |
| 960 Oct 6 | 71 976.8474 | 198.6167 | 198.3521 | 88.16 | + 21.30 ± 12.72 |
| 1016 Jun 15 | 92 318.0036 | 23.6015[a] | 23.5657[a] | 0.43 | +113.88 ±824.42 |
| 1016 Sep 2 | 92 396.8316 | 165.1833 | 165.1188 | 76.44 | + 5.99 ± 4.65 |
| 1016 Oct 2 | 92 426.8252 | 194.8500 | 194.7541 | 77.77 | + 8.75 ± 4.56 |
| 1016 Dec 7 | 92 492.8947 | - 23.2257[a] | - 23.2992[a] | - 5.07 | - 25.20 ± 69.99 |
| 1016 Dec 7 | 92 492.9419 | - 23.2898[a] | - 23.3062[a] | - 5.04 | + 47.74 ± 70.41 |

[a]These positions are declinations; the others are longitudes.

The results of the analysis are summarized in Table XI.5. The first column gives the date in the Julian calendar and the second column gives the day and time by means of the Julian day number and fraction; the day number serves to tie this table and Table VII.3 together. The following three columns give $\theta_{ei}$, $\theta_{Oi}$, and $\dot{\theta}_{Oi}\tau^2$, respectively, and the last column gives the value of $\nu_S{}'$ along with its standard deviation. The standard deviation of $\nu_S{}'$ equals 7097 (from Equation XI.13) times the standard deviation of $\alpha$, and the standard deviation of $\alpha$ (by Equation XI.12) is the standard deviation of the measurement divided by $\dot{\theta}_{Oi}\tau^2$. The variability of $\dot{\theta}_{Oi}\tau^2$ explains the variability of the standard deviation of $\nu_S'$.

For the sake of uniformity with earlier results, I have kept two decimal places in the values of $\nu_S'$ even though this precision is absurd for most values in Table XI.5.

The value of $\nu_S'$ inferred from Table XI.5 is

$$\nu_S' = 1.38 \pm 1.29 \quad ''/\text{cy}^2 \; . \qquad \text{(XI.18)}$$

## 6. Equations of Condition from the Planetary and Solar Observations

Some of the observations of Mercury in Chapter VII give its position relative to Venus, some of the observations of Venus give its position relative to Mercury and some give it relative to Mars, and some of the observations of Mars give its position relative to Venus. Further, the observed positions of all the planets are geocentric and thus they depend upon the position of the sun. As a result, we cannot analyze the observations of, say, Mercury independently of observations of the other planets or of the sun. What we must do is to derive equations of condition from each observation and then solve all of the equations simultaneously for all of the accelerations.

The best way to explain how the equations of condition are obtained is by example. As examples I shall use the second and third observations of Mercury in Table VII.4. One gives the position of Mercury relative to $\alpha$ Scorpii and the other gives it relative to Venus.

The date of the second observation is equivalent to 918 December 24. At 6.24 hours on that day in local mean time, which is equivalent to 3.01 hours in Greenwich mean solar time, Mercury was 19°49′ = 19°.8167 east of $\alpha$ Scorpii. By means of the ephemeris programs described in Chapter IX, I calculated three positions of Mercury at $3^h.01$ on 918 December 24. For the first position, I assumed that the accelerations of both the sun and Mercury are zero. For the second position, I assumed an acceleration of the sun but not of Mercury, and for the third position, I assumed an acceleration of Mercury but not of the sun. It is reasonable to assume that the position is a linear function of both accelerations for small values of the accelerations. Thus the three ephemeris calculations of Mercury give its geocentric longitude at the stated time to be

$$\lambda_{☿} = 254°.5123 + 0°.01418\nu_S' - 0°.00036\nu_{☿}'. \qquad \text{(XI.19)}$$

In this, $\nu_S'$ and $\nu_{☿}'$ are the accelerations of the sun and of Mercury, respectively, in the customary units.

The calculated longitude of α Scorpii at the time of the observation is 234°.7004, and Mercury was 19°.8167 to the east. Thus the observed longitude of Mercury was 254°.5171. Equating this to the value given in Equation XI.19 yields

$$0°.01418\nu_S' - 0°.00036\nu_{☿}' = 0°.0048. \qquad \text{(XI.20)}$$

The date of the third observation is equivalent to 1000 May 19. At 19.91 hours on that day in local mean time, which is equal to 17.83 hours in Greenwich mean time, Mercury and Venus had the same longitude. In the manner just described, I calculate the longitude of Mercury to be

$$\lambda_{☿} = 84°.5094 + 0°.00978\nu_S' + 0°.00114\nu_{☿}'. \qquad \text{(XI.21)}$$

A similar calculation yields the longitude of Venus to be

$$\lambda_{♀} = 84°.4809 + 0°.03218\nu_S' - 0°.02206\nu_{♀}'. \qquad \text{(XI.22)}$$

$\nu_{♀}'$ is the acceleration of Venus. Since the two longitudes were equal, according to the observation, the combination of Equations XI.21 and XI.22 yields

## TABLE XI.6

## EQUATIONS OF CONDITION FROM OBSERVATIONS OF MERCURY

| Date | Julian Day Number -2 000 000 | $1000A_i$ | $1000B_i$ | $1000C_i$ | $1000Z_i$[a] |
|---|---|---|---|---|---|
| 858 Sep 22 | 34 706.5882 | 5.76 | 0.38 | - 6.10 | 69.2 |
| 918 Dec 24 | 56 714.6253 | 14.18 | -0.36 | 0 | 4.8 |
| 1000 May 19 | 86 447.2428 | -22.40 | 1.14 | 22.06 | -28.5 |

[a]The standard deviation of the $Z_i$ will be taken as 0°.15.

## TABLE XI.7

## EQUATIONS OF CONDITION FROM OBSERVATIONS OF VENUS

| Date | Julian Day Number -2 000 000 | $1000A_i$ | $1000C_i$ | $1000D_i$ | $1000Z_i$[a] |
|---|---|---|---|---|---|
| 830 Oct 26 | 24 514.1245 | 21.47 | - 4.83 | | -167.8 |
| 858 Aug 28 | 34 681.5787 | 9.07 | 5.84 | | 79.8 |
| 864 Feb 13 | 36 677.1482 | 3.93 | 5.34 | -10.32 | - 61.0 |
| 864 Oct 22 | 36 928.6037 | 6.82 | 3.38 | - 8.52 | 44.0 |
| 885 Sep 10 | 44 556.5506 | 9.46 | 4.70 | | 23.4 |
| 896 Oct 25 | 48 619.5823 | 5.67 | 3.84 | - 8.01 | -109.6 |
| 901 Oct 13 | 50 433.6052 | 6.71 | 5.29 | | 407.0 |
| 918 Dec 24 | 56 714.6253 | 27.43 | -13.00 | | -117.2 |
| 987 Jun 18 | 81 728.2461 | 12.90 | - 1.76 | | -624.5 |
| 988 Jan 20 | 81 943.6799 | 7.66 | 4.14 | | - 40.8 |
| 990 Jun 23 | 82 829.2499 | 7.84 | 3.42 | | 358.2 |
| 992 Sep 17 | 83 645.6132 | 8.62 | 2.70 | | - 52.2 |
| 995 Jun 11 | 84 643.2036 | 10.56 | - 0.91 | | 48.9 |
| 995 Jun 18 | 84 650.2324 | 13.11 | - 2.18 | | -329.3 |
| 996 Aug 15 | 85 074.2099 | 5.04 | 4.56 | | 74.4 |
| 997 May 24 | 85 355.5936 | 12.66 | - 1.65 | | - 35.3 |
| 998 Jun 23 | 85 751.2499 | 7.76 | 3.29 | | - 34.8 |
| 998 Jul 4 | 85 762.2486 | 4.44 | 2.92 | - 6.29 | 190.8 |
| 999 Apr 10 | 86 041.6003 | 2.67 | 4.10 | - 8.11 | 30.2 |
| 1002 Apr 18 | 87 145.5765 | 7.46 | 2.26 | | 90.1 |
| 1002 Jul 14 | 87 233.2495 | 6.67 | 4.28 | | -232.4 |
| 1003 Jan 7 | 87 410.4207 | 2.04 | 4.62 | - 7.98 | - 26.0 |
| 1003 Feb 18 | 87 452.1736 | 5.19 | 4.31 | | -126.8 |
| 1003 Feb 19 | 87 453.1745 | 5.19 | 4.30 | | -125.0 |
| 1003 Jun 18 | 87 572.4961 | 13.41 | - 2.68 | | -201.8 |

[a]The standard deviation of the $Z_i$ will be taken as 0°.15.

$$-0^\circ.02240\nu_S' + 0^\circ.00114\nu_{☿}' + 0^\circ.02206\nu_{♀}' = -0^\circ.0285. \qquad \text{(XI.23)}$$

Let any particular planetary observation be identified by a subscript i. Each observation yields an equation analogous to Equations XI.20 or XI.23, and the equation thus has the type form

$$A_i\nu_S' + B_i\nu_{☿}' + C_i\nu_{♀}' + D_i\nu_{♂}' = Z_i \,. \qquad \text{(XI.24)}$$

The equations that have this form are called the equations of condition.

TABLE XI.8

EQUATIONS OF CONDITION FROM OBSERVATIONS OF MARS

| Date | Julian Day Number -2 000 000 | $1000A_i$ | $1000D_i$ | $1000Z_i$ [a] |
|---|---|---|---|---|
| 903 Feb 28 | 50 936.5881 | 5.44 | 10.15 | 364.6 |
| 919 Jan 2 | 56 724.1770 | -13.76 | 24.40 | -259.5 |
| 983 May 10 | 80 228.2265 | 1.72 | 7.46 | 68.2 |
| 987 Oct 10 | 81 842.1953 | 2.92 | 8.31 | 127.8 |
| 988 Sep 18 | 82 186.4078 | 3.65 | 6.53 | -140.6 |
| 989 Dec 15 | 82 638.9132 | 2.84 | 8.94 | 60.6 |
| 990 Aug 30 | 82 896.5911 | 4.09 | 6.11 | 2.4 |
| 991 Nov 2 | 83 325.5270 | 1.99 | 11.82 | 308.2 |
| 993 Oct 19 | 84 043.1949 | - 6.97 | 22.14 | 156.7 |
| 994 Jun 1 | 84 267.9107 | 2.37 | 6.18 | 0.9 |
| 994 Jun 1 | 84 268.2249 | 2.37 | 6.17 | -131.2 |
| 995 Jun 11 | 84 642.4511 | 0.86 | 12.77 | 299.4 |
| 1001 Jul 19 | 86 872.9161 | 3.84 | 6.87 | -435.8 |
| 1002 Mar 14 | 87 111.2520 | -11.26 | 18.61 | -171.9 |

[a]The standard deviation of the $Z_i$ will be taken as $0^\circ.15$.

The coefficients in the equations of condition derived from the observations of Mercury, Venus, and Mars are listed in Tables XI.6, XI.7, and XI.8, respectively. The equations in these tables correspond to the observations listed in Tables VII.4, VII.5, and VII.6, respectively. Observations that involve more than one planet are listed under the inner planet involved. Tables XI.6, XI.7, and XI.8 should be self-explanatory.

TABLE XI.9

EQUATIONS OF CONDITION FROM SOLAR OBSERVATIONS[a]

| Date | Julian Day Number -2 000 000 | $Z_i$ "/cy$^2$ | $\sigma(Z_i)$ "/cy$^2$ |
|---|---|---|---|
| 832 Sep 18 | 25 207.3645 | 2.40 | 1.71 |
| 859 Sep 19 | 35 068.8727 | 6.32 | 1.71 |
| 882 Sep 18 | 43 469.4381 | 4.66 | 1.71 |
| 1000 Nov 30 | 86 641.9132 | 3.30 | 1.71 |
| 829 Sep 19 | 24 111.6545 | - 0.09 | 5.13 |
| 830 Sep 19 | 24 476.9127 | - 1.05 | 3.63 |
| 831 Mar 17 | 24 655.4653 | 5.38 | 3.63 |
| 831 Sep 19 | 24 842.1627 | - 1.73 | 3.63 |
| 832 Mar 16 | 25 020.7152 | 4.65 | 3.63 |
| 832 Jun 18 | 25 114.3759 | - 2.28 | 3.42 |
| 888 Sep 18 | 45 660.8755 | 3.90 | 3.42 |
| 970 Sep 18 | 75 610.8068 | - 1.27 | 3.42 |
| 971 Sep 18 | 75 976.0984 | - 5.04 | 3.42 |
| 1019 Sep 18 | 93 507.5960 | 7.05 | 3.42 |
| 934[b] | ——— | 6.75 | 2.13 |
| 832 Jan 31 | 24 975.4794 | 5.00 | 3.09 |
| 832 May 1 | 25 067.2480 | - 3.40 | 3.24 |
| 832 Aug 4 | 25 161.4138 | - 5.07 | 3.23 |
| 960 Oct 6 | 71 976.8474 | 21.30 | 12.72 |
| 1016 Jun 15 | 92 318.0036 | 113.88 | 824.42 |
| 1016 Sep 2 | 92 396.8316 | 5.99 | 4.65 |
| 1016 Oct 2 | 92 426.8252 | 8.75 | 4.56 |
| 1016 Dec 7 | 92 492.8947 | - 25.20 | 69.99 |
| 1016 Dec 7 | 92 492.9419 | 47.74 | 70.41 |

[a] $A_i = 1$, $B_i = C_i = D_i = 0$, for all entries in this table.

[b] This is the resultant of the observations in Table XI.3, for which the mean epoch is approximately 934.

The *a priori* estimate of the standard deviation of an observation is taken as 0°.15 for each planetary observation (Sections VII.4, VII.5, and VII.6).

Each solar observation that was discussed in the earlier sections of this chapter also yields an equation having the form of Equation XI.24, with $B_i = C_i = D_i = 0$, with the exception of the meridian observations listed in Table XI.3. The observations in that table involve both $\nu_S'$ and errors in the direction of the local vertical. In order to avoid the trouble of carrying the vertical errors throughout the analysis, I solved those observations for $\nu_S'$ and for the errors in the vertical, and the result for $\nu_S'$ is given in Equation XI.14.

The individual solar observations from Tables XI.1, XI.2, and XI.5, along with the composite observation from Equation XI.14, are summarized in Table XI.9. Since $A_i = 1$ and since $B_i = C_i = D_i = 0$ for each observation, it is not necessary to list those coefficients. Only the quantity $Z_i$ (which is the value of $\nu_S'$ already inferred for each observation), along with its estimated standard deviation $\sigma(Z_i)$, is listed in the table.

Some of the observations involve the positions of Jupiter and Saturn. The geocentric longitude of either of these planets involves the accelerations of both the sun and the appropriate planet. We expect the accelerations of both Jupiter and Saturn to be small, and I accordingly neglected them in deriving the equations of condition. The coefficient of $\nu_S'$ that appears in the geocentric longitude is inversely proportional to the orbital radius, approximately. Since the orbital radius of Saturn is about 10 astronomical units, I neglected the coefficient of $\nu_S'$ when dealing with Saturn. I retained the coefficient when dealing with Jupiter, because its orbital radius is only about 5.

The equations of condition listed in Tables XI.6 through XI.9 are now to be solved for the accelerations, using the method of weighted least-squares. This is a standard method that does not need to be described. The solution will be discussed in the next section.

7. The Accelerations That Are Derived from the Islamic Data

The solution of the equations of condition is

$$\nu_S' = 2.80, \quad \nu_☿' = -93.57, \quad \nu_♀' = 9.59, \quad \nu_♂' = 2.32. \qquad \text{(XI.25)}$$

The ratios of the accelerations that are required by the spin fluctuation hypothesis are

$$\frac{\nu_☿'}{\nu_S'} = 4.1520, \quad \frac{\nu_♀'}{\nu_S'} = 1.6255, \quad \frac{\nu_♂'}{\nu_S'} = 0.5317 \ . \qquad \text{(XI.26)}$$

The accelerations in Equation XI.25 are far from satisfying the ratios in Equation XI.26. Before we put any stress upon this point, we must examine the statistical significance of the solution in Equation XI.25.

One way to examine the statistical significance is to see what happens when portions of the data sample are omitted. Since the solution in Equation XI.25 is based upon the data in Tables XI.6 through XI.9, it is natural to study the effect of omitting one or more tables. Accordingly, I tried four cases, as follows:

Case 1: All data.

Case 2: Tables XI.7, XI.8, and XI.9 only.

Case 3: Tables XI.8 and XI.9 only.

Case 4: Table XI.9 only.

Between Cases 1 and 2, we must drop the acceleration of Mercury from the solution because there are no Mercury observations in Case 2, and so on until we have only $\nu_S'$ left in Case 4. The results

TABLE XI.10

THE EFFECT OF CHANGING THE DATA SAMPLE USED IN INFERRING THE ACCELERATIONS

| Case | $\nu_S'$ "/cy$^2$ | $\nu_☿'$ "/cy$^2$ | $\nu_♀'$ "/cy$^2$ | $\nu_♂'$ "/cy$^2$ |
|---|---|---|---|---|
| 1 | 2.80 | -93.57 | 9.59 | 2.32 |
| 2 | 2.79 | | 14.39 | 2.68 |
| 3 | 3.16 | | | 1.73 |
| 4 | 3.08 | | | |

are shown in Table XI.10. The large changes shown in the table suggest that there is a problem with the data.

All previous experience has shown that there are two classes of error in ancient data. There are errors of observation and there are errors of recording. Errors of recording are surprisingly common, and they are usually larger than the genuine errors of observation.

Errors of recording have already been rather well eliminated from Table XI.9, but Tables XI.6, XI.7, and XI.8 still contain the data from all records that could be read unambiguously. Since the values of the $Z_i$ in Tables XI.6, XI.7, and XI.8 are dominated by errors in the recorded values,† we can get a good idea of the error distribution by looking at the distribution of the $Z_i$.

If the reader will prepare a histogram of the $Z_i$, he will see that they seem to contain two populations. The values out to about 0°.25 seem to belong to a population with a standard deviation of about 0°.1 or 0°.15, but there are many more values larger than 0°.25 than we expect from a normal

†We expect the accelerations to be a few seconds per century per century, and the data are about 10 centuries old. Thus the effect of the accelerations is about 200″, or about 0°.05, say, which is smaller than most of the $Z_i$. Thus the residuals in the data will be close to the $Z_i$.

distribution. There is no way to establish a rigorous division between the two populations, but we shall not be far wrong if we say that errors less than 0°.25 are errors of observation and that larger errors are errors of recording. The apparent standard deviation of the errors of observation is close to the *a priori* estimate that I made in Sections VII.4, VII.5, and VII.6.

TABLE XI.11

THE EFFECT OF ELIMINATING RECORDING ERRORS UPON THE INFERENCE OF THE ACCELERATIONS

| Case | $\nu_S'$ ″/cy² | $\nu_☿'$ ″/cy² | $\nu_♀'$ ″/cy² | $\nu_♂'$ ″/cy² |
|---|---|---|---|---|
| 1 | 2.82 | 12.51 | 2.29 | 1.40 |
| 2 | 2.82 | | 4.60 | 1.69 |
| 3 | 3.09 | | | 1.14 |
| 4 | 3.08 | | | |

Accordingly, I repeated the calculations summarized in Table XI.10 but with the omission of data that seem to include recording errors ($Z_i >$ 0°.25). No values were omitted from Table XI.6, four values were omitted from Table XI.7, and five values were omitted from Table XI.8. The new calculations are summarized in Table XI.11. The cases have the same meaning in this table that they had in Table XI.10.

The solutions in Table XI.11 are much more stable than the solutions in Table XI.10, so that we seem to be justified in omitting the values of $Z_i$ that are greater than 0°.25. Accordingly, I propose to take Case 1 in Table XI.11 as the best estimate of the accelerations that we can make from the Islamic data. We still need to attach an estimated standard deviation to each acceleration, which we do by calculating the scatter of the residuals and applying the standard statistical formulae. The results are

$$\nu_S' = 2.82 \pm 0.60 , \qquad \nu_{☿}' = 12.51 \pm 148.56,$$

(XI.27)

$$\nu_{♀}' = 2.29 \pm 5.99 , \qquad \nu_{♂}' = 1.40 \pm 3.86.$$

The epoch at which these values apply can be taken as 920. The standard deviation of the residuals is 0°.11.

For each planet, the standard deviation of the estimated acceleration is greater than the acceleration. Thus we cannot test the spin fluctuation hypothesis by means of the values listed in Equations XI.27. There is a more delicate test that we can apply; the penalty that we pay for increasing the delicacy is the inability to estimate the planetary accelerations. Since the estimates in Equations XI.27 are not significant anyway, the loss is small.

Consider the standard form of the equation of condition in Equation XI.24 in the preceding section. If we assume that the spin fluctuation hypothesis is valid, we may replace each planetary acceleration by its value derived from Equations XI.26. The result is

$$(A_i + 4.1520B_i + 1.6255C_i + 0.5317D_i)\nu_S' = Z_i.$$

(XI.28)

TABLE XI.12

VALUES OF $\nu_S'$ DERIVED FROM OBSERVATIONS OF THE SUN AND OF SEVERAL PLANETS

| Body Observed | $\nu_S'$ ″/cy$^2$ | $\sigma(\nu_S')$ ″/cy$^2$ |
|---|---|---|
| Sun | 3.08 | 0.77 |
| Mercury | -1.28 | 2.75 |
| Venus | -3.10 | 1.80 |
| Mars | 2.10 | 4.51 |

By means of this relation, we can derive a value of $\nu_S'$ from each body of data. If the spin fluctuation hypothesis is valid, the values of $\nu_S'$ should be equal within observational error. The values obtained in this way are listed in Table XI.12.

In Table XI.12, the first column identifies the body of data used. For the sun and for Mercury, the bodies of data are those in Tables XI.9 and XI.6, respectively. For Venus, the body of data is that in Table XI.7 plus the two observations in Table XI.6 that involve Venus, and for Mars the body is that in Table XI.8 plus the six observations in Table XI.7 that involve Mars. The second and third columns in the table give the inferred values of $\nu_S'$ and of its standard deviation. The standard deviation, as is usual in these investigations, is based only upon the apparent precision of the data and does not include an allowance for possible systematic error.

The accelerations derived from the solar data and from the Mars data are not significantly different. The accelerations derived from the Venus and the Mercury data, however, do differ significantly. The difference between the sun and Mercury data is 4.36, and the standard deviation of the difference is $[(0.77)^2 + (2.75)^2]^{\frac{1}{2}} = 2.86$. The odds that the difference did not happen by chance are about 7 to 1. The difference between the Venus and the sun data is 6.18, the standard deviation of the difference is 1.96, and the odds that the difference did not happen by chance are about 640 to 1.

A few other comments are needed. Mercury and Venus both yield negative values of $\nu_S'$ in Table XI.12. Since Mercury and Venus are interior to the earth while Mars is exterior, this fact is interesting, but we should not give it too much significance until we have analyzed the Hellenistic and Babylonian data.

We should also note that the Mars data give the smallest estimated error for a planetary acceleration in Equations XI.27 and the largest error in Table XI.12. This is a result of the geometry of the orbits. The Venus and Mercury observations tend to be made near the points of maximum elongation, where the geocentric position depends strongly upon the solar acceleration and weakly upon the planetary acceleration. Thus they give poorly determined planetary accelerations in Equations XI.27 and a well-determined solar acceleration in Table XI.12. The Mars observations, on the other hand, tend to be made at all points around the orbit. This gives a relatively strong determination of the planetary acceleration and a relatively weak determination of the solar one.

I studied the observations of Venus in an earlier work [Newton, 1970, Section VIII.3]. There I assumed the validity of the spin fluctuation hypothesis without question, and the result obtained, in the notation of this work, is

$$\nu_S{}' = -2.60 \pm 3.12 \ . \qquad \text{(XI.29)}$$

In the earlier study, I also made the simplifying assumption that the heliocentric orbits of the planets are ellipses, instead of using accurate ephemerides as I have done here. The estimate of $\nu_S{}'$ is not changed significantly by the new analysis, but the error limits are tightened considerably.

## 8. Summary of the Inferences from Islamic Data

The best estimates of the solar and planetary accelerations, as deduced from the body of Islamic data, are given by Equations XI.27, which will be repeated here for convenience:

$$\nu_S{}' = 2.82 \pm 0.60, \qquad \nu_{☿}{}' = 12.51 \pm 148.56,$$
$$\text{(XI.30)}$$
$$\nu_{♀}{}' = 2.29 \pm 5.99, \qquad \nu_{♂}{}' = 1.40 \pm 3.86.$$

Since the error estimates in the planetary accelerations exceed the estimated values, we cannot test the spin fluctuation hypothesis by means of Equations XI.30.

If we estimate the solar acceleration by means of solar data alone, the result is taken from the first line in Table XI.12:

$$\nu_S' = 3.08 \pm 0.77 \quad ''/\text{cy}^2 \ . \qquad \text{(XI.31)}$$

This does not differ significantly from the value in Equations XI.30, and we may take the solar acceleration as being reasonably well established. Since the value in Equations XI.30 is based upon more data, we should probably prefer it.

We can test the spin fluctuation hypothesis by a more sensitive method, but one that does not allow us to infer planetary accelerations. The results of the test are summarized in Table XI.12. The odds that Mercury does not satisfy the hypothesis are about 7 to 1, and the odds that Venus does not are about 640 to 1. The bodies of data that enter into these estimates are not independent, since some of the observations are conjunctions of Mercury with Venus, and there is no simple way to estimate the combined probability. It is probably conservative to say that the overall odds, on the basis of the Islamic data only, are more than 1000 to 1 against the hypothesis.

We should not put much stress on this conclusion until we have analyzed the Hellenistic and Babylonian data. The conclusion depends critically upon the estimated standard deviations listed in Table XI.12. These standard deviations are based only upon the apparent scatter in the data, and it is a commonplace in the observational and experimental sciences that such estimates are too small. One role of the other bodies of data is to supply error estimates that rest upon a broader base.

## CHAPTER XII

## ANALYSIS OF THE HELLENISTIC DATA

### 1. The Solar Data

The body of Hellenistic solar data contains two main parts. There are the equinox observations made by Hipparchus, and there are the observations of the summer solstice made by Aristarchus and Hipparchus.

I studied Hipparchus's equinox observations in an earlier work [Newton, 1970] and summarized the analysis in Section V.3 of this work. Here it is sufficient to repeat the value of the solar acceleration that is derived from the equinoxes:

$$\nu_S' = 2.98 \pm 0.71 \quad ''/\mathrm{cy}^2 . \qquad \text{(XII.1)}$$

This is the same as Equation V.2 in Section V.3.

The solstice observations made by Aristarchus and Hipparchus are described immediately after Equation V.2 in Section V.3. They found that summer solstices occurred at 18 hours on -279 June 26 and at 12 hours on -134 June 26, respectively. Since Aristarchus is known as Aristarchus of Samos, I assume that his observation was made there. Hipparchus's observation was presumably made on Rhodes. We do not change the results appreciably if we use any place in the Hellenistic world for either observation.

TABLE XII.1

THE SOLSTICES OF ARISTARCHUS AND HIPPARCHUS

| Date | Greenwich Mean Time | Julian Day Number -1 600 000 | Longitude of Sun, degrees | | $\nu_S'$ |
|---|---|---|---|---|---|
| | | | Newcomb's theory | observed | $''/\mathrm{cy}^2$ |
| -279 Jun 26 | 16.15 | 19 330 | 89.4079 | 90 | 8.98 |
| -134 Jun 26 | 10.06 | 72 291 | 90.0558 | 90 | -0.97 |

The analysis of these two solstices is summarized in Table XII.1, which should be self-explanatory. The acceleration inferred from the two solstices is

$$\nu_S' = 4.01 \pm 4.98 \quad ''/\text{cy}^2 . \qquad \text{(XII.2)}$$

The accuracy of the results does not warrant the number of decimal places used in writing Equation XII.2. I have kept two decimal places only to maintain parallelism with other stated results.

When I come to making an inference using all the Hellenistic data, I shall use the observations in Table XII.1 individually. However, I shall use Equation XII.1 rather than the individual observations in Table V.2 because the individual observations are highly correlated and should not be used directly in a statistical analysis.

## 2. Observations of Mercury

Each planetary observation furnishes an equation of condition that connects the solar and planetary accelerations. I shall continue to use the notation of Equation XI.24 in writing the equations of condition, and I repeat the equation here for the convenience of the reader:

$$A_i \nu_S' + B_i \nu_{☿}' + C_i \nu_{♀}' + D_i \nu_{♂}' = Z_i . \qquad \text{(XII.3)}$$

The coefficients $A_i$, $B_i$, $C_i$, $D_i$, and $Z_i$ are found by the method described in Section XI.6.

Some of the Islamic observations of Mercury, for example, involved the accelerations $\nu_{♀}'$ and $\nu_{♂}'$ of Venus and Mars as well as the acceleration $\nu_{☿}'$ of Mercury and the acceleration $\nu_S'$ of the sun. As a result, we could not analyze the observations of one planet separately from those of another. Here we do not have that complication. Each observation of Mercury involves only the accelerations $\nu_S'$ and $\nu_{☿}'$, and we can analyze the Mercury observations independently of other planetary observations.

Before looking at the longitude data, it is interesting to look at the latitudes. For seven of the observations in Table V.5, the difference in latitude between Mercury and the reference star has been preserved. Since we can calculate the latitude of the reference star quite accurately, these records give us a measured latitude of Mercury at the time of an observation. Further, since the latitude of Mercury calculated from the ephemeris program is almost independent of the accelerations of Mercury and the sun, a comparison of the calculated and observed latitudes gives us an idea of the accuracy of the observations.

TABLE XII.2

OBSERVATIONS OF THE LATITUDE OF MERCURY

| Date | Observed Latitude, deg. | Calculated Latitude, deg. | Error,[a] deg. |
|---|---|---|---|
| -264 Nov 15 | 2.289 | 2.770 | -0.481 |
| -264 Nov 19 | 2.289 | 2.784 | -0.495 |
| -261 Feb 12 | -0.693 | 0.158 | -0.851 |
| -261 Apr 25 | 3.352 | 2.435 | +0.917 |
| -256 May 28 | 0.009 | 0.698 | -0.689 |
| -244 Nov 19 | 2.537[b] | 2.486 | +0.051 |
| -236 Oct 30 | 1.854[b] | 2.278 | -0.424 |

[a] In the sense observed minus calculated.

[b] Using 1 ammat = 2½ degrees.

The comparison is presented in Table XII.2, which should be largely self-explanatory. In the observations of -244 November 19 and -236 October 30, the latitude difference was given in terms of the Babylonian ammat. We decided in Section X.4 that the ammat was equal to 2½ degrees at this time in history, and I have used this value in Table XII.2.

The standard deviation of the error in the observed values is 0°.82. This compares unfavorably with the estimate of 0°.11 that I made for the

Islamic observations in Section XI.7. It is comparable, however, with the accuracy estimated for Babylonian observations in Section X.5. It is likely that there are recording errors in Table XII.2, but there is no obvious way to separate errors of recording from errors of observation.

TABLE XII.3

EQUATIONS OF CONDITION FROM OBSERVATIONS OF MERCURY

| Date | Julian Day Number -1 000 000 | $100A_i$ | $100B_i$ | $100Z_i$ | Reliability |
|---|---|---|---|---|---|
| -264 Nov 15 | 624 950.650 | 8.121 | -1.344 | - 35.76 | 1 |
| -264 Nov 19 | 624 954.652 | 7.152 | -0.410 | - 57.40 | 1 |
| -261 Feb 12 | 625 769.668 | 6.784 | -0.162 | + 2.34 | 1 |
| -261 Apr 25 | 625 842.220 | 6.570 | -0.301 | + 80.03 | 1 |
| -261 Aug 23 | 625 962.225 | 5.719 | +0.753 | -179.78 | 1 |
| -256 May 28 | 627 702.234 | 7.029 | -0.831 | +105.53 | 1 |
| -244 Nov 19 | 632 259.652 | 6.212 | +0.371 | - 20.14 | 1 |
| -236 Oct 30 | 635 161.641 | 5.652 | +0.836 | + 26.02 | 1 |
| 130 Jul 4 | 768 725.243 | 4.349 | -0.128 | + 44.68 | 1 |
| 132 Feb 2 | 769 303.182 | 4.599 | -0.219 | +166.29 | 0 |
| 134 Jun 4 | 770 155.593 | 5.464 | -1.324 | -120.52 | 0 |
| 134 Oct 3 | 770 276.631 | 3.873 | +0.509 | +116.05 | 0 |
| 135 Apr 5 | 770 461.212 | 4.889 | -0.681 | +116.15 | 0 |
| 138 Jun 4 | 771 617.238 | 4.216 | -0.052 | +140.68 | 0 |
| 139 May 17 | 771 964.227 | 4.308 | -0.150 | + 69.60 | 0 |
| 139 Jul 5 | 772 012.597 | 4.926 | -0.794 | - 22.71 | 0 |
| 141 Feb 2 | 772 590.671 | 4.042 | +0.356 | -105.35 | 0 |

The information needed for estimating the accelerations is summarized in Table XII.3, which resembles Tables XI.6, XI.7, and XI.8. The observations are now put into chronological order. Instead of tabulating 1000 times the coefficients, I have tabulated 100 times the coefficients in Table XII.3. I have also included a column for the reliability assigned to each observation in Table XII.3. Eight observations have a reliability of zero and nine have a reliability of unity.

The nine observations whose reliability is unity lead to

$$\nu_S' = -\ 2.06 \pm\ 4.53,$$
$$\nu_☿' = -45.73 \pm 42.98. \qquad \text{(XII.4)}$$

Plausible values of the accelerations lie within $1\frac{1}{2}$ standard deviations of the central estimates in Equations XII.4, and the reader may wish to accept these equations. However, inspection of the values of $Z_i$ in Table XII.3 suggests that we have, as usual, a mixture of observing and recording errors. If we confine our attention to the nine observations with a reliability of unity, we see that the values of $Z_i$ are greater than 1° for -261 August 23 and -256 May 28. All other values of $Z_i$ are much smaller except for the observation of -261 April 25, which is a borderline case. I believe that a safe course is to omit the three observations just mentioned, leaving six observations to be used.

The accelerations inferred from the 'best six' observations are

$$\nu_S' = -\ 1.02 \pm\ 2.39,$$
$$\nu_☿' = 24.94 \pm 22.48. \qquad \text{(XII.5)}$$

The estimated standard deviations are halved by omitting three observations, in spite of the fact that we have only six observations left. Accordingly, I shall assume that Equations XII.5 furnish the best estimates that we can make from the Mercury observations.

For all nine observations, the standard deviation of the residuals is 0°.84, which is almost exactly the error that we inferred from the latitudes in Table XII.2. For the 'best six', the standard deviation is only 0°.36. Further, when we omit the records for which we suspected a recording error on the basis of the Z's, the standard deviation in latitude also falls, to 0°.53. This increases the strength of the case for omitting the three observations, and it suggests that 0°.45 is a better estimate than 0°.84 for the accuracy of the early Hellenistic observations, per coordinate.

The standard deviations of the parameters in both Equations XII.4 and XII.5 are so large that we cannot infer anything about the spin fluctuation hypothesis from them alone. However, we can apply the more delicate test that we used with the Islamic data in Section XI.7 by assuming that $\nu_{☿}' = 4.1520\nu_S'$ (Equations XI.26) and estimating $\nu_S'$ alone from the data in Table XII.3. The result is

$$\nu_S' = -2.64 \pm 4.43 \quad ''/\text{cy}^2 \qquad \text{(XII.6)}$$

if we use all nine observations, and

$$\nu_S' = -0.60 \pm 2.55 \quad ''/\text{cy}^2 \qquad \text{(XII.7)}$$

if we use only the 'best six'. I shall defer the discussion of these results to Section XII.5, after the analysis of the other planetary data.

The residuals in longitude show some tendency to be large when the separation in latitude between Mercury and the reference star is large. This may merely indicate that it is more difficult to estimate the separation in longitude, or to estimate when it is zero, when the stars are widely apart in latitude. However, there is also a chance that the observers unconsciously assessed the separation in right ascension and declination even though they were trying to use latitude and longitude.

In order to test this possibility, I repeated the analysis of the Mercury data on the assumption that the separations were in right ascension and declination. The residuals were much larger in this analysis than in the analysis based upon latitude and longitude. Therefore it seems that the observers did indeed use the latter pair of coordinates for Mercury, and I shall assume that they did the same for the other planets.

The eight observations with a reliability of 0

in Table XII.3 are those that Ptolemy claims to have made himself. I shall defer discussion of them to Section XIII.8. Here I shall remark only that the Z's give a good indication of the errors in the observations, and that the Z's for the other observations and the Z's for Ptolemy's claimed observations exhibit quite different properties.

TABLE XII.4

OBSERVATIONS OF VENUS BY PROCLUS AND HELIODORUS

| Date | Other Body | Calculated Longitudes, degrees Venus | Other Body | Stated Difference in Longitude[a] |
|---|---|---|---|---|
| 475 Nov 18 | Sun | 284.905 | 237.773 | 48° |
| | Mean sun | 284.905 | 235.887[b] | 48°? |
| 510 Aug 13 | Jupiter | 160.537 | 168.060 | - 8 digits? |
| 510 Aug 21 | Jupiter | 170.401 | 169.720 | +10 digits |

[a] In the sense Venus minus other body.

[b] Calculated from Ptolemy's theory, which is the one that Proclus would have used. The value from Newcomb's theory is 238°.432.

## 3. Observations of Venus

Before turning to the observations of Venus found in the Syntaxis, it is convenient to dispose of the observations of Venus made by Proclus and Heliodorus that are listed in Table V.8. The analysis of the observations is summarized in Table XII.4.

The record of the observation made by Proclus on 475 November 18 says that the moon occulted Venus when Venus was 48° from the sun. I shall take up the occultation along with some other miscellaneous observations in Section XII.6. From Table XII.4, we see that Venus was only about 47°.13 from the true sun. The longitude of the mean sun that Proclus would have calculated from the Syntaxis is 235°.887, so that Venus was about 49°.02 from where Proclus thought the mean sun was. Thus the observation is of poor quality whether the true or the mean sun is meant.

The record of the two observations in August of 510 is badly garbled. I suggested a tentative restoration of the text in Section V.9, with the intention of giving half weight to each observation if the restoration proved to be plausible. In the suggested restoration, Venus was 8 digits (40′) west of Jupiter on 510 August 13, while calculation shows that it was about 7°.52 west. Also in the restoration, Venus was 10 digits (50′) east of Jupiter on 510 August 21, while calculation shows that it was about 0°.68, or about 41′, to the east. This suggests an emendation in the tentative restoration: Venus was 8 degrees, not 8 digits, west on 510 August 13, and it was 10 digits (50′ rather than 41′) east on 510 August 21.

It is clear that we cannot restore the text with enough confidence to let us use the record in the study of the accelerations. We can now turn to the observations preserved by Ptolemy in the Syntaxis, which were tabulated in Table V.6.

Among the observations in Table V.6 that have a reliability greater than zero, two give information about the difference in latitude between Venus

TABLE XII.5

EQUATIONS OF CONDITION FROM OBSERVATIONS OF VENUS

| Date | Julian Day Number -1 000 000 | $100A_i$ | $100C_i$ | $100Z_i$ | Reliability |
|---|---|---|---|---|---|
| -271 Oct 12 | 622 359.633 | 4.755 | +1.853 | + 15.35 | 1 |
| -271 Oct 16 | 622 363.635 | 4.690 | +1.931 | + 8.91 | 1/2 |
| 127 Oct 12 | 767 728.635 | 3.684 | +0.711 | - 2.82[a] | 1/4 |
| 129 May 20 | 768 314.597 | 3.620 | +0.639 | - 4.94 | 1 |
| 132 Mar 8 | 769 338.200 | 5.142 | -0.937 | + 81.15[b] | 1/4 |
| 134 Feb 18 | 770 049.664 | 4.406 | +0.041 | +112.09 | 0 |
| 136 Nov 18 | 771 054.159 | 3.672 | +0.783 | +149.48 | 0 |
| 136 Dec 25 | 771 091.161 | 5.371 | -0.939 | + 91.10 | 0 |
| 138 Dec 16 | 771 811.611 | 3.926 | +0.540 | + 21.64 | 0 |
| 140 Feb 18 | 772 241.191 | 4.115 | +0.128 | + 50.88 | 0 |
| 140 Jul 30 | 772 403.605 | 3.454 | +0.726 | -108.50 | 0 |

[a]This should be changed to 47.18; see the main text.
[b]This should be changed to 31.15; see the main text.

and the reference star. Venus was described as touching $\eta$ Virginis on -271 October 12, while calculation shows that it was about 13′ to the south. Since Venus appears to the naked eye to have a finite diameter, the word "touching" implies a small difference in latitude, and this seems to be an accurate observation. We concluded from the description of the observation of 129 May 20 that Venus was 3°.35 farther north (in latitude) than $\mu$ Ceti, while calculation puts it 3°.39 farther north. This is also an accurate observation.

The information concerning the longitudes is summarized in Table XII.5, which has the same format as Table XII.3 for Mercury. Unfortunately, there are only five observations whose reliability is greater than zero. These five lead to

$$\begin{aligned} \nu_S' &= 7.46 \pm 3.35, \\ \nu_♀' &= -16.23 \pm 10.27. \end{aligned} \qquad \text{(XII.8)}$$

The value of $\nu_S'$ is only about one standard deviation away from a plausible value, but the value of $\nu_♀'$ is about $1\frac{1}{2}$ standard deviations less than zero, and it is about two standard deviations away from a plausible value. Thus the Venus observations seem to conflict with the spin fluctuation hypothesis, as have all other observations of Mercury and Venus that we have studied.

If we assume the spin fluctuation hypothesis, we set $\nu_♀' = 1.6255\nu_S'$. This leads to

$$\nu_S' = 1.65 \pm 1.33, \qquad \text{(XII.9)}$$

if we use all the observations in Table XII.5 whose reliability is not zero. The value obtained from solar data alone is probably between 3 and 4. Thus Equation XII.9 conflicts with the hypothesis at only about the '$1\sigma$' level. However, the value in Equation XII.9 is certainly small, as have been all other estimates of $\nu_S'$ formed from Mercury and Venus data with the aid of the hypothesis.

The value of Z for the observation of 132 March 8 is many times bigger than the others. If we omit it, we get $\nu_S' = 1.08 \pm 0.68$. However, I do not believe that this is the correct thing to do. The observations of 127 October 12 and 132 March 8 both involved the width of the Pleiades as a measuring unit. If the reader will refer back to the discussion of these observations in Section V.6, he will see that it was not clear whether the width should be taken as 1 or as $1\frac{1}{2}$ degrees. Since Venus was west of the reference star in one observation and east in the other, an error in the unit should not affect the central estimate much, but it can affect the size of the residuals. The width of the Pleiades was taken as 1° in preparing Table XII.5. If we change it to $1\frac{1}{2}$ degrees, all values of Z become reasonable, and the estimate of $\nu_S'$ becomes

$$\nu_S' = 1.78 \pm 1.21 \quad ''/\text{cy}^2. \qquad \text{(XII.10)}$$

This is the estimate that I shall prefer. The reader is justified in adopting either of the other two estimates if he prefers.

The observations with a weight of zero in Table XII.5 are those that Ptolemy claims to have made

TABLE XII.6

EQUATIONS OF CONDITION FROM OBSERVATIONS OF MARS

| Date | Julian Day Number -1 000 000 | $100A_i$ | $100D_i$ | $100Z_i$ | Reliability |
|---|---|---|---|---|---|
| -271 Jan 18[a] | 622 092.675 | + 0.709 | + 5.804 | -85.39 | 1 |
| 130 Dec 15 | 768 888.455 | - 6.882 | + 9.631 | +47 | 0 |
| 135 Feb 21 | 770 418.302 | - 6.732 | + 9.456 | -49 | 0 |
| 139 May 27 | 771 974.332 | - 9.988 | +16.311 | +54 | 0 |
| 139 May 30 | 771 977.290 | -10.012 | +16.371 | +68.80 | 0 |
| 498 May 1 | 903 073.250 | - 0.767 | + 3.168 | + 6.04 | 1 |
| 509 Jun 13 | 907 134.243 | + 0.534 | + 1.534 | + 5.26 | 1 |

[a] Ptolemy uses the Egyptian equivalent of this date, but the correct date is probably -271 January 16. The value listed for $100Z_i$ corresponds to the date -271 January 18.

himself. I shall discuss these 'observations' in Section XIII.5. Here I shall remark only that the values of Z for these observations seem to differ in their properties from the other values, just as they did for Mercury.

## 4. Observations of Mars

We have altogether seven Hellenistic observations of Mars. Five were preserved by Ptolemy and are discussed in Section V.7. Two were made by Heliodorus and preserved in notes that Heliodorus made in his copy of Ptolemy [ca. 142]; these are discussed in Section V.9. Of the five that Ptolemy preserved, four were made by himself and therefore receive a reliability of zero. Information concerning the seven observations of Mars is summarized in Table XII.6, which has the same format as Table XII.5.

In the record that Ptolemy dates as the equivalent of -271 January 18, Mars is described as touching β Scorpii. The latitude of β Scorpii was 1°.2900 and the calculated latitude of Mars was 1°.3186, in excellent agreement. However, the longitude of β Scorpii was 211°.6290 while the longitude of Mars, calculated with both $\nu_{\text{☋}}'$ and $\nu_{\text{♂}}'$ equal to zero, was 212°.4829, about 0°.85 away from the star. I cannot believe that the bodies would have been described as touching if they were this far apart, and I therefore believe that there is an error in the record. The only place where an error is plausible is in the date. The date was recorded in an old Greek calendar† that we do not know much about, and we can only take the date of this observation from its equivalent in the Egyptian calendar as given by Ptolemy. The calculations that Ptolemy makes with this observation, as well as its statement in the Egyptian calendar, show that he took the date to be the equivalent of -271 January 18.

The only date that is consistent with the astronomical circumstances is -271 January 16, and I believe that this is the correct date. The date

†See Section V.5.

appears in the Greek text as day 25 of the month Agon. If my deduction about the date is correct, this was originally day 23. I gather that we cannot be sure about the forms of the numerals that were used in Ptolemy's sources. However, it is possible that 23 was written as KΓ, and it is possible that Ptolemy misread this as KE(25), this making the date come two days later than it was in the original.

Because there is undoubtedly a serious error in this record, whether in the date or somewhere else, I shall not use it. Thus none of the observations of Mars found in the Syntaxis are usable.

The observations of Venus that Heliodorus preserved proved not to be usable. The observation of 475 November 18 (Table XII.4), which was actually made by Proclus, is of poor quality unless there is an error in the date. The record of the observations dated 510 August 13 and 21 in Table XII.4 is badly garbled and we do not even know whether the listed dates are correct. Luckily, Heliodorus's observations of Mars are of good quality and seem to be recorded correctly.

Thus Heliodorus made the only usable Hellenistic observations of Mars that are left to us.

If we solve Heliodorus's two observations for $\nu_S'$ and $\nu_{♂}'$, we get

$$\nu_S' = 2.58 \pm 18.41,$$
$$\nu_{♂}' = 2.53 \pm 8.15. \qquad \text{(XII.11)}$$

Since we have estimated two parameters from only two observations, we cannot estimate the errors from the residuals of the observations, which are necessarily zero. In deriving the error estimates in Equations XII.11, I have assumed that the standard error of a single observation is 0°.25.

If we assume that the spin fluctuation hypothesis is valid, we have $\nu_S' = 0.5317\nu_{♂}'$ from Equations XI.26. The observations then give

$$\nu_S' = 4.75 \pm 9.19. \qquad \text{(XII.12)}$$

This does not tell us anything significant about the spin fluctuation hypothesis.

The best way to use the Mars observations is probably to make a single determination of all accelerations from all the solar and planetary observations. If we do this, the solar acceleration will be determined almost entirely by the observations that do not involve Mars. The acceleration of Mars will then be the only parameter determined by the pair of useful observations in Table XII.6. We can still not expect to get a useful result from only two observations.

## 5. Summary of the Solar and Planetary Data

The total body of usable Hellenistic solar and planetary data is represented by Equation XII.1 and by Tables XII.1, XII.3, XII.5, and XII.6. In Table XII.5, we should add 0°.50 to the value of $Z_i$ for 127 October 12 and we should subtract 0°.50 from the value for 132 March 8, in accordance with the discussion in Section XII.3. I take 0.71 $''/cy^2$ for the standard deviation of the value in Equation XII.1, 4.98 $''/cy^2$ for the standard deviation of the values in Table XII.1, and 0°.25 for the standard deviation of all other values. The latter is intermediate between the accuracy of the Islamic planetary observations that I estimated in Section XI.7 and the accuracy of the Hellenistic observations of Mercury that I estimated in Section XII.2.

In doing this, I am assigning the same relative weights to the first five observations in Table XII.5, although I originally assigned various weights. I have arbitrarily changed one weight from $\frac{1}{2}$ to 1 because the difference is not really enough to matter. The weights of $\frac{1}{4}$ were assigned because the observations depended upon the originally uncertain width that Hellenistic astronomers used for the Pleiades. We are now rather sure what this width is, and we also know that an error in assigning the width cancels out of the final inference. Hence I have changed these weights to 1 also, after changing the width used for the Pleiades.

The solution using simultaneously all of the observations previously used in smaller groups is

$$\nu_S{}' = 2.53 \pm 0.63, \quad \nu_{☿}{}' = 35.50 \pm 14.92, \tag{XII.13}$$
$$\nu_{♀}{}' = 0.14 \pm 8.45, \quad \nu_{♂}{}' = 2.52 \pm 7.10.$$

The only one of these values that is significant is $\nu_S{}'$, which is slightly less than the value that we would infer from solar data alone.

TABLE XII.7

VALUES OF $\nu_S{}'$ DERIVED FROM HELLENISTIC OBSERVATIONS OF THE SUN AND OF SEVERAL PLANETS

| Body Observed | $\nu_S{}'$ "/cy$^2$ | $\sigma(\nu_S{}')$ "/cy$^2$ |
|---|---|---|
| Sun | 3.02 | 0.70 |
| Mercury | -0.60 | 2.55 |
| Venus | 1.78 | 1.21 |
| Mars | 4.75 | 9.19 |

When we use the ratios $\nu_{☿}{}'/\nu_S{}'$, and so on, that are implied by the spin fluctuation hypothesis (Equations XI.26), we can find a value of $\nu_S{}'$ from the solar observations alone, from the Mercury observations alone, and so on. The resulting values are summarized in Table XII.7, which should be compared with Table XI.12. Both tables show a similar pattern. The values of $\nu_S{}'$ inferred from observations of Mercury and Venus are significantly smaller than the value inferred from solar observations, in both tables. The observations of Mars do not yield a significant estimate of $\nu_S{}'$.

In spite of the great Hellenistic contributions to astronomy, the surviving Hellenistic data do not let us infer meaningful values of the planetary accelerations. This is enough to make one weep at the damage that Ptolemy has done to the science of astronomy. The real damage does not come directly from the fraudulent data that he produced; we can detect and eliminate them fairly easily. The real damage comes from an indirect effect: Because Ptolemy based his Syntaxis upon fraudulent data, it could pretend to a universality that was denied to works based upon honest data. As a result of its meretricious universality, it displaced the genuine works of Hellenistic astronomy. In consequence, Ptolemy has probably caused us to lose almost all of the vast body of accurate Hellenistic observations.

## 6. Some Miscellaneous Observations

Proclus and Heliodorus made three observations of the moon, one of which involves Saturn as well, and one observation that involves Jupiter. In addition, one of the observations that Ptolemy claims to have made of each planet also involves the moon. It is interesting to analyze these observations, even though they are not directly involved in finding the planetary accelerations. We shall need Ptolemy's observations of the moon when we come to study the authenticity of his claimed planetary observations in the next chapter.

TABLE XII.8

SOME MISCELLANEOUS OBSERVATIONS OF THE MOON AND PLANETS

| Date | Julian Day Number -1 000 000 | Primary Body | Secondary Body | $\lambda_1$, degrees | |
|---|---|---|---|---|---|
| | | | | Implied by observation | Calculated |
| 139 May 17 | 771 964.227 | Moon | Mercury | 77.205 | 78.178 |
| 138 Dec 16 | 771 811.611 | Moon | Venus | 217.499 | 216.859 |
| 139 May 30 | 771 977.290 | Moon | Mars | 240.631 | 241.304 |
| 139 Jul 11 | 772 018.628 | Moon | Jupiter | 76.525 | 76.714 |
| 138 Dec 22 | 771 818.248 | Moon | Saturn | 309.636 | 310.399 |
| 475 Nov 18 | 894 873.181 | Moon | Venus | 284.905 | 285.400 |
| 503 Feb 21 | 904 830.388 | Moon | Saturn | 95.645 | 95.792 |
| 508 Sep 27 | 906 875.417[a] | Jupiter | $\alpha$ Leo | 129.163 | 129.193 |
| 509 Mar 11 | 907 040.202 | Moon | $\alpha$ Tau | $<$ 49.812 | 50.069 |

[a]Time assumed to be midnight at Alexandria.

The circumstances of the observations by Proclus and Heliodorus are summarized in Table V.8. The circumstances of Ptolemy's claimed observations are summarized in the tables in Chapter V and Appendix VI that are devoted to the planets involved.

The analysis of the observations is summarized in Table XII.8. In the table, Ptolemy's claimed observations are listed in the order of the planet involved; the other observations are in chronological order. The first group of five are Ptolemy's observations, the observation on 475 November 18 was made by Proclus, and the others were made by Heliodorus. The observation by Proclus was made in Athens and the others were made in Alexandria.

When the time of an observation is not stated in the record, I take it to be 45 minutes from sunrise or sunset, whichever is appropriate, except for the observation of Jupiter and $\alpha$ Leonis on 508 September 27. Since there is no clue to the time of this observation, I take it to be local midnight.

The first two columns in Table XII.8 give the date and the Julian day number, the latter being in terms of Greenwich mean solar time. The next two columns identify the primary and secondary bodies involved in the observation, with the faster body being taken as the primary one. The fifth column gives the longitude $\lambda_1$ of the primary body that is implied by the observation. In order to find this value, I calculated the longitude of the secondary body and added the longitude difference that is found in the record. The accelerations used in calculating the longitude of the secondary body are those that were useful in the analysis of earlier sections in this chapter. For Mercury, Venus, and Mars, the accelerations of both the sun and the planet are taken to be zero. For Jupiter and Saturn, the acceleration of the sun is taken to be 3.60† and that of the planet is taken to be zero. The calculated positions include the effect of aberration.

† I intended to use 3.0, but I accidentally punched 3.6 in the input to the ephemeris program. Since the error has a trivial effect upon the over-all results, I did not bother to correct it after I discovered it.

The last column gives the longitude $\lambda_1$ of the primary body as calculated from the ephemeris program for that body. The lunar longitude includes the effect of parallax. That is, it is the apparent longitude of the moon among the stars as seen from the place of observation. The lunar acceleration used is $\nu_M' = 8.18$. I chose this acceleration by using the mean epoch of the observations in the formula for $D''(= \nu_M' - \nu_S')$ that I have derived elsewhere [Newton, 1972a]; this value is 5.18 $''/\text{cy}^2$. I then added $\nu_S' = 3.00$ to get $\nu_M'$.

In two instances, we have a latitude difference as well as a longitude difference between the two bodies. Heliodorus says that Jupiter was 3 digits (0°.25) north of $\alpha$ Leonis in the night of 508 September 27, while I calculate that it was 0°.32. Ptolemy says that the moon was 2°.0 north of Venus at the time of the observation on 138 December 16, while the calculated difference is 1°.86.

Proclus says that the moon occulted Venus on the night of 475 November 18, but he does not give the hour. I took the time to be 45 minutes after sunset in Table XII.8. The error in the calculated position of the moon that comes from an uncertainty in its acceleration is probably no more than 0°.1 in Table XII.8. According to the calculations, the moon was 0°.5 east of the moon, so that the moon could not have been occulting Venus unless the error in the lunar ephemeris is much greater than 0°.1. 45 minutes earlier, at sunset, the calculated position of the moon is only 0°.1 east of Venus. Since the latitude difference between them is only about 0°.09, Venus was certainly occulted at sunset if the calculations are accurate.

Thus Proclus's observation is correct so far as it goes. However, he is not as careful as Heliodorus in his recording. Heliodorus, and his brother Ammonius, observed an occultation of Saturn by the moon on 503 February 21. They first observed a time when the moon was west of Saturn. They then waited until the moon had passed Saturn and Saturn was again visible, and they measured this time. From the two times, they estimated the time when the center of the moon passed the planet; their estimate was used in preparing Table XII.8. At this time, the calculated position of the moon is about $0°.15$ (= 9′) east of Saturn. This much error could be the result of naked-eye observations, of an error in measuring the times, and of an error in the lunar acceleration used. The observation is clearly of high quality.

The conjunction of Jupiter and $\alpha$ Leonis is also of high quality, except that Heliodorus does not give the time. Since Jupiter moves only about $2\frac{1}{2}$ minutes of arc during a night, he may have thought that it was not necessary to give the time. With the naked eye, he could hardly have determined the time of the conjunction to better than a day.

In the last observation in Table XII.8, Heliodorus says that $\alpha$ Tauri was west of the bright arc of the moon by at most 6 digits after sunset. 6 digits is $0°.5$, the semi-diameter of the moon is about $0°.25$, and the bright edge of the moon appears to the naked eye to be about 3′ (= $0°.05$) from its true position.† Thus $\alpha$ Tauri, according to the observation, was no more than $0°.8$ from the center of the moon. The longitude of the star was $49°.012$, so the longitude of the moon was no more than $49°.812$. At 45 minutes after sunset, the calculated position of the moon was $50°.069$, and it would have been about $49°.7$ at sunset. Thus the observation is valid, but Heliodorus should have given the time a little more accurately. Perhaps he did; perhaps "after sunset" to him meant immediately after sunset. Immediately after sunset, $\alpha$ Tauri was about 5 digits west of the bright edge of the moon, and Heliodorus says that it was no more than 6 digits west.

In sum, the observations of Heliodorus seem to be of good quality, except for the observation of Venus and Jupiter in August of 510. The record of this observation has clearly been garbled in transmission. One garbled record out of six is

---

†Fotheringham [1915] uses this amount for the optical effect of the moon's bright limb, but he does not give an explicit source for it. Since the value seems plausible, I have adopted it. The exact amount is of no consequence for the present discussion.

about what we have always found in the records of ancient and early medieval astronomers.

I shall discuss Ptolemy's claimed observations in the next chapter. Here I remark only that the errors shown in Table XII.8 seem unreasonably large, except for the one involving Jupiter.

# CHAPTER XIII

## THE USE OF PLANETARY TABLES AND THE AUTHENTICITY OF PTOLEMY'S PLANETARY DATA

### 1. Preliminaries; Two Uses of Epicycles

The main purpose of this chapter, from the viewpoint of the goals of this book, is to determine whether ancient and medieval tables of planetary motion are useful in studying the accelerations within the solar system. In order to determine this, we must look at Ptolemy's planetary models in detail in order to know their characteristics and limitations. This process is an intimate part of studying the authenticity of the planetary observations that Ptolemy claims to have made himself. Hence a secondary purpose of this chapter, and one that will actually take more space, is to study the authenticity of these observations.

We know from Chapter VIII and from earlier work [Newton, 1970, Section II.4] that solar and lunar tables are quite useful in studying the accelerations, under suitable conditions.

The epicycle model appears in theories of the sun, the moon, and the planets. Its role in the theory of the sun was studied in Section VIII.2, and its role in the theory of the moon is the same in fundamentals† but more complex in detail. If the reader will refer to Equations VIII.1 and VIII.7, he will see that the epicycle model for the sun leads to the following equation for the longitude $\lambda_S$ of the sun:

$$\lambda_S = L_S + \tan^{-1}[r \sin M/(60 - r \cos M)]. \quad \text{(XIII.1)}$$

†Ptolemy uses an oval rather than a circular deferent for the moon. See Appendix I of Neugebauer [1957], or the discussion by Newton [1974].

The number 60 appears in this because Hellenistic and Islamic astronomers frequently chose 60 rather than 1 for a reference unit. $L_S$ is the mean longitude of the sun and M is its mean anomaly. If we expand the inverse tangent and keep terms through the second degree in the parameter r, we find

$$\lambda_S = L_S + (r/60) \sin M + \tfrac{1}{2}(r/60)^2 \sin 2M, \qquad \text{(XIII.2)}$$

when we express angles in radians rather than degrees.

For an elliptic Kepler orbit of eccentricity e, we have (Equation VIII.2)

$$\lambda_S = L_S + 2e \sin M + (5/4)e^2 \sin 2M, \qquad \text{(XIII.3)}$$

again with angles in radians. If we identify r/60 with 2e, Equation XIII.2 becomes

$$\lambda_S = L_S + 2e \sin M + 2e^2 \sin 2M. \qquad \text{(XIII.4)}$$

The difference between Equations XIII.3 and XIII.4 has a maximum value of $(3/4)e^2$. For the sun, e is about 0.017 and $(3/4)e^2$ is about 0.0002 radians, the equivalent of about 0.7 minutes of arc. Thus the epicycle model is capable of representing the solar motion with a maximum error of less than 1′ if the parameters are chosen accurately.

In general, errors in the solar parameters were more important than defects in the epicycle model at all times covered in this work. Hipparchus (see Section VIII.3 above) had an error of about 20′ in his value of 2e, and Ptolemy preserved this error. ibn Yunis (Section VIII.10) had a 'maximum equation' of 2;0,30 degrees, which is equivalent to 0.03504 for 2e. The epoch of his observations is about 1000, and Newcomb's theory [<u>Explanatory Supplement</u>, 1961, p. 98] gives 0.03423 for the value of 2e at this epoch. Although ibn Yunis's parameter is much more accurate than Hipparchus's, the error in it is still about 0.00081 radians or 2′.8, which exceeds the defect in the model.

Thus, in modern terms, the function of the epicycle model in solar theory is to provide an accurate approximation to an elliptical orbit. Its function in lunar theory is similar but requires a more complex statement. Its primary function in planetary theory, however, is quite different. For the planets, its primary function is to provide a transformation from heliocentric coordinates to geocentric coordinates.

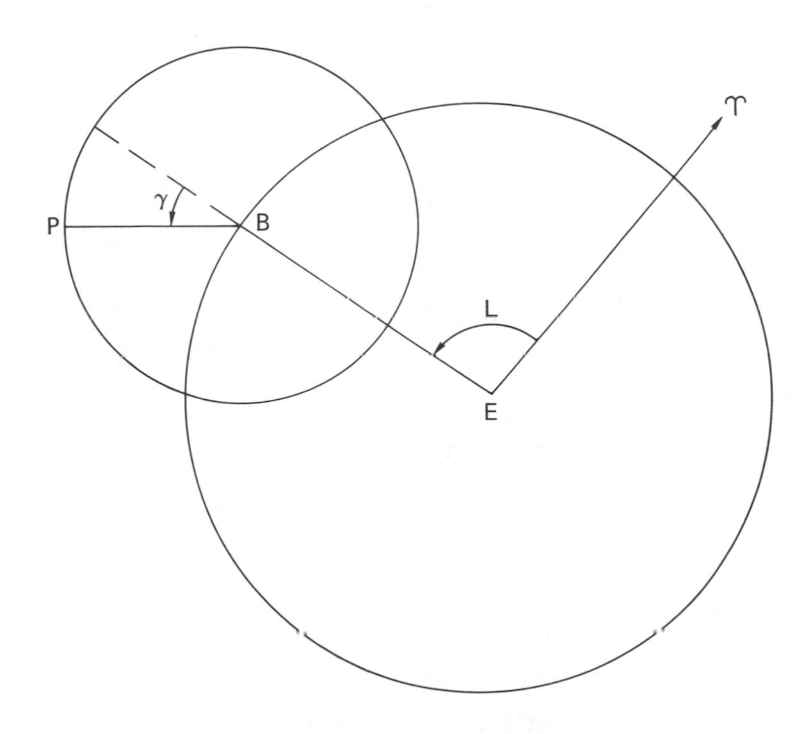


Figure XIII.1. The simplest form of the epicycle model for a planet. The circle whose center is E is called the deferent. The circle whose center is B is called the epicycle. P is the position of the planet. The position of P is fixed by giving the angles L and $\gamma$. L is the mean longitude of the planet and $\gamma$ is called its anomaly.

As we saw in Section III.2, the epicycle model would be exact for the planets if the heliocentric orbits of the earth and planets were circular. It is useful to review the situation briefly with the aid of Figure XIII.1. In this figure, E is the earth and the line E♈ points to the vernal equinox. The point B travels around the deferent circle, and L is its longitude measured from the equinox. B is also the center of the epicycle, and the planet P moves uniformly around the epicycle. The angle $\gamma$ shown in the figure is called the anomaly. Note that $\gamma$ is measured from the extension of the deferent radius EB rather than from a fixed line; this differs from the modern definition of the anomaly. At the level of accuracy embodied in Figure XIII.1, we can physically identify point B with the sun.

Thus the deferent circle is simply the orbit of the sun around the earth and the epicycle is the orbit of the planet around the sun.

For an outer planet, this interpretation makes the epicycle larger than the deferent circle. As I have remarked in Section III.2 and in Appendix IV, the early astronomers chose instead to keep the deferent always larger than the epicycle. For an outer planet, we can still keep a pictorial interpretation by taking E to be the outer planet and P to be the earth. B is still the sun in this picture. Ancient and medieval astronomers, so far as I know, did not use this pictorial interpretation for the outer planets, although they were aware of some of its implications. Instead, they regarded E as being always the earth and P as being always the planet. This did not give them a simple interpretation of B.

The earth, sun, and planet return to the same relative configuration when the anomaly $\gamma$ increases from any value $\gamma_0$ to the value $360° + \gamma_0$, although their orientation in space may change. The time required for $\gamma$ to increase by 360° is therefore the synodic period (Section III.1), whose value is given in Equation III.1. From this equation, we can immediately derive

$$\gamma' = |\nu_S - \nu_P| \qquad \text{(XIII.5)}$$

for the time derivative of $\gamma$ with respect to solar time. Ptolemy, and presumably all other astronomers in Hellenistic times and later, knew† this relation.

For an outer planet, the pictorial interpretation described above identifies the line PB with the earth-sun line. Even if we give up the pictorial interpretation, Equation XIII.5 still means that the line PB is always parallel to the earth-sun line.

## 2. The Eccentric Model

The eccentric model was pictured in Figure VIII.2 and discussed in Section VIII.2. In this model, the celestial body moves uniformly around a

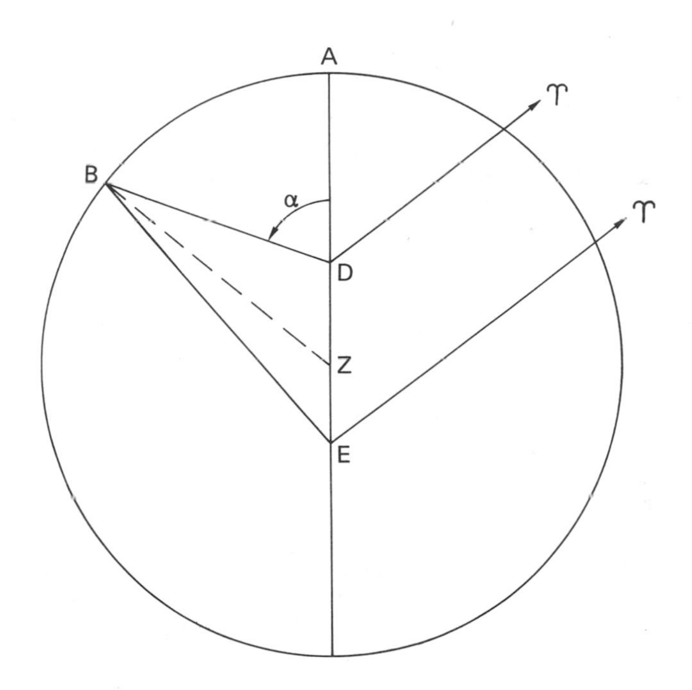


Figure XIII.2. The double eccentric model. B, the body being observed, always stays on the circle whose center is Z, but B moves uniformly about D rather than Z. That is, the angle ♈DB increases linearly with time. The observer is at the point E.

†See Ptolemy [ca. 142, Chapter IX.3].

fixed circle, but the observer is eccentric to the circle. If the radius of the circle is 60 and if the distance from the center to the observer is r, Equation XIII.2 again gives the longitude. That is, the epicyclic and eccentric models give exactly the same observed results, if the corresponding dimensions are the same.

In the rest of this chapter, I shall follow modern practice and take the reference distance to be 1 rather than 60. For an epicycle or eccentric of size r, with this convention, we omit "60" where-ever it occurs in Equation XIII.2.

An interesting variant of the eccentric model is shown in Figure XIII.2. This variant may be called the double eccentric model. In the double eccentric model, the body B moves around the circle whose center is Z and whose radius is unity. The observer is at E, which is eccentric to the circle; let $e_1$ denote the distance ZE (the first eccentric). The center of rotation of B is also eccentric to the circle. Although B always stays on the circle, its rotation is uniform with respect to the point D. That is, the angle ♈DB varies linearly with time, as does the angle $\alpha$. $\pi + \alpha$ is the mean anomaly M, as this term is used in modern astronomy.†

The distance DZ will be denoted by $e_2$ (the second eccentric). The angle ♈EB is the longitude $\lambda$ and the angle ♈DB is the mean longitude L.

From elementary geometry and trigonometry, we have‡

$$\lambda = L - \angle DBE = L - \angle DBZ - \angle ZBE, \quad \sin \angle DBZ = e_2 \sin \alpha,$$

$$\angle ZEB = \alpha - \angle DBE = \alpha - \angle DBZ - \angle ZBE, \quad \sin \angle ZBE = e_1 \sin \angle ZEB.$$

If we expand these relations through terms of the second degree in the eccentrics $e_1$ and $e_2$, and if we

---

†Here, $\pi$ denotes the numerical constant 3.14159..., not a position of perigee.

‡An angle is denoted by a triad of letters, in which the middle letter denotes the vertex.

replace $\alpha$ by $\pi + M$, we get

$$\lambda = L + (e_1+e_2) \sin M + e_1(e_1+e_2) \sin M \cos M.$$

In order to make this parallel to Equations XIII.3 and XIII.4, we let $2e = e_1 + e_2$. That is, $2e$ is the total distance between D and E in Figure XIII.2. This gives us

$$\lambda = L + 2e \sin M + e_1 e \sin 2M. \qquad \text{(XIII.6)}$$

The coefficient of sin M is simply the total distance, and thus it does not depend upon where the center Z is with respect to D and E.

There are two cases of immediate interest. In the simple eccentric model, D is at Z, $e_2 = 0$, and $e_1 = 2e$. For this case, the coefficient of sin 2M in Equation XIII.6 is $2e^2$, in agreement with Equation XIII.4.

We may also choose to make Equations XIII.3 and XIII.6 identical. This requires $e_1 = (5/4)e$ and hence $e_2 = (3/4)e$. That is

$$e_1/e_2 = 5/3. \qquad \text{(XIII.7)}$$

This case is quite interesting. With the ratio of eccentrics given in Equation XIII.7, the double eccentric model gives the same orbit as an ellipse, through terms of the second degree in the eccentricities. Ancient astronomers, of course, had no conscious reason to duplicate an ellipse. However, many orbits are ellipses within the accuracy of naked-eye measurements and, for these orbits, the choice in Equation XIII.7 should give the best fit of the model to the observations.

Ptolemy uses a double eccentric as a part of his planetary model (Section XIII.4 below), but with the ratio $e_1/e_2 = 1$ rather than 5/3.

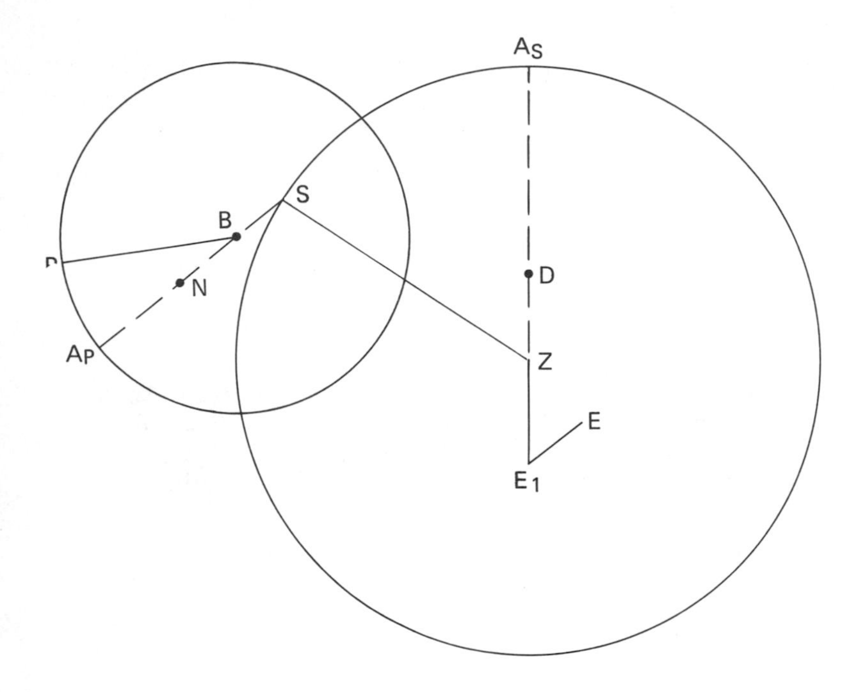


Figure XIII.3. A possible epicycle model that was apparently never used. S represents the sun, and the line SZ rotates uniformly around the point Z. The earth $E_1$ is eccentric to the deferent circle generated by SZ. The point $A_S$ is the solar apogee, and the distance $ZE_1$ is called the solar eccentric. P represents a planet, which moves uniformly around the epicycle whose center is B rather than S. The point $A_P$ is the aphelion of the heliocentric planetary orbit. The distance BS, which was not used by early astronomers, so far as I know, can be called the planetary eccentric. The line $EE_1$ is a construction line used in a discussion in the text. The line $SBA_P$ always remains parallel to itself, except for the slow precession of aphelion.

## 3. The Problem of Two Ellipticities

Determining the geocentric coordinates of a planet involves two heliocentric orbits, one of the earth and one of the planet. We have seen that the epicycle model would give exact results if both orbits were circular. It would give highly accurate, but not exact, results if one orbit were elliptical and one circular. To see this, we recall from Section VIII.2 that the eccentric and epicyclic models give identical results. Thus we could use circular deferent and epicycle circles, but we could either move the earth away from the center of the deferent or make some non-central point within the epicycle be the point that moves uniformly around the deferent.†

In either case, the only error would arise from the fact that an epicycle or a single eccentric cannot represent an elliptical orbit exactly. The error is $(3/4)e^2$. Of the planets that were known in pre-telescopic times, the one with the largest eccentricity is Mercury. Its eccentricity is about 0.2, and the defect in the model is about 0.03 radians, or about 1°.7. The defect is much smaller for the other planets.

In the actual case, both heliocentric orbits are elliptical. It would be possible to handle this situation by the device shown in Figure XIII.3, which uses eccentric circles for both the epicycle and deferent. The caption should supply all the information needed about the figure except the line $EE_1$, which will be discussed in a moment. This model would supply accurate geocentric positions for the planets, except for those with highly elliptical orbits.

However, ancient and medieval astronomers never used the model of Figure XIII.3, so far as I know. They always seem to have required the center B of the epicycle to lie on the deferent. They could still have gotten the equivalent of Figure XIII.3 if they had allowed an "epicycle upon an epicycle"

†Figure XIII.3, which will be discussed in a moment, will help clarify the latter possibility.

or a double epicycle. That is, they could have used an epicycle whose center moves uniformly around the deferent. Instead of putting the planet upon the circumference of this epicycle, they could then have put the planet on a second epicycle whose center moves uniformly around the circumference of the first epicycle. Since a single epicycle is equivalent to an eccentric, this would give the same geocentric motion as Figure XIII.3.

The double epicycle requires the use of three circles. If early astronomers had insisted upon using only two circles, we could see why they never used the double epicycle model. Actually, Ptolemy uses three circles† in his models of the moon and of Mercury and, at least in the case of the moon, he uses fraudulent data [Newton, 1973, 1974] to 'prove' the accuracy of his model. His third circle, however, is usually regarded as the path of a crank handle‡ rather than a second epicycle. Copernicus was the first western astronomer to use a double epicycle for the motion of the moon, and an even more complex model for Mercury, and Copernicus is about fourteen centuries later than Ptolemy.‡

---

†It might be better to say that he uses three uniform rotations. The number of circles obtained depends upon how one draws the figure. In Section XIII.1, I described his lunar model by referring to an oval deferent. See Neugebauer [1957, p. 194] for further discussion.

‡This terminology must not be pushed too far. It is possible to regard the crank as a small deferent, with what is usually called the deferent being regarded as a much larger epicycle, making what is usually called the epicycle into a second epicycle. In any case, Ptolemy's model of Mercury is not a double epicycle in the sense that I have used it, with the center of the first epicycle being the sun.

‡Some non-western astronomers used a double epicycle before Copernicus. It is not known whether Copernicus drew upon the earlier use or whether he invented the model independently. See Neugebauer [1957, p. 197].

The double epicycle is superior to the 'crank' model of Ptolemy, for reasons that will be described in Section XIII.7 below. Hence it seems hard to understand why astronomers were so slow to adopt it when its need is so obvious to us. Perhaps the answer lies in our backgrounds. The need for the double epicycle or the equivalent model in Figure XIII.3 is obvious to us because we come to astronomy through the thought of Kepler and his successors. The modes of thought that lead us to the double epicycle are not so obvious to a pre-Keplerian astronomer, and they are probably not the modes that led Copernicus to it.

If we insist, as Ptolemy and his contemporaries did, upon using a single epicycle whose center is on the deferent, we can still find a partial solution to the problem of two ellipticities. Let the lines BS and $E_1E$ in Figure XIII.3 be parallel and of equal length. If we move the center of the epicycle from B to S, carrying the planet and the entire epicycle with it, and if we also move the earth from $E_1$ to E, we do not change the relation of the earth to the planet. However, this is not a full solution to the problem, for two reasons. First, the point B (or S) can no longer be identified with the sun. Second, the motion around the epicycle is now taken to be uniform about a point lying on the deferent, whereas it should be uniform about a point not on the deferent.

In modern parlance, the line from Z to E is the vector sum of $ZE_1$, the eccentric of the solar orbit, and $E_1E$ (or BS), the eccentric of the (heliocentric) planetary orbit. Thus, except for the two points just made, the problem of two ellipticities can be reduced to a single ellipticity. In the new picture, a point moves uniformly around a deferent whose center is Z, while the earth is at an eccentric point E. The planet then moves uniformly around the point lying on the deferent.

In this discussion, I have been using a single eccentric for each of the two orbits involved. It would be possible to use a double eccentric for each orbit, as the points labelled D and N in Figure XIII.3 suggest. However, we cannot combine the second eccentricities into a single eccentric point applying to the deferent, as we could for the first eccentricity, because the second eccentric points refer to centers of rotation rather than to observing positions. Thus, we should ignore the point N in Figure XIII.3 and leave point D undisturbed. This means that the points D, Z, and E do not lie on a straight line, and we no longer have a proper picture for the double eccentric model.

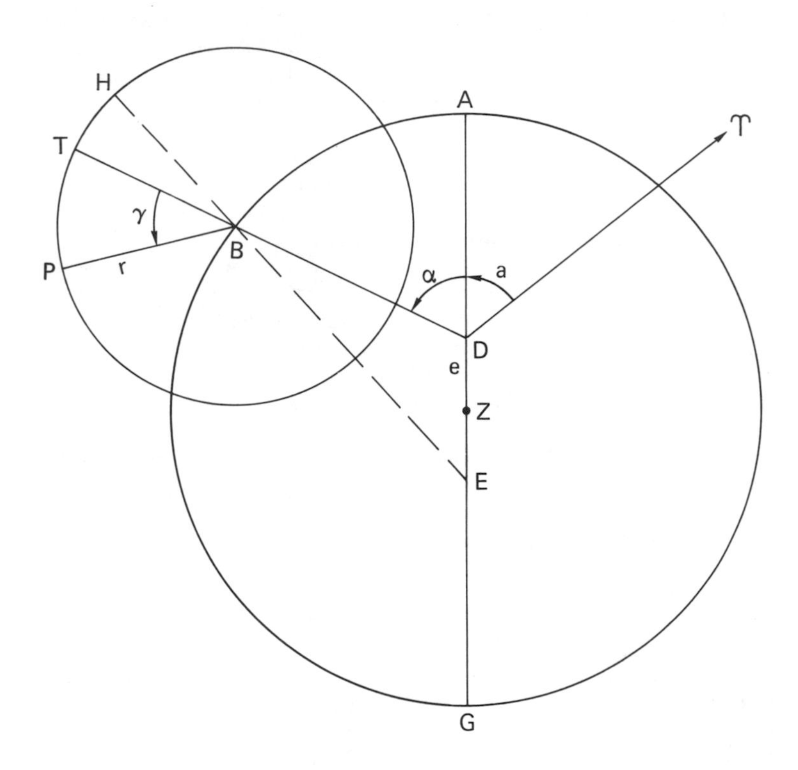


Figure XIII.4. Ptolemy's model for Venus and the outer planets. The planet P moves uniformly around the epicycle of radius r and center B. B always lies on the deferent circle whose center is Z. The rotation of B is not uniform around Z, however; instead, the angle $\alpha (= \angle ADB)$ increases uniformly with time. The distances DZ and ZE are both equal to the eccentric distance e. ♈ is the first line of Aries.

In spite of this, Ptolemy uses a double eccentric in generating the deferent in his models for all the planets except Mercury. I shall next describe the model that he uses for Venus and the outer planets, because it is simpler than the model that he uses for Mercury.

## 4. Ptolemy's Model of Venus and the Outer Planets

The model that Ptolemy uses for Venus and the outer planets is pictured in Figure XIII.4. I shall use the lettering of points given in Figure XIII.4 in all discussions of this model.†

In Figure XIII.4, the planet P moves uniformly around the epicycle whose center is B and whose radius is r. The angle TBP is the anomaly $\gamma$. The center B always remains on the deferent circle ABG whose center is Z. I shall take the radius AZ (= ZG) of the deferent to be unity, although early astronomers, who worked in sexagesimal notation, usually took it to be 60. The rotation of B is uniform about the point D rather than about the point Z. That is, the angle $\alpha$(= angle ADB) is a linear function of time. The observer is at neither D nor Z; he is at the point E, which lies on the diameter ADZEG, with the distance DZ equal to the distance ZE. The line D♈ points to the vernal equinox or the first line of Aries.

Thus, so far as the motion of point B is concerned, Ptolemy uses the double eccentric model of Section XIII.2. However, he uses the ratio $e_1/e_2 = 1$ rather than the value (Equation XIII.7) that gives the closest approximation to elliptical motion.

In much of the writing about Ptolemy's model, the distance DZ, or sometimes the point D itself, is called the equant, and the model of Figure XIII.4 is often called the equant model.

The point G is the point on the deferent that is closest to the observer E, and the point A is the

†Ptolemy is not so kind to his readers. He uses new lettering in almost every separate discussion of the model.

point that is farthest away. When Ptolemy uses apogee in reference to a planet, he usually means A rather than H, which would be the actual apogee position of the planet. Similarly, perigee usually means the point G. I shall adopt these usages in this chapter except in circumstances that will be specifically noted. Ptolemy usually calls E the center of the zodiac, but I shall call it the observer. He calls Z the center of the deferent and he calls D the point about which the motion of the epicycle is uniform. I shall call it the center of rotation.

The angle $\underline{a}$ (= AD♈) is the longitude of apogee. Ptolemy often uses "longitude from the apogee" or some similar phrase to denote the angle $\alpha$; I shall call it the mean distance. Finally, I shall call the distance e (= DZ or ZE) the eccentricity.

Modern writing about the Ptolemaic theory of Venus often refers to B as the mean sun, but this is correct only in a special sense. B is not the position of the mean sun as seen by the observer, and it can be called the mean sun only in the sense that the sum $\underline{a} + \alpha$ is identical with the longitude $L_S$ of the mean sun:

$$L_S = \underline{a} + \alpha \qquad \text{for Venus.} \qquad \text{(XIII.8)}$$

This is only a formal and not a geometric relation, because $L_S$ is measured about the point E while $\underline{a} + \alpha$ is measured about D.

For an outer planet, the line PB in the simplified model of Figure XIII.1 is parallel to the mean sun, and it keeps this property in the fully developed Ptolemaic model of Figure XIII.4. That is,

$$L_S = \underline{a} + \alpha + \gamma \quad \text{for an outer planet.} \quad \text{(XIII.9)}$$

Perhaps the most convincing way to see that Ptolemy uses Equation XIII.9 for the outer planets is to inspect his tables of $\alpha$ and $\gamma$ for the outer planets [<u>Ptolemy</u>, ca. 142, Chapter IX.4] and his table of

$L_S$ [Chapter III.4]. If we calculate the quantity $L_S - \alpha - \gamma$ from these tables as a function of time, we find that it is a constant for a given planet, and this constant is equal to the value that Ptolemy lists for the apogee a of that planet.

As we saw in the preceding section, the points D, Z, and E should not lie on the same straight line. The point D should be the undisturbed point of uniform rotation for generating the deferent, while the line ZE should be the vector sum of the eccentricities of the geocentric solar orbit and the heliocentric planetary orbit. In spite of this, Ptolemy had fair success with the double eccentric in Figure XIII.4 for the following reason: For Venus, the deferent is the orbit of the sun around the earth and the epicycle is the idealized circular orbit of Venus around the sun. The eccentricity† of Venus is about 0.0068 while that of the sun is about 0.017. Thus the point E in Figure XIII.3 is close to $E_1$, and we do not make a large error in making the small compromise necessary to keep D, Z, and E collinear.

A similar thing happens for all the outer planets. For them, the deferent in Figure XIII.4 is the orbit of the sun about the planet and the epicycle is the orbit of the earth about the sun. As it happens, the eccentricities of Mars, Jupiter, and Saturn are all considerably larger than the eccentricity of the earth's orbit.

The model of Figure XIII.4 has five unknown constants for each planet. They are the longitude a of apogee, the eccentricity e, the radius r of the epicycle, the value $\gamma_0$ of $\gamma$ at some epoch, and the time derivative $\gamma'$ of $\gamma$. Ptolemy determines the parameters by analyzing a set of observations. He uses some observations made by other astronomers and he uses some observations that he claims to have made himself. In order to study the authenticity of the latter observations, we must study his use of them in fixing the planetary parameters. I shall do this for Venus and Mars in the next two sections.

---

†In this immediate passage, I am using eccentricity in its modern meaning.

## 5. Ptolemy's Parameters for Venus

The observations that Ptolemy uses in fixing the parameters of Venus are listed in Table V.6. All the observations except the last three were made at times of maximum elongation. Ptolemy uses elongation, in the case of a planet, in a special sense. Let $D_{♀}$ and $\lambda_{♀}$ denote the elongation and the geocentric longitude of Venus, respectively, and let $L_S$ be the calculated longitude of the mean sun. As Ptolemy defines elongation, we have

$$D_{♀} = \lambda_{♀} - L_S. \qquad \text{(XIII.10)}$$

In spite of its definition, $D_{♀}$ is not the apparent angle between Venus and the position of the mean sun, for reasons that were explained in the preceding section.

The places in Ptolemy's text where the observations can be found were given in Section V.6, and I shall not repeat them here.

Ptolemy first finds the position of apogee by considerations of symmetry. Theon had observed a maximum elongation on 132 March 8, when $L_S$ was $344\frac{1}{4}$ degrees and $\lambda_{♀}$ was $31\frac{1}{2}$. Thus the elongation was $47\frac{1}{4}$ degrees to the east.† Ptolemy then makes an observation of Venus when it had the same maximum elongation to the west.‡ This happened on 140 July 30, when $L_S$ was 125 3/4 degrees and $\lambda_{♀}$ was $78\frac{1}{2}$, making $D_{♀} = -47\frac{1}{4}$. Hence, he concludes, the apogee-perigee line bisects the two values of $L_S$, and hence it is the line that passes through longitudes 55° and 235°. He cannot tell yet which end is apogee and which is perigee.

---

†When I refer to an east elongation, I mean that Venus is east of the sun. Since it is east of the sun, it appears in the evening and hence in the west. An east elongation means a positive value of $D_{♀}$.

‡In writing that Ptolemy "makes an observation", I am not implying that the observation is authentic. I am merely avoiding continuous use of the circumlocution "claims to have made an observation".

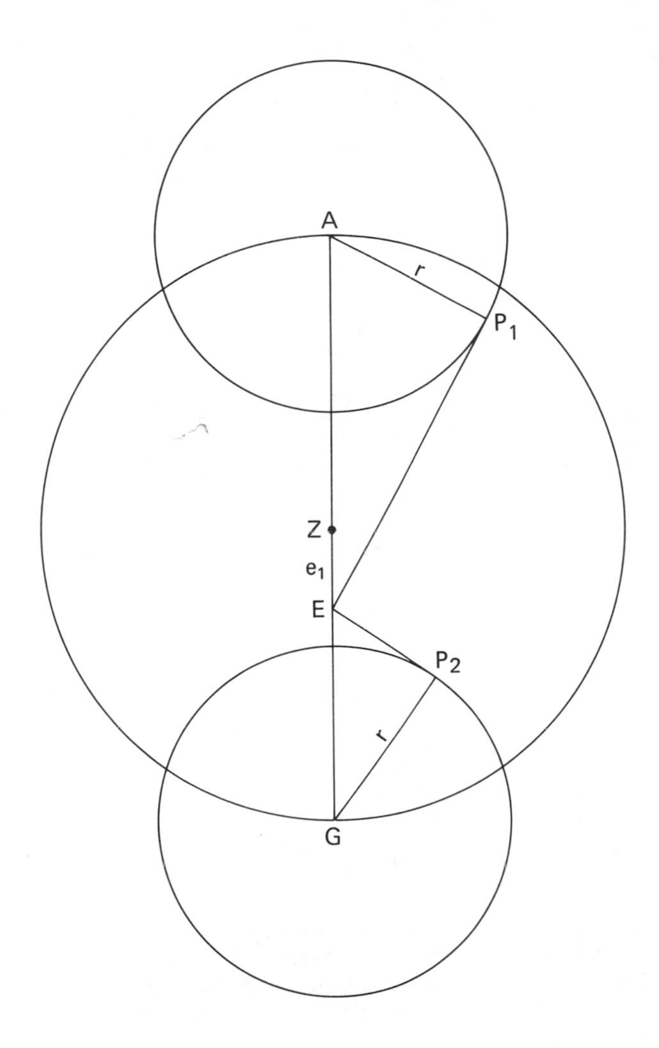


Figure XIII.5. The extremes of elongation for Venus. On 129 May 20, according to Ptolemy, the mean sun was at A, Venus was at $P_1$, and the angle $AEP_1$ was 44°48′. On 136 November 18, the mean sun was at G, Venus was at $P_2$, and the angle $GEP_2$ was 47°20′.

Ptolemy confirms the position of apogee by another pair of observations. On 127 October 12, Theon had observed another maximum elongation of Venus. He found $L_S = 197°52'$, $\lambda_♀ = 150°20'$, and $D_♀ = -47°32'$. Ptolemy then observes that $D_♀ = +47°32'$ on 136 December 25, when $L_S = 272°4'$ and $\lambda_♀ = 319°36'$. The average of the two values of $L_S$ goes through the longitudes 54°58′ and 234°58′. Ptolemy takes these to be equivalent to 55° and 235°.

So far, only two values of the elongation have occurred. The sun is closer to 235° than to 55° at the time of the larger value. Hence 235° must be the perigee and 55° the apogee.

Ptolemy does not assume that the distances DZ and ZE in Figure XIII.4 are equal. Instead, he takes the distances to be independent parameters and finds each one independently by means of observations. Hence we need two eccentricities in discussing his model for Venus. I shall denote the distance ZE by $e_1$ and call it the first eccentricity. I shall denote DZ by $e_2$ and call it the second eccentricity.

Ptolemy finds the radius r of the epicycle and the first eccentricity $e_1$ by means of observations made when $L_S$ is 55° and 235°, as closely as possible. Theon made one observation on 129 May 20, when he found $L_S$ = 55 2/5 degrees,† $\lambda_♀$ = 10 3/5 degrees, and $D_♀$ = -44 4/5 degres. Ptolemy made the other on 136 November 18, when he found $L_S = 235\frac{1}{2}$ degrees, $\lambda_♀$ = 282 5/6, and $D_♀$ = 47 1/3 degrees.

The picture given by these observations is shown in Figure XIII.5. At the time of the first observation, the mean sun was at A and Venus was at maximum elongation. Hence the line of sight from the observer E was tangent to the epicycle

---

†This appears as "55° plus two thirds" in Halma's text, with the fraction being written out in words. However, Ptolemy's use of the observation shows that 55 2/5 is meant.

and Venus must have been at point $P_1$,† since it was west of the sun. Similarly, the sun was at G‡ and Venus was at $P_2$ at the time of the second observation. The radius AZ = ZG of the deferent will be taken as unity. The angles $AEP_1$ and $GEP_2$ are 44°48′ and 47°20′, respectively.

Thus we have

$$r = (1+e_1) \sin 44°.8 \text{ and } r = (1-e_1) \sin 47°.333$$

from the two triangles in Figure XIII.5. The solution is

$$e_1 = 0.021\ 303, \qquad r = 0.719\ 645. \qquad \text{(XIII.11)}$$

These agree closely with the solution that Ptolemy gives (Chapter X.2):

$$e_1 = 1\tfrac{1}{4}/60 = 0.020\ 833, \quad r = 43\tfrac{1}{6}/60 = 0.719\ 444. \qquad \text{(XIII.12)}$$

These observations locate the point E in Figure XIII.4, but they do not tell us where the point D is. Ptolemy uses the next two observations to locate D, and he uses observations that he had made himself. On 134 February 18, he found $L_S$ = 325°30′, $\lambda_♀$ = 281°55′, and $D_♀$ = -43°35′. On 140 February 18, he found $L_S$ = 325°30′ again, but he found $\lambda_♀$ = 13°50′ and $D_♀$ = 48°20′. Venus was a morning star on the first occasion and an evening star on the second.

†This construction corresponds to measuring the longitude $L_S$ of the mean sun about the point E rather than the point D, which is not shown in Figure XIII.5. However, this is the construction that Ptolemy uses.

‡On both occasions, the mean sun was 30′ beyond points A and G, respectively. If we assume that it was exactly at A and G, we make an error that is proportional to the difference between cos 30′ and unity. This difference is about 0.00004, or 0.004 percent.

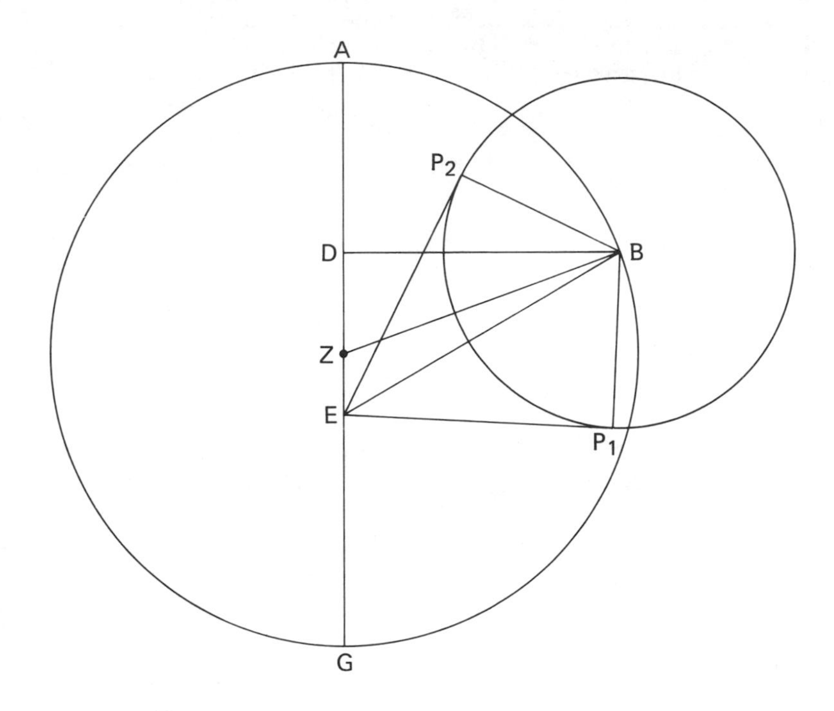


Figure XIII.6. Locating the center of rotation D in Ptolemy's model of Venus. On 134 February 18, according to Ptolemy, Venus was at $P_1$, where its longitude was 281°55′. On 140 February 18, Venus was at $P_2$, where its longitude was 13°50′. On both occasions, the line from the center B of the epicycle to the center of rotation D was perpendicular to the diameter AG of the deferent.

The situation is illustrated in Figure XIII.6. On both occasions, $L_S$ had the same value, which is 90° less† than the longitude of the apogee point A.

†There is a trifling error in calling the angle 90° when it is really 89°30′.

Hence DB is perpendicular to the diameter AG of the deferent. Venus was at $P_1$ on 134 February 18 and at $P_2$ on 140 February 18. The problem is to find the distance DZ, which is the second eccentricity $e_2$.

First we find the distance EB, which is the hypotenuse of the right triangle EDB. The angle $P_2EP_1$ is the difference between the longitudes of Venus on the two occasions, namely $91°55' = 91°.917$. Since the elongation angle $BEP_1$ is half of this, and since $BP_1E$ is a right angle, we can immediately find BE from the angle and from the epicycle radius, which equals 0.719 645 (Equation XIII.11). The result is EB = 1.001 1274. Ptolemy, using the values in Equation XIII.12, gets 1.000 8333.

Now, using the exact values, angle ADB is $89°30'$ and angle $AEP_2$ is $41°10'$. Since angle $BEP_2$ is $45°52\frac{1}{2}'$, angle DBE is $2°22\frac{1}{2}'$. Solving triangle DBE gives $DE = e_1 + e_2$ and using Equation XIII.12 then gives $e_2$. My result is $e_2 = 0.020\ 173$ while Ptolemy's result is

$$e_2 = e_1 = 0.020\ 833. \qquad \text{(XIII.13)}$$

We now have all of the parameters in Ptolemy's model except $\gamma_0$ and $\gamma'$, the parameters needed to calculate the anomaly $\gamma$. Ptolemy needs two observations in order to find $\gamma_0$ and $\gamma'$. He observed Venus, he says, at 4 3/4 equal hours after the midnight that begins 138 December 16, when he found Venus at longitude $216°30'$, and latitude $+2°40'$. The mean sun was at $262°9'$.

He uses an observation by Timocharis for the other observation needed. Timocharis had observed Venus in the morning of -271 October 12, and Ptolemy concludes from the data that $\lambda_♀ = 154°10'$ and $L_S = 197°3'$. Ptolemy does not state the value of the time explicitly, but he clearly takes it to be sunrise in his calculations. This is $45^m$ later than the time that I used in Table V.6.

Ptolemy needs the position of apogee at the time of Timocharis's observation in order to perform his calculations. By means of some observations of Mercury that I shall discuss later, Ptolemy has already 'proved' that the longitude of Mercury's apogee increases by 1° per century; this is exactly his precession of the equinoxes. From this, he concludes that the apogees of all the planets remain fixed with regard to the stars and hence that they increase in longitude just as the stars do.

Ptolemy also needs to know whether the observation was made before or after maximum elongation. He uses the observation made by Timocharis on -271 October 16 to show that it was after, but he makes no other use of the latter observation.†

For each observation, we now know the longitude of P as seen from E, and we know all other angles and distances in Figure XIII.4. Even with modern computing aids, it is tedious, although straightforward, to calculate the anomaly $\gamma$. The calculation must have been extremely tedious in Ptolemy's time, and we can see why ancient astronomers did not tend to use large volumes of data. The results are

$\gamma = 252°11'$ on -271 October 12,

$\gamma = 230°40'$ on 138 December 16,

rounded to the nearest minute. Ptolemy gives

$\gamma = 252°7'$ on -271 October 12,

$\gamma = 230°32'$ on 138 December 16. (XIII.14)

The agreement is good, particularly in view of the fact that an important step is finding an angle near 90° from its sine.

The time between the two observations is within about $30^m$ of being 149 452 days. If we use 149 452

†We can see from Figure XIII.4 that two different values of $\gamma$, one before and one after maximum elongation, give the same geocentric longitude. When Ptolemy wants to infer $\gamma$, therefore, he must know whether Venus is before or after maximum elongation in order to know which value of $\gamma$ to use.

days and the anomalies from Equations XIII.14, we get $\gamma' = 0.616\ 508\ 4219$ degrees per day,† or

$$\gamma' = 0;36,59,25,49,8,49 \quad \text{degrees per day}$$

in sexagesimal notation. Ptolemy says that the value obtained this way is close to the value that he has given earlier (in the tables in Chapter IX.4); this value is

$$\gamma' = 0;36,59,25,53,11,28 \text{ degrees per day.} \quad \text{(XIII.15)}$$

The value from Equation XIII.15 is the one that Ptolemy adopts.

Finally, Ptolemy calculates the anomaly at the epoch of Nabonassar, which is noon, apparent time at Alexandria, on -746 February 26, using the value of $\gamma'$ from Equation XIII.15 and the value of $\gamma$ on -271 October 12. The value calculated this way is 71°8′, but Ptolemy gives 71°7′. If he had used the value that is actually derived from Equations XIII.14, he would have gotten 71°12′, so it is clear that he did not use this value.

It is useful to close this discussion with two general remarks.

Although the detailed study of the authenticity of Ptolemy's observations is being deferred to Section XIII.9, this is a good place to ask about possible methods of fudging the data. In fraudulent work that has already been analyzed, Ptolemy uses observations made by himself and by others, and the number of observations made by others is enough to fix all the needed parameters, with one exception: Ptolemy does not refer to Aristarchus's measurement of the ratio between the mean solar and lunar

---

†I shall depart from the conventional set of units in this discussion.

distances† nor to any equivalent measurement that he could have used in fudging the solar and lunar parallaxes. However, it is clear [Newton, 1973] that Ptolemy used Aristarchus's result in fudging his fraudulent lunar observation of 135 October 1, or some result that is very close to Aristarchus's.

Thus, when Ptolemy fudges data, he does not always give us the authentic observations that formed the basis for his fudging.

The other general remark concerns the accuracy of Ptolemy's model. I have programmed Ptolemy's model of Venus and have compared geocentric longitudes calculated from it with those calculated by the accurate ephemeris programs of Chapter IX. Since Ptolemy's equinox is 1°.1 too far to the east, I added 1°.1 to his longitude of apogee. I also used Newcomb's expression for the mean longitude of the sun instead of Ptolemy's. I did the comparisons for 51 positions calculated for times 44 days apart. I chose this time span so that both Venus and the sun would return to nearly the same positions at the end of the total time of 2200 days. The total interval is centered on the epoch 137 July 20,‡ which is nearly in the middle of Ptolemy's career. The maximum discrepancy that I found is 4°.29, on 137 March 10, and the root-mean-square discrepancy is 1°.01.

I then allowed all six parameters ($a$, $e_1$, $e_2$, $r$, $\gamma_0$, and $\gamma'$) to vary until the model gave the

---

†Aristarchus found that the ratio lies between 18 and 20, a result that is too small but that is nonetheless remarkable in view of the time (about -280) when the measurement was made. The work in which Aristarchus gives the method and the result has been preserved, although I have not consulted it. The title is usually given as "On the Magnitudes and Distances of the Sun and Moon" in English. T.L. Heath published an edition with a parallel English translation in 1913; I do not know the publisher.

‡This is the first day of the first year of the emperor Antoninus Pius; it is an epoch that Ptolemy frequently uses.

TABLE XIII.1

A COMPARISON OF PTOLEMY'S PARAMETERS FOR VENUS WITH THOSE THAT GIVE THE BEST FIT

| Parameter | Ptolemy's Value | Best Fit |
|---|---|---|
| Apogee, degrees | 56.1[b] | 60.202 |
| First eccentricity ($e_1$) | 0.020 833 | 0.012 883 |
| Second eccentricity ($e_2$) | 0.020 833 | 0.014 711 |
| Epicycle radius | 0.719 444 | 0.722 804 |
| Anomaly at the epoch,[a] degrees | 273.825 | 274.675 |
| Anomalistic rate, degrees per day | 0.616 508 734 | 0.616 595 195 |

[a]Noon, Alexandria time, 137 July 20.

[b]Ptolemy's value is 55°, but this is referred to an erroneous equinox. I added 1°.1 in order to refer to an accurate position of the equinox.

best fit to the ephemeris in a least-squares sense. The values that give the best fit are compared with Ptolemy's values in Table XIII.1, which should be self-explanatory.

The biggest relative changes in the parameters are those in $e_1$ and $e_2$. In absolute terms, however, these changes are not particularly important. They amount to about 0.008 and 0.006, which are approximately equal to the heliocentric eccentricity of Venus. I have not tested the matter explicitly, but the most important changes are perhaps those in $\gamma_0$, the anomaly at the epoch, and in apogee. Apogee should be about 4° larger than Ptolemy's value, after we refer it to the correct equinox.

The fitting decreased the greatest discrepancy from 4°.29 to 0°.32, and it decreased the root-mean-square discrepancy from 1°.01 to 0°.14. Thus we see that Ptolemy does a rather poor job of finding the parameters of Venus's orbit. His model is capable of representing the orbit, at least in longitude, with a root-mean-square error of 0°.14, which is about equal to the accuracy of observations in Ptolemy's time. However, the root-mean-square error when Ptolemy's parameters are used is more than 1° and the maximum error is greater than 4°. A few observations of mediocre quality would have shown that his model is seriously in error. It is hard to tell how his trouble arises. Part of it undoubtedly comes from the observations that he uses in finding the parameters. However, some trouble seems to come from small errors in his calculations.

## 6. Ptolemy's Parameters for Mars

The qualitative relation between the solar orbit and the planet is the same for Mars, Jupiter, and Saturn, and Ptolemy uses identical methods of finding the parameters in the models for them. Thus, if we can discover what he did for Mars, we know almost surely what he did for the others.

For each outer planet, Ptolemy first uses three observations of mean opposition, all made by himself; I shall explain what he means by mean opposition in a moment. From these observations, he finds the apogee a and the eccentricity e,† and he finds the anomaly $\gamma$ at the epoch of one of the observations. He then uses an observation, also made by himself, when the planet is not at opposition; this gives him the epicycle radius r. For all three planets, he makes this observation when the planet is close to the moon, for reasons that I do not understand.‡ Finally, he uses an observation made in the -3rd century to find another value of $\gamma$. The two values of $\gamma$ then give him $\gamma_0$ and $\gamma'$.

By mean opposition, Ptolemy means the configuration in which the points B, P, and E of Figure

†He has shown that the two eccentricities $e_1$ and $e_2$ are equal for Venus, and he assumes $e_1 = e_2 = e$ for the outer planets without discussion.

‡Several writers have said that he did this in order to improve his accuracy. Ptolemy may have thought that this process improves the accuracy, but I do not see how it does. His accuracy depends upon the difference between two positions, and the accuracy is governed by the poorer observation, not the better one. Even if referring one position to the moon made that position more accurate, which is itself a doubtful matter, having only one accurate position is like measuring all but the last centimeter of a line with a meterstick and measuring the last centimeter with a micrometer.

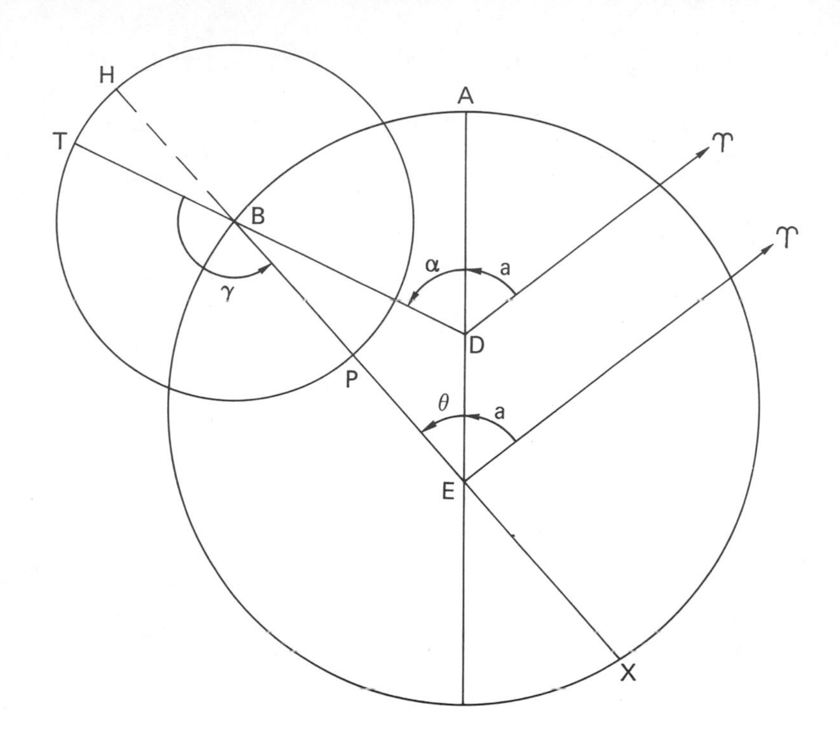


Figure XIII.7. Ptolemy's picture of mean opposition and mean conjunction of an outer planet. When the planet is at the point **P**, it is a theorem with Ptolemy to prove that the mean position of the sun, as seen from the earth **E**, is at the point X; this is mean opposition. When the planet is at H, it is also a theorem to prove that the mean sun, as seen from earth, is at the point B; this is mean conjunction.

XIII.4 are collinear, with **P** lying between B and E. To us, **E** is the planet, **P** is earth, and B is the sun, so that the earth is between the sun and the planet, and the planet is at opposition. The picture that Ptolemy visualizes, however, is quite different; his picture is shown in Figure XIII.7. To him, **E** is the earth and **P** is the planet. He must then prove as a theorem that the mean sun, as seen from earth, is at the point X, in opposition to **P**. That is, he must prove that $L_S$, the longitude of the mean sun, is equal to $\underline{a} + \theta + 180°$.

The proof is simple. The angle $\theta$ is equal to the mean distance $\alpha$ minus the angle EBD. The latter angle is equal to 180° minus the anomaly $\gamma$. Hence,

$$\theta = \alpha - (180° - \gamma).$$

The longitude of point X from the equinox ♈ is clearly equal to $\underline{a} + \theta + 180°$, in which $\underline{a}$ is the longitude of the apogee point A. Hence,

$$\underline{a} + \theta + 180° = \underline{a} + \alpha + \gamma = L_S. \qquad \text{(XIII.16)}$$

The last step follows from Equation XIII.9 in Section XIII.4. Ptolemy proves this theorem in his Chapter X.6.

For Ptolemy, mean conjunction occurs when the planet is at the point H in Figure XIII.7. That is, he proves as a theorem that $L_S = \underline{a} + \theta$ when the planet is at H, so that B is the position of the mean sun as seen from earth. The proof follows the method just used.

Let us now confine our attention to Mars, and let $\lambda_{♂}$ denote its longitude, which is the angle $\underline{a} + \theta$ in Figure XIII.7. That is,

$$\lambda_{♂} = \underline{a} + \theta. \qquad \text{(XIII.17)}$$

Ptolemy presumably finds mean opposition by allegedly measuring $\lambda_{♂}$, by calculating $L_S$, and finding when they differ by 180°. He finds $\lambda_{♂} = 81°$ and $L_S = 261°$ at 01 hours on 130 December 15 (Table V.7), he finds $\lambda_{♂} = 148°50'$ and $L_S = 328°50'$ at 21 hours on 135 February 21, and he finds $\lambda_{♂} = 242°34'$ and $L_S = 62°34'$ at 22 hours on 139 May 27. He uses Equation XIII.17 to find the changes in $\theta$ between the observations. He also needs the changes in $\alpha$, but he cannot find them from Equation XIII.16 because he does not know the values of the anomaly $\gamma$. Hence he calculates the changes in $\alpha$ from the differences in time and from $\alpha'$, the rate of change of $\alpha$, which he takes from his table of planetary mean motions in his Chapter IX.4.†

It is a difficult job to find the desired quantities from the three observations. Ptolemy needs many pages to do so, and I made no attempt to follow his method. Instead, I took the three unknowns to be $\alpha_1$, the value of $\alpha$ at the first observation, $\theta_1$, the value of $\theta$ at the first observation, and the value e of the eccentricity. From the increments in $\alpha$ and $\theta$, I could then express $\alpha$ and $\theta$ at the other observations in terms of $\alpha_1$ and $\theta_1$. If we draw in the radius of the deferent circle in Figure XIII.7, we easily find that

$$\alpha - \sin^{-1}(e \sin \alpha) = \theta + \sin^{-1}(e \sin \theta) \qquad \text{(XIII.18)}$$

at the time of each observation. I next solved the three resulting equations for e, $\alpha_1$, and $\theta_1$. Equation XIII.17 then supplies the value of $\underline{a}$. Finally, I calculated $\alpha_3$, the value of $\alpha$ at the third observation, from $\alpha_1$ and the known increments, and used Equation XIII.16 to find $\gamma_3$, the value of $\gamma$ at the third observation. The results are

$$\underline{a} = 115° \ 29' \ 37'',$$
$$e = 0.100\ 003,$$
$$\gamma_3 = 171° \ 24' \ 57''.$$

The errors in $\underline{a}$ and $\gamma_3$, assuming that the data are exact, are no more than 2″.

Ptolemy does an extremely good job of solving the observations, much better than he does for

†Ptolemy's goal in using these observations is to find apogee $\underline{a}$, eccentricity e, and anomaly $\gamma$ at the time of the third observation. However, if he could find $\gamma$ at the third observation, he could also find it at the first observation, by symmetry. This would give him four quantities from three measurements, which is impossible. Hence he assumes $\alpha'$, and does not find it from any observations that he tells us about.

Venus. He finds

$$a = 115^\circ\ 30',$$
$$e = 0.1, \qquad \text{(XIII.19)}$$
$$\gamma_3 = 171^\circ\ 25',$$

and I shall use his values in the remaining discussion.

Since Ptolemy has already assumed the value of $\alpha'$, he also knowsthe value of $\gamma'$; it is 0.461 575 5672 degrees per day, or 0;27,41,40,19,20,58 in sexagesimal notation. From the value of $\gamma_3$, we can get $\gamma_0$, the value of $\gamma$ at the epoch of Nabonassar. I get 327°13'.6, while Ptolemy gives 327°13'.

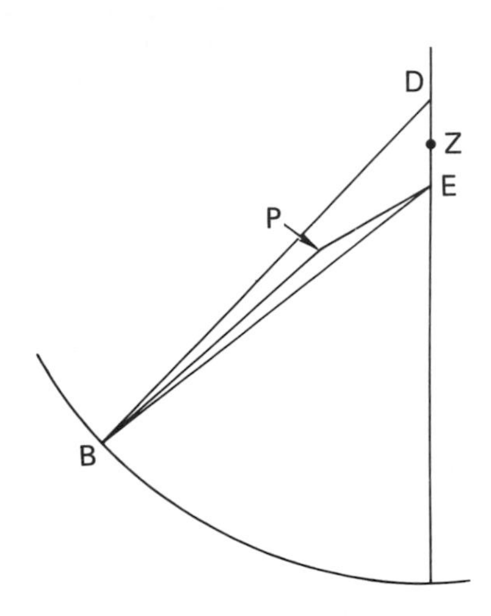


Figure XIII.8. The configuration of Ptolemy's model for Mars, on the basis of an observation that he claims to have made at 21 hours, Alexandria time, on 139 May 30. Mars was at P. According to the observation, angle BDE is 180° - 137°.2, angle DEB is 128°.8366, and angle DEP is the longitude of Mars with respect to the apogee point, or 126°.1. The drawing is to scale except that angle BEP has been enlarged so that it may appear clearly.

The radius r of the epicycle is the only parameter that has not been found, and Ptolemy uses an observation made by himself at 21 hours on 139 May 30 in order to find it; this is 2 days and 23 hours after the third mean opposition that he has just used. He observed $\lambda_{♂} = 241°36'$ at a time when $L_S$ was 65°27'. The configuration is shown in Figure XIII.8. The notation in Figure XIII.8 is the same as in Figure XIII.4, but I have omitted the epicycle and most of the deferent for clarity. The figure is approximately to scale, except that I have enlarged the angle BEP slightly in order that it may appear clearly. The distance BP is the epicycle radius r, which is what we want to find.

The angle BDE is the supplement of $\alpha$, which, from the earlier results, we find to be 137°.2. Angle DEB, which is called $\theta$ in Figure XIII.7, is then equal to 128°.8366 (Equation XIII.18). Angle DBE is 8°.3634, the difference between $\alpha$ and $\theta$. We also calculate from the earlier results that $\gamma$ equals 172°.7821, and angle DBP is the supplement of this. From angles DBE and DBP, we get angle PBE.

Since P is the position of Mars as seen from E, angle DEP equals $\lambda_{♂} - \underline{a}$, or 126°.1, and angle PEB is $\theta$ minus this value. We now have two angles in triangle BPE and can immediately get the third. We can also find the distance BE by solving the triangle ZBE, in which we know that side BZ is unity, that side ZE is the eccentricity e, (Equation XIII.19), and in which $\theta$ is the angle at E. Finally, we solve the triangle BPE for r, getting

$$r = 0.6598. \qquad \text{(XIII.20)}$$

Ptolemy gives $r = 39\frac{1}{2}$ if the radius BZ is taken as 60; thus his value of r is 0.6583. In modern terms, r is the ratio of the earth's semi-major axis to that of Mars, equal to 0.6563. Ptolemy's value is accurate to about 1 part in 300.

In fact, Ptolemy did a remarkably good job in finding the parameters of Mars, much better than he did for Venus. After I corrected his position of apogee so that it would refer to the correct equinox, I compared values of the longitude from his model

TABLE XIII.2

A COMPARISON OF PTOLEMY'S PARAMETERS FOR MARS WITH THOSE THAT GIVE THE BEST FIT

| Parameter | Ptolemy's Value | Best Fit |
|---|---|---|
| Apogee, degrees | 116.6[b] | 117.977 |
| First eccentricity ($e_1$) | 0.1 | 0.097 536 |
| Second eccentricity ($e_2$) | 0.1 | 0.099 953 |
| Epicycle radius | 0.658 333 | 0.657 767 |
| Anomaly at the epoch,[a] degrees | 219.189 | 219.140 |
| Anomalistic rate, degrees per day | 0.461 575 567 | 0.461 504 019 |

[a]Noon, Alexandria time, 137 July 20.

[b]Ptolemy's value is 115°.5, but this is referred to an erroneous equinox. I added 1°.1 in order to refer to an accurate position of the equinox.

with those from modern ephemerides. The maximum discrepancy that I found was 0°.89, and the root-mean-square discrepancy was 0°.44.

I then found the parameters that give the best fit, without requiring $e_1 = e_2$. The best-fitting parameters are compared with Ptolemy's values in Table XIII.2. After the fit, the maximum discrepancy actually rose slightly to 0°.91, while the root-mean-square discrepancy fell slightly to 0°.36.

The values of the eccentricities are worthy of note. Ptolemy 'proved' from observations that the two eccentricities are equal for Venus, but, as we see from Table XIII.1, he made a moderately serious error in doing so. The ratio $e_1/e_2$ should be about 0.875 rather than 1, and both eccentricities should be about 2/3 of the value that Ptolemy finds. Ptolemy then takes it for granted that the eccentricities will be equal for Mars. In fact, they are about equal for Mars and they are nearly equal to the value that Ptolemy derives.

Except for the formality of finding the anomaly at the epoch of Nabonassar, Ptolemy concludes his treatment of Mars by quoting a measurement that he dates to the equivalent of -271 January 18,† but he

†In Section XII.4, I concluded that the correct date was probably -271 January 16.

does not name the person who made the observation. He calculates that the anomaly at the time of the observation was 109° 42′. When we use the same date and hour for this observation that Ptolemy does, and when we use $\gamma_3$ from Equations XIII.19, which Ptolemy also uses, we find $\gamma' = 0.461\ 575\ 576$ degrees per day. However, the value of $\gamma'$ that Ptolemy adopts is 0.461 575 567 degrees per day. Thus, with Mars just as with Venus, Ptolemy does not adopt the value of $\gamma'$ that his data lead to, and he does not tell us the basis for the value that he does adopt.

In sum, Ptolemy quotes only four observations of Mars that he actually uses in constructing his model of its motion. All four are observations that he claims to have made himself. If it turns out that these observations were fudged, we do not seem to have any way to reconstruct the process by which he found the model.

## 7. Ptolemy's Model for Mercury

We should remember that the orbit which naturally forms the deferent in a planetary model is the sun's apparent orbit around the earth when we are dealing with Mercury or Venus, and that it is the sun's apparent orbit around the planet when we are dealing with an outer planet. The natural epicycle is the planet's heliocentric orbit for Mercury or Venus, and the earth's heliocentric orbit when we have to do with an outer planet.

Mercury poses a problem that is unique among the planets known in classical antiquity. For all the other planets, it happens that the eccentricity of the natural epicycle is small compared with the eccentricity of the natural deferent. Thus it makes a relatively small error to take the epicycle to be circular, as it is in Figure XIII.4. With Mercury, however, the eccentricity of the natural epicycle is about 0.2, and this fact would cause large errors if we used the same model for Mercury that we use for the other planets.

Ptolemy did not word the problem in this way, of course. The problem as he saw it concerns the maximum elongations of Mercury, and it can be

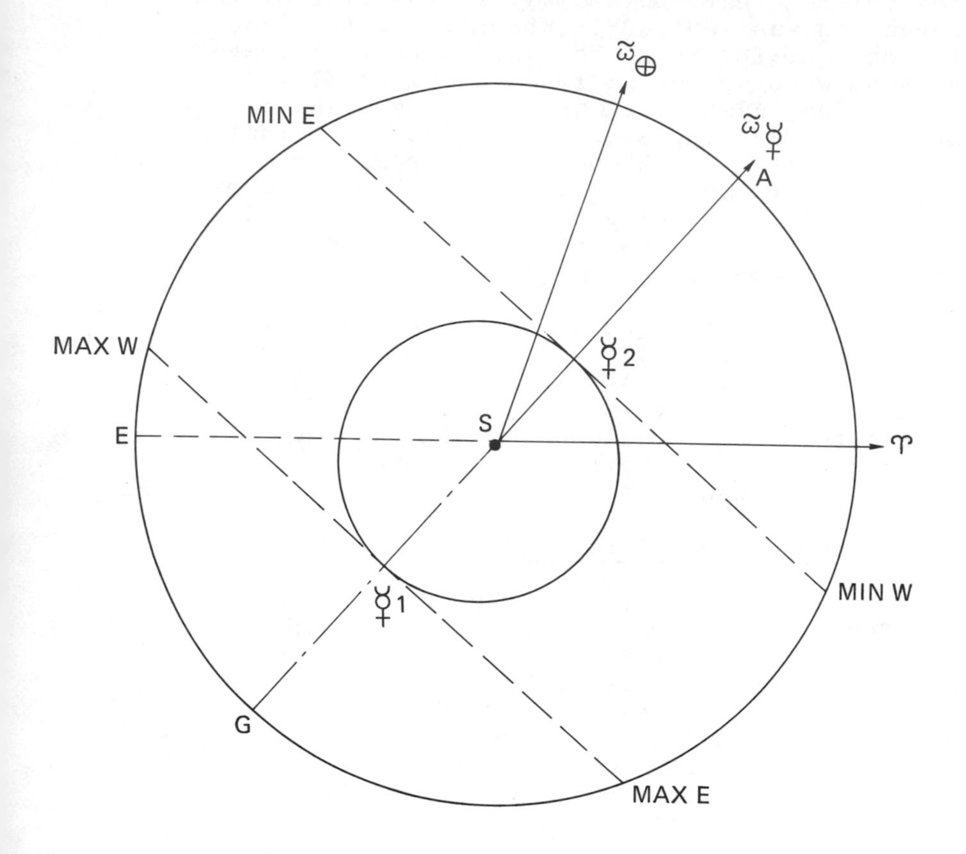


Figure XIII.9. The heliocentric orbits of Mercury and the earth, drawn approximately to scale. The line AG is the line of apsides of Mercury, and the perihelion of Mercury was at longitude 48°.39, approximately, in the time of Ptolemy. The perihelion of the earth, marked $\tilde{\omega}_\oplus$, was at about 71°.04. When Mercury is at ☿1, its maximum possible elongation is about 28°, and when it is at ☿2, its maximum possible elongation is about 18°. The line from S to ♈ points to the first line of Aries, so that the earth is at E at the vernal equinox, when the longitude of the sun is zero.

understood in modern terms by means of Figure XIII.9. In Figure XIII.9, I have drawn the heliocentric orbits of Mercury and the earth approximately to scale, using the values that the orbital parameters had in Ptolemy's time.

Suppose that the earth is at the point marked "Max E" in the figure and that Mercury is at the point ☿1. Then the elongation of Mercury is a relative maximum and, since Mercury appears to the east of the sun as seen from earth, Mercury is at a maximum elongation east. When the earth is at the point "Min W" and Mercury is at ☿2, Mercury is at a maximum elongation west. The points marked "Min E" and "Max W" have analogous meanings. The points ☿1 and ☿2 are the ends of the semi-major axis of Mercury's orbit.

The difficulty with Mercury is that the maximum possible elongation when Mercury is at ☿1 is about 28°, while it is only about 18° when Mercury is at ☿2. For Venus, on the other hand, whose eccentricity is about 0.0068, the variation of the maximum elongation is almost negligible for observations made with the naked eye.†

Thus the main problems presented by the moon and by Mercury are similar. The maximum difference in longitude between the true and mean moons is about 7°.5 when the moon is at the quarters, but it is only about 5°.0 at new and full moons. Similarly, the longitude of the sun is close to the mean longitude of Mercury. The phenomenon illustrated in Figure XIII.9, when put in geocentric terms, makes the separation between the true and mean positions of Mercury about 28° in some positions and only about 18° in others. The ratio of separations is about 3/2 for both the moon and Mercury.

---

†Ptolemy's definition of the elongation of Mercury, which will be given in Equation XIII.22 below, is analogous to his definition of the elongation of Venus in Equation XIII.10 and it differs slightly from the geometric definition that I have tacitly used in this discussion. The difference in definition is not important in the immediate discussion.

For either body, the maximum separation is equal to $\sin^{-1}(r/R)$, in which r is the epicycle radius and R is the deferent radius. In order to vary the separation, then, we must cause either r or R to vary with longitude.

It is possibly to Ptolemy's credit that he realizes this point.† Having realized it, however, he has to decide whether to vary r or R, and he makes the wrong choice. He chooses to vary R. When I say that Ptolemy makes the wrong choice, I mean that he is wrong in terms of the information available to him, not merely that he is wrong in modern terms.

In order to make $\sin^{-1}(r/R)$ vary by about 3/2, either r or R must vary by about 3/2, and this fact has implications about the distance from the earth to the body in question. Suppose first that r varies from $r_1$ to $r_2$, with $r_2/r_1 = 3/2$. Then the distance can vary from $R - r_2$ to $R + r_2$. On the other hand, if R varies from $R_1$ to $R_2$, with $R_2/R_1 = 3/2$, the distance can vary from $R_1 - r$ to $R_2 + r$. Thus the ratios of greatest to least distance are, in the two cases,

$$(R + r_2)/(R - r_2) \quad \text{and} \quad (R_2 + r)/(R_1 - r),$$

respectively. Since either r or $r_2$ is small compared with R, $R_1$, or $R_2$, the first ratio is much closer to 1 than is the second ratio.

For the moon, the second ratio, which corresponds to Ptolemy's choice, is nearly 2/1, and in consequence the lunar distance varies by almost 2/1 in Ptolemy's lunar model [Neugebauer, 1957, Appendix I or Newton, 1974]. There were two ways by which Ptolemy should have known that this variation is not correct.

First, a variation in distance means a variation in apparent size by approximately the same ratio, almost 2/1 in this case. It is manifest that the size of the moon does not vary by more than about

†Since we have lost so much of pre-Ptolemaic astronomy, we do not know whether to credit him with this point or not.

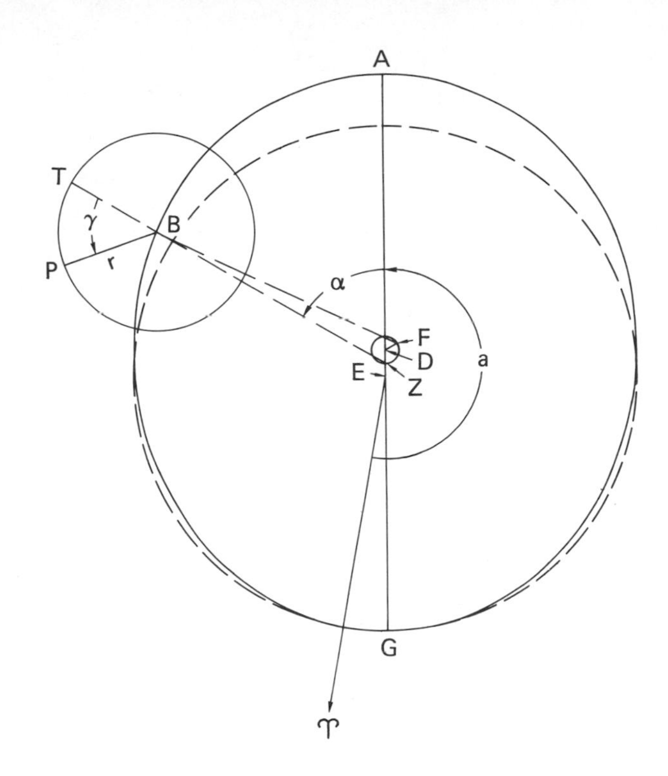


Figure XIII.10. Ptolemy's model for Mercury, drawn approximately to scale. ♈ is the vernal equinox and A is the point that Ptolemy usually calls apogee. The earth is at E. The mean sun is taken to rotate uniformly around the point Z, and $\alpha$ is the longitude of the mean sun minus the longitude a of apogee. As usual in Ptolemy's models, the planet P rotates uniformly around the epicycle whose center is B; $\gamma$ is the angle that Ptolemy calls the anomaly. The point F rotates clockwise (in the opposite sense to the sun) around a small circle whose center is D, at the same rate as the mean sun. That is, the angle ADF is also equal to $\alpha$. The distance BF is fixed. The distances EZ, ZE, and DF are all equal.

The solid curve on which B moves is not a circle. This is shown by comparison with the circle shown by the broken line whose center is E. The points at which B is closest to the earth E lie about 120° from A; Ptolemy calls these the perigee points.

11/10. Second, the distance can be measured by means of the parallax. I have shown elsewhere [Newton, 1974] that Ptolemy's treatment of the apparent size is misleading and, probably, deliberately deceptive, and that his treatment of the parallax is demonstrably fraudulent. In other words, Ptolemy knew that he had made a serious error in his lunar model and he attempted to cover it up by the fabrication of data.

Ptolemy may not have known that his model of Mercury gives the wrong variation of distances, however, since neither the apparent size nor the parallax of Mercury can be measured by naked-eye observations. We know that the sun is the center of Mercury's epicycle and hence that varying R implies varying the apparent size of the sun, but it is not clear that Ptolemy realized this point. He takes the deferent radius R to be less than the solar distance for both Mercury and Venus, but, it seems to me, he does not attach much importance to this action. He merely points out (Chapter IX.1) that we cannot prove anything about the planetary distances except that the planets are much farther away than the moon, because we cannot measure their parallaxes.†

In order to meet the problem caused by the maximum elongations of Mercury, Ptolemy introduces a 'crank' into his Mercury model, just as he does in his lunar model. His model is shown in Figure XIII.10. In contrast to earlier practice, I have drawn the figure to scale. This gives the reader

---

†Some source that I cannot locate says that Galileo observed that Venus has both gibbous and crescent phases, that this proves that Venus can be both nearer and farther than the sun, and hence that Galileo had disproved the geocentric model of the solar system. This is a misreading of Ptolemy. The parameters of his model for Venus are such that the orbit of Venus encloses the sun, whether the sun is at the geometric center of Venus's epicycle or not. A quantitative analysis of Galileo's observations might have shown that the sun is at the center of the epicycle, but this does not contradict the Ptolemaic (geocentric) model.

a good idea of some of the quantities involved, although it makes a few details hard to follow.

In the figure, ♈ is the vernal equinox and A is the point that Ptolemy calls apogee. $\underline{a}$ is the longitude of apogee, which Ptolemy takes to be 190° in his own time. He makes $\underline{a}$ increase with time at the rate of 1° per century; that is, he takes A to be fixed with respect to the stars. The deferent curve on which B moves, which is shown by the solid curve ABG, is not a circle. In order to generate the deferent curve, we start by letting the mean distance $\alpha$ be

$$\alpha = L_S - \underline{a}, \tag{XIII.21}$$

just as we did for Venus (Equation XIII.8). The line on which B lies thus rotates eastward (counter-clockwise) around the point Z, which is somewhere between the earth E and apogee A. Let $e_1$ denote the distance ZE; I call $e_1$ the first eccentricity.

D is a point that lies between Z and A; let $e_2$ denote the distance DZ, which I call the second eccentricity. The point F rotates clockwise (in the opposite sense to the sun) on a circle whose center is D and whose radius will be denoted by $e_3$. I call this the third eccentricity or the crank radius. The rotation of F is such that the angle ADF, measured clockwise from the line DA, is always equal to $\alpha$. During the rotation, the distance BF remains constant. When $\alpha = 180°$, we see that F is at the point Z and that B is at the point G.† Hence DF = ZG. When the constants $e_1$, $e_2$, $e_3$, and BF have been assigned, we can locate B as a function of the angle $\alpha$.

When we consider the model for the value $\alpha = 0°$, we see that BF is also equal to the distance from A to the unlabelled point at the top of the circle that is centered on D. Hence D is the center of the line AG, whatever may be the values of $e_1$, $e_2$, $e_3$, and BF.

†In contrast to the model in Figure XIII.4, G is not the point that Ptolemy calls perigee, as we shall see in a moment.

The broken curve in Figure XIII.10 is a circle whose center is E. When we compare the circle with the deferent curve ABG, we see that B is farthest from E (and hence that the local deferent radius R is a maximum) when B coincides with A, and that B is closest to E when it is about 120° from A. Ptolemy introduced the 'crank' in order to create the latter property. This means that the greatest possible elongations can occur when the sun is about 120° from the point A. The points where B is closest to E are the points that Ptolemy calls perigee, and he describes the situation by saying (Chapter IX.8) that "Mercury is perigeed twice in each revolution."†

The epicyclic part of Figure XIII.10 is standard. Mercury, whose position is denoted by P, rotates uniformly around the circle whose radius is r and whose center is B. The point T lies on the extension of the line ZB, and the anomaly $\gamma$ is the angle TBP. $\gamma$ is taken as positive‡ when it is measured in the direction shown.

By accident, it happened that the perihelia of Mercury and of the earth differed by only about 22°.65 in the time of Ptolemy. Since the eccentricity of the earth's orbit is much less than Mercury's eccentricity, the sun-earth-Mercury picture is nearly symmetrical about the perihelion-aphelion line of Mercury, as Figure XIII.9 shows us. Thus it is a good approximation for Ptolemy to use a symmetrical model, as he does in Figure XIII.10, in spite of the fact that Mercury's eccentricity is about 0.206. At points that are symmetrically located in Figure XIII.10, the differences between the correct elongations are typically about 1°.

---

†In more accurate terms, Mercury goes through a perigee once in each synodic period of about 116 days (Table III.1). The center of its epicycle has two perigees per year, not two for each revolution of Mercury.

‡Ptolemy did not explicitly use negative numbers, so far as I know, but he often did the equivalent. For example, he might say that Mercury in its anomaly was west of T by, say, 100° instead of saying that $\gamma$ was 260°.

As we shall see, the root-mean-square error in Ptolemy's model is almost 3°, so the main defect in his theory does not come from the lack of symmetry.

Altogether, there are eight parameters in Ptolemy's model of Mercury, which he determines by means of observations. They are:

$\underline{a}$, the longitude of A,

$\underline{a}'$, the rate of change of $\underline{a}$,

$e_1$, the distance ZE,

$e_2$, the distance DZ,

$e_3$, the crank radius DF,

r, the epicycle radius r,

$\gamma_0$, the value of $\gamma$ at some specific epoch,

$\gamma'$, the rate of change of $\gamma$.

I shall give Ptolemy's values of these parameters in Table XIII.4 below.

Since Ptolemy does not assign a distance to Mercury, the distance scale in Figure XIII.10 is arbitrary. Ptolemy chooses 120 for the length of the line AE. I prefer to use unity for the arm BF.

Ptolemy uses the 17 observations listed in Table V.5 in order to find the values of his 8 parameters.† We are now ready to study the process by which he does this.

---

†More accurately, Ptolemy uses the last two observations only in order to 'confirm' a value of $\gamma'$ that he has found elsewhere, but he does not use the value that actually results from these two observations. He does not tell us how he found the value of $\gamma'$ that he adopts. Thus he does the same thing for Mercury that he does for Venus and Mars (Sections XIII.5 and XIII.6). In reality, then, Ptolemy uses 15 observations in order to find 7 parameters.

TABLE XIII.3

MAXIMUM ELONGATIONS OF MERCURY USED BY PTOLEMY

| Identifying number | Date | Reliability[a] | $L_S$[b], degrees | Elongation, degrees | |
|---|---|---|---|---|---|
| | | | | Stated by Ptolemy | Calculated from modern theory[c] |
| 1 | 132 Feb 2 | 0 | 309.75 | +21.25 | +19.78 |
| 2 | 134 Jun 4 | 0 | 70.00 | -21.25 | -19.77 |
| 3 | 138 Jun 4 | 0 | 70.50 | +26.50 | +25.58 |
| 4 | 141 Feb 2 | 0 | 310.00 | -26.50 | -25.26 |
| 5 | -261 Feb 12 | 1 | 318.17 | -25.83 | -25.36 |
| 6 | -261 Apr 25 | 1 | 29.50 | +24.17 | +23.70 |
| 7 | -256 May 28 | 1 | 62.83 | +26.50 | +25.51 |
| 8 | -261 Aug 23 | 1 | 147.83 | +21.67 | +23.94 |
| 9 | -236 Oct 30 | 1 | 215.17 | -21.00 | -20.71 |
| 10 | -244 Nov 19 | 1 | 234.83 | -22.50 | -21.97 |
| 11 | 134 Oct 3 | 0 | 189.25 | -19.05 | -19.66 |
| 12 | 135 Apr 5 | 0 | 11.08 | +23.25 | +22.19 |
| 13 | 130 Jul 4 | 1 | 100.08 | +26.25 | +26.25 |
| 14 | 139 Jul 5 | 0 | 100.33 | -20.25 | -19.73 |

[a] This column partially identifies the observer. The reliability is 0 when Ptolemy claims to be the observer and it is 1 otherwise.

[b] These are the values stated by Ptolemy, not the correct values.

[c] Using zero accelerations.

## 8. Ptolemy's Parameters for Mercury

Of the 15 observations that Ptolemy actually uses, 14 were made when Mercury was at maximum elongation. We noted in Section XIII.5 that Ptolemy does not use "elongation" to mean the angle as seen from earth between the planet and either the true or the mean sun. His definition of the elongation of Venus is given in Equation XIII.10, and his definition of the elongation $D_☿$ of Mercury is exactly analogous:

$$D_☿ = \lambda_☿ - L_S. \qquad \text{(XIII.22)}$$

In this, $\lambda_☿$ denotes the (geocentric) longitude of Mercury.

With so many observations of maximum elongation to be discussed, it is convenient to start by tabulating them. In Table XIII.3, the first column gives an integer that will be used to identify an observation; the observations in Table XIII.3 are the first 14 observations in Table V.5 and they are listed in the same order. The second column gives the date and the third column gives the reliability that was assigned in Table V.5. The main use of the reliability here will be to distinguish the observations that Ptolemy claims to have made himself from those made by others; Ptolemy's claimed observations have a reliability of 0 and all others in Table XIII.3 have a reliability of 1.

The next two columns in Table XIII.3 give the longitude of the mean sun and the elongation of Mercury as stated by Ptolemy, but changed into decimal form.† If the reader wants the longitude of Mercury as stated by Ptolemy, he can find it by using values from these columns in Equation XIII.22. The final column gives the values of $D_{☿}$, as defined by the equation, but calculated using the values from the ephemeris programs described in Chapter IX, using zero for the accelerations of Mercury and the sun.

Ptolemy chooses the first two observations because the elongations are equal but opposite. Reference to Figure XIII.10 then shows us that the symmetry axis of the figure must lie along the line bisecting the two values of $L_S$; that is, the symmetry axis lies along the line connecting longitudes 9 7/8 and 189 7/8 degrees. He confirms this by means of observations 3 and 4, which make the line run between longitudes $10\frac{1}{4}$ and $190\frac{1}{4}$ degrees.‡ He settles for putting the ends of the symmetry axis at 10° and 190°. We can in fact see already that point G in Figure XIII.10 must be the end that is

---

†I use a negative sign for the elongation when $D_{☿}$, as defined by Equation XIII.22, has a negative value, whereas Ptolemy refers to maximum elongations east or west. In modern terms, $D_{☿}$ has a local minimum when it is negative in Table XIII.3. I hope it will not confuse the reader if I continue to use "maximum elongation" in these cases.

‡Ptolemy has a specific reason for repeating the values of $L_S$ in the first four observations as closely as possible and for using values as close to 70° and 310° as he can. This reason does not come out until we come to observations 11 and 12.

at 10° while point A is the end that is at 190°, but Ptolemy does not draw this conclusion yet.

So far, Ptolemy has used four observations in order to find a. He uses the next six observations in combination with these four, to find a′, the rate at which a changes.

He first quotes observation 5. He may have chosen it because it gives the largest negative elongation that he could find among the observations from early Alexandria, where he lived.† If this be his reason, he does not state it. He then says that he cannot find, among the ancient observations,‡ an elongation of exactly the same size but in the opposite direction, and he has to use an indirect method in order to find the symmetry axis in earlier times.‡‡ The elongations in observations 6 and 7 span the value needed, which is +25°.83. He interpolates linearly between observations 6 and 7 in order to find the value of $L_S$ when $D_☿$ = 25°.83, and he finds $L_S$ = 53°.5. He should have found $L_S$ = 53°.3, so his interpolation is not very accurate.

---

†We should remember that Alexandria was founded after Alexander's conquest of Egypt in -331. Ptolemy [ca. 142, Chapter VII.3] uses a conjunction of the moon with $\beta^1$ Scorpii observed in Alexandria on -294 December 21. So far as I know, this is the oldest surviving Alexandrian observation that can be dated.

‡That is, among those already ancient to him.

‡‡If Ptolemy had been in the habit of fabricating old data as well as his own data, he could simply have fudged the needed ancient observation without resorting to an indirect device. This is one of the minor reasons which make me believe in the validity of most of the data that Ptolemy attributes to others. As I said in Section V.2, the strongest reason is that the ancient sources were still widely available in Ptolemy's time, and he would have run a grave risk of immediate detection if he had fudged many old observations. He could run the risk of fudging a few.

He now has values of $L_S$ for which the elongation $D_☿$ has equal values but with opposite signs. Thus he finds that the symmetry axis ran from longitudes 5 5/6 to 185 5/6 degrees at a time near -260, four centuries before his own observations. He should have found longitudes 5°.73 and 185°.73. Since the longitudes of the axis are 10° and 190° in his own time, he concludes that $\underline{a}'$ = 1° per century.

Ptolemy confirms this value with the aid of observations 8, 9, and 10. He first quotes observation 8, with its positive elongation that he cannot match exactly with an ancient negative value. Accordingly, he interpolates between observations 9 and 10 and finds that $L_S$ should be 224° 10′ when the elongation is -21.67. He should have found $L_S$ = 223°54′, again his calculations are not very accurate.

In this way, he finds that the symmetry axis ran between the 6° and the 186° points four centuries before his own time, which he takes as an adequate confirmation of the result that he found from observations 5, 6, and 7. Hence he is sure that $\underline{a}'$ = 1° per century.

We should note carefully what Ptolemy has really done here. The position of Mercury in observations 5 through 10 was given with reference to stated stars. Ptolemy calculates the longitudes of these stars at the times of the observations by using his own star tables and correcting the stellar longitudes for precession at the rate of 1° per century. That is, he 'knows' that longitudes of stars increase at the rate of 1° per century and he has now 'proved' that the symmetry points A and G in Figure XIII.10 increase in longitude at the same rate. Hence he has 'proved' that Figure XIII.10 remains fixed with respect to the stars, and he applies this conclusion to all the other planets without further examination.

The correct procession rate is about 5020″ per century. The perihelion of Mercury increases in longitude at the rate of about 5600″ per century, about 580″ per century greater than the precession. The perihelion of earth moves about 1160″ per century eastward with respect to the stars. Thus Figure XIII.10 should not remain rigid, but we should probably not expect Ptolemy to have discovered this, since the eccentricity of the earth's orbit is fairly small. To the extent that it may be considered rigid, Figure XIII.10 probably rotates at a rate that is dominated by the orbit of Mercury, say at a rate of 600″ per century with respect to the stars. Thus the figure should have rotated by about 40′ with respect to the stars during the four centuries involved, and a careful analysis of careful observations, even ones made with the naked eye, should have revealed this motion.

Up to this point, Ptolemy has employed a high degree of redundancy, having used 10 observations in order to find only two parameters. He drops his redundancy here, and uses only five more observations to find the remaining five parameters.

He uses observations 11 and 12 to find r, the radius of the epicycle, and the distance DE in Figure XIII.10, which is the sum $e_1 + e_2$. The positions of the mean sun are close to the points A and G. The picture resembles Figure XIII.5, except that point D rather than Z is the midpoint between A and G. The solution† is:

$$r = 39.16718, \quad e_1 + e_2 = 10.38912. \quad \text{(XIII.23)}$$

Ptolemy's values are:

$$r = 39;9 = 39.15, \quad e_1 + e_2 = 10;25 = 10.41667. \quad \text{(XIII.24)}$$

Ptolemy's solution of the figure is thus rather accurate.

Ptolemy now turns back to observations 1 through 4. So far, he says, he has not shown that

---

†For the moment, I am still using Ptolemy's convention that the distance AE in Figure XIII.10 equals 120.

Figure XIII.4, which works well for the other planets, does not work well for Mercury. If Figure XIII.4 were correct for Mercury, the maximum elongation would have its greatest possible value when the mean sun is at point G, where $L_S = 10°$. However, observation 12 shows that the maximum elongation is only about $23\frac{1}{4}$ degrees at this point. The greatest elongations occur when $L_S$ is 60° from this value, at 70° and at 310°. Hence he has devised the model of Figure XIII.10 for Mercury, which makes the deferent come closest to earth, and hence allows the greatest possible elongations, for these values of $L_S$.

We now see why Ptolemy chooses values of $L_S$ as close to 70° or 310° as he can in observations 1 through 4. He needs symmetry, and he also needs the values of $L_S$ for which Mercury has its greatest possible elongations.

Ptolemy says, at the beginning of Chapter IX.8, that he cannot find any old observations of Mercury made when the sun is at points A and G ($L_S = 10°$ and 190°) because there are not any stars bright enough in the right parts of the sky. It takes an astrolabe or a similar instrument, and therefore he has to use observations that he has made himself for the observations numbered 11 and 12 in Table XIII.3. What he means by this, I think, is that the old observers could measure only small celestial angles. Hence they could locate Mercury only when it was within a few degrees of a reference star, and, since Mercury is visible only near sunrise or sunset, the star has to be a bright one. The astrolabe, however, can measure large angles and therefore Ptolemy can use any star bright enough to be seen.†

At point A, where $L_S = 190°$, the maximum elongation is about 19°, and hence we need stars whose longitudes were near, say, either 170° or

---

†I think that the reference star would also have to be fairly close to the ecliptic, although Ptolemy does not explicitly say so. It must of course be in the appropriate celestial hemisphere.

210° in Ptolemy's time.† At point G, we have $L_S$ = 10°, the maximum elongation is about 23°, and hence we need stars whose longitudes were near 33° or 347°.

It is simply not true that older observers could not have located Mercury accurately at maximum elongation when the mean sun is at A or G. At the time of observation 12, which was the evening of 135 April 5, Mercury set at about $1^h.65$ after sunset at Alexandria, its longitude was about 34°.3, and its latitude was about 2°.8. The sun was at longitude 13°.8. Mercury in fact set after the beginning of astronomical twilight‡ and even faint stars could have been seen. Even if we assume that Mercury could not be seen all the way down to the horizon, because of haze or other kinds of interference, it is certainly safe to assume that stars of the third magnitude could already be seen before Mercury became invisible. The Babylonian "normal stars" (see Table III.7) include η Tauri, whose magnitude is 2.96.♯ At the time of observation 12, this star was separated from Mercury by only about 0°.2 in longitude and about 1°.1 in latitude. Configurations close to this one must have occurred many times within the period of Babylonian and Hellenistic observations, and the older observers must have had many opportunities to make the measurement that Ptolemy wants.

Further, the star that Ptolemy actually uses on this occasion is α Tauri, which is also one of

---

†Since point A precesses at the same rate as the stars in Ptolemy's model, the stars needed would be the same at any epoch. Hence we use here the longitudes that apply at Ptolemy's time. The longitudes themselves would have been smaller at an older epoch.

‡This is the time when the indirect light from the sun is less than the light from the stars. See Explanatory Supplement [1961, p. 399] or the discussion in Section IV.7 above.

♯The stars in the preserved Babylonian conjunctions of Mercury, which are listed in Table IV.6, range in magnitude from 1.21 to 4.17.

the normal stars. It was higher in the sky than Mercury by about 11°. The Babylonian observations tabulated in Chapter IV show that they measured angles of this size on several occasions. Thus older observers could have used the very star that Ptolemy claims to have used.

In the morning of 134 October 3, which is the time of observation 11, Mercury was at longitude 170°.5, and latitude 2°.1, while the sun was at longitude 188°.4. Mercury rose $1^h.40$ before sunrise, when the sun was still 18°.0 below the horizon, at the very end of astronomical twilight. The very bright star $\alpha$ Virginis was at longitude 178°.0, and it is possible that it could not be seen before it was obscured by the rising sun. However, the normal star $\gamma$ Virginis, of magnitude 2.91, was about 6° above Mercury, and it could have been seen under ordinary circumstances at the same time as Mercury. Thus again older observers certainly had many opportunities to make the measurement that Ptolemy wants.

In sum, there were many occasions when Babylonian and Hellenistic observers could have made the observations that Ptolemy needs for his observations 11 and 12, and the astrolabe was not needed in order to make them accurately. We also know that the archives available to Ptolemy contained many observations made by the Babylonians and by Hellenistic astronomers anterior to Ptolemy. In spite of this, it is possible that the archives by accident did not have the observations that Ptolemy needs. Nonetheless, Ptolemy's wrong and unnecessary statement about the need to use the astrolabe, which 'proved' the need to use observations that he claims to have made himself, arouses our suspicions.

The sum $e_1 + e_2$ in Equations XIII.24 is the distance from D to E in Figure XIII.10, but we have not yet located the point Z. Ptolemy locates Z, and also finds the crank radius $e_3$, from observations 13 and 14. For these observations, he wants $L_S$ to be midway between points A and G in Figure XIII.10, that is, at either 100° or 280°, and he wants values of the maximum elongation in both directions at whichever of these points he uses. Observation 13 was made by Theon and observation 14 is one that he claims to have made himself.

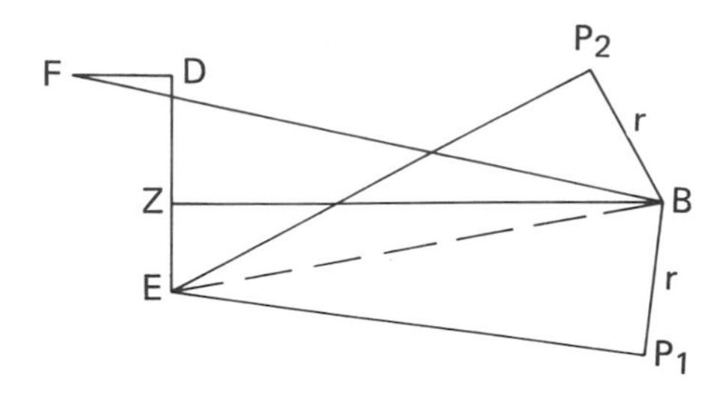


Figure XIII.11. Locating the point Z, and finding the crank radius DF, in Ptolemy's model of Mercury. On 130 July 4, according to Theon as quoted by Ptolemy, Mercury was at $P_2$, where its longitude was 126°20′. On 139 July 5, it was at $P_1$, where its longitude was 80°05′. On both occasions, the line from the center B of the epicycle to the point Z was nearly perpendicular to the line DE, as was the crank radius DF. The figure is not to scale.

The configuration represented by the observations is shown in Figure XIII.11. We can assume that both values of $L_S$ equal 100° provided that we keep the observed values of $D_☿$. Finding the distance ZE is just like finding the distance DE in Figure XIII.6. We find that the angle ZBE equals 3° and that ZE = 5;12 in sexagesimal notation. Ptolemy alters the sum $e_1 + e_2$ in Equation XIII.24 to 10;24, so that

$$e_1 = e_2 = 5.2 = 5;12. \qquad \text{(XIII.26)}$$

We also find the distance BE in the same way that we did before, and Ptolemy's value for it is 99;9. BZ equals BE cos 3°, so BZ = 99, very closely. In Figure XIII.11, the constant distance BF, which we have represented by $\rho$, is the hypotenuse of a right triangle. One leg is the distance DZ, which is the second eccentricity $e_2$. The other is the sum of BZ, which we have just found, and DF, which is the crank radius $e_3$. Since $e_2$ is now known, this right triangle gives us one relation between $\rho$ and $e_3$. From the discussion that immediately follows Equation XIII.21 in the preceding section, we can also express the distance AE in terms of $\rho$ and $e_3$. That is:

$$\rho + e_1 + e_2 + e_3 = AE,$$
$$(99 + e_3)^2 + e_2^2 = \rho^2. \quad \text{(XIII.27)}$$

The easiest way to solve Equations XIII.27 is probably to use successive approximations, noting that the second equation is almost the same as $\rho - e_3 = 99$. Since AE = 120 by Ptolemy's convention, the first equation is the same as $\rho + e_3 = 109;36$, if we use Equation XIII.26. The solution gives $e_3 = 5;14$, so that all three e's are substantially equal. Ptolemy then adopts the final values:

$$\rho = 104;24, \quad e_1 = e_2 = e_3 = 5;12. \quad \text{(XIII.28)}$$

All distances in Figure XIII.10 have been given using Ptolemy's convention that AE = 120. The distance $\rho$ seems to me to be a more important property of the model than AE is, and I shall shortly change to the convention that $\rho = 1$. This change will also reveal some interesting features of the parameters that are not obvious in sexagesimal notation; at least they are not obvious to a modern reader who works almost exclusively in decimal notation. Before I summarize the parameters and change them to the convention that $\rho = 1$, however, I shall complete Ptolemy's analysis. The remaining task is that of finding the anomaly $\gamma$ at some epoch.

From the observation of 139 May 17 (Table V.5), which Ptolemy claims to have made himself, he finds that the longitude of Mercury was $77\frac{1}{2}$ degrees. From this, an obvious but tedius calculation yields $99°.5779 = 99°34'40''$ for the value of the anomaly $\gamma$, while Ptolemy finds $99°27'$. He then finds the anomaly at the epoch of Nabonassar, using the value of $\gamma'$ that he has already given in his Chapter IX.3.

TABLE XIII.4

A COMPARISON OF PTOLEMY'S PARAMETERS FOR MERCURY WITH THOSE THAT GIVE THE BEST FIT

| Parameter | Ptolemy's Value | Best Fit |
|---|---|---|
| Apogee, degrees | 191.1[b] | 219.013 |
| Apogee rate, degrees per century | 1.00 | ~ 1.6[c] |
| First eccentricity ($e_1$) | 0.049 900 | 0.085 738 |
| Second eccentricity ($e_2$) | 0.049 900 | -0.015 284 |
| Crank radius ($e_3$) | 0.049 900 | 0.012 025 |
| Epicycle radius | 0.375 090 | 0.376 704 |
| Anomaly at the epoch,[a] degrees | 189.430 | 187.195 |
| Anomalistic rate, degrees per day | 3.106 699 043 | 3.106 404 735 |

[a]Noon, Alexandria time, 137 July 20.

[b]Ptolemy's value is 190° when it is referred to his erroneous equinox. I added 1°.1 in order to refer to a more accurate equinox.

[c]Not studied by the fitting process.

Although he does not really use them, we should note what Ptolemy says about the observations of -264 November 15 and -264 November 19 in Table V.5. He uses the second of these only to show that Mercury had not reached maximum elongation at the time of the first one; he needs to know this in order to discriminate between two possible values of $\gamma$. He then finds that $\gamma$ equals $212°34'$ at the time of the first observation.† When he combines this with the

†The value of $\gamma$ at maximum westward elongation, according to Ptolemy's model, is about 250°. The value of $\gamma$ that he finds is so far from this that he could use an approximate value of $\gamma'$ in order to discriminate between the possible values of $\gamma$. However, we may be thankful that Ptolemy uses this unnecessary observation; it preserves an ancient observation that would otherwise be lost.

value of $\gamma$ on 139 May 17, he says that the resulting value of $\gamma'$ agrees closely with the value that he has already adopted. He continues to leave us in the dark about the sources of his values of $\gamma'$.

The 'connecting rod' BF, which connects the center B of the epicycle to the end F of the crank radius, is a more fundamental part of Ptolemy's model than is the distance AE in Figure XIII.10, and I have chosen to take its length $\rho$ as unity. Ptolemy's parameters, adjusted to make $\rho = 1$, are summarized in the central column of Table XIII.4. With this convention, the eccentricities are all nearly equal to 0.05 = 1/20 and the epicycle radius is nearly equal to 0.375 = 3/8. I do not know whether Ptolemy planned matters to make them come out this way or not.

We expect the quantitative errors in Ptolemy's model of Mercury to be large because the model suffers from an obvious defect. From Figure XIII.9 we see that the maximum elongations of Mercury occur when the longitude of the mean sun is about 100° or 340°. However, Ptolemy has the maximum elongations occur at 70° and 310°, about 30° from their correct values.

As I did with Mars and Venus, I programmed Ptolemy's model of Mercury and calculated values of its geocentric longitude at 51 times. I took the times to be separated by 80 days, with the central time being noon on 137 July 20. In the total span of 4000 days, both Mercury and the earth make approximately an integral number of revolutions around the sun. I then compared the values from Ptolemy's model with accurate values. The largest error that I found is 7°.84, on the central date itself, and the root-mean-square error is 2°.99. These errors are far larger than those found for Venus and Mars.

I then varied the parameters in Ptolemy's model until it gave the best fit, in a least-squares sense, to the accurate ephemeris. The best-fitting parameters are listed in the third column of Table XIII.4. I did not bother to vary $\underline{a}'$, the rate of change of the apogee position in the model; we know from Figure XIII.9 that the value of $\underline{a}$ must be dominated by the major axis of Mercury's heliocentric orbit and hence we know that $\bar{a}'$ must be about 1°.6 per century. After the fitting, the largest error that I found is still 3°.26, on 134 January 17, and the root-mean-square error is 1°.70. There are some interesting features about the best-fitting parameters.

Apogee moved by a large amount from Ptolemy's value, but not as much as I expected. I expected that it would move close to 228°, which was the position of Mercury's aphelion in Ptolemy's time, but it moved only to 219°.† We should note that Ptolemy did not depend upon the detailed properties of his model in finding apogee. The line AG in Figure XIII.10 is the symmetry line, and symmetry was the only property that Ptolemy used in finding the apogee point A. I shall take up the significance of Ptolemy's error in the next section.

The value of $e_2$ became negative. This means that the point Z in Figures XIII.10 and XIII.11 should lie above, not below, the point D. Put another way, Z should lie outside the line DE, but Ptolemy locates it at the center of DE.

Finally, we note that both $e_2$ and $e_3$ become rather small in magnitude. Ptolemy would have done much better, in a quantitative sense, if he had used the same model for Mercury that he used for the other planets, but simplified by the elimination of the equant. That is, he should have used the model of Figure XIII.4, modified by moving point D into coincidence with Z. I tested this matter by repeating the calculations with Ptolemy's model, with $e_2 = e_3 = 0$, but with all other parameters having the values listed in the last column of Table XIII.4. This change in parameters hardly affected the accuracy. The greatest error that I found came on the same date as before, and it was 3°.31 rather than 3°.26. The root-mean-square error rose from 1°.70 to 1°.73. These are far better than the errors that occur with Ptolemy's original model.

---

†The eccentricity of the earth's orbit has some influence upon $\bar{a}$. I have not tried to find out if it is responsible for the change from 228° to 219°.

Thus Ptolemy's model for Mercury should have only a simple eccentric and an epicycle. This simple model gives better overall accuracy, it preserves the double maximum and the double minimum in the elongation, and it puts them in the right places. The doubling is a result of the eccentric; it does not come from the elaborate crank mechanism that Ptolemy devised.

Dreyer [1905, p. 200], says, with reference to Ptolemy's system for the motions of the sun, moon, and planets: "Nearly in every detail (except the variation of distance of the moon) it represented geometrically these movements almost as closely as the simple instruments then in use enabled observers to follow them, ..." I do not understand Dreyer's basis for this statement. It is far from being correct, even if it is restricted to Ptolemy's own time, before the inevitable errors in his values of the mean motions had a chance to produce large errors in position.

The standard deviation of the error in a Hellenistic measurement of a planetary longitude, as we have seen in Chapter XII, is something like 0°.25. The standard deviation that I found for Ptolemy's model of Mars, for times centered on 137 July 20, was 0°.44; it was 1°.01 for Venus and it was 2°.99 for Mercury. There is no question that the standard deviations for Venus and Mercury exceed the standard deviation of observation, and this is probably true for Mars also.

It is conceivable that we cannot judge Ptolemy's models on the basis of the standard deviations alone. It is conceivable that the standard deviations are dominated by large errors that occur only when the planet is too close to the sun to be seen. Investigation shows that this is not the case. The largest error in my sample of 51 longitudes of Mars was about 0°.9, and it came when Mars was about 100° from the sun. The largest error for Venus was about 4°.3, and it came when the true elongation was more than 20°. The largest error for Mercury was about 7°.8, but it came when the elongation was only about 8° and hence when Mercury may not have been visible. However, many errors ranging from 3°.8 to 7°.1 came when the elongation was more than 20° and hence when Mercury was readily visible.

Thus Ptolemy's models for the planets are far from meeting the normal standards of observational accuracy, even in his own time. The idea that his models met the needs of naked-eye observation seems to be a modern one; we find Heliodorus (Section V.9 above) already criticizing his accuracy in the year 510, and such criticism was commonplace with some early Islamic astronomers.

## 9. The Authenticity of the Planetary Observations that Ptolemy Claims to Have Made Himself

The most important purpose of this chapter is to study the usefulness of planetary tables in studying the accelerations. Before I conclude that topic, I shall take up the interesting question of the authenticity of the planetary observations that Ptolemy claims to have made himself.

In three earlier papers [Newton, 1973, 1974, and 1974b], I have studied the authenticity of Ptolemy's alleged observations of the equinoxes and solstices, of lunar positions, of eclipses, and of the stellar positions that he uses in finding the precession of the equinoxes. The proof is overwhelming that all of these alleged observations are fraudulent and that Ptolemy is the person guilty of the fraud. I use the word 'fraudulent' advisedly. The false observations cannot be taken as illustrative examples that Ptolemy calculates in order to explain his theories. Ptolemy uses them explicitly in his 'proofs' that his theories are correct, and he often emphasizes the care that he took in making the measurements. Further, he is not the victim of a trusted associate who fudged the data and passed them off as genuine. It is clear that Ptolemy knew what he was doing.

The stellar data are particularly revealing. In order to find the rate of precession, Ptolemy [ca. 142, Chapter VII.2] first 'measures' the longitude of Regulus ($\alpha$ Leonis) and compares his value with the value that Hipparchus measured 2 2/3 centuries earlier; he finds a rate of change of 36″ per year. He also enters the 'measured' value for the longitude of Regulus in his star table. His claimed measurement is clearly fudged, and thus at least one of the entries in his star table is known to be fudged. However, we cannot conclude from this alone that Ptolemy is responsible for the fudging; this proof comes from Ptolemy's next step.

Ptolemy now (Chapter VII.3) lists the declinations of 18 stars that he claims to have measured himself, and he finds the precession rate by comparing the declinations of six of the stars with earlier measurements of their declinations. He finds that the value is 36″ per year, in exact agreement with the rate that he finds from the fudged observation of Regulus. There is no question that these six declinations have been fudged, but, interestingly, the twelve declinations that he does not use turn out to be genuine. We can see this in several ways. One way is to calculate the precession from the declinations that Ptolemy does not use. The rates calculated from the data for individual stars agree amazingly well, and none gives a value less than 45″ per year. The mean rate [Newton, 1974b] is 52″.8 ± 2″.0 per year, in excellent agreement with the value from modern observations and theory.

In summary, the data sample that Ptolemy starts with contains exactly 6 fraudulent observations and 12 genuine ones. Ptolemy chooses to use exactly 6 observations in his work, and the 6 that he chooses to use are the fraudulent ones. The probability that this happened by chance, with Ptolemy being ignorant that he is using fudged data, is negligible. In other words, Ptolemy knowingly uses false data for the purpose of proving an erroneous proposition, and such a use is scientific fraud.

The study of the fraudulent solar and lunar data was simplified by the fact that Ptolemy preserves a number of genuine solar and lunar observations made by earlier astronomers. These observations are the basis of Ptolemy's solar and lunar theories. It is straightforward to calculate the circumstances of Ptolemy's alleged observations and to compare with the stated circumstances. The agreement is often exact, to every digit that Ptolemy wrote down. When it is not exact, the discrepancy is of a size that can be attributed to minor computational matters, such as differences between Ptolemy and us in rounding numbers or in interpolating in tables. There is one interesting exception.

Ptolemy claims (Chapter IV.6) to have observed the lunar eclipses of 133 May 6, 134 October 20, and 136 March 6 in Alexandria. Let us calculate the longitude $\lambda_S$ of the sun and the longitude $\lambda_M$ of the moon from Ptolemy's tables for the time that he states, and form the quantity $\lambda_M - \lambda_S \pm 180°$ from these calculated values. If the stated time is correct, this quantity, which can be called the phase, should be negative because of the errors in Ptolemy's tables. However, the phase turns out to be positive or nearly zero for each of the three eclipses [Newton, 1974b], which shows that the times stated by Ptolemy have been fudged.

In the cited paper, I made an error in quoting from Ptolemy. He says that the longitude of the sun was 14 1/12 degrees in the sign of Pisces (344°5′) at the time of the eclipse of 136 March 6, but I quoted this as 344°12′.† The calculated longitude of the moon turns out to be 344°12′ also. Thus the phase is +7′, but I took it to be zero, saying that Ptolemy made a large error in calculating the longitude of the sun.

Correcting the quotation does not affect any major conclusion that I made. In fact, it strengthens the argument that Ptolemy fudged the observation and made a mistake in calculation in doing so. The only change needed is to say that the mistake was in calculating the position of the moon rather than of the sun.

When we come to the precession of the equinoxes, we have lost the basis on which Ptolemy chose 36″ per year, although it is clear that he fudged the data in order to make them yield this value. The only clue to the origin of this value that I have found is in Chapter VII.2 of Ptolemy. There Ptolemy says that Hipparchus measured 174° for the longitude of $\alpha$ Virginis while Timocharis had measured 172°, and he says that Hipparchus had made many comparisons of this sort. Later in the same chapter, Ptolemy quotes Hipparchus as concluding

---

†The fraction 1/12 is written as ιβ′, which I misread as ιβ (12) minutes.

that the precession "is not less than a hundredth of a degree in a year."

The time of Hipparchus's observation is around the year -128 and that of Timocharis is around -290. A motion of 2° in 162 years gives a rate of 44″.4 per year. It is plausible that Hipparchus found a value of this size or larger from all the comparisons that he made, but that he realized the likelihood of measurement error. To be safe, he stated only that the rate was surely "not less than" 36″ per year. Ptolemy for some reason chose to use Hipparchus's lower limit. This limit is the same as 1° per century, and I suspect that Ptolemy chose 1° per century because it is a nice round number. However, this is only a suspicion; we apparently have no rigorous way to recover Ptolemy's basis of choice.

When we come to the planetary theories, we have only a few pre-Ptolemaic observations that could have served as his basis for Mercury and Venus, and we have only one old observation each for Mars, Jupiter, and Saturn. In studying the authenticity of Ptolemy's alleged planetary observations, we cannot proceed by deriving theories from pre-Ptolemaic observations and thence calculating the circumstances of the alleged observations. Thus we shall not be able to prove that all of them have been fudged; we can only prove that many of them have been.

We can begin by looking at the errors in Ptolemy's alleged observations of Mercury, Venus, and Mars. I have not analyzed most of his alleged observations of Jupiter and Saturn. These errors are called $Z_i$ in the preceding chapter,† and the values of $100Z_i$ are listed in Tables XII.3, XII.5, and XII.6. Ptolemy's alleged observations are the

†Actually the values of $Z_i$ are the differences between the observed values and the values calculated with zero accelerations of the sun and planets. Since the changes in position that result from the accelerations are smaller than the errors of observation, the values of $Z_i$ give a good idea of the measurement errors.

ones with a reliability of 0 in the tables.

The root-mean-square value of $Z_i$ for Ptolemy's observations is 1°.15 for Mercury, 0°.98 for Venus, and 0°.56 for Mars. With the possible exception of Mars, the errors in Ptolemy's observations are much larger than we expect on the basis of other Hellenistic observations. This alone does not prove that the observations are fudged. We must admit the possibility that Ptolemy was simply a poor observer, but we are entitled to be suspicious.

We are on firmer ground when we consider the values of anomaly $\gamma$ that Ptolemy discusses. For each of the five planets known in antiquity, Ptolemy deduces a value of the anomaly from an observation that he claims to have made himself. He also deduces the anomaly from an observation that had been made about four centuries before. As I said in the sections that deal with Mercury, Venus, and Mars, the values that Ptolemy adopts for $\gamma'$, the rate of change of $\gamma$, are not the values that we deduce from these observations, and Ptolemy in fact does not state the values of $\gamma'$ that come from them. He merely says that the change in anomaly between two observation equals, or nearly equals, the value found from the tables that he has given earlier.

In Chapter IX.3, Ptolemy says that Hipparchus had measured the periods of the planets. Apparently Hipparchus started with the resonant periods listed in Table IV.1, which are due to the Babylonians,† and then found the amount by which N synodic periods fail to equal exactly the appropriate number of years. Ptolemy says that he has corrected Hipparchus's results slightly by means of measurements that he has made himself, and he proceeds to construct tables that give changes in anomaly as functions of time for each planet.

I think that Ptolemy tries to give the impression at this point that his tables are based upon

†Ptolemy mentions only one resonance period for each of Mars and Jupiter, whereas the Babylonians used two periods for each of these planets. Ptolemy mentions only 37 synodic periods for Mars and 65 synodic periods for Jupiter.

the observations that he will present in later chapters, but I do not insist upon this impression and I may well be misreading him. In any case, I have shown in earlier sections that his 'observations' are not the basis for his tables of anomaly for Mercury, Venus, and Mars, and I have verified that the same situation exists for Jupiter and Saturn.

If this had happened for only one of the planets, we could accept some sort of accident for that one case. However, it happens for all five, and we must conclude that the situation is not accidental. Thus we are forced to the conclusion that Ptolemy uses his own observation, in combination with an old observation, only to 'verify' a value of $\gamma'$ that he has already decided to adopt.

The observations used to 'verify' the anomaly tables therefore give us an exact parallel to Ptolemy's alleged equinox and solstice observations [Newton, 1973]. Ptolemy took, for example, Hipparchus's observation of the vernal equinox at 06 hours on -145 March 24, and he adopted Hipparchus's value of 365.2466 6667 days for the length of the tropical year. He added 285 multiples of this length to Hipparchus's observed time, rounded to the nearest hour, and claimed that he had observed the vernal equinox of 140 at this exact hour (1 hour after noon on 140 March 22).

The simplest method of fudging the planetary observations is as follows: (1) Select a plausible time of observation when the planet is readily visible and when, in the case of an inner planet, it is not too near maximum elongation. (2) Select an old observation not too near maximum elongation and calculate the anomaly. (3) Multiply the time interval by the adopted value of $\gamma'$ and add the product to the old value of the anomaly $\gamma$. (4) Using this fudged value of the anomaly, and the parameters $\underline{a}$, $\underline{e}$, and $\underline{r}$ of the planetary model, calculate the planetary longitude.

TABLE XIII.5

PLANETARY OBSERVATIONS THAT PTOLEMY USES TO FIND THE ANOMALY

| Planet: | Mercury | Venus | Mars | Jupiter | Saturn |
|---|---|---|---|---|---|
| $\gamma'$, degrees/day | 3.106 699 0430 | 0.616 508 7338 | 0.461 575 5672 | 0.902 512 8420 | 0.952 146 7383 |
| Old observation | | | | | |
| Date | -264 Nov 15 | -271 Oct 12 | -271 Jan 18[a] | -240 Sep 4 | -228 Mar 1 |
| Hour | 6 | 4.75 | 6 | 6 | 18 |
| Anomaly | 212°34′ | 252°7′ | 109°42′ | 77°2′ | 183°17′ |
| Ptolemy's 'observation' | | | | | |
| Date | 139 May 17 | 138 Dec 16 | 139 May 27 | 137 Oct 8 | 136 Jul 8 |
| Hour | 19.5 | 4.75 | 22 | 5 | 12 |
| Anomaly | 99°27′ | 230°32′ | 171°25′ | 182°47′ | 174°44′ |
| Longitude | 77°30′ | 216°30′ | 242°34′ | 14°23′ | 284°14′ |
| Calculated anomaly | 99°27′.6 | 230°34′.8 | 171°24′.9 | 182°47′.8 | 174°44′.0 |
| Calculated longitude | 77°29′.2 | 216°29′.2 | 242°33′.6 | 14°22′.9 | 284°14′.0 |

[a]The correct date is probably -271 January 16, but -271 January 18 is the date that Ptolemy uses.

The test of this method of fudging is summarized in Table XIII.5. For each planet, the table first gives the value of $\gamma'$ that Ptolemy uses in his tables. Next come the date, hour, and anomaly that Ptolemy states for the old observation,† and the corresponding quantities for the observation that Ptolemy claims to have made himself, as well as his 'observed' longitude. The calculated anomaly and the calculated longitude are the values calculated by the method outlined in the preceding paragraph.

When the calculated longitude is rounded to the nearest minute, it agrees exactly with the 'observed' longitude for Mars, Jupiter, and Saturn. For Mercury and Venus, as it happens, the calculated longitude ends with 29′.2; this is apparently rounded to the powerfully attractive value of 30′ rather than to the nearest minute.

It is quite unlikely that the agreement shown in Table XIII.5 happened by chance. For simplicity, let us say that each observed longitude lies within a

†I believe that the observation of Mars listed for Mars on -271 January 18 was really made on -271 January 16, but here it is essential to use the date that Ptolemy uses.

pre-assigned zone whose width is 1′.5, and that the standard error of a single observation of longitude is 15′. The probability that a single value would lie in this zone by chance is less than 0.04, and the probability that all five values would do so is less than $10^{-7}$.

With odds of at least $10^7$ to 1, then, we may say that Ptolemy fudged the observations in Table XIII.5 that he claims to have made.

The calculated values of the anomaly show larger deviations from those stated by Ptolemy than do the values of the longitude, on the average. This is to be expected. After he fudged the longitude, Ptolemy had to round it in order to preserve verisimilitude. He then had to derive a value of the anomaly from the rounded longitude. The anomaly does not have the same sensitivity to a change in longitude for all observations. It is most sensitive for the observation of Venus, and the difference between the calculated anomaly and the anomaly that Ptolemy states is 2′.8. For the other planets, the difference is less than 1′.

For each planet, Ptolemy claims to have made an observation in which he also observed the position of the moon. In these observations, he measures the separation of the planet from a reference star and he also measures the distance of the planet from the moon. He first finds the longitude of the planet from the stellar observation. Then he calculates the apparent position of the moon from his tables, including the effects of parallax and of the equation of time. The longitude of the planet inferred from the position of the moon agrees exactly with the longitude inferred from the star in every case.

For Mars, Jupiter, and Saturn, the observation that involves the moon is the one that Ptolemy uses to find the radius r of the epicycle. For Mercury and Venus, the observation is the one that he uses to 'verify' the value of $\gamma'$. The planetary part of these observations has been analyzed in Table XIII.5. The lunar observation is independent of the planetary observation if it is genuine.

TABLE XIII.6

LUNAR OBSERVATIONS THAT PTOLEMY CLAIMS TO HAVE MADE HIMSELF IN CONNECTION WITH PLANETARY OBSERVATIONS

| Date | Planet Used With | Apparent Longitude[a] of the Moon | | | | | | | |
|---|---|---|---|---|---|---|---|---|---|
| | | 'Observed' | | Calculated from modern theory | | Calculated by me from Ptolemy's theory[b] | | Calculated by Ptolemy | |
| | | ° | ' | ° | ' | ° | ' | ° | ' |
| 139 May 17 | Mercury | 76 | 20 | 78 | 11 | 76 | 17 | 76 | 20 |
| 138 Dec 16 | Venus | 216 | 45 | 216 | 52 | 216 | 46 | 216 | 45 |
| 139 May 30 | Mars | 240 | 00 | 241 | 18 | 239 | 58[c] | 240 | 00 |
| 139 Jul 11 | Jupiter | 75 | 45 | 76 | 43 | 75 | 28 | 75 | 45 |
| 138 Dec 22 | Saturn | 308 | 34 | 310 | 24 | 308 | 26 | 308 | 34 |

[a]As seen from Alexandria.

[b]Based upon my calculation of the geocentric longitude from Ptolemy's theory and his value of the parallax, except for Mars.

[c]Based upon my calculation of the geocentric longitude from Ptolemy's theory and the parallax calculated from modern theory.

The lunar observations are summarized in Table XIII.6. The 'observed' longitude of the moon is the value that Ptolemy claims to have measured with the astrolabe; he claims to find it by using the measured separation of the moon from the planet, which in turn is referred to a star. The table then shows the apparent longitude of the moon calculated from the ephemeris program described in Section IX.6, using $\nu_M' = 8''.18$ per century per century. This is the value that I used in Section XII.6. It is not quite equal to the best estimate that one can make for the epoch of the observations in Table XIII.6, but it is good enough for the purposes of this section. Finally, the table gives the values of the longitude that I calculate from Ptolemy's tables and the value that Ptolemy calculates from his tables. Ptolemy finds that the 'observed' value agrees exactly with his calculated value.

When I calculated the lunar position from Ptolemy's tables, I verified that Ptolemy's calculation of the parallax is reasonably accurate,† and I adopted his value of the parallax in finding the positions listed in Table XIII.6. For Mars, Ptolemy

†That is, that the calculation agrees well with Ptolemy's model; it need not agree well with the values from an accurate model.

does not state the parallax. For this observation, the moon was within about 1° of being full, and Ptolemy's parallax is quite accurate when the moon is this close to syzygy. Thus I used the parallax calculated from modern theory for this observation and added it to the geocentric position calculated from Ptolemy's tables.

We cannot expect my calculations to agree exactly with Ptolemy's since we do not know how he may have handled rounding, since he worked in sexagesimal arithmetic while I work basically in decimal arithmetic, and, most importantly, since we do not know how he interpolated in tables. In some instances, he seems to have made some effort at careful interpolation and, in others, he seems merely to have used the closest tabular value without attempting to interpolate. The agreement of the calculations is close for Mercury, Venus, and Mars and rather poor for Jupiter and Saturn. Perhaps Ptolemy was getting bored by the time he came to these planets.

Except for Saturn, the longitudes are all multiples of 5′. It is conservative to say that Ptolemy intended to show agreement between the observed and calculated longitudes only after rounding to the nearest multiple of 5′ and that he forgot to round for Saturn. The observed value then lies in a pre-assigned zone whose width is 1/3 of a standard deviation. The probability that this could happen by chance is less than 0.132 for a single observation, and it is less than $4 \times 10^{-5}$ for the entire set of five observations.

It is safe to conclude that Ptolemy fudged the observations listed in Table XIII.6.

In the case of Saturn, there is additional evidence that is even more impressive, at least as it seems to me. Ptolemy (Chapter XI.6) says that the longitude of Saturn was greater than the longitude of the center of the moon by ½ degree, since Saturn was ½ degree east of the northern horn of the crescent. According to the modern ephemeris programs, the longitude of Saturn was 310°.136 and that of the moon was 310°.399; this is the apparent longitude after correction for parallax. Thus Saturn was almost surely west of the center of the moon at the time stated by Ptolemy. However, the calculated positions are slightly uncertain and, further, perhaps Ptolemy made an error in stating the time. Instead of using the calculated longitudes, let us look at the configuration of Saturn and the moon when Saturn was ½ degree east of the moon in longitude. The calculated configuration for this condition depends only upon the calculated latitudes, which are highly accurate.

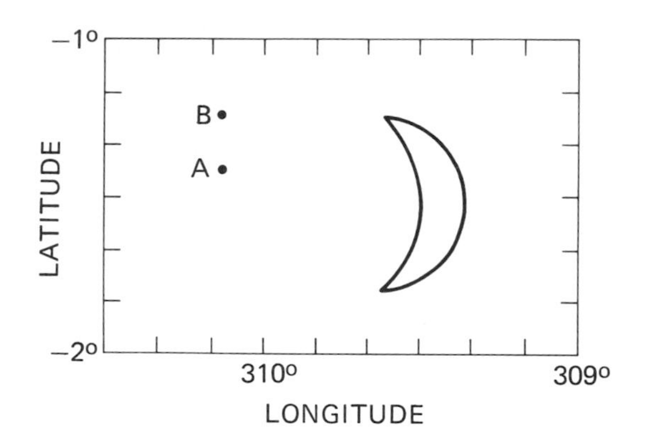


Figure XIII.12. Configuration of Saturn and the moon on the night of 138 December 22, at a time when Saturn was ½ degree east of the moon. Saturn was really at point A but Ptolemy says that he observed it at point B.

The configuration is shown in Figure XIII.12. Saturn is at longitude 310°.136 and latitude -1°.411, while the center of the moon is at longitude 309°.636 and latitude -1°.524; this happened about 1½ hours before the time that Ptolemy states and about 1½ hours after sunset. The moon was about halfway to the first quarter. Saturn was at the position marked A, but Ptolemy says that it was at the position marked B. I do not believe that a person could look at the planet when it was in position A and assert that it was east of the northern tip of the crescent, especially not an observer using an astrolabe, as Ptolemy claims that he was doing.

In sum, at the time of Ptolemy's claimed observation, Saturn was either occulted by the moon or it had just emerged from behind the bright limb. The best observation that Ptolemy could have made was to measure the times when the occultation began and ended, but he does not do this, and he does not seem to have realized that the time of the observation was near the time of an occultation. He says that Saturn was east of the northern tip when in fact it was closer to the center of the moon than it was to the tip. I do not believe that Ptolemy looked at Saturn and the moon at all on the night of 138 December 22, but I do not see how to estimate the confidence level of this conclusion.

We cannot do any more direct testing for the outer planets Mars, Jupiter, and Saturn, but we can do considerably more for Mercury and Venus. For one thing, there is the matter of the eccentricities. Ptolemy uses three 'eccentricities' in his model of Mercury, and we saw in Section XIII.8 how he uses 'observations' to show that all three are equal.† He uses only two eccentricities for Venus, and we saw in Section XIII.5 how he uses his own 'observations' to show that they are equal. He also uses two eccentricities for the outer planets, but he tacitly takes them to be equal without investigation. So much equality of eccentricities needs to be studied, since there is no *a priori* physical reason for the equality.

†I used the term "crank radius" for one of the eccentricities in earlier discussion.

Ptolemy finds the three eccentricities of Mercury and its epicycle radius by using the observations numbered 11 through 14 in Table XIII.3. The necessary analysis was summarized in Section XIII.8. Let us repeat the analysis using the values in Table XIII.3 that were calculated from modern theory rather than the values that Ptolemy states. The result is

$$r = 40;22, \qquad e_1 = 5;53$$

$$e_2 = 0;40, \qquad e_3 = 5;7.$$

These should be compared with the values $r = 39;9$ and $e_1 = e_2 = e_3 = 5;12$ (Equations XIII.24 and XIII.28) which Ptolemy finds. The value just found for $e_3$ is about the same as before, the value found for $e_1$ is considerably different, and the value found for $e_2$ is drastically different.

Of course Ptolemy did not have access to the values found from modern theory, and we must allow him the benefit of reasonable errors of observation. In order to test the effect of observational error, I altered the value of the elongation for the last observation in Table XIII.3 by 0°.25, from -19°.73 to -19°.98, and repeated the analysis. Since r is not affected by the change, the three eccentricities become

$$e_1 = 5;37, \qquad e_2 = 0;56, \qquad e_3 = 5;23.$$

The values of $e_1$ and $e_3$ have come closer together, although both now differ substantially from Ptolemy's value of 5;12 for both. The value of $e_2$ has increased somewhat, but it is still far from being equal to Ptolemy's value or to the other two.

It is not feasible in the scope of this work to study exhaustively the effect of all possible errors. However, if we take 0°.25 as the standard deviation of a single observation, and if we make a reasonable allowance for the fact that four observations altogether are involved, we find that the error in Ptolemy's value of $e_2$ is something like 10 standard deviations. The probability that this happened by chance is about $10^{-23}$.

Hence there is no substantial doubt that the observations have been fabricated in order to make $e_2$ equal the other two eccentricities. It is likely that reasonable errors of observation could make $e_3$ equal to $e_1$. However, the observations that yield $e_3$ are also those that yield $e_2$. Hence, if the value of $e_2$ comes from fabricated observations, so does the value of $e_3$.

Moderate changes in observations 11 and 12 can change r by 1 part in 40, and this changes $e_3$ by 1;15, about 1 part in 4. Hence, in order to make $e_3$ equal to $e_1$, Ptolemy has to have a precise value of r. This indicates strongly that observations 11 and 12 have been fabricated also, although I have not seen a way to assign a confidence level to this conclusion. This conclusion receives strong confirmation from Ptolemy's false statement that observations made under the necessary conditions were beyond the capacity of earlier observers, so that he had to use observations made by himself. His reason for this claim was undoubtedly to justify his own observations, which he could adjust to yield the value of r that he wanted.

To summarize the eccentricities of Mercury, they undoubtedly come from fudged observations, and their equality has been intentionally created by Ptolemy. However, we do not know why he chose to make all of them equal to 5;12. In view of the results shown in Table XIII.4, I should point out that this is not a particularly accurate value.

Matters are not as striking with Venus, mostly because the two eccentricities should be nearly equal, as Table XIII.1 shows, although they should be considerably smaller than the values that Ptolemy finds. Let us suppose, however, that Ptolemy's value of $e_1$ in Equation XIII.12 is correct, and let us look at the solution for $e_2$ using Figure XIII.6.

In order to test the effect of error, let us change the elongation of Venus in the first observation by 15′, from -43°35′ to -43°20′, and repeat the solution. In doing so, let us use Ptolemy's values of r and $e_1$ from Equations XIII.12. The result is that $e_2$ = 1;22,30 in sexagesimal notation if we use 60 for the radius of the deferent. Changing one observation by 15′, which we may use as the standard deviation of a single observation, thus changes $e_2$ by 0;7,30. Since both observations affect $e_2$ about equally, we may say that the standard deviation of the inferred value of $e_2$ is 0;10.

Ptolemy's value of $e_2$ is 1;15, and he has shown that he is willing to round from 1;14 to this value. However, his discussion implies that he does not accept a greater amount of rounding. Thus the value of $e_2$ lies within a preassigned range centered on 1;15 and with a total width of 0;2. Hence the value lies within a preassigned range whose width is 0.2 standard deviations. The probability that this would happen by chance is less than 0.08.

We could accept an occasional result with a probability of this size, but what we cannot accept is the fact that all of Ptolemy's results are improbable. Hence, in spite of the relatively large probability just derived, I have no doubt that Ptolemy's observations used in Figure XIII.6 are fudged. Further, the true probability is considerably smaller than 0.08, but I have not tried to determine by how much. The reason is that I accepted Ptolemy's value of $e_1$ as being substantially correct when I derived the probability above. Actually, as Table XIII.1 shows, Ptolemy should have found a much smaller value.

Finally, we can turn to the interesting matter of the apogees. The eccentricity of Mercury is far greater than that of Venus or the earth, and it should be much easier to locate the apogee of Mercury than of Venus. Yet Ptolemy misses the apogee of Mercury (Table XIII.4) by nearly 30°, while he misses the apogee of Venus (Table XIII.1) by only about 4°. We must enquire how he came to be so far out for Mercury.

First, let us look briefly again at how he found the apogee of Venus. Theon measured the maximum elongation of Venus when it was $+47\frac{1}{4}$ degrees and Ptolemy 'measured' it when it was $-47\frac{1}{4}$ degrees. Ptolemy averages the values of $L_S$ at the two measurements and finds 55°0′ for the longitude of apogee. He confirms this by two more measurements. Theon measured a maximum elongation of -47°32′ and

Ptolemy measured one of +47°32′. Ptolemy again averages the values of $L_S$ and finds that apogee is at 54°58′, an excellent confirmation.†

Let us look at the last observation. The elongation matches that measured by Theon to the minute of arc. Hence the observed longitude of Venus falls within a zone that is within ½ minute of a pre-assigned value, while the standard error of observation is about 15′. The probability that this happened by chance is about 0.027.

When I studied the text of this observation in Section V.6, I said that this observation, out of all of Ptolemy's claimed observations, had the best chance of being genuine when judged by the text alone. I said this because of Ptolemy's remark that Venus paled the reference star φ Aquarii by its brightness. A remark of this sort suggests that the writer may well have observed the situation. However, we conclude, with odds of about 36 to 1, that the observation was fudged. It is possible, in spite of this, that Ptolemy looked at the stars on the occasion stated and noted the paling of φ Aquarii, without actually making the measurement. It is also possible that Ptolemy made up the remark out of his imagination.

Let us now turn to Mercury. In Ptolemy's time, the greatest positive elongation of Mercury was about 26°16′, and it occurred when $L_S$ was about 100°. The greatest negative elongation was about -25°47′, when $L_S$ was about 339°. Ptolemy knows that the greatest positive and negative elongations are about 26° (he uses 26½ degrees) and that the corre-

†If Ptolemy is fudging his data, we naturally ask why the confirmation is not exact. The answer comes from the fact that Ptolemy must use $L_S$ and the elongation in locating apogee. For the last observation, for example, he must have +47°32′ for the elongation. Since the elongation is positive, the observation must be made in the evening, and Ptolemy does not have full control over $L_S$; he must use its value at the necessary time and find the longitude of Venus by adding the elongation to $L_S$. He was lucky to find 'confirmation' within 2′, but he was not so lucky with Mercury.

sponding values of $L_S$ differ from each other, but this seems to be all that he knows on the subject. He thinks that the correct values of $L_S$ are 70° and 310°, respectively, and he uses observations 3 and 4 in Table XIII.3 to prove these values. The errors in his 'observations' are so gross that we do not need to calculate the probability of their chance occurrence; we can say instantly that the observations are fraudulent.

Ptolemy averages 70° and 310° and finds 190° for the position of apogee.† Ostensibly, he uses observations 3 and 4 to 'confirm' the position of apogee that he finds from observations 1 and 2, which were also made with $L_S$ near 70° and 310°.‡ It is clear that these observations are also fraudulent, since they confirm a grossly wrong result. Since the values of $L_S$ are chosen because they are the ones that correspond to the greatest possible elongations, it is clear that observations 3 and 4 were chosen before 1 and 2, although they occur later in Ptolemy's discussion.

We naturally ask how Ptolemy came to make such an enormous error, which he could not have made if he had studied a body of Mercury observations with even moderate care. As a preliminary to answering this question, let us look at Figure XIII.9 again. Let us mark any point on the earth's orbit and calculate the value of $L_S$ at that point. Let us then draw the pair of tangents to Mercury's orbit from the point and calculate the geocentric longitudes of the points of tangency. The differences between the longitudes and the value of $L_S$ are the maximum elongations that can occur for the given value of $L_S$. In this way, we can calculate the maximum elongation of Mercury as a function of $L_S$, for both positive (east) and negative (west)

---

†More accurately, Ptolemy does not know for a short time whether 10° or 190° is the position of apogee, but we can ignore this minor matter here.

‡The values of $L_S$ are not exactly 70° and 310°, and the two values of apogee differ by 3/8 degrees. This illustrates the limitation in fudging maximum elongations that was mentioned in the second footnote before this one.

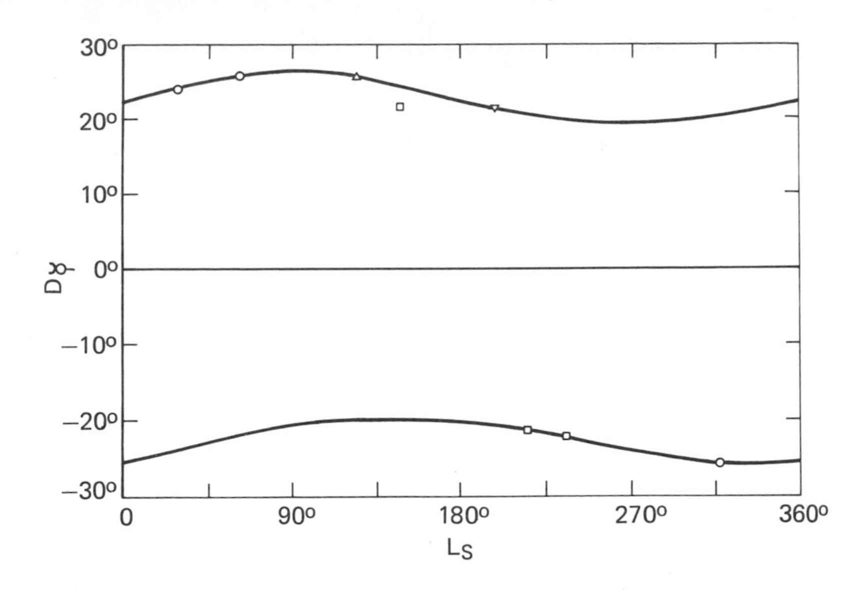


Figure XIII.13. The maximum elongation D☿ of Mercury from the mean sun, plotted as a function of $L_S$, the longitude of the mean sun. The curves apply for epochs near -260. Ptolemy uses six observations near -260 to find the apogee of Mercury. He uses an observation near the point ○ on the lower curve together with the observations marked ○ on the upper curve. He does not realize that the two positive elongations are on the wrong side of the maximum; he should have used observations near the point marked △. He also uses an observation near the point □ below the upper curve together with observations near the points marked □ on the lower curve. These points are on the correct sides of the maxima, but the error in the measured elongation for the point below the upper curve is greater than 2°, an essentially impossible error. Ptolemy may have fudged this observation. A valid observation would have been near the point marked ▽ on the upper curve.

elongations. The results, for an epoch near -260, are plotted in Figure XIII.13.

Ptolemy makes his error in Mercury's apogee because of observations 5, 6, and 7 in Table XIII.3, which have dates near the year -260. From observation 5, he finds an elongation of -25°.83 for $L_S$ = 318°.17. This point is shown by the circle on the lower curve in Figure XIII.13. Observations 6 and 7 are shown by the circles on the upper curve. Ptolemy interpolates between observations 6 and 7 to find the value of $L_S$ when the elongation is +25.83, and finds $L_S$ = 53°.5. He averages this with 318°.17 and finds 185°.83 for the position of apogee in -260, four centuries before his own time.

We now see what Ptolemy has done. He does not realize that there are two values of $L_S$ at which the maximum elongation is the same,† and he has used the wrong value of $L_S$ from the upper curve to go with his value from the lower curve. He should have used an observation, or a pair of observations if necessary, near the point marked Δ on the upper curve. The value of $L_S$ at this point is about 124°, and this would have given him about 221°, rather than about 186°, for the position of apogee. The value 221° is close to being correct.

Unfortunately, this does not end Ptolemy's misuse of observations from the period near -260. Observation 8, with an elongation of 21°.67, is shown by the point □ just below the upper curve. We see from Table XIII.3 that the error in elongation for this observation is more than 2°. Ptolemy interpolates between observations 9 and 10, shown by the points marked □ on the lower curve, to find that $L_S$ = 224°.17 when the elongation is -21°.67. Since $L_S$ = 147°.83 for observation 8, this gives him exactly 186° for apogee. He should have had a point near the one marked ∇ on the upper curve to average with 224°.17 from the lower curve. This would have given him about 209° for apogee, which is too small but is still much closer to the correct value.

An error of more than 2° in the elongation is essentially impossible except by a recording error.

---

†Except, of course, for the two extrema of each curve in Figure XIII.13.

The fact that Ptolemy uses an observation with such an error and finds accurate agreement with an earlier but grossly wrong result shows that one of two things happened: (1) Ptolemy fudged observation 8, or (2) Ptolemy searched through a body of observations until he found one that gave him the result he wanted, without mentioning that most observations gave quite different results. It does not matter greatly which possibility is correct; the actions are equally dishonest.

Neither action can be charged to a dishonest colleague who hoodwinked an innocent and trusting Ptolemy. Ptolemy alone must have been responsible for the misuse of the old observations.

After he found 186° for apogee in -260, Ptolemy decided to keep a constant relation between Mercury's orbit and the stars. This requires the longitude of apogee to increase by 1° per century, since 1° per century is Ptolemy's value for the precession of the equinoxes. In Ptolemy's time, then, which is 4 centuries later than -260, apogee was to be 190°. Ptolemy then decided that the points of greatest possible elongation should be 120° from apogee, namely at 70° and 310°, and he proceeded to fudge observations 1 through 4 accordingly. We do not know why he chose 120° for the separation. We may speculate that he read this value somewhere and adopted it, probably because it is a neat round number; it makes apogee and the two maxima exactly trisect a circle. Regardless of why he chose 120°, he applied it to a seriously wrong position of apogee, as we have seen.

I do not see how to test the authenticity of any more of Ptolemy's claimed planetary observations, and so I shall summarize the situation.

1. For each of Mars, Jupiter, and Saturn, Ptolemy quotes four observations that he claims to have made himself. The first three observations were made at opposition while the fourth was not. We have shown that the third oppositions and the observations not at opposition were fudged.

2. Ptolemy says that Saturn was even with the northern tip of the lunar crescent at a stated time

on 138 December 22 when in fact it was closer to the center of the moon (Figure XIII.12). Such an error in observation is essentially impossible.

3. For each of Mercury and Venus, Ptolemy claims to have made an observation in which he simultaneously measured the planetary position and the lunar position. Both of these observations were fudged.

4. In proving that the eccentricities of Mercury are equal, Ptolemy uses one observation made by Theon and one by himself. The one made by himself was fudged.

5. In proving that the eccentricities of Venus are equal, Ptolemy uses two observations that he claims to have made himself. We can prove that at least one was fudged, and I have no doubt that both were.

6. In finding the radius of Mercury's epicycle, Ptolemy uses two observations that he claims to have made himself. He explains that he has to use his own observations because earlier observers did not have the necessary instruments. This explanation is false. It is likely that Ptolemy, from his earlier fudged observations, was committed to a specific value of the radius; if so, he had to use his own 'observations' because genuine observations would not yield the radius that he needed.

7. In finding the apogee of Venus, Ptolemy uses two observations by Theon and two of his own. We can show that at least one of his own observations was fudged.

8. In finding the apogee of Mercury in his own time, Ptolemy uses four observations that he claims to have made himself. All four were fudged.

9. In finding the apogee of Mercury four centuries earlier, Ptolemy first uses three observations made near -260. He seriously misinterprets the theoretical significance of these observations and finds a grossly wrong position of apogee. He then uses another triad of observations from near -260. One of them is seriously in error, but the

resulting position of apogee nonetheless agrees closely with the position already found. Either this observation was fudged or Ptolemy chose it carefully from an ancient compilation of data, without mentioning that most older observations were in conflict with the result.

In this section, and in three earlier papers [Newton, 1973, 1974, and 1974b], I have examined all of the observations that Ptolemy claims to have made that can be tested. As a result of this detailed examination, we can reach these conclusions:

1. All of his own 'observations' that Ptolemy actually uses, and that are subject to test, prove to be fraudulent. A few stellar observations that he does not use seem to be genuine.

2. Ptolemy develops models for the motions of the sun, moon, the planets, and the equinoxes, and he presents a table of stellar positions. All of these models or tables, without exception, depend intimately upon fraudulent data.

3. Ptolemy is responsible for the fraud. Logically, it is possible that Ptolemy had an assistant who committed some of the fraud unknown to Ptolemy. That is, we cannot categorically deny the possibility that Ptolemy and an associate worked closely together, each committing fraud unknown to the other. However, it is clear that much of the fraud could have been committed only by the person who wrote the *Syntaxis*, and probably all of it was.

4. We cannot accept, without serious reservations, the idea that Ptolemy was a gifted though dishonest astronomer. His treatment of Mercury and the moon is not even competent.

## 10. The Usefulness of Ancient and Medieval Planetary Tables in Studying the Accelerations

In Chapter VIII, I concluded that ancient and medieval solar tables are quite valuable in studying the acceleration of the sun. Here we shall find, in contrast, that planetary tables are not useful. This conclusion needs to be confined in scope to the time period covered in this work. It is possible that planetary tables belonging to the extreme end of the medieval period may be useful; I have not examined any tables from the end of the Middle Ages in enough detail to decide about their value for the purposes of this work.

The distinction between solar and planetary tables rests upon their respective accuracies. In Section 1 of this chapter, we saw that the eccentric or epicyclic model is capable of representing solar position within a maximum error of less than 1′ of arc. Even though most early astronomers did not find the parameters of the model well enough to achieve this accuracy,† we still know their procedures and the structure of their errors well enough to make their tables quite valuable.

All the astronomers whose work is covered by this study used the planetary models developed by Ptolemy; these are the 'crank' model for Mercury and the 'equant' model for the other planets. I found the limiting accuracy of Ptolemy's models for Mercury, Venus, and Mars by making a least-squares fit of the parameters in his models to a

TABLE XIII.7

THE BEST POSSIBLE ACCURACY OF PTOLEMY'S PLANETARY MODELS

| Planet | Root-Mean-Square Error in Best-Fitting Model, degrees | Root-Mean-Square Error in Ptolemy's Model, degrees |
|---|---|---|
| Mercury | 1.70 | 2.99 |
| Venus | 0.14 | 1.01 |
| Mars | 0.36 | 0.44 |

set of accurate ephemeris positions. The limiting accuracies are listed in Table XIII.7. The model for Venus has the greatest potential for accuracy; it is capable of representing the longitude of Venus with a root-mean-square error of $0°.14 = 8'.4$.

†For example, we found that the maximum error in ibn Yunis's solar table is probably about 3′.

Ptolemy did not come close to realizing the potential of his model for Mercury and Venus, as the last column in the table shows.

In using a planetary table to study the accelerations, the quantity that we use is the mean longitude at a stated epoch for the outer planets and the anomaly at an epoch for the inner planets. We should remember from the earlier discussion that the mean longitude for an outer planet is, in modern terms, the mean heliocentric longitude. The anomaly for an inner planet is the heliocentric mean longitude minus the mean longitude of the sun. Thus we can compare a tabular value directly with a value calculated from modern theory and hence infer an acceleration.

If a table is to be useful, the error in an acceleration calculated from it should not exceed, say, 2 $''/cy^2$. For the time of ibn Yunis, the error

TABLE XIII.8

PLANETARY PARAMETERS FROM THE TABLES OF PTOLEMY, YAHYA, AND IBN YUNIS

| | Mercury | Venus | Mars | Jupiter | Saturn |
|---|---|---|---|---|---|
| Ptolemy[a] | | | | | |
| apogee, deg. | 190 | 55 | 115;30 | 161 | 233 |
| eq. of center | 3;2 | 2;24 | 11;25 | 5;15 | 6;31 |
| eq. of epicycle | 22;2 | 45;57 | 41;9 | 11;3 | 6;13 |
| rate, deg./year[d,g] | 53;56,42,33[e] | 225;1,32,29 | 191;16,54,28 | 30;20,22,53 | 12;13,23,57 |
| Yahya[b] | | | | | |
| apogee, deg. | 201 | 82;39 | 93;33[f] | 172;32 | 244;30 |
| eq. of center | 3;2 | 1;59 | 11;25 | 5;15 | 6;31 |
| eq. of epicycle | 22;2 | 45;49 | 41;9 | 11;3 | 6;13 |
| rate, deg./year[d] | 53;56,42,33[e] | 225;2,0,2 | 191;17,17,27 | 30;20,38,12 | 12;13,39,33 |
| ibn Yunis[c] | | | | | |
| apogee, deg. | 202;3 | 86;10 | 125;36 | 173;35 | 244;30 |
| eq. of center | 4;2 | 2;0,30 | 11;25 | 5;15 | 6;31 |
| eq. of epicycle | 22;24 | 46;25 | 41;9 | 11;3 | 6;13 |
| rate, deg./year[d] | 53;56,50[e] | 225;2,24,20 | 191;17,9,46,2 | 30;20,33 | 12;13,36 |

[a]For the year 137.

[b]For the year 830.

[c]For the year 1003.

[d]Rate of change of anomaly for Mercury and Venus, of the mean longitude for the others. The year is 365 days exactly.

[e]The excess over three full revolutions.

[f]This is written as $3^s3°;33$, meaning 3;33 degrees past the end of the third zodiacal sign. It is probably an error for $4^s3°;33$ = 123;33.

[g]I have rounded Ptolemy's numbers. He states the values to a precision of the sixth sexagesimal position, which is clearly absurd.

in the angle used should thus not exceed about 81″, less than 1′.4. For Venus, the error in the mean must be this small in the face of a root-mean-square error of at least 8′.4. In order to achieve this accuracy, the astronomer responsible for the table would have had to find the parameters in the table by fitting to at least 36 observations of position. In view of the difficulty of making the necessary calculations with the means available, we doubt that any early astronomer used this many observations.

Before I look more carefully at the matter of accuracy, it is interesting to look at the parameters in a few tables. We have used Ptolemy's parameters for Mercury, Venus, and Mars in earlier parts of this chapter, and we can easily find his parameters for Jupiter and Saturn in Chapters XI.1 through XI.8 of his work. ibn Yunis [1008, pp. 216 ff] compares his parameters (in the Hakémite tables) with those used by Yahya nearly two centuries earlier. Four important parameters for each planet, from the tables of Ptolemy, Yahya, and ibn Yunis, are compared in Table XIII.8. The comparison allows us to be sure that we understand the tables of ibn Yunis and Yahya.

Instead of using the eccentricities and epicycle radii of the models (Figure XIII.10 for Mercury and XIII.4 for the other planets), ibn Yunis uses the "equation of the center" and the "equation of the epicycle". This usage compresses several parameters into two. For Mercury, the equation of the center means the maximum angle subtended by the eccentric distance ZE at the point B in Figure XIII.10. For the other planets, the equation of the center is the maximum angle subtended by the distance DE at the point B in Figure XIII.4. In each case, Ptolemy's value of the equation of the center† is found by adding the third and fourth columns in his planetary tables (Chapter XI.11) and finding the maximum value of the sum. The equation of the epicycle is the maximum value of the quantity in the sixth column of Ptolemy's planetary tables. In principle, it is equal to $\sin^{-1}(r/R)$,

†We should emphasize that Ptolemy did not use this term.

in which r is the epicycle radius and R is the mean distance. In practice, ibn Yunis seems to mean the largest value that appears in the table in the form that Ptolemy actually wrote down. Because Ptolemy uses an interval of either 3° or 6° in the argument, the value of $\sin^{-1}(r/R)$ does not necessarily appear in the table.

For all of the planets except Venus, Yahya uses the same values of the equations that Ptolemy uses; apparently Yahya uses post-Ptolemaic observations only to update the position of apogee and to refine the mean motion slightly. For Mercury, Yahya's model is identical with Ptolemy's except for the position of apogee. ibn Yunis keeps the same equations as Ptolemy for the outer planets, but he is independent for the inner ones. Both ibn Yunis and Yahya use the same apogee and equation of the center for Venus that they use for the sun, as we can see by comparing Table XIII.8 with the values given in Sections VIII.5 and VIII.10.† Finally, we should notice that neither Yahya nor ibn Yunis made an independent determination of Mercury's apogee. If they had, they would have found that apogee was near 230°. Instead, it is clear that they have preserved Ptolemy's curious and enormous error, and

TABLE XIII.9

PLANETARY PARAMETERS AT THE EPOCH[a] FROM THE TABLES OF IBN YUNIS

| Planet | Anomaly or Mean Longitude,[b] degrees | | Rate degrees/day | |
|---|---|---|---|---|
| | ibn Yunis | from modern theory | ibn Yunis | from modern theory |
| Mercury | 307.756 | 316.500 | 3.106 704 7 | 3.106 729 7 |
| Venus | 292.602 | 290.933 | 0.616 548 2 | 0.616 521 4 |
| Mars | 280.722 | 280.284 | 0.524 071 4 | 0.524 071 0 |

[a]Cairo mean noon, 1000 November 30.

[b]Mean longitude for Mars, anomaly for the others.

†This makes us naturally wonder whether Yahya and ibn Yunis thought that the sun was the physical center of Venus's orbit.

that they have merely added some amount of precession to the apogee that Ptolemy gives.

It is also clear from Table XIII.8 that the planetary models of Yahya and ibn Yunis are identical in structure with those used by Ptolemy.

ibn Yunis does not give positions at the epoch from Yahya's tables, although we could doubtless find them from other sources. Here, it is sufficient to examine positions from ibn Yunis's own tables. Positions at the epoch of mean noon at Cairo on 1000 November 30, along with the mean motions, are compared with values derived from modern theory in Table XIII.9, for Mercury, Venus, and Mars. There is no ready way to make the comparison for Jupiter and Saturn, because modern mean elements of high accuracy are not available for these planets.†

The excellent agreement of the rates in Table XIII.9 shows us that we are interpreting the quantities in ibn Yunis's tables correctly. Thus we can

TABLE XIII.10

A TEST OF IBN YUNIS'S MODEL OF MARS

| Date | Julian Day Number -2 000 000 | Longitude from Modern Theory ° | Longitude from Model ° | Error in Model ° |
|---|---|---|---|---|
| 983 May 10 | 80.228.2265 | 121.181 | 119.819 | -1.362 |
| 987 Oct 10 | 81 842.1953 | 248.155 | 249.678 | +1.523 |
| 988 Sep 18 | 82 186.4078 | 135.938 | 135.617 | -0.321 |
| 989 Dec 15 | 82 638.9132 | 316.431 | 318.043 | +1.612 |
| 990 Aug 30 | 82 896.5911 | 135.821 | 135.477 | -0.344 |
| 991 Nov 2 | 83 325.5278 | 303.412 | 305.795 | +2.383 |
| 993 Oct 19 | 84 043.1949 | 326.999 | 330.439 | +3.440 |
| 994 Jun 1 | 84 267.9107 | 102.971 | 102.220 | -0.751 |
| 994 Jun 1 | 84 268.2249 | 103.169 | 102.417 | —— |
| 995 Jun 11 | 84 642.4511 | 358.517 | 360.957 | +2.440 |
| 998 Jul 4 | 85 762.2486 | 148.488 | 147.667 | -0.821 |
| 999 Apr 10 | 86 041.6003 | 351.140 | 352.301 | +1.161 |
| 1001 Jul 19 | 86 872.9161 | 78.731 | 79.114 | +0.383 |
| 1002 Mar 14 | 87 111.2520 | 138.164 | 135.955 | -2.209 |
| 1003 Jan 7 | 87 410.4207 | 311.296 | 312.672 | +1.376 |

†See the discussion on page 114 of the Explanatory Supplement [1961].

be sure that ibn Yunis's positions are seriously in error, as the table shows. The error in position ranges from more than 8°.7 for Mercury down to about 26′ for Mars.

For Mars, I have confirmed that the error shown in Table XIII.9 is real and not a matter of interpretation. I did so by using ibn Yunis's parameters to calculate the geocentric longitude of Mars at 15 epochs. The epochs are those of the observations of Mars that ibn Yunis made himself, from Table VII.6, combined with the epochs at which ibn Yunis observed conjunctions of Venus and Mars, from Table VII.5. The results of the calculations are shown in Table XIII.10 which should be nearly self-explanatory. It is necessary to point out that the value calculated from modern theory is calculated using the accelerations $\nu_S' = \nu_♂' = 0$.

The last column in Table XIII.10 is the error† in ibn Yunis's model, in the sense of ibn Yunis's value minus that from modern theory. The algebraic mean error‡ is +0°.608, which is far in excess of any amount that we can attribute to the accelerations. We see that ibn Yunis did a poorer job of finding the parameters than Ptolemy did, in spite of the fact that Ptolemy used fraudulent data. The error of +0°.608 should be compared with the error of +0°.438 in the mean longitude from Table XIII.9. In view of the small sample size in Table XIII.10, there is no significant difference between the two errors.

We saw near the beginning of this section that the error must be less than 1′.4 if it is to be useful in finding the accelerations. If ibn Yunis's error is typical, we need about 400 independent tables in order to form a useful estimate of the accelerations from them.

---

†More accurately, it is the apparent error before we correct the third column for the effect of the accelerations. The effect of the accelerations should be less than 0°.05.

‡Two of the observations are on 994 June 1. In order not to give this date an undue weight, I use only the first observation on this date in calculating the mean error.

We should close with a brief reference to the Babylonian planetary tables, which we know a great deal about [Neugebauer, 1955]. Unfortunately, we are ignorant of one crucial fact. In order to use tables properly in the study of the accelerations, we must know the epoch of the underlying observations with reasonable accuracy. Even for a table as accurate as the solar table of Hipparchus, we must know the epoch of the basic observations within about a decade, as we saw in Section VIII.3. We do not know the epoch of the underlying Babylonian observations even to the century.

# CHAPTER XIV

# ANALYSIS OF THE BABYLONIAN PLANETARY DATA

## 1. The Value of the Ammat and the Dating of Observations

One of the first tasks in using the Babylonian planetary data is to decide upon the value of the angular unit called the ammat. Since the latitude of a planet is an insensitive function of the solar and planetary accelerations, we can find the value of the ammat from measurements of the difference in latitude between a planet and a star.

I studied the value of the ammat in Section X.4 by using Babylonian lunar observations. The conclusion (Equations X.6 and X.7) was that the ammat equalled 2° in the records from the Seleucid year -256 (-567/-566 before the common era), that it equalled 2°.5 in the records from the Seleucid year 38 and after (-273 before the common era and after), and that we cannot tell its value from the data in the records from the Seleucid year -67 (-378/-377 before the common era). It is desirable to test this conclusion by using the planetary data.

64 observations from Tables IV.6, IV.8, and IV.10 give the difference in latitude between a planet and a star in terms of the ammat. A number of other observations give the latitude difference by using a mixture of ammats and ubans, but I have ignored these for simplicity. Two of the observations in Table IV.8 are from the year SE -256. These give 3°.24 for the ammat instead of 2°, but the difference is not significant in view of the small sample size. Further, we hardly need the ammat for this year for the purposes of this work, because only one observation from this year gives a longitude difference between a planet and a reference star. I shall assume that the ammat equalled 2° in the year SE -256.

This leaves 62 observations, all from years SE 38 or after, from which we can estimate the value of the ammat. The largest value deduced from an individual observation is 4°.74 and the smallest is 1°.34. The mean value inferred from the observations is

$$1 \text{ ammat} = 2^\circ.54 \pm 0^\circ.08 \text{ for the year SE 38 and after.} \qquad \text{(XIV.1)}$$

This confirms the earlier conclusion. The standard error of a single observation is about 0°.67. This is considerably better than the error of 1°.4 that we found for these years by using the lunar data.

There is one observation of latitude from the year SE -67 that involves the ammat. The observation of Mars in the evening of SE -67 VIII 1 gives a difference of 2 am. 10 u. between Mars and β Arietis, while the calculated difference is 9°.368. Since the uban equals 5′, this gives 4°.27 for the ammat, which is certainly wrong. Thus we still cannot decide upon the value of the ammat for this year. Fortunately, we do not need it for the analysis of the planetary longitudes.

The average daily motion of Mercury and Venus is about 1° per day. I had originally hoped that the standard error of the Babylonian observations would be considerably less than 0°.5. Such accuracy would allow us to determine the dates of most planetary observations, at least of Mercury and Venus, on the basis of the observed positions, with considerable confidence. However, we have found, both here and in Chapter X, that the standard error is considerably greater than 0°.5.† This means that we cannot test the dates by means of single planetary observations. When we cannot verify the dates by means of lunar observations within the same Babylonian month, we must therefore use the tentative dates assigned in Chapter IV, with one type of exception. If we have several planetary

---

†I analyzed the Islamic and Hellenistic data first in order to set limits to the accelerations. On the basis of these results, we may be sure that the effects of the accelerations upon the planetary positions are less than the standard errors of observation.

TABLE XIV.1

FIRST AND LAST APPEARANCES OF MERCURY

| Date | Greenwich Mean Time, hours | Event | $\lambda_S$[a] degrees | $\lambda_☿$[a] degrees |
|---|---|---|---|---|
| -567 May 31 | 16.77 | Missing | 62.384 | 72.474 |
| -522 Jul 3 | 16.99 | See Table IV.5[b] | 93.979 | 116.479 |
| -378 Nov 17[c] | 14.79 | First, evening[c] | 230.764 | 248.615 |
| -278 Nov 9 | 14.89 | First, evening | 223.334 | 242.121 |
| -278 Nov 14 | 14.84 | Last, evening | 228.428 | 241.091 |
| -277 Jan 7 | 3.34 | Last, morning | 282.967 | 268.880 |
| -276 May 6 | 16.47 | Missing | 41.081 | 51.598 |
| -266 Apr 12 | 16.17 | Last, evening | 17.683 | 31.376 |
| -266 Aug 10 | 16.71 | Last, evening | 132.624 | 155.077 |
| -266 Sep 24 | 2.06 | Last, morning | 176.404 | 165.616 |
| -266 Nov 16 | 14.82 | First, evening | 230.555 | 249.461 |
| -266 Dec 1 | 14.71 | Last, evening | 245.864 | 260.520 |
| -265 Mar 25 | 15.99 | Last, evening | 0.042 | 11.941 |
| -239 Mar 28 | 16.02 | Last, evening | 3.643 | 18.287 |
| -239 May 10 | 1.44 | First, morning | 44.357 | 21.032 |
| -239 May 23 | 1.29 | Last, morning | 56.736 | 41.064 |
| -239 Jun 21 | 16.94 | First, evening | 84.972 | 101.331 |
| -239 Jul 25 | 16.89 | Last, evening | 117.574 | 142.631 |
| -239 Aug 23 | 1.74 | First, morning | 145.165 | 129.447 |
| -239 Sep 15 | 1.97 | Last, morning | 167.876 | 159.809 |

[a] Calculated with $\nu_S' = \nu_☿' = 0$. $\lambda_☿$ is the geocentric longitude.

[b] The text says that Mercury was 3 am. east of the moon. This corresponds to a plausible position of the moon, but I have not checked it accurately.

[c] This appears as first, morning, on -378 Nov 18 in Table IV.5, but Mercury was clearly an evening star then.

TABLE XIV.1 (Continued)

| Date | Greenwich Mean Time, hours | Event | $\lambda_S$[a] degrees | $\lambda_☿$[a] degrees |
|---|---|---|---|---|
| -239 Nov 8 | 14.92 | First, evening | 222.852 | 243.469 |
| -239 Nov 17 | 14.81 | Last, evening | 232.026 | 249.937 |
| -239 Dec 5 | 3.01 | First, morning | 249.899 | 234.560 |
| -238 Jan 14 | 3.34 | Last, morning | 290.588 | 275.320 |
| -238 Feb 16 | 15.49 | First, evening | 324.157 | 336.261 |
| -238 Mar 17 | 15.89 | Last, evening | 352.706 | 356.792 |
| -238 Apr 9 | 2.02 | Last, morning | 14.440 | 350.594 |
| -217 Apr 9 | 2.02 | Last, morning | 14.357 | 352.223 |
| -217 May 11 | 16.51 | Missing | 45.599 | 57.304 |
| -217 Jul 18 | 1.32 | First, morning | 109.896 | 92.437 |
| -217 Sep 26 | 15.76 | First, evening | 179.090 | 202.144 |
| -217 Oct 30 | 2.49 | First, morning[d] | 212.825 | 198.495 |
| -217 Dec 7 | 3.04 | Last, morning | 251.584 | 241.063 |
| -216 Jan 11 | 14.96 | First, evening | 287.698 | 300.862 |
| -216 Mar 23 | 2.37 | Last, morning | 358.654 | 336.698 |
| -193 May 12 | 1.44 | First, morning | 46.145 | 23.064 |
| -193 May 25 | 1.24 | Last, morning | 58.521 | 43.588 |
| -192 Apr 10 | 1.87 | Last, morning | 16.238 | 352.095 |
| -170 Feb 1 | 15.27 | Last, evening | 309.694 | 320.423 |
| -132 May 20 | 16.66 | First, evening | 55.569 | 68.980 |

[a]Calculated with $\nu_S' = \nu_♀' = 0$. $\lambda_☿$ is the geocentric longitude.
[d]There is also a conjunction with $\alpha$ Librae. See Tables IV.5, IV.6, and XIV.3.

TABLE XIV.1 (Continued)

| Date | Greenwich Mean Time, hours | Event | $\lambda_S$[a] degrees | $\lambda_☿$[a] degrees |
|---|---|---|---|---|
| -121 May 18 | 1.32 | First, morning | 52.428 | 29.758 |
| -121 Jun 1 | 1.19 | Last, morning | 65.749 | 52.195 |
| -121 Jun 29 | 16.99 | First, evening | 93.062 | 110.124 |
| -121 Aug 1 | 16.81 | Last, evening | 124.772 | 150.052 |
| -121 Sep 2 | 1.84 | First, morning | 155.407 | 137.932 |
| -121 Nov 12 | 14.86 | First, evening | 227.273 | 246.857 |
| -121 Nov 27 | 14.72 | Last, evening | 242.568 | 256.745 |
| -121 Dec 12 | 3.12 | First, morning | 257.391 | 241.843 |
| -120 Jan 16 | 3.34 | Last, morning[e] | 292.964 | 274.323 |
| -120 Feb 24 | 15.62 | First, evening | 332.470 | 345.544 |
| -120 Mar 19 | 15.92 | Last, evening | 356.043 | 7.246 |
| - 75 May 18 | 1.31 | First, morning | 53.267 | 30.564 |
| - 75 Jun 1 | 1.19 | Last, morning | 66.589 | 52.843 |

[a]Calculated with $\nu_S' = \nu_☿' = 0$. $\lambda_☿$ is the geocentric longitude.
[e]The text says first, but calculation shows that last is correct.

observations within the same Babylonian month, we can see whether the observations have correlated errors that can be reduced by changing all the dates by the same amount. If we can, then it is safe to use the dates that minimize the root-mean-square error of all the correlated observations.

In Section IV.3, I said that I would attach low weights to the Babylonian observations because of difficulties in interpreting goal-year texts, and I said that I would do so by assigning low values of the reliability unless the residuals of the observations are so large that they automatically result in low weights. We see now that the errors of observation, and hence the residuals, will be large, and thus I do not need to take any action about the reliability in order to assure low weights.

## 2. The Synodic Phenomena Involving Mercury

The Babylonian observations of synodic phenomena involving Mercury, which are tabulated in Table IV.5, can be divided into two broad classes. One class gives the interval of visibility of Mercury, presumably on the occasions of a first or last visibility. The other class simply records the dates of first and last visibilities, usually but not always accompanied by a statement about the constellation in which Mercury was found at the time.†

The records that do not give an interval of visibility have no clear value in the study of the accelerations, but they may have interest in other types of study. Therefore I have summarized the calculations concerning them in Table XIV.1. The first column of Table XIV.1 gives the date of the event in the Julian calendar, using the dates that were assigned in Table IV.5, with one exception. Table IV.5 lists a first, morning visibility on -378 November 18, but the position of Mercury shows that it was an evening star at that time. Therefore I have changed this to a first, evening visibility on -378 November 17, which comes on the same Babylonian date as the morning of -378 November 18.

The second column of the table gives the Greenwich mean time used in the calculations; this time is 45 minutes from sunrise or sunset at Babylon, whichever is appropriate. The third column repeats the type of event, but with the constellation omitted. The final two columns give the longitude of the sun and the (geocentric) longitude of Mercury, calculated using $\nu_S' = \nu_☿' = 0$.

The differences in longitude between the sun and Mercury vary widely, which is to be expected. However, all of the differences look plausible except for the last, evening visibility on -238 March

†It is conceivable that the constellation applies to the sun's position rather than to Mercury's. I have not inspected this point, although it would be easy to do so, because it is not important for this study.

TABLE XIV.2

ANALYSIS OF THE INTERVALS OF VISIBILITY OF MERCURY

| Date | Greenwich Mean Time, hours | Observed $\Delta H$, degrees | $\Delta H_0$, degrees | 100 $A_i$ | 100 $B_i$ |
|---|---|---|---|---|---|
| -277 Feb 20 | 15.57 | 1/2 | 22.35 | -0.056 | -0.139 |
| -220 Jun 30 | 16.99 | 15 | 15.68 | -1.944 | +1.361 |
| -216 Jan 31 | 15.26 | 15 | 15.35 | +2.528 | -2.694 |
| -216 Feb 19 | 3.02 | - 6 2/3 | -17.30 | +2.694 | -1.556 |
| -193 Jun 21 | 16.96 | 15 | 18.09 | -1.889 | +1.306 |
| -193 Aug 23 | 1.72 | -17 2/3 | -17.18 | +1.833 | -2.000 |
| -171 May 10 | 16.54 | 16 | 13.34 | -1.472 | +1.556 |
| -171 Jul 16 | 1.32 | - 15 | -16.31 | +1.361 | -2.083 |
| -171 Oct 30 | 2.51 | - 17 | -17.16 | +1.750 | -1.861 |
| -170 Feb 21 | 2.96 | -14 1/2 | -18.12 | +2.361 | -1.194 |
| - 86 May 20 | 16.67 | 15 1/2 | 15.71 | -1.306 | +1.417 |
| - 75 Jun 27 | 16.99 | 15 | 17.27 | -1.667 | +1.139 |
| - 75 Aug 31 | 1.82 | -16 2/3 | -19.49 | +1.139 | -1.167 |
| - 74 Feb 24 | 15.64 | 15 | 15.31 | -1.222 | +1.167 |

17. On this date, Mercury was only 4° in longitude from the sun, and it is doubtful that it could have been seen.

We now turn to the records in which the interval of visibility is given. The analysis of these goes just like the analysis of the lunar rise and set times in Section X.3, and the analysis of the lunar data confirms the interpretation of the stated times that was given in Section IV.7. We can use the same description and most of the same notation that we used earlier, merely replacing "moon" wherever it occurred by "Mercury". The data are summarized in Table XIV.2. The first two columns give the date in the Julian calendar and the Greenwich mean time of the observation. The third column gives the observed value of $\Delta H$. This is the observed time of the Mercury event (rise or set) minus the observed time of the solar event, so that it is positive when Mercury was an evening star and negative when it was a morning star. The interval is given in units of degrees rather than of hours or minutes. The fourth column, called $\Delta H_o$, is the value of $\Delta H$ calculated from the ephemerides of Mercury and the sun, when the accelerations $\nu_S' = \nu_☿' = 0$ are used. The calculated value of $\Delta H$ for other values of the accelerations can be written as

$$\Delta H = \Delta H_o + A_i \nu_S' + B_i \nu_☿'. \qquad \text{(XIV.2)}$$

The last two columns give the values of $A_i$ and $B_i$ for each observation, multiplied by 100 for convenience.

When we compare the values of $\Delta H_o$ with the observed values, we find the usual mixture of small differences and large differences. It is plausible that differences of more than 1° represent recording errors of some sort. The disturbing feature of the table is the large number of large errors: eight of the fourteen entries show discrepancies of more than 1°. I shall neglect the discrepancies of more than 1° in the analysis, on the assumption that they come from recording errors and not errors of observation, but the large number of records that must be neglected decreases our confidence in the results.

One possible type of recording error could come from misinterpreting the records in the astronomical diaries and other sources. The reader may remember from the discussion in Section IV.3 that students of the subject often misunderstood the year involved for some time after the texts were discovered. In other words, the Babylonian year that I have assigned may be wrong. If it is, the error in the Babylonian year is almost surely 46, because this is the interval listed for Mercury in Table IV.1. In several instances, I tried changing the Babylonian year by 46 in each direction, recalculating the Julian calendar date, and then repeating all the steps of the analysis. In no case could I make an appreciable improvement by doing so. Thus the recorded year is not the main source of the trouble.†

†Of course, this does not prove that the years listed in Table XIV.2 are necessarily correct.

Another possible type of recording error concerns the dates. None of the dates in Table XIV.2 can be tested by means of lunar observations, and hence all of the dates are the tentative dates assigned in Chapter IV. It is possible that some of the dates are in error by a day, or even by two days in a few extreme cases. Errors of this sort are not errors in the original record, and hence it is perhaps not fair to call them recording errors. They arise from errors in our interpretation of the record, that is, in knowing how to correlate the recorded date with the Julian calendar. For simplicity, I shall call them recording errors.

The information needed in the analysis is given in Table XIV.2 for all of the records. Thus the reader can make his own analysis, using as many or as few of the records as he wishes.

It quickly turns out that $\Delta H$ depends more upon the difference $\nu_S' - \nu_☿'$ than upon either individual acceleration or upon any linear combination that is independent of the difference. Instead of using $\nu_S'$ and $\nu_☿'$, I introduce the parameters

$$x = \nu_S' - \nu_☿' ,$$
$$y = \nu_S' + \nu_☿' . \qquad \text{(XIV.3)}$$

The parameter y defined in Equation XIV.3 has nothing to do with the parameter $10^9(\dot{\omega}_e/\omega_e)$ that I have denoted by y in other discussions.

Since $A_i = \partial(\Delta H)/\partial\nu_S'$ and $B_i = \partial(\Delta H)/\partial\nu_☿'$ from Equation XIV.2, we have

$$\partial(\Delta H)/\partial x = \tfrac{1}{2}(A_i - B_i), \quad \partial(\Delta H)/\partial y = \tfrac{1}{2}(A_i + B_i). \qquad \text{(XIV.4)}$$

From the data in Table XIV.2, we readily find

$$x = 1.56 \pm 6.79,$$
$$y = 244.08 \pm 92.80, \qquad \text{(XIV.5)}$$

in the customary units. When we omit y entirely from the analysis, assuming that ΔH is a function of x only, we find

$$x = 1.07 \pm 10.03. \qquad \text{(XIV.6)}$$

It clearly does not matter much whether we take the value of x from Equation XIV.5 or from Equation XIV.6. Rather arbitrarily, I shall take x from Equation XIV.5, but without using y from the same equation.

We may also test the spin fluctuation hypothesis by the method used in Chapters XI and XII. That is, we assume that $\nu_{☿}' = 4.1520\nu_S'$, as demanded by the hypothesis (Equation XI.26), and use the data from Table XIV.2 to estimate $\nu_S'$. The result is

$$\nu_S' = 0.33 \pm 3.15 \ . \qquad \text{(XIV.7)}$$

Here, as in all previous tests of this sort that use Mercury or Venus data, we have a value of $\nu_S'$ that is considerably smaller than any plausible estimate of $\nu_S'$ from solar data. That is, the rising and setting times of Mercury continue the trend that we have found from other data with regard to the spin fluctuation hypothesis.

### 3. Conjunctions Involving Mercury

The Babylonian conjunctions of Mercury with various stars are listed in Table IV.6. In several cases, there are two or more conjunctions that occur within the same Babylonian month and, in spite of the relatively large errors of measurement, we can estimate the likelihood that the tentative Julian calendar dates listed in Table IV.6 are correct. The only cases in which the tentative date seems questionable occur for the last two observations, which came in month XII of the Seleucid year 190. When we use the dates listed in Table IV.6, the calculated positions of Mercury are too far west by amounts between 1° and 2° for both observations. Therefore I made the dates in the Julian calendar 1 day later, changing them to -120

TABLE XIV.3

ANALYSIS OF THE CONJUNCTIONS INVOLVING MERCURY

| Date | Julian Day Number -1 000 000 | Observed $\lambda_{☿}$, degrees | $100A_i$ | $100B_i$ | $100Z_i$ |
|---|---|---|---|---|---|
| -567 Jun 20[a] | 514 132.206 | 106.347 | 7.300 | +0.925 | - 9.8 |
| -265 Mar 8 | 624 333.157 | 2.495 | 6.127 | +0.296 | + 69.9 |
| -239 Jul 2 | 633 946.208 | 118.859 | 5.208 | +1.003 | - 30.0 |
| -232 Dec 10[a] | 636 663.629 | 230.385 | 5.997 | +0.513 | -156.1 |
| -217 May 23 | 641 941.195 | 79.506 | 5.346 | +0.691 | + 67.1 |
| -217 May 26 | 641 944.197 | 82.768 | 5.555 | +0.475 | - 55.5 |
| -217 Jun 8 | 641 957.204 | 97.884 | 6.910 | -0.936 | - 25.3 |
| -217 Oct 30 | 642 100.604 | 198.075 | 8.042 | -1.578 | - 42.0 |
| -217 Nov 17 | 642 118.614 | 212.382 | 5.128 | +1.248 | + 68.3 |
| -132 May 26 | 672 991.197 | 80.679 | 4.636 | +0.940 | + 63.2 |
| -132 May 29 | 672 994.199 | 83.929 | 4.771 | +0.801 | -115.3 |
| -121 Jul 5 | 677 048.207 | 120.483 | 4.459 | +1.096 | + 42.6 |
| -120 Mar 8[b] | 677 295.158 | 4.500 | 6.203 | -0.621 | + 20.9 |
| -120 Mar 12[b] | 677 299.161 | 8.152 | 7.022 | -1.458 | +128.7 |

[a]These dates are confirmed by lunar observations. See Table X.1.

[b]These are one day later than the dates in Table IV.6.

March 8 and -120 March 12, respectively.†

The analysis proceeds along familiar lines. Let $\lambda_{obs}$ denote the longitude of Mercury that is inferred from an observation and let $\lambda_{oo}$ denote the longitude that is calculated‡ from the ephemeris

†Neither of these dates comes in a month for which we have lunar data. The only dates in Table IV.6 that can be confirmed by lunar observations are -567 June 20 and -232 December 10. The change in dates corresponds to starting the month one day later. In contrast, all the changes required by the lunar data (Section X.1) corresponded to starting the month one day earlier than the date given by Parker and Dubberstein [1956].

‡The times used in the calculations are 45 minutes from sunrise or sunset, whichever is appropriate. The analysis of the lunar data in Section X.5 shows that this assignment of times is reasonably accurate.

program using $\nu_S' = \nu_☿' = 0$. Then we write

$$\lambda_{obs} = \lambda_{oo} + A_i\nu_S' + B_i\nu_☿'. \qquad \text{(XIV.8)}$$

The information needed to infer $\nu_S'$ and $\nu_☿'$ is summarized in Table XIV.3. This table is like Table XII.3, for example, which applies to Hellenistic observations of Mercury, except that we do not need a column for the reliability of an observation. Accordingly, I have omitted this column from Table XIV.3, but I have included a column giving the observed longitudes. The values in this column are found by taking the calculated longitudes of the stars and adding the relative positions listed in Table IV.6. $Z_i$, as before, means the observed longitude minus $\lambda_{oo}$.

Of the 14 values of $Z_i$ in the table, 11 are less than 1° and 3 are greater. This is a familiar distribution, and I assume that the small values represent errors of measurement and that the large values come from recording errors. I neglect the latter in the analysis.

The accelerations of the sun and Mercury that are inferred from Table XIV.3 are

$$\begin{aligned} \nu_S' &= 0.94 \pm 2.25, \\ \nu_☿' &= 25.65 \pm 14.11. \end{aligned} \qquad \text{(XIV.9)}$$

When we infer $\nu_S'$ alone by using $\nu_☿' = 4.1520\nu_S'$, we find

$$\nu_S' = 2.70 \pm 1.80. \qquad \text{(XIV.10)}$$

This value is reasonably consistent with the spin fluctuation hypothesis, but it is the first such value that we have found from either Mercury or Venus data. Even so, the central value in Equation XIV.10 is somewhat below the most probable value of $\nu_S'$, and in this sense it continues the trend.

## TABLE XIV.4

## SYNODIC PHENOMENA INVOLVING VENUS

| Date | Greenwich Mean Time, hours | Event | $\lambda_S$[a] degrees | $\lambda_♀$[a] degrees |
|---|---|---|---|---|
| -567 May 22[b] | 16.66 | Max. elongation | 53.817 | 94.895 |
| -522 Jun 12 | 16.91 | Last, evening | 73.933 | 84.845 |
| -522 Jun 30 | 1.19 | First, morning | 90.482 | 74.953 |
| -522 Sep 24[c] | 2.04 | Max. elongation | 174.494 | 129.892 |
| -521 Mar 4 | 2.86 | Last, morning | 337.063 | 326.640 |
| -521 Apr 7[d] | 16.12 | First, evening | 10.650 | 9.235 |
| -521 May 6 | 16.46 | First, evening | 38.447 | 44.883 |
| -281 Nov 4[e] | 14.96 | Last, evening | 217.971 | 235.102 |
| -281 Nov 17[e] | 2.77 | First, morning | 230.700 | 228.387 |
| -273 Nov 15[e] | 2.72 | Visible for 10° | 228.722 | 225.614 |
| -240 Nov 27 | 14.72 | First, evening | 242.470 | 267.638 |
| -238 Apr 26 | 16.37 | First, evening | 31.339 | 38.085 |
| -237 Jan 15 | 15.01 | Last, evening | 291.851 | 293.193 |
| -201 Oct 9 | 15.46 | Last, evening | 192.248 | 210.615 |
| -201 Nov 6 | 2.57 | First, morning | 220.066 | 197.302 |
| -179 Apr 9 | 1.94 | Last, morning | 15.130 | 3.043 |
| -179 Jul 20 | 16.91 | First, evening | 113.215 | 129.181 |
| -178 Mar 6 | 15.76 | Last, evening | 342.389 | 351.201 |
| -178 Mar 14 | 2.56 | First, morning | 349.691 | 346.590 |
| - 94 Nov 28 | 14.71 | Last, evening | 244.044 | 251.335 |
| - 94 Dec 3 | 3.02 | First, morning | 248.653 | 248.636 |
| - 83 May 21 | 16.69 | First, evening | 56.660 | 60.450 |
| - 82 Feb 10 | 15.44 | Last, evening | 319.351 | 321.372 |
| - 82 Feb 16 | 3.04 | First, morning | 324.806 | 317.958 |

[a]Calculated using $\nu_S' = \nu_♀' = 0$.

[b]Correct date may be one Babylonian month later. The elongation was 43°.774 on -567 June 8.

[c]Calculation shows that the elongation was 44°.653 on -522 September 23.

[d]This entry is wrong, and the next one should probably be substituted for it.

[e]These entries, which are closely correlated by the Babylonian text, seem inexplicable.

## 4. The Synodic Phenomena Involving Venus

The synodic observations of Mercury were all observations of first or last visibility, with a measured value of the rise or set time accompanying the observation in a few instances. The synodic observations of Venus include these kinds of observation and also involve a few measurements of maximum elongation.

Only one synodic observation of Mercury resulted in calculated values that seem unreasonable, which suggests that the basic method of interpreting the records is probably correct. However, the Venus phenomena pose a number of problems. The information concerning the phenomena is summarized in Table XIV.4, which has the same form as Table XIV.1 for the Mercury phenomena.†

The first observation in the table is a maximum elongation dated SE -256 II 1 (Table IV.7). Since calculation shows that Venus was an evening star at the time, the date has been equated to the evening of -567 May 22, and the elongation on this date is seen from the table to be 41°.078. On the other hand, Table IV.8 lists a conjunction of Venus with $\alpha$ Leonis on SE -256 II 18, 17 days later, and calculation shows that the elongation was 43°.774 at this time. Therefore the entry for SE -256 II 1 is in error. The most plausible error is a month, which would make the date -SE 256 III 1, but I have not tested this possibility by calculation.

Table IV.7 shows another maximum elongation on SE -211 VI 24, but it does not give the time of day. The observation must have been in the night between -522 September 23 and 24, and I calculated the positions for the evening of the earlier day and the

---

†One of the only two dates in Table XIV.4 whose conversion from the Babylonian to the Julian calendar is confirmed by lunar observations is the first one, which is equivalent to SE -256 II 1. It is ironic that the Babylonian date seems to be wrong in this case. The other confirmed date is -273 November 15, which is equivalent to SE 38 VIII 11.

morning of the later day. Calculation shows that the observation was made in the morning and that it was close to maximum elongation. As a note in the table shows, the elongation changed by only 0°.05 between the two times.

Table IV.7 lists a first, evening visibility on SE -210 I 23, and a note indicates that the month is probably wrong. Calculation confirms this. This date is equated to -521 April 7, but Venus was still a morning star then and so close to the sun that it was not visible. If we change the date to SE -210 II 23, which is equated to -521 May 6, Venus is an evening star, certainly close to first visibility, and it is also in the right part of the sky (Taurus). It may be that the wrong month was put down in the record, but there may be another possibility. Parker and Dubberstein [1956] show an intercalary month at the end of this year; we do not need a recording error if the intercalary month came at the end of the preceding year. This would require the intercalary month to begin on or near March 27. Such a late beginning of an intercalary month is rare but not unknown, according to the tables of Parker and Dubberstein.

In Table IV.7, the successive records dated SE 30 VIII 1, SE 30 VIII 13, and SE 38 VIII 11 pose a problem that I have not been able to answer by postulating a single error. The Seleucid dates have been equated to -281 November 4 (evening), -281 November 17 (morning), and -273 November 15 (morning). According to the records, a last, evening visibility occurred on the first date and a first, morning visibility occurred on the second. Calculation shows that Venus did in fact change from an evening to a morning star between these two dates. This means that both the year and the month are probably correct. However, the elongation on the first occasion was 17°.131, which seems too large for a last visibility, and it was 2°.313 on the second occasion, which is certainly too small.

The daily rate of change of the elongation when Venus is near perigee is about 1°.6. In order to make the results plausible, we need to make both dates at least two days later in the Julian calendar. Since it is rare but possible for the tables of

*Parker and Dubberstein* [1956] to be in error by two days,† we are at first tempted to say that their conversion tables of Babylonian dates happen to be wrong by two days for month VIII of the Seleucid year 30, but the entry for SE 38 VIII 11 seems to remove even this possibility. The record in the diary‡ says that Venus was seen on the morning of day 11 of month VIII, presumably in the diary year (SE 38) itself, for 10°. Calculation shows, however, that Venus was only 3°.108 from the sun on -273 November 15, the date that I have equated to SE 38 VIII 11. Thus we need to postulate that the table for converting dates also happens to be in error by the rare amount of two days for this date also, which is in a different year. This is extremely unlikely. Further, as I noted above, the date (SE 38 VIII 11) is confirmed by lunar observations. For the present, at least, I leave these three records as an unexplained mystery.

The elongation on -240 November 27, which is recorded as a first, evening visibility is more than 25°, which seems excessive. Perhaps a combination of the weather with the inclination of the ecliptic to the horizonǂ can explain this. The elongation on -237 January 15 is only 1°.342, and visibility on this date seems impossible. The elongation on -178 March 14 was 3°.101, which again seems impossible. The elongation on -94 December 3 was only 0°.017, which is certainly impossible. Finally, the elongation on -82 February 10 was 2°.021.

Altogether, we find 7 unreasonably small values of the elongation out of 22 observations of first or last visibility. This contrasts strongly with the Mercury records, for which we found only 1 unreasonable value out of 53. We can perhaps explain 5 of the 7 small values by postulating an error in writing the day of the month, or an error in the tables

---

†We should remember that these tables are based upon the theoretical calculations of Schoch (Section II.2).

‡See Table IV.2 for an identification of the text.

ǂSee Section III.5. Also compare with the entry for -232 November 26 in Table XIV.5.

for changing Babylonian dates to dates in the Julian calendar. The other unreasonable values, however, seem inexplicable by single errors.

I do not believe that we can explain the errors by saying that my basic method of interpreting the records† of the synodic phenomena is wrong. If it were wrong, we should have to shift the Seleucid year by 8. I tried making such a shift in each direction in several instances, but in no case could I change an unreasonable first or last visibility into a reasonable one.

TABLE XIV.5

ANALYSIS OF THE INTERVALS OF VISIBILITY OF VENUS

| Date | Greenwich Mean Time, hours | Observed ΔH, degrees | $\Delta H_o$, degrees | $100A_i$ | $100C_i$ |
|---|---|---|---|---|---|
| -273 Nov 15[a] | 2.72 | -10 | - 3.54 | +14.500 | -12.944 |
| -232 Nov 26 | 14.74 | +17 | +23.86 | - 1.556 | + 2.611 |
| -228 Aug 1[b] | 16.81 | + 8 | + 9.51 | + 9.389 | -10.750 |
| -220 Aug 2[b] | 16.79 | + 8 | + 5.66 | +10.361 | -11.667 |
| -201 Oct 5[c] | 15.54 | + 9 | + 9.08 | +10.361 | - 9.417 |
| -193 Oct 6[c] | 15.49 | + 9 | + 5.94 | +10.889 | - 9.917 |
| -193 Nov 5 | 2.57 | -12 | -18.47 | + 7.611 | - 6.944 |
| -170 Mar 15 | 2.57 | - 8 | - 8.32 | +16.806 | -17.056 |
| - 74 Feb 14 | 3.07 | -12 | - 6.00 | +15.611 | -15.694 |

[a]This entry also appears in Table XIV.4, where it was used in a test of dates.

[b]These are not separate observations. I tried using dates separated by 8 Babylonian years in order to see whether the results were appreciably affected.

[c]These are not separate observations. I tried using dates separated by 8 Babylonian years in order to see whether the results were appreciably affected.

We can now turn to the records in which an interval of visibility of Venus is recorded. The information concerning these records is summarized in Table XIV.5. This table has the same form as Table XIV.2, except that I use $C_i$ rather than $B_i$ for the coefficient of $\nu_{♀}'$. The first observation in the table, dated -273 November 15, also appears

---

†See Section IV.3.

in Table XIV.4, where it was used in studying the possible need of changing the dates -281 November 4 and -281 November 17.

The visibility intervals for Venus show even more problems that those for Mercury. A comparison of $\Delta H$ and $\Delta H_o$ shows that only three values agree within a plausible amount. These are the observations dated -228 August 1, -201 October 5, and -170 March 15. For the first two of these observations, I have also listed the conditions that would exist if the correct Babylonian date had been 8 Babylonian years later than the dates listed in Table IV.7. We see that such a change in date can change the interval of visibility by about 3° or 4°. Changes this small obviously cannot bring the other values of $\Delta H$ and $\Delta H_o$ into reasonable agreement.

The difficulty with the visibility intervals for Mercury and Venus is probably not a matter of interpretation. The corresponding data for the moon, which were studied in Section X.3, yielded a reasonable value for the acceleration of the moon. The problem may simply lie in the nature of the object being observed. The thin lunar crescent does not resemble any other object in the heavens, and there should be no problem in identifying it. Mercury and Venus, however, on the occasions of their first and last appearances, look like stars to the naked eye. Further, I suspect, attempts to see them at first and last visibility are likely to be made when the sky is still relatively bright and thus when there are few other objects to aid in their identification. If this is so, the seriously wrong observations may be matters of wrong identification; that is, a star may have been mistaken for the planet.

Whether this explanation is correct or not, it seems clear that the data in Table XIV.5 will not furnish a useful estimate of the accelerations.

## 5. Conjunctions Involving Venus

The Babylonian conjunctions of Venus with various stars are listed in Table IV.8. The calculations needed to estimate the accelerations of

TABLE XIV.6

ANALYSIS OF THE CONJUNCTIONS INVOLVING VENUS

| Date | Julian Day Number -1 000 000 | Observed λ♀, degrees | 100$A_i$ | 100$C_i$ | 100$Z_i$ |
|---|---|---|---|---|---|
| -567 Jun 8[b] | 514 120.202 | 114.347 | 6.664 | +1.563 | + 57.1 |
| -566 Feb 2[b] | 514 358.636 | 268.295 | 5.773 | +2.820 | -114.0 |
| -273 Oct 8[b] | 621 625.145 | 232.317 | 12.611 | -5.759 | +194.4 |
| -238 Jul 6 | 634 315.208 | 125.292 | 3.820 | +2.437 | + 33.5 |
| -238 Jul 21 | 634 330.204 | 145.571 | 3.924 | +2.364 | +239.9 |
| -238 Oct 3 | 634 404.149 | 230.300 | 5.052 | +1.442 | - 59.1 |
| -238 Nov 28 | 634 460.113 | 290.581 | 9.063 | -2.449 | +258.3 |
| -232 Dec 2[b] | 636 656.113 | 273.256 | 3.940 | +2.520 | -123.3 |
| -232 Dec 15[b] | 636 669.113 | 290.664 | 3.991 | +2.441 | + 7.5 |
| -232 Dec 17[b] | 636 671.113 | 292.429 | 4.000 | +2.430 | - 62.9 |
| -231 Feb 17[b] | 636 733.146 | 6.614 | 4.591 | +1.657 | - 81.9 |
| -228 May 19 | 637 920.194 | 97.732 | 5.202 | +0.880 | - 52.1 |
| -228 Jun 15 | 637 947.205 | 125.431 | 6.989 | -0.938 | + 75.0 |
| -228 Oct 8[a] | 638 061.592 | 145.815 | 8.120 | -1.731 | - 6.3 |
| -228 Nov 24 | 638 108.619 | 194.175 | 5.093 | +1.378 | - 18.9 |
| -228 Nov 29 | 638 113.622 | 198.441 | 4.958 | +1.514 | -170.7 |
| -227 Jan 30 | 638 175.636 | 272.981 | 4.051 | +2.336 | -173.2 |
| -201 Apr 10 | 647 742.175 | 39.165 | 3.631 | +2.386 | - 45.5 |
| -201 Apr 19 | 647 751.179 | 51.978 | 3.663 | +2.340 | +144.1 |
| -201 Apr 29 | 647 761.184 | 62.886 | 3.711 | +2.281 | + 27.3 |
| -201 May 1 | 647 763.185 | 64.675 | 3.722 | +2.268 | - 34.5 |
| -201 May 4 | 647 766.186 | 68.491 | 3.740 | +2.247 | - 13.2 |
| -201 May 13 | 647 775.191 | 70.727 | 3.804 | +2.180 | + 33.3 |
| -201 May 17 | 647 779.193 | 82.986 | 3.837 | +2.146 | -117.4 |
| -201 May 28[c] | 647 790.197 | 98.106 | 3.945 | +2.040 | + 91.4 |
| -201 Jun 7[c] | 647 800.204 | 110.101 | 4.069 | +1.922 | +116.9 |
| -201 Jun 15[c] | 647 808.204 | 119.382 | 4.191 | +1.808 | +115.5 |
| -201 Jun 22[c] | 647 815.207 | 125.804 | 4.317 | +1.691 | - 48.3 |
| -201 Jul 1 | 647 824.207 | 141.220 | 4.512 | +1.509 | +469.4 |
| -201 Jul 9 | 647 832.207 | 146.091 | 4.724 | +1.311 | + 60.8 |

[a]The text says that Venus was "slightly east" of β Virginis. I have taken this to be 0°.1 east.

[b]These dates are confirmed by lunar observations. See Table X.1.

[c]I tried adding 1 day to each of these well-correlated dates, with the result of enlarging the root-mean-square error slightly. Therefore I have kept the original dates.

TABLE XIV.6 (Continued)

| Date | Julian Day Number -1 000 000 | Observed λ♀, degrees | $100A_i$ | $100C_i$ | $100Z_i$ |
|---|---|---|---|---|---|
| -201 Jul 22 | 647 845.204 | 159.861 | 5.177 | +0.886 | + 18.6 |
| -201 Aug 3 | 647 857.200 | 173.280 | 5.767 | +0.328 | +103.2 |
| -201 Aug 26 | 647 880.184 | 194.545 | 7.763 | -1.587 | + 58.1 |
| -201 Sep 1 | 647 886.179 | 198.811 | 8.609 | -2.406 | - 0.4 |
| -201 Nov 23 | 647 968.618 | 198.814 | 12.724 | -6.053 | - 62.4 |
| -201 Dec 14 | 647 989.631 | 212.606 | 8.234 | -1.738 | - 83.6 |
| -201 Dec 20 | 647 995.633 | 219.190 | 7.491 | -1.028 | + 43.0 |
| -200 Jan 1 | 648 007.637 | 230.816 | 6.406 | -0.003 | + 40.5 |
| -200 Feb 8 | 648 045.632 | 273.355 | 4.726 | +1.527 | +128.3 |
| -200 Feb 24 | 648 061.622 | 291.098 | 4.377 | +1.825 | + 47.1 |
| -200 Feb 26 | 648 063.620 | 292.862 | 4.341 | +1.855 | - 10.4 |
| -179 Aug 25 | 655 915.184 | 173.584 | 3.661 | +2.353 | + 6.2 |
| -179 Sep 16 | 655 937.163 | 199.115 | 3.797 | +2.267 | -139.2 |
| -179 Sep 26 | 655 947.155 | 212.630 | 3.871 | +2.216 | - 10.0 |
| -179 Oct 2 | 655 953.150 | 219.491 | 3.920 | +2.179 | - 55.3 |
| -179 Oct 11 | 655 962.141 | 231.116 | 4.001 | +2.117 | + 13.4 |
| -179 Nov 16 | 655 998.117 | 273.657 | 4.466 | +1.698 | - 48.7 |
| -179 Nov 30 | 656 012.112 | 291.400 | 4.758 | +1.403 | + 95.5 |
| -179 Dec 3 | 656 015.112 | 293.665 | 4.835 | +1.324 | - 21.3 |
| - 94 May 12 | 686 856.191 | 69.973 | 3.238 | +2.162 | + 32.1 |
| - 94 May 21 | 686 865.195 | 81.203 | 3.270 | +2.129 | + 61.7 |
| - 94 May 25 | 686 869.197 | 84.449 | 3.287 | +2.112 | - 98.6 |
| - 94 Jun 6 | 686 881.202 | 99.587 | 3.348 | +2.056 | - 34.8 |
| - 94 Jun 22 | 686 897.207 | 120.855 | 3.458 | +1.962 | +170.0 |
| - 94 Jun 29 | 686 904.207 | 127.284 | 3.518 | +1.911 | - 22.5 |
| - 94 Jul 16 | 686 921.206 | 147.595 | 3.702 | +1.755 | - 4.2 |
| - 94 Jul 28 | 686 933.201 | 161.326 | 3.875 | +1.608 | - 33.7 |
| - 94 Aug 8 | 686 944.194 | 174.759 | 4.074 | +1.434 | + 40.7 |
| - 94 Aug 27 | 686 963.182 | 196.023 | 4.560 | +1.004 | + 26.8 |
| - 94 Aug 31 | 686 967.178 | 200.290 | 4.694 | +0.882 | + 14.1 |
| - 94 Sep 19 | 686 986.161 | 220.667 | 5.577 | +0.069 | + 42.7 |
| - 94 Oct 1 | 686 998.149 | 232.292 | 6.469 | -0.771 | + 36.3 |
| - 93 Feb 13 | 687 132.625 | 274.837 | 5.631 | +0.047 | - 14.2 |
| - 93 Mar 2 | 687 149.616 | 292.584 | 4.772 | +0.830 | - 14.9 |
| - 93 Mar 4 | 687 151.614 | 294.348 | 4.696 | +0.898 | - 55.0 |

TABLE XIV.6 (Continued)

| Date | Julian Day Number -1 000 000 | Observed $\lambda_♀$, degrees | $100A_i$ | $100C_i$ | $100Z_i$ |
|---|---|---|---|---|---|
| - 83 Aug 9 | 690 963.195 | 161.477 | 3.362 | +2.079 | - 4.2 |
| - 83 Aug 18 | 690 972.188 | 174.911 | 3.419 | +2.041 | +243.1 |
| - 83 Sep 6 | 690 991.173 | 196.175 | 3.567 | +1.936 | + 66.0 |
| - 83 Sep 10 | 690 995.170 | 200.442 | 3.603 | +1.909 | + 9.7 |
| - 83 Sep 21 | 691 006.160 | 213.625 | 3.714 | +1.823 | + 5.2 |
| - 83 Sep 27 | 691 012.152 | 220.818 | 3.784 | +1.768 | + 6.9 |
| - 83 Oct 6 | 691 021.146 | 232.444 | 3.902 | +1.670 | + 97.8 |
| - 83 Nov 14 | 691 060.118 | 274.985 | 4.757 | +0.878 | -152.5 |
| - 83 Nov 29 | 691 075.113 | 292.733 | 5.398 | +0.240 | + 9.5 |
| - 83 Dec 1 | 691 077.112 | 294.497 | 5.508 | +0.129 | - 20.2 |

the sun and Venus from these observations are summarized in Table XIV.6, which has the same format as Table XIV.3, except for the use of $C_i$ rather than $B_i$ for the coefficient of the planetary acceleration. Several points about Table XIV.6 need preliminary discussion.

Table IV.8 contains an entry for SE -256 XI 4 in which I could not tell whether the reference star was intended to be $\gamma$ Capricorni or $\delta$ Capricorni, and which I therefore listed as $\gamma\delta$ Capricorni. This observation is useless for present purposes and I have therefore omitted it from Table XIV.6.

The entry for SE 83 VII 19 (equated to -228 October 8) says that Venus was slightly east of β Virginis. I have arbitrarily taken this to be 0°.1 east.

The four entries for SE 110 III 3 through SE 110 III 28 come within the same Babylonian month, although it is not one for which we have a confirming lunar observation. In Table IV.8, I equated these dates to -201 May 28, -201 June 7, -201 June 15, and -201 June 22. Table XIV.6 shows us that the value of $Z_i$ is about 1° for the first three dates and about -0°.5 for the fourth. Therefore I tried adding one day to the Julian date for each of the dates in this Babylonian month in order to see whether I could decrease the root-mean-square value of the $Z_i$ for the four observations by doing so. The change decreased the values of $Z_i$ for the first three observations, but it increased the value for the fourth observation by so much that the root-mean-square value for all four observations actually increased. Therefore I keptthe Julian dates listed in Table IV.8.

The condition of the text for the observation SE 83 III 23 (-228 June 15) left open the possibility that a longitude separation between Venus and ρ Leonis might be missing. However, the value of $Z_i$ for this observation shown in Table XIV.6 is plausible, and thus it is likely that no longitude separation was originally recorded.

The value of $Z_i$ for -238 July 21 (SE 73 IV 9 in Table IV.8) in Table XIV.6 is greater than 2°. It would be necessary to change the date to -238 July 23 in order to obtain a reasonable value of $Z_i$. As I noted in Table IV.8, the month of the observation was missing from the record and was supplied by the editor of the text. It is possible but not likely that some other month would fit the observation better, but I have not tried to test the matter. I have simply listed the observation in Table XIV.6 under the date originally assigned, but I shall not use it in the inference of the accelerations.

The days of the month listed under SE 217 III 11 or 12 (-94 June 22 or 23) and SE 217 IV 6 (-94 July 16 or 17) were hard to read. Table XIV.6 suggests that I should perhaps have used -94 June 23 rather than 22, but that the date -94 July 16 is probably correct. Instead of bothering to repeat the calculations using the date -94 June 23, I shall simply omit this observation.

In addition to the two observations that I have decided not to use, Table XIV.6 contains 73 observations. Of the values of the $Z_i$, 16 are greater than 1° and 57 are less. I assume that the values greater than 1° come from recording errors and ignore them. The accelerations of the sun and of Venus that are inferred from the 57 observations are

$$\nu_S' = 0.16 \pm 1.27,$$
$$\nu_♀' = 1.50 \pm 3.47. \qquad \text{(XIV.11)}$$

When we infer $\nu_S'$ alone by using the condition $\nu_♀' = 1.6255\nu_S'$ (Equation XI.26), we find

$$\nu_S' = 0.40 \pm 0.93. \qquad \text{(XIV.12)}$$

Thus we continue the trend of finding values of $\nu_S'$ from observations of Mercury and Venus that are smaller than the values found from solar observations.

TABLE XIV.7

SYNODIC PHENOMENA INVOLVING MARS

| Date | Greenwich Mean Time, hours | Event | $\lambda_S$[a] degrees | $\lambda_♂$[a] degrees |
|---|---|---|---|---|
| -522 Jun 1 | 16.76 | Last, evening | 63.441 | 81.045 |
| -522 Sep 13 | 1.92 | First, morning | 163.507 | 147.181 |
| -521 Aug 2[b] | 21.04 | Turning point | 122.812 | 346.656 |
| -520 May 20 | 16.64 | Last, evening | 52.513 | 86.090 |
| -378 Nov 17[b] | 21.04 | Turning point | 231.032 | 2.950 |
| -299 Oct 8[b] | 21.04 | Opposition | 191.252 | 14.241 |
| -299 Nov 13[b] | 21.04 | Turning point[c] | 227.774 | 6.227 |
| -250 Oct 4[b] | 21.04 | Turning point[c] | 187.343 | 61.698 |
| -250 Nov 16[b] | 21.04 | Opposition | 230.940 | 52.405 |
| -250 Dec 17[b] | 21.04 | Turning point | 262.577 | 43.760 |

[a] Calculated using $\nu_S' = \nu_♂' = 0$.

[b] Since the record gives no indication of the hour of the observation, I used local midnight at Babylon.

[c] The position at the turning point is also recorded. See Table XIV.8.

## 6. The Synodic Phenomena Involving Mars

The small number of recorded synodic phenomena involving Mars include first and last visibilities, turning points, and oppositions. These phenomena are listed in Table XIV.7. The position of Mars is

TABLE XIV.8

ANALYSIS OF THE CONJUNCTIONS INVOLVING MARS

| Date | Julian Day Number -1 000 000 | Observed $\lambda_{♂}$, degrees | $100A_i$ | $100D_i$ | $100Z_i$ |
|---|---|---|---|---|---|
| -567 May 24[d] | 514 105.195 | a | + 2.856 | + 4.595 | b |
| -567 May 26[d] | 514 107.197 | a | + 2.867 | + 4.583 | b |
| -567 Jun 20[d] | 514 132.206 | 106.347 | + 2.944 | + 4.491 | - 51.0 |
| -378 Oct 27[d] | 583 293.129 | 0.952 | -11.775 | +19.659 | -305.4 |
| -299 Nov 13[e] | 612 165.377 | 1.044 | - 7.799 | +14.822 | -518.3 |
| -273 Oct 27[d] | 621 644.129 | 272.357 | + 2.040 | + 5.652 | - 7.4 |
| -267 Jun 14 | 623 700.548 | 355.327 | c | c | c |
| -267 Jun 24 | 623 710.549 | 2.476 | + 0.451 | + 6.932 | + 81.7 |
| -266 Feb 28 | 623 960.152 | 51.081 | + 0.971 | + 4.849 | - 18.4 |
| -266 Mar 20 | 623 980.164 | 61.989 | + 1.390 | + 4.370 | - 25.1 |
| -268 Mar 28 | 623 988.168 | 67.593 | + 1.519 | + 4.223 | + 83.5 |
| -250 Oct 4[e] | 630 022.377 | 62.223 | - 5.416 | +10.853 | + 52.5 |
| -240 Sep 10 | 633 650.580 | 140.685 | + 2.259 | + 3.436 | +488.5 |
| -218 May 21 | 641 573.556 | 356.001 | + 1.960 | + 4.900 | + 53.1 |
| -154 May 14 | 664 942.556 | 356.885 | + 2.052 | + 4.321 | -165.5 |
| -154 May 24 | 664 952.552 | 4.035 | + 1.991 | + 4.311 | -176.1 |
| -154 May 29 | 664 957.550 | 7.686 | + 1.956 | + 4.308 | -169.4 |
| -154 Jun 24 | 664 983.548 | 30.060 | + 1.733 | + 4.316 | +256.6 |
| -154 Aug 1 | 665 021.561 | 52.636 | + 1.180 | + 4.516 | + 70.0 |
| -154 Aug 6 | 665 026.564 | 54.849 | + 1.074 | + 4.572 | - 7.3 |

[a]The mean of the observed values of $\lambda_{♂}$ for these two observations is 91°.727, but neither individual value is known.

[b]The mean value of $100Z_i$ for these observations is +106.5, but neither individual value is known.

[c]These values are omitted because the editor calculated the date from the observed position. Therefore this is not a true observation.

[d]These dates are confirmed by lunar observations. See Table X.1.

[e]These are also turning points. See Table XIV.7.

given for two of the turning points, and these observations are also listed in the later table of conjunctions. The format of the table is the same as that of Table XIV.4.

Most of the observations do not tell us whether the observation was made in the morning or evening. I have listed the (Greenwich mean) time of $21^h.04$, which corresponds to local midnight at Babylon, for these observations.

In view of the small number of Mars synodic phenomena, and of the large uncertainty in time of most of the observations, it is clear that they cannot contribute significantly to the inference of the accelerations. They are tabulated only as a matter of record.

## 7. Conjunctions Involving Mars

The conjunctions involving Mars are listed in Table XIV.8, which includes the two observations of turning points for which the position of the planet was given. The format of the table is the same as that of Table XIV.6, except that I have used $D_i$ rather than $C_i$ for the coefficient of the planetary acceleration. A few comments about the table are needed.

Table IV.10 gives the dates (SE -256 II 3 and SE -256 II 5) when Mars was seen to enter and to leave the galactic cluster NGC 2632, which is identified in Table IV.10 by the old term ε Cancri. We have no way to know exactly what the observers meant by entering and leaving the cluster. In the calculations, therefore, I shall use the average time of the two observations, which has Julian day number 1 514 106.196, and I shall use the longitude of the center of ε Cancri, which was 91°.727 at this time.

Table IV.10 lists a conjunction of Mars with η Piscium on the date SE 44 III 11, which is equated to -267 June 14. However, the translator had supplied the day of the month by calculation. Since Mars moves too slowly to let us infer the day, in view of the expected errors of observation, the day is clearly not reliable. I have included a line for this observation in Table XIV.8, but I have omitted the values of $A_i$, $D_i$, and $Z_i$, which would clearly be meaningless.

The three successive entries in Table XIV.8 for the dates -154 May 14, -154 May 24, and -154 May 29 have almost identical and large values of $Z_i$, which suggests that the assigned dates are uniformly in error. However, we must subtract about $2^d.5$ from the assigned times in order to find reasonable values. This would imply that the calendar is wrong by two days and also that the recorder put all the observations in the morning instead of the evening. Such a combination of errors is quite unlikely. Further, the succeeding entry, for -154 June 24, requires adding about four days to the assigned date, and this is certainly incompatible with the earlier requirement for subtracting from the dates. The most likely conclusion, therefore, is that the observer for these four observations was rather careless, and I have left the originally assigned dates in Table XIV.8.

If we count the two observations involving ε Cancri (-567 May 24 and -567 May 26) as a single observation, there are 18 observations in Table XIV.8. Of these, 10 show values of $Z_i$ less than $1°$ and 8 show values that are greater than $1°$. This distribution is distinctly inferior to that shown by the conjunctions involving Mercury and Venus and the inferred accelerations are correspondingly inferior.

When we infer the accelerations of both Mars and the sun from the conjunctions, we find

$$\begin{aligned} \nu_S' &= -3.49 \pm 6.82, \\ \nu_♂' &= 4.61 \pm 2.70. \end{aligned} \qquad \text{(XIV.13)}$$

When we infer only the acceleration of the sun, using the relation $\nu_♂' = 0.5317\nu_S'$ that is required by the spin fluctuation hypothesis, we find

$$\nu_S' = 4.25 \pm 4.19. \qquad \text{(XIV.14)}$$

In view of the large standard deviations attached to these estimates, they are hardly significant.

## 8. A Simultaneous Inference of All the Accelerations

The sources discussed in Chapter IV contain several observations of conjunctions of planets with each other. Because there are ambiguities in interpreting the records of these conjunctions,† I decided that it would be unsafe to use them.

As a result, there are no direct connections of planetary accelerations with each other in the conjunctions that have been used in this chapter. However, each individual set of conjunctions involves the acceleration of the sun as well as that of the planet, and thus a simultaneous inference of all the relevant accelerations makes sense.

The results are

$$\begin{aligned} \nu_S' &= 0.22 \pm 1.10, \\ \nu_☿' &= 26.57 \pm 14.85, \\ \nu_♀' &= 1.45 \pm 3.39, \\ \nu_♂' &= 4.54 \pm 2.53. \end{aligned} \qquad \text{(XIV.15)}$$

These do not differ significantly from the values inferred from the individual sets of accelerations.

---

†See Sections IV.4 and Appendix V.

# CHAPTER XV

## DISCUSSION AND CONCLUSIONS; APPARENT FAILURE OF THE SPIN FLUCTUATION HYPOTHESIS

### 1. The Goals of the Work

The primary goals of this work that were stated in Section I.6 are threefold:

a. to find the secular acceleration of the sun with respect to solar time,

b. to find the secular accelerations of the planets with respect to solar time, and

c. to determine whether the secular accelerations of the sun and planets are consistent with the spin fluctuation hypothesis.

Since it is necessary to analyze some of the Babylonian lunar data in order to correlate the Babylonian and Julian calendars, I adopted a secondary goal in Section IV.8. The secondary goal is to estimate the lunar acceleration with respect to solar time from the Babylonian lunar data given in Chapter IV.

The work has also produced some useful by-products.

The main purpose of this chapter is to discuss the extent to which this work has met its primary goals. Before I take up this matter, however, it is useful to dispose of the discussion relating to the secondary goal and to the by-products.

### 2. The Secondary Goal and Some Useful By-products of the Work

In Equation X.5 of Section X.3, I estimated the quantity $D''$, which is the difference between the lunar and solar accelerations $\nu_M'$ and $\nu_S'$, by using Babylonian measurements of the rising and setting times of the moon. This estimate is

$$D'' = 5.6 \pm 8.0 \quad ''/\text{cy}^2. \tag{XV.1}$$

It applies at the epoch -370. In Equation X.10 of Section X.5, I estimated $\nu_M'$ from Babylonian observations of lunar conjunctions. This estimate is

$$\nu_M' = 3.4 \pm 1.7. \tag{XV.2}$$

It applies at the epoch -280.

In Section X.6, I combined the estimates in Equations XV.1 and XV.2, with the aid of a tentative estimate of $\nu_S'$, in order to obtain an estimate of $\nu_M'$ from all the Babylonian lunar data. This estimate is

$$\nu_M' = 3.6 \pm 1.7. \tag{XV.3}$$

I adopted this as the best estimate of $\nu_M'$ that we can form from the lunar data; we can take the epoch to be -300. As we saw in Section X.6, it is considerably smaller than the estimate formed from other data.

Four by-products of this work are useful enough to warrant mention.

The first by-product concerns solar tables based upon the epicycle-eccentric model that was developed by Hellenistic astronomers and that became the standard solar model in medieval Islamic and European astronomy. We saw in Section XIII.4 that this model is capable of representing the solar position to an accuracy that is better than 1′, provided that the parameters of the model are adequately chosen. This means that we can calculate a position from a solar table and use it in inferring the solar acceleration $\nu_S'$, provided that we know approximately the epochs of the observations used in determining the tables. I used several solar tables in Chapter VIII.

The second by-product concerns planetary tables based upon the Hellenistic models. We saw in Section XIII.10 that these tables are not capable of good accuracy, even if the parameters are chosen as well as possible, and that planetary tables, at least during the period covered by this study, are not useful in studying the accelerations.

The third by-product concerns the time of the summer solstice that was observed by the Athenian astronomers Meton and Euctemon in the year -431. Ptolemy [ca. 142, Chapter III.1] gives the time as 06 hours on the Egyptian date that is equivalent to -431 June 27. An inscription called the Milesian parapegm† gives the same Egyptian date, and it also gives the date in the Athenian calendar, namely the date called Skirophorion 13 in the year -431. As a result, all chronologists who have studied the Athenian calendar have taken Skirophorion 13 to be equivalent to June 27 in -431, and they have encountered insuperable difficulties in trying to reconcile this correspondence with other evidence.

I showed in Section V.3 that the date and time given by Ptolemy are not those observed by Meton and Euctemon. Instead, they were calculated from Hipparchus's solar theory, and the calculations were probably made about the year -110. However, the date Skirophorion 13 is probably the date observed by Meton and Euctemon. Thus Skirophorion 13 should not be equated to June 27 in -431. Elimination of this false correspondence should make it easier to construct a self-consistent theory of the Athenian calendar. Skirophorion 13 should be equated to either June 28 or June 29, probably the former.

The fourth by-product concerns the authenticity of the planetary observations that Ptolemy claims to have measured himself. I studied this matter in detail in Chapter XIII. We do not have enough information to test all of Ptolemy's claimed planetary observations, but we can test many of them. All those that we can test prove to be fraudulent. In earlier works [Newton, 1973, 1974, 1974b], I have studied the authenticity of all of the other observations that Ptolemy claims to have made himself.

†See Section V.3 for a discussion of this inscription.

As a result of these combined studies, we can say that all of Ptolemy's claimed observations that he uses and that can be tested prove to be fraudulent. That is, they were calculated, not observed, and they were calculated for the purpose of 'proving' the accuracy of Ptolemy's theories.

There are also strong grounds for questioning Ptolemy's competence as a theoretical astronomer. He makes more fundamental errors in theoretical matters than we expect from one whom we would consider to be a competent theoretician.

## 3. A Recapitulation of the Inferred Planetary Accelerations

In Chapters XI, XII, and XIV, we have inferred accelerations of the sun, Mercury, Venus, and Mars, with respect to solar time, by using observations of solar and planetary positions. Most of the planetary data come from observations of conjunctions of the planets with stars. Most of the solar observations are measurements of the times of equinoxes and solstices, or measurements of the meridian elevation of the sun at various seasons. In addition, in Chapter VIII, we used several values of the mean longitude of the sun at the epoch from solar tables.

TABLE XV.1

A SUMMARY OF THE ACCELERATIONS

| Provenance | Type of Data | Approximate Epoch | Source Equation | $\nu_S'$ | $\nu_☿'$ | $\nu_♀'$ | $\nu_♂'$ |
|---|---|---|---|---|---|---|---|
| Babylonian | Planetary conjunctions | -195 | XIV.15 | 0.22 ±1.10 | -26.57 ±14.85 | 1.45 ±3.39 | 4.54 ±2.53 |
| Hellenistic | Solar positions | -140 | XII.1, XII.2 | 3.02 ±0.70 | — | — | — |
| | Planetary conjunctions | - 40 | XII.13 | 2.53 ±0.63 | 35.50 ±14.92 | 0.14 ±8.45 | 2.52 ±7.10 |
| Islamic | Solar positions | +925 | a | 3.08 ±0.77 | — | — | — |
| | Solar positions plus planetary conjunctions | +940 | XI.27 | 2.82 ±0.60 | 12.51 ±148.56 | 2.29 ±5.99 | 1.40 ±3.86 |

[a]From Table XI.9.

The accelerations that have been inferred are summarized in Table XV.1. The first column in the table gives us the provenance of the observations used, which are either late Babylonian, Hellenistic, or early Islamic (through about 1020). The second and third columns are obvious. The fourth column gives the numbers of the equations in earlier chapters in which the accelerations are stated, except for the Islamic observations of the sun. These data are listed in Table XI.9, but I have not previously given the acceleration of the sun that is inferred from this table alone. Finally, the last four columns give the values of the inferred accelerations, along with the estimated standard deviations of the inferred values.

In addition to the estimates of the accelerations that are listed in Table XV.1, I estimated the difference $\nu_S' - \nu_☿'$ in Equation XIV.5 of Section XIV.2 from Babylonian measurements of the rising and setting times of Mercury. This estimate, which I repeat here for convenience, is

$$\nu_S' - \nu_☿' = 1.56 \pm 6.79. \qquad \text{(XV.4)}$$

Table XV.1 shows that we have met the first primary goal, that of finding $\nu_S'$, rather well. All of the values of $\nu_S'$ except the value from Babylonian planetary conjunctions agree well with each other and with the tentative estimate 3.0 that I have used in discussion. The estimate from the Babylonian data is less than the mean of the other estimates by somewhat more than two standard deviations. Since the epoch of the Babylonian estimates is the earliest epoch involved, it is conceivable that this reflects a real change in $\nu_S'$. However, the epoch is only about a century earlier than the mean epoch of the Hellenistic data, and it is unlikely that $\nu_S'$ changed by such an amount during a time as short as a century. It is more likely that the small value of $\nu_S'$ from the Babylonian data is a statistical accident.

If we assume that $\nu_S'$ should be constant, we can find a composite estimate from all the values listed in Table XV.1. This estimate is

$$\nu_S' = 2.52 \pm 0.35.$$

The error estimate is found by assuming that the variance of the composite estimate should equal the sum of the variances of the individual estimates. Since there is no reason to assume that $\nu_S'$ should be constant over the time span covered by Table XV.1, I do not propose this estimate as the best estimate to be formed from the table. However, I shall use it as a 'talking value' in the discussion of the spin fluctuation hypothesis.

The large errors attached to the estimates of the planetary accelerations in Table XV.1 show that we have not met the second primary goal, that of finding the planetary accelerations, very well. About all that we can say is that the accelerations have the magnitude that we expect, but their values are not accurate enough to let us test the spin fluctuation hypothesis by direct comparison with the solar acceleration.

We cannot find the planetary accelerations with accuracy because rather large errors of observation are combined with small numbers of usable observations. The errors in observing planetary positions run from about $\frac{1}{4}$ degree for the Islamic and Hellenistic observations to about 3/4 degree for the Babylonian observations.

We can presumably overcome the effect of observing errors, provided that they are not systematic, by using enough observations. There is little hope of finding enough more Hellenistic observations to be useful. However, there is a large body of extant but unpublished Islamic literature, and we may hope for eventual improvement in the volume of Islamic observations that can be used for the purposes of this work. The same is true of Babylonian observations, and it is my understanding that Professor A. Sachs (see Section IV.3) is readying a large body of Babylonian observations for publication. If these observations include enough planetary data, we may be able to make accurate estimates of the planetary accelerations from Babylonian observations within a few years.

Although we cannot meet the third primary goal, that of testing the spin fluctuation hypothesis, by direct use of the planetary accelerations, we can meet it by a method that I have used in Chapters XI, XII, and XIV. This method involves the use of what we may call 'planetary estimates of the solar acceleration'.

## 4. A Test of the Spin Fluctuation Hypothesis

This section will be devoted to discussing a test of the spin fluctuation hypothesis by using the planetary estimates of the solar acceleration. The method, which is described in Section XI.7, will be summarized here for the convenience of the reader.

Let $\lambda_{e,i}$ denote the geocentric longitude of a planet that is inferred from the ith observation, and let $\lambda_{oo,i}$ denote the geocentric longitude that is calculated for the time of the observation, with the solar and planetary accelerations being taken as zero in the calculations. Further, let

$$\lambda_{e,i} - \lambda_{oo,i} = Z_i \pm \sigma(Z_i). \qquad \text{(XV.5)}$$

In this, $\sigma(Z_i)$ is the standard error that goes with the observed position $\lambda_{e,i}$.

The calculated position $\lambda_{oo,i}$ is not the correctly calculated position because the accelerations are taken to be zero in the calculation. If the observation is that of a conjunction of a planet with a star, the correctly calculated position depends upon both the solar acceleration and the planetary acceleration. If the observation is the conjunction of two planets with each other, the correctly calculated position depends upon the solar acceleration and both planetary accelerations. If we assume that the dependence is linear, and if we confine ourselves to the planets Mercury, Venus, and Mars, we may write the correctly calculated position in the type form

$$\lambda_{oo,i} + A_i \nu_S' + B_i \nu_{☿}' + C_i \nu_{♀}' + D_i \nu_{♂}'.$$

Using this, we may rewrite Equation XV.5 in the form

$$A_i \nu_S' + B_i \nu_☿' + C_i \nu_♀' + D_i \nu_♂' = Z_i \pm \sigma(Z_i). \quad \text{(XV.6)}$$

In order to test the spin fluctuation hypothesis, we assume that it is valid, apply it to Equation XV.6, and examine the consequences. If the hypothesis is valid, the planetary accelerations have the same ratio to the solar acceleration as the corresponding mean motions; the ratios are given in Equation XI.26 of Section XI.7. Use of these ratios in Equation XV.6 gives

$$(A_i + 4.1520B_i + 1.6255C_i + 0.5317D_i)\nu_S' = Z_i \pm \sigma(Z_i). \quad \text{(XV.7)}$$

A value of $\nu_S'$ calculated from Equation XV.7 is what I mean by a planetary estimate of the solar acceleration.

TABLE XV.2

PLANETARY ESTIMATES OF THE SOLAR ACCELERATION

| Provenance | Approximate Epoch | Planetary Estimates of $\nu_S'$ as Inferred from Observations of | | |
|---|---|---|---|---|
| | | Mercury | Venus | Mars |
| Babylonian[a] | -200 | 2.12 ± 1.56 | 0.40 ± 0.93 | 4.25 ± 4.19 |
| Hellenistic[b] | - 40 | -0.60 ± 2.55 | 1.78 ± 1.21 | 4.75 ± 9.19 |
| Islamic[c] | +940 | -1.28 ± 2.75 | -3.10 ± 1.80 | 2.10 ± 4.51 |

[a]From Equations XIV.7, XIV.10, XIV.12, and XIV.14.

[b]From Equations XII.7, XII.10, and XIV.12.

[c]From Table XI.12.

We can make planetary estimates of $\nu_S'$ by using Babylonian, Hellenistic, and Islamic data separately, and by using observations of Mercury, Venus, and Mars separately within each body of data. This gives us the nine planetary estimates listed in Table XV.2, whose form should be self-explanatory. I have used both the value obtained from the Babylonian conjunctions in Equation XIV.10 and the value obtained from the intervals of visibility in Equation XIV.7 in obtaining the value 2.12 ± 1.56 listed for the Babylonian observations of Mercury.

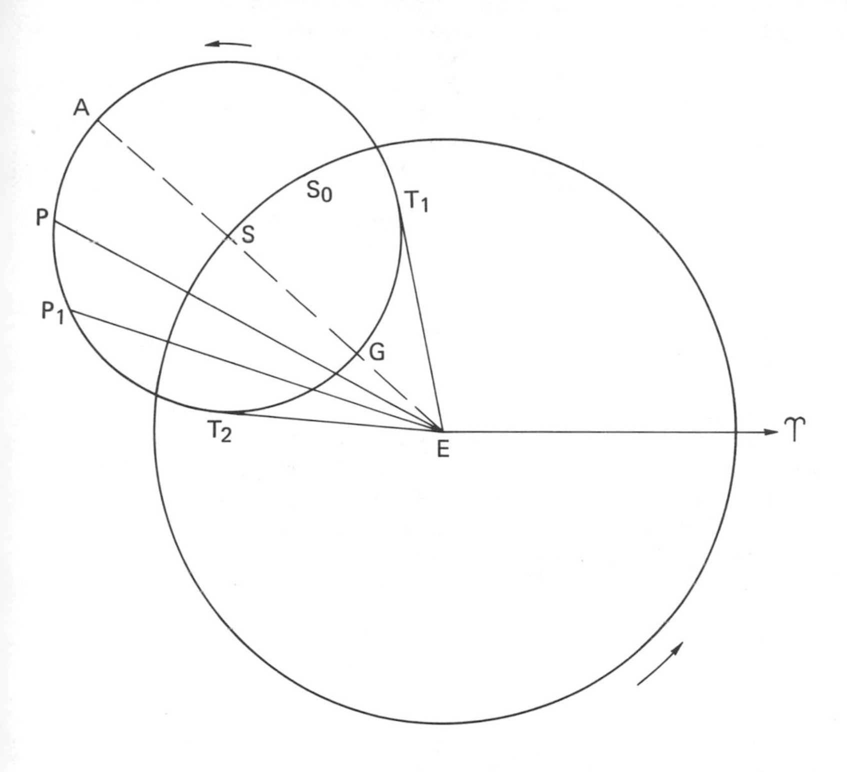


Figure XV.1. The significance of the values of $\nu_S'$ deduced from the observations of Mercury and Venus. E is the earth and the line E♈ points toward the vernal equinox. The point $S_o$ represents the position that the sun would have at some value $\tau_o$ of solar time if its acceleration $\nu_S'$ were zero, while S represents its actual position at that time. S is the center of the circle on which the points A, P, $P_1$, $T_2$, G, and $T_1$ lie. P denotes the point where the planet would be if its own acceleration satisfied the spin fluctuation hypothesis, and $P_1$ denotes the point where it would be if its acceleration is greater than the value demanded by the hypothesis. The motions of the sun and planet are in the directions shown by the arrows. $T_1$ and $T_2$ are the points of maximum elongation.

If the spin fluctuation hypothesis is valid, the estimates obtained from the different planets in Table XV.2 should agree with each other and with the estimate of $\nu_S'$ obtained from solar data, within an error that can be reasonably attributed to observation. Further, the estimates from the Babylonian and Hellenistic data should agree rather well, because the epochs are close together, although there is no requirement for agreement with the Islamic data, which differ in epoch by about a milenium.

If we use 2.52 as the 'talking value' of $\nu_S'$ obtained from solar data, we see that the planetary estimates of $\nu_S'$ obtained from Mars data agree reasonably well with the spin fluctuation hypothesis. There are planetary estimates that are both larger and smaller than 2.52, and all estimates are within one standard deviation of the value 2.52. However, the standard deviations of the estimates are so large that the test is not highly significant. The most that we can say with safety is that the Mars data are consistent with the hypothesis, but that a more extensive body of data might reveal an inconsistency.

The situation with regard to the observations of Mercury and Venus is quite different. Each planetary estimate obtained from either Mercury or Venus is smaller than 2.52, some by amounts that are not statistically significant and some by amounts that are. Before I make an estimate of the overall statistical significance of the estimates in Table XV.2, it is useful to study the physical interpretation of having a planetary estimate that is smaller than $\nu_S'$, in terms of planetary energy and angular momentum. We can do this with the aid of Figure XV.1.

In the figure, E is the earth and the line E♈ points toward the vernal equinox. Suppose that the sun would be at some point $S_o$ at some value $\tau_o$ of solar time, if its acceleration $\nu_S'$ with respect to solar time were zero. Since $\nu_S'$ is positive, however, the sun is farther advanced in longitude than the point $S_o$ at time $\tau_o$. Suppose that it is actually at the point S at the time $\tau_o$. S is then the center of the heliocentric orbit of the inner planet at time $\tau_o$.†

Let P denote the point where the planet would be if its own acceleration, $\nu_P'$, say, satisfied the spin fluctuation hypothesis. That is, the planet would be at P if the ratio $\nu_P'/\nu_S'$ were equal to the ratio of the corresponding mean motions. Its geocentric longitude would then be the angle PE♈. Suppose, however, that the planet's acceleration exceeds the value demanded by the spin fluctuation hypothesis. The planet will then be farther advanced than P in heliocentric longitude at time $\tau_o$, and it will be, say, at the point $P_1$. Its geocentric longitude will then be the angle $P_1$E♈.

The points $T_1$ and $T_2$ are the points of maximum elongation of the planet. If the point P is anywhere along the portion $T_1AT_2$ of its heliocentric orbit, $P_1$E♈ will exceed PE♈. On the other hand, if P is anywhere along the portion $T_2GT_1$ of the orbit, $P_1$E♈ will be less than PE♈ Thus an excess of $\nu_P'$ over the value demanded by the spin fluctuation hypothesis may either increase or decrease the geocentric longitude of the planet, which is the quantity that is directly observed.

However, it is clear that the planet is visible more of the time on the arc $T_1AT_2$ than it is on the arc $T_2GT_1$. If we use the average time intervals listed in Table III.6, Mercury is visible on $T_1AT_2$ about 1.5 times as long as on $T_2GT_1$, and Venus is visible about 3.0 times as long. Since the number of observations should be proportional to the periods of visibility, an excess value of $\nu_P'$ therefore increases the geocentric longitude on the average.

For Mercury, for example, if the spin fluctuation hypothesis is correct, and if no other planet is involved, Equation XV.7 reduces to

---

†For simplicity, I am taking the orbits involved to be circular. The true situation is more complex, but none of the qualitative considerations of this discussion would be altered if we used more accurate descriptions of the orbits.

$$(A_i + 4.1520B_i)\nu_S' = Z_i \pm \sigma(Z_i).$$

On the average, $A_i$ and $B_i$ are positive, and the value of $\nu_S'$ that we get from a specific observation is

$$\nu_S' = Z_i/(A_i + 4.1520B_i). \qquad \text{(XV.8)}$$

If the actual acceleration $\nu_P'$ of Mercury is greater than $4.1520\nu_S'$, $Z_i$ is increased on the average, as we have just seen, and Equation XV.8 shows us that the value of $\nu_S'$ inferred from the observations is likewise increased on the average. In other words, if the actual acceleration $\nu_P'$ of a planet is greater than the value demanded by the spin fluctuation hypothesis, the planetary estimate of the solar acceleration $\nu_S'$ is greater than the correct value of $\nu_S'$.

If the conclusions implied by Table XV.2 are correct, the actual accelerations of Mercury and Venus are less than the values demanded by the spin fluctuation hypothesis. That is, the values of $\nu_☿'$ and $\nu_♀'$ are less than $4.1520\nu_S'$ and $1.6255\nu_S'$, respectively.

We should like to know whether this conclusion means that Mercury and Venus are gaining or losing angular momentum or energy. For this purpose, we use Equation I.14 in Section I.7, which we can write specifically as

$$\dot{n}_P = \nu_P' + (n_P/K)(\dot{\omega}_e - \dot{n}_S). \qquad \text{(XV.9)}$$

In this, a subscript P denotes either Mercury or Venus, $\dot{n}_P$ is the acceleration of either planet with respect to atomic time t, $\dot{n}_S$ is the acceleration of the sun with respect to t, $\dot{\omega}_e$ is the acceleration of the earth's spin with respect to t, and K is a numerical constant introduced in Equation I.13 of Section I.7. We get $\dot{n}_S$ if we replace the subscript P by the subscript S in Equation XV.9.

Straightforward manipulation of the expressions for $\dot{n}_P$ and $\dot{n}_S$ leads to the form

$$\dot{n}_P = (n_P/n_S)\dot{n}_S + \nu_P' - (n_P/n_S)\nu_S'. \qquad \text{(XV.10)}$$

We have found from the data that $\nu_P' < (n_P/n_S)\nu_S'$. If we assume that $\dot{n}_S = 0$,† Equation XV.10 then leads to

$$\dot{n}_P < 0 \qquad \text{(XV.11)}$$

for both Mercury and Venus.

A negative angular acceleration $\dot{n}_P$ means that the inner planets are gaining angular momentum and energy.

TABLE XV.3

PROBABILITY OF CHANCE OCCURRENCE OF THE PLANETARY ESTIMATES

| Provenance | Planetary Estimate of $\nu_S'$ | Probability of Chance Occurrence |
|---|---|---|
| A. Observations of Mercury | | |
| Babylonian | 2.12 ± 1.56 | 0.399 |
| Hellenistic | -0.60 ± 2.55 | 0.111 |
| Islamic | -1.28 ± 2.75 | 0.084 |
| Combined | | $3.7 \times 10^{-3}$ |
| B. Observations of Venus | | |
| Babylonian | 0.40 ± 0.93 | 0.011 |
| Hellenistic | 1.78 ± 1.21 | 0.270 |
| Islamic | -3.10 ± 1.80 | 0.0009 |
| Combined | | $2.75 \times 10^{-6}$ |

†There is an objection to this assumption that will be discussed in the next section.

Therefore Table XV.2 leads to the conclusion that Mercury and Venus do not obey the spin fluctuation hypothesis. Instead, both inner planets seem to be gaining orbital energy and angular momentum. Table XV.3 shows the probability that this result could have been obtained by chance. According to the table, the probability that the result could happen by chance is about $3.7 \times 10^{-3}$ for Mercury and about $2.75 \times 10^{-6}$ for Venus. The probability that the same result would happen by chance for both planets is about $1.02 \times 10^{-8}$.

## 5. Discussion; Possible Reasons for the Apparent Failure of the Hypothesis

As we have just seen, the spin fluctuation hypothesis apparently fails for the inner planets Mercury and Venus, but not for the outer planet Mars. The odds in favor of this conclusion are of the order of $10^8$ to 1.

I believe that the tests of my ephemeris programs for the sun, Mercury, and Venus that were described in Chapter IX rule out the possibility that the results can come from an error in my programming the orbital theories involved. If this is so, there remain two basic possibilities.

The first possibility concerns a possible error in the theories that were programmed. These theories come from Newcomb's solutions [Newcomb, 1895, 1895a, and 1895b] of the gravitational equations of motion of the earth, Mercury, and Venus. Various workers have thoroughly checked various aspects of Newcomb's solutions,† but I do not recall seeing a check that is specifically directed at the long-term behavior of the longitudes. Thus, we cannot categorically rule out the possibility that terms have been omitted from Newcomb's solutions that affect the long-term behavior of the longitudes, although such an omission seems unlikely.

†For example, it has been suggested that the rate of change of the obliquity does not agree with Newcomb's theory, and those aspects of his theory that deal with the rotation of the ecliptic have been extensively studied.

If such terms have been omitted, they probably deal with the secular behavior of the mean orbital elements.

If there are no errors in the solution of the gravitational equations of motion, the remaining possibility is that there are significant forces acting on the planets that are non-gravitational in origin. With regard to this possibility, the change in energy and angular momentum that is implied is important. We found from relation XV.11 of the preceding section, by making the assumption $\dot{n}_S = 0$, that Mercury and Venus are gaining energy and angular momentum.

Some results by van Flandern [1974] indicate that this is not a good assumption. van Flandern finds that the present acceleration of the moon with respect to atomic time is $-83 \pm 10$ ″/cy$^2$. Morrison [1973] and Oesterwinter and Cohen [1972], on the other hand, find that the lunar acceleration is about $-40$ for a time scale on which $\dot{n}_S = 0$.† van Flandern suggests that the difference arises from a secular change in the gravitational 'constant' G, or from a secular change in some other 'constant' that produces similar changes in the gravitational equations of motion. If we adopt the assumption of a changing G for simplicity in the discussion, and if we let $\dot{G}$ be its rate of change with respect to atomic time, any body in the solar system has an acceleration $\dot{n}$ with respect to atomic time that is given by‡

---

†This is the time scale that I have called ephemeris time in a limited sense, being based only upon the yearly motion of the sun without implying the validity of the spin fluctuation hypothesis. Oesterwinter and Cohen say that their value of the acceleration is with respect to atomic time. However, the scale that they call atomic time is really a kind of ephemeris time. It is a time scale based principally upon the motion of the sun and moon, adjusted in rate and epoch to agree with cesium time over the period from about 1955 to 1969.

‡This relation has been known for a long time; I do not know where it originated.

$$\dot{n}/n = 2\dot{G}/G. \tag{XV.12}$$

Thus a decrease in G makes a negative contribution to an acceleration with respect to atomic time, in accordance with van Flandern's results for the moon. His results require $\dot{G}/G = (-12 \pm 3) \times 10^{-9}$ $cy^{-1}$.

The time scale with respect to which the acceleration of the sun is zero is the ephemeris time scale based upon the yearly motion of the sun. If the acceleration of the sun with respect to ephemeris time is zero, Equation XV.12 shows us that $\dot{n}_S$ is about $-2.5$ $''/cy^2$ with respect to atomic time.

In the remaining discussion, let T denote the ephemeris time scale defined by the yearly motion of the sun, and let $dn/dT$ denote an acceleration with respect to T; a particular acceleration will be specified by attaching an appropriate subscript to n. The definition of T requires $d\dot{n}_S/dT = 0$. If we use this instead of requiring $\dot{n}_S = 0$, the argument of the preceding section leads to

$$dn_P/dT < 0 \tag{XV.13}$$

rather than to $\dot{n}_P < 0$.

Relation XV.13 requires only a minor change in the wording of the earlier conclusion. We should say that Mercury and Venus are gaining angular momentum and energy if we use a time scale in which $dn_S/dT$ is zero. This is a time scale in which the earth is not changing its orbital angular momentum or energy. If the earth is gaining momentum and energy with respect to an atomic time scale, Mercury and Venus are gaining more rapidly. Conversely, if the earth is losing, Mercury and Venus are not losing as rapidly.

If Mercury and Venus were losing angular momentum and energy (in the time scale with $dn_S/dT = 0$), we might try to speculate that the solar atmosphere is extensive enough that Mercury and

Venus are affected by its drag.† However, since Mercury and Venus are gaining energy and angular momentum, drag cannot be the cause. Various students of the accelerations have tried to find ways of changing the momentum and energy of the earth other than drag. No effect that has been studied is within orders of magnitude of being large enough to matter for the purposes of this work. The same effects, calculated for Mercury and Venus, are still orders of magnitude too small to account for the difference between $\nu_P{}'$ and $(n_P/n_S)\nu_S{}'$ for the inner planets.

Thus we have a dilemma. The data indicate with statistical odds of the order of $10^8$ to 1, that Mercury and Venus are gaining energy and angular momentum. We would normally take a conclusion with such odds as being well established. However, there is no physical explanation for the conclusion, and we should be cautious about accepting a conclusion that has no physical explanation. In view of this dilemma, the best way to word the results of this work is probably to pose a series of choices. We may pose the choices by saying that at least one of the following statements is correct:

(1) There are errors in the orbital theories used. This includes the possibility that the theories themselves are correct but that a mistake was made in programming them for the computer.

(2) There are forces causing Mercury and Venus to gain energy and angular momentum with respect to the earth and Mars.

(3) The data used are not valid.

With regard to the last possibility, I thank R.L. Duncombe and T.C. van Flandern, of the U. S.

†It may be that the tracking of the artificial space probes that have passed close to or landed on Mercury or Venus would prevent such speculation. An atmosphere that can produce a measureable drag on a planet should produce a large effect on a small probe, and one that is easily found by tracking.

Naval Observatory, for pointing out a peculiar property of one data sample to me. This sample is the collection of Islamic observations of Venus in Table XI.7. There are 13 observations for which 1000 $|Z_i|$ is less than 100. Of these, 7 have positive values and 6 have negative values; this is a well-balanced distribution. There are 8 observations for which 1000$|Z_i|$ lies between 100 and 250, which was the largest value that I used in the analysis. Of these, only 1 is positive while 7 are negative, so that this range of 1000$|Z_i|$ is largely responsible for the negative value of $\nu_S'$ found, namely -3.10 ± 1.80.

This suggests that the sample used may still contain some invalid observations, and the true significance of the results may not be as large as Table XV.3 indicates. However, even if we restrict the Islamic observations of Venus to those with 1000$|Z_i|$ less than 100, there is still a problem. The planetary estimate of $\nu_S'$ based upon this sample is 1.16 ± 1.26. This is but slightly more than one standard deviation away from 2.52, but it is still smaller than 2.52, just as all the other estimates in Table XV.3 are.

Thus we still have an unlikely situation. The probability that the new result for the Venus data happened by chance is 0.140 rather than 0.0009. The combined probability for all the Venus observations becomes $4.2 \times 10^{-4}$ rather than $2.75 \times 10^{-6}$, and the combined probability for both Mercury and Venus becomes $1.5 \times 10^{-6}$. This is two orders of magnitude larger than the earlier result, but it is still quite small.

It is clearly important to find additional ancient and medieval observations with which we may test the spin fluctuation hypothesis before we decide whether or not the hypothesis is valid.

A recent result by Morrison [1973] should be mentioned in closing this work. The work of Spencer Jones [1939], which was discussed in Chapter I, leads to the value $dn_M/dT$ = -22, approximately, for the lunar acceleration, but all other recent estimates give values near -40. These estimates are based upon both ancient and medieval data. Morrison points out that Spencer Jones took $\nu_S'$ but not $\nu_M'$ as a parameter to be determined from the data. For $\nu_M'$, he used a value 10.44 taken from ancient data. He also assumed that the planetary accelerations $\nu_P'$ are related to $\nu_S'$ by the spin fluctuation hypothesis. That is, Spencer Jones allowed $\nu_S'$ to be determined by a combination of solar and planetary data. Morrison has analysed Spencer Jones's fitting process and finds that his result for $\nu_S'$, and hence his result for $dn_M/dT$ (Section I.7), is strongly dependent upon the observations of Mercury. He finds that Spencer Jones's results can be reconciled with the other results if Mercury has a small negative acceleration in a time scale in which $dn_S/dT = 0$. This is just what the results of this work suggest.

Note added in proof: I should repeat here part of a note that I added in proof at the end of Chapter I. Duncombe (Astronomical Papers, XVI, Part I, 1972) has made a careful reduction of the modern observations of Venus, and he finds that they fit the spin fluctuation hypothesis to high accuracy, in contradiction to the results found by using the reduction of Spencer Jones [1939]. Thus the results found in this work by the use of ancient data seem to conflict with the results from modern data, and this conflict increases the importance of finding more ancient data to analyze.

APPENDICES

# APPENDIX I

## STARTING CONDITIONS FOR THE BACKWARD INTEGRATION OF THE FIVE OUTER PLANETS

As we saw in Section IX.4, we need twelve quantities for each coordinate in order to start the numerical integration of the equations of motion of the five outer planets. Ten of these quantities are the values of an acceleration at ten successive time steps; the accelerations are, of course, the same whether the integration is to proceed forward or backward in time. The other two quantities were denoted by $''X(t_o)$ and $''X(t_o-1)$, for an X-coordinate, in Section IX.4, and they do depend upon the direction of the integration.

The starting conditions needed to integrate backward in time from the Julian day 2 430 000.5 are given in Tables A.I-1 through A.I-5, which have been reproduced directly from the computer listing. Planet 5 denotes Jupiter, planet 6 denotes Saturn, and so on. The first part of each table lists the X, Y, and Z components of acceleration. The second part lists the values of $''X(t_o)$, $''X(t_o-1)$, and the corresponding values for the Y and Z coordinates; they are called "X SUM", "Y SUM", and "Z SUM" in the tables.

If the values in the tables are used in the program described in Section IX.4, the first set of coordinates listed will be for the Julian day 2 429 960.5.

**TABLE A.I-1**

STARTING CONDITIONS FOR PLANET 5

JULIAN DATE = 2430000.5

| X ACCEL. | Y ACCEL. | Z ACCEL. |
|---|---|---|
| -0.00365516441054 | -0.01655124624800 | -0.00701140174172 |
| -0.00476621272909 | -0.01641010746960 | -0.00692370081775 |
| -0.00587545304976 | -0.01620165976647 | -0.00680716754638 |
| -0.00697755224719 | -0.01592491890713 | -0.00666150954247 |
| -0.00806695246540 | -0.01557930699543 | -0.00648661414143 |
| -0.00913791252663 | -0.01516468872744 | -0.00628256296393 |
| -0.01018455587568 | -0.01468140363576 | -0.00604964461633 |
| -0.01120092466950 | -0.01413029317862 | -0.00578836504517 |
| -0.01218103941778 | -0.01351272158502 | -0.00549945508872 |
| -0.01311896336242 | -0.01283058948349 | -0.00518387481919 |

| X SUM | Y SUM | Z SUM |
|---|---|---|
| 3.20122013661802 | 3.55074301652669 | 1.44499321318640 |
| 3.43056759900037 | 3.35493907199817 | 1.35538111299905 |

TABLE A.I-2

STARTING CONDITIONS FOR PLANET 6

JULIAN DATE = 2430000.5

| X ACCEL. | Y ACCEL. | Z ACCEL. |
|---|---|---|
| -0.00313568256436 | -0.00447517857140 | -0.00171490455226 |
| -0.00325284651806 | -0.00439541275633 | -0.00167683609014 |
| -0.00336712342591 | -0.00431293072181 | -0.00163776732229 |
| -0.00347841793302 | -0.00422782194884 | -0.00159774038192 |
| -0.00358663818411 | -0.00414017390524 | -0.00155679667383 |
| -0.00369169717152 | -0.00405007161852 | -0.00151497659557 |
| -0.00379351444313 | -0.00395759769215 | -0.00147231938542 |
| -0.00389201800670 | -0.00386283289375 | -0.00142886315790 |
| -0.00398714615968 | -0.00376585735935 | -0.00138464516727 |
| -0.00407884890650 | -0.00366675232572 | -0.00133970229840 |

| X SUM | Y SUM | Z SUM |
|---|---|---|
| 6.47353510709432 | 6.11630651767276 | 2.24941030170064 |
| 6.64179534400071 | 5.97187515000108 | 2.18242664200024 |

## TABLE A.I-3

STARTING CONDITIONS FOR PLANET 7

JULIAN DATE = 2430000.5

| X ACCEL. | Y ACCEL. | Z ACCEL. |
|---|---|---|
| -0.00064988945413 | -0.00100063585552 | -0.00042905873408 |
| -0.00065897176121 | -0.00099513524800 | -0.00042651049775 |
| -0.00066799086755 | -0.00098951878653 | -0.00042391292914 |
| -0.00067694149209 | -0.00098378561534 | -0.00042126578150 |
| -0.00068581823585 | -0.00097793529428 | -0.00041856899180 |
| -0.00069461562336 | -0.00097196785205 | -0.00041582270310 |
| -0.00070332814527 | -0.00096588383480 | -0.00041302728477 |
| -0.00071195030196 | -0.00095968434770 | -0.00041018334951 |
| -0.00072047664766 | -0.00095337108700 | -0.00040729176615 |
| -0.00072890183522 | -0.00094694636060 | -0.00040435366721 |

| X SUM | Y SUM | Z SUM |
|---|---|---|
| 11.13261044216518 | 14.77074498763965 | 6.31452713633275 |
| 11.26310446299979 | 14.69533659200007 | 6.27963894700004 |

## TABLE A.I-4

STARTING CONDITIONS FOR PLANET 8

JULIAN DATE = 2430000.5

| X ACCEL. | Y ACCEL. | Z ACCEL. |
|---|---|---|
| 0.000513377716544 | -0.00002763130239 | -0.00002453980821 |
| 0.000512241056391 | -0.00002945454515 | -0.00002525558435 |
| 0.000511100011169 | -0.00003121276089 | -0.00002594333595 |
| 0.000509957590064 | -0.00003290499750 | -0.00002660277921 |
| 0.000508819599521 | -0.00003453069128 | -0.00002723380218 |
| 0.000507691492850 | -0.00003608970135 | -0.00002783647851 |
| 0.000506578878070 | -0.00003758234036 | -0.00002841107953 |
| 0.000505487420560 | -0.00003900940039 | -0.00002895808408 |
| 0.000504422808070 | -0.00004037217317 | -0.00002947818580 |
| 0.000503390684790 | -0.00004167246346 | -0.00002997229738 |

| X SUM | Y SUM | Z SUM |
|---|---|---|
| -30.16463609531522 | 1.54191859353643 | 1.39095726170256 |
| -30.15526882500011 | 1.65700313700012 | 1.43786002500012 |

## TABLE A.I-5

STARTING CONDITIONS FOR PLANET 9

JULIAN DATE = 2430000.5

| X ACCEL. | Y ACCEL. | Z ACCEL. |
|---|---|---|
| 0.00017810259775 | -0.00024902523265 | -0.00013574934779 |
| 0.00017603339329 | -0.00024910542504 | -0.00013543420981 |
| 0.00017396902896 | -0.00024911729656 | -0.00013509096749 |
| 0.00017191463992 | -0.00024905996356 | -0.00013471936708 |
| 0.00016987557380 | -0.00024893292699 | -0.00013431932450 |
| 0.00016785735053 | -0.00024873610683 | -0.00013389093915 |
| 0.00016586561588 | -0.00024846987303 | -0.00013343450606 |
| 0.00016390608916 | -0.00024813507155 | -0.00013295052585 |
| 0.00016198450566 | -0.00024773304482 | -0.00013243971208 |
| 0.00016010655474 | -0.00024726564531 | -0.00013190299565 |

| X SUM | Y SUM | Z SUM |
|---|---|---|
| -21.19451277339816 | 28.35984936429058 | 15.38226504694539 |
| -21.12384867995640 | 28.44653041993920 | 15.38827695994407 |

# APPENDIX II

## CONTINUING CONDITIONS FOR THE BACKWARD INTEGRATION OF THE OUTER PLANETS BEFORE JULIAN DAY 1 720 960.5

I have carried the numerical integration of the equations of motion of the five outer planets backward in time to the Julian day 1 720 960.5, with the results that are tabulated in the next appendix for Jupiter and Saturn. If the reader should wish to carry the integration still farther back in time, he may do so by using the 'continuing conditions' that are tabulated in this appendix. That is, if he uses the values listed in this appendix, along with the program described in Section IX.4, the first day for which he will generate coordinates will be Julian day 1 720 920.5.

The continuing conditions are listed in Tables A.II-1 through A.II-5, which have the same format as the corresponding tables in Appendix A.I.

## TABLE A.II-1

CONTINUING CONDITIONS FOR PLANET 5

JULIAN DATE = 1720960.5

| X ACCEL. | Y ACCEL. | Z ACCEL. |
|---|---|---|
| 0.01524730317146 | 0.00467220819886 | 0.00162012488699 |
| 0.01547807189799 | 0.00392315744177 | 0.00129188381667 |
| 0.01566919973288 | 0.00316517236208 | 0.00096081310186 |
| 0.01582043998589 | 0.00239958563594 | 0.00062749253413 |
| 0.01593154650657 | 0.00162772496128 | 0.00029249974898 |
| 0.01600227222716 | 0.00085091967726 | -0.00004358688836 |
| 0.01603236859032 | 0.00007050751951 | -0.00038018536137 |
| 0.01602158587270 | -0.00071215853860 | -0.00071670706665 |
| 0.01596967441851 | -0.00149570340952 | -0.00105255388638 |
| 0.01587638679918 | -0.00227872303288 | -0.00138711529231 |

| X SUM | Y SUM | Z SUM |
|---|---|---|
| -5.39604473206009 | 0.50519989366165 | 0.35557376179042 |
| -5.35210467228409 | 0.76800692138572 | 0.46753989681217 |

## TABLE A.II-2

CONTINUING CONDITIONS FOR PLANET 6

JULIAN DATE = 1720960.5

| X ACCEL. | Y ACCEL. | Z ACCEL. |
|---|---|---|
| 0.00077763095979 | -0.00537271119335 | -0.00223464861854 |
| 0.00062480724122 | -0.00539633579133 | -0.00223819587091 |
| 0.00047114136679 | -0.00541584260827 | -0.00224002108641 |
| 0.00031678567585 | -0.00543119536713 | -0.00224011548204 |
| 0.00016189436109 | -0.00544236367298 | -0.00223847276081 |
| 0.00000662314105 | -0.00544932311760 | -0.00223508914131 |
| -0.00014887107476 | -0.00545205536593 | -0.00222996337947 |
| -0.00030443052653 | -0.00545054822376 | -0.00222309678248 |
| -0.00045989694938 | -0.00544479568625 | -0.00221449321457 |
| -0.00061511192162 | -0.00543479796703 | -0.00220415909466 |

| X SUM | Y SUM | Z SUM |
|---|---|---|
| 0.71927814661477 | 8.27890117376716 | 3.36662186121987 |
| 0.95494370339470 | 8.26135520639744 | 3.34993830524773 |

## TABLE A.II-3

CONTINUING CONDITIONS FOR PLANET 7

JULIAN DATE = 1720960.5

| X ACCEL. | Y ACCEL. | Z ACCEL. |
|---|---|---|
| -0.00106914451363 | -0.00046464976666 | -0.00018836696133 |
| -0.00107233861753 | -0.00045765746824 | -0.00018524733549 |
| -0.00107550838490 | -0.00045065658757 | -0.00018212384493 |
| -0.00107865405297 | -0.00044364618554 | -0.00017899608432 |
| -0.00108177586172 | -0.00043662532847 | -0.00017586365062 |
| -0.00108487405548 | -0.00042959308189 | -0.00017272614040 |
| -0.00108794888351 | -0.00042254850401 | -0.00016958314700 |
| -0.00109100059986 | -0.00041549063924 | -0.00016643425775 |
| -0.00109402946231 | -0.00040841851164 | -0.00016327905115 |
| -0.00109703573045 | -0.00040133111837 | -0.00016011709411 |

| X SUM | Y SUM | Z SUM |
|---|---|---|
| 18.55346588414594 | 6.78680563690413 | 2.70591406489885 |
| 18.61161720100804 | 6.65958175435201 | 2.64923648865352 |

## TABLE A.II-4

CONTINUING CONDITIONS FOR PLANET 8

JULIAN DATE = 1720960.5

| X ACCEL. | Y ACCEL. | Z ACCEL. |
|---|---|---|
| 0.00014872723825 | 0.00046537037435 | 0.00018698764849 |
| 0.00015094750818 | 0.00046413720383 | 0.00018641259383 |
| 0.00015312750127 | 0.00046288929739 | 0.00018583235492 |
| 0.00015526697229 | 0.00046162796528 | 0.00018524750035 |
| 0.00015736567650 | 0.00046035450707 | 0.00018465859424 |
| 0.00015942336852 | 0.00045907021808 | 0.00018406619912 |
| 0.00016143980200 | 0.00045777639611 | 0.00018347087878 |
| 0.00016341473003 | 0.00045647434819 | 0.00018287320114 |
| 0.00016534790644 | 0.00045516539743 | 0.00018227374119 |
| 0.00016723908791 | 0.00045385088982 | 0.00018167308381 |

| X SUM | Y SUM | Z SUM |
|---|---|---|
| -8.83193501608836 | -26.90554404505308 | -10.80029551257760 |
| -8.95099760441995 | -26.87292019293444 | -10.78393901650774 |

## TABLE A.II-5

CONTINUING CONDITIONS FOR PLANET 9

JULIAN DATE = 1720960.5

| X ACCEL. | Y ACCEL. | Z ACCEL. |
|---|---|---|
| 0.00048428875966 | 0.00021702817532 | -0.00007381160700 |
| 0.00048490428734 | 0.00021406997177 | -0.00007496320212 |
| 0.00048546637158 | 0.00021110297112 | -0.00007611266109 |
| 0.00048597488061 | 0.00020812855608 | -0.00007725942053 |
| 0.00048642969023 | 0.00020514810837 | -0.00007840292244 |
| 0.00048683068155 | 0.00020216301529 | -0.00007954261086 |
| 0.00048717773943 | 0.00019917467621 | -0.00008067792850 |
| 0.00048747075169 | 0.00019618450905 | -0.00008180831331 |
| 0.00048770960928 | 0.00019319395657 | -0.00008293319514 |
| 0.00048789420722 | 0.00019020449254 | -0.00008405199244 |

| X SUM | Y SUM | Z SUM |
|---|---|---|
| -27.59671383934741 | -11.44913715237731 | 4.74596141429683 |
| -27.64941583178092 | -11.33176867311408 | 4.79894559942457 |

## APPENDIX III

## COORDINATES OF JUPITER AND SATURN FROM JULIAN DAY 1 720 960.5 TO DAY 2 325 760.5

Table A.III-1 lists the rectangular coordinates of Jupiter at intervals of 160 days from Julian day 1 720 960.5 to Julian day 2 325 760.5. Table X.III-2 lists the coordinates of Saturn at intervals of 320 days between the same dates. The tables should be self-explanatory.

## TABLE A.III-1

HELIOCENTRIC COORDINATES OF JUPITER

REFERENCE AXES OF 1950.0

| JULIAN DATE | X | Y | Z |
|---|---|---|---|
| 1720960.5 | -5.350781467 | 0.767817020 | 0.467424296 |
| 1721120.5 | -5.430546656 | -0.288991184 | 0.014700465 |
| 1721280.5 | -5.255142259 | -1.332261474 | -0.438731611 |
| 1721440.5 | -4.832934355 | -2.312951952 | -0.871553960 |
| 1721600.5 | -4.182434576 | -3.184315288 | -1.263176236 |
| 1721760.5 | -3.332018973 | -3.903322995 | -1.594355815 |
| 1721920.5 | -2.319548219 | -4.432222321 | -1.847877477 |
| 1722080.5 | -1.191761098 | -4.740282467 | -2.009320365 |
| 1722240.5 | -0.003264537 | -4.805731859 | -2.067917494 |
| 1722400.5 | 1.185088064 | -4.617805582 | -2.017478321 |
| 1722560.5 | 2.308632646 | -4.178710987 | -1.857296953 |
| 1722720.5 | 3.302276255 | -3.505192305 | -1.592912025 |
| 1722880.5 | 4.104796376 | -2.629265579 | -1.236533038 |
| 1723040.5 | 4.663777900 | -1.597660159 | -0.806926438 |
| 1723200.5 | 4.940610612 | -0.469606943 | -0.328591815 |
| 1723360.5 | 4.914730876 | 0.687109067 | 0.169828149 |
| 1723520.5 | 4.586242789 | 1.801504770 | 0.657785361 |
| 1723680.5 | 3.976284771 | 2.805427511 | 1.105411124 |
| 1723840.5 | 3.124983857 | 3.639287367 | 1.486035243 |
| 1724000.5 | 2.087393695 | 4.256584784 | 1.778238364 |
| 1724160.5 | 0.928225366 | 4.626657664 | 1.967169016 |
| 1724320.5 | -0.283689453 | 4.735528138 | 2.045049777 |
| 1724480.5 | -1.480412292 | 4.585126933 | 2.010971884 |
| 1724640.5 | -2.598907746 | 4.191396426 | 1.870181615 |
| 1724800.5 | -3.583894983 | 3.581807959 | 1.633084964 |
| 1724960.5 | -4.389631771 | 2.792732483 | 1.314161854 |
| 1725120.5 | -4.980854379 | 1.866952708 | 0.930919642 |
| 1725280.5 | -5.333132789 | 0.851458031 | 0.502953328 |
| 1725440.5 | -5.432874699 | -0.204450391 | 0.051130242 |
| 1725600.5 | -5.277155757 | -1.250791829 | -0.403116657 |
| 1725760.5 | -4.873488431 | -2.238364468 | -0.838422311 |
| 1725920.5 | -4.239577759 | -3.120136540 | -1.234094429 |
| 1726080.5 | -3.403051651 | -3.852676129 | -1.570735673 |
| 1726240.5 | -2.401095472 | -4.397710267 | -1.830923951 |
| 1726400.5 | -1.279864804 | -4.723871255 | -1.999979037 |
| 1726560.5 | -0.093500184 | -4.808633326 | -2.066821394 |
| 1726720.5 | 1.097464936 | -4.640361164 | -2.024894687 |
| 1726880.5 | 2.228504184 | -4.220283579 | -1.873076579 |
| 1727040.5 | 3.234433903 | -3.564081057 | -1.616447036 |
| 1727200.5 | 4.053656260 | -2.702667199 | -1.266732888 |

HELIOCENTRIC COORDINATES OF JUPITER

REFERENCE AXES OF 1950.0

| JULIAN DATE | X | Y | Z |
|---|---|---|---|
| 1727360.5 | 4.633043340 | -1.681708389 | -0.842225957 |
| 1727520.5 | 4.932910654 | -0.559523715 | -0.367006602 |
| 1727680.5 | 4.931291486 | 0.596730391 | 0.130588678 |
| 1727840.5 | 4.626663264 | 1.716293880 | 0.620149602 |
| 1728000.5 | 4.038483986 | 2.730757393 | 1.071740577 |
| 1728160.5 | 3.205346782 | 3.579797497 | 1.458415870 |
| 1728320.5 | 2.181100014 | 4.215789734 | 1.758303949 |
| 1728480.5 | 1.029699163 | 4.606701931 | 1.955984252 |
| 1728640.5 | -0.180290728 | 4.737109734 | 2.043064864 |
| 1728800.5 | -1.380745809 | 4.607573999 | 2.018043846 |
| 1728960.5 | -2.508082541 | 4.232855569 | 1.885645923 |
| 1729120.5 | -3.506221184 | 3.639494331 | 1.655856540 |
| 1729280.5 | -4.328479252 | 2.863196531 | 1.342846048 |
| 1729440.5 | -4.938602254 | 1.946330725 | 0.963918536 |
| 1729600.5 | -5.311184107 | 0.935687436 | 0.538558544 |
| 1729760.5 | -5.431710343 | -0.119458060 | 0.087598550 |
| 1729920.5 | -5.296405100 | -1.169018165 | -0.367504622 |
| 1730080.5 | -4.911999567 | -2.163583444 | -0.805317099 |
| 1730240.5 | -4.295476199 | -3.055833464 | -1.205041621 |
| 1730400.5 | -3.473782874 | -3.801971753 | -1.547142334 |
| 1730560.5 | -2.483454090 | -4.363274942 | -1.814021108 |
| 1730720.5 | -1.370021778 | -4.707816181 | -1.990774221 |
| 1730880.5 | -0.187049619 | -4.812372049 | -2.066037646 |
| 1731040.5 | 1.005408723 | -4.664445042 | -2.032896811 |
| 1731200.5 | 2.143126784 | -4.264229007 | -1.889791641 |
| 1731360.5 | 3.160937908 | -3.626222469 | -1.641293521 |
| 1731520.5 | 3.996880083 | -2.780084269 | -1.298579551 |
| 1731680.5 | 4.597017092 | -1.770278930 | -0.879403301 |
| 1731840.5 | 4.920414227 | -0.654137870 | -0.407387854 |
| 1732000.5 | 4.943499561 | 0.501793336 | 0.089434427 |
| 1732160.5 | 4.662959166 | 1.626868661 | 0.580748315 |
| 1732320.5 | 4.096493145 | 2.652256248 | 1.036481522 |
| 1732480.5 | 3.281188848 | 3.516729415 | 1.429338280 |
| 1732640.5 | 2.269806131 | 4.171415627 | 1.736937952 |
| 1732800.5 | 1.125708154 | 4.582867228 | 1.943266139 |
| 1732960.5 | -0.082646682 | 4.734251027 | 2.039327527 |
| 1733120.5 | -1.286833940 | 4.624862033 | 2.023067917 |
| 1733280.5 | -2.422676052 | 4.268421341 | 1.898749631 |
| 1733440.5 | -3.433324222 | 3.690686555 | 1.676002160 |
| 1733600.5 | -4.271253291 | 2.926830110 | 1.368745125 |

HELIOCENTRIC COORDINATES OF JUPITER

REFERENCE AXES OF 1950.0

| JULIAN DATE | X | Y | Z |
|---|---|---|---|
| 1733760.5 | -4.899365089 | 2.018899488 | 0.994123583 |
| 1733920.5 | -5.291449463 | 1.013526066 | 0.571533718 |
| 1734080.5 | -5.432239805 | -0.040067017 | 0.121766666 |
| 1734240.5 | -5.317251000 | -1.091782929 | -0.333735157 |
| 1734400.5 | -4.952523512 | -2.092165628 | -0.773536251 |
| 1734560.5 | -4.354335168 | -2.993808328 | -1.176821249 |
| 1734720.5 | -3.548886083 | -3.752761807 | -1.524008795 |
| 1734880.5 | -2.571906563 | -4.330037849 | -1.797408407 |
| 1735040.5 | -1.468081857 | -4.693275382 | -1.981952105 |
| 1735200.5 | -0.290136057 | -4.818589496 | -2.066013472 |
| 1735360.5 | 0.902620456 | -4.692550889 | -2.042296579 |
| 1735520.5 | 2.046483118 | -4.314142594 | -1.908734054 |
| 1735680.5 | 3.076440561 | -3.696419377 | -1.669280120 |
| 1735840.5 | 3.930184569 | -2.867478568 | -1.334430940 |
| 1736000.5 | 4.552849340 | -1.870288137 | -0.921271896 |
| 1736160.5 | 4.902001722 | -0.760971842 | -0.452867368 |
| 1736320.5 | 4.952138368 | 0.394629941 | 0.043104074 |
| 1736480.5 | 4.697825691 | 1.525933665 | 0.536409404 |
| 1736640.5 | 4.154754561 | 2.563508939 | 0.996771335 |
| 1736800.5 | 3.358386229 | 3.444982765 | 1.396443554 |
| 1736960.5 | 2.360413052 | 4.119996945 | 1.712434434 |
| 1737120.5 | 1.223734827 | 4.553516994 | 1.928059390 |
| 1737280.5 | 0.016880646 | 4.727225120 | 2.033682992 |
| 1737440.5 | -1.191259525 | 4.639163247 | 2.026700147 |
| 1737600.5 | -2.335802743 | 4.302079226 | 1.910936231 |
| 1737760.5 | -3.359085545 | 3.741022201 | 1.695694564 |
| 1737920.5 | -4.212745889 | 2.990670692 | 1.394659956 |
| 1738080.5 | -4.858869964 | 2.092734989 | 1.024810295 |
| 1738240.5 | -5.270463629 | 1.093621240 | 0.605423472 |
| 1738400.5 | -5.431498129 | 0.042417239 | 0.157211834 |
| 1738560.5 | -5.336729703 | -1.010827931 | -0.298422701 |
| 1738720.5 | -4.991428134 | -2.016609693 | -0.740043530 |
| 1738880.5 | -4.411083735 | -2.927362390 | -1.146786880 |
| 1739040.5 | -3.621100547 | -3.698849876 | -1.498968030 |
| 1739200.5 | -2.656425516 | -4.291649071 | -1.778729171 |
| 1739360.5 | -1.561008132 | -4.672797157 | -1.970762022 |
| 1739520.5 | -0.386934075 | -4.817627690 | -2.063120077 |
| 1739680.5 | 0.806962689 | -4.711750380 | -2.048106020 |
| 1739840.5 | 1.957216083 | -4.353030278 | -1.923177521 |
| 1740000.5 | 2.998737660 | -3.753300508 | -1.691761244 |

## HELIOCENTRIC COORDINATES OF JUPITER

## REFERENCE AXES OF 1950.0

| JULIAN DATE | X | Y | Z |
|---|---|---|---|
| 1740160.5 | 3.868763971 | -2.939422393 | -1.363810041 |
| 1740320.5 | 4.511570521 | -1.953235213 | -0.955901983 |
| 1740480.5 | 4.883495459 | -0.849979026 | -0.490690290 |
| 1740640.5 | 4.957539038 | 0.305019790 | 0.004405169 |
| 1740800.5 | 4.726662624 | 1.441244232 | 0.499220204 |
| 1740960.5 | 4.205022523 | 2.488877771 | 0.963350089 |
| 1741120.5 | 3.426768944 | 3.384779324 | 1.368751893 |
| 1741280.5 | 2.442603603 | 4.077561865 | 1.692018015 |
| 1741440.5 | 1.314792488 | 4.531030529 | 1.915982721 |
| 1741600.5 | 0.111576462 | 4.725681164 | 2.030509376 |
| 1741760.5 | -1.098136168 | 4.658417163 | 2.032504057 |
| 1741920.5 | -2.249186108 | 4.340943066 | 1.925337217 |
| 1742080.5 | -3.283422786 | 3.797389277 | 1.717905018 |
| 1742240.5 | -4.151828284 | 3.061658503 | 1.423541894 |
| 1742400.5 | -4.815700805 | 2.174840324 | 1.058938377 |
| 1742560.5 | -5.247158843 | 1.182884061 | 0.643152687 |
| 1742720.5 | -5.429217405 | 0.134591337 | 0.196748937 |
| 1742880.5 | -5.355637402 | -0.920098966 | -0.258944665 |
| 1743040.5 | -5.030684575 | -1.931612707 | -0.702488254 |
| 1743200.5 | -4.468868572 | -2.852133112 | -1.112936192 |
| 1743360.5 | -3.694670837 | -3.636992574 | -1.470442885 |
| 1743520.5 | -2.742211575 | -4.246161904 | -1.756912094 |
| 1743680.5 | -1.654750101 | -4.645909719 | -1.956723704 |
| 1743840.5 | -0.483860870 | -4.810659489 | -2.057553845 |
| 1744000.5 | 0.711912898 | -4.725001201 | -2.051274893 |
| 1744160.5 | 1.869138753 | -4.385714795 | -1.934879282 |
| 1744320.5 | 2.922478126 | -3.803539452 | -1.711317081 |
| 1744480.5 | 3.808622534 | -3.004300414 | -1.390081485 |
| 1744640.5 | 4.471016459 | -2.028931152 | -0.987339126 |
| 1744800.5 | 4.864912911 | -0.931968117 | -0.525411673 |
| 1744960.5 | 4.962028819 | 0.221699430 | -0.031496392 |
| 1745120.5 | 4.753924642 | 1.361675896 | 0.464332770 |
| 1745280.5 | 4.253343586 | 2.417923351 | 0.931614093 |
| 1745440.5 | 3.493134361 | 3.326738248 | 1.342097290 |
| 1745600.5 | 2.522930225 | 4.035868793 | 1.672030325 |
| 1745760.5 | 1.404253327 | 4.508033446 | 1.903798542 |
| 1745920.5 | 0.204969191 | 4.722529734 | 2.026758704 |
| 1746080.5 | -1.006026957 | 4.675064832 | 2.037301981 |
| 1746240.5 | -2.163272595 | 4.376242694 | 1.938317097 |
| 1746400.5 | -3.208065712 | 3.849250910 | 1.738280135 |

HELIOCENTRIC COORDINATES OF JUPITER

REFERENCE AXES OF 1950.0

| JULIAN DATE | X | Y | Z |
|---|---|---|---|
| 1746560.5 | -4.090661756 | 3.127238264 | 1.450182333 |
| 1746720.5 | -4.771522870 | 2.250736338 | 1.090452600 |
| 1746880.5 | -5.221874723 | 1.265324220 | 0.677966934 |
| 1747040.5 | -5.423821234 | 0.219605693 | 0.233180980 |
| 1747200.5 | -5.370220381 | -0.836522527 | -0.222618277 |
| 1747360.5 | -5.064460889 | -1.853387164 | -0.667970680 |
| 1747520.5 | -4.520214270 | -2.782941307 | -1.081852886 |
| 1747680.5 | -3.761174622 | -3.580162156 | -1.444286427 |
| 1747840.5 | -2.820739760 | -4.204542732 | -1.736987718 |
| 1748000.5 | -1.741530711 | -4.621751833 | -1.944094626 |
| 1748160.5 | -0.574593519 | -4.805491759 | -2.052986320 |
| 1748320.5 | 0.621914839 | -4.739512776 | -2.055183850 |
| 1748480.5 | 1.784768824 | -4.419644617 | -1.947276682 |
| 1748640.5 | 2.848616181 | -3.855585148 | -1.731767544 |
| 1748800.5 | 3.749867612 | -3.072066265 | -1.417673282 |
| 1748960.5 | 4.431372103 | -2.108942199 | -1.020682088 |
| 1749120.5 | 4.847441063 | -1.019780785 | -0.562675634 |
| 1749280.5 | 4.968514321 | 0.131266062 | -0.070501739 |
| 1749440.5 | 4.784613781 | 1.274185472 | 0.425969576 |
| 1749600.5 | 4.306823321 | 2.338875176 | 0.896291603 |
| 1749760.5 | 3.566396819 | 3.261091393 | 1.312022951 |
| 1749920.5 | 2.611625198 | 3.987616607 | 1.649031179 |
| 1750080.5 | 1.503086390 | 4.479897838 | 1.889179644 |
| 1750240.5 | 0.308173136 | 4.715813579 | 2.021227620 |
| 1750400.5 | -0.904225131 | 4.689662926 | 2.040961534 |
| 1750560.5 | -2.068250738 | 4.410781159 | 1.950715017 |
| 1750720.5 | -3.124506388 | 3.901309517 | 1.758496907 |
| 1750880.5 | -4.022361153 | 3.193609453 | 1.476937605 |
| 1751040.5 | -4.721297661 | 2.327683608 | 1.122213804 |
| 1751200.5 | -5.191545259 | 1.348815224 | 0.713048964 |
| 1751360.5 | -5.414246329 | 0.305507654 | 0.269830983 |
| 1751520.5 | -5.381360606 | -0.752287594 | -0.186152680 |
| 1751680.5 | -5.095451424 | -1.774722812 | -0.633388243 |
| 1751840.5 | -4.569433608 | -2.713483185 | -1.050756943 |
| 1752000.5 | -3.826301179 | -3.523198176 | -1.418149289 |
| 1752160.5 | -2.898794820 | -4.162932699 | -1.717115171 |
| 1752320.5 | -1.828913488 | -4.597832530 | -1.931584325 |
| 1752480.5 | -0.667122836 | -4.800956753 | -2.048675020 |
| 1752640.5 | 0.528928141 | -4.755264143 | -2.059581548 |
| 1752800.5 | 1.696387059 | -4.455626662 | -1.960491177 |

HELIOCENTRIC COORDINATES OF JUPITER

REFERENCE AXES OF 1950.0

| JULIAN DATE | X | Y | Z |
|---|---|---|---|
| 1752960.5 | 2.770047528 | -3.910625828 | -1.753429910 |
| 1753120.5 | 3.686134209 | -3.143767477 | -1.446881890 |
| 1753280.5 | 4.386892192 | -2.193670120 | -1.055987734 |
| 1753440.5 | 4.825577451 | -1.112802530 | -0.602128883 |
| 1753600.5 | 4.971167986 | 0.035478826 | -0.111773275 |
| 1753760.5 | 4.811950319 | 1.181506903 | 0.385402417 |
| 1753920.5 | 4.357201620 | 2.254994254 | 0.858918141 |
| 1754080.5 | 3.636524888 | 3.190992022 | 1.280071467 |
| 1754240.5 | 2.696912677 | 3.935160278 | 1.624280322 |
| 1754400.5 | 1.598119028 | 4.447570579 | 1.872837417 |
| 1754560.5 | 0.407217396 | 4.704654303 | 2.013887659 |
| 1754720.5 | -0.806767654 | 4.699349287 | 2.042628447 |
| 1754880.5 | -1.977490028 | 4.439824437 | 1.960879254 |
| 1755040.5 | -3.044830040 | 3.947304481 | 1.776236378 |
| 1755200.5 | -3.957312118 | 3.253493300 | 1.501026112 |
| 1755360.5 | -4.673540547 | 2.397972126 | 1.151221985 |
| 1755520.5 | -5.162891270 | 1.425797905 | 0.745429018 |
| 1755680.5 | -5.405706391 | 0.385393519 | 0.303980678 |
| 1755840.5 | -5.393200862 | -0.673273595 | -0.151847582 |
| 1756000.5 | -5.127233189 | -1.700282800 | -0.600529115 |
| 1756160.5 | -4.620027068 | -2.647207760 | -1.020915904 |
| 1756320.5 | -3.893869185 | -3.468520113 | -1.392849956 |
| 1756480.5 | -2.980753312 | -4.123049032 | -1.697800167 |
| 1756640.5 | -1.921887406 | -4.575577407 | -1.919561571 |
| 1756800.5 | -0.766927573 | -4.798616552 | -2.045038476 |
| 1756960.5 | 0.427241931 | -4.774339382 | -2.065107386 |
| 1757120.5 | 1.598385704 | -4.496562831 | -1.975517936 |
| 1757280.5 | 2.681632246 | -3.972556218 | -1.777740330 |
| 1757440.5 | 3.613115395 | -3.224330277 | -1.479613012 |
| 1757600.5 | 4.334447806 | -2.288971009 | -1.095600620 |
| 1757760.5 | 4.797668634 | -1.217582121 | -0.646465392 |
| 1757920.5 | 4.970023092 | -0.072548136 | -0.158211731 |
| 1758080.5 | 4.837738441 | 1.076855818 | 0.339706220 |
| 1758240.5 | 4.407984189 | 2.160066090 | 0.816744821 |
| 1758400.5 | 3.708504201 | 3.111233370 | 1.243863102 |
| 1758560.5 | 2.784916062 | 3.874656927 | 1.595929879 |
| 1758720.5 | 1.696201144 | 4.408818630 | 1.853574653 |
| 1758880.5 | 0.509252117 | 4.688562103 | 2.004264525 |
| 1759040.5 | -0.706603391 | 4.705414616 | 2.042582620 |
| 1759200.5 | -1.884386270 | 4.466406635 | 1.969842414 |

## HELIOCENTRIC COORDINATES OF JUPITER

## REFERENCE AXES OF 1950.0

| JULIAN DATE | X | Y | Z |
|---|---|---|---|
| 1759360.5 | -2.963164724 | 3.991915959 | 1.793254675 |
| 1759520.5 | -3.890588250 | 3.313057753 | 1.524869258 |
| 1759680.5 | -4.624393857 | 2.469024560 | 1.180469107 |
| 1759840.5 | -5.133127178 | 1.504625761 | 0.778530028 |
| 1760000.5 | -5.396337846 | 0.468135187 | 0.339299598 |
| 1760160.5 | -5.404471819 | -0.590548746 | -0.115997204 |
| 1760320.5 | -5.158622813 | -1.621491603 | -0.565847832 |
| 1760480.5 | -4.670239006 | -2.576155670 | -0.989076119 |
| 1760640.5 | -3.960817838 | -3.408784029 | -1.365445587 |
| 1760800.5 | -3.061562339 | -4.077838351 | -1.676286841 |
| 1760960.5 | -2.012916243 | -4.547573162 | -1.905186798 |
| 1761120.5 | -0.863842192 | -4.789792497 | -2.038762937 |
| 1761280.5 | 0.329337603 | -4.785775353 | -2.067521371 |
| 1761440.5 | 1.504743439 | -4.528269404 | -1.986760703 |
| 1761600.5 | 2.597571200 | -4.023339173 | -1.797434510 |
| 1761760.5 | 3.543659057 | -3.291730864 | -1.506829717 |
| 1761920.5 | 4.283914561 | -2.369318653 | -1.128871771 |
| 1762080.5 | 4.769268011 | -1.306185162 | -0.683855620 |
| 1762240.5 | 4.965527075 | -0.164023172 | -0.197451891 |
| 1762400.5 | 4.857296651 | 0.988149849 | 0.301036770 |
| 1762560.5 | 4.450125581 | 2.079554692 | 0.781014466 |
| 1762720.5 | 3.770320357 | 3.043693946 | 1.213197537 |
| 1762880.5 | 2.862373451 | 3.823895086 | 1.572065853 |
| 1763040.5 | 1.784509413 | 4.377485882 | 1.837773266 |
| 1763200.5 | 0.603227158 | 4.678103469 | 1.997280022 |
| 1763360.5 | -0.612218077 | 4.716108004 | 2.044671068 |
| 1763520.5 | -1.794652999 | 4.497455425 | 1.980794304 |
| 1763680.5 | -2.882737122 | 4.041568808 | 1.812440468 |
| 1763840.5 | -3.823539513 | 3.378742097 | 1.551291932 |
| 1764000.5 | -4.574078723 | 2.547488559 | 1.212821271 |
| 1764160.5 | -5.102071540 | 1.592087716 | 0.815255007 |
| 1764320.5 | -5.386153299 | 0.560441009 | 0.378656689 |
| 1764480.5 | -5.415795103 | -0.497759343 | -0.075863035 |
| 1764640.5 | -5.191082055 | -1.532610136 | -0.526831752 |
| 1764800.5 | -4.722450159 | -2.495422116 | -0.953035057 |
| 1764960.5 | -4.030415753 | -3.340098650 | -1.334116608 |
| 1765120.5 | -3.145271617 | -4.024572401 | -1.651203993 |
| 1765280.5 | -2.106667007 | -4.512385288 | -1.887599209 |
| 1765440.5 | -0.962934832 | -4.774460623 | -2.029558272 |
| 1765600.5 | 0.230017370 | -4.791056963 | -2.067160151 |

## HELIOCENTRIC COORDINATES OF JUPITER

### REFERENCE AXES OF 1950.0

| JULIAN DATE | X | Y | Z |
|---|---|---|---|
| 1765760.5 | 1.410462562 | -4.553805631 | -1.995228202 |
| 1765920.5 | 2.513450404 | -4.067618931 | -1.814217433 |
| 1766080.5 | 3.474354922 | -3.352129126 | -1.530924194 |
| 1766240.5 | 4.233307584 | -2.442217195 | -1.158826858 |
| 1766400.5 | 4.740185752 | -1.387174921 | -0.717852697 |
| 1766560.5 | 4.959530670 | -0.248176600 | -0.233415697 |
| 1766720.5 | 4.874553404 | 0.905957712 | 0.265303959 |
| 1766880.5 | 4.489382988 | 2.004304153 | 0.747686836 |
| 1767040.5 | 3.828990789 | 2.979887650 | 1.184280223 |
| 1767200.5 | 2.936730592 | 3.775262793 | 1.549262036 |
| 1767360.5 | 1.869979652 | 4.346754131 | 1.822376448 |
| 1767520.5 | 0.694737658 | 4.666855192 | 1.990103950 |
| 1767680.5 | -0.519881884 | 4.724745443 | 2.046024417 |
| 1767840.5 | -1.706521603 | 4.525258685 | 1.990497632 |
| 1768000.5 | -2.803389455 | 4.086825636 | 1.829875370 |
| 1768160.5 | -3.756904155 | 3.438919296 | 1.575469857 |
| 1768320.5 | -4.523300894 | 2.619419399 | 1.242460861 |
| 1768480.5 | -5.069435230 | 1.672157576 | 0.848860062 |
| 1768640.5 | -5.373045312 | 0.644761632 | 0.414590277 |
| 1768800.5 | -5.422698778 | -0.413190161 | -0.039309977 |
| 1768960.5 | -5.217591385 | -1.451762269 | -0.491377081 |
| 1769120.5 | -4.767298650 | -2.422096830 | -0.920346004 |
| 1769280.5 | -4.091518075 | -3.277798163 | -1.305752552 |
| 1769440.5 | -3.219779525 | -3.976371733 | -1.628559844 |
| 1769600.5 | -2.191043976 | -4.480804416 | -1.871848455 |
| 1769760.5 | -1.053056070 | -4.761337327 | -2.021595714 |
| 1769920.5 | 0.138731871 | -4.797423434 | -2.067545513 |
| 1770080.5 | 1.322875483 | 4.570774484 | -2.004132953 |
| 1770240.5 | 2.434491995 | -4.112287600 | -1.831378595 |
| 1770400.5 | 3.408765545 | -3.413517623 | -1.555611380 |
| 1770560.5 | 4.185340036 | -2.517261813 | -1.189832091 |
| 1770720.5 | 4.713276720 | -1.471804994 | -0.753514896 |
| 1770880.5 | 4.955969183 | -0.337498896 | -0.271691238 |
| 1771040.5 | 4.895196777 | 0.817356189 | 0.226718565 |
| 1771200.5 | 4.533482896 | 1.921950822 | 0.711185332 |
| 1771360.5 | 3.894182484 | 2.908964493 | 1.152149324 |
| 1771520.5 | 3.019206141 | 3.720163020 | 1.523486514 |
| 1771680.5 | 1.964822976 | 4.310728040 | 1.804478386 |
| 1771840.5 | 0.796362810 | 4.651796412 | 1.981039676 |
| 1772000.5 | -0.417263714 | 4.731134160 | 2.046148181 |

HELIOCENTRIC COORDINATES OF JUPITER

REFERENCE AXES OF 1950.0

| JULIAN DATE | X | Y | Z |
|---|---|---|---|
| 1772160.5 | -1.608474072 | 4.552245086 | 1.999586530 |
| 1772320.5 | -2.714907724 | 4.132415212 | 1.847200221 |
| 1772480.5 | -3.682175665 | 3.500214859 | 1.599893091 |
| 1772640.5 | -4.465567776 | 2.692879386 | 1.272544000 |
| 1772800.5 | -5.030941925 | 1.753841010 | 0.882967618 |
| 1772960.5 | -5.355048844 | 0.730542156 | 0.450981948 |
| 1773120.5 | -5.425518705 | -0.327448221 | -0.002402444 |
| 1773280.5 | -5.240679291 | -1.370064917 | -0.455688474 |
| 1773440.5 | -4.809311523 | -2.348211956 | -0.887529279 |
| 1773600.5 | -4.150385070 | -3.215167431 | -1.277339376 |
| 1773760.5 | -3.292757265 | -3.928030201 | -1.605924765 |
| 1773920.5 | -2.274761937 | -4.449294328 | -1.856170799 |
| 1774080.5 | -1.143561077 | -4.748603606 | -2.013813589 |
| 1774240.5 | 0.045914515 | -4.804683169 | -2.068298082 |
| 1774400.5 | 1.232638172 | -4.607362805 | -2.013691250 |
| 1774560.5 | 2.351967296 | -4.159497462 | -1.849571650 |
| 1774720.5 | 3.339050362 | -3.478467440 | -1.581761800 |
| 1774880.5 | 4.133120792 | -2.596837632 | -1.222721393 |
| 1775040.5 | 4.682390722 | -1.561727008 | -0.791400920 |
| 1775200.5 | 4.948970638 | -0.432547345 | -0.312394342 |
| 1775360.5 | 4.913014625 | 0.722955456 | 0.185658270 |
| 1775520.5 | 4.575251872 | 1.834025733 | 0.672296856 |
| 1775680.5 | 3.957295097 | 2.832862981 | 1.117799802 |
| 1775840.5 | 3.099580613 | 3.660281090 | 1.495671303 |
| 1776000.5 | 2.057330781 | 4.270177232 | 1.784667080 |
| 1776160.5 | 0.895329127 | 4.632248721 | 1.970096607 |
| 1776320.5 | -0.317574201 | 4.732838314 | 2.044327030 |
| 1776480.5 | -1.513451102 | 4.574175486 | 2.006584064 |
| 1776640.5 | -2.629301993 | 4.172500928 | 1.862247374 |
| 1776800.5 | -3.609921644 | 3.555598623 | 1.621861968 |
| 1776960.5 | -4.409706821 | 2.760168610 | 1.300054053 |
| 1777120.5 | -4.993619917 | 1.829327356 | 0.914480237 |
| 1777280.5 | -5.337559709 | 0.810378440 | 0.484877785 |
| 1777440.5 | -5.428366203 | -0.247116817 | 0.032234403 |
| 1777600.5 | -5.263634087 | -1.293012619 | -0.421936813 |
| 1777760.5 | -4.851448419 | -2.278077019 | -0.856248965 |
| 1777920.5 | -4.210093375 | -3.155409852 | -1.250058695 |
| 1778080.5 | -3.367724624 | -3.881877269 | -1.584091375 |
| 1778240.5 | -2.361941525 | -4.419650734 | -1.841113872 |
| 1778400.5 | -1.239143953 | -4.737910183 | -2.006683561 |

HELIOCENTRIC COORDINATES OF JUPITER

REFERENCE AXES OF 1950.0

| JULIAN DATE | X | Y | Z |
|---|---|---|---|
| 1778560.5 | -0.053510840 | -4.814719358 | -2.069981153 |
| 1778720.5 | 1.134595549 | -4.639003319 | -2.024702162 |
| 1778880.5 | 2.260994348 | -4.212453441 | -1.869936818 |
| 1779040.5 | 3.260953579 | -3.551063900 | -1.610914461 |
| 1779200.5 | 4.073349404 | -2.685895357 | -1.259438227 |
| 1779360.5 | 4.645472699 | -1.662617976 | -0.833811100 |
| 1779520.5 | 4.937953879 | -0.539468844 | -0.358082836 |
| 1779680.5 | 4.929033449 | 0.616492945 | 0.139452175 |
| 1779840.5 | 4.617329828 | 1.734564982 | 0.628412206 |
| 1780000.5 | 4.022443152 | 2.746339428 | 1.078867337 |
| 1780160.5 | 3.183171175 | 3.591466703 | 1.463864572 |
| 1780320.5 | 2.153655782 | 4.222327494 | 1.761536113 |
| 1780480.5 | 0.998214547 | 4.606988404 | 1.956502786 |
| 1780640.5 | -0.214214856 | 4.730257422 | 2.040469864 |
| 1780800.5 | -1.415198905 | 4.593065370 | 2.012091712 |
| 1780960.5 | -2.540972825 | 4.210650317 | 1.876296953 |
| 1781120.5 | -3.535445736 | 3.610083402 | 1.643300732 |
| 1781280.5 | -4.352106500 | 2.827594680 | 1.327502893 |
| 1781440.5 | -4.955040260 | 1.906011646 | 0.946411935 |
| 1781600.5 | -5.319314963 | 0.892473319 | 0.519671453 |
| 1781760.5 | -5.430978728 | -0.163537237 | 0.068213633 |
| 1781920.5 | -5.286855751 | -1.211879805 | -0.386471670 |
| 1782080.5 | -4.894263563 | -2.203247151 | -0.822985230 |
| 1782240.5 | -4.270709753 | -3.090562334 | -1.220627317 |
| 1782400.5 | -3.443565467 | -3.830388549 | -1.560012570 |
| 1782560.5 | -2.449655746 | -4.384443387 | -1.823731898 |
| 1782720.5 | -1.334652496 | -4.721280004 | -1.997091230 |
| 1782880.5 | -0.152107125 | -4.818140108 | -2.068936821 |
| 1783040.5 | 1.038074510 | -4.662972461 | -2.032545338 |
| 1783200.5 | 2.171914577 | -4.256283757 | -1.886512797 |
| 1783360.5 | 3.184548198 | -3.612814976 | -1.635522597 |
| 1783520.5 | 4.014318534 | -2.762363502 | -1.290820927 |
| 1783680.5 | 4.607561621 | -1.749465852 | -0.870198944 |
| 1783840.5 | 4.923574138 | -0.631500818 | -0.397303521 |
| 1784000.5 | 4.939001118 | 0.524927935 | 0.099806976 |
| 1784160.5 | 4.650781762 | 1.649097374 | 0.590783940 |
| 1784320.5 | 4.076953208 | 2.672108717 | 1.045525160 |
| 1784480.5 | 3.255035133 | 3.532735662 | 1.436732488 |
| 1784640.5 | 2.238270479 | 4.182233741 | 1.742077468 |
| 1784800.5 | 1.090469142 | 4.587444787 | 1.945668667 |

HELIOCENTRIC COORDINATES OF JUPITER

REFERENCE AXES OF 1950.0

| JULIAN DATE | X | Y | Z |
|---|---|---|---|
| 1784960.5 | -0.119590891 | 4.731972574 | 2.038699738 |
| 1785120.5 | -1.323352661 | 4.615636680 | 2.019346130 |
| 1785280.5 | -2.456707855 | 4.252688494 | 1.892105062 |
| 1785440.5 | -3.463051729 | 3.669345750 | 1.666812699 |
| 1785600.5 | -4.295223809 | 2.901119128 | 1.357543566 |
| 1785760.5 | -4.916548227 | 1.990252086 | 0.981535755 |
| 1785920.5 | -5.301238429 | 0.983440139 | 0.558220119 |
| 1786080.5 | -5.434411901 | -0.070127210 | 0.108378335 |
| 1786240.5 | -5.311906106 | -1.120440895 | -0.346582105 |
| 1786400.5 | -4.940020820 | -2.118148727 | -0.785269600 |
| 1786560.5 | -4.335246606 | -3.015939923 | -1.186910968 |
| 1786720.5 | -3.523977700 | -3.769945650 | -1.531961266 |
| 1786880.5 | -2.542149660 | -4.341250596 | -1.802763283 |
| 1787040.5 | -1.434686698 | -4.697581740 | -1.984288128 |
| 1787200.5 | -0.254593069 | -4.815191341 | -2.064968433 |
| 1787360.5 | 0.938513869 | -4.680875560 | -2.037604143 |
| 1787520.5 | 2.080636483 | -4.293965691 | -1.900276054 |
| 1787680.5 | 3.106550525 | -3.668003947 | -1.657147115 |
| 1787840.5 | 3.953897455 | -2.831698037 | -1.318977408 |
| 1788000.5 | 4.568002020 | -1.828690217 | -0.903147615 |
| 1788160.5 | 4.906908574 | -0.715742129 | -0.433006541 |
| 1788320.5 | 4.945865004 | 0.440827234 | 0.063546931 |
| 1788480.5 | 4.680378021 | 1.570226272 | 0.556173366 |
| 1788640.5 | 4.127125312 | 2.603149007 | 1.014632708 |
| 1788800.5 | 3.322439228 | 3.477667755 | 1.411357913 |
| 1788960.5 | 2.318633248 | 4.144107763 | 1.723645633 |
| 1789120.5 | 1.178907953 | 4.568229892 | 1.935155250 |
| 1789280.5 | -0.028223299 | 4.732493061 | 2.036594075 |
| 1789440.5 | -1.234139832 | 4.635595770 | 2.025652827 |
| 1789600.5 | -2.374388086 | 4.290766659 | 1.906379214 |
| 1789760.5 | -3.391798796 | 3.723349758 | 1.688218595 |
| 1789920.5 | -4.238491838 | 2.968156597 | 1.384925997 |
| 1790080.5 | -4.876973283 | 2.066911672 | 1.013495137 |
| 1790240.5 | -5.280585057 | 1.065965024 | 0.593185861 |
| 1790400.5 | -5.433552486 | 0.014322059 | 0.144677803 |
| 1790560.5 | -5.330823060 | -1.038043379 | -0.310659433 |
| 1790720.5 | -4.977824949 | -2.041675265 | -0.751411790 |
| 1790880.5 | -4.390208680 | -2.949025674 | -1.156725441 |
| 1791040.5 | -3.593573875 | -3.715857166 | -1.506917039 |
| 1791200.5 | -2.623124671 | -4.302753546 | -1.784132416 |

## HELIOCENTRIC COORDINATES OF JUPITER

### REFERENCE AXES OF 1950.0

| JULIAN DATE | X | Y | Z |
|---|---|---|---|
| 1791360.5 | -1.523141430 | -4.676807698 | -1.973086159 |
| 1791520.5 | -0.346104788 | -4.813506837 | -2.061894989 |
| 1791680.5 | 0.848730482 | -4.698760234 | -2.042986529 |
| 1791840.5 | 1.997522267 | -4.330909481 | -1.914019998 |
| 1792000.5 | 3.034949480 | -3.722436272 | -1.678700248 |
| 1792160.5 | 3.898266144 | -2.900966798 | -1.347313113 |
| 1792320.5 | 4.532092620 | -1.909113564 | -0.936779267 |
| 1792480.5 | 4.893446144 | -0.802755934 | -0.470042534 |
| 1792640.5 | 4.956256142 | 0.352412504 | 0.025300404 |
| 1792800.5 | 4.714503566 | 1.485865310 | 0.519062933 |
| 1792960.5 | 4.183270238 | 2.528131625 | 0.980974883 |
| 1793120.5 | 3.397392090 | 3.416677830 | 1.383247063 |
| 1793280.5 | 2.407951049 | 4.100840183 | 1.702782620 |
| 1793440.5 | 1.277304773 | 4.545123388 | 1.922722700 |
| 1793600.5 | 0.073570788 | 4.730608961 | 2.033192877 |
| 1793760.5 | -1.134588513 | 4.654640206 | 2.031300518 |
| 1793920.5 | -2.282302932 | 4.329228300 | 1.920559621 |
| 1794080.5 | -3.311704037 | 3.778712644 | 1.709966106 |
| 1794240.5 | -4.174029545 | 3.037147359 | 1.412927130 |
| 1794400.5 | -4.830810587 | 2.145748104 | 1.046192991 |
| 1794560.5 | -5.254393138 | 1.150583715 | 0.628877374 |
| 1794720.5 | -5.428035955 | 0.100573279 | 0.181598205 |
| 1794880.5 | -5.345779464 | -0.954236799 | -0.274266881 |
| 1795040.5 | -5.012217447 | -1.964191974 | -0.717239892 |
| 1795200.5 | -4.442239265 | -2.881445325 | -1.126358105 |
| 1795360.5 | -3.660749443 | -3.661374955 | -1.481790962 |
| 1795520.5 | -2.702313760 | -4.264097578 | -1.765500217 |
| 1795680.5 | -1.610625912 | -4.656146570 | -1.961975417 |
| 1795840.5 | 0.437638340 | -4.812336951 | -2.059058339 |
| 1796000.5 | 0.757837133 | -4.717768688 | -2.048840456 |
| 1796160.5 | 1.912262334 | -4.369820684 | -1.928575085 |
| 1796320.5 | 2.960398810 | -3.779867567 | -1.701493040 |
| 1796480.5 | 3.839262893 | -2.974334524 | -1.377357196 |
| 1796640.5 | 4.492831228 | -1.994639215 | -0.972557408 |
| 1796800.5 | 4.877037890 | -0.895616262 | -0.509560082 |
| 1796960.5 | 4.964338141 | 0.257772728 | -0.015609148 |
| 1797120.5 | 4.746984776 | 1.395280945 | 0.479273334 |
| 1797280.5 | 4.238281037 | 2.447191572 | 0.944756891 |
| 1797440.5 | 3.471457475 | 3.350218352 | 1.352769620 |
| 1797600.5 | 2.496352860 | 4.052545415 | 1.679750855 |

HELIOCENTRIC COORDINATES OF JUPITER

REFERENCE AXES OF 1950.0

| JULIAN DATE | X | Y | Z |
|---|---|---|---|
| 1797760.5 | 1.374557026 | 4.517290008 | 1.908264574 |
| 1797920.5 | 0.173918439 | 4.724087986 | 2.027821432 |
| 1798080.5 | -1.036723621 | 4.668930093 | 2.034942575 |
| 1798240.5 | -2.191975808 | 4.362671421 | 1.932631505 |
| 1798400.5 | -3.233215595 | 3.828743262 | 1.729474495 |
| 1798560.5 | -4.110802887 | 3.100547914 | 1.438576258 |
| 1798720.5 | -4.785353125 | 2.218884626 | 1.076485099 |
| 1798880.5 | -5.228318453 | 1.229604866 | 0.662199063 |
| 1799040.5 | -5.422119406 | 0.181567872 | 0.216289701 |
| 1799200.5 | -5.360023691 | -0.875122608 | -0.239859831 |
| 1799360.5 | -5.045908073 | -1.890672911 | -0.684730012 |
| 1799520.5 | -4.493978497 | -2.817040567 | -1.097292265 |
| 1799680.5 | -3.728463892 | -3.609358981 | -1.457629484 |
| 1799840.5 | -2.783240708 | -4.227434668 | -1.747589318 |
| 1800000.5 | -1.701291479 | -4.637387285 | -1.951503170 |
| 1800160.5 | -0.533852242 | -4.813463319 | -2.056987981 |
| 1800320.5 | 0.660935535 | -4.739987919 | -2.055820159 |
| 1800480.5 | 1.820067575 | -4.413324909 | -1.944830764 |
| 1800640.5 | 2.878578972 | -3.843598319 | -1.726720009 |
| 1800800.5 | 3.773361935 | -3.055811140 | -1.410635784 |
| 1800960.5 | 4.447756197 | -2.089921205 | -1.012323258 |
| 1801120.5 | 4.856498538 | -0.999460379 | -0.553656999 |
| 1801280.5 | 4.970345827 | 0.151536225 | -0.061437805 |
| 1801440.5 | 4.779530016 | 1.293186428 | 0.434522202 |
| 1801600.5 | 4.295279695 | 2.355484227 | 0.903822759 |
| 1801760.5 | 3.548988736 | 3.274236746 | 1.318050510 |
| 1801920.5 | 2.589133429 | 3.996259752 | 1.653092475 |
| 1802080.5 | 1.476539255 | 4.483066173 | 1.890843303 |
| 1802240.5 | 0.278886322 | 4.712681556 | 2.020126190 |
| 1802400.5 | -0.934659263 | 4.679659999 | 2.036836501 |
| 1802560.5 | -2.098033001 | 4.393696500 | 1.943461716 |
| 1802720.5 | -3.151750110 | 3.877366928 | 1.748197846 |
| 1802880.5 | -4.045240339 | 3.163496186 | 1.463876786 |
| 1803040.5 | -4.738200361 | 2.292528368 | 1.106869495 |
| 1803200.5 | -5.201211354 | 1.310119815 | 0.696066348 |
| 1803360.5 | -5.415874542 | 0.265042340 | 0.251978620 |
| 1803520.5 | -5.374675406 | -0.792612480 | -0.204037424 |
| 1803680.5 | -5.080725590 | -1.812996060 | -0.650459348 |
| 1803840.5 | -4.547466167 | -2.747930309 | -1.066220171 |
| 1804000.5 | -3.798353605 | -3.552305979 | -1.431317892 |

HELIOCENTRIC COORDINATES OF JUPITER

REFERENCE AXES OF 1950.0

| JULIAN DATE | X | Y | Z |
|---|---|---|---|
| 1804160.5 | -2.866491268 | -4.185550353 | -1.727456223 |
| 1804320.5 | -1.794113900 | -4.613240139 | -1.938751544 |
| 1804480.5 | -0.631781057 | -4.808894528 | -2.052524760 |
| 1804640.5 | 0.562905912 | -4.755919073 | -2.060170180 |
| 1804800.5 | 1.727266216 | -4.449579213 | -1.958054458 |
| 1804960.5 | 2.796355210 | -3.898766548 | -1.748348115 |
| 1805120.5 | 3.706705276 | -3.127200705 | -1.439637852 |
| 1805280.5 | 4.400871800 | -2.173626226 | -1.047127067 |
| 1805440.5 | 4.832395237 | -1.090577297 | -0.592230930 |
| 1805600.5 | 4.970510340 | 0.058550711 | -0.101436669 |
| 1805760.5 | 4.803758527 | 1.204055139 | 0.395563268 |
| 1805920.5 | 4.341710242 | 2.275619447 | 0.868276418 |
| 1806080.5 | 3.614329604 | 3.208304837 | 1.288004005 |
| 1806240.5 | 2.669030845 | 3.947868435 | 1.630204117 |
| 1806400.5 | 1.565995577 | 4.454604775 | 1.876264163 |
| 1806560.5 | 0.372646899 | 4.705303095 | 2.014483836 |
| 1806720.5 | -0.841793643 | 4.693363179 | 2.040263025 |
| 1806880.5 | -2.010972404 | 4.427456478 | 1.955643973 |
| 1807040.5 | -3.074944541 | 3.929276889 | 1.768433511 |
| 1807200.5 | -3.982550527 | 3.230902671 | 1.491128674 |
| 1807360.5 | -4.692792593 | 2.372158731 | 1.139817146 |
| 1807520.5 | -5.175465673 | 1.398207698 | 0.733157170 |
| 1807680.5 | -5.411302587 | 0.357458909 | 0.291481833 |
| 1807840.5 | -5.391849711 | -0.700214769 | -0.163971100 |
| 1808000.5 | -5.119225907 | -1.725022678 | -0.611730200 |
| 1808160.5 | -4.605850944 | -2.668665139 | -1.030703260 |
| 1808320.5 | -3.874166387 | -3.485714456 | -1.400778417 |
| 1808480.5 | -2.956312668 | 4.133070292 | -1.703458400 |
| 1808640.5 | -1.893669574 | -4.581569608 | -1.922564728 |
| 1808800.5 | -0.736114197 | -4.797791179 | -2.045032806 |
| 1808960.5 | 0.459193644 | -4.766030532 | -2.061792347 |
| 1809120.5 | 1.629709374 | -4.480328281 | -1.968688159 |
| 1809280.5 | 2.710270364 | -3.948316054 | -1.767344585 |
| 1809440.5 | 3.636816792 | -3.192518510 | -1.465820253 |
| 1809600.5 | 4.350958179 | -2.250659457 | -1.078855973 |
| 1809760.5 | 4.805001513 | -1.174527320 | -0.627514690 |
| 1809920.5 | 4.966765738 | -0.027119382 | -0.138075607 |
| 1810080.5 | 4.823320668 | 1.121881799 | 0.359816470 |
| 1810240.5 | 4.382833507 | 2.201817191 | 0.835559771 |
| 1810400.5 | 3.674036108 | 3.147095720 | 1.260209131 |

HELIOCENTRIC COORDINATES OF JUPITER

REFERENCE AXES OF 1950.0

| JULIAN DATE | X | Y | Z |
|---|---|---|---|
| 1810560.5 | 2.743356164 | 3.902588105 | 1.608869152 |
| 1810720.5 | 1.650285308 | 4.427551964 | 1.862500468 |
| 1810880.5 | 0.461880394 | 4.697670107 | 2.008936458 |
| 1811040.5 | -0.752683026 | 4.705241167 | 2.043103863 |
| 1811200.5 | -1.926811192 | 4.457907399 | 1.966595180 |
| 1811360.5 | -3.000086660 | 3.976457605 | 1.786815071 |
| 1811520.5 | -3.920704465 | 3.292221229 | 1.515921440 |
| 1811680.5 | -4.646904277 | 2.444443645 | 1.169732866 |
| 1811840.5 | -5.147647601 | 1.477876356 | 0.766709259 |
| 1812000.5 | -5.402798168 | 0.440677629 | 0.327053418 |
| 1812160.5 | -5.403022842 | -0.617381200 | -0.128063049 |
| 1812320.5 | -5.149568578 | -1.646470527 | -0.577174196 |
| 1812480.5 | -4.654004971 | -2.598116923 | -0.999134712 |
| 1812640.5 | -3.937958896 | -3.426585935 | -1.373721480 |
| 1812800.5 | -3.032808290 | -4.090336219 | -1.682266864 |
| 1812960.5 | -1.979245258 | -4.553627236 | -1.908361381 |
| 1813120.5 | -0.826564688 | -4.788323818 | -2.038648044 |
| 1813280.5 | 0.368510859 | -4.775877621 | -2.063704629 |
| 1813440.5 | 1.543682369 | -4.509372668 | -1.978970763 |
| 1813600.5 | 2.633792042 | -3.995402171 | -1.785624516 |
| 1813760.5 | 3.574501159 | -3.255417046 | -1.491256135 |
| 1813920.5 | 4.306828558 | -2.326096882 | -1.110143009 |
| 1814080.5 | 4.782172021 | -1.258300615 | -0.662925863 |
| 1814240.5 | 4.967145629 | -0.114311527 | -0.175546634 |
| 1814400.5 | 4.847383514 | 1.036584825 | 0.322556933 |
| 1814560.5 | 4.429499903 | 2.123730077 | 0.800822896 |
| 1814720.5 | 3.740708305 | 3.081105021 | 1.230160807 |
| 1814880.5 | 2.826114807 | 3.852753538 | 1.585353644 |
| 1815040.5 | 1.744222792 | 4.396798180 | 1.846899466 |
| 1815200.5 | 0.561516916 | 4.687608985 | 2.002082106 |
| 1815360.5 | -0.652965542 | 4.716135023 | 2.045251902 |
| 1815520.5 | -1.832374632 | 4.488756076 | 1.977452333 |
| 1815680.5 | -2.915719715 | 4.025175320 | 1.805606984 |
| 1815840.5 | -3.850399081 | 3.355864116 | 1.541484785 |
| 1816000.5 | -4.593723358 | 2.519453237 | 1.200616603 |
| 1816160.5 | -5.113670243 | 1.560313329 | 0.801273690 |
| 1816320.5 | -5.389123575 | 0.526430163 | 0.363559799 |
| 1816480.5 | -5.409813540 | -0.532421354 | -0.091375782 |
| 1816640.5 | -5.176115761 | -1.566264481 | -0.542026113 |
| 1816800.5 | -4.698802328 | -2.526363023 | -0.967153380 |

HELIOCENTRIC COORDINATES OF JUPITER

REFERENCE AXES OF 1950.0

| JULIAN DATE | X | Y | Z |
|---|---|---|---|
| 1816960.5 | -3.998775435 | -3.366624233 | -1.346399125 |
| 1817120.5 | -3.106755291 | -4.045063012 | -1.660922007 |
| 1817280.5 | -2.062837820 | -4.525408463 | -1.894100113 |
| 1817440.5 | -0.915784710 | -4.778896471 | -2.032320088 |
| 1817600.5 | 0.278136174 | -4.786231525 | -2.065850514 |
| 1817760.5 | 1.456965242 | -4.539612941 | -1.989759636 |
| 1817920.5 | 2.555706246 | -4.044606817 | -1.804788130 |
| 1818080.5 | 3.509918563 | -3.321522156 | -1.518032424 |
| 1818240.5 | 4.260164458 | -2.405855821 | -1.143249235 |
| 1818400.5 | 4.756968111 | -1.347364897 | -0.700582711 |
| 1818560.5 | 4.965657707 | -0.207465252 | -0.215568613 |
| 1818720.5 | 4.870261991 | 0.945038899 | 0.282603470 |
| 1818880.5 | 4.475643202 | 2.039478455 | 0.763411753 |
| 1819040.5 | 3.807332776 | 3.009306571 | 1.197581790 |
| 1819200.5 | 2.909032051 | 3.797589581 | 1.559511961 |
| 1819360.5 | 1.838266259 | 4.361162158 | 1.829171373 |
| 1819520.5 | 0.661032803 | 4.672968931 | 1.993243250 |
| 1819680.5 | -0.553646440 | 4.722559934 | 2.045476660 |
| 1819840.5 | -1.738544416 | 4.515067401 | 1.986368996 |
| 1820000.5 | -2.832009458 | 4.069172979 | 1.822387783 |
| 1820160.5 | -3.780605459 | 3.414579926 | 1.564950553 |
| 1820320.5 | -4.540730160 | 2.589396621 | 1.229340569 |
| 1820480.5 | -5.079441171 | 1.637688135 | 0.833674704 |
| 1820640.5 | -5.374741608 | 0.607313374 | 0.397980363 |
| 1820800.5 | -5.415544063 | -0.451939793 | -0.056607690 |
| 1820960.5 | -5.201473162 | -1.489976657 | -0.508550392 |
| 1821120.5 | -4.742603421 | -2.457866267 | -0.936544025 |
| 1821280.5 | -4.059169835 | -3.309262367 | -1.320139023 |
| 1821440.5 | -3.181232061 | -4.001868890 | -1.640378721 |
| 1821600.5 | -2.148216562 | -4.499025981 | -1.880492828 |
| 1821760.5 | -1.008208540 | -4.771462609 | -2.026668968 |
| 1821920.5 | 0.183177897 | -4.799208970 | -2.068903858 |
| 1822080.5 | 1.364548394 | -4.573579353 | -2.001900624 |
| 1822240.5 | 2.471280731 | -4.099026619 | -1.825931392 |
| 1822400.5 | 3.438989006 | -3.394548344 | -1.547530219 |
| 1822560.5 | 4.207844473 | -2.494226265 | -1.179835030 |
| 1822720.5 | 4.727450666 | -1.446459926 | -0.742381988 |
| 1822880.5 | 4.961688004 | -0.311571577 | -0.260198523 |
| 1823040.5 | 4.892727235 | 0.842260420 | 0.237843725 |
| 1823200.5 | 4.523386305 | 1.944384351 | 0.721284209 |

HELIOCENTRIC COORDINATES OF JUPITER

REFERENCE AXES OF 1950.0

| JULIAN DATE | X | Y | Z |
|---|---|---|---|
| 1823360.5 | 3.877248332 | 2.927633975 | 1.160633488 |
| 1823520.5 | 2.996428282 | 3.733916926 | 1.529833721 |
| 1823680.5 | 1.937407361 | 4.318563175 | 1.808234967 |
| 1823840.5 | 0.765740778 | 4.652896717 | 1.981835910 |
| 1824000.5 | -0.449443633 | 4.724936918 | 2.043725430 |
| 1824160.5 | -1.640392485 | 4.538516772 | 1.993829154 |
| 1824320.5 | -2.744660956 | 4.111315603 | 1.838163322 |
| 1824480.5 | -3.707892631 | 3.472332140 | 1.587818599 |
| 1824640.5 | -4.485540851 | 2.659226936 | 1.257860765 |
| 1824800.5 | -5.043754612 | 1.715814534 | 0.866274567 |
| 1824960.5 | -5.359685643 | 0.689841739 | 0.433015917 |
| 1825120.5 | -5.421446574 | -0.368923019 | -0.020810985 |
| 1825280.5 | -5.227894105 | -1.410337259 | -0.473667818 |
| 1825440.5 | -4.788346293 | -2.385357169 | -0.904222786 |
| 1825600.5 | -4.122279473 | -3.247439766 | -1.291961425 |
| 1825760.5 | -3.258990466 | -3.953977309 | -1.617812170 |
| 1825920.5 | -2.237153236 | -4.467850304 | -1.864825616 |
| 1826080.5 | -1.104145821 | -4.759152355 | -2.018933527 |
| 1826240.5 | 0.085023048 | -4.807085710 | -2.069791462 |
| 1826400.5 | 1.269387174 | -4.601945594 | -2.011674826 |
| 1826560.5 | 2.384487719 | -4.147004119 | -1.844351874 |
| 1826720.5 | 3.365751150 | -3.459982852 | -1.573803004 |
| 1826880.5 | 4.152746036 | -2.573699370 | -1.212607117 |
| 1827040.5 | 4.694041876 | -1.535441262 | -0.779795944 |
| 1827200.5 | 4.952107272 | -0.404722217 | -0.300013487 |
| 1827360.5 | 4.907453995 | 0.750657655 | 0.198073093 |
| 1827520.5 | 4.561184469 | 1.859929456 | 0.683995857 |
| 1827680.5 | 3.935319208 | 2.855334512 | 1.128049442 |
| 1827840.5 | 3.070739258 | 3.677814551 | 1.503790977 |
| 1828000.5 | 2.023118671 | 4.281511543 | 1.790079481 |
| 1828160.5 | 0.857635841 | 4.636495357 | 1.972384268 |
| 1828320.5 | -0.356598133 | 4.729588798 | 2.043281759 |
| 1828480.5 | -1.551583950 | 4.563558861 | 2.002235340 |
| 1828640.5 | -2.664452903 | 4.155172931 | 1.854860703 |
| 1828800.5 | -3.640303169 | 3.532663713 | 1.611907130 |
| 1828960.5 | -4.433950902 | 2.733053707 | 1.288151111 |
| 1829120.5 | -5.010827506 | 1.799634312 | 0.901334908 |
| 1829280.5 | -5.347289188 | 0.779741857 | 0.471217929 |
| 1829440.5 | -5.430576823 | -0.277144197 | 0.018757772 |
| 1829600.5 | -5.258605885 | -1.321031294 | -0.434596168 |

## HELIOCENTRIC COORDINATES OF JUPITER

## REFERENCE AXES OF 1950.0

| JULIAN DATE | X | Y | Z |
|---|---|---|---|
| 1829760.5 | -4.839699500 | -2.302863675 | -0.867533076 |
| 1829920.5 | -4.192312648 | -3.175900211 | -1.259480317 |
| 1830080.5 | -3.344734740 | -3.897127122 | -1.591218612 |
| 1830240.5 | -2.334700245 | -4.428796799 | -1.845553775 |
| 1830400.5 | -1.208782173 | -4.740151655 | -2.008073559 |
| 1830560.5 | -0.021394233 | -4.809338210 | -2.067996270 |
| 1830720.5 | 1.166805140 | -4.625435558 | -2.019083767 |
| 1830880.5 | 2.291310377 | -4.190412747 | -1.860544252 |
| 1831040.5 | 3.287104912 | -3.520702766 | -1.597794010 |
| 1831200.5 | 4.092917686 | -2.647969177 | -1.242895271 |
| 1831360.5 | 4.656134232 | -1.618600676 | -0.814463581 |
| 1831520.5 | 4.937799066 | -0.491564354 | -0.336871554 |
| 1831680.5 | 4.916899882 | 0.665482279 | 0.161314549 |
| 1831840.5 | 4.593062570 | 1.781513510 | 0.649555476 |
| 1832000.5 | 3.987003253 | 2.788169563 | 1.097922844 |
| 1832160.5 | 3.138553541 | 3.625533568 | 1.479634019 |
| 1832320.5 | 2.102634333 | 4.246727989 | 1.773131366 |
| 1832480.5 | 0.943980890 | 4.620732207 | 1.963428118 |
| 1832640.5 | -0.268432623 | 4.733281906 | 2.042636853 |
| 1832800.5 | -1.466461137 | 4.586118438 | 2.009775383 |
| 1832960.5 | -2.586850911 | 4.195090245 | 1.870053362 |
| 1833120.5 | -3.574129914 | 3.587649853 | 1.633869150 |
| 1833280.5 | -4.382412680 | 2.800195484 | 1.315712266 |
| 1833440.5 | -4.976343990 | 1.875554025 | 0.933104771 |
| 1833600.5 | -5.331444808 | 0.860752279 | 0.505651448 |
| 1833760.5 | -5.434098729 | -0.194893190 | 0.054217887 |
| 1833920.5 | -5.281360867 | -1.241414048 | -0.399778884 |
| 1834080.5 | -4.880705290 | -2.229645225 | -0.835002638 |
| 1834240.5 | -4.249762465 | -3.112604072 | -1.230797804 |
| 1834400.5 | -3.416037340 | -3.846900979 | -1.567803348 |
| 1834560.5 | -2.416540871 | -4.394274293 | -1.828622315 |
| 1834720.5 | -1.297212314 | -4.723311480 | -1.998576795 |
| 1834880.5 | -0.111959367 | -4.811366025 | -2.066556801 |
| 1835040.5 | 1.078891354 | -4.646599267 | -2.025937876 |
| 1835200.5 | 2.210940683 | -4.229966211 | -1.875494893 |
| 1835360.5 | 3.219004615 | -3.576835079 | -1.620182247 |
| 1835520.5 | 4.041336042 | -2.717816135 | -1.271597593 |
| 1835680.5 | 4.624525133 | -1.698334324 | -0.847922389 |
| 1835840.5 | 4.928521078 | -0.576574338 | -0.373165418 |
| 1836000.5 | 4.930970622 | 0.580311221 | 0.124358005 |

HELIOCENTRIC COORDINATES OF JUPITER

REFERENCE AXES OF 1950.0

| JULIAN DATE | X | Y | Z |
|---|---|---|---|
| 1836160.5 | 4.630007831 | 1.701441653 | 0.614208860 |
| 1836320.5 | 4.044838509 | 2.718195559 | 1.066384763 |
| 1836480.5 | 3.213917028 | 3.569994401 | 1.453853793 |
| 1836640.5 | 2.191060900 | 4.208963543 | 1.754657901 |
| 1836800.5 | 1.040276267 | 4.602860402 | 1.953303734 |
| 1836960.5 | -0.169774765 | 4.736105912 | 2.041346793 |
| 1837120.5 | -1.370857080 | 4.609160559 | 2.017252146 |
| 1837280.5 | -2.499283824 | 4.236726760 | 1.885725932 |
| 1837440.5 | -3.498890119 | 3.645311573 | 1.656742585 |
| 1837600.5 | -4.322926761 | 2.870599176 | 1.344462814 |
| 1837760.5 | -4.935084772 | 1.954936475 | 0.966178167 |
| 1837920.5 | -5.309906790 | 0.945087777 | 0.541355742 |
| 1838080.5 | -5.432821129 | -0.109702502 | 0.090805811 |
| 1838240.5 | -5.299981323 | -1.159378472 | -0.364039948 |
| 1838400.5 | -4.918029877 | -2.154556720 | -0.801772708 |
| 1838560.5 | -4.303841137 | -3.047929001 | -1.201616333 |
| 1838720.5 | -3.484237858 | -3.795688472 | -1.544048008 |
| 1838880.5 | -2.495618756 | -4.359070619 | -1.811470459 |
| 1839040.5 | -1.383379701 | -4.706070263 | -1.988965150 |
| 1839200.5 | -0.200962030 | -4.813345356 | -2.065135283 |
| 1839360.5 | 0.991672617 | -4.668241334 | -2.033015219 |
| 1839520.5 | 2.130342417 | -4.270764866 | -1.890977810 |
| 1839680.5 | 3.149863632 | -3.635212629 | -1.643516681 |
| 1839840.5 | 3.988187857 | -2.791048610 | -1.301728628 |
| 1840000.5 | 4.591225981 | -1.782574155 | -0.883294222 |
| 1840160.5 | 4.917838773 | -0.667011705 | -0.411780351 |
| 1840320.5 | 4.944223176 | 0.489133509 | 0.084813234 |
| 1840480.5 | 4.666837060 | 1.615184911 | 0.576178104 |
| 1840640.5 | 4.103181718 | 2.642219413 | 1.032225585 |
| 1840800.5 | 3.290192205 | 3.508877018 | 1.425626381 |
| 1840960.5 | 2.280528157 | 4.166130370 | 1.733956187 |
| 1841120.5 | 1.137500848 | 4.580372232 | 1.941152473 |
| 1841280.5 | -0.070442940 | 4.734615086 | 2.038170533 |
| 1841440.5 | -1.274857610 | 4.628008619 | 2.022906416 |
| 1841600.5 | -2.411515681 | 4.274137634 | 1.899571963 |
| 1841760.5 | -3.423490065 | 3.698631927 | 1.677745089 |
| 1841920.5 | -4.263147884 | 2.936546305 | 1.371293590 |
| 1842080.5 | -4.893255213 | 2.029827260 | 0.997313625 |
| 1842240.5 | -5.287445347 | 1.025034832 | 0.575161807 |
| 1842400.5 | -5.430287823 | -0.028631627 | 0.125607831 |

## HELIOCENTRIC COORDINATES OF JUPITER

## REFERENCE AXES OF 1950.0

| JULIAN DATE | X | Y | Z |
|---|---|---|---|
| 1842560.5 | -5.317143809 | -1.081036739 | -0.329901900 |
| 1842720.5 | -4.953929074 | -2.082621883 | -0.769901230 |
| 1842880.5 | -4.356842372 | -2.985829210 | -1.173524025 |
| 1843040.5 | -3.552059970 | -3.746539217 | -1.521130465 |
| 1843200.5 | -2.575343088 | -4.325607734 | -1.794977899 |
| 1843360.5 | -1.471450594 | -4.690559399 | -1.979963715 |
| 1843520.5 | -0.293199848 | -4.817451503 | -2.064450481 |
| 1843680.5 | 0.900015608 | -4.692853613 | -2.041154843 |
| 1843840.5 | 2.044447313 | -4.315789564 | -1.908039110 |
| 1844000.5 | 3.075098961 | -3.699368770 | -1.669091746 |
| 1844160.5 | 3.929738010 | -2.871720850 | -1.334831198 |
| 1844320.5 | 4.553613161 | -1.875789213 | -0.922336895 |
| 1844480.5 | 4.904401044 | -0.767596394 | -0.454628795 |
| 1844640.5 | 4.956649253 | 0.387188742 | 0.040696205 |
| 1844800.5 | 4.704870689 | 1.518188388 | 0.533509188 |
| 1844960.5 | 4.164580699 | 2.556155069 | 0.993634243 |
| 1845120.5 | 3.370963771 | 3.438816979 | 1.393396277 |
| 1845280.5 | 2.375388295 | 4.115795871 | 1.709824580 |
| 1845440.5 | 1.240451920 | 4.551909243 | 1.926198797 |
| 1845600.5 | 0.034465089 | 4.728589467 | 2.032799303 |
| 1845760.5 | -1.173779660 | 4.643574678 | 2.026907046 |
| 1845920.5 | -2.319367592 | 4.309308828 | 1.912226105 |
| 1846080.5 | -3.344491911 | 3.750583304 | 1.697951395 |
| 1846240.5 | -4.200569691 | 3.001895832 | 1.397686493 |
| 1846400.5 | -4.849428828 | 2.104866280 | 1.028362349 |
| 1846560.5 | -5.263817751 | 1.105898776 | 0.609244748 |
| 1846720.5 | -5.427482469 | 0.054154049 | 0.161064206 |
| 1846880.5 | -5.335000100 | -1.000191649 | -0.294736862 |
| 1847040.5 | -4.991559211 | -2.007476676 | -0.736668815 |
| 1847200.5 | -4.412588569 | -2.919973320 | -1.143812599 |
| 1847360.5 | -3.623522915 | -3.693301324 | -1.496435665 |
| 1847520.5 | -2.659376291 | -4.287929986 | -1.776647802 |
| 1847680.5 | -1.564187056 | -4.670837029 | -1.969128887 |
| 1847840.5 | -0.390124596 | -4.817346933 | -2.061942190 |
| 1848000.5 | 0.803925640 | -4.713101329 | -2.047417724 |
| 1848160.5 | 1.954499409 | -4.356016432 | -1.923048361 |
| 1848320.5 | 2.996570905 | -3.757963863 | -1.692289311 |
| 1848480.5 | 3.867486840 | -2.945795355 | -1.365098700 |
| 1848640.5 | 4.511643158 | -1.961264876 | -0.958022806 |
| 1848800.5 | 4.885452747 | -0.859445100 | -0.493641126 |

HELIOCENTRIC COORDINATES OF JUPITER

REFERENCE AXES OF 1950.0

| JULIAN DATE | X | Y | Z |
|---|---|---|---|
| 1848960.5 | 4.961886417 | 0.294562268 | 0.000733148 |
| 1849120.5 | 4.733736130 | 1.430467207 | 0.495052151 |
| 1849280.5 | 4.214860434 | 2.478612360 | 0.959005169 |
| 1849440.5 | 3.439049566 | 3.375888247 | 1.364591442 |
| 1849600.5 | 2.456679758 | 4.070789477 | 1.688379803 |
| 1849760.5 | 1.329809251 | 4.526881198 | 1.913124462 |
| 1849920.5 | 0.126623033 | 4.724362930 | 2.028579254 |
| 1850080.5 | -1.083895684 | 4.659856629 | 2.031541998 |
| 1850240.5 | -2.236433849 | 4.344850942 | 1.925298287 |
| 1850400.5 | -3.272660852 | 3.803344142 | 1.718691060 |
| 1850560.5 | -4.143390356 | 3.069181225 | 1.425029764 |
| 1850720.5 | -4.809784165 | 2.183445247 | 1.060998870 |
| 1850880.5 | -5.243861699 | 1.192104995 | 0.645659160 |
| 1851040.5 | -5.428570571 | 0.143987091 | 0.199577439 |
| 1851200.5 | -5.357623868 | -0.910952133 | -0.255920758 |
| 1851360.5 | -5.035244370 | -1.923133909 | -0.699405853 |
| 1851520.5 | -4.475889421 | -2.844747767 | -1.109949290 |
| 1851680.5 | -3.703967599 | -3.631135374 | -1.467725857 |
| 1851840.5 | -2.753499545 | -4.242267193 | -1.754657339 |
| 1852000.5 | -1.667619114 | -4.644386489 | -1.955132205 |
| 1852160.5 | -0.497758727 | -4.811850821 | -2.056818226 |
| 1852320.5 | 0.697678507 | -4.729131492 | -2.051556047 |
| 1852480.5 | 1.855371551 | -4.392832507 | -1.936279537 |
| 1852640.5 | 2.910033187 | -3.813467655 | -1.713854898 |
| 1852800.5 | 3.798317221 | -3.016610431 | -1.393674290 |
| 1852960.5 | 4.463524381 | -2.042953864 | -0.991800887 |
| 1853120.5 | 4.860662467 | -0.946850955 | -0.530469471 |
| 1853280.5 | 4.961132724 | 0.206893948 | -0.036825329 |
| 1853440.5 | 4.756162614 | 1.347848522 | 0.459063097 |
| 1853600.5 | 4.258206369 | 2.405821828 | 0.926692931 |
| 1853760.5 | 3.499918236 | 3.316874908 | 1.337737885 |
| 1853920.5 | 2.530855732 | 4.028489644 | 1.668354826 |
| 1854080.5 | 1.412576897 | 4.503139453 | 1.900843748 |
| 1854240.5 | 0.213063163 | 4.719935948 | 2.024497375 |
| 1854400.5 | -0.998636931 | 4.674475576 | 2.035671016 |
| 1854560.5 | -2.156909981 | 4.377322091 | 1.937244092 |
| 1854720.5 | -3.202933579 | 3.851674714 | 1.737702370 |
| 1854880.5 | -4.086886327 | 3.130721206 | 1.450055639 |
| 1855040.5 | -4.769196569 | 2.255036043 | 1.090750400 |
| 1855200.5 | -5.221088758 | 1.270227080 | 0.678671630 |

## HELIOCENTRIC COORDINATES OF JUPITER

### REFERENCE AXES OF 1950.0

| JULIAN DATE | X | Y | Z |
|---|---|---|---|
| 1855360.5 | -5.424681977 | 0.224902119 | 0.234270749 |
| 1855520.5 | -5.372847445 | -0.831065670 | -0.221183975 |
| 1855680.5 | -5.068967761 | -1.848049581 | -0.666263734 |
| 1855840.5 | -4.526674709 | -2.778061273 | -1.079984442 |
| 1856000.5 | -3.769580104 | -3.576132170 | -1.442407658 |
| 1856160.5 | -2.830955290 | -4.201785127 | -1.735281640 |
| 1856320.5 | -1.753258360 | -4.620673343 | -1.942758185 |
| 1856480.5 | -0.587354910 | -4.806420582 | -2.052203820 |
| 1856640.5 | 0.608764797 | -4.742628410 | -2.055095604 |
| 1856800.5 | 1.771986474 | -4.424917793 | -1.947949932 |
| 1856960.5 | 2.836970539 | -3.862748024 | -1.733178368 |
| 1857120.5 | 3.740015770 | -3.080628029 | -1.419706608 |
| 1857280.5 | 4.423744181 | -2.118253257 | -1.023152097 |
| 1857440.5 | 4.842173409 | -1.029133733 | -0.565362450 |
| 1857600.5 | 4.965452193 | 0.122522928 | -0.073194841 |
| 1857760.5 | 4.783378438 | 1.266552281 | 0.423433692 |
| 1857920.5 | 4.306918240 | 2.332647583 | 0.894008354 |
| 1858080.5 | 3.567319147 | 3.256360749 | 1.310019853 |
| 1858240.5 | 2.612951896 | 3.984314886 | 1.647286187 |
| 1858400.5 | 1.504517473 | 4.477866033 | 1.887648941 |
| 1858560.5 | 0.309531004 | 4.714869397 | 2.019872726 |
| 1858720.5 | -0.903028151 | 4.689649662 | 2.039768040 |
| 1858880.5 | -2.067258797 | 4.411589911 | 1.949699558 |
| 1859040.5 | -3.123763966 | 3.902876409 | 1.757703094 |
| 1859200.5 | -4.021943325 | 3.195895408 | 1.476424162 |
| 1859360.5 | -4.721322065 | 2.330646165 | 1.122039085 |
| 1859520.5 | -5.192166223 | 1.352381048 | 0.713255673 |
| 1859680.5 | -5.415634879 | 0.309552137 | 0.270433984 |
| 1859840.5 | -5.383675663 | -0.747949723 | -0.185173655 |
| 1860000.5 | -5.098807936 | -1.770334240 | -0.632090162 |
| 1860160.5 | -4.573873146 | -2.709328306 | -1.049229270 |
| 1860320.5 | -3.831769990 | -3.519576653 | -1.416504623 |
| 1860480.5 | -2.905133574 | -4.160125340 | -1.715475903 |
| 1860640.5 | -1.835861940 | -4.596063873 | -1.930067061 |
| 1860800.5 | -0.674341036 | -4.800359943 | -2.047375024 |
| 1860960.5 | 0.521822785 | -4.755854653 | -2.058559907 |
| 1861120.5 | 1.689770786 | -4.457291478 | -1.959767497 |
| 1861280.5 | 2.764237006 | -3.913133323 | -1.752982837 |
| 1861440.5 | 3.681340100 | -3.146799574 | -1.446658367 |
| 1861600.5 | 4.383191637 | -2.196872687 | -1.055919592 |

HELIOCENTRIC COORDINATES OF JUPITER

REFERENCE AXES OF 1950.0

| JULIAN DATE | X | Y | Z |
|---|---|---|---|
| 1861760.5 | 4.822914812 | -1.115843453 | -0.602152700 |
| 1861920.5 | 4.969385813 | 0.032858940 | -0.111847478 |
| 1862080.5 | 4.810842390 | 1.179466053 | 0.385289144 |
| 1862240.5 | 4.356571338 | 2.253590420 | 0.858750400 |
| 1862400.5 | 3.636229946 | 3.190211847 | 1.279821396 |
| 1862560.5 | 2.696882582 | 3.934962368 | 1.623927128 |
| 1862720.5 | 1.598343146 | 4.447926398 | 1.872384879 |
| 1862880.5 | 0.407712165 | 4.705572512 | 2.013373512 |
| 1863040.5 | -0.805997526 | 4.700877094 | 2.042122891 |
| 1863200.5 | -1.976481614 | 4.442027967 | 1.960473514 |
| 1863360.5 | -3.043673417 | 3.950236093 | 1.776024872 |
| 1863520.5 | -3.956139182 | 3.257156084 | 1.501086840 |
| 1863680.5 | -4.672494372 | 2.402291368 | 1.151599651 |
| 1863840.5 | -5.162084530 | 1.430607830 | 0.746124894 |
| 1864000.5 | -5.405181167 | 0.390446164 | 0.304952679 |
| 1864160.5 | -5.392902051 | -0.668273493 | -0.150672516 |
| 1864320.5 | -5.127003610 | -1.695620656 | -0.599232038 |
| 1864480.5 | -4.619627296 | -2.643093795 | -1.019558765 |
| 1864640.5 | -3.893019999 | -3.465035446 | -1.391453181 |
| 1864800.5 | -2.979193551 | -4.120120394 | -1.696332230 |
| 1864960.5 | -1.919436433 | -4.572988658 | -1.917943790 |
| 1865120.5 | -0.763539178 | -4.796054347 | -2.043163810 |
| 1865280.5 | 0.431448292 | -4.771462838 | -2.062869168 |
| 1865440.5 | 1.603127926 | -4.493083174 | -1.972842799 |
| 1865600.5 | 2.686508202 | -3.968309685 | -1.774618246 |
| 1865760.5 | 3.617681154 | -3.219326249 | -1.476116551 |
| 1865920.5 | 4.338316391 | -2.283399742 | -1.091886349 |
| 1866080.5 | 4.800604512 | -1.211774710 | -0.642753802 |
| 1866240.5 | 4.972003845 | -0.066895019 | -0.154748079 |
| 1866400.5 | 4.838961526 | 1.082006424 | 0.342700969 |
| 1866560.5 | 4.408813343 | 2.164499044 | 0.819119194 |
| 1866720.5 | 3.709371098 | 3.114917943 | 1.245563109 |
| 1866880.5 | 2.786208485 | 3.877743692 | 1.597002196 |
| 1867040.5 | 1.698171885 | 4.411586393 | 1.854145971 |
| 1867200.5 | 0.511972475 | 4.691335299 | 2.004505436 |
| 1867360.5 | -0.703238869 | 4.708477725 | 2.042667677 |
| 1867520.5 | -1.880613025 | 4.469938212 | 1.969916791 |
| 1867680.5 | -2.959277730 | 3.995955028 | 1.793413801 |
| 1867840.5 | -3.886867662 | 3.317504652 | 1.525153178 |
| 1868000.5 | -4.621042764 | 2.473670440 | 1.180869716 |

## HELIOCENTRIC COORDINATES OF JUPITER

### REFERENCE AXES OF 1950.0

| JULIAN DATE | X | Y | Z |
|---|---|---|---|
| 1868160.5 | -5.130231361 | 1.509200783 | 0.779007266 |
| 1868320.5 | -5.393851555 | 0.472364501 | 0.339802065 |
| 1868480.5 | -5.402228885 | -0.586891141 | -0.115511484 |
| 1868640.5 | -5.156369238 | -1.618539117 | -0.565394074 |
| 1868800.5 | -4.667681148 | -2.573921403 | -0.988631560 |
| 1868960.5 | -3.957678943 | -3.407152922 | -1.364946240 |
| 1869120.5 | -3.057639254 | -4.076580973 | -1.675633182 |
| 1869280.5 | -2.008127433 | -4.546380564 | -1.904258287 |
| 1869440.5 | -0.858258029 | -4.788328691 | -2.037440041 |
| 1869600.5 | 0.335490388 | -4.783740497 | -2.065712186 |
| 1869760.5 | 1.511108471 | -4.525462993 | -1.984427554 |
| 1869920.5 | 2.603720243 | -4.019709527 | -1.794614471 |
| 1870080.5 | 3.549172711 | -3.287396565 | -1.503642590 |
| 1870240.5 | 4.288468495 | -2.364551505 | -1.125510577 |
| 1870400.5 | 4.772700421 | -1.301353633 | -0.680558720 |
| 1870560.5 | 4.967863137 | -0.159508714 | -0.194461814 |
| 1870720.5 | 4.858716680 | 0.992038869 | 0.303516888 |
| 1870880.5 | 4.450884916 | 2.082642779 | 0.782853476 |
| 1871040.5 | 3.770653379 | 3.045951985 | 1.214347540 |
| 1871200.5 | 2.862425895 | 3.825406180 | 1.572549870 |
| 1871360.5 | 1.784327622 | 4.378383103 | 1.837657516 |
| 1871520.5 | 0.602798463 | 4.678511449 | 1.996645845 |
| 1871680.5 | -0.612906910 | 4.716112571 | 2.043598444 |
| 1871840.5 | -1.795575362 | 4.497105351 | 1.979359418 |
| 1872000.5 | -2.883819905 | 4.040898125 | 1.810722211 |
| 1872160.5 | -3.824682777 | 3.377795448 | 1.549379610 |
| 1872320.5 | -4.575186130 | 2.546335782 | 1.210817705 |
| 1872480.5 | -5.103070003 | 1.590822965 | 0.813273139 |
| 1872640.5 | -5.387046990 | 0.559169138 | 0.376810866 |
| 1872800.5 | -5.416618400 | -0.498943171 | -0.077468585 |
| 1872960.5 | -5.191927222 | -1.533641311 | -0.528114578 |
| 1873120.5 | -4.723439824 | -2.496282593 | -0.953943961 |
| 1873280.5 | -4.031676609 | -3.340824145 | -1.334637181 |
| 1873440.5 | -3.146905322 | -4.025248363 | -1.651359017 |
| 1873600.5 | -2.108723847 | -4.513130941 | -1.887443445 |
| 1873760.5 | -0.965396111 | -4.775402384 | -2.029167958 |
| 1873920.5 | 0.227242544 | -4.792295041 | -2.066618552 |
| 1874080.5 | 1.407522021 | -4.555380313 | -1.994609762 |
| 1874240.5 | 2.510514457 | -4.069485391 | -1.813574143 |
| 1874400.5 | 3.471567224 | -3.354150431 | -1.530278333 |

HELIOCENTRIC COORDINATES OF JUPITER

REFERENCE AXES OF 1950.0

| JULIAN DATE | X | Y | Z |
|---|---|---|---|
| 1874560.5 | 4.230732211 | -2.444182975 | -1.158173576 |
| 1874720.5 | 4.737766465 | -1.388847062 | -0.717173577 |
| 1874880.5 | 4.957079045 | -0.249354028 | -0.232700658 |
| 1875040.5 | 4.871776277 | 0.905372362 | 0.266033004 |
| 1875200.5 | 4.485947038 | 2.004257586 | 0.748359260 |
| 1875360.5 | 3.824607917 | 2.980166209 | 1.184774231 |
| 1875520.5 | 2.931238846 | 3.775525555 | 1.549419067 |
| 1875680.5 | 1.863395636 | 4.346599773 | 1.822029004 |
| 1875840.5 | 0.687266690 | 4.665903181 | 1.989108956 |
| 1876000.5 | -0.527877533 | 4.722709904 | 2.044293165 |
| 1876160.5 | -1.714584657 | 4.521998584 | 1.988015416 |
| 1876320.5 | -2.811042721 | 4.082363842 | 1.826707627 |
| 1876480.5 | -3.763722389 | 3.433431602 | 1.571755222 |
| 1876640.5 | -4.528967401 | 2.613200244 | 1.238393981 |
| 1876800.5 | -5.073776560 | 1.665571744 | 0.844668131 |
| 1876960.5 | -5.376042182 | 0.638191212 | 0.410507305 |
| 1877120.5 | -5.424474819 | -0.419395447 | -0.043067643 |
| 1877280.5 | -5.218383522 | -1.457325027 | -0.494631153 |
| 1877440.5 | -4.767414300 | -2.426837498 | -0.922970446 |
| 1877600.5 | -4.091285322 | -3.281642875 | -1.307680023 |
| 1877760.5 | -3.219497287 | -3.979340890 | -1.629779518 |
| 1877920.5 | -2.190939963 | -4.482982312 | -1.872395075 |
| 1878080.5 | -1.053260854 | -4.762826179 | -2.021531015 |
| 1878240.5 | 0.138186679 | -4.798289803 | -2.066935045 |
| 1878400.5 | 1.322028808 | -4.580002243 | -2.003024002 |
| 1878560.5 | 2.433393371 | -4.111756423 | -1.829787219 |
| 1878720.5 | 3.407393510 | -3.412023799 | -1.553526099 |
| 1878880.5 | 4.183525587 | -2.514585473 | -1.187236834 |
| 1879040.5 | 4.710669925 | -1.467808741 | -0.750425750 |
| 1879200.5 | 4.952077506 | -0.332226770 | -0.268194509 |
| 1879360.5 | 4.889493315 | 0.823615243 | 0.230441930 |
| 1879520.5 | 4.525550533 | 1.928662104 | 0.714860764 |
| 1879680.5 | 3.883845259 | 2.915415979 | 1.155438556 |
| 1879840.5 | 3.006605807 | 3.725585102 | 1.526039304 |
| 1880000.5 | 1.950420296 | 4.314429538 | 1.805991359 |
| 1880160.5 | 0.780869074 | 4.653279739 | 1.981306313 |
| 1880320.5 | -0.432999925 | 4.730163968 | 2.045088614 |
| 1880480.5 | -1.623592624 | 4.548862139 | 1.997253231 |
| 1880640.5 | -2.728647098 | 4.126902704 | 1.843762391 |
| 1880800.5 | -3.693950031 | 3.493033320 | 1.595606627 |

HELIOCENTRIC COORDINATES OF JUPITER

REFERENCE AXES OF 1950.0

| JULIAN DATE | X | Y | Z |
|---|---|---|---|
| 1880960.5 | -4.475004888 | 2.684588477 | 1.267714983 |
| 1881120.5 | -5.037884654 | 1.745023927 | 0.877916314 |
| 1881280.5 | -5.359528471 | 0.721743678 | 0.446012334 |
| 1881440.5 | -5.427710245 | -0.335763885 | -0.007024716 |
| 1881600.5 | -5.240849207 | -1.377534442 | -0.459748544 |
| 1881760.5 | -4.807767186 | -2.354572503 | -0.890866245 |
| 1881920.5 | -4.147433791 | -3.220239043 | -1.279841477 |
| 1882080.5 | -3.288680981 | -3.931686393 | -1.607518738 |
| 1882240.5 | -2.269811509 | -4.451427931 | -1.856808288 |
| 1882400.5 | -1.137971365 | -4.749096308 | -2.013458123 |
| 1882560.5 | 0.051891365 | -4.803387921 | -2.066916508 |
| 1882720.5 | 1.238688567 | -4.604107670 | -2.011253342 |
| 1882880.5 | 2.357673706 | -4.154116518 | -1.846060473 |
| 1883040.5 | 3.343867177 | -3.470858285 | -1.577194946 |
| 1883200.5 | 4.136385690 | -2.587036024 | -1.217179417 |
| 1883360.5 | 4.683382917 | -1.549979356 | -0.785055122 |
| 1883520.5 | 4.947014341 | -0.419355451 | -0.305523154 |
| 1883680.5 | 4.907608630 | 0.736840939 | 0.192674385 |
| 1883840.5 | 4.566191583 | 1.847673832 | 0.679006817 |
| 1884000.5 | 3.944749766 | 2.845283398 | 1.123735887 |
| 1884160.5 | 3.084099978 | 3.670569120 | 1.500413613 |
| 1884320.5 | 2.039777488 | 4.277645466 | 1.787902421 |
| 1884480.5 | 0.876756182 | 4.636513984 | 1.971657843 |
| 1884640.5 | -0.336068004 | 4.733848228 | 2.044205664 |
| 1884800.5 | -1.530853887 | 4.572177276 | 2.004916757 |
| 1884960.5 | -2.644784928 | 4.167968484 | 1.859283074 |
| 1885120.5 | -3.622886961 | 3.549139455 | 1.617919034 |
| 1885280.5 | -4.419700479 | 2.752431293 | 1.295476496 |
| 1885440.5 | -5.000646628 | 1.820925243 | 0.909600297 |
| 1885600.5 | -5.341495872 | 0.801839926 | 0.479989845 |
| 1885760.5 | -5.429249391 | -0.255365593 | 0.027583433 |
| 1885920.5 | -5.261526829 | -1.300630860 | -0.426150978 |
| 1886080.5 | -4.846416372 | -2.284763490 | -0.859853812 |
| 1886240.5 | -4.202214401 | -3.160843729 | -1.252884759 |
| 1886400.5 | -3.357126294 | -3.885667826 | -1.585951422 |
| 1886560.5 | -2.348858591 | -4.421316719 | -1.841794395 |
| 1886720.5 | -1.223982959 | -4.736900920 | -2.005953492 |
| 1886880.5 | -0.036905329 | -4.810477936 | -2.067617950 |
| 1887040.5 | 1.151758336 | -4.631062918 | -2.020531819 |
| 1887200.5 | 2.277593318 | -4.200550952 | -1.863881446 |

HELIOCENTRIC COORDINATES OF JUPITER

REFERENCE AXES OF 1950.0

| JULIAN DATE | X | Y | Z |
|---|---|---|---|
| 1887360.5 | 3.275711706 | -3.535248622 | -1.603038154 |
| 1887520.5 | 4.084969047 | -2.666605474 | -1.249979462 |
| 1887680.5 | 4.652810958 | -1.640697811 | -0.823189705 |
| 1887840.5 | 4.940203597 | -0.516109311 | -0.346873405 |
| 1888000.5 | 4.925865479 | 0.639889178 | 0.150580424 |
| 1888160.5 | 4.608956516 | 1.756568580 | 0.638778751 |
| 1888320.5 | 4.009583307 | 2.765686633 | 1.087868177 |
| 1888480.5 | 3.166926398 | 3.607212400 | 1.471043885 |
| 1888640.5 | 2.135333900 | 4.233926585 | 1.766626618 |
| 1888800.5 | 0.979147001 | 4.614297271 | 1.959430132 |
| 1888960.5 | -0.232821211 | 4.733470598 | 2.041326831 |
| 1889120.5 | -1.432353042 | 4.592616116 | 2.011094412 |
| 1889280.5 | -2.555931190 | 4.207105249 | 1.873736398 |
| 1889440.5 | -3.547696244 | 3.604054034 | 1.639500290 |
| 1889600.5 | -4.361323670 | 2.819678099 | 1.322787640 |
| 1889760.5 | -4.961030429 | 1.896762777 | 0.941091637 |
| 1889920.5 | -5.321966837 | 0.882404382 | 0.514035739 |
| 1890080.5 | -5.430229459 | -0.173938206 | 0.062536159 |
| 1890240.5 | -5.282677789 | -1.222120899 | -0.391924052 |
| 1890400.5 | -4.886673402 | -2.212804495 | -0.827942358 |
| 1890560.5 | -4.259796306 | -3.098860627 | -1.224808489 |
| 1890720.5 | -3.429532865 | -3.836797703 | -1.563125497 |
| 1890880.5 | -2.432872712 | -4.388297005 | -1.825480493 |
| 1891040.5 | -1.315697824 | -4.721918966 | -1.997193816 |
| 1891200.5 | -0.131798656 | -4.814988907 | -2.067154804 |
| 1891360.5 | 1.058680983 | -4.655594792 | -2.028720176 |
| 1891520.5 | 2.191568471 | -4.244525427 | -1.880606271 |
| 1891680.5 | 3.201895689 | -3.596855659 | -1.627652472 |
| 1891840.5 | 4.028038269 | -2.742774583 | -1.281282019 |
| 1892000.5 | 4.616526037 | -1.727205050 | -0.859460063 |
| 1892160.5 | 4.927004059 | -0.607840170 | -0.385974805 |
| 1892320.5 | 4.936574943 | 0.548534433 | 0.111034139 |
| 1892480.5 | 4.642668104 | 1.671180322 | 0.601213271 |
| 1892640.5 | 4.063758801 | 2.691345731 | 1.054531035 |
| 1892800.5 | 3.237690978 | 3.548074268 | 1.443817994 |
| 1892960.5 | 2.217901331 | 4.192956053 | 1.746905354 |
| 1893120.5 | 1.068286234 | 4.593177342 | 1.948066285 |
| 1893280.5 | -0.142362274 | 4.732661541 | 2.038647656 |
| 1893440.5 | -1.345524705 | 4.611509653 | 2.016960644 |
| 1893600.5 | -2.477180794 | 4.244213217 | 1.887619327 |

HELIOCENTRIC COORDINATES OF JUPITER

REFERENCE AXES OF 1950.0

| JULIAN DATE | X | Y | Z |
|---|---|---|---|
| 1893760.5 | -3.480870341 | 3.657187253 | 1.660556749 |
| 1893920.5 | -4.309630877 | 2.886092799 | 1.349921907 |
| 1894080.5 | -4.927022913 | 1.973270669 | 0.973001803 |
| 1894240.5 | -5.307513371 | 0.965456584 | 0.549248560 |
| 1894400.5 | -5.436468763 | -0.088175939 | 0.099437730 |
| 1894560.5 | -5.309952808 | -1.137683467 | -0.355053949 |
| 1894720.5 | -4.934455402 | -2.133818169 | -0.792885529 |
| 1894880.5 | -4.326613404 | -3.029398832 | -1.193348414 |
| 1895040.5 | -3.512921761 | -3.780697142 | -1.536969752 |
| 1895200.5 | -2.529375184 | -4.348936073 | -1.806165782 |
| 1895360.5 | -1.420924716 | -4.701968501 | -1.985977373 |
| 1895520.5 | -0.240582279 | -4.816155103 | -2.064900146 |
| 1895680.5 | 0.952031697 | -4.678384806 | -2.035789967 |
| 1895840.5 | 2.092909031 | -4.288075740 | -1.896779476 |
| 1896000.5 | 3.116802994 | -3.658867627 | -1.652085383 |
| 1896160.5 | 3.961328831 | -2.819595733 | -1.312530876 |
| 1896320.5 | 4.571804651 | -1.814074464 | -0.895574870 |
| 1896480.5 | 4.906323037 | -0.699282516 | -0.424659743 |
| 1896640.5 | 4.940271879 | 0.458220401 | 0.072215270 |
| 1896800.5 | 4.669412677 | 1.587421225 | 0.564621227 |
| 1896960.5 | 4.110787497 | 2.618866500 | 1.022262174 |
| 1897120.5 | 3.301164142 | 3.490604109 | 1.417569403 |
| 1897280.5 | 2.293298806 | 4.153083865 | 1.727903163 |
| 1897440.5 | 1.150771113 | 4.572333783 | 1.937048963 |
| 1897600.5 | -0.057648405 | 4.731186407 | 2.035887212 |
| 1897760.5 | -1.263235937 | 4.628766412 | 2.022305211 |
| 1897920.5 | -2.401591738 | 4.278724215 | 1.900545664 |
| 1898080.5 | -3.415736164 | 3.706773332 | 1.600225905 |
| 1898240.5 | -4.258078605 | 2.948009356 | 1.375235006 |
| 1898400.5 | -4.891473192 | 2.044340383 | 1.002655085 |
| 1898560.5 | -5.289628768 | 1.042196991 | 0.581787376 |
| 1898720.5 | -5.437122673 | -0.009428192 | 0.133309356 |
| 1898880.5 | -5.329213705 | -1.060651019 | -0.321448269 |
| 1899040.5 | -4.971583322 | -2.062161370 | -0.761138297 |
| 1899200.5 | -4.380068427 | -2.966590864 | -1.164991913 |
| 1899360.5 | -3.580387335 | -3.729892682 | -1.513418941 |
| 1899520.5 | -2.607802258 | -4.312831647 | -1.788657118 |
| 1899680.5 | -1.506605382 | -4.682655459 | -1.975503726 |
| 1899840.5 | -0.329263358 | -4.814970502 | -2.062143823 |
| 1900000.5 | 0.864984992 | -4.695771444 | -2.041058428 |

## HELIOCENTRIC COORDINATES OF JUPITER

## REFERENCE AXES OF 1950.0

| JULIAN DATE | X | Y | Z |
|---|---|---|---|
| 1900160.5 | 2.012292888 | -4.323469298 | -1.909951303 |
| 1900320.5 | 3.047299811 | -3.710634254 | -1.672576601 |
| 1900480.5 | 3.907190736 | -2.885044030 | -1.339292239 |
| 1900640.5 | 4.536523649 | -1.889558308 | -0.927127821 |
| 1900800.5 | 4.892327834 | -0.780397272 | -0.459173034 |
| 1900960.5 | 4.948694946 | 0.376358869 | 0.036811756 |
| 1901120.5 | 4.699962366 | 1.509835673 | 0.530493430 |
| 1901280.5 | 4.161745279 | 2.550347778 | 0.991512789 |
| 1901440.5 | 3.369511275 | 3.435353002 | 1.392080750 |
| 1901600.5 | 2.374958902 | 4.114409086 | 1.709201182 |
| 1901760.5 | 1.240936906 | 4.552441034 | 1.926204205 |
| 1901920.5 | 0.035852300 | 4.731076442 | 2.033461632 |
| 1902080.5 | -1.171571659 | 4.648240614 | 2.028345304 |
| 1902240.5 | -2.316613769 | 4.316480486 | 1.914615458 |
| 1902400.5 | -3.341714680 | 3.760566306 | 1.701469273 |
| 1902560.5 | -4.198518299 | 3.014848221 | 1.402455776 |
| 1902720.5 | -4.848997640 | 2.120699124 | 1.034404701 |
| 1902880.5 | -5.265925726 | 1.124221446 | 0.616452305 |
| 1903040.5 | -5.432938124 | 0.074270813 | 0.169193119 |
| 1903200.5 | -5.344382092 | -0.979235549 | -0.286051245 |
| 1903360.5 | -5.005082784 | -1.986807038 | -0.727877946 |
| 1903520.5 | -4.430091890 | -2.900768589 | -1.135406488 |
| 1903680.5 | -3.644422817 | -3.676658071 | -1.488886294 |
| 1903840.5 | -2.682718533 | -4.274728864 | -1.770351200 |
| 1904000.5 | -1.588742457 | -4.661627636 | -1.964353761 |
| 1904160.5 | -0.414531275 | -4.812270756 | -2.058793508 |
| 1904320.5 | 0.780990666 | -4.711870227 | -2.045823091 |
| 1904480.5 | 1.934145630 | -4.357958256 | -1.922772762 |
| 1904640.5 | 2.979550709 | -3.762135398 | -1.692979075 |
| 1904800.5 | 3.854117941 | -2.951146228 | -1.366348467 |
| 1904960.5 | 4.501822058 | -1.966818106 | -0.959451372 |
| 1905120.5 | 4.878761268 | -0.864448044 | -0.494956411 |
| 1905280.5 | 4.957762418 | 0.290560048 | -0.000300749 |
| 1905440.5 | 4.731653265 | 1.427630835 | 0.494349490 |
| 1905600.5 | 4.214454961 | 2.476920058 | 0.958607168 |
| 1905760.5 | 3.440152222 | 3.375265828 | 1.364454252 |
| 1905920.5 | 2.459250621 | 4.071223624 | 1.688493821 |
| 1906080.5 | 1.333814823 | 4.528469893 | 1.913538960 |
| 1906240.5 | 0.131917472 | 4.727293926 | 2.029394494 |
| 1906400.5 | -1.077632823 | 4.664333965 | 2.032876699 |

HELICCENTRIC COORDINATES OF JUPITER

REFERENCE AXES OF 1950.0

| JULIAN DATE | X | Y | Z |
|---|---|---|---|
| 1906560.5 | -2.229686921 | 4.351004027 | 1.927247882 |
| 1906720.5 | -3.266011795 | 3.811153986 | 1.721292492 |
| 1906880.5 | -4.137426519 | 3.078445627 | 1.428242799 |
| 1907040.5 | -4.805007390 | 2.193787303 | 1.064706258 |
| 1907200.5 | -5.240617964 | 1.203016693 | 0.649682509 |
| 1907360.5 | -5.427011217 | 0.154894130 | 0.203703117 |
| 1907520.5 | -5.357702260 | -0.900619040 | -0.251911933 |
| 1907680.5 | -5.036745913 | -1.913874488 | -0.695710902 |
| 1907840.5 | -4.478487573 | -2.836945018 | -1.106722216 |
| 1908000.5 | -3.707293062 | -3.625033376 | -1.465067361 |
| 1908160.5 | -2.757205394 | -4.237978036 | -1.752617400 |
| 1908320.5 | -1.671424609 | -4.641925581 | -1.953725370 |
| 1908480.5 | -0.501460148 | -4.811191424 | -2.056047549 |
| 1908640.5 | 0.694233464 | -4.730260287 | -2.051437681 |
| 1908800.5 | 1.852340453 | -4.395784380 | -1.936858506 |
| 1908960.5 | 2.907650984 | -3.818322019 | -1.715204141 |
| 1909120.5 | 3.796958595 | -3.023441839 | -1.395873260 |
| 1909280.5 | 4.463726906 | -2.051744924 | -0.994896465 |
| 1909440.5 | 4.863087283 | -0.957392364 | -0.534429354 |
| 1909600.5 | 4.966457699 | 0.195079523 | -0.041499763 |
| 1909760.5 | 4.764928861 | 1.335522587 | 0.453954899 |
| 1909920.5 | 4.270661610 | 2.393966814 | 0.921541929 |
| 1910080.5 | 3.515907562 | 3.306559920 | 1.332991949 |
| 1910240.5 | 2.549796350 | 4.020699844 | 1.664446878 |
| 1910400.5 | 1.433524895 | 4.498615922 | 1.898121201 |
| 1910560.5 | 0.234850843 | 4.719066316 | 2.023170838 |
| 1910720.5 | -0.977232171 | 4.677256038 | 2.035792320 |
| 1910880.5 | -2.137005810 | 4.383389089 | 1.938714122 |
| 1911040.5 | -3.185421827 | 3.860388263 | 1.740301921 |
| 1911200.5 | -4.072365471 | 3.141274908 | 1.453489091 |
| 1911360.5 | -4.757958016 | 2.266569830 | 1.094692028 |
| 1911520.5 | -5.213147817 | 1.281924315 | 0.682806451 |
| 1911680.5 | -5.419839548 | 0.236059560 | 0.238324668 |
| 1911840.5 | -5.370767161 | -0.821000117 | -0.217427319 |
| 1912000.5 | -5.069254092 | -1.839471292 | -0.662959765 |
| 1912160.5 | -4.528939326 | -2.771231481 | -1.077236686 |
| 1912320.5 | -3.773486111 | -3.571221350 | -1.440286137 |
| 1912480.5 | -2.836231857 | -4.198926912 | -1.733846368 |
| 1912640.5 | -1.759682252 | -4.620015981 | -1.942082234 |
| 1912800.5 | -0.594701054 | -4.808160944 | -2.052389405 |

HELIOCENTRIC COORDINATES OF JUPITER

REFERENCE AXES OF 1950.0

| JULIAN DATE | X | Y | Z |
|---|---|---|---|
| 1912960.5 | 0.600793953 | -4.747013468 | -2.056276374 |
| 1913120.5 | 1.763834079 | -4.432202352 | -1.950274055 |
| 1913280.5 | 2.829274227 | -3.873107618 | -1.736771172 |
| 1913440.5 | 3.733602733 | -3.094043179 | -1.424618962 |
| 1913600.5 | 4.419550453 | -2.134400511 | -1.029309306 |
| 1913760.5 | 4.841088214 | -1.047328063 | -0.572533058 |
| 1913920.5 | 4.968122795 | 0.103299697 | -0.080995359 |
| 1914080.5 | 4.790037858 | 1.247529015 | 0.415492965 |
| 1914240.5 | 4.317297532 | 2.315073786 | 0.886445962 |
| 1914400.5 | 3.580677000 | 3.241303165 | 1.303296497 |
| 1914560.5 | 2.628211891 | 3.972501749 | 1.641735004 |
| 1914720.5 | 1.520467237 | 4.469621812 | 1.883442316 |
| 1914880.5 | 0.325022118 | 4.710148916 | 2.017031201 |
| 1915040.5 | -0.888932797 | 4.688145878 | 2.038201987 |
| 1915200.5 | -2.055221071 | 4.412868444 | 1.949264536 |
| 1915360.5 | -3.114192874 | 3.906490909 | 1.758249041 |
| 1915520.5 | -4.015078602 | 3.201456390 | 1.477825661 |
| 1915680.5 | -4.717337254 | 2.337836321 | 1.124202604 |
| 1915840.5 | -5.191253109 | 1.360924368 | 0.716106329 |
| 1916000.5 | -5.418046773 | 0.319153503 | 0.273888704 |
| 1916160.5 | -5.389722255 | -0.737673845 | -0.181237778 |
| 1916320.5 | -5.108805230 | -1.759914084 | -0.627863671 |
| 1916480.5 | -4.588059099 | -2.699471705 | -1.044987022 |
| 1916640.5 | -3.850201345 | -3.511160851 | -1.412605942 |
| 1916800.5 | -2.927582993 | -4.154143376 | -1.712345001 |
| 1916960.5 | -1.861737124 | -4.593525562 | -1.928153153 |
| 1917120.5 | -0.702648501 | -4.802155369 | -2.047094925 |
| 1917280.5 | 0.492448726 | -4.762597575 | -2.060230004 |
| 1917440.5 | 1.660960461 | -4.469170222 | -1.963536510 |
| 1917600.5 | 2.737700029 | -3.929804761 | -1.758780615 |
| 1917760.5 | 3.658623893 | -3.167363270 | -1.454177141 |
| 1917920.5 | 4.365429136 | -2.219951911 | -1.064640296 |
| 1918080.5 | 4.810623967 | -1.139775181 | -0.611418921 |
| 1918240.5 | 4.962382931 | 0.009757618 | -0.120974809 |
| 1918400.5 | 4.808305574 | 1.158619586 | 0.376894893 |
| 1918560.5 | 4.357243955 | 2.235951391 | 0.851499064 |
| 1918720.5 | 3.638705745 | 3.176171739 | 1.273895028 |
| 1918880.5 | 2.699887173 | 3.924403381 | 1.619296443 |
| 1919040.5 | 1.600933933 | 4.440383599 | 1.868874254 |
| 1919200.5 | 0.409354625 | 4.700444542 | 2.010749096 |

## HELIOCENTRIC COORDINATES OF JUPITER

### REFERENCE AXES OF 1950.0

| JULIAN DATE | X | Y | Z |
|---|---|---|---|
| 1919360.5 | -0.805474629 | 4.697619617 | 2.040175649 |
| 1919520.5 | -1.977015451 | 4.440282525 | 1.959076139 |
| 1919680.5 | -3.045129874 | 3.949874980 | 1.775153165 |
| 1919840.5 | -3.958462042 | 3.258249032 | 1.500807038 |
| 1920000.5 | -4.675807578 | 2.405015913 | 1.152030382 |
| 1920160.5 | -5.166730615 | 1.435131645 | 0.747385569 |
| 1920320.5 | -5.411693171 | 0.396809200 | 0.307110260 |
| 1920480.5 | -5.401916997 | -0.660253438 | -0.147647977 |
| 1920640.5 | -5.139133808 | -1.686400928 | -0.595494817 |
| 1920800.5 | -4.635312586 | -2.633406084 | -1.015391798 |
| 1920960.5 | -3.912389207 | -3.455824571 | -1.387245630 |
| 1921120.5 | -3.001964206 | -4.112423791 | -1.692530203 |
| 1921280.5 | -1.944877525 | -4.567770168 | -1.914979822 |
| 1921440.5 | -0.790514624 | -4.794020373 | -2.041376894 |
| 1921600.5 | 0.404348317 | -4.772899249 | -2.062431535 |
| 1921760.5 | 1.577377476 | -4.497754638 | -1.973711351 |
| 1921920.5 | 2.663389920 | -3.975458348 | -1.776527162 |
| 1922080.5 | 3.598039242 | -3.227790519 | -1.478621243 |
| 1922240.5 | 4.322386255 | -2.291843476 | -1.094456251 |
| 1922400.5 | 4.787980083 | -1.218976581 | -0.644893881 |
| 1922560.5 | 4.961769754 | -0.072014626 | -0.156113893 |
| 1922720.5 | 4.829961781 | 1.079271221 | 0.342231826 |
| 1922880.5 | 4.399972849 | 2.163912798 | 0.819442049 |
| 1923040.5 | 3.699963871 | 3.115858575 | 1.246406510 |
| 1923200.5 | 2.775991632 | 3.879444248 | 1.598029823 |
| 1923360.5 | 1.687360506 | 4.413383165 | 1.855064439 |
| 1923520.5 | 0.501087428 | 4.692847879 | 2.005145231 |
| 1923680.5 | -0.713584583 | 4.709001407 | 2.043018696 |
| 1923840.5 | -1.889919906 | 4.471129081 | 1.970117077 |
| 1924000.5 | -2.967306829 | 3.997637039 | 1.793702994 |
| 1924160.5 | -3.893705635 | 3.320239937 | 1.525807977 |
| 1924320.5 | -4.627085744 | 2.477931758 | 1.182138243 |
| 1924480.5 | -5.136104310 | 1.515254914 | 0.781055200 |
| 1924640.5 | -5.400287050 | 0.480203228 | 0.342678350 |
| 1924800.5 | -5.409930351 | -0.577566540 | -0.111886163 |
| 1924960.5 | -5.165881305 | -1.608282032 | -0.561216414 |
| 1925120.5 | -4.679286020 | -2.563460601 | -0.984185281 |
| 1925280.5 | -3.971333570 | -3.397281835 | -1.360557041 |
| 1925440.5 | -3.072963641 | -4.068030931 | -1.671615354 |
| 1925600.5 | -2.024449262 | -4.539696268 | -1.900860682 |

HELIOCENTRIC COORDINATES OF JUPITER

REFERENCE AXES OF 1950.0

| JULIAN DATE | X | Y | Z |
|---|---|---|---|
| 1925760.5 | -0.874712010 | -4.783766458 | -2.034800243 |
| 1925920.5 | 0.319818901 | -4.781210321 | -2.063828096 |
| 1926080.5 | 1.497016832 | -4.524532482 | -1.983154642 |
| 1926240.5 | 2.591730855 | -4.019679233 | -1.793693496 |
| 1926400.5 | 3.539423183 | -3.287441174 | -1.502756308 |
| 1926560.5 | 4.280685748 | -2.363901340 | -1.124359356 |
| 1926720.5 | 4.766278120 | -1.299472069 | -0.678936478 |
| 1926880.5 | 4.962031570 | -0.156212876 | -0.192309532 |
| 1927040.5 | 4.852759615 | 0.996555599 | 0.306098849 |
| 1927200.5 | 4.444335166 | 2.087896491 | 0.785639993 |
| 1927360.5 | 3.763402293 | 3.051331970 | 1.217058625 |
| 1927520.5 | 2.854713660 | 3.830356686 | 1.574930620 |
| 1927680.5 | 1.776627229 | 4.382540138 | 1.839540466 |
| 1927840.5 | 0.595645147 | 4.681756838 | 1.997977891 |
| 1928000.5 | -0.619073523 | 4.718545303 | 2.044430611 |
| 1928160.5 | -1.800504265 | 4.498958305 | 1.979810815 |
| 1928320.5 | -2.887462330 | 4.042437489 | 1.810934778 |
| 1928480.5 | -3.827145780 | 3.379235988 | 1.549478803 |
| 1928640.5 | -4.576657245 | 2.547792166 | 1.210888410 |
| 1928800.5 | -5.103748862 | 1.592301432 | 0.813353099 |
| 1928960.5 | -5.387049816 | 0.560592148 | 0.376898811 |
| 1929120.5 | -5.415998088 | -0.497692072 | -0.077395319 |
| 1929280.5 | -5.190637000 | -1.532666284 | -0.528078816 |
| 1929440.5 | -4.721370822 | -2.495630708 | -0.953949410 |
| 1929600.5 | -4.028709501 | -3.340457466 | -1.334655865 |
| 1929760.5 | -3.142971941 | -4.025040335 | -1.651329104 |
| 1929920.5 | -2.103861063 | -4.512890817 | -1.887279509 |
| 1930080.5 | -0.959774118 | -4.774926059 | -2.028783159 |
| 1930240.5 | 0.233329282 | -4.791431955 | -2.065954156 |
| 1930400.5 | 1.413703920 | -4.554096978 | -1.993663485 |
| 1930560.5 | 2.516429047 | -4.067902797 | -1.812417438 |
| 1930720.5 | 3.476956632 | -3.352537352 | -1.529054518 |
| 1930880.5 | 4.235528556 | -2.442898730 | -1.157073419 |
| 1931040.5 | 4.742134733 | -1.388243479 | -0.716391839 |
| 1931200.5 | 4.961395031 | -0.249663121 | -0.232384575 |
| 1931360.5 | 4.876536871 | 0.904128195 | 0.265827314 |
| 1931520.5 | 4.491635360 | 2.002299281 | 0.747687159 |
| 1931680.5 | 3.831555053 | 2.977920228 | 1.183791728 |
| 1931840.5 | 2.939526712 | 3.773525201 | 1.548344011 |
| 1932000.5 | 1.872829776 | 4.345357803 | 1.821087329 |

HELIOCENTRIC COORDINATES OF JUPITER

REFERENCE AXES OF 1950.0

| JULIAN DATE | X | Y | Z |
|---|---|---|---|
| 1932160.5 | 0.697421869 | 4.665795113 | 1.988482898 |
| 1932320.5 | -0.517559502 | 4.723898977 | 2.044085424 |
| 1932480.5 | -1.704677204 | 4.524415872 | 1.988236088 |
| 1932640.5 | -2.802028093 | 4.085739476 | 1.827283165 |
| 1932800.5 | -3.755917910 | 3.437361439 | 1.572553397 |
| 1932960.5 | -4.522493589 | 2.617228209 | 1.239256690 |
| 1933120.5 | -5.068565801 | 1.669269584 | 0.845444249 |
| 1933280.5 | -5.371879817 | 0.641222013 | 0.411079524 |
| 1933440.5 | -5.421061043 | -0.417238371 | -0.042765865 |
| 1933600.5 | -5.215403503 | -1.456107354 | -0.494610371 |
| 1933760.5 | -4.764604350 | -2.426500670 | -0.923191792 |
| 1933920.5 | -4.088484146 | -3.282044464 | -1.308072152 |
| 1934080.5 | -3.216673093 | -3.980311454 | -1.630263092 |
| 1934240.5 | -2.188186842 | -4.484388626 | -1.872910587 |
| 1934400.5 | -1.050761811 | -4.764627851 | -2.022064384 |
| 1934560.5 | 0.140226673 | -4.800574322 | -2.067534394 |
| 1934720.5 | 1.323468280 | -4.582983880 | -2.003800986 |
| 1934880.5 | 2.434237895 | -4.115729367 | -1.830898142 |
| 1935040.5 | 3.407850051 | -3.417271548 | -1.555132932 |
| 1935200.5 | 4.184000403 | -2.521266813 | -1.189456215 |
| 1935360.5 | 4.711694299 | -1.475858366 | -0.753282093 |
| 1935520.5 | 4.954170765 | -0.341312397 | -0.271595675 |
| 1935680.5 | 4.893004277 | 0.814054426 | 0.226694512 |
| 1935840.5 | 4.530536238 | 1.919301076 | 0.711028203 |
| 1936000.5 | 3.890034254 | 2.906889296 | 1.151780724 |
| 1936160.5 | 3.013459028 | 3.718344594 | 1.522753200 |
| 1936320.5 | 1.957264868 | 4.308661642 | 1.803172475 |
| 1936480.5 | 0.787056494 | 4.640905834 | 1.978944344 |
| 1936640.5 | -0.427961928 | 4.726916219 | 2.043095806 |
| 1936800.5 | -1.619969727 | 4.546402082 | 1.995512507 |
| 1936960.5 | -2.726482381 | 4.124938416 | 1.842176826 |
| 1937120.5 | -3.693133778 | 3.491398034 | 1.594133920 |
| 1937280.5 | -4.475382741 | 2.683264004 | 1.266378768 |
| 1937440.5 | -5.039366193 | 1.744111609 | 0.876794563 |
| 1937600.5 | -5.362163725 | 0.721396521 | 0.445208185 |
| 1937760.5 | -5.431716839 | -0.335428886 | -0.007421711 |
| 1937920.5 | -5.246587172 | -1.376523265 | -0.459701952 |
| 1938080.5 | -4.815668158 | -2.353081620 | -0.890424857 |
| 1938240.5 | -4.157897639 | -3.218688498 | -1.279157228 |
| 1938400.5 | -3.301957319 | -3.930708164 | -1.606845598 |

HELIOCENTRIC COORDINATES OF JUPITER

REFERENCE AXES OF 1950.0

| JULIAN DATE | X | Y | Z |
|---|---|---|---|
| 1938560.5 | -2.285887036 | -4.451804919 | -1.856479581 |
| 1938720.5 | -1.156486389 | -4.751654308 | -2.013842666 |
| 1938880.5 | 0.031673560 | -4.808851883 | -2.068357767 |
| 1939040.5 | 1.217839644 | -4.612944108 | -2.014000429 |
| 1939200.5 | 2.337475864 | -4.166394418 | -1.850204550 |
| 1939360.5 | 3.325614773 | -3.486167310 | -1.582627862 |
| 1939520.5 | 4.121145399 | -2.604497770 | -1.223590728 |
| 1939680.5 | 4.671768254 | -1.568370067 | -0.791976839 |
| 1939840.5 | 4.939041487 | -0.437326987 | -0.312418380 |
| 1940000.5 | 4.902688024 | 0.720491960 | 0.186298863 |
| 1940160.5 | 4.563271140 | 1.833764334 | 0.673495766 |
| 1940320.5 | 3.942573473 | 2.834103375 | 1.119221010 |
| 1940480.5 | 3.081503754 | 3.661883392 | 1.496808410 |
| 1940640.5 | 2.035933531 | 4.270828983 | 1.784954884 |
| 1940800.5 | 0.871301435 | 4.630768833 | 1.969040335 |
| 1940960.5 | -0.343038909 | 4.728432057 | 2.041612372 |
| 1941120.5 | -1.538907931 | 4.566581727 | 2.002141943 |
| 1941280.5 | -2.653333338 | 4.162012223 | 1.856263865 |
| 1941440.5 | -3.631377454 | 3.542969320 | 1.614738808 |
| 1941600.5 | -4.427863402 | 2.746447074 | 1.292334564 |
| 1941760.5 | -5.008239778 | 1.815655373 | 0.906759005 |
| 1941920.5 | -5.348857009 | 0.797798301 | 0.477711853 |
| 1942080.5 | -5.436898376 | -0.257813987 | 0.026071148 |
| 1942240.5 | -5.270148676 | -1.301372914 | -0.426803786 |
| 1942400.5 | -4.856715720 | -2.283992809 | -0.859690454 |
| 1942560.5 | -4.214754352 | -3.159056532 | -1.252088691 |
| 1942720.5 | -3.372181765 | -3.883595209 | -1.584821046 |
| 1942880.5 | -2.366315064 | -4.419796940 | -1.840690585 |
| 1943040.5 | -1.243306306 | -4.736708390 | -2.005226717 |
| 1943200.5 | -0.057199202 | -4.812140825 | -2.067528454 |
| 1943360.5 | 1.131604549 | -4.634711761 | -2.021180862 |
| 1943520.5 | 2.258683025 | -4.205842588 | -1.865172828 |
| 1943680.5 | 3.258888142 | -3.541400992 | -1.604686212 |
| 1943840.5 | 4.070595055 | -2.672556257 | -1.251570630 |
| 1944000.5 | 4.640653182 | -1.645362679 | -0.824288985 |
| 1944160.5 | 4.929474066 | -0.518672621 | -0.347148138 |
| 1944320.5 | 4.915427160 | 0.639745199 | 0.151259599 |
| 1944480.5 | 4.597638675 | 1.758576876 | 0.640295400 |
| 1944640.5 | 3.996505156 | 2.769079597 | 1.089890514 |
| 1944800.5 | 3.151723201 | 3.610943712 | 1.473116045 |

## HELIOCENTRIC COORDINATES OF JUPITER

### REFERENCE AXES OF 1950.0

| JULIAN DATE | X | Y | Z |
|---|---|---|---|
| 1944960.5 | 2.118222131 | 4.236952341 | 1.768288613 |
| 1945120.5 | 0.960827936 | 4.615827958 | 1.960328537 |
| 1945280.5 | -0.251368638 | 4.733128789 | 2.041287053 |
| 1945440.5 | -1.450117013 | 4.590475194 | 2.010142257 |
| 1945600.5 | -2.572082998 | 4.203622173 | 1.872073053 |
| 1945760.5 | -3.561734891 | 3.599932118 | 1.637445270 |
| 1945920.5 | -4.373133552 | 2.815701970 | 1.320707585 |
| 1946080.5 | -4.970858093 | 1.893642219 | 0.939330957 |
| 1946240.5 | -5.330335594 | 0.880652803 | 0.512860703 |
| 1946400.5 | -5.437815861 | -0.174076382 | 0.062099206 |
| 1946560.5 | -5.290174206 | -1.220689244 | -0.391597177 |
| 1946720.5 | -4.894658609 | -2.210102555 | -0.826943184 |
| 1946880.5 | -4.268632694 | -3.095368778 | -1.223317051 |
| 1947040.5 | -3.439305212 | -3.833071052 | -1.561367080 |
| 1947200.5 | -2.443379124 | -4.384844381 | -1.823674491 |
| 1947360.5 | -1.326497857 | -4.719086541 | -1.995502948 |
| 1947520.5 | -0.142315755 | -4.812870581 | -2.065644216 |
| 1947680.5 | 1.049015995 | -4.653991593 | -2.027336275 |
| 1947840.5 | 2.183158453 | -4.242971604 | -1.879183433 |
| 1948000.5 | 3.194839638 | -3.594717717 | -1.625951185 |
| 1948160.5 | 4.022056948 | -2.739410919 | -1.279054163 |
| 1948320.5 | 4.610984913 | -1.722151505 | -0.856527269 |
| 1948480.5 | 4.921043834 | -0.600969128 | -0.382297760 |
| 1948640.5 | 4.929323997 | 0.556935693 | 0.115319266 |
| 1948800.5 | 4.633476669 | 1.680445582 | 0.605806959 |
| 1948960.5 | 4.052379855 | 2.700574364 | 1.059030242 |
| 1949120.5 | 3.224348819 | 3.556337515 | 1.447804793 |
| 1949280.5 | 2.203330164 | 4.199495959 | 1.750034938 |
| 1949440.5 | 1.053168215 | 4.597537268 | 1.950127167 |
| 1949600.5 | -0.156998413 | 4.734720467 | 2.039580440 |
| 1949760.5 | -1.358877596 | 4.611431957 | 2.016838726 |
| 1949920.5 | -2.488676494 | 4.242349867 | 1.886606367 |
| 1950080.5 | -3.490179346 | 3.653967007 | 1.658858093 |
| 1950240.5 | -4.316624853 | 2.881933172 | 1.347743664 |
| 1950400.5 | -4.931703869 | 1.968525162 | 0.970525270 |
| 1950560.5 | -5.309942302 | 0.960400484 | 0.546621419 |
| 1950720.5 | -5.436711517 | -0.093329131 | 0.096778957 |
| 1950880.5 | -5.308052768 | -1.142747667 | -0.357640713 |
| 1951040.5 | -4.930436099 | -2.138596065 | -0.795295925 |
| 1951200.5 | -4.320508639 | -3.033651476 | -1.195464131 |

## HELIOCENTRIC COORDINATES OF JUPITER

## REFERENCE AXES OF 1950.0

| JULIAN DATE | X | Y | Z |
|---|---|---|---|
| 1951360.5 | -3.504825308 | -3.784131921 | -1.538652743 |
| 1951520.5 | -2.519496100 | -4.351219709 | -1.807263283 |
| 1951680.5 | -1.409629410 | -4.702767421 | -1.986339167 |
| 1951840.5 | -0.228404933 | -4.815199363 | -2.064406589 |
| 1952000.5 | 0.964421379 | -4.675542709 | -2.034385116 |
| 1952160.5 | 2.104781170 | -4.283417503 | -1.894499858 |
| 1952320.5 | 3.127475158 | -3.652694357 | -1.649074002 |
| 1952480.5 | 3.970283612 | -2.812416704 | -1.309028697 |
| 1952640.5 | 4.578786223 | -1.806529599 | -0.891887767 |
| 1952800.5 | 4.911381460 | -0.692024886 | -0.421106614 |
| 1952960.5 | 4.943735877 | 0.464650903 | 0.075360112 |
| 1953120.5 | 4.671794255 | 1.592694846 | 0.567174618 |
| 1953280.5 | 4.112646438 | 2.622900900 | 1.024153591 |
| 1953440.5 | 3.302975178 | 3.493532693 | 1.418832089 |
| 1953600.5 | 2.295361383 | 4.155170752 | 1.728640440 |
| 1953760.5 | 1.153183089 | 4.573867755 | 1.937389114 |
| 1953920.5 | -0.054955917 | 4.732388327 | 2.035941929 |
| 1954080.5 | -1.260423827 | 4.629734002 | 2.022142618 |
| 1954240.5 | -2.398826518 | 4.279422109 | 1.900182214 |
| 1954400.5 | -3.413116732 | 3.707062190 | 1.679635339 |
| 1954560.5 | -4.255591741 | 2.947700385 | 1.374368764 |
| 1954720.5 | -4.888983312 | 2.043257346 | 1.001467329 |
| 1954880.5 | -5.286899422 | 1.040233253 | 0.580258399 |
| 1955040.5 | -5.433862241 | -0.012269641 | 0.131462663 |
| 1955200.5 | -5.325135161 | -1.064240267 | -0.323538188 |
| 1955360.5 | -4.966467184 | -2.066249571 | -0.763348841 |
| 1955520.5 | -4.373818096 | -2.970842578 | -1.167165748 |
| 1955680.5 | -3.573066844 | -3.733938509 | -1.515386417 |
| 1955840.5 | -2.599647851 | -4.316333958 | -1.790264808 |
| 1956000.5 | -1.498004910 | -4.685376738 | -1.976644699 |
| 1956160.5 | -0.320701784 | -4.816830727 | -2.062783808 |
| 1956320.5 | 0.873008987 | -4.696879054 | -2.041251016 |
| 1956480.5 | 2.019365330 | -4.324111392 | -1.909835589 |
| 1956640.5 | 3.053181810 | -3.711218188 | -1.672353216 |
| 1956800.5 | 3.911872246 | -2.885996302 | -1.339179581 |
| 1956960.5 | 4.540215552 | -1.891206199 | -0.927308886 |
| 1957120.5 | 4.895383255 | -0.782874666 | -0.459750661 |
| 1957280.5 | 4.951478073 | 0.373142200 | 0.035833680 |
| 1957440.5 | 4.702711329 | 1.506145895 | 0.529194859 |
| 1957600.5 | 4.164484125 | 2.546516926 | 0.990014268 |

## HELIOCENTRIC COORDINATES OF JUPITER

### REFERENCE AXES OF 1950.0

| JULIAN DATE | X | Y | Z |
|---|---|---|---|
| 1957760.5 | 3.372045552 | 3.431649218 | 1.390489054 |
| 1957920.5 | 2.376952620 | 4.110936382 | 1.707564890 |
| 1958080.5 | 1.242036688 | 4.549104080 | 1.924495297 |
| 1958240.5 | 0.035810923 | 4.727619435 | 2.031588339 |
| 1958400.5 | -1.172813385 | 4.644342699 | 2.026189953 |
| 1958560.5 | -2.318908531 | 4.311872459 | 1.912083008 |
| 1958720.5 | -3.344754498 | 3.755128450 | 1.698529433 |
| 1958880.5 | -4.201929532 | 3.008659037 | 1.399165156 |
| 1959040.5 | -4.852457708 | 2.114025381 | 1.030903962 |
| 1959200.5 | -5.269267694 | 1.117458222 | 0.612941264 |
| 1959360.5 | -5.436215082 | 0.067848976 | 0.165890875 |
| 1959520.5 | -5.347877497 | -0.984954071 | -0.288952601 |
| 1959680.5 | -5.009266310 | -1.991625856 | -0.730257249 |
| 1959840.5 | -4.435529563 | -2.904727176 | -1.137247080 |
| 1960000.5 | -3.651656332 | -3.680061436 | -1.490292714 |
| 1960160.5 | -2.692133245 | -4.278127062 | -1.771544414 |
| 1960320.5 | -1.600447313 | -4.665741777 | -1.965642882 |
| 1960480.5 | -0.428276304 | -4.817870342 | -2.060525401 |
| 1960640.5 | 0.765834266 | -4.719617096 | -2.048315721 |
| 1960800.5 | 1.918523894 | -4.368244294 | -1.926243415 |
| 1960960.5 | 2.964578893 | -3.774951721 | -1.697483447 |
| 1961120.5 | 3.840863567 | -2.966025428 | -1.371748901 |
| 1961280.5 | 4.491063240 | -1.982882410 | -0.965429472 |
| 1961440.5 | 4.870785714 | -0.880571108 | -0.501075104 |
| 1961600.5 | 4.952274970 | 0.275507322 | -0.006104946 |
| 1961760.5 | 4.727834421 | 1.414516001 | 0.489220464 |
| 1961920.5 | 4.211163945 | 2.466147492 | 0.954330106 |
| 1962080.5 | 3.436215700 | 3.360702131 | 1.360984438 |
| 1962240.5 | 2.453747598 | 4.064270210 | 1.685589999 |
| 1962400.5 | 1.326274324 | 4.522252071 | 1.910839729 |
| 1962560.5 | 0.122382391 | 4.720901716 | 2.026519575 |
| 1962720.5 | -1.088675130 | 4.657044129 | 2.029522898 |
| 1962880.5 | -2.241469935 | 4.342431103 | 1.923256561 |
| 1963040.5 | -3.277697235 | 3.801307020 | 1.716677378 |
| 1963200.5 | -4.148302460 | 3.067694048 | 1.423178452 |
| 1963360.5 | -4.814639157 | 2.182757086 | 1.059485433 |
| 1963520.5 | -5.248930483 | 1.192445875 | 0.644654428 |
| 1963680.5 | -5.434296244 | 0.145477420 | 0.199205281 |
| 1963840.5 | -5.364555089 | -0.908370166 | -0.255616908 |
| 1964000.5 | -5.043944327 | -1.919734783 | -0.698483876 |

HELIOCENTRIC COORDINATES OF JUPITER

REFERENCE AXES OF 1950.0

| JULIAN DATE | X | Y | Z |
|---|---|---|---|
| 1964160.5 | -4.486832510 | -2.841025690 | -1.108574111 |
| 1964320.5 | -3.717437159 | -3.627769491 | -1.466158622 |
| 1964480.5 | -2.769501200 | -4.240049698 | -1.753227975 |
| 1964640.5 | -1.685825201 | -4.644118603 | -1.954197032 |
| 1964800.5 | -0.517498868 | -4.814217311 | -2.056706758 |
| 1964960.5 | 0.677368713 | -4.734571336 | -2.052514105 |
| 1965120.5 | 1.835639455 | -4.401430335 | -1.938419013 |
| 1965280.5 | 2.892044604 | -3.824894824 | -1.717122665 |
| 1965440.5 | 3.783065109 | -3.030141846 | -1.397852057 |
| 1965600.5 | 4.451653816 | -2.057572792 | -0.996542319 |
| 1965760.5 | 4.852361211 | -0.961428001 | -0.535369925 |
| 1965920.5 | 4.956124158 | 0.193392664 | -0.041505970 |
| 1966080.5 | 4.753807186 | 1.336186971 | 0.454883250 |
| 1966240.5 | 4.257679362 | 2.396404437 | 0.923158920 |
| 1966400.5 | 3.500403763 | 3.309757116 | 1.334862005 |
| 1966560.5 | 2.531696600 | 4.023468182 | 1.666053458 |
| 1966720.5 | 1.413345833 | 4.499880735 | 1.898990824 |
| 1966880.5 | 0.213554230 | 4.718098610 | 2.022975855 |
| 1967040.5 | -0.998473392 | 4.673795922 | 2.034408747 |
| 1967200.5 | -2.157056242 | 4.377650143 | 1.936228324 |
| 1967360.5 | -3.203388085 | 3.852965000 | 1.736974371 |
| 1967520.5 | -4.087722957 | 3.132992036 | 1.449691113 |
| 1967680.5 | -4.770591623 | 2.258313713 | 1.090832791 |
| 1967840.5 | -5.223316235 | 1.274489603 | 0.679264683 |
| 1968000.5 | -5.428082808 | 0.230033197 | 0.235395144 |
| 1968160.5 | -5.377780626 | -0.825305728 | -0.219566854 |
| 1968320.5 | -5.075751793 | -1.842032617 | -0.664259228 |
| 1968480.5 | -4.535527391 | -2.772278714 | -1.077762636 |
| 1968640.5 | -3.780564170 | -3.571161828 | -1.440192101 |
| 1968800.5 | -2.843936995 | -4.198239427 | -1.733329435 |
| 1968960.5 | -1.767885860 | -4.619132479 | -1.941332731 |
| 1969120.5 | -0.603062900 | -4.807355990 | -2.051542248 |
| 1969280.5 | 0.592718840 | -4.746324463 | -2.055374037 |
| 1969440.5 | 1.756451734 | -4.431402889 | -1.949251116 |
| 1969600.5 | 2.822800455 | -3.871750109 | -1.735468024 |
| 1969760.5 | 3.727944025 | -3.091572843 | -1.422826553 |
| 1969920.5 | 4.414258920 | -2.130321809 | -1.026837813 |
| 1970080.5 | 4.835419436 | -1.041382064 | -0.569287455 |
| 1970240.5 | 4.961196203 | 0.111002509 | -0.077034152 |
| 1970400.5 | 4.781061472 | 1.256470493 | 0.419938053 |

HELIOCENTRIC COORDINATES OF JUPITER

REFERENCE AXES OF 1950.0

| JULIAN DATE | X | Y | Z |
|---|---|---|---|
| 1970560.5 | 4.305789293 | 2.324405983 | 0.891000066 |
| 1970720.5 | 3.566610528 | 3.250021841 | 1.307514217 |
| 1970880.5 | 2.612040693 | 3.979659028 | 1.645192236 |
| 1971040.5 | 1.503026157 | 4.474509466 | 1.885817787 |
| 1971200.5 | 0.307346600 | 4.712400778 | 2.018154494 |
| 1971360.5 | -0.905807614 | 4.687744089 | 2.038059854 |
| 1971520.5 | -2.070422201 | 4.410074556 | 1.947973147 |
| 1971680.5 | -3.127095497 | 3.901740997 | 1.756008611 |
| 1971840.5 | -4.025313147 | 3.195256174 | 1.474873517 |
| 1972000.5 | -4.724742125 | 2.330680195 | 1.120775832 |
| 1972160.5 | -5.195805255 | 1.353249783 | 0.712419474 |
| 1972320.5 | -5.419795434 | 0.311330430 | 0.270126827 |
| 1972480.5 | -5.388741572 | -0.745326859 | -0.184914056 |
| 1972640.5 | -5.105173545 | -1.767099671 | -0.631306266 |
| 1972800.5 | -4.581866749 | -2.705882303 | -1.048047223 |
| 1972960.5 | -3.841583494 | -3.516458086 | -1.415125198 |
| 1973120.5 | -2.916768559 | -4.157957591 | -1.714154020 |
| 1973280.5 | -1.849098647 | -4.595480938 | -1.929082130 |
| 1973440.5 | -0.688733110 | -4.801921056 | -2.046995519 |
| 1973600.5 | 0.506925712 | -4.759958407 | -2.059006417 |
| 1973760.5 | 1.675168657 | -4.464101227 | -1.961178757 |
| 1973920.5 | 2.750790582 | -3.922525760 | -1.755389587 |
| 1974080.5 | 3.669853413 | -3.158349732 | -1.449970921 |
| 1974240.5 | 4.374283089 | -2.209888020 | -1.059935130 |
| 1974400.5 | 4.816905146 | -1.129452093 | -0.606585394 |
| 1974560.5 | 4.966235064 | 0.019575847 | -0.116378306 |
| 1974720.5 | 4.810159778 | 1.167325405 | 0.380953598 |
| 1974880.5 | 4.357701594 | 2.243179872 | 0.854825893 |
| 1975040.5 | 3.638394152 | 3.181819138 | 1.276415737 |
| 1975200.5 | 2.699332946 | 3.928578441 | 1.621039672 |
| 1975360.5 | 1.600488381 | 4.443315048 | 1.869933658 |
| 1975520.5 | 0.409183638 | 4.702379454 | 2.011239226 |
| 1975680.5 | -0.805346788 | 4.698743734 | 2.040195397 |
| 1975840.5 | -1.976632826 | 4.440680237 | 1.958688144 |
| 1976000.5 | -3.044528829 | 3.949531199 | 1.774381022 |
| 1976160.5 | -3.957613427 | 3.257083676 | 1.499646969 |
| 1976320.5 | -4.674588342 | 2.402936216 | 1.150471405 |
| 1976480.5 | -5.164927168 | 1.432087160 | 0.745431759 |
| 1976640.5 | -5.409032160 | 0.392837038 | 0.304799661 |
| 1976800.5 | -5.398115348 | -0.665001711 | -0.150231675 |

HELIOCENTRIC COORDINATES OF JUPITER

REFERENCE AXES OF 1950.0

| JULIAN DATE | X | Y | Z |
|---|---|---|---|
| 1976960.5 | -5.133958024 | -1.691655448 | -0.598220134 |
| 1977120.5 | -4.628636826 | -2.638799140 | -1.018087657 |
| 1977280.5 | -3.904241905 | -3.460933484 | -1.389719332 |
| 1977440.5 | -2.992553742 | -4.116831271 | -1.694593307 |
| 1977600.5 | -1.934589087 | -4.571133963 | -1.916478551 |
| 1977760.5 | -0.779873412 | -4.796140614 | -2.042222184 |
| 1977920.5 | 0.414747222 | -4.773768947 | -2.062622386 |
| 1978080.5 | 1.586964463 | -4.497577604 | -1.973344443 |
| 1978240.5 | 2.671723692 | -3.974621533 | -1.775786608 |
| 1978400.5 | 3.604890748 | -3.226786724 | -1.477746029 |
| 1978560.5 | 4.327774738 | -2.291156641 | -1.093689927 |
| 1978720.5 | 4.792141003 | -1.218961994 | -0.644430061 |
| 1978880.5 | 4.965055549 | -0.072818584 | -0.156057037 |
| 1979040.5 | 4.832704941 | 1.077717849 | 0.341874659 |
| 1979200.5 | 4.402363085 | 2.161823608 | 0.818735573 |
| 1979360.5 | 3.701989084 | 3.113472713 | 1.245438977 |
| 1979520.5 | 2.777463498 | 3.876907267 | 1.596861981 |
| 1979680.5 | 1.688008508 | 4.410672663 | 1.853695312 |
| 1979840.5 | 0.500680082 | 4.689768411 | 2.003506356 |
| 1980000.5 | -0.715137004 | 4.705926333 | 2.040997125 |
| 1980160.5 | -1.892513734 | 4.466385223 | 1.967597835 |
| 1980320.5 | -2.970656092 | 3.991696594 | 1.790616237 |
| 1980480.5 | -3.897408795 | 3.313083676 | 1.522165272 |
| 1980640.5 | -4.630729258 | 2.469758481 | 1.178046559 |
| 1980800.5 | -5.139376804 | 1.506451824 | 0.776705508 |
| 1980960.5 | -5.403074188 | 0.471269137 | 0.338313353 |
| 1981120.5 | -5.412365837 | -0.586125675 | -0.116017188 |
| 1981280.5 | -5.168342155 | -1.616064323 | -0.564906827 |
| 1981440.5 | -4.682332849 | -2.570266014 | -0.987315674 |
| 1981600.5 | -3.975605631 | -3.403178619 | -1.363127291 |
| 1981760.5 | -3.079044847 | -4.073376301 | -1.673757555 |
| 1981920.5 | -2.032726694 | -4.545101949 | -1.902828385 |
| 1982080.5 | -0.885252869 | -4.790006634 | -2.036932241 |
| 1982240.5 | 0.307343354 | -4.789079246 | -2.066489128 |
| 1982400.5 | 1.483333417 | -4.534671442 | -1.986660701 |
| 1982560.5 | 2.577874625 | -4.032406147 | -1.798235539 |
| 1982720.5 | 3.526556046 | -3.302626597 | -1.508341138 |
| 1982880.5 | 4.269851236 | -2.380935452 | -1.130787507 |
| 1983040.5 | 4.758149041 | -1.317351808 | -0.685831140 |
| 1983200.5 | 4.956725092 | -0.173745300 | -0.199196844 |

HELIOCENTRIC COORDINATES OF JUPITER

REFERENCE AXES OF 1950.0

| JULIAN DATE | X | Y | Z |
|---|---|---|---|
| 1983360.5 | 4.849786810 | 0.980479823 | 0.299676023 |
| 1983520.5 | 4.442714076 | 2.074038464 | 0.780007755 |
| 1983680.5 | 3.761906968 | 3.039932210 | 1.212333469 |
| 1983840.5 | 2.852182161 | 3.821108135 | 1.571001143 |
| 1984000.5 | 1.772233623 | 4.374707720 | 1.836112195 |
| 1984160.5 | 0.589060655 | 4.674396400 | 1.994662876 |
| 1984320.5 | -0.627660578 | 4.710750442 | 2.040851959 |
| 1984480.5 | -1.810494211 | 4.490068093 | 1.975693172 |
| 1984640.5 | -2.898030205 | 4.032162063 | 1.806161296 |
| 1984800.5 | -3.837453222 | 3.367686104 | 1.544107420 |
| 1984960.5 | -4.586042727 | 2.535422544 | 1.205130562 |
| 1985120.5 | -5.111863556 | 1.579791766 | 0.807524373 |
| 1985280.5 | -5.393922141 | 0.548694960 | 0.371353683 |
| 1985440.5 | -5.422023192 | -0.508308074 | -0.082332440 |
| 1985600.5 | -5.196497263 | -1.541552147 | -0.532175240 |
| 1985760.5 | -4.727900966 | -2.502651875 | -0.957109286 |
| 1985920.5 | -4.036726622 | -3.345831712 | -1.336941531 |
| 1986080.5 | -3.153097972 | -4.029309542 | -1.652953746 |
| 1986240.5 | -2.116371381 | -4.516824428 | -1.888569301 |
| 1986400.5 | -0.974506929 | -4.779364385 | -2.030111712 |
| 1986560.5 | 0.216975296 | -4.797092782 | -2.067659008 |
| 1986720.5 | 1.396669894 | -4.561385731 | -1.995961653 |
| 1986880.5 | 2.499800260 | -4.076776691 | -1.815341681 |
| 1987040.5 | 3.461703835 | -3.362471513 | -1.532431573 |
| 1987200.5 | 4.222247907 | -2.452983989 | -1.160558426 |
| 1987360.5 | 4.730865575 | -1.397410004 | -0.719558410 |
| 1987520.5 | 4.951585827 | -0.256977184 | -0.234849598 |
| 1987680.5 | 4.067107217 | 0.099107041 | 0.264283300 |
| 1987840.5 | 4.481579012 | 1.999679348 | 0.747047092 |
| 1988000.5 | 3.819793275 | 2.977013183 | 1.183802399 |
| 1988160.5 | 2.925501358 | 3.773342529 | 1.548586262 |
| 1988320.5 | 1.856554067 | 4.344798958 | 1.821088294 |
| 1988480.5 | 0.679446338 | 4.663917996 | 1.987832611 |
| 1988640.5 | -0.536311633 | 4.720118787 | 2.042525369 |
| 1988800.5 | -1.723139906 | 4.518594530 | 1.985701993 |
| 1988960.5 | -2.819220560 | 4.078165646 | 1.823900556 |
| 1989120.5 | -3.771121388 | 3.428647471 | 1.568596116 |
| 1989280.5 | -4.535352008 | 2.608161313 | 1.235083893 |
| 1989440.5 | -5.079106716 | 1.660653439 | 0.841431413 |
| 1989600.5 | -5.380466708 | 0.633737260 | 0.407555994 |

HELIOCENTRIC COORDINATES OF JUPITER

REFERENCE AXES OF 1950.0

| JULIAN DATE | X | Y | Z |
|---|---|---|---|
| 1989760.5 | -5.428295692 | -0.423135917 | -0.045564214 |
| 1989920.5 | -5.221999790 | -1.460240833 | -0.496568556 |
| 1990080.5 | -4.771254592 | -2.428975875 | -0.924321515 |
| 1990240.5 | -4.095737554 | -3.283205984 | -1.308496713 |
| 1990400.5 | -3.224843807 | -3.980658965 | -1.630184276 |
| 1990560.5 | -2.197306211 | -4.484465664 | -1.872562460 |
| 1990720.5 | -1.060585460 | -4.764900161 | -2.021661022 |
| 1990880.5 | 0.130153024 | -4.801319316 | -2.067220670 |
| 1991040.5 | 1.313690159 | -4.584215492 | -2.003616931 |
| 1991200.5 | 2.425239528 | -4.117181226 | -1.830767152 |
| 1991360.5 | 3.399900832 | -3.418453126 | -1.554880896 |
| 1991520.5 | 4.177043054 | -2.521590450 | -1.188861726 |
| 1991680.5 | 4.705315914 | -1.474812017 | -0.752148009 |
| 1991840.5 | 4.947679681 | -0.338631069 | -0.269822522 |
| 1992000.5 | 4.885602235 | 0.818274662 | 0.229057485 |
| 1992160.5 | 4.521538186 | 1.924589757 | 0.713773853 |
| 1992320.5 | 3.879065744 | 2.912498235 | 1.154582257 |
| 1992480.5 | 3.000567052 | 3.723422869 | 1.525237633 |
| 1992640.5 | 1.942910950 | 4.312449090 | 1.805004309 |
| 1992800.5 | 0.772004355 | 4.650883136 | 1.979892519 |
| 1992960.5 | -0.442818519 | 4.726872518 | 2.043066028 |
| 1993120.5 | -1.633781886 | 4.544416305 | 1.994541891 |
| 1993280.5 | -2.738575798 | 4.121301115 | 1.840401323 |
| 1993440.5 | -3.703067172 | 3.486512497 | 1.591745010 |
| 1993600.5 | -4.482937366 | 2.677557637 | 1.263583344 |
| 1993760.5 | -5.044491263 | 1.737974600 | 0.873786042 |
| 1993920.5 | -5.364907908 | 0.715153555 | 0.442151850 |
| 1994080.5 | -5.432168342 | -0.341516881 | -0.010390091 |
| 1994240.5 | -5.244835266 | -1.382236438 | -0.462467878 |
| 1994400.5 | -4.811791556 | -2.358209398 | -0.892882196 |
| 1994560.5 | -4.151979698 | -3.223000742 | -1.281195832 |
| 1994720.5 | -3.294123271 | -3.933940032 | -1.608344218 |
| 1994880.5 | -2.276353185 | -4.453663341 | -1.857307814 |
| 1995040.5 | -1.145605326 | -4.751849573 | -2.013873911 |
| 1995200.5 | 0.043391677 | -4.807152117 | -2.067492628 |
| 1995360.5 | 1.229746713 | -4.609243306 | -2.012196645 |
| 1995520.5 | 2.348852018 | -4.160775983 | -1.847505495 |
| 1995680.5 | 3.335772322 | -3.478939200 | -1.579179063 |
| 1995840.5 | 4.129548416 | -2.596178644 | -1.219635141 |
| 1996000.5 | 4.678136018 | -1.559619127 | -0.787825109 |

HELIOCENTRIC COORDINATES OF JUPITER

REFERENCE AXES OF 1950.0

| JULIAN DATE | X | Y | Z |
|---|---|---|---|
| 1996160.5 | 4.943402474 | -0.428829475 | -0.308398575 |
| 1996320.5 | 4.905362239 | 0.728153456 | 0.189899419 |
| 1996480.5 | 4.564781212 | 1.840213788 | 0.676479547 |
| 1996640.5 | 3.943510526 | 2.839217933 | 1.121505328 |
| 1996800.5 | 3.082390370 | 3.665770672 | 1.498420052 |
| 1996960.5 | 2.037124815 | 4.273748065 | 1.785998815 |
| 1997120.5 | 0.872947815 | 4.633025633 | 1.969655130 |
| 1997280.5 | -0.340967125 | 4.730283272 | 2.041927213 |
| 1997440.5 | -1.536553024 | 4.568172266 | 2.002246557 |
| 1997600.5 | -2.650866497 | 4.163355959 | 1.856196489 |
| 1997760.5 | -3.628923531 | 3.543969228 | 1.614491271 |
| 1997920.5 | -4.425450604 | 2.746942307 | 1.291869648 |
| 1998080.5 | -5.005781343 | 1.815479963 | 0.906033744 |
| 1998240.5 | -5.346164083 | 0.796837358 | 0.476700710 |
| 1998400.5 | -5.433718058 | -0.259581477 | 0.024784221 |
| 1998560.5 | -5.266218300 | -1.303851822 | -0.428310810 |
| 1998720.5 | -4.851823108 | -2.286973674 | -0.861316404 |
| 1998880.5 | -4.208793461 | -3.162241680 | -1.253697434 |
| 1999040.5 | -3.365193399 | -3.886646280 | -1.586261453 |
| 1999200.5 | -2.358503326 | -4.422397588 | -1.841823449 |
| 1999360.5 | -1.235023572 | -4.738632338 | -2.005954573 |
| 1999520.5 | -0.048896498 | -4.813311880 | -2.067822548 |
| 1999680.5 | 1.139458256 | -4.635239896 | -2.021098283 |
| 1999840.5 | 2.265699791 | -4.206020496 | -1.864856700 |
| 2000000.5 | 3.264855212 | -3.541651003 | -1.604343964 |
| 2000160.5 | 4.075534126 | -2.673330093 | -1.251430626 |
| 2000320.5 | 4.644817133 | -1.647015624 | -0.824544228 |
| 2000480.5 | 4.933369783 | -0.521353146 | -0.347904674 |
| 2000640.5 | 4.919280380 | 0.636143591 | 0.150008766 |
| 2000800.5 | 4.601844795 | 1.754375648 | 0.638658536 |
| 2000960.5 | 4.001123173 | 2.764699784 | 1.088031565 |
| 2001120.5 | 3.156554786 | 3.606759816 | 1.471193255 |
| 2001280.5 | 2.122880502 | 4.233169693 | 1.766400366 |
| 2001440.5 | 0.964864859 | 4.612424180 | 1.958484670 |
| 2001600.5 | -0.248326880 | 4.729873000 | 2.039412822 |
| 2001760.5 | -1.448269334 | 4.587011628 | 2.008110573 |
| 2001920.5 | -2.571414714 | 4.199587648 | 1.869752701 |
| 2002080.5 | -3.562043842 | 3.595070943 | 1.634749712 |
| 2002240.5 | -4.374107381 | 2.809945207 | 1.317629764 |
| 2002400.5 | -4.972180351 | 1.887129850 | 0.935954145 |

HELIOCENTRIC COORDINATES OF JUPITER

REFERENCE AXES OF 1950.0

| JULIAN DATE | X | Y | Z |
|---|---|---|---|
| 2002560.5 | -5.331794885 | 0.873696062 | 0.509343511 |
| 2002720.5 | -5.439390972 | -0.181078519 | 0.058640269 |
| 2002880.5 | -5.292073148 | -1.227358131 | -0.394806010 |
| 2003040.5 | -4.897300982 | -2.216190522 | -0.829765922 |
| 2003200.5 | -4.272576915 | -3.100851652 | -1.225715980 |
| 2003360.5 | -3.445131468 | -3.838204214 | -1.563430427 |
| 2003520.5 | -2.451546070 | -4.390167623 | -1.825622548 |
| 2003680.5 | -1.337197279 | -4.725368538 | -1.997668088 |
| 2003840.5 | -0.155354548 | -4.820992008 | -2.068423606 |
| 2004000.5 | 1.034274750 | -4.664777613 | -2.031121022 |
| 2004160.5 | 2.167766802 | -4.256999175 | -1.884274342 |
| 2004320.5 | 3.180132532 | -3.612138314 | -1.632478094 |
| 2004480.5 | 4.009424389 | -2.759838895 | -1.286920640 |
| 2004640.5 | 4.601588488 | -1.744664607 | -0.865401703 |
| 2004800.5 | 4.915547844 | -0.624244442 | -0.391662728 |
| 2004960.5 | 4.927723647 | 0.534368198 | 0.106064729 |
| 2005120.5 | 4.635086810 | 1.659894798 | 0.597217380 |
| 2005280.5 | 4.055998056 | 2.682915221 | 1.051494129 |
| 2005440.5 | 3.228545503 | 3.541856053 | 1.441471887 |
| 2005600.5 | 2.206675268 | 4.187893032 | 1.744809972 |
| 2005760.5 | 1.054898323 | 4.588076062 | 1.945726207 |
| 2005920.5 | -0.157437706 | 4.726459702 | 2.035626569 |
| 2006080.5 | -1.361431584 | 4.603471293 | 2.012966127 |
| 2006240.5 | -2.492898781 | 4.234025572 | 1.882545734 |
| 2006400.5 | -3.495418339 | 3.644962239 | 1.654487003 |
| 2006560.5 | -4.322226765 | 2.872294844 | 1.343097247 |
| 2006720.5 | -4.937188661 | 1.958600053 | 0.965771596 |
| 2006880.5 | -5.315123513 | 0.950714828 | 0.542011327 |
| 2007040.5 | -5.441745305 | -0.102218309 | 0.092582092 |
| 2007200.5 | -5.313413398 | -1.150401861 | -0.361201838 |
| 2007360.5 | -4.936824217 | -2.144820543 | -0.798103086 |
| 2007520.5 | -4.328706240 | -3.038575781 | -1.197542469 |
| 2007680.5 | -3.515518708 | -3.788229839 | -1.540184333 |
| 2007840.5 | -2.533096442 | -4.355259322 | -1.808570257 |
| 2008000.5 | -1.426127242 | -4.707690311 | -1.987835045 |
| 2008160.5 | -0.247296216 | -4.821940149 | -2.066519888 |
| 2008320.5 | 0.944109230 | -4.684817185 | -2.037466979 |
| 2008480.5 | 2.084351374 | -4.295525453 | -1.898735233 |
| 2008640.5 | 3.108324153 | -3.667392485 | -1.654420342 |
| 2008800.5 | 3.953607383 | -2.828911195 | -1.315204393 |

HELIOCENTRIC COORDINATES OF JUPITER

REFERENCE AXES OF 1950.0

| JULIAN DATE | X | Y | Z |
|---|---|---|---|
| 2008960.5 | 4.565303427 | -1.823613166 | -0.898422780 |
| 2009120.5 | 4.901159437 | -0.708340648 | -0.427449910 |
| 2009280.5 | 4.936186078 | 0.450286191 | 0.069703676 |
| 2009440.5 | 4.665849517 | 1.581010539 | 0.562521627 |
| 2009600.5 | 4.107065652 | 2.614030886 | 1.020575794 |
| 2009760.5 | 3.296673898 | 3.487053411 | 1.416166007 |
| 2009920.5 | 2.287662470 | 4.150288095 | 1.726561946 |
| 2010080.5 | 1.143921267 | 4.569675046 | 1.935521278 |
| 2010240.5 | -0.065481841 | 4.728109910 | 2.033959841 |
| 2010400.5 | -1.271608822 | 4.624892230 | 2.019847655 |
| 2010560.5 | -2.409964095 | 4.273901902 | 1.897533277 |
| 2010720.5 | -3.423591959 | 3.701080149 | 1.676738738 |
| 2010880.5 | -4.265022648 | 2.941705580 | 1.371437517 |
| 2011040.5 | -4.897292566 | 2.037797664 | 0.998764057 |
| 2011200.5 | -5.294315888 | 1.035818500 | 0.578035227 |
| 2011360.5 | -5.440863488 | -0.015283919 | 0.129908720 |
| 2011520.5 | -5.332347845 | -1.065732480 | -0.324334482 |
| 2011680.5 | -4.974544337 | -2.066366796 | -0.763418573 |
| 2011840.5 | -4.383312481 | -2.969987754 | -1.166657845 |
| 2012000.5 | -3.584318779 | -3.732712551 | -1.514546086 |
| 2012160.5 | -2.612704666 | -4.315439956 | -1.789394728 |
| 2012320.5 | -1.512587056 | -4.685498458 | -1.976053804 |
| 2012480.5 | -0.336226623 | -4.818503128 | -2.062730961 |
| 2012640.5 | 0.857341774 | -4.700373759 | -2.041893273 |
| 2012800.5 | 2.004433948 | -4.329363078 | -1.911192337 |
| 2012960.5 | 3.039769953 | -3.717815824 | -1.674296980 |
| 2013120.5 | 3.900500823 | -2.893252999 | -1.341460304 |
| 2013280.5 | 4.531024142 | -1.898302143 | -0.929610507 |
| 2013440.5 | 4.888102227 | -0.789042858 | -0.461768099 |
| 2013600.5 | 4.945509659 | 0.368438435 | 0.034318405 |
| 2013760.5 | 4.697301797 | 1.503099347 | 0.528261550 |
| 2013920.5 | 4.158934215 | 2.544964166 | 0.989595604 |
| 2014080.5 | 3.365890650 | 3.431160312 | 1.390405657 |
| 2014240.5 | 2.370057242 | 4.110969689 | 1.707589824 |
| 2014400.5 | 1.234583276 | 4.549168440 | 1.924423689 |
| 2014560.5 | 0.028202091 | 4.727390563 | 2.031288893 |
| 2014720.5 | -1.180094736 | 4.643705216 | 2.025625021 |
| 2014880.5 | -2.325432508 | 4.310890600 | 1.911297409 |
| 2015040.5 | -3.350233929 | 3.753973402 | 1.697619273 |
| 2015200.5 | -4.206248611 | 3.007528343 | 1.398241878 |

## HELIOCENTRIC COORDINATES OF JUPITER

## REFERENCE AXES OF 1950.0

| JULIAN DATE | X | Y | Z |
|---|---|---|---|
| 2015360.5 | -4.855644684 | 2.113079850 | 1.030065064 |
| 2015520.5 | -5.271434707 | 1.116787318 | 0.612252973 |
| 2015680.5 | -5.437492142 | 0.067465929 | 0.165383756 |
| 2015840.5 | -5.348360189 | -0.985092497 | -0.289276978 |
| 2016000.5 | -5.008987532 | -1.991584218 | -0.730412361 |
| 2016160.5 | -4.434458747 | -2.904552003 | -1.137244924 |
| 2016320.5 | -3.649724611 | -3.679748865 | -1.490129550 |
| 2016480.5 | -2.689278563 | -4.277607151 | -1.771193128 |
| 2016640.5 | -1.596668407 | -4.664886030 | -1.965055106 |
| 2016800.5 | -0.423678941 | -4.816523885 | -2.059644481 |
| 2016960.5 | 0.771018459 | -4.717650577 | -2.047098788 |
| 2017120.5 | 1.923960711 | -4.365613464 | -1.924687385 |
| 2017280.5 | 2.969897915 | -3.771745265 | -1.695646366 |
| 2017440.5 | 3.845752642 | -2.962478809 | -1.369756705 |
| 2017600.5 | 4.495363637 | -1.979342837 | -0.963461250 |
| 2017760.5 | 4.874555029 | -0.877415551 | -0.499327092 |
| 2017920.5 | 4.955790362 | 0.277979435 | -0.004742080 |
| 2018080.5 | 4.731526993 | 1.416179943 | 0.490110162 |
| 2018240.5 | 4.215500513 | 2.467104555 | 0.954762308 |
| 2018400.5 | 3.441565325 | 3.367264915 | 1.361076637 |
| 2018560.5 | 2.460276102 | 4.064886314 | 1.685531681 |
| 2018720.5 | 1.333897866 | 4.523392401 | 1.910845131 |
| 2018880.5 | 0.130788212 | 4.722948783 | 2.026778131 |
| 2019040.5 | -1.079954713 | 4.660210479 | 2.030160878 |
| 2019200.5 | -2.232956782 | 4.346723782 | 1.924317612 |
| 2019360.5 | -3.269867605 | 3.806538605 | 1.718123037 |
| 2019520.5 | -4.141509195 | 3.073529687 | 1.424904765 |
| 2019680.5 | -4.809067131 | 2.188781950 | 1.061349322 |
| 2019840.5 | -5.244587631 | 1.198237267 | 0.646503121 |
| 2020000.5 | -5.431036036 | 0.150668525 | 0.200903123 |
| 2020160.5 | -5.362122419 | -0.904044121 | -0.254168451 |
| 2020320.5 | -5.042034896 | -1.916413419 | -0.697335621 |
| 2020480.5 | -4.485155116 | -2.838725439 | -1.107729125 |
| 2020640.5 | -3.715768299 | -3.626409543 | -1.465582537 |
| 2020800.5 | -2.767721971 | -4.239497642 | -1.752868056 |
| 2020960.5 | -1.683931921 | -4.644248081 | -1.954006877 |
| 2021120.5 | -0.515581792 | -4.814966336 | -2.056671839 |
| 2021280.5 | 0.679179000 | -4.735986709 | -2.052671857 |
| 2021440.5 | 1.837250295 | -4.403682182 | -1.938866311 |
| 2021600.5 | 2.893486306 | -3.828247826 | -1.718004427 |

## HELIOCENTRIC COORDINATES OF JUPITER

### REFERENCE AXES OF 1950.0

| JULIAN DATE | X | Y | Z |
|---|---|---|---|
| 2021760.5 | 3.784555333 | -3.034878543 | -1.399328246 |
| 2021920.5 | 4.453610483 | -2.063880217 | -0.998737469 |
| 2022080.5 | 4.855342488 | -0.969279159 | -0.538318413 |
| 2022240.5 | 4.960698331 | 0.184315215 | -0.045113776 |
| 2022400.5 | 4.760386112 | 1.326482095 | 0.450840528 |
| 2022560.5 | 4.266371981 | 2.386846708 | 0.918995528 |
| 2022720.5 | 3.510946331 | 3.301126762 | 1.330911533 |
| 2022880.5 | 2.543487895 | 4.016375501 | 1.662592851 |
| 2023040.5 | 1.425572980 | 4.494641779 | 1.896183581 |
| 2023200.5 | 0.225362790 | 4.714695055 | 2.020849770 |
| 2023360.5 | -0.987819758 | 4.671920736 | 2.032870389 |
| 2023520.5 | -2.148068581 | 4.376815610 | 1.935105464 |
| 2023680.5 | -3.196316625 | 3.852637373 | 1.736071635 |
| 2023840.5 | -4.082589845 | 3.132716161 | 1.448843143 |
| 2024000.5 | -4.767272891 | 2.257797092 | 1.089941195 |
| 2024160.5 | -5.221647716 | 1.273627047 | 0.678309898 |
| 2024320.5 | -5.427964143 | 0.228870623 | 0.234421932 |
| 2024480.5 | -5.379252319 | -0.826653298 | -0.220484182 |
| 2024640.5 | -5.079024885 | -1.843486041 | -0.665062359 |
| 2024800.5 | -4.540955827 | -2.773899198 | -1.078455700 |
| 2024960.5 | -3.788561883 | -3.573232919 | -1.440879586 |
| 2025120.5 | -2.854851882 | -4.201307022 | -1.734236930 |
| 2025280.5 | -1.781851251 | -4.623987373 | -1.942803330 |
| 2025440.5 | -0.619854646 | -4.814950008 | -2.054004042 |
| 2025600.5 | 0.573785636 | -4.757621540 | -2.059278363 |
| 2025760.5 | 1.736550442 | -4.447181071 | -1.954987469 |
| 2025920.5 | 2.803515363 | -3.892388161 | -1.743269318 |
| 2026080.5 | 3.711075934 | -3.116871737 | -1.432686001 |
| 2026240.5 | 4.401533242 | -2.159418785 | -1.038462752 |
| 2026400.5 | 4.828153737 | -1.072804174 | -0.582111797 |
| 2026560.5 | 4.960013688 | 0.079132199 | -0.090297479 |
| 2026720.5 | 4.785740107 | 1.226110723 | 0.407058017 |
| 2026880.5 | 4.315302530 | 2.297241937 | 0.879235511 |
| 2027040.5 | 3.579353603 | 3.227177550 | 1.297379158 |
| 2027200.5 | 2.626173359 | 3.961556426 | 1.636913789 |
| 2027360.5 | 1.516828531 | 4.460900304 | 1.879340659 |
| 2027520.5 | 0.319492753 | 4.702540472 | 2.013207908 |
| 2027680.5 | -0.896113498 | 4.680639783 | 2.034259365 |
| 2027840.5 | -2.063453019 | 4.404733094 | 1.944925746 |
| 2028000.5 | -3.122719689 | 3.897359811 | 1.753396008 |

HELIOCENTRIC COORDINATES OF JUPITER

REFERENCE AXES OF 1950.0

| JULIAN DATE | X | Y | Z |
|---|---|---|---|
| 2028160.5 | -4.023174255 | 3.191327643 | 1.472499800 |
| 2028320.5 | -4.724448955 | 2.327005862 | 1.118576767 |
| 2028480.5 | -5.197092459 | 1.349878867 | 0.710438690 |
| 2028640.5 | -5.422628235 | 0.308446691 | 0.268468556 |
| 2028800.5 | -5.393351521 | -0.747542663 | -0.186144299 |
| 2028960.5 | -5.112023016 | -1.768605005 | -0.632061656 |
| 2029120.5 | -4.591548552 | -2.706881320 | -1.048389574 |
| 2029280.5 | -3.854669781 | -3.517462280 | -1.415255439 |
| 2029440.5 | -2.933631587 | -4.159782293 | -1.714415486 |
| 2029600.5 | -1.869735407 | -4.599168298 | -1.929931690 |
| 2029760.5 | -0.712633781 | -4.808590083 | -2.048941766 |
| 2029920.5 | 0.480825013 | -4.770596902 | -2.062520518 |
| 2030080.5 | 1.648419157 | -4.479333747 | -1.966592136 |
| 2030240.5 | 2.725241088 | -3.942410634 | -1.762801034 |
| 2030400.5 | 3.647357663 | -3.182266322 | -1.459189744 |
| 2030560.5 | 4.356359636 | -2.236564996 | -1.070483485 |
| 2030720.5 | 4.804435622 | -1.157155483 | -0.617770022 |
| 2030880.5 | 4.959293874 | -0.007274267 | -0.127420780 |
| 2031040.5 | 4.808037095 | 1.142990850 | 0.370762077 |
| 2031200.5 | 4.359116447 | 2.222507301 | 0.845992900 |
| 2031360.5 | 3.641825681 | 3.165297509 | 1.269180284 |
| 2031520.5 | 2.703360640 | 3.916082491 | 1.615383248 |
| 2031680.5 | 1.604041014 | 4.434290018 | 1.865653243 |
| 2031840.5 | 0.411643360 | 4.696085339 | 2.008048624 |
| 2032000.5 | -0.804181238 | 4.694478668 | 2.037820609 |
| 2032160.5 | -1.976678990 | 4.437929966 | 1.956933155 |
| 2032320.5 | -3.045590717 | 3.948026228 | 1.773154806 |
| 2032480.5 | -3.959537878 | 3.256773821 | 1.498954755 |
| 2032640.5 | -4.677374405 | 2.403908130 | 1.150380411 |
| 2032800.5 | -5.168774287 | 1.434454807 | 0.746023968 |
| 2032960.5 | -5.414327795 | 0.396632152 | 0.306123270 |
| 2033120.5 | -5.405372081 | -0.659918424 | -0.148202969 |
| 2033280.5 | -5.143715244 | -1.685648375 | -0.595613633 |
| 2033440.5 | -4.641344285 | -2.632469163 | -1.015140118 |
| 2033600.5 | -3.920143159 | -3.455083989 | -1.386765757 |
| 2033760.5 | -3.011587605 | -4.112390846 | -1.692036498 |
| 2033920.5 | -1.956326679 | -4.569042946 | -1.914742044 |
| 2034080.5 | -0.803506539 | -4.797214451 | -2.041691700 |
| 2034240.5 | 0.390354877 | -4.778557730 | -2.063583785 |
| 2034400.5 | 1.563160121 | -4.506247070 | -1.975928904 |

## HELIOCENTRIC COORDINATES OF JUPITER

## REFERENCE AXES OF 1950.0

| JULIAN DATE | X | Y | Z |
|---|---|---|---|
| 2034560.5 | 2.649891837 | -3.986883804 | -1.779935405 |
| 2034720.5 | 3.586249058 | -3.241905787 | -1.483206253 |
| 2034880.5 | 4.313181187 | -2.308040876 | -1.100047530 |
| 2035040.5 | 4.781961202 | -1.236333940 | -0.651177345 |
| 2035200.5 | 4.959130915 | -0.089426057 | -0.162680338 |
| 2035360.5 | 4.830435447 | 1.062904164 | 0.335810068 |
| 2035520.5 | 4.402880638 | 2.149481040 | 0.813527351 |
| 2035680.5 | 3.704365769 | 3.103898067 | 1.241231318 |
| 2035840.5 | 2.780889463 | 3.870084097 | 1.593670423 |
| 2036000.5 | 1.691892017 | 4.406396249 | 1.851457578 |
| 2036160.5 | 0.504656863 | 4.687775990 | 2.002137083 |
| 2036320.5 | -0.711264542 | 4.705988298 | 2.040427100 |
| 2036480.5 | -1.888868481 | 4.468336286 | 1.967787844 |
| 2036640.5 | -2.967370332 | 3.995412559 | 1.791547118 |
| 2036800.5 | -3.894673139 | 3.318429179 | 1.523814583 |
| 2036960.5 | -4.628794962 | 2.476535863 | 1.180364898 |
| 2037120.5 | -5.138516985 | 1.514371052 | 0.779601418 |
| 2037280.5 | -5.403523469 | 0.479944811 | 0.341649787 |
| 2037440.5 | -5.414261308 | -0.577152320 | -0.112415405 |
| 2037600.5 | -5.171682080 | -1.607284641 | -0.561236943 |
| 2037760.5 | -4.686961225 | -2.562153239 | -0.983775971 |
| 2037920.5 | -3.981224574 | -3.396137371 | -1.359895358 |
| 2038080.5 | -3.085252327 | -4.067702403 | -1.670972255 |
| 2038240.5 | -2.039073049 | -4.540961226 | -1.900579425 |
| 2038400.5 | -0.891304709 | -4.787438024 | -2.035259973 |
| 2038560.5 | 0.301948354 | -4.788022019 | -2.065394927 |
| 2038720.5 | 1.478859419 | -4.535009695 | -1.986125299 |
| 2038880.5 | 2.574500170 | -4.034017180 | -1.798239531 |
| 2039040.5 | 3.524422859 | -3.305414457 | -1.508879317 |
| 2039200.5 | 4.269129741 | -2.384831374 | -1.131868770 |
| 2039360.5 | 4.759091567 | -1.322273989 | -0.687460215 |
| 2039520.5 | 4.959677601 | -0.179528641 | -0.201341935 |
| 2039680.5 | 4.855142423 | 0.974156265 | 0.297117897 |
| 2039840.5 | 4.450814096 | 2.067691726 | 0.777232450 |
| 2040000.5 | 3.772918630 | 3.034257566 | 1.209626384 |
| 2040160.5 | 2.865994934 | 3.816898914 | 1.568705871 |
| 2040320.5 | 1.788410485 | 4.372732572 | 1.834578718 |
| 2040480.5 | 0.606860685 | 4.675268589 | 1.994189884 |
| 2040640.5 | -0.609196038 | 4.714823635 | 2.041638470 |
| 2040800.5 | -1.792416818 | 4.497381097 | 1.977809946 |

HELIOCENTRIC COORDINATES OF JUPITER

REFERENCE AXES OF 1950.0

| JULIAN DATE | X | Y | Z |
|---|---|---|---|
| 2040960.5 | -2.881351879 | 4.042438490 | 1.809545579 |
| 2041120.5 | -3.823031567 | 3.380381187 | 1.548578447 |
| 2041280.5 | -4.574501122 | 2.549801431 | 1.210419615 |
| 2041440.5 | -5.103552244 | 1.595021671 | 0.813311999 |
| 2041600.5 | -5.388918586 | 0.563935450 | 0.377307298 |
| 2041760.5 | -5.420164712 | -0.493828620 | -0.076525896 |
| 2041920.5 | -5.197436205 | -1.528481760 | -0.526785228 |
| 2042080.5 | -4.731171497 | -2.491485934 | -0.952348193 |
| 2042240.5 | -4.041812071 | -3.336910260 | -1.332962365 |
| 2042400.5 | -3.159488027 | -4.022838418 | -1.649858161 |
| 2042560.5 | -2.123599554 | -4.512914354 | -1.886423647 |
| 2042720.5 | -0.982160899 | -4.778078653 | -2.028967242 |
| 2042880.5 | 0.209273755 | -4.798490752 | -2.067570757 |
| 2043040.5 | 1.389312031 | -4.565547036 | -1.996999358 |
| 2043200.5 | 2.493251903 | -4.083791562 | -1.817586083 |
| 2043360.5 | 3.456558566 | -3.372388403 | -1.535951689 |
| 2043520.5 | 4.219241972 | -2.465718103 | -1.165370603 |
| 2043680.5 | 4.730827311 | -1.412634606 | -0.725575999 |
| 2043840.5 | 4.955310380 | -0.274038038 | -0.241840679 |
| 2044000.5 | 4.875258598 | 0.881285797 | 0.256708397 |
| 2044160.5 | 4.494191548 | 1.982185786 | 0.739403105 |
| 2044320.5 | 3.836635919 | 2.961244666 | 1.176654133 |
| 2044480.5 | 2.945752060 | 3.760462708 | 1.542453199 |
| 2044640.5 | 1.878991479 | 4.335623724 | 1.816358240 |
| 2044800.5 | 0.702643288 | 4.658803533 | 1.984708077 |
| 2044960.5 | -0.513771751 | 4.718952017 | 2.041012956 |
| 2045120.5 | -1.702480163 | 4.520873679 | 1.985641103 |
| 2045280.5 | -2.801357289 | 4.083137928 | 1.825018329 |
| 2045440.5 | -3.756631994 | 3.435464558 | 1.570572127 |
| 2045600.5 | -4.524513604 | 2.616017066 | 1.237609111 |
| 2045760.5 | -5.071982035 | 1.668876778 | 0.844249032 |
| 2045920.5 | -5.377009563 | 0.641824328 | 0.410476482 |
| 2046080.5 | -5.428450171 | -0.415550615 | -0.042674595 |
| 2046240.5 | -5.225770145 | -1.453458771 | -0.493819345 |
| 2046400.5 | -4.778727400 | -2.423330331 | -0.921839702 |
| 2046560.5 | -4.107047117 | -3.279154322 | -1.306467917 |
| 2046720.5 | -3.240083617 | -3.978848112 | -1.628883475 |
| 2046880.5 | -2.216400311 | -4.485750252 | -1.872365149 |
| 2047040.5 | -1.083143373 | -4.770297448 | -2.023027389 |
| 2047200.5 | 0.104965363 | -4.811893084 | -2.070648571 |

## HELIOCENTRIC COORDINATES OF JUPITER

### REFERENCE AXES OF 1950.0

| JULIAN DATE | X | Y | Z |
|---|---|---|---|
| 2047360.5 | 1.287224631 | -4.600893945 | -2.009564813 |
| 2047520.5 | 2.399345967 | -4.140531843 | -1.839555725 |
| 2047680.5 | 3.376784216 | -3.448458060 | -1.566591553 |
| 2047840.5 | 4.158994079 | -2.557482899 | -1.203257588 |
| 2048000.5 | 4.694349892 | -1.515037046 | -0.768645833 |
| 2048160.5 | 4.945160331 | -0.380973875 | -0.287536047 |
| 2048320.5 | 4.891952403 | 0.776395430 | 0.211201002 |
| 2048480.5 | 4.536129682 | 1.885723020 | 0.696866850 |
| 2048640.5 | 3.900330503 | 2.878758402 | 1.139562593 |
| 2048800.5 | 3.026299702 | 3.696192669 | 1.512753856 |
| 2048960.5 | 1.970671929 | 4.292258180 | 1.795353173 |
| 2049120.5 | 0.799510354 | 4.637472128 | 1.973035806 |
| 2049280.5 | -0.417408810 | 4.719396145 | 2.038709296 |
| 2049440.5 | -1.611731384 | 4.541710339 | 1.992244054 |
| 2049600.5 | -2.720577608 | 4.122138949 | 1.839683321 |
| 2049760.5 | -3.689360217 | 3.489798997 | 1.592174237 |
| 2049920.5 | -4.473477825 | 2.682435760 | 1.264822114 |
| 2050080.5 | -5.039131090 | 1.743844380 | 0.875602752 |
| 2050240.5 | -5.363542645 | 0.721619062 | 0.444400610 |
| 2050400.5 | -5.434829801 | -0.334748702 | -0.007811715 |
| 2050560.5 | -5.251716883 | -1.375478288 | -0.459671827 |
| 2050720.5 | -4.823204456 | -2.351910952 | -0.890042228 |
| 2050880.5 | -4.168246263 | -3.217835977 | -1.278588352 |
| 2051040.5 | -3.315423467 | -3.930842752 | -1.606367803 |
| 2051200.5 | -2.302548530 | -4.453794130 | -1.856473488 |
| 2051360.5 | -1.176076092 | -4.756485614 | -2.014761439 |
| 2051520.5 | 0.009851937 | -4.817503442 | -2.070673276 |
| 2051680.5 | 1.194935888 | -4.626216026 | -2.018130376 |
| 2051840.5 | 2.315034236 | -4.184724613 | -1.856430198 |
| 2052000.5 | 3.305419612 | -3.509469403 | -1.591018087 |
| 2052160.5 | 4.104981294 | -2.632066823 | -1.233952351 |
| 2052320.5 | 4.661134273 | -1.598897972 | -0.803850330 |
| 2052480.5 | 4.934886008 | -0.469047259 | -0.325129366 |
| 2052640.5 | 4.905239489 | 0.689548615 | 0.173538084 |
| 2052800.5 | 4.572017387 | 1.805464164 | 0.661459137 |
| 2052960.5 | 3.956392265 | 2.809938358 | 1.108552237 |
| 2053120.5 | 3.098896074 | 3.642807900 | 1.487947252 |
| 2053280.5 | 2.055285352 | 4.257230904 | 1.778118781 |
| 2053440.5 | 0.891104814 | 4.622550715 | 1.964252451 |
| 2053600.5 | -0.324046634 | 4.725151759 | 2.038755892 |

HELIOCENTRIC COORDINATES OF JUPITER

REFERENCE AXES OF 1950.0

| JULIAN DATE | X | Y | Z |
|---|---|---|---|
| 2053760.5 | -1.521693458 | 4.567602499 | 2.001019865 |
| 2053920.5 | -2.638586568 | 4.166626865 | 1.856649155 |
| 2054080.5 | -3.619571255 | 3.550481435 | 1.616407194 |
| 2054240.5 | -4.419325920 | 2.756207741 | 1.295078763 |
| 2054400.5 | -5.003218325 | 1.827063171 | 0.910387362 |
| 2054560.5 | -5.347560977 | 0.810274822 | 0.482036320 |
| 2054720.5 | -5.439515794 | -0.244860484 | 0.030889894 |
| 2054880.5 | -5.276836173 | -1.288584357 | -0.421724290 |
| 2055040.5 | -4.867565011 | -2.272089269 | -0.854629780 |
| 2055200.5 | -4.229740668 | -3.148847624 | -1.247379831 |
| 2055360.5 | -3.391101518 | -3.875968927 | -1.580848035 |
| 2055520.5 | -2.388722064 | -4.415681536 | -1.837874214 |
| 2055680.5 | -1.268458113 | -4.737005138 | -2.003997048 |
| 2055840.5 | -0.084027593 | -4.817628912 | -2.068284223 |
| 2056000.5 | 1.104483487 | -4.645931789 | -2.024238018 |
| 2056160.5 | 2.232904479 | -4.222972274 | -1.870707579 |
| 2056320.5 | 3.236210519 | -3.564144893 | -1.612681611 |
| 2056480.5 | 4.052711027 | -2.700078485 | -1.261779230 |
| 2056640.5 | 4.628961614 | -1.676294126 | -0.836225868 |
| 2056800.5 | 4.924856137 | -0.551217298 | -0.360126416 |
| 2056960.5 | 4.918083532 | 0.607602576 | 0.138047916 |
| 2057120.5 | 4.607039805 | 1.728797719 | 0.627666810 |
| 2057280.5 | 4.011479787 | 2.743301675 | 1.078554484 |
| 2057440.5 | 3.170665785 | 3.590287603 | 1.463588006 |
| 2057600.5 | 2.139359031 | 4.221948220 | 1.760851798 |
| 2057760.5 | 0.982468003 | 4.606463436 | 1.955046894 |
| 2057920.5 | -0.230656720 | 4.728982022 | 2.038055584 |
| 2058080.5 | -1.431424466 | 4.590881255 | 2.008753413 |
| 2058240.5 | -2.556169109 | 4.207825424 | 1.872279034 |
| 2058400.5 | -3.549094887 | 3.607194353 | 1.639002827 |
| 2058560.5 | -4.364091795 | 2.825355622 | 1.323400060 |
| 2058720.5 | -4.965656746 | 1.905093124 | 0.942968857 |
| 2058880.5 | -5.329198131 | 0.893346015 | 0.517266129 |
| 2059040.5 | -5.440974570 | -0.160710864 | 0.067081107 |
| 2059200.5 | -5.297859063 | -1.207294090 | -0.386269811 |
| 2059360.5 | -4.907051089 | -2.197439361 | -0.821563809 |
| 2059520.5 | -4.285792696 | -3.084341009 | -1.218254414 |
| 2059680.5 | -3.461082905 | -3.824714904 | -1.557064352 |
| 2059840.5 | -2.469326554 | -4.380282262 | -1.820632017 |
| 2060000.5 | -1.355797440 | -4.719441955 | -1.994243660 |

## HELIOCENTRIC COORDINATES OF JUPITER

## REFERENCE AXES OF 1950.0

| JULIAN DATE | X | Y | Z |
|---|---|---|---|
| 2060160.5 | -0.173743424 | -4.819147837 | -2.066662718 |
| 2060320.5 | 1.017077420 | -4.666929511 | -2.031035071 |
| 2060480.5 | 2.152644438 | -4.262888465 | -1.885803223 |
| 2060640.5 | 3.167860906 | -3.621374377 | -1.635506254 |
| 2060800.5 | 4.000689054 | -2.771925995 | -1.291287325 |
| 2060960.5 | 4.597008656 | -1.759002566 | -0.870899589 |
| 2061120.5 | 4.915679657 | -0.640103505 | -0.398024691 |
| 2061280.5 | 4.933022410 | 0.517880958 | 0.099183465 |
| 2061440.5 | 4.645825388 | 1.643848835 | 0.590249149 |
| 2061600.5 | 4.072161193 | 2.668522939 | 1.044948571 |
| 2061760.5 | 3.249736413 | 3.530381908 | 1.435899976 |
| 2061920.5 | 2.232071937 | 4.180518063 | 1.740746824 |
| 2062080.5 | 1.083281716 | 4.585748811 | 1.943625477 |
| 2062240.5 | -0.127584794 | 4.729769467 | 2.035802120 |
| 2062400.5 | -1.331776565 | 4.612571114 | 2.015553480 |
| 2062560.5 | -2.465093486 | 4.248616475 | 1.887489262 |
| 2062720.5 | -3.470945344 | 3.664334759 | 1.661552072 |
| 2062880.5 | -4.302279251 | 2.895413479 | 1.351901776 |
| 2063040.5 | -4.922594600 | 1.984212213 | 0.975829460 |
| 2063200.5 | -5.306315683 | 0.977466506 | 0.552779995 |
| 2063360.5 | -5.438772144 | -0.075674599 | 0.103508188 |
| 2063520.5 | -5.315981970 | -1.125317151 | -0.350643728 |
| 2063680.5 | -4.944364814 | -2.122283780 | -0.788381335 |
| 2063840.5 | -4.340447593 | -3.019475290 | -1.189050530 |
| 2064000.5 | -3.530562912 | -3.773241541 | -1.533234213 |
| 2064160.5 | -2.550483290 | -4.344858305 | -1.803396098 |
| 2064320.5 | -1.444879063 | -4.702179033 | -1.984604125 |
| 2064480.5 | -0.266437070 | 4.021403007 | -2.065346768 |
| 2064640.5 | 0.925561069 | -4.689468607 | -2.038425017 |
| 2064800.5 | 2.067410994 | -4.305235722 | -1.901858986 |
| 2064960.5 | 3.094071021 | -3.681980960 | -1.659693235 |
| 2065120.5 | 3.943193895 | -2.848002603 | -1.322529650 |
| 2065280.5 | 4.559911776 | -1.846549710 | -0.907582225 |
| 2065440.5 | 4.901887130 | -0.734100801 | -0.438063649 |
| 2065600.5 | 4.943866422 | 0.423117261 | 0.058196817 |
| 2065760.5 | 4.680848451 | 1.554175958 | 0.550842905 |
| 2065920.5 | 4.129120561 | 2.589427224 | 1.009536220 |
| 2066080.5 | 3.324836433 | 3.466478540 | 1.406559591 |
| 2066240.5 | 2.320375604 | 4.135182140 | 1.719052218 |
| 2066400.5 | 1.179214397 | 4.560926913 | 1.930552636 |

## HELIOCENTRIC COORDINATES OF JUPITER

## REFERENCE AXES OF 1950.0

| JULIAN DATE | X | Y | Z |
|---|---|---|---|
| 2066560.5 | -0.029733909 | 4.725973312 | 2.031713951 |
| 2066720.5 | -1.237429454 | 4.629016702 | 2.020249020 |
| 2066880.5 | -2.379074959 | 4.283453492 | 1.900294639 |
| 2067040.5 | -3.397298517 | 3.714911299 | 1.681429795 |
| 2067200.5 | -4.244188260 | 2.958532997 | 1.377559433 |
| 2067360.5 | -4.882385097 | 2.056354690 | 1.005814947 |
| 2067520.5 | -5.285499627 | 1.054958583 | 0.585555718 |
| 2067680.5 | -5.438106614 | 0.003454638 | 0.137501200 |
| 2067840.5 | -5.335519372 | -1.048236950 | -0.317013114 |
| 2068000.5 | -4.983479195 | -2.050876544 | -0.756684058 |
| 2068160.5 | -4.397828478 | -2.957269442 | -1.160834999 |
| 2068320.5 | -3.604177906 | -3.723614029 | -1.510003236 |
| 2068480.5 | -2.637523078 | -4.310935391 | -1.786565520 |
| 2068640.5 | -1.541710133 | -4.686680531 | -1.975436879 |
| 2068800.5 | -0.368596767 | -4.826506951 | -2.064860195 |
| 2068960.5 | 0.823287305 | -4.716227431 | -2.047275435 |
| 2069120.5 | 1.970792046 | -4.353775905 | -1.920215761 |
| 2069280.5 | 3.009087168 | -3.750936527 | -1.687124080 |
| 2069440.5 | 3.875551783 | -2.934452618 | -1.357926863 |
| 2069600.5 | 4.514462792 | -1.946050835 | -0.949161775 |
| 2069760.5 | 4.882041103 | -0.840944704 | -0.483460795 |
| 2069920.5 | 4.951123938 | 0.315421371 | 0.011732129 |
| 2070080.5 | 4.714570101 | 1.452258840 | 0.506175880 |
| 2070240.5 | 4.186600870 | 2.499378046 | 0.969352515 |
| 2070400.5 | 3.401668369 | 3.393263077 | 1.373106089 |
| 2070560.5 | 2.411012709 | 4.082268097 | 1.693961735 |
| 2070720.5 | 1.277596058 | 4.530157284 | 1.914773554 |
| 2070880.5 | 0.070380485 | 4.717658586 | 2.025535385 |
| 2071040.5 | -1.141123945 | 4.642170999 | 2.023391346 |
| 2071200.5 | -2.291392063 | 4.316088747 | 1.912029044 |
| 2071360.5 | -3.322207635 | 3.764320072 | 1.700696451 |
| 2071520.5 | -4.184794317 | 3.021519586 | 1.403065861 |
| 2071680.5 | -4.840957833 | 2.129403235 | 1.036108620 |
| 2071840.5 | -5.263516234 | 1.134349105 | 0.619074892 |
| 2072000.5 | -5.436285485 | 0.085350000 | 0.172613465 |
| 2072160.5 | -5.353834634 | -0.967710535 | -0.281975206 |
| 2072320.5 | -5.021156416 | -1.975541542 | -0.723384669 |
| 2072480.5 | -4.453330824 | -2.890795223 | -1.130891254 |
| 2072640.5 | -3.675195926 | -3.669397421 | -1.484934062 |
| 2072800.5 | -2.720981674 | -4.271960874 | -1.767731426 |

## HELIOCENTRIC COORDINATES OF JUPITER

### REFERENCE AXES OF 1950.0

| JULIAN DATE | X | Y | Z |
|---|---|---|---|
| 2072960.5 | -1.633806779 | -4.665359903 | -1.963970191 |
| 2073120.5 | -0.464886379 | -4.824503077 | -2.061587282 |
| 2073280.5 | 0.727744851 | -4.734273789 | -2.052635565 |
| 2073440.5 | 1.881208430 | -4.391512743 | -1.934185182 |
| 2073600.5 | 2.930641394 | -3.806794088 | -1.709158321 |
| 2073760.5 | 3.813016489 | -3.005618543 | -1.386937481 |
| 2073920.5 | 4.471795112 | -2.028557443 | -0.983543764 |
| 2074080.5 | 4.861995831 | -0.929897371 | -0.521181234 |
| 2074240.5 | 4.954960954 | 0.225482066 | -0.027014449 |
| 2074400.5 | 4.741920565 | 1.366904638 | 0.468803620 |
| 2074560.5 | 4.235522160 | 2.423823235 | 0.935640124 |
| 2074720.5 | 3.468867434 | 3.331958758 | 1.345045059 |
| 2074880.5 | 2.492178786 | 4.038620623 | 1.673121284 |
| 2075040.5 | 1.367774665 | 4.506387540 | 1.902231539 |
| 2075200.5 | 0.164334095 | 4.714779525 | 2.021859711 |
| 2075360.5 | -1.048600733 | 4.660046310 | 2.028656952 |
| 2075520.5 | -2.205213665 | 4.353540148 | 1.925857042 |
| 2075680.5 | -3.246792103 | 3.819261089 | 1.722305991 |
| 2075840.5 | -4.123903350 | 3.091103045 | 1.431333080 |
| 2076000.5 | -4.797574091 | 2.210176172 | 1.069629776 |
| 2076160.5 | -5.239751944 | 1.222412586 | 0.656232448 |
| 2076320.5 | -5.433318903 | 0.176523091 | 0.211644712 |
| 2076480.5 | -5.371873068 | -0.877722787 | -0.242907095 |
| 2076640.5 | -5.059426337 | -1.890976477 | -0.686117294 |
| 2076800.5 | -4.510097007 | -2.815658304 | -1.097187749 |
| 2076960.5 | -3.747812610 | -3.607287525 | -1.456406824 |
| 2077120.5 | -2.805979394 | -4.225898924 | -1.745766352 |
| 2077280.5 | -1.727018543 | 4.607626721 | -1.949653837 |
| 2077440.5 | -0.561615933 | -4.816494738 | -2.055640445 |
| 2077600.5 | 0.632512105 | -4.746385250 | -2.055357876 |
| 2077760.5 | 1.792555040 | -4.423062539 | -1.945416284 |
| 2077920.5 | 2.853436395 | -3.856003620 | -1.728261627 |
| 2078080.5 | 3.751617422 | -3.069658155 | -1.412812126 |
| 2078240.5 | 4.429754070 | -2.103667484 | -1.014668766 |
| 2078400.5 | 4.841794406 | -1.011585836 | -0.555693183 |
| 2078560.5 | 4.957800809 | 0.142172229 | -0.062816482 |
| 2078720.5 | 4.767596048 | 1.287074709 | 0.433907688 |
| 2078880.5 | 4.282399681 | 2.352369239 | 0.903796248 |
| 2079040.5 | 3.533990186 | 3.273224894 | 1.318196792 |
| 2079200.5 | 2.571496879 | 3.996092412 | 1.652871083 |

HELIOCENTRIC COORDINATES OF JUPITER

REFERENCE AXES OF 1950.0

| JULIAN DATE | X | Y | Z |
|---|---|---|---|
| 2079360.5 | 1.456472230 | 4.482464456 | 1.889734007 |
| 2079520.5 | 0.257204075 | 4.710657740 | 2.017761515 |
| 2079680.5 | -0.956786892 | 4.675717655 | 2.033087157 |
| 2079840.5 | -2.119383403 | 4.387888071 | 1.938460414 |
| 2080000.5 | -3.171319820 | 3.870221258 | 1.742305851 |
| 2080160.5 | -4.062434443 | 3.155855879 | 1.457609739 |
| 2080320.5 | -4.752917197 | 2.285345753 | 1.100803401 |
| 2080480.5 | -5.213825280 | 1.304255914 | 0.690743105 |
| 2080640.5 | -5.427135492 | 0.261102203 | 0.247826015 |
| 2080800.5 | -5.385551896 | -0.794386978 | -0.206761558 |
| 2080960.5 | -5.092220135 | -1.812765905 | -0.651693793 |
| 2081120.5 | -4.560431473 | -2.746230408 | -1.066096678 |
| 2081280.5 | -3.813335797 | -3.549953964 | -1.430129294 |
| 2081440.5 | -2.883623529 | -4.183506304 | -1.725601358 |
| 2081600.5 | -1.813079469 | -4.612437114 | -1.936664914 |
| 2081760.5 | -0.651857383 | -4.810066041 | -2.050601707 |
| 2081920.5 | 0.542721786 | -4.759453835 | -2.058700372 |
| 2082080.5 | 1.708067810 | -4.455433387 | -1.957176774 |
| 2082240.5 | 2.779107533 | -3.906458134 | -1.748037960 |
| 2082400.5 | 3.692043619 | -3.135897907 | -1.439732739 |
| 2082560.5 | 4.388965878 | -2.182320114 | -1.047387472 |
| 2082720.5 | 4.822915739 | -1.098303317 | -0.592422624 |
| 2082880.5 | 4.962712695 | 0.052497376 | -0.101412304 |
| 2083040.5 | 4.796665336 | 1.199968584 | 0.395807493 |
| 2083200.5 | 4.334340673 | 2.273337044 | 0.868586243 |
| 2083360.5 | 3.605908922 | 3.207273078 | 1.288097911 |
| 2083520.5 | 2.659127167 | 3.947292253 | 1.629737393 |
| 2083680.5 | 1.554565616 | 4.453630334 | 1.874904837 |
| 2083840.5 | 0.360002306 | 4.703183643 | 2.011979434 |
| 2084000.5 | -0.855076992 | 4.689576329 | 2.036485712 |
| 2084160.5 | -2.024177997 | 4.421764372 | 1.950611061 |
| 2084320.5 | -3.087347215 | 3.921733755 | 1.762305288 |
| 2084480.5 | -3.993535692 | 3.221818870 | 1.484187373 |
| 2084640.5 | -4.701947844 | 2.362032591 | 1.132431889 |
| 2084800.5 | -5.182636526 | 1.387637488 | 0.725740206 |
| 2084960.5 | -5.416609562 | 0.347047228 | 0.284441382 |
| 2085120.5 | -5.395667894 | -0.709953022 | -0.170276587 |
| 2085280.5 | -5.122130926 | -1.733738984 | -0.617032094 |
| 2085440.5 | -4.608537341 | -2.676234257 | -1.034853387 |
| 2085600.5 | -3.877346315 | -3.492262349 | -1.403767012 |

HELIOCENTRIC COORDINATES OF JUPITER

REFERENCE AXES OF 1950.0

| JULIAN DATE | X | Y | Z |
|---|---|---|---|
| 2085760.5 | -2.960605157 | -4.140967324 | -1.705416565 |
| 2085920.5 | -1.899493343 | -4.587385385 | -1.923749397 |
| 2086080.5 | -0.743599922 | -4.804211037 | -2.045793587 |
| 2086240.5 | 0.450253589 | -4.773739409 | -2.062521962 |
| 2086400.5 | 1.619856125 | -4.489875703 | -1.969761600 |
| 2086560.5 | 2.700306433 | -3.959985239 | -1.769056217 |
| 2086720.5 | 3.627666309 | -3.206230256 | -1.468331853 |
| 2086880.5 | 4.343477852 | -2.265943048 | -1.082170036 |
| 2087040.5 | 4.799781522 | -1.190581122 | -0.631484165 |
| 2087200.5 | 4.963974187 | -0.042961507 | -0.142452430 |
| 2087360.5 | 4.822642855 | 1.107202518 | 0.355304298 |
| 2087520.5 | 4.383533249 | 2.189005285 | 0.831124256 |
| 2087680.5 | 3.675127209 | 3.136450888 | 1.255937451 |
| 2087840.5 | 2.743834622 | 3.893950155 | 1.604697863 |
| 2088000.5 | 1.649354601 | 4.420359738 | 1.858234931 |
| 2088160.5 | 0.459110103 | 4.691109455 | 2.004306701 |
| 2088320.5 | -0.757297952 | 4.698435191 | 2.037838454 |
| 2088480.5 | -1.932891668 | 4.450099429 | 1.960492871 |
| 2088640.5 | -3.006988525 | 3.967149424 | 1.779798486 |
| 2088800.5 | -3.927681671 | 3.281246975 | 1.508064304 |
| 2088960.5 | -4.653281981 | 2.431976039 | 1.161258931 |
| 2089120.5 | -5.152975921 | 1.464372436 | 0.757965477 |
| 2089280.5 | -5.406963545 | 0.426775627 | 0.318461656 |
| 2089440.5 | -5.406300849 | -0.631001018 | -0.136070813 |
| 2089600.5 | -5.152609921 | -1.659244797 | -0.584230344 |
| 2089760.5 | -4.657753380 | -2.609754185 | -1.005009625 |
| 2089920.5 | -3.943504925 | -3.437188634 | -1.378383492 |
| 2090080.5 | -3.041190941 | 4.100461750 | -1.685916206 |
| 2090240.5 | -1.991224939 | -4.564265874 | -1.911428011 |
| 2090400.5 | -0.842403528 | -4.800780335 | -2.041746669 |
| 2090560.5 | 0.349214981 | -4.791557239 | -2.067545755 |
| 2090720.5 | 1.522057021 | -4.529492784 | -1.984235324 |
| 2090880.5 | 2.611601543 | -4.020678685 | -1.792821575 |
| 2091040.5 | 3.553885995 | -3.285800931 | -1.500595436 |
| 2091200.5 | 4.289896106 | -2.360648001 | -1.121460583 |
| 2091360.5 | 4.770522993 | -1.295267226 | -0.675694758 |
| 2091520.5 | 4.961469991 | -0.151431560 | -0.188983854 |
| 2091680.5 | 4.847262546 | 1.001619653 | 0.309308033 |
| 2091840.5 | 4.433494443 | 2.092774406 | 0.788485991 |
| 2092000.5 | 3.746715767 | 3.055175753 | 1.219160033 |

## HELIOCENTRIC COORDINATES OF JUPITER

## REFERENCE AXES OF 1950.0

| JULIAN DATE | X | Y | Z |
|---|---|---|---|
| 2092160.5 | 2.831882537 | 3.831875436 | 1.575738024 |
| 2092320.5 | 1.747857246 | 4.380109997 | 1.838375398 |
| 2092480.5 | 0.561845841 | 4.673673211 | 1.994133477 |
| 2092640.5 | -0.656254318 | 4.703340408 | 2.037302949 |
| 2092800.5 | -1.838815957 | 4.475696423 | 1.969024177 |
| 2092960.5 | -2.924315667 | 4.010919577 | 1.796430784 |
| 2093120.5 | -3.859940861 | 3.340078903 | 1.531553433 |
| 2093280.5 | -4.603110080 | 2.502380291 | 1.190174530 |
| 2093440.5 | -5.122164340 | 1.542631579 | 0.790756300 |
| 2093600.5 | -5.396507164 | 0.509044742 | 0.353499935 |
| 2093760.5 | -5.416432057 | -0.548618856 | -0.100455664 |
| 2093920.5 | -5.182811927 | -1.580624133 | -0.549718669 |
| 2094080.5 | -4.706755583 | -2.538658827 | -0.973252971 |
| 2094240.5 | -4.009270985 | -3.377163170 | -1.350957917 |
| 2094400.5 | -3.120914652 | -4.054700703 | -1.664266082 |
| 2094560.5 | -2.081350124 | -4.535460074 | -1.896799583 |
| 2094720.5 | -0.938684545 | -4.790943418 | -2.035112110 |
| 2094880.5 | 0.251595126 | -4.801841110 | -2.069521039 |
| 2095040.5 | 1.428295386 | -4.560009371 | -1.994998959 |
| 2095200.5 | 2.527001230 | -4.070355155 | -1.812045637 |
| 2095360.5 | 3.483502928 | -3.352304208 | -1.527404410 |
| 2095520.5 | 4.238130779 | -2.440417196 | -1.154435204 |
| 2095680.5 | 4.740706247 | -1.383684017 | -0.712935014 |
| 2095840.5 | 4.955512432 | -0.243135221 | -0.228235776 |
| 2096000.5 | 4.865441344 | 0.912300959 | 0.270472150 |
| 2096160.5 | 4.474429446 | 2.011339649 | 0.752460085 |
| 2096320.5 | 3.807540205 | 2.986497108 | 1.188105184 |
| 2096480.5 | 2.908560367 | 3.779849503 | 1.551425511 |
| 2096640.5 | 1.835569174 | 4.347455886 | 1.822091481 |
| 2096800.5 | 0.655373069 | 4.661873332 | 1.986645147 |
| 2096960.5 | -0.562192433 | 4.712685830 | 2.038869056 |
| 2097120.5 | -1.749287111 | 4.505392368 | 1.979433508 |
| 2097280.5 | -2.843960369 | 4.059218673 | 1.815053426 |
| 2097440.5 | -3.792807485 | 3.404424556 | 1.557397734 |
| 2097600.5 | -4.552527048 | 2.579553232 | 1.221946418 |
| 2097760.5 | -5.090640419 | 1.628898630 | 0.826920321 |
| 2097920.5 | -5.385654896 | 0.600311396 | 0.392342575 |
| 2098080.5 | -5.426912908 | -0.456649955 | -0.060759519 |
| 2098240.5 | -5.214304408 | -1.492287298 | -0.511035469 |
| 2098400.5 | -4.757949603 | -2.458151657 | -0.937416200 |

## HELIOCENTRIC COORDINATES OF JUPITER

## REFERENCE AXES OF 1950.0

| JULIAN DATE | X | Y | Z |
|---|---|---|---|
| 2098560.5 | -4.077893130 | -3.308366301 | -1.319690148 |
| 2098720.5 | -3.203790086 | -4.000994212 | -1.639098643 |
| 2098880.5 | -2.174507121 | -4.499542381 | -1.878992189 |
| 2099040.5 | -1.037506983 | -4.774661697 | -2.025576189 |
| 2099200.5 | 0.152164291 | -4.806045715 | -2.068752038 |
| 2099360.5 | 1.333501009 | -4.584447927 | -2.003024382 |
| 2099520.5 | 2.441981273 | -4.113623688 | -1.828396455 |
| 2099680.5 | 3.412968610 | -3.411871607 | -1.551117245 |
| 2099840.5 | 4.186046510 | -2.512734204 | -1.184090711 |
| 2100000.5 | 4.710000246 | -1.464378689 | -0.746734067 |
| 2100160.5 | 4.947849646 | -0.327284813 | -0.264109949 |
| 2100320.5 | 4.881084954 | 0.829843395 | 0.234724416 |
| 2100480.5 | 4.512207950 | 1.935594121 | 0.719023993 |
| 2100640.5 | 3.864928005 | 2.922014561 | 1.159000655 |
| 2100800.5 | 2.981871336 | 3.730410458 | 1.528372164 |
| 2100960.5 | 1.920247620 | 4.315836889 | 1.806399820 |
| 2101120.5 | 0.746345308 | 4.649703721 | 1.979145294 |
| 2101280.5 | -0.470154499 | 4.720407665 | 2.039884112 |
| 2101440.5 | -1.661233778 | 4.532317505 | 1.988795821 |
| 2101600.5 | -2.764491611 | 4.103659435 | 1.832154097 |
| 2101760.5 | -3.725871347 | 3.463867845 | 1.581256552 |
| 2101920.5 | -4.501286244 | 2.650851213 | 1.251290580 |
| 2102080.5 | -5.057396100 | 1.708457081 | 0.860263548 |
| 2102240.5 | -5.371813614 | 0.684265327 | 0.428059809 |
| 2102400.5 | -5.432983690 | -0.372276983 | -0.024363673 |
| 2102560.5 | -5.239915772 | -1.411438988 | -0.475664329 |
| 2102720.5 | -4.801880090 | -2.384606189 | -0.904723141 |
| 2102880.5 | -4.138112831 | -3.245610181 | -1.291223330 |
| 2103040.5 | -3.277514290 | -3.952098414 | -1.616246418 |
| 2103200.5 | -2.258266473 | -4.467037973 | -1.862927882 |
| 2103360.5 | -1.127241172 | -4.760414647 | -2.017201971 |
| 2103520.5 | 0.060980786 | -4.811133725 | -2.068642941 |
| 2103680.5 | 1.245693121 | -4.609045258 | -2.011374863 |
| 2103840.5 | 2.362462754 | -4.156903193 | -1.844973374 |
| 2104000.5 | 3.346486791 | -3.471938236 | -1.575225246 |
| 2104160.5 | 4.136886344 | -2.586611754 | -1.214559276 |
| 2104320.5 | 4.681657413 | -1.548077082 | -0.781938142 |
| 2104480.5 | 4.942698089 | -0.415973504 | -0.302035898 |
| 2104640.5 | 4.900086473 | 0.741543032 | 0.196360733 |
| 2104800.5 | 4.554726049 | 1.853210129 | 0.682610855 |

HELIOCENTRIC COORDINATES OF JUPITER

REFERENCE AXES OF 1950.0

| JULIAN DATE | X | Y | Z |
|---|---|---|---|
| 2104960.5 | 3.928701740 | 2.850764700 | 1.126833862 |
| 2105120.5 | 3.063174594 | 3.674751947 | 1.502457798 |
| 2105280.5 | 2.014215636 | 4.279100424 | 1.788288981 |
| 2105440.5 | 0.847411446 | 4.633866709 | 1.969825839 |
| 2105600.5 | -0.367787258 | 4.726026696 | 2.039739875 |
| 2105760.5 | -1.563156976 | 4.558605106 | 1.997628674 |
| 2105920.5 | -2.675739454 | 4.148666888 | 1.849253632 |
| 2106080.5 | -3.650674161 | 3.524727922 | 1.605498116 |
| 2106240.5 | -4.442927554 | 2.724039140 | 1.281243711 |
| 2106400.5 | -5.018157043 | 1.790039042 | 0.894297229 |
| 2106560.5 | -5.352981925 | 0.770116213 | 0.464436125 |
| 2106720.5 | -5.434904024 | -0.286294061 | 0.012588715 |
| 2106880.5 | -5.262061076 | -1.329330855 | -0.439868476 |
| 2107040.5 | -4.842928235 | -2.310139342 | -0.871733311 |
| 2107200.5 | -4.196017696 | -3.182222437 | -1.262566279 |
| 2107360.5 | -3.349565853 | -3.902820534 | -1.593291107 |
| 2107520.5 | -2.341140348 | -4.434412145 | -1.846849320 |
| 2107680.5 | -1.217044946 | -4.746397240 | -2.008938343 |
| 2107840.5 | -0.031351003 | -4.816973261 | -2.068841697 |
| 2108000.5 | 1.155649958 | -4.635135146 | -2.020326921 |
| 2108160.5 | 2.279778508 | -4.202623244 | -1.862540685 |
| 2108320.5 | 3.276230082 | -3.535516368 | -1.600774648 |
| 2108480.5 | 4.083766308 | -2.665051747 | -1.246922310 |
| 2108640.5 | 4.649590717 | -1.637203819 | -0.819419675 |
| 2108800.5 | 4.934365672 | -0.510636103 | -0.342490414 |
| 2108960.5 | 4.916568832 | 0.647102190 | 0.155379594 |
| 2109120.5 | 4.595304936 | 1.764858761 | 0.643641812 |
| 2109280.5 | 3.990885112 | 2.773936645 | 1.092273622 |
| 2109440.5 | 3.142947922 | 3.613954792 | 1.474343222 |
| 2109600.5 | 2.106466755 | 4.237559665 | 1.768133590 |
| 2109760.5 | 0.946440376 | 4.613351690 | 1.958532306 |
| 2109920.5 | -0.267766554 | 4.726856295 | 2.037589964 |
| 2110080.5 | -1.467593899 | 4.579794753 | 2.004335884 |
| 2110240.5 | -2.589440160 | 4.188161929 | 1.864055876 |
| 2110400.5 | -3.577616351 | 3.579670065 | 1.627267677 |
| 2110560.5 | -4.386180494 | 2.791025305 | 1.308596620 |
| 2110720.5 | -4.979880040 | 1.865347599 | 0.925690624 |
| 2110880.5 | -5.334471743 | 0.849889069 | 0.498247799 |
| 2111040.5 | -5.436662936 | -0.205915988 | 0.047176899 |
| 2111200.5 | -5.283861383 | -1.252115801 | -0.406124601 |

HELIOCENTRIC COORDINATES OF JUPITER

REFERENCE AXES OF 1950.0

| JULIAN DATE | X | Y | Z |
|---|---|---|---|
| 2111360.5 | -4.883856640 | -2.239707728 | -0.840408895 |
| 2111520.5 | -4.254491744 | -3.122007245 | -1.235174136 |
| 2111680.5 | -3.423322287 | -3.856020412 | -1.571262224 |
| 2111840.5 | -2.427203877 | -4.403912724 | -1.831499827 |
| 2112000.5 | -1.311697169 | -4.734642600 | -2.001414848 |
| 2112160.5 | -0.130130658 | -4.825780016 | -2.070041090 |
| 2112320.5 | 1.057876357 | -4.665456378 | -2.030792823 |
| 2112480.5 | 2.188635903 | -4.254288916 | -1.882347010 |
| 2112640.5 | 3.197515257 | -3.606999024 | -1.629416514 |
| 2112800.5 | 4.022987380 | -2.753325001 | -1.283243286 |
| 2112960.5 | 4.611405536 | -1.737766327 | -0.861607587 |
| 2113120.5 | 4.922006446 | -0.617754441 | -0.388166898 |
| 2113280.5 | 4.931374452 | 0.539925890 | 0.108970209 |
| 2113440.5 | 4.636481133 | 1.664253896 | 0.599366800 |
| 2113600.5 | 4.055565368 | 2.685983036 | 1.052817782 |
| 2113760.5 | 3.226547911 | 3.543591834 | 1.441946085 |
| 2113920.5 | 2.203245366 | 4.188196633 | 1.744407738 |
| 2114080.5 | 1.050135586 | 4.586726325 | 1.944386374 |
| 2114240.5 | -0.163354222 | 4.723119097 | 2.033249841 |
| 2114400.5 | -1.368155102 | 4.597748557 | 2.009436030 |
| 2114560.5 | -2.499877069 | 4.225571732 | 1.877764958 |
| 2114720.5 | -3.501913232 | 3.633570929 | 1.648417149 |
| 2114880.5 | -4.327382413 | 2.857982607 | 1.335791528 |
| 2115040.5 | -4.940131429 | 1.941651409 | 0.957393637 |
| 2115200.5 | -5.315074108 | 0.931686071 | 0.532837008 |
| 2115360.5 | -5.438125308 | -0.122535443 | 0.082984464 |
| 2115520.5 | -5.305931699 | -1.171050716 | -0.370782419 |
| 2115680.5 | -4.925532450 | -2.164775449 | -0.807200136 |
| 2115840.5 | -4.314015623 | -3.056852694 | -1.205711494 |
| 2116000.5 | -3.498173827 | -3.803999575 | -1.547052787 |
| 2116160.5 | -2.514104997 | -4.367949863 | -1.813883529 |
| 2116320.5 | -1.406649282 | -4.717063447 | -1.991491965 |
| 2116480.5 | -0.228502488 | -4.828127516 | -2.068591032 |
| 2116640.5 | 0.961192393 | -4.688302839 | -2.038189656 |
| 2116800.5 | 2.098984961 | -4.297068129 | -1.898481485 |
| 2116960.5 | 3.120155351 | -3.667891920 | -1.653638944 |
| 2117120.5 | 3.962693203 | -2.829238741 | -1.314344772 |
| 2117280.5 | 4.572039915 | -1.824443673 | -0.897856425 |
| 2117440.5 | 4.906123270 | -0.710032894 | -0.427410145 |
| 2117600.5 | 4.939930315 | 0.447717731 | 0.069143453 |

HELIOCENTRIC COORDINATES OF JUPITER

REFERENCE AXES OF 1950.0

| JULIAN DATE | X | Y | Z |
|---|---|---|---|
| 2117760.5 | 4.668723620 | 1.577813865 | 0.561420669 |
| 2117920.5 | 4.109129294 | 2.610547146 | 1.019057333 |
| 2118080.5 | 3.297736190 | 3.483524811 | 1.414334596 |
| 2118240.5 | 2.287421487 | 4.146721450 | 1.724441591 |
| 2118400.5 | 1.142137848 | 4.565810895 | 1.933037270 |
| 2118560.5 | -0.068843140 | 4.723495575 | 2.030964696 |
| 2118720.5 | -1.276311289 | 4.619015808 | 2.016173663 |
| 2118880.5 | -2.415510779 | 4.266334056 | 1.893049675 |
| 2119040.5 | -3.429301936 | 3.691584364 | 1.671396865 |
| 2119200.5 | -4.270143855 | 2.930297683 | 1.365295383 |
| 2119360.5 | -4.901122734 | 2.024751919 | 0.991988625 |
| 2119520.5 | -5.296308838 | 1.021628106 | 0.570885235 |
| 2119680.5 | -5.440705877 | -0.029978994 | 0.122703208 |
| 2119840.5 | -5.329998621 | -1.080235977 | -0.331256011 |
| 2120000.5 | -4.970234797 | -2.080021151 | -0.769738530 |
| 2120160.5 | -4.377508993 | -2.982261165 | -1.172120103 |
| 2120320.5 | -3.577653502 | -3.743267271 | -1.518987898 |
| 2120480.5 | -2.605882599 | -4.324172598 | -1.792766821 |
| 2120640.5 | -1.506280973 | -4.692545236 | -1.978425259 |
| 2120800.5 | -0.330974787 | -4.824206241 | -2.064276760 |
| 2120960.5 | 0.861217419 | -4.705210236 | -2.042864534 |
| 2121120.5 | 2.006870610 | -4.333844819 | -1.911872395 |
| 2121280.5 | 3.040952835 | -3.722378458 | -1.674950419 |
| 2121440.5 | 3.900792579 | -2.898159230 | -1.342285834 |
| 2121600.5 | 4.530845489 | -1.903588162 | -0.930710354 |
| 2121760.5 | 4.887792730 | -0.794529757 | -0.463144371 |
| 2121920.5 | 4.945210954 | 0.363064267 | 0.032737134 |
| 2122080.5 | 4.696904437 | 1.498157533 | 0.526570703 |
| 2122240.5 | 4.158110198 | 2.540644741 | 0.987858781 |
| 2122400.5 | 3.364202562 | 3.427427812 | 1.388612361 |
| 2122560.5 | 2.367106516 | 4.107539096 | 1.705641339 |
| 2122720.5 | 1.230147972 | 4.545558006 | 1.922150656 |
| 2122880.5 | 0.022323116 | 4.723030298 | 2.028492846 |
| 2123040.5 | -1.187098180 | 4.638065571 | 2.022129597 |
| 2123200.5 | -2.333013950 | 4.303596443 | 1.906994564 |
| 2123360.5 | -3.357716678 | 3.744879232 | 1.692500047 |
| 2123520.5 | -4.212942420 | 2.996744786 | 1.392406984 |
| 2123680.5 | -4.860957628 | 2.100953458 | 1.023715713 |
| 2123840.5 | -5.274962101 | 1.103843386 | 0.605665690 |
| 2124000.5 | -5.439071510 | 0.054327577 | 0.158874587 |

## HELIOCENTRIC COORDINATES OF JUPITER

## REFERENCE AXES OF 1950.0

| JULIAN DATE | X | Y | Z |
|---|---|---|---|
| 2124160.5 | -5.348087535 | -0.997795007 | -0.295393007 |
| 2124320.5 | -5.007195094 | -2.003300720 | -0.735860912 |
| 2124480.5 | -4.431658180 | -2.914884413 | -1.141825968 |
| 2124640.5 | -3.646522909 | -3.688497829 | -1.493740429 |
| 2124800.5 | -2.686277680 | -4.284785829 | -1.773837442 |
| 2124960.5 | -1.594364009 | -4.670697602 | -1.966835440 |
| 2125120.5 | -0.422378485 | -4.821305384 | -2.060739390 |
| 2125280.5 | 0.771236485 | -4.721792347 | -2.047727496 |
| 2125440.5 | 1.923231941 | -4.369469531 | -1.925068608 |
| 2125600.5 | 2.968504165 | -3.775554444 | -1.695959967 |
| 2125760.5 | 3.844020811 | -2.966317536 | -1.370117533 |
| 2125920.5 | 4.493559617 | -1.983125472 | -0.963912466 |
| 2126080.5 | 4.872799481 | -0.880945250 | -0.499855159 |
| 2126240.5 | 4.954018122 | 0.274922682 | -0.005308727 |
| 2126400.5 | 4.729502922 | 1.413740749 | 0.489529496 |
| 2126560.5 | 4.212889226 | 2.465274342 | 0.954146489 |
| 2126720.5 | 3.438034468 | 3.365849732 | 1.360344089 |
| 2126880.5 | 2.455601500 | 4.063529386 | 1.684547448 |
| 2127040.5 | 1.328039793 | 4.521642141 | 1.909446634 |
| 2127200.5 | 0.123920116 | 4.720349994 | 2.024810871 |
| 2127360.5 | -1.087467327 | 4.656393194 | 2.027513598 |
| 2127520.5 | -2.240616032 | 4.341469629 | 1.920948414 |
| 2127680.5 | -3.277123465 | 3.799814269 | 1.714071916 |
| 2127840.5 | -4.147842146 | 3.065486511 | 1.420291897 |
| 2128000.5 | -4.814059080 | 2.179727248 | 1.056361314 |
| 2128160.5 | -5.247972365 | 1.188586155 | 0.641370824 |
| 2128320.5 | -5.432725404 | 0.140885616 | 0.195874920 |
| 2128480.5 | -5.362208135 | -0.913504690 | 0.250054207 |
| 2128640.5 | -5.040768233 | -1.925162625 | -0.701476356 |
| 2128800.5 | -4.482908279 | -2.846483109 | -1.111178615 |
| 2128960.5 | -3.712982276 | -3.633032890 | -1.468265542 |
| 2129120.5 | -2.764846647 | -4.244991424 | -1.754786565 |
| 2129280.5 | -1.681365266 | -4.648754544 | -1.955237326 |
| 2129440.5 | -0.513616017 | -4.818734883 | -2.057352911 |
| 2129600.5 | 0.680395703 | -4.739325876 | -2.052983649 |
| 2129760.5 | 1.837728220 | -4.406899976 | -1.939003160 |
| 2129920.5 | 2.893379128 | -3.831590417 | -1.718145441 |
| 2130080.5 | 3.784114834 | -3.038480385 | -1.399611142 |
| 2130240.5 | 4.453118235 | -2.067739140 | -0.999241818 |
| 2130400.5 | 4.855033986 | -0.973265816 | -0.539062985 |

HELIOCENTRIC COORDINATES OF JUPITER

REFERENCE AXES OF 1950.0

| JULIAN DATE | X | Y | Z |
|---|---|---|---|
| 2130560.5 | 4.960700356 | 0.180413163 | -0.046069103 |
| 2130720.5 | 4.760686612 | 1.322892203 | 0.449728118 |
| 2130880.5 | 4.266832117 | 2.383740553 | 0.917776301 |
| 2131040.5 | 3.511349872 | 3.298567083 | 1.329610195 |
| 2131200.5 | 2.543610530 | 4.014296134 | 1.661198332 |
| 2131360.5 | 1.425249369 | 4.492862265 | 1.894652203 |
| 2131520.5 | 0.224533601 | 4.712960104 | 2.019118330 |
| 2131680.5 | -0.989088447 | 4.669948910 | 2.030873066 |
| 2131840.5 | -2.149591122 | 4.374342679 | 1.932788640 |
| 2132000.5 | -3.197810845 | 3.849445955 | 1.733403199 |
| 2132160.5 | -4.083709087 | 3.128651601 | 1.445816322 |
| 2132320.5 | -4.767641897 | 2.252776959 | 1.086576296 |
| 2132480.5 | -5.220899661 | 1.267653436 | 0.674658057 |
| 2132640.5 | -5.425777846 | 0.222047213 | 0.230571607 |
| 2132800.5 | -5.375390430 | -0.834107883 | -0.224402058 |
| 2132960.5 | -5.073372634 | -1.851241619 | -0.668876616 |
| 2133120.5 | -4.533555516 | -2.781544180 | -1.081968936 |
| 2133280.5 | -3.779634836 | -3.580332128 | -1.443894062 |
| 2133440.5 | -2.844797830 | -4.207477563 | -1.736588062 |
| 2133600.5 | -1.771216211 | -4.628976874 | -1.944394209 |
| 2133760.5 | -0.609263714 | -4.818700332 | -2.054833867 |
| 2133920.5 | 0.583725677 | -4.760301877 | -2.059457783 |
| 2134080.5 | 1.745364377 | -4.449177089 | -1.954734680 |
| 2134240.5 | 2.810965481 | -3.894240131 | -1.742883217 |
| 2134400.5 | 3.717231831 | -3.119165609 | -1.432497351 |
| 2134560.5 | 4.406778096 | -2.162649082 | -1.038770886 |
| 2134720.5 | 4.833111700 | -1.077242150 | -0.583121102 |
| 2134880.5 | 4.965405760 | 0.073524667 | -0.092071677 |
| 2135040.5 | 4.792202628 | 1.219688541 | 0.404608204 |
| 2135200.5 | 4.323221125 | 2.290595763 | 0.876325226 |
| 2135360.5 | 3.588764545 | 3.220985599 | 1.294290174 |
| 2135520.5 | 2.636761696 | 3.956414697 | 1.633921825 |
| 2135680.5 | 1.528018145 | 4.457182430 | 1.876651441 |
| 2135840.5 | 0.330594154 | 4.700323903 | 2.010918927 |
| 2136000.5 | -0.885744409 | 4.679709466 | 2.032353764 |
| 2136160.5 | -2.054286338 | 4.404649910 | 1.943294379 |
| 2136320.5 | -3.114977485 | 3.897567055 | 1.751877186 |
| 2136480.5 | -4.016817010 | 3.191266210 | 1.470924567 |
| 2136640.5 | -4.719214222 | 2.326214169 | 1.116809354 |
| 2136800.5 | -5.192569795 | 1.348062201 | 0.708405225 |

HELICCENTRIC COORDINATES OF JUPITER

REFERENCE AXES OF 1950.0

| JULIAN DATE | X | Y | Z |
|---|---|---|---|
| 2136960.5 | -5.418352271 | 0.305509147 | 0.266170647 |
| 2137120.5 | -5.388896200 | -0.751503819 | -0.188631645 |
| 2137280.5 | -5.107082734 | -1.773338123 | -0.634606270 |
| 2137440.5 | -4.585994960 | -2.712044830 | -1.050829631 |
| 2137600.5 | -3.848575982 | -3.522705147 | -1.417434276 |
| 2137760.5 | -2.927258526 | -4.164828834 | -1.716218854 |
| 2137920.5 | -1.863479013 | -4.603893936 | -1.931321649 |
| 2138080.5 | -0.706936521 | -4.813073214 | -2.049980675 |
| 2138240.5 | 0.485587567 | -4.775132093 | -2.063378388 |
| 2138400.5 | 1.652057643 | -4.484394482 | -1.967531688 |
| 2138560.5 | 2.727850112 | -3.948557491 | -1.764137886 |
| 2138720.5 | 3.649359808 | -3.190009692 | -1.461230811 |
| 2138880.5 | 4.358467464 | -2.246208611 | -1.073454249 |
| 2139040.5 | 4.807520381 | -1.168662718 | -0.621748676 |
| 2139200.5 | 4.964179941 | -0.020204733 | -0.132300695 |
| 2139360.5 | 4.815276850 | 1.129439104 | 0.365264254 |
| 2139520.5 | 4.368822855 | 2.209345937 | 0.840279014 |
| 2139680.5 | 3.653627915 | 3.153524876 | 1.263677023 |
| 2139840.5 | 2.716499481 | 3.906462851 | 1.610444795 |
| 2140000.5 | 1.617561628 | 4.427209910 | 1.861494084 |
| 2140160.5 | 0.424608876 | 4.691525679 | 2.004724934 |
| 2140320.5 | -0.792522453 | 4.692091464 | 2.035256572 |
| 2140480.5 | -1.966804590 | 4.437173565 | 1.954973213 |
| 2140640.5 | -3.037694622 | 3.948307481 | 1.771620292 |
| 2140800.5 | -3.953580603 | 3.257561281 | 1.497689621 |
| 2140960.5 | -4.673167010 | 2.404800112 | 1.149278816 |
| 2141120.5 | -5.166070569 | 1.435201084 | 0.745038656 |
| 2141280.5 | -5.412905152 | 0.397116838 | 0.305257504 |
| 2141440.5 | -5.405088367 | -0.659721181 | -0.148915270 |
| 2141600.5 | -5.144535140 | -1.685733168 | -0.596134323 |
| 2141760.5 | -4.643333590 | -2.632860298 | -1.015454596 |
| 2141920.5 | -3.923433302 | -3.455887392 | -1.386907233 |
| 2142080.5 | -3.016316049 | -4.113826975 | -1.692102304 |
| 2142240.5 | -1.962562244 | -4.571452463 | -1.914898181 |
| 2142400.5 | -0.811172239 | -4.801031040 | -2.042162803 |
| 2142560.5 | 0.381545565 | -4.784246664 | -2.064627384 |
| 2142720.5 | 1.553735532 | -4.514214533 | -1.977795553 |
| 2142880.5 | 2.640606071 | -3.997370463 | -1.782819932 |
| 2143040.5 | 3.578007285 | -3.254887132 | -1.487199500 |
| 2143200.5 | 4.306909269 | -2.323162989 | -1.105101033 |

HELIOCENTRIC COORDINATES OF JUPITER

REFERENCE AXES OF 1950.0

| JULIAN DATE | X | Y | Z |
|---|---|---|---|
| 2143360.5 | 4.778440956 | -1.252911944 | -0.657093428 |
| 2143520.5 | 4.958839968 | -0.106519763 | -0.169135262 |
| 2143680.5 | 4.833438190 | 1.046345794 | 0.329211768 |
| 2143840.5 | 4.408806974 | 2.134440254 | 0.807180244 |
| 2144000.5 | 3.712487540 | 3.091122526 | 1.235458475 |
| 2144160.5 | 2.790271688 | 3.859980413 | 1.588674648 |
| 2144320.5 | 1.701572702 | 4.399006860 | 1.847305964 |
| 2144480.5 | 0.513805378 | 4.682834799 | 1.998778459 |
| 2144640.5 | -0.703246013 | 4.703028080 | 2.037732102 |
| 2144800.5 | -1.882312763 | 4.466812200 | 1.965597102 |
| 2144960.5 | -2.962373994 | 3.994810570 | 1.789714347 |
| 2145120.5 | -3.891165522 | 3.318339830 | 1.522235998 |
| 2145280.5 | -4.626623730 | 2.476686609 | 1.178991611 |
| 2145440.5 | -5.137529475 | 1.514620178 | 0.778435682 |
| 2145600.5 | -5.403628955 | 0.480245926 | 0.340728594 |
| 2145760.5 | -5.415465431 | -0.576803609 | -0.113044634 |
| 2145920.5 | -5.174089922 | -1.606909746 | -0.561543505 |
| 2146080.5 | -4.690749744 | -2.561843722 | -0.983770896 |
| 2146240.5 | -3.986588646 | -3.396091651 | -1.359649672 |
| 2146400.5 | -3.092332107 | -4.068237133 | -1.670624427 |
| 2146560.5 | -2.047874232 | -4.542491185 | -1.900328975 |
| 2146720.5 | -0.901627814 | -4.790423289 | -2.035346119 |
| 2146880.5 | 0.290549244 | -4.792888401 | -2.066062159 |
| 2147040.5 | 1.467067873 | -4.542056118 | -1.987581244 |
| 2147200.5 | 2.563167917 | -4.043329425 | -1.800613362 |
| 2147360.5 | 3.514448632 | -3.316807032 | -1.512190376 |
| 2147520.5 | 4.261314572 | -2.397836692 | -1.136014228 |
| 2147680.5 | 4.754005581 | -1.336189151 | -0.692227058 |
| 2147840.5 | 4.957568674 | -0.193515070 | -0.206442353 |
| 2148000.5 | 4.855906613 | 0.960941922 | 0.291995866 |
| 2148160.5 | 4.454034014 | 2.055967813 | 0.772372128 |
| 2148320.5 | 3.777958446 | 3.024520893 | 1.205241087 |
| 2148480.5 | 2.872125067 | 3.809382268 | 1.564918295 |
| 2148640.5 | 1.794928778 | 4.367419234 | 1.831423194 |
| 2148800.5 | 0.613182639 | 4.671950230 | 1.991632594 |
| 2148960.5 | -0.603491598 | 4.713178940 | 2.039606666 |
| 2149120.5 | -1.787585494 | 4.497052584 | 1.976220939 |
| 2149280.5 | -2.877513412 | 4.043091628 | 1.808328782 |
| 2149440.5 | -3.820216145 | 3.381738595 | 1.547687322 |
| 2149600.5 | -4.572698027 | 2.551652390 | 1.209833422 |

HELIOCENTRIC COORDINATES OF JUPITER

REFERENCE AXES OF 1950.0

| JULIAN DATE | X | Y | Z |
|---|---|---|---|
| 2149760.5 | -5.102750388 | 1.597211840 | 0.813029304 |
| 2149920.5 | -5.389132832 | 0.566343741 | 0.377334064 |
| 2150080.5 | -5.421445627 | -0.491318869 | -0.076190631 |
| 2150240.5 | -5.199863597 | -1.526010918 | -0.526163376 |
| 2150400.5 | -4.734833983 | -2.489240239 | -0.951494056 |
| 2150560.5 | -4.046776136 | -3.335133852 | -1.331969618 |
| 2150720.5 | -3.165761112 | -4.021832198 | -1.648860950 |
| 2150880.5 | -2.131092652 | -4.513019791 | -1.885590804 |
| 2151040.5 | -0.990657134 | -4.779646168 | -2.028489220 |
| 2151200.5 | 0.200137360 | -4.801834430 | -2.067639736 |
| 2151360.5 | 1.380042392 | -4.570890259 | -1.997784207 |
| 2151520.5 | 2.484471654 | -4.091211592 | -1.819205354 |
| 2151680.5 | 3.448949854 | -3.381771090 | -1.538449362 |
| 2151840.5 | 4.213466640 | -2.476735398 | -1.168700175 |
| 2152000.5 | 4.727435520 | -1.424755554 | -0.729598031 |
| 2152160.5 | 4.954656303 | -0.286577465 | -0.246336448 |
| 2152320.5 | 4.877443647 | 0.869086819 | 0.252008621 |
| 2152480.5 | 4.499050234 | 1.971063793 | 0.734783095 |
| 2152640.5 | 3.843767696 | 2.951823665 | 1.172374792 |
| 2152800.5 | 2.954589082 | 3.753189412 | 1.538723936 |
| 2152960.5 | 1.888883158 | 4.330738344 | 1.813321486 |
| 2153120.5 | 0.712937523 | 4.656344678 | 1.982437055 |
| 2153280.5 | -0.503664727 | 4.718785470 | 2.039519065 |
| 2153440.5 | -1.693046276 | 4.522731707 | 1.984885776 |
| 2153600.5 | -2.792956928 | 4.086657909 | 1.824924888 |
| 2153760.5 | -3.749492235 | 3.440220931 | 1.571035232 |
| 2153920.5 | -4.518729575 | 2.621546886 | 1.238502075 |
| 2154000.5 | -5.067525997 | 1.674704765 | 0.845431774 |
| 2154240.5 | -5.373748250 | 0.647494784 | 0.411807484 |
| 2154400.5 | -5.426170757 | -0.410433729 | -0.041321936 |
| 2154560.5 | -5.224214873 | -1.449192376 | -0.492540262 |
| 2154720.5 | -4.777635045 | -2.420087405 | -0.920688004 |
| 2154880.5 | -4.106195380 | -3.276985420 | -1.305457429 |
| 2155040.5 | -3.239323664 | -3.977712412 | -1.628002051 |
| 2155200.5 | -2.215673681 | -4.485568518 | -1.871597768 |
| 2155360.5 | -1.082472534 | -4.771012307 | -2.022382172 |
| 2155520.5 | 0.105516777 | -4.813520141 | -2.070178810 |
| 2155680.5 | 1.287617874 | -4.603548233 | -2.009380245 |
| 2155840.5 | 2.399646264 | -4.144417136 | -1.839817326 |
| 2156000.5 | 3.377231156 | -3.453813250 | -1.567486748 |

HELIOCENTRIC COORDINATES OF JUPITER

REFERENCE AXES OF 1950.0

| JULIAN DATE | X | Y | Z |
|---|---|---|---|
| 2156160.5 | 4.160036902 | -2.564490834 | -1.204957569 |
| 2156320.5 | 4.696624136 | -1.523715589 | -0.771253093 |
| 2156480.5 | 4.949395845 | -0.391081638 | -0.291036177 |
| 2156640.5 | 4.898825349 | 0.765402384 | 0.206966791 |
| 2156800.5 | 4.546095144 | 1.874657899 | 0.692195935 |
| 2156960.5 | 3.913485659 | 2.868592278 | 1.134849417 |
| 2157120.5 | 3.042321401 | 3.687891394 | 1.508422572 |
| 2157280.5 | 1.988848151 | 4.286612490 | 1.791783192 |
| 2157440.5 | 0.818853976 | 4.634966181 | 1.970500207 |
| 2157600.5 | -0.397998978 | 4.720143594 | 2.037339889 |
| 2157760.5 | -1.593304921 | 4.545463884 | 1.992027236 |
| 2157920.5 | -2.703997530 | 4.128359631 | 1.840482147 |
| 2158080.5 | -3.675218950 | 3.497760915 | 1.593767131 |
| 2158240.5 | -4.462068723 | 2.691340680 | 1.266947587 |
| 2158400.5 | -5.030468600 | 1.752925448 | 0.878001122 |
| 2158560.5 | -5.357417991 | 0.730219984 | 0.446846583 |
| 2158720.5 | -5.430887067 | -0.327125778 | -0.005488928 |
| 2158880.5 | -5.249535003 | -1.369157494 | -0.457581809 |
| 2159040.5 | -4.822372462 | -2.347057353 | -0.888239282 |
| 2159200.5 | -4.168421275 | -3.214494625 | -1.277087962 |
| 2159360.5 | -3.316363266 | -3.928994266 | -1.605170439 |
| 2159520.5 | -2.304113154 | -4.453416781 | -1.855591555 |
| 2159680.5 | -1.178196430 | -4.757611812 | -2.014244132 |
| 2159840.5 | 0.007235519 | -4.820257052 | -2.070623337 |
| 2160000.5 | 1.191955010 | -4.630817054 | -2.018706573 |
| 2160160.5 | 2.311975911 | -4.191451158 | -1.857830625 |
| 2160320.5 | 3.302785895 | -3.518574887 | -1.593442690 |
| 2160480.5 | 4.103496325 | -2.643663819 | -1.237548848 |
| 2160640.5 | 4.661676015 | -1.612836038 | -0.808654894 |
| 2160800.5 | 4.938338808 | -0.484824760 | -0.331019552 |
| 2160960.5 | 4.912293373 | 0.672798021 | 0.166859415 |
| 2161120.5 | 4.582971213 | 1.788886723 | 0.654436400 |
| 2161280.5 | 3.971032477 | 2.794785482 | 1.101707230 |
| 2161440.5 | 3.116502707 | 3.630216358 | 1.481784728 |
| 2161600.5 | 2.074765684 | 4.248026578 | 1.773037769 |
| 2161760.5 | 0.911202234 | 4.617136208 | 1.960491293 |
| 2161920.5 | -0.304543865 | 4.723500582 | 2.036382638 |
| 2162080.5 | -1.503800746 | 4.569341145 | 1.999960941 |
| 2162240.5 | -2.623048291 | 4.171161217 | 1.856738010 |
| 2162400.5 | -3.606854278 | 3.557123389 | 1.617434318 |

HELIOCENTRIC COORDINATES OF JUPITER

REFERENCE AXES OF 1950.0

| JULIAN DATE | X | Y | Z |
|---|---|---|---|
| 2162560.5 | -4.409660267 | 2.764273750 | 1.296830972 |
| 2162720.5 | -4.996661541 | 1.835934037 | 0.912671463 |
| 2162880.5 | -5.344065871 | 0.819418124 | 0.484687343 |
| 2163040.5 | -5.438986872 | -0.235897521 | 0.033766850 |
| 2163200.5 | -5.279165856 | -1.280201207 | -0.418750952 |
| 2163360.5 | -4.872645673 | -2.264666822 | -0.851694510 |
| 2163520.5 | -4.237452530 | -3.142781028 | -1.244637726 |
| 2163680.5 | -3.401279753 | -3.871689315 | -1.578485955 |
| 2163840.5 | -2.401109591 | -4.413658812 | -1.836113696 |
| 2164000.5 | -1.282653723 | -4.737726056 | -2.003086228 |
| 2164160.5 | -0.099442103 | -4.821547935 | -2.068477252 |
| 2164320.5 | 1.088648070 | -4.653397198 | -2.025762533 |
| 2164480.5 | 2.217637695 | -4.234136882 | -1.873723715 |
| 2164640.5 | 3.222627036 | -3.578878642 | -1.617239620 |
| 2164800.5 | 4.041933620 | -2.717906827 | -1.267785925 |
| 2164960.5 | 4.621967154 | -1.696387572 | -0.843431288 |
| 2165120.5 | 4.922314838 | -0.572449925 | -0.368139469 |
| 2165280.5 | 4.920232658 | 0.586525453 | 0.129713197 |
| 2165440.5 | 4.613630466 | 1.709165513 | 0.619522592 |
| 2165600.5 | 4.021816506 | 2.726219470 | 1.071064674 |
| 2165760.5 | 3.183729489 | 3.576537377 | 1.457107298 |
| 2165920.5 | 2.153978052 | 4.211922100 | 1.755592594 |
| 2166080.5 | 0.997490521 | 4.600176043 | 1.951078477 |
| 2166240.5 | -0.216228515 | 4.726145216 | 2.035329743 |
| 2166400.5 | -1.418359533 | 4.591011176 | 2.007144624 |
| 2166560.5 | -2.544992744 | 4.210350968 | 1.871626422 |
| 2166720.5 | -3.540120103 | 3.611544420 | 1.639144212 |
| 2166880.5 | -4.357476848 | 2.831013761 | 1.324192942 |
| 2167040.5 | -4.961469214 | 1.911618250 | 0.944297423 |
| 2167200.5 | -5.327470210 | 0.900364636 | 0.519036277 |
| 2167360.5 | -5.441739588 | -0.153531882 | 0.069207088 |
| 2167520.5 | -5.301164196 | -1.200284010 | -0.383882881 |
| 2167680.5 | -4.912945365 | -2.190959151 | -0.819037293 |
| 2167840.5 | -4.294292720 | -3.078808597 | -1.215749251 |
| 2168000.5 | -3.472121790 | -3.820611792 | -1.554786198 |
| 2168160.5 | -2.482694646 | -4.378135016 | -1.818825374 |
| 2168320.5 | -1.371066958 | -4.719775369 | -1.993173181 |
| 2168480.5 | -0.190312824 | -4.822412370 | -2.066582233 |
| 2168640.5 | 1.000100881 | -4.673413794 | -2.032148337 |
| 2168800.5 | 2.136307960 | -4.272635417 | -1.888223950 |

HELIOCENTRIC COORDINATES OF JUPITER

REFERENCE AXES OF 1950.0

| JULIAN DATE | X | Y | Z |
|---|---|---|---|
| 2168960.5 | 3.153272536 | -3.634123073 | -1.639227025 |
| 2169120.5 | 3.988866570 | -2.787098873 | -1.296166361 |
| 2169280.5 | 4.588728271 | -1.775748905 | -0.876670952 |
| 2169440.5 | 4.911363672 | -0.657396805 | -0.404332048 |
| 2169600.5 | 4.932691508 | 0.501109525 | 0.092736008 |
| 2169760.5 | 4.649122782 | 1.628571137 | 0.584041254 |
| 2169920.5 | 4.078438078 | 2.655501353 | 1.039296559 |
| 2170080.5 | 3.258173508 | 3.520106664 | 1.431028557 |
| 2170240.5 | 2.241805220 | 4.173200768 | 1.736783035 |
| 2170400.5 | 1.093504534 | 4.581361869 | 1.940611284 |
| 2170560.5 | -0.117558701 | 4.728111114 | 2.033717712 |
| 2170720.5 | -1.322490444 | 4.613334618 | 2.014342579 |
| 2170880.5 | -2.456956245 | 4.251447629 | 1.887079850 |
| 2171040.5 | -3.464257947 | 3.668869844 | 1.661869214 |
| 2171200.5 | -4.297265882 | 2.901297069 | 1.352872194 |
| 2171360.5 | -4.919429067 | 1.991098775 | 0.977379451 |
| 2171520.5 | -5.305137597 | 0.985011781 | 0.554829180 |
| 2171680.5 | -5.439690991 | -0.067826288 | 0.105961663 |
| 2171840.5 | -5.319069692 | -1.117544343 | -0.347902554 |
| 2172000.5 | -4.949639211 | -2.114993883 | -0.785495402 |
| 2172160.5 | -4.347849856 | -3.013102146 | -1.186189947 |
| 2172320.5 | -3.539933352 | -3.768232685 | -1.530592311 |
| 2172480.5 | -2.561539870 | -4.341651187 | -1.801180269 |
| 2172640.5 | -1.457203809 | -4.701166867 | -1.983021526 |
| 2172800.5 | -0.279475584 | -4.822972708 | -2.064585674 |
| 2172960.5 | 0.912482308 | -4.693634859 | -2.038633535 |
| 2173120.5 | 2.055046477 | -4.312079204 | -1.903123833 |
| 2173280.5 | 3.083197391 | -3.691297854 | -1.662022259 |
| 2173440.5 | 3.934535359 | -2.859375738 | -1.325842135 |
| 2173600.5 | 4.554060127 | -1.859367931 | -0.911710408 |
| 2173760.5 | 4.899231002 | -0.747608428 | -0.442766904 |
| 2173920.5 | 4.944545046 | 0.409746673 | 0.053206739 |
| 2174080.5 | 4.684740631 | 1.541754715 | 0.545870351 |
| 2174240.5 | 4.135870530 | 2.578674299 | 1.004870433 |
| 2174400.5 | 3.333908261 | 3.457960772 | 1.402449912 |
| 2174560.5 | 2.331121322 | 4.129283279 | 1.715693093 |
| 2174720.5 | 1.190940621 | 4.557843116 | 1.928078346 |
| 2174880.5 | -0.017711103 | 4.725727249 | 2.030200309 |
| 2175040.5 | -1.225744593 | 4.631480366 | 2.019718321 |
| 2175200.5 | -2.368285528 | 4.288371727 | 1.900720830 |

## HELIOCENTRIC COORDINATES OF JUPITER

### REFERENCE AXES OF 1950.0

| JULIAN DATE | X | Y | Z |
|---|---|---|---|
| 2175360.5 | -3.387864358 | 3.721921906 | 1.682742325 |
| 2175520.5 | -4.236453534 | 2.967183366 | 1.379645912 |
| 2175680.5 | -4.876561882 | 2.066118408 | 1.008524326 |
| 2175840.5 | -5.281656508 | 1.065258303 | 0.588704999 |
| 2176000.5 | -5.436167635 | 0.013697789 | 0.140888805 |
| | | | |
| 2176160.5 | -5.335277098 | -1.038608708 | -0.313587067 |
| 2176320.5 | -4.984623010 | -2.042333139 | -0.753394972 |
| 2176480.5 | -4.399986144 | -2.950148865 | -1.157815611 |
| 2176640.5 | -3.606964129 | -3.718103574 | -1.507336116 |
| 2176800.5 | -2.640585944 | -4.307083948 | -1.784288649 |
| | | | |
| 2176960.5 | -1.544765354 | -4.684437413 | -1.973560099 |
| 2177120.5 | -0.371440230 | -4.825774845 | -2.063387319 |
| 2177280.5 | 0.820794578 | -4.716914346 | -2.046225692 |
| 2177440.5 | 1.968765453 | -4.355831151 | -1.919638138 |
| 2177600.5 | 3.007676874 | -3.754358022 | -1.687098817 |
| | | | |
| 2177760.5 | 3.874999752 | -2.939258373 | -1.358550568 |
| 2177920.5 | 4.515135396 | -1.952217343 | -0.950516856 |
| 2178080.5 | 4.884416183 | -0.848327853 | -0.485575693 |
| 2178240.5 | 4.955722818 | 0.307155931 | 0.008920102 |
| 2178400.5 | 4.721844561 | 1.443665761 | 0.502841288 |
| | | | |
| 2178560.5 | 4.196805072 | 2.491201993 | 0.965775830 |
| 2178720.5 | 3.414755953 | 3.386347501 | 1.369639253 |
| 2178880.5 | 2.426593270 | 4.077424766 | 1.690973519 |
| 2179040.5 | 1.294965237 | 4.528031982 | 1.912589648 |
| 2179200.5 | 0.088613909 | 4.718627660 | 2.024388709 |
| | | | |
| 2179360.5 | -1.123039923 | 4.646290600 | 2.023393134 |
| 2179520.5 | -2.274425025 | 4.323101310 | 1.913163944 |
| 2179680.5 | -3.307165927 | 3.773708141 | 1.702838481 |
| 2179840.5 | -4.172251909 | 3.032589381 | 1.406008345 |
| 2180000.5 | -4.831221981 | 2.141379119 | 1.039600299 |
| | | | |
| 2180160.5 | -5.256634316 | 1.146463098 | 0.622855360 |
| 2180320.5 | -5.432082560 | 0.096915144 | 0.176442931 |
| 2180480.5 | -5.351971226 | -0.957249415 | -0.278294909 |
| 2180640.5 | -5.021195861 | -1.966582326 | -0.719999264 |
| 2180800.5 | -4.454806173 | -2.883577834 | -1.127893506 |
| | | | |
| 2180960.5 | -3.677667372 | -3.664025291 | -1.482372612 |
| 2181120.5 | -2.724076661 | -4.268439570 | -1.765627025 |
| 2181280.5 | -1.637236581 | -4.663645689 | -1.962336310 |
| 2181440.5 | -0.468435543 | -4.824551782 | -2.060450955 |
| 2181600.5 | 0.724255618 | -4.736078354 | -2.052053314 |

HELIOCENTRIC COORDINATES OF JUPITER

REFERENCE AXES OF 1950.0

| JULIAN DATE | X | Y | Z |
|---|---|---|---|
| 2181760.5 | 1.877979194 | -4.395117230 | -1.934248484 |
| 2181920.5 | 2.927955409 | -3.812274337 | -1.709983906 |
| 2182080.5 | 3.811286581 | -3.013026561 | -1.388640639 |
| 2182240.5 | 4.471569477 | -2.037838291 | -0.986197898 |
| 2182400.5 | 4.863904133 | -0.940801307 | -0.524773221 |
| 2182560.5 | 4.959598243 | 0.213459136 | -0.031410457 |
| 2182720.5 | 4.749692139 | 1.354520313 | 0.463866771 |
| 2182880.5 | 4.246498119 | 2.412011680 | 0.930529161 |
| 2183040.5 | 3.482705668 | 3.321689283 | 1.340172792 |
| 2183200.5 | 2.508154370 | 4.030730935 | 1.668873498 |
| 2183360.5 | 1.384907533 | 4.501445361 | 1.898902482 |
| 2183520.5 | 0.181564545 | 4.713014706 | 2.019616426 |
| 2183680.5 | -1.032252084 | 4.661363312 | 2.027538309 |
| 2183840.5 | -2.190543344 | 4.357590773 | 1.925799406 |
| 2184000.5 | -3.234375418 | 3.825542653 | 1.723180642 |
| 2184160.5 | -4.114105000 | 3.099049650 | 1.432981585 |
| 2184320.5 | -4.790587055 | 2.219224452 | 1.071889307 |
| 2184480.5 | -5.235646124 | 1.232037249 | 0.658948839 |
| 2184640.5 | -5.432085604 | 0.186246292 | 0.214674676 |
| 2184800.5 | -5.373456497 | -0.868340369 | -0.239701447 |
| 2184960.5 | -5.063737789 | -1.882354000 | -0.682877985 |
| 2185120.5 | -4.517011030 | -2.808214018 | -1.094070926 |
| 2185280.5 | -3.757147419 | -3.601451397 | -1.453589903 |
| 2185440.5 | -2.817466285 | -4.222112326 | -1.743449496 |
| 2185600.5 | -1.740267492 | -4.636324650 | -1.948053564 |
| 2185760.5 | -0.576086742 | -4.818068505 | -2.054974158 |
| 2185920.5 | 0.617522101 | -4.751127257 | -2.055819591 |
| 2186080.5 | 1.777895596 | -4.431099706 | -1.947145954 |
| 2186240.5 | 2.840052846 | -3.867231790 | -1.731313024 |
| 2186400.5 | 3.740461230 | -3.083694872 | -1.417126619 |
| 2186560.5 | 4.421664276 | -2.119842214 | -1.020063503 |
| 2186720.5 | 4.837371329 | -1.028982100 | -0.561870656 |
| 2186880.5 | 4.957302624 | 0.124620524 | -0.069397683 |
| 2187040.5 | 4.770888925 | 1.270447705 | 0.427331575 |
| 2187200.5 | 4.288979709 | 2.337616706 | 0.897605547 |
| 2187360.5 | 3.543073617 | 3.261048859 | 1.312693399 |
| 2187520.5 | 2.582153578 | 3.986884443 | 1.648249073 |
| 2187680.5 | 1.467765835 | 4.476306706 | 1.886075634 |
| 2187840.5 | 0.268305294 | 4.707377081 | 2.015055144 |
| 2188000.5 | -0.946533504 | 4.674969578 | 2.031258075 |

HELIOCENTRIC COORDINATES OF JUPITER

REFERENCE AXES OF 1950.0

| JULIAN DATE | X | Y | Z |
|---|---|---|---|
| 2188160.5 | -2.110442052 | 4.389244746 | 1.937404081 |
| 2188320.5 | -3.163985354 | 3.873243974 | 1.741915282 |
| 2188480.5 | -4.056877194 | 3.160141359 | 1.457792410 |
| 2188640.5 | -4.749234274 | 2.290544533 | 1.101486675 |
| 2188800.5 | -5.212085168 | 1.310067018 | 0.691869762 |
| 2188960.5 | -5.427407429 | 0.267251599 | 0.249342885 |
| 2189120.5 | -5.387915811 | -0.788175635 | -0.204918311 |
| 2189280.5 | -5.096756516 | -1.806799835 | -0.649613212 |
| 2189440.5 | -4.567193301 | -2.740867953 | -1.063904048 |
| 2189600.5 | -3.822307099 | -3.545609839 | -1.427990731 |
| 2189760.5 | -2.894671403 | -4.180636576 | -1.723719458 |
| 2189920.5 | -1.825908890 | -4.611500805 | -1.935263836 |
| 2190080.5 | -0.665980512 | -4.811464247 | -2.049901628 |
| 2190240.5 | 0.527988191 | -4.763453727 | -2.058883946 |
| 2190400.5 | 1.693563409 | -4.462092839 | -1.958353859 |
| 2190560.5 | 2.765740344 | -3.915572885 | -1.750217837 |
| 2190720.5 | 3.680661430 | -3.146991966 | -1.442813060 |
| 2190880.5 | 4.380217485 | -2.194690142 | -1.051165373 |
| 2191040.5 | 4.817144562 | -1.111112154 | -0.596627053 |
| 2191200.5 | 4.959916395 | 0.040097986 | -0.105750165 |
| 2191360.5 | 4.796530068 | 1.188716188 | 0.391603406 |
| 2191520.5 | 4.336333720 | 2.263768665 | 0.864720572 |
| 2191680.5 | 3.609398294 | 3.199687565 | 1.284696233 |
| 2191840.5 | 2.663493678 | 3.941766529 | 1.626851231 |
| 2192000.5 | 1.559281291 | 4.450074353 | 1.872532335 |
| 2192160.5 | 0.364664424 | 4.701410421 | 2.010092107 |
| 2192320.5 | -0.850751451 | 4.689365083 | 2.035052175 |
| 2192400.5 | -2.020383176 | 4.422903092 | 1.949611981 |
| 2192640.5 | -3.084227845 | 3.924037331 | 1.761737920 |
| 2192800.5 | -3.991221746 | 3.225126387 | 1.484061199 |
| 2192960.5 | -4.700576356 | 2.366191200 | 1.132758532 |
| 2193120.5 | -5.182357802 | 1.392480980 | 0.726521132 |
| 2193280.5 | -5.417578821 | 0.352376215 | 0.285656391 |
| 2193440.5 | -5.398025689 | -0.704384274 | -0.168677847 |
| 2193600.5 | -5.125976772 | -1.728225937 | -0.615134179 |
| 2193760.5 | -4.613901361 | -2.671114228 | -1.032773952 |
| 2193920.5 | -3.884163750 | -3.487895339 | -1.401650406 |
| 2194080.5 | -2.968698435 | -4.137706834 | -1.703422554 |
| 2194240.5 | -1.908566040 | -4.585541915 | -1.922038000 |
| 2194400.5 | -0.753246162 | -4.804012513 | -2.044507672 |

## HELIOCENTRIC COORDINATES OF JUPITER

## REFERENCE AXES OF 1950.0

| JULIAN DATE | X | Y | Z |
|---|---|---|---|
| 2194560.5 | 0.440522279 | -4.775294356 | -2.061769914 |
| 2194720.5 | 1.610567052 | -4.493146227 | -1.969602974 |
| 2194880.5 | 2.691969287 | -3.964777072 | -1.769493477 |
| 2195040.5 | 3.620712508 | -3.212205457 | -1.469310733 |
| 2195200.5 | 4.338207653 | -2.272656760 | -1.083589500 |
| 2195360.5 | 4.796331664 | -1.197537483 | -0.633214395 |
| 2195520.5 | 4.962315139 | -0.049677196 | -0.144356548 |
| 2195680.5 | 4.822606566 | 1.101142432 | 0.353350745 |
| 2195840.5 | 4.384862388 | 2.183912871 | 0.829221524 |
| 2196000.5 | 3.677529738 | 3.132528504 | 1.254160068 |
| 2196160.5 | 2.747027123 | 3.891307435 | 1.603101826 |
| 2196320.5 | 1.653084224 | 4.419042736 | 1.856868843 |
| 2196480.5 | 0.463154789 | 4.691128175 | 2.003221219 |
| 2196640.5 | -0.753143067 | 4.699779292 | 2.037089885 |
| 2196800.5 | -1.928830124 | 4.452738496 | 1.960139576 |
| 2196960.5 | -3.003230329 | 3.971020407 | 1.779891358 |
| 2197120.5 | -3.924436828 | 3.286235218 | 1.508634253 |
| 2197280.5 | -4.650739020 | 2.437897607 | 1.162304883 |
| 2197440.5 | -5.151270135 | 1.470965184 | 0.759446395 |
| 2197600.5 | -5.406144416 | 0.433705857 | 0.320295977 |
| 2197760.5 | -5.406309788 | -0.624110902 | -0.133995587 |
| 2197920.5 | -5.153276658 | -1.652767701 | -0.582037962 |
| 2198080.5 | -4.658814310 | -2.603998593 | -1.002810020 |
| 2198240.5 | -3.944641772 | -3.432344977 | -1.376250213 |
| 2198400.5 | -3.042083161 | -4.096572127 | -1.683873778 |
| 2198560.5 | -1.991605913 | -4.561227238 | -1.909453235 |
| 2198720.5 | -0.842109139 | -4.798380368 | -2.039783281 |
| 2198880.5 | 0.350217291 | -4.789533265 | -2.065528438 |
| 2199040.5 | 1.523668591 | -4.527599770 | -1.982117358 |
| 2199200.5 | 2.613626425 | -4.018750636 | -1.790599491 |
| 2199360.5 | 3.556094189 | -3.283789811 | -1.498323746 |
| 2199520.5 | 4.292102040 | -2.358627271 | -1.119250727 |
| 2199680.5 | 4.772657023 | -1.293394980 | -0.673695759 |
| 2199840.5 | 4.963618236 | -0.149880685 | -0.187347644 |
| 2200000.5 | 4.849657237 | 1.002743713 | 0.310467638 |
| 2200160.5 | 4.436452935 | 2.093500145 | 0.789127994 |
| 2200320.5 | 3.750545676 | 3.055691623 | 1.219332981 |
| 2200480.5 | 2.836783860 | 3.832504902 | 1.575573113 |
| 2200640.5 | 1.753852942 | 4.381243844 | 1.838058642 |
| 2200800.5 | 0.568760449 | 4.675681510 | 1.993866296 |

## HELIOCENTRIC COORDINATES OF JUPITER

## REFERENCE AXES OF 1950.0

| JULIAN DATE | X | Y | Z |
|---|---|---|---|
| 2200960.5 | -0.648766100 | 4.706490511 | 2.037262883 |
| 2201120.5 | -1.831203755 | 4.480097444 | 1.969335349 |
| 2201280.5 | -2.917051312 | 4.016502859 | 1.797149135 |
| 2201440.5 | -3.853439518 | 3.346613282 | 1.532667095 |
| 2201600.5 | -4.597669129 | 2.509513885 | 1.191616351 |
| 2201760.5 | -5.117927733 | 1.549948800 | 0.792423988 |
| 2201920.5 | -5.393456976 | 0.516127403 | 0.355279100 |
| 2202080.5 | -5.414404750 | -0.542135666 | -0.098669633 |
| 2202240.5 | -5.181534254 | -1.575008322 | -0.548003035 |
| 2202400.5 | -4.705893702 | -2.534054364 | -0.971646580 |
| 2202560.5 | -4.008485607 | -3.373582512 | -1.349457916 |
| 2202720.5 | -3.119914628 | -4.052037264 | -1.662832901 |
| 2202880.5 | -2.079936614 | -4.533519339 | -1.895369721 |
| 2203040.5 | -0.936778131 | -4.789488840 | -2.033617175 |
| 2203200.5 | 0.253950926 | -4.800648003 | -2.067910154 |
| 2203360.5 | 1.430957933 | -4.558916864 | -1.993260223 |
| 2203520.5 | 2.529778545 | -4.069303971 | -1.810221205 |
| 2203680.5 | 3.486220700 | -3.351348069 | -1.525593640 |
| 2203840.5 | 4.240699720 | -2.439699894 | -1.152782042 |
| 2204000.5 | 4.743167265 | -1.383383520 | -0.711599808 |
| 2204160.5 | 4.958038430 | -0.243388910 | -0.227357436 |
| 2204320.5 | 4.868285908 | 0.911464811 | 0.270812534 |
| 2204480.5 | 4.477833314 | 2.010036795 | 0.752261340 |
| 2204640.5 | 3.811629251 | 2.984972035 | 1.187445008 |
| 2204800.5 | 2.913280667 | 3.778413532 | 1.550436036 |
| 2204960.5 | 1.840690941 | 4.346402570 | 1.820921232 |
| 2205120.5 | 0.660558977 | 4.661405876 | 1.985423572 |
| 2205280.5 | -0.557201040 | 4.712883424 | 2.037687041 |
| 2205440.5 | -1.744954560 | 4.506219819 | 1.978342934 |
| 2205600.5 | -2.840378358 | 4.060565160 | 1.814079844 |
| 2205760.5 | -3.790063629 | 3.406150312 | 1.556556878 |
| 2205920.5 | -4.550643851 | 2.581525920 | 1.221256677 |
| 2206080.5 | -5.089611188 | 1.631010700 | 0.826407454 |
| 2206240.5 | -5.385474271 | 0.602478976 | 0.392036818 |
| 2206400.5 | -5.427595417 | -0.454501019 | -0.060831687 |
| 2206560.5 | -5.215888449 | -1.490241161 | -0.510861947 |
| 2206720.5 | -4.760487221 | -2.456321253 | -0.937009440 |
| 2206880.5 | -4.081428140 | -3.306905515 | -1.319094419 |
| 2207040.5 | -3.208328843 | -4.000099163 | -1.638392425 |
| 2207200.5 | -2.179987669 | -4.499439098 | -1.878284350 |

## HELIOCENTRIC COORDINATES OF JUPITER

## REFERENCE AXES OF 1950.0

| JULIAN DATE | X | Y | Z |
|---|---|---|---|
| 2207360.5 | -1.043773550 | -4.775579520 | -2.024995407 |
| 2207520.5 | 0.145373807 | -4.808178472 | -2.068430028 |
| 2207680.5 | 1.326547380 | -4.587908325 | -2.003074938 |
| 2207840.5 | 2.435291296 | -4.118400130 | -1.828894002 |
| 2208000.5 | 3.406977739 | -3.417799546 | -1.552079986 |
| 2208160.5 | 4.181124565 | -2.519494522 | -1.185473932 |
| 2208320.5 | 4.706376279 | -1.471531394 | -0.748437696 |
| 2208480.5 | 4.945557527 | -0.334337061 | -0.266000849 |
| 2208640.5 | 4.879949631 | 0.823348272 | 0.232780110 |
| 2208800.5 | 4.511881768 | 1.929990035 | 0.717127675 |
| 2208960.5 | 3.864969830 | 2.917454139 | 1.157197734 |
| 2209120.5 | 2.981845791 | 3.726851796 | 1.526645784 |
| 2209280.5 | 1.919815292 | 4.313077320 | 1.804683127 |
| 2209440.5 | 0.745319935 | 4.647447057 | 1.977347215 |
| 2209600.5 | -0.471792565 | 4.718343004 | 2.037919875 |
| 2209760.5 | -1.663364229 | 4.530188339 | 1.986613495 |
| 2209920.5 | -2.766907572 | 4.101308833 | 1.829751169 |
| 2210080.5 | -3.728342845 | 3.461254514 | 1.578684259 |
| 2210240.5 | -4.503618971 | 2.648038797 | 1.248646894 |
| 2210400.5 | -5.059475197 | 1.705583627 | 0.857677657 |
| 2210560.5 | -5.373627118 | 0.681501561 | 0.425671732 |
| 2210720.5 | -5.434624353 | -0.374772011 | -0.026424725 |
| 2210880.5 | -5.241562937 | -1.413557472 | -0.477299625 |
| 2211040.5 | -4.803765323 | -2.386320250 | -0.905879471 |
| 2211200.5 | -4.140474490 | -3.246984961 | -1.291901287 |
| 2211360.5 | -3.280547518 | -3.953286251 | -1.616500307 |
| 2211520.5 | -2.262076451 | -4.468252025 | -1.862856325 |
| 2211680.5 | -1.131808667 | -4.761882996 | -2.016929317 |
| 2211840.5 | 0.055811864 | -4.813039259 | -2.068293360 |
| 2212000.5 | 1.240197164 | -4.611463122 | -2.011043768 |
| 2212160.5 | 2.356977317 | -4.159754694 | -1.844703995 |
| 2212320.5 | 3.341324112 | -3.474982596 | -1.574999637 |
| 2212480.5 | 4.132231246 | -2.589489002 | -1.214310075 |
| 2212640.5 | 4.677488389 | -1.550397297 | -0.781580024 |
| 2212800.5 | 4.938769730 | -0.417427945 | -0.301507339 |
| 2212960.5 | 4.895988872 | 0.741084230 | 0.197059624 |
| 2213120.5 | 4.550002780 | 1.853646618 | 0.683398266 |
| 2213280.5 | 3.922984496 | 2.851779042 | 1.127552118 |
| 2213440.5 | 3.056289686 | 3.675882521 | 1.502903830 |
| 2213600.5 | 2.006232136 | 4.279848920 | 1.788259247 |

## HELIOCENTRIC COORDINATES OF JUPITER

## REFERENCE AXES OF 1950.0

| JULIAN DATE | X | Y | Z |
|---|---|---|---|
| 2213760.5 | 0.838623968 | 4.633806279 | 1.969161475 |
| 2213920.5 | -0.376926947 | 4.724881984 | 2.038359762 |
| 2214080.5 | -1.572133649 | 4.556289291 | 1.995543862 |
| 2214240.5 | -2.684067041 | 4.145275709 | 1.846562926 |
| 2214400.5 | -3.657968038 | 3.520501108 | 1.602369171 |
| 2214560.5 | -4.448946999 | 2.719303928 | 1.277885960 |
| 2214720.5 | -5.022816828 | 1.785149907 | 0.890932819 |
| 2214880.5 | -5.356337735 | 0.765401741 | 0.461273707 |
| 2215040.5 | -5.437119223 | -0.290570163 | 0.009803287 |
| 2215200.5 | -5.263363195 | -1.332991190 | -0.442148037 |
| 2215360.5 | -4.843563226 | -2.313096095 | -0.873428596 |
| 2215520.5 | -4.196209183 | -3.184464065 | -1.263646093 |
| 2215680.5 | -3.349485554 | -3.904385232 | -1.593761891 |
| 2215840.5 | -2.340894553 | -4.435353102 | -1.846741244 |
| 2216000.5 | -1.216680142 | -4.746746302 | -2.008289742 |
| 2216160.5 | -0.030879047 | -4.816710996 | -2.067685180 |
| 2216320.5 | 1.156210578 | -4.634175970 | -2.018680793 |
| 2216480.5 | 2.280353643 | -4.200823190 | -1.860409012 |
| 2216640.5 | 3.276646408 | -3.532707346 | -1.598157844 |
| 2216800.5 | 4.083729513 | -2.661099223 | -1.243837309 |
| 2216960.5 | 4.648699948 | -1.632075779 | -0.815924988 |
| 2217120.5 | 4.932170381 | -0.504463019 | -0.338707838 |
| 2217280.5 | 4.912660051 | 0.653999246 | 0.159256104 |
| 2217440.5 | 4.589417969 | 1.771988046 | 0.647357004 |
| 2217600.5 | 3.982982313 | 2.780704709 | 1.095542129 |
| 2217760.5 | 3.133254308 | 3.619767710 | 1.476894101 |
| 2217920.5 | 2.095446770 | 4.241928094 | 1.769756136 |
| 2218080.5 | 0.934723918 | 4.615971114 | 1.959110767 |
| 2218240.5 | -0.279485905 | 4.727644181 | 2.037119109 |
| 2218400.5 | -1.478660785 | 4.578881192 | 2.002915737 |
| 2218560.5 | -2.599316603 | 4.185843009 | 1.861869463 |
| 2218720.5 | -3.585925114 | 3.576340086 | 1.624548766 |
| 2218880.5 | -4.392709935 | 2.787104919 | 1.305594958 |
| 2219040.5 | -4.984556060 | 1.861222009 | 0.922641058 |
| 2219200.5 | -5.337308368 | 0.845866165 | 0.495347962 |
| 2219360.5 | -5.437707052 | -0.209621365 | 0.044576689 |
| 2219520.5 | -5.283148103 | -1.255368340 | -0.408319853 |
| 2219680.5 | -4.881387907 | -2.242410855 | -0.842122366 |
| 2219840.5 | -4.250243615 | -3.124046134 | -1.236333741 |
| 2220000.5 | -3.417279288 | -3.857208004 | -1.571777878 |

HELIOCENTRIC COORDINATES OF JUPITER

REFERENCE AXES OF 1950.0

| JULIAN DATE | X | Y | Z |
|---|---|---|---|
| 2220160.5 | -2.419412730 | -4.403959213 | -1.831250541 |
| 2220320.5 | -1.302328165 | -4.733162291 | -2.000251301 |
| 2220480.5 | -0.119532414 | -4.822338199 | -2.067804639 |
| 2220640.5 | 1.069144902 | -4.659648558 | -2.027347885 |
| 2220800.5 | 2.199811937 | -4.245837950 | -1.877621016 |
| 2220960.5 | 3.207687521 | -3.595850582 | -1.623439328 |
| 2221120.5 | 4.031200884 | -2.739718010 | -1.276176066 |
| 2221280.5 | 4.616801709 | -1.722255165 | -0.853750910 |
| 2221440.5 | 4.923970221 | -0.601167375 | -0.379941057 |
| 2221600.5 | 4.929650774 | 0.556591537 | 0.117072969 |
| 2221760.5 | 4.631226509 | 1.679979118 | 0.606849431 |
| 2221920.5 | 4.047317371 | 2.699880389 | 1.059249013 |
| 2222080.5 | 3.216121286 | 3.555024776 | 1.447016642 |
| 2222240.5 | 2.191586187 | 4.196837362 | 1.747959343 |
| 2222400.5 | 1.038174447 | 4.592546179 | 1.946410046 |
| 2222560.5 | -0.174815348 | 4.726326262 | 2.033859293 |
| 2222720.5 | -1.378504336 | 4.598699399 | 2.008827493 |
| 2222880.5 | -2.508702459 | 4.224682417 | 1.876175355 |
| 2223040.5 | -3.508973646 | 3.631248524 | 1.646089583 |
| 2223200.5 | -4.332557399 | 2.854584758 | 1.332952719 |
| 2223360.5 | -4.943367121 | 1.937474869 | 0.954244016 |
| 2223520.5 | -5.316338112 | 0.926978591 | 0.529551494 |
| 2223680.5 | -5.437378518 | -0.127548090 | 0.079720617 |
| 2223840.5 | -5.303120441 | -1.176131437 | -0.373872643 |
| 2224000.5 | -4.920600850 | -2.169644529 | -0.809958593 |
| 2224160.5 | -4.306935352 | -3.061165708 | -1.207965224 |
| 2224320.5 | -3.488986247 | -3.807340448 | -1.548611229 |
| 2224480.5 | -2.502968700 | -4.369843505 | -1.814543600 |
| 2224640.5 | -1.393884122 | -4.717010322 | -1.991052197 |
| 2224800.5 | -0.214619564 | -4.825659566 | -2.066874635 |
| 2224960.5 | 0.975486030 | -4.683055966 | -2.035074614 |
| 2225120.5 | 2.112816677 | -4.288861218 | -1.893933448 |
| 2225280.5 | 3.132557453 | -3.656796966 | -1.647740309 |
| 2225440.5 | 3.972709828 | -2.815623220 | -1.307311675 |
| 2225600.5 | 4.578854545 | -1.808967635 | -0.890035835 |
| 2225760.5 | 4.909179611 | -0.693591765 | -0.419252763 |
| 2225920.5 | 4.939014197 | 0.464098409 | 0.077132152 |
| 2226080.5 | 4.663980309 | 1.593108630 | 0.568741298 |
| 2226240.5 | 4.101010511 | 2.623852581 | 1.025274484 |
| 2226400.5 | 3.286898155 | 3.494144099 | 1.419117893 |

HELIOCENTRIC COORDINATES OF JUPITER

REFERENCE AXES OF 1950.0

| JULIAN DATE | X | Y | Z |
|---|---|---|---|
| 2226560.5 | 2.274615859 | 4.154198854 | 1.727582906 |
| 2226720.5 | 1.128134430 | 4.569923409 | 1.934446689 |
| 2226880.5 | -0.083290915 | 4.724222737 | 2.030655246 |
| 2227040.5 | -1.290476018 | 4.616508488 | 2.014245136 |
| 2227200.5 | -2.428703059 | 4.260890222 | 1.889675702 |
| 2227360.5 | -3.440899540 | 3.683629066 | 1.666813937 |
| 2227520.5 | -4.279632849 | 2.920360997 | 1.359790886 |
| 2227680.5 | -4.908135962 | 2.013439589 | 0.985885576 |
| 2227840.5 | -5.300650826 | 1.009583979 | 0.564523446 |
| 2228000.5 | -5.442359823 | -0.042115070 | 0.116418497 |
| 2228160.5 | -5.329115800 | -1.091869165 | -0.337152146 |
| 2228320.5 | -4.967109268 | -2.090636301 | -0.774976462 |
| 2228480.5 | -4.372541913 | -2.991447185 | -1.176485107 |
| 2228640.5 | -3.571312112 | -3.750728807 | -1.522327510 |
| 2228800.5 | -2.598659440 | -4.329728271 | -1.794992048 |
| 2228960.5 | -1.498659094 | -4.696113476 | -1.979505869 |
| 2229120.5 | -0.323404411 | -4.825781879 | -2.064231788 |
| 2229280.5 | 0.868326572 | -4.704837416 | -2.041750139 |
| 2229440.5 | 2.013139931 | -4.331593992 | -1.909770213 |
| 2229600.5 | 3.046011572 | -3.718336314 | -1.671960156 |
| 2229760.5 | 3.904249838 | -2.892438913 | -1.338525416 |
| 2229920.5 | 4.532271924 | -1.896361889 | -0.926324958 |
| 2230080.5 | 4.886727969 | -0.786077112 | -0.458322105 |
| 2230240.5 | 4.941207629 | 0.372308799 | 0.037749261 |
| 2230400.5 | 4.689607208 | 1.507579976 | 0.531459460 |
| 2230560.5 | 4.147352503 | 2.549469099 | 0.992254049 |
| 2230720.5 | 3.350092600 | 3.434782136 | 1.392116102 |
| 2230880.5 | 2.350073790 | 4.112559543 | 1.707872078 |
| 2231040.5 | 1.210933289 | 4.547506771 | 1.922793564 |
| 2231200.5 | 0.001912543 | 4.721399478 | 2.027343698 |
| 2231360.5 | -1.207579579 | 4.632645248 | 2.019123334 |
| 2231520.5 | -2.352425117 | 4.294497093 | 1.902214084 |
| 2231680.5 | -3.375017492 | 3.732510285 | 1.686165868 |
| 2231840.5 | -4.227287824 | 2.981759198 | 1.384851879 |
| 2232000.5 | -4.871759991 | 2.084171481 | 1.015349937 |
| 2232160.5 | -5.281919728 | 1.086168521 | 0.596938411 |
| 2232320.5 | -5.442165923 | 0.036665487 | 0.150236224 |
| 2232480.5 | -5.347554930 | -1.014606613 | -0.303523993 |
| 2232640.5 | -5.003478634 | -2.018545817 | -0.743124293 |
| 2232800.5 | -4.425348310 | -2.928002635 | -1.147937520 |

HELIOCENTRIC COORDINATES OF JUPITER

REFERENCE AXES OF 1950.0

| JULIAN DATE | X | Y | Z |
|---|---|---|---|
| 2232960.5 | -3.638293169 | -3.699097089 | -1.498500306 |
| 2233120.5 | -2.676823746 | -4.292633585 | -1.777128950 |
| 2233280.5 | -1.584353916 | -4.675692864 | -1.968614825 |
| 2233440.5 | -0.412421703 | -4.823436421 | -2.061018481 |
| 2233600.5 | 0.780595010 | -4.721089339 | -2.046552836 |
| 2233760.5 | 1.931490337 | -4.365966869 | -1.922503155 |
| 2233920.5 | 2.975158703 | -3.769281822 | -1.692075555 |
| 2234080.5 | 3.848511534 | -2.957336093 | -1.365005464 |
| 2234240.5 | 4.495233925 | -1.971609335 | -0.957713185 |
| 2234400.5 | 4.870928796 | -0.867284276 | -0.492797591 |
| 2234560.5 | 4.947890016 | 0.290041790 | 0.002258238 |
| 2234720.5 | 4.718576308 | 1.429317080 | 0.497131072 |
| 2234880.5 | 4.196964284 | 2.480061297 | 0.961211512 |
| 2235040.5 | 3.417376845 | 3.378501456 | 1.366267524 |
| 2235200.5 | 2.430976994 | 4.072789183 | 1.688771313 |
| 2235360.5 | 1.300648620 | 4.526523528 | 1.911538495 |
| 2235520.5 | 0.095251977 | 4.720267260 | 2.024521158 |
| 2235680.5 | -1.115817189 | 4.651218358 | 2.024801257 |
| 2235840.5 | -2.267127602 | 4.331515628 | 1.915974260 |
| 2236000.5 | -3.300497873 | 3.785761038 | 1.707166437 |
| 2236160.5 | -4.167096457 | 3.048273986 | 1.411907750 |
| 2236320.5 | -4.828567669 | 2.160438761 | 1.047020674 |
| 2236480.5 | -5.257462088 | 1.168340486 | 0.631615727 |
| 2236640.5 | -5.437236594 | 0.120749494 | 0.186225953 |
| 2236800.5 | -5.362034698 | -0.932577836 | -0.267928112 |
| 2236960.5 | -5.036389425 | -1.942366492 | -0.709575894 |
| 2237120.5 | -4.474923016 | -2.861167190 | -1.117981463 |
| 2237280.5 | -3.702054888 | -3.644690723 | -1.473523713 |
| 2237440.5 | -2.751670778 | -4.253235308 | -1.758317439 |
| 2237600.5 | -1.666649964 | -4.653285390 | -1.956911536 |
| 2237760.5 | -0.498094524 | -4.819317443 | -2.057083452 |
| 2237920.5 | 0.695938819 | -4.735780168 | -2.050720813 |
| 2238080.5 | 1.852422052 | -4.399118375 | -1.934739483 |
| 2238240.5 | 2.906240349 | -3.819585196 | -1.711932997 |
| 2238400.5 | 3.794051533 | -3.022457722 | -1.391589930 |
| 2238560.5 | 4.458983144 | -2.048183489 | -0.989672869 |
| 2238720.5 | 4.855736085 | -0.951006845 | -0.528354697 |
| 2238880.5 | 4.955364520 | 0.204183997 | -0.034782858 |
| 2239040.5 | 4.748828398 | 1.346680840 | 0.460904219 |
| 2239200.5 | 4.248499739 | 2.405889611 | 0.928087973 |

HELIOCENTRIC COORDINATES OF JUPITER

REFERENCE AXES OF 1950.0

| JULIAN DATE | X | Y | Z |
|---|---|---|---|
| 2239360.5 | 3.487187673 | 3.317452024 | 1.338322429 |
| 2239520.5 | 2.514819601 | 4.028533820 | 1.667686959 |
| 2239680.5 | 1.393452326 | 4.501482738 | 1.898480296 |
| 2239840.5 | 0.191577623 | 4.715505786 | 2.020081464 |
| 2240000.5 | -1.021344078 | 4.666488843 | 2.029007409 |
| 2240160.5 | -2.179460900 | 4.365412048 | 1.928345686 |
| 2240320.5 | -3.223914924 | 3.835935288 | 1.726801349 |
| 2240480.5 | -4.105040924 | 3.111674118 | 1.437581581 |
| 2240640.5 | -4.783575438 | 2.233539138 | 1.077283131 |
| 2240800.5 | -5.231148985 | 1.247346758 | 0.664878563 |
| 2240960.5 | -5.430327206 | 0.201771614 | 0.220838072 |
| 2241120.5 | -5.374415269 | -0.853384321 | -0.233618567 |
| 2241280.5 | -5.067172186 | -1.868686430 | -0.677170883 |
| 2241440.5 | -4.522512987 | -2.796432310 | -1.088991227 |
| 2241600.5 | -3.764212294 | -3.591998538 | -1.449330623 |
| 2241760.5 | -2.825563510 | -4.215271826 | -1.740142081 |
| 2241920.5 | -1.748894352 | -4.632243091 | -1.945777290 |
| 2242080.5 | -0.584791950 | -4.816799304 | -2.053775038 |
| 2242240.5 | 0.609149921 | -4.752680236 | -2.055731965 |
| 2242400.5 | 1.770270162 | -4.435475446 | -1.948207595 |
| 2242560.5 | 2.833649017 | -3.874422048 | -1.733565478 |
| 2242720.5 | 3.735866504 | -3.093640005 | -1.420596394 |
| 2242880.5 | 4.419594007 | -2.132351636 | -1.024726034 |
| 2243040.5 | 4.838623356 | -1.043641077 | -0.567606568 |
| 2243200.5 | 4.962644270 | 0.108521136 | -0.075957076 |
| 2243360.5 | 4.780895618 | 1.253920992 | 0.420339833 |
| 2243520.5 | 4.303869142 | 2.321906396 | 0.890688631 |
| 2243680.5 | 3.562588968 | 3.247479656 | 1.300414037 |
| 2243840.5 | 2.605536696 | 3.976671326 | 1.643147347 |
| 2244000.5 | 1.493831532 | 4.470373401 | 1.882585344 |
| 2244160.5 | 0.295595623 | 4.706228690 | 2.013446698 |
| 2244320.5 | -0.919556785 | 4.678643812 | 2.031610633 |
| 2244480.5 | -2.085211740 | 4.397341660 | 1.939611868 |
| 2244640.5 | -3.141683903 | 3.885015512 | 1.745720470 |
| 2244800.5 | -4.038350017 | 3.174613557 | 1.462833520 |
| 2244960.5 | -4.734962938 | 2.306645787 | 1.107351200 |
| 2245120.5 | -5.202211287 | 1.326741663 | 0.698141931 |
| 2245280.5 | -5.421792438 | 0.283545681 | 0.255640575 |
| 2245440.5 | -5.386220237 | -0.773062462 | -0.198920530 |
| 2245600.5 | -5.098523193 | -1.793496674 | -0.644175464 |

## HELIOCENTRIC COORDINATES OF JUPITER

## REFERENCE AXES OF 1950.0

| JULIAN DATE | X | Y | Z |
|---|---|---|---|
| 2245760.5 | -4.571920979 | -2.729845408 | -1.059225656 |
| 2245920.5 | -3.829503431 | -3.537215874 | -1.424224987 |
| 2246080.5 | -2.903877258 | -4.175145325 | -1.720994491 |
| 2246240.5 | -1.836689848 | -4.609161498 | -1.933704131 |
| 2246400.5 | -0.677886820 | -4.812535278 | -2.049643687 |
| 2246560.5 | 0.515484487 | -4.768206407 | -2.060079164 |
| 2246720.5 | 1.681136344 | -4.470772317 | -1.961152711 |
| 2246880.5 | 2.754256405 | -3.928312656 | -1.754733092 |
| 2247040.5 | 3.671173726 | -3.163694827 | -1.449066794 |
| 2247200.5 | 4.373882746 | -2.214909495 | -1.059033791 |
| 2247360.5 | 4.815058526 | -1.133981812 | -0.605803907 |
| 2247520.5 | 4.962897905 | 0.015842933 | -0.115747949 |
| 2247680.5 | 4.804917590 | 1.164611659 | 0.381405990 |
| 2247840.5 | 4.349865635 | 2.241409326 | 0.854992444 |
| 2248000.5 | 3.627222263 | 3.180486238 | 1.276051798 |
| 2248160.5 | 2.684304148 | 3.926754495 | 1.619761835 |
| 2248320.5 | 1.581540480 | 4.439796609 | 1.867275597 |
| 2248480.5 | 0.386839780 | 4.695934483 | 2.006748960 |
| 2248640.5 | -0.829995771 | 4.688382801 | 2.033544658 |
| 2248800.5 | -2.002079879 | 4.425877851 | 1.949762633 |
| 2248960.5 | -3.069094671 | 3.930344122 | 1.763328435 |
| 2249120.5 | -3.979724856 | 3.234149004 | 1.486874234 |
| 2249280.5 | -4.693028947 | 2.377364184 | 1.136596282 |
| 2249440.5 | -5.179012124 | 1.405281026 | 0.731201935 |
| 2249600.5 | -5.418683629 | 0.366278110 | 0.290995414 |
| 2249760.5 | -5.403837618 | -0.689967108 | -0.162895400 |
| 2249920.5 | -5.136726344 | -1.713996055 | -0.609178652 |
| 2250080.5 | -4.629722255 | -2.657922682 | -1.026988012 |
| 2250240.5 | -3.904999236 | -3.476739137 | -1.396451127 |
| 2250400.5 | -2.994202739 | -4.129683295 | -1.699284932 |
| 2250560.5 | -1.938021485 | -4.581757688 | -1.919458386 |
| 2250720.5 | -0.785518763 | -4.805452264 | -2.043948677 |
| 2250880.5 | 0.406966609 | -4.782663320 | -2.063592276 |
| 2251040.5 | 1.577568701 | -4.506711371 | -1.973994497 |
| 2251200.5 | 2.661498473 | -3.984241405 | -1.776409679 |
| 2251360.5 | 3.594624748 | -3.236654745 | -1.478444355 |
| 2251520.5 | 4.317970562 | -2.300611898 | -1.094384123 |
| 2251680.5 | 4.782784367 | -1.227124867 | -0.644929056 |
| 2251840.5 | 4.955528519 | -0.078895252 | -0.156173867 |
| 2252000.5 | 4.821894181 | 1.074118903 | 0.342196492 |

HELIOCENTRIC COORDINATES OF JUPITER

REFERENCE AXES OF 1950.0

| JULIAN DATE | X | Y | Z |
|---|---|---|---|
| 2252160.5 | 4.388948458 | 2.160470127 | 0.819330034 |
| 2252320.5 | 3.684828047 | 3.113464183 | 1.245895728 |
| 2252480.5 | 2.755947695 | 3.876826883 | 1.596585561 |
| 2252640.5 | 1.662281490 | 4.408884295 | 1.852025853 |
| 2252800.5 | 0.471660898 | 4.684761686 | 1.999862137 |
| 2252960.5 | -0.745910964 | 4.696609916 | 2.034996676 |
| 2253120.5 | -1.923160881 | 4.452261222 | 1.959132257 |
| 2253280.5 | -2.999260292 | 3.972897311 | 1.779862605 |
| 2253440.5 | -3.922293291 | 3.290294164 | 1.509550284 |
| 2253600.5 | -4.650648716 | 2.444069587 | 1.164179225 |
| 2253760.5 | -5.153611610 | 1.479186286 | 0.762297211 |
| 2253920.5 | -5.411437645 | 0.443811063 | 0.324098508 |
| 2254080.5 | -5.415147649 | -0.612478017 | -0.129350996 |
| 2254240.5 | -5.166211314 | -1.640210280 | -0.576773669 |
| 2254400.5 | -4.676218693 | -2.591373137 | -0.997268194 |
| 2254560.5 | -3.966572537 | -3.420711022 | -1.370874896 |
| 2254720.5 | -3.068174383 | -4.087083148 | -1.679166469 |
| 2254880.5 | -2.021020176 | -4.554971602 | -1.905905502 |
| 2255040.5 | -0.873566082 | -4.796195752 | -2.037796253 |
| 2255200.5 | 0.318324244 | -4.791830638 | -2.065334350 |
| 2255360.5 | 1.493071404 | -4.534239054 | -1.983721076 |
| 2255520.5 | 2.585923031 | -4.029008261 | -1.793757726 |
| 2255680.5 | 3.532479950 | -3.296445861 | -1.502574945 |
| 2255840.5 | 4.273159392 | -2.372173294 | -1.123997695 |
| 2256000.5 | 4.758270346 | -1.306324066 | -0.678326124 |
| 2256160.5 | 4.953053913 | -0.160982287 | -0.191359862 |
| 2256320.5 | 4.841797090 | 0.994175763 | 0.307372065 |
| 2256480.5 | 4.430110375 | 2.007610770 | 0.787014898 |
| 2256640.5 | 3.744759400 | 3.052166902 | 1.218074142 |
| 2256800.5 | 2.830997034 | 3.830785025 | 1.574934284 |
| 2256960.5 | 1.747937575 | 4.380765049 | 1.837802971 |
| 2257120.5 | 0.562906449 | 4.676068081 | 1.993838875 |
| 2257280.5 | -0.654237457 | 4.707649039 | 2.037434620 |
| 2257440.5 | -1.836032647 | 4.482205877 | 1.969801429 |
| 2257600.5 | -2.921184465 | 4.019918497 | 1.798091320 |
| 2257760.5 | -3.857099923 | 3.351738685 | 1.534296057 |
| 2257920.5 | -4.601347547 | 2.516659389 | 1.194110713 |
| 2258080.5 | -5.122310136 | 1.559221234 | 0.795879166 |
| 2258240.5 | -5.399310277 | 0.527362744 | 0.359674164 |
| 2258400.5 | -5.422444852 | -0.529388632 | -0.093483459 |

HELIOCENTRIC COORDINATES OF JUPITER

REFERENCE AXES OF 1950.0

| JULIAN DATE | X | Y | Z |
|---|---|---|---|
| 2258560.5 | -5.192298469 | -1.561454394 | -0.542292165 |
| 2258720.5 | -4.719636075 | -2.520575085 | -0.965765555 |
| 2258880.5 | -4.025109852 | -3.361126809 | -1.343804867 |
| 2259040.5 | -3.138955493 | -4.041493467 | -1.657795649 |
| 2259200.5 | -2.100596046 | -4.525585373 | -1.891269573 |
| 2259360.5 | -0.958016879 | -4.784559941 | -2.030658652 |
| 2259520.5 | 0.233273502 | -4.798742713 | -2.066145697 |
| 2259680.5 | 1.411913241 | -4.559660710 | -1.992578558 |
| 2259840.5 | 2.513195972 | -4.071983915 | -1.810365633 |
| 2260000.5 | 3.472550235 | -3.355037831 | -1.526211157 |
| 2260160.5 | 4.229948773 | -2.443437204 | -1.153496097 |
| 2260320.5 | 4.734938336 | -1.386359821 | -0.712090474 |
| 2260480.5 | 4.951667197 | -0.245099271 | -0.227426663 |
| 2260640.5 | 4.863044096 | 0.911158817 | 0.271211488 |
| 2260800.5 | 4.473132630 | 2.010951856 | 0.753040477 |
| 2260960.5 | 3.807158331 | 2.986736227 | 1.188437257 |
| 2261120.5 | 2.909032296 | 3.780634265 | 1.551467877 |
| 2261280.5 | 1.836880910 | 4.348805580 | 1.821874667 |
| 2261440.5 | 0.657479466 | 4.663900899 | 1.986267178 |
| 2261600.5 | -0.559420457 | 4.715549068 | 2.038471267 |
| 2261760.5 | -1.746078897 | 4.509231722 | 1.979169008 |
| 2261920.5 | -2.840628275 | 4.064103523 | 1.815058064 |
| 2262080.5 | -3.789707247 | 3.410319560 | 1.557769177 |
| 2262240.5 | -4.550009021 | 2.586307868 | 1.222733308 |
| 2262400.5 | -5.089003164 | 1.636257553 | 0.828120968 |
| 2262560.5 | -5.385109269 | 0.607940303 | 0.393911153 |
| 2262720.5 | -5.427561423 | -0.449129412 | -0.058901913 |
| 2262880.5 | -5.216137853 | -1.485260056 | -0.508987695 |
| 2263040.5 | -4.760860440 | -2.451975257 | -0.935284543 |
| 2263200.5 | -4.081702568 | -3.303345337 | -1.317578708 |
| 2263360.5 | -3.208282196 | -3.997366611 | -1.637104368 |
| 2263520.5 | -2.179459482 | -4.497478999 | -1.877205488 |
| 2263680.5 | -1.042706315 | -4.774278277 | -2.024086739 |
| 2263840.5 | 0.146927992 | -4.807421848 | -2.067657111 |
| 2264000.5 | 1.328459620 | -4.587641795 | -2.002434355 |
| 2264160.5 | 2.437422846 | -4.118669507 | -1.828431504 |
| 2264320.5 | 3.409265014 | -3.418752295 | -1.551891621 |
| 2264480.5 | 4.183656177 | -2.521332838 | -1.185685794 |
| 2264640.5 | 4.709431650 | -1.474424302 | -0.749166756 |
| 2264800.5 | 4.949583018 | -0.338316426 | -0.267307119 |

HELICCENTRIC COORDINATES OF JUPITER

REFERENCE AXES OF 1950.0

| JULIAN DATE | X | Y | Z |
|---|---|---|---|
| 2264960.5 | 4.885467114 | 0.818471015 | 0.230934171 |
| 2265120.5 | 4.519350574 | 1.924651117 | 0.714894537 |
| 2265280.5 | 3.874644395 | 2.912290876 | 1.154830009 |
| 2265440.5 | 2.993676157 | 3.722591593 | 1.524451931 |
| 2265600.5 | 1.933422477 | 4.310395408 | 1.802967051 |
| 2265760.5 | 0.760050814 | 4.646835247 | 1.976349843 |
| 2265920.5 | -0.456752039 | 4.720023703 | 2.037778129 |
| 2266080.5 | -1.648852356 | 4.534089525 | 1.987344479 |
| 2266240.5 | -2.753661122 | 4.107096406 | 1.831261079 |
| 2266400.5 | -3.716906186 | 3.468405418 | 1.580795667 |
| 2266560.5 | -4.494298763 | 2.655933469 | 1.251135555 |
| 2266720.5 | -5.052339819 | 1.713597871 | 0.860310359 |
| 2266880.5 | -5.368542460 | 0.689082912 | 0.428238564 |
| 2267040.5 | -5.431313898 | -0.368052661 | -0.024087924 |
| 2267200.5 | -5.239679083 | -1.407982974 | -0.475300695 |
| 2267360.5 | -4.802959446 | -2.382032828 | -0.904271485 |
| 2267520.5 | -4.140454552 | -3.244016767 | -1.290694945 |
| 2267680.5 | -3.281113806 | -3.951608249 | -1.615684588 |
| 2267840.5 | -2.263128414 | -4.467831710 | -1.862423453 |
| 2268000.5 | -1.133319481 | -4.762738627 | -2.016898812 |
| 2268160.5 | 0.053850920 | -4.815276096 | -2.068728811 |
| 2268320.5 | 1.237852722 | -4.615276856 | -2.012055665 |
| 2268480.5 | 2.354451443 | -4.165390964 | -1.846433290 |
| 2268640.5 | 3.339006610 | -3.482651503 | -1.577580869 |
| 2268800.5 | 4.130698297 | -2.599250847 | -1.217820347 |
| 2268960.5 | 4.677428389 | -1.562055718 | -0.785989084 |
| 2269120.5 | 4.940840199 | -0.430476307 | -0.306648750 |
| 2269280.5 | 4.900640881 | 0.727430039 | 0.191480948 |
| 2269440.5 | 4.557333940 | 1.840327141 | 0.677759736 |
| 2269600.5 | 3.932677562 | 2.839714918 | 1.122240875 |
| 2269760.5 | 3.067660960 | 3.665798697 | 1.498240994 |
| 2269920.5 | 2.018371322 | 4.272155084 | 1.784445267 |
| 2270080.5 | 0.850576005 | 4.628566623 | 1.966258071 |
| 2270240.5 | -0.365995276 | 4.721874199 | 2.036310042 |
| 2270400.5 | -1.562825760 | 4.555123287 | 1.994219341 |
| 2270560.5 | -2.676726491 | 4.145526504 | 1.845818582 |
| 2270720.5 | -3.652722282 | 3.521812536 | 1.602087781 |
| 2270880.5 | -4.445801527 | 2.721438712 | 1.278000397 |
| 2271040.5 | -5.021763211 | 1.787983978 | 0.891424493 |
| 2271200.5 | -5.357439921 | 0.768874615 | 0.462151604 |

HELIOCENTRIC COORDINATES OF JUPITER

REFERENCE AXES OF 1950.0

| JULIAN DATE | X | Y | Z |
|---|---|---|---|
| 2271360.5 | -5.440555803 | -0.286530781 | 0.011071060 |
| 2271520.5 | -5.269418642 | -1.328550265 | -0.440528313 |
| 2271680.5 | -4.852572189 | -2.308578323 | -0.871567666 |
| 2271840.5 | -4.208465014 | -3.180391107 | -1.261746637 |
| 2272000.5 | -3.365129116 | -3.901471386 | -1.592120095 |
| 2272160.5 | -2.359803228 | -4.434452873 | -1.845727510 |
| 2272320.5 | -1.238379960 | -4.748753807 | -2.008307941 |
| 2272480.5 | -0.054504242 | -4.822419641 | -2.069113379 |
| 2272640.5 | 1.131888426 | -4.644117360 | -2.021801013 |
| 2272800.5 | 2.256812918 | -4.215115587 | -1.865338594 |
| 2272960.5 | 3.255421473 | -3.550952724 | -1.604799175 |
| 2273120.5 | 4.066161580 | -2.682368581 | -1.251863573 |
| 2273280.5 | 4.635684851 | -1.655008964 | -0.824814518 |
| 2273440.5 | 4.923972243 | -0.527479957 | -0.347827594 |
| 2273600.5 | 4.908854650 | 0.632417795 | 0.150534112 |
| 2273760.5 | 4.588996074 | 1.753024271 | 0.639537091 |
| 2273920.5 | 3.984590286 | 2.765013073 | 1.088919137 |
| 2274080.5 | 3.135500126 | 3.607424417 | 1.471525398 |
| 2274240.5 | 2.097179024 | 4.232525913 | 1.765495819 |
| 2274400.5 | 0.935208340 | 4.608816229 | 1.955689360 |
| 2274560.5 | -0.280524858 | 4.721985735 | 2.034239934 |
| 2274720.5 | -1.481119771 | 4.574107268 | 2.000339390 |
| 2274880.5 | -2.602870465 | 4.181600358 | 1.859467207 |
| 2275040.5 | -3.590207807 | 3.572564386 | 1.622318016 |
| 2275200.5 | -4.397473354 | 2.783972170 | 1.303640417 |
| 2275360.5 | -4.989776290 | 1.859045463 | 0.921131298 |
| 2275520.5 | -5.343225070 | 0.844967698 | 0.494459534 |
| 2275680.5 | -5.444794914 | -0.209038028 | 0.044437526 |
| 2275840.5 | -5.292028869 | -1.253319536 | -0.407677656 |
| 2276000.5 | -4.892698621 | -2.239192443 | -0.840792365 |
| 2276160.5 | -4.264483116 | -3.120238484 | -1.234541724 |
| 2276320.5 | -3.434660985 | -3.853620219 | -1.569861663 |
| 2276480.5 | -2.439754708 | -4.401513295 | -1.829612615 |
| 2276640.5 | -1.325008543 | -4.732728986 | -1.999289018 |
| 2276800.5 | -0.143529628 | -4.824551484 | -2.067829305 |
| 2276960.5 | 1.045119373 | -4.664738886 | -2.028509711 |
| 2277120.5 | 2.177098981 | -4.253529734 | -1.879859481 |
| 2277280.5 | 3.187418962 | -3.605364324 | -1.626476972 |
| 2277440.5 | 4.014051190 | -2.749899018 | -1.279566526 |
| 2277600.5 | 4.602826558 | -1.731816922 | -0.856977853 |

## HELIOCENTRIC COORDINATES OF JUPITER

### REFERENCE AXES OF 1950.0

| JULIAN DATE | X | Y | Z |
|---|---|---|---|
| 2277760.5 | 4.912594913 | -0.608993556 | -0.382545453 |
| 2277920.5 | 4.919830743 | 0.551178044 | 0.115375966 |
| 2278080.5 | 4.621741750 | 1.677072341 | 0.606104125 |
| 2278240.5 | 4.037107706 | 2.699020237 | 1.059265500 |
| 2278400.5 | 3.204551428 | 3.555380706 | 1.447444407 |
| 2278560.5 | 2.178561811 | 4.197475541 | 1.748399462 |
| 2278720.5 | 1.024092505 | 4.592689705 | 1.946527690 |
| 2278880.5 | -0.189241328 | 4.725535195 | 2.033464280 |
| 2279040.5 | -1.392473398 | 4.596936290 | 2.007906595 |
| 2279200.5 | -2.521540422 | 4.222269087 | 1.874877344 |
| 2279360.5 | -3.520284489 | 3.628746399 | 1.644675841 |
| 2279520.5 | -4.342291365 | 2.852640361 | 1.331730911 |
| 2279680.5 | -4.951808018 | 1.936668317 | 0.953501187 |
| 2279840.5 | -5.324026624 | 0.927702574 | 0.529498855 |
| 2280000.5 | -5.444994674 | -0.125162967 | 0.080457391 |
| 2280160.5 | -5.311349210 | -1.172238996 | -0.372373154 |
| 2280320.5 | -4.930002739 | -2.164654618 | -0.807840879 |
| 2280480.5 | -4.317840706 | -3.055671976 | -1.205463896 |
| 2280640.5 | -3.501428425 | -3.802015454 | -1.546008205 |
| 2280800.5 | -2.516667422 | -4.365315912 | -1.812116213 |
| 2280960.5 | -1.408285362 | -4.713742546 | -1.989019900 |
| 2281120.5 | -0.228991462 | -4.823847150 | -2.065353418 |
| 2281280.5 | 0.961911353 | -4.682571136 | -2.034048956 |
| 2281440.5 | 2.100677505 | -4.289260219 | -1.893254703 |
| 2281600.5 | 3.122207348 | -3.657401523 | -1.647156741 |
| 2281760.5 | 3.964114849 | -2.815669592 | -1.306528366 |
| 2281920.5 | 4.571581128 | -1.807794461 | -0.888793054 |
| 2282080.5 | 4.902491771 | -0.690817540 | -0.417402268 |
| 2282240.5 | 4.932064494 | 0.468464642 | 0.079577287 |
| 2282400.5 | 4.656041965 | 1.598664823 | 0.571602189 |
| 2282560.5 | 4.091676661 | 2.629914666 | 1.028251051 |
| 2282720.5 | 3.276182746 | 3.499932686 | 1.421867789 |
| 2282880.5 | 2.262927856 | 4.159035707 | 1.729807018 |
| 2283040.5 | 1.116144705 | 4.573372902 | 1.935952824 |
| 2283200.5 | -0.094834964 | 4.726142460 | 2.031383216 |
| 2283360.5 | -1.300922360 | 4.617013316 | 2.014253502 |
| 2283520.5 | -2.437602567 | 4.260263035 | 1.889103570 |
| 2283680.5 | -3.448031734 | 3.682217303 | 1.665834929 |
| 2283840.5 | -4.284966814 | 2.918489462 | 1.358573135 |
| 2284000.5 | -4.911758765 | 2.011357111 | 0.984565979 |

HELIOCENTRIC COORDINATES OF JUPITER

REFERENCE AXES OF 1950.0

| JULIAN DATE | X | Y | Z |
|---|---|---|---|
| 2284160.5 | -5.302692206 | 1.007448088 | 0.563198409 |
| 2284320.5 | -5.442932258 | -0.044221827 | 0.115148772 |
| 2284480.5 | -5.328280381 | -1.093903022 | -0.338327342 |
| 2284640.5 | -4.964871286 | -2.092549281 | -0.776021597 |
| 2284800.5 | -4.368874913 | -2.993149649 | -1.177352079 |
| 2284960.5 | -3.566204363 | -3.752067853 | -1.522945644 |
| 2285120.5 | -2.592171313 | -4.330489256 | -1.795267859 |
| 2285280.5 | -1.490974192 | -4.696047999 | -1.979334493 |
| 2285440.5 | -0.314855195 | -4.824660461 | -2.063519577 |
| 2285600.5 | 0.877272823 | -4.702517495 | -2.040444344 |
| 2285760.5 | 2.021938484 | -4.328084561 | -1.907887584 |
| 2285920.5 | 3.054132932 | -3.713835589 | -1.669604149 |
| 2286080.5 | 3.911288127 | -2.887325482 | -1.335883283 |
| 2286240.5 | 4.538039572 | -1.891132280 | -0.923641089 |
| 2286400.5 | 4.891304712 | -0.781240496 | -0.455851576 |
| 2286560.5 | 4.944921233 | 0.376350340 | 0.039795069 |
| 2286720.5 | 4.692944773 | 1.510627369 | 0.532958534 |
| 2286880.5 | 4.150827068 | 2.551565195 | 0.993195558 |
| 2287040.5 | 3.354107671 | 3.436180283 | 1.392591363 |
| 2287200.5 | 2.354828611 | 4.113634522 | 1.708038960 |
| 2287360.5 | 1.216394304 | 4.548640310 | 1.922827567 |
| 2287520.5 | 0.007851153 | 4.722877179 | 2.027391651 |
| 2287680.5 | -1.201503677 | 4.634592050 | 2.019272076 |
| 2287840.5 | -2.346564841 | 4.296861702 | 1.902479564 |
| 2288000.5 | -3.369650833 | 3.735093557 | 1.686501345 |
| 2288160.5 | -4.222559768 | 2.984271957 | 1.385169575 |
| 2288320.5 | -4.867661028 | 2.086304360 | 1.015548911 |
| 2288480.5 | -5.278299246 | 1.087658160 | 0.596931916 |
| 2288640.5 | -5.438773149 | 0.037345417 | 0.149973362 |
| 2288800.5 | -5.344097221 | -1.014777576 | -0.304045660 |
| 2288960.5 | -4.999685982 | -2.019480807 | -0.743856818 |
| 2289120.5 | -4.421033052 | -2.929509253 | -1.148791464 |
| 2289280.5 | -3.633394632 | -3.700921739 | -1.499363211 |
| 2289440.5 | -2.671428878 | -4.294521078 | -1.777890794 |
| 2289600.5 | -1.578687275 | -4.677451487 | -1.969196384 |
| 2289760.5 | -0.406802110 | -4.824995544 | -2.061397052 |
| 2289920.5 | 0.785827649 | -4.722534395 | -2.046778668 |
| 2290080.5 | 1.936065948 | -4.367536886 | -1.922699840 |
| 2290240.5 | 2.978964164 | -3.771319820 | -1.692419627 |
| 2290400.5 | 3.851645699 | -2.960194674 | -1.365685501 |

## HELIOCENTRIC COORDINATES OF JUPITER

## REFERENCE AXES OF 1950.0

| JULIAN DATE | X | Y | Z |
|---|---|---|---|
| 2290560.5 | 4.498003793 | -1.975532597 | -0.958876929 |
| 2290720.5 | 4.873773980 | -0.872305936 | -0.494504156 |
| 2290880.5 | 4.951248524 | 0.284139873 | 0.000061572 |
| 2291040.5 | 4.722734467 | 1.422962120 | 0.494596081 |
| 2291200.5 | 4.201948934 | 2.473773535 | 0.958543476 |
| 2291360.5 | 3.422933291 | 3.372746242 | 1.363663367 |
| 2291520.5 | 2.436639162 | 4.067850574 | 1.686364016 |
| 2291680.5 | 1.305870844 | 4.522442121 | 1.909366852 |
| 2291840.5 | 0.099552515 | 4.716858397 | 2.022533145 |
| 2292000.5 | -1.112747622 | 4.648157853 | 2.022886905 |
| 2292160.5 | -2.265380060 | 4.328459576 | 1.914014618 |
| 2292320.5 | -3.299967134 | 3.782458426 | 1.705081173 |
| 2292480.5 | -4.167558049 | 3.044637106 | 1.409685947 |
| 2292640.5 | -4.829785402 | 2.156555317 | 1.044727134 |
| 2292800.5 | -5.259293968 | 1.164432108 | 0.629374069 |
| 2292960.5 | -5.439709064 | 0.117092967 | 0.184185032 |
| 2293120.5 | -5.365367383 | -0.935746671 | -0.269636195 |
| 2293280.5 | -5.040966516 | -1.944946014 | -0.710877611 |
| 2293440.5 | -4.481217674 | -2.863262084 | -1.118895521 |
| 2293600.5 | -3.710519718 | -3.646647835 | -1.474179799 |
| 2293760.5 | -2.762612282 | -4.255632496 | -1.758955004 |
| 2293920.5 | -1.680108901 | -4.656868186 | -1.957855820 |
| 2294080.5 | -0.513757137 | -4.824885695 | -2.058699036 |
| 2294240.5 | 0.678773057 | -4.744036669 | -2.053347153 |
| 2294400.5 | 1.834794590 | -4.410503342 | -1.938618071 |
| 2294560.5 | 2.889399576 | -3.834132208 | -1.717141131 |
| 2294720.5 | 3.779241215 | -3.039713597 | -1.397999540 |
| 2294880.5 | 4.447190911 | -2.067230091 | -0.996951660 |
| 2295040.5 | 4.847464694 | -0.970601770 | -0.536018542 |
| 2295200.5 | 4.950498084 | 0.185367179 | -0.042293090 |
| 2295360.5 | 4.746647663 | 1.329771930 | 0.454022367 |
| 2295520.5 | 4.247854726 | 2.391583537 | 0.922140871 |
| 2295680.5 | 3.486779747 | 3.305887530 | 1.333390389 |
| 2295840.5 | 2.513505441 | 4.019323205 | 1.663630084 |
| 2296000.5 | 1.390479089 | 4.493873404 | 1.895001785 |
| 2296160.5 | 0.186690604 | 4.708613397 | 2.016824107 |
| 2296320.5 | -1.027931861 | 4.659529270 | 2.025652449 |
| 2296480.5 | -2.187206525 | 4.357873588 | 1.924688395 |
| 2296640.5 | -3.232138830 | 3.827660978 | 1.722789259 |
| 2296800.5 | -4.113122706 | 3.102852255 | 1.433313412 |

HELIOCENTRIC COORDINATES OF JUPITER

REFERENCE AXES OF 1950.0

| JULIAN DATE | X | Y | Z |
|---|---|---|---|
| 2296960.5 | -4.791111264 | 2.224619934 | 1.072975170 |
| 2297120.5 | -5.238046216 | 1.238912741 | 0.660809799 |
| 2297280.5 | -5.436826990 | 0.194391096 | 0.217286452 |
| 2297440.5 | -5.381045514 | -0.859294128 | -0.236437297 |
| 2297600.5 | -5.074641592 | -1.872964761 | -0.679152080 |
| 2297760.5 | -4.531563280 | -2.799232020 | -1.090170352 |
| 2297920.5 | -3.775452111 | -3.593784053 | -1.449886892 |
| 2298080.5 | -2.839313724 | -4.216752966 | -1.740373783 |
| 2298240.5 | -1.765077851 | -4.634247756 | -1.946049343 |
| 2298400.5 | -0.602897238 | -4.820100010 | -2.054444830 |
| 2298560.5 | 0.590012249 | -4.757805591 | -2.057066575 |
| 2298720.5 | 1.751214365 | -4.442551114 | -1.950311341 |
| 2298880.5 | 2.815782730 | -3.883089647 | -1.736338808 |
| 2299040.5 | 3.720026390 | -3.103093084 | -1.423743901 |
| 2299200.5 | 4.406118512 | -2.141502574 | -1.027821201 |
| 2299360.5 | 4.827239712 | -1.051391963 | -0.570203560 |
| 2299520.5 | 4.952523543 | 0.102976534 | -0.077719584 |
| 2299680.5 | 4.770876078 | 1.250864959 | 0.419535864 |
| 2299840.5 | 4.292784314 | 2.321019586 | 0.890715922 |
| 2300000.5 | 3.549592587 | 3.247943256 | 1.306932020 |
| 2300160.5 | 2.590318501 | 3.977405408 | 1.643697367 |
| 2300320.5 | 1.476666002 | 4.470325122 | 1.882716483 |
| 2300480.5 | 0.277231029 | 4.704620725 | 2.012821780 |
| 2300640.5 | -0.938112607 | 4.675122403 | 2.030074234 |
| 2300800.5 | -2.102935152 | 4.392005581 | 1.937207096 |
| 2300960.5 | -3.157745535 | 3.878342984 | 1.742661039 |
| 2301120.5 | -4.052252687 | 3.167323903 | 1.459446093 |
| 2301280.5 | -4.746594934 | 2.299536569 | 1.104005256 |
| 2301440.5 | -5.211821999 | 1.320535699 | 0.695182141 |
| 2301600.5 | -5.429908534 | 0.278771780 | 0.253332370 |
| 2301760.5 | -5.393524932 | -0.776141750 | -0.200425942 |
| 2301920.5 | -5.105721307 | -1.794905665 | -0.644854098 |
| 2302080.5 | -4.579610557 | -2.729866130 | -1.059172076 |
| 2302240.5 | -3.838072044 | -3.536316014 | -1.423624461 |
| 2302400.5 | -2.913438148 | -4.173874399 | -1.720080364 |
| 2302560.5 | -1.847069388 | -4.608030515 | -1.932706675 |
| 2302720.5 | -0.688669499 | -4.811899311 | -2.048738563 |
| 2302880.5 | 0.504855217 | -4.768172796 | -2.059345235 |
| 2303040.5 | 1.671214153 | -4.471157998 | -1.960549566 |
| 2303200.5 | 2.745431313 | -3.928669701 | -1.754107050 |

HELIOCENTRIC COORDINATES OF JUPITER

REFERENCE AXES OF 1950.0

| JULIAN DATE | X | Y | Z |
|---|---|---|---|
| 2303360.5 | 3.663533144 | -3.163411161 | -1.448189300 |
| 2303520.5 | 4.367138356 | -2.213371593 | -1.057667653 |
| 2303680.5 | 4.808574265 | -1.130762109 | -0.603782314 |
| 2303840.5 | 4.955827856 | 0.020829288 | -0.113042892 |
| 2304000.5 | 4.796424015 | 1.171036333 | 0.384650277 |
| 2304160.5 | 4.339352585 | 2.248577998 | 0.858475828 |
| 2304320.5 | 3.614508444 | 3.187492729 | 1.279381434 |
| 2304480.5 | 2.669679369 | 3.932690376 | 1.622541321 |
| 2304640.5 | 1.565694220 | 4.443945961 | 1.869190168 |
| 2304800.5 | 0.370695912 | 4.697895262 | 2.007620754 |
| 2304960.5 | -0.845477517 | 4.688090806 | 2.033346301 |
| 2305120.5 | -2.016074270 | 4.423548330 | 1.948594178 |
| 2305280.5 | -3.081011064 | 3.926369867 | 1.761373696 |
| 2305440.5 | -3.989226437 | 3.228994047 | 1.484353621 |
| 2305600.5 | -4.699990744 | 2.371479306 | 1.133728142 |
| 2305760.5 | -5.183450364 | 1.399053827 | 0.728178489 |
| 2305920.5 | -5.420684374 | 0.360018922 | 0.287974285 |
| 2306080.5 | -5.403502388 | -0.696010782 | -0.165786882 |
| 2306240.5 | -5.134143769 | -1.719608337 | -0.611831209 |
| 2306400.5 | -4.624968492 | -2.662884191 | -1.029295703 |
| 2306560.5 | -3.898164881 | -3.480799959 | -1.398299387 |
| 2306720.5 | -2.985439072 | -4.132553966 | -1.700545851 |
| 2306880.5 | -1.927593170 | -4.583124586 | -1.919996400 |
| 2307040.5 | -0.773844550 | -4.805019022 | -2.043637818 |
| 2307200.5 | 0.419305111 | -4.780213878 | -2.062343340 |
| 2307360.5 | 1.589863286 | -4.502182098 | -1.971786707 |
| 2307520.5 | 2.672996112 | -3.977780915 | -1.773317453 |
| 2307680.5 | 3.604641750 | -3.228646898 | -1.474648440 |
| 2307840.5 | 4.326013654 | -2.291644480 | -1.090159107 |
| 2308000.5 | 4.788645078 | -1.217902204 | -0.640605858 |
| 2308160.5 | 4.959318861 | -0.070112873 | -0.152084371 |
| 2308320.5 | 4.824006665 | 1.081902835 | 0.345777911 |
| 2308480.5 | 4.389948725 | 2.166926918 | 0.822230599 |
| 2308640.5 | 3.685313217 | 3.118521978 | 1.248060762 |
| 2308800.5 | 2.756415437 | 3.880628845 | 1.598064254 |
| 2308960.5 | 1.663047687 | 4.411697867 | 1.852934045 |
| 2309120.5 | 0.472842407 | 4.686873241 | 2.000336548 |
| 2309280.5 | -0.744355854 | 4.698237781 | 2.035155130 |
| 2309440.5 | -1.921357631 | 4.453507528 | 1.959049446 |
| 2309600.5 | -2.997336417 | 3.973743243 | 1.779564248 |

HELIOCENTRIC COORDINATES OF JUPITER

REFERENCE AXES OF 1950.0

| JULIAN DATE | X | Y | Z |
|---|---|---|---|
| 2309760.5 | -3.920312104 | 3.290630754 | 1.509022976 |
| 2309920.5 | -4.648571490 | 2.443749790 | 1.163390277 |
| 2310080.5 | -5.151291912 | 1.478083047 | 0.761218040 |
| 2310240.5 | -5.408644672 | 0.441868412 | 0.322725948 |
| 2310400.5 | -5.411611199 | -0.615210184 | -0.130979667 |
| 2310560.5 | -5.161678984 | -1.643563296 | -0.578574623 |
| 2310720.5 | -4.670515110 | -2.595070736 | -0.999115073 |
| 2310880.5 | -3.959650437 | -3.424403687 | -1.372612832 |
| 2311040.5 | -3.060147835 | -4.090401773 | -1.680634673 |
| 2311200.5 | -2.012171458 | -4.557593355 | -1.906965563 |
| 2311360.5 | -0.864320173 | -4.797911574 | -2.038361705 |
| 2311520.5 | 0.327459044 | -4.792600481 | -2.065395524 |
| 2311680.5 | 1.501591074 | -4.534219008 | -1.983358235 |
| 2311840.5 | 2.593428401 | -4.028535062 | -1.793135785 |
| 2312000.5 | 3.538765318 | -3.295971414 | -1.501915985 |
| 2312160.5 | 4.278259211 | -2.372156356 | -1.123533632 |
| 2312320.5 | 4.762439767 | -1.307106436 | -0.678242889 |
| 2312480.5 | 4.956678194 | -0.162693721 | -0.191752263 |
| 2312640.5 | 4.845253271 | 0.991642073 | 0.306516064 |
| 2312800.5 | 4.433630516 | 2.084538274 | 0.785793311 |
| 2312960.5 | 3.748350201 | 3.048894657 | 1.216622093 |
| 2313120.5 | 2.834446806 | 3.827573653 | 1.573366862 |
| 2313280.5 | 1.750902612 | 4.377700006 | 1.836171472 |
| 2313440.5 | 0.565037995 | 4.673032114 | 1.992115141 |
| 2313600.5 | -0.653174278 | 4.704367767 | 2.035526521 |
| 2313760.5 | -1.836086140 | 4.478345882 | 1.967591821 |
| 2313920.5 | -2.922207000 | 4.015203381 | 1.795486285 |
| 2314080.5 | -3.858798032 | 3.346045918 | 1.531266420 |
| 2314240.5 | -4.603374061 | 2.510068743 | 1.190713731 |
| 2314400.5 | -5.124373869 | 1.552003522 | 0.792254832 |
| 2314560.5 | -5.401274744 | 0.519917019 | 0.356019063 |
| 2314720.5 | -5.424389953 | -0.536630633 | -0.096957061 |
| 2314880.5 | -5.194529849 | -1.568135447 | -0.545403213 |
| 2315040.5 | -4.722639828 | -2.526510977 | -0.968408439 |
| 2315200.5 | -4.029460204 | -3.366377880 | -1.345983018 |
| 2315360.5 | -3.145190025 | -4.046394161 | -1.659637729 |
| 2315520.5 | -2.109078459 | -4.530721292 | -1.893023238 |
| 2315680.5 | -0.968811937 | -4.790688170 | -2.032659450 |
| 2315840.5 | 0.220485006 | -4.806660026 | -2.068762331 |
| 2316000.5 | 1.397851856 | -4.570035996 | -1.996140805 |

## HELIOCENTRIC COORDINATES OF JUPITER

### REFERENCE AXES OF 1950.0

| JULIAN DATE | X | Y | Z |
|---|---|---|---|
| 2316160.5 | 2.498912296 | -4.085184764 | -1.815087170 |
| 2316320.5 | 3.459259968 | -3.370993182 | -1.532125130 |
| 2316480.5 | 4.218794723 | -2.461581967 | -1.160423653 |
| 2316640.5 | 4.726730083 | -1.405696365 | -0.719658549 |
| 2316800.5 | 4.946671476 | -0.264381070 | -0.235139266 |
| 2316960.5 | 4.860898514 | 0.893158308 | 0.263862327 |
| 2317120.5 | 4.472923390 | 1.995159723 | 0.746454846 |
| 2317280.5 | 3.807647151 | 2.973576964 | 1.182816015 |
| 2317440.5 | 2.908963136 | 3.769966154 | 1.546778476 |
| 2317600.5 | 1.835270682 | 4.340009518 | 1.817882010 |
| 2317760.5 | 0.653810132 | 4.656082484 | 1.982614818 |
| 2317920.5 | -0.565148838 | 4.707784303 | 2.034784209 |
| 2318080.5 | -1.753426706 | 4.500786183 | 1.975148040 |
| 2318240.5 | -2.848885105 | 4.054576548 | 1.810543989 |
| 2318400.5 | -3.798096943 | 3.399694441 | 1.552767612 |
| 2318560.5 | -4.557881924 | 2.574914175 | 1.217401565 |
| 2318720.5 | -5.095980585 | 1.624667318 | 0.822725528 |
| 2318880.5 | -5.391160990 | 0.596826189 | 0.388767357 |
| 2319040.5 | -5.433010534 | -0.459145738 | -0.063496006 |
| 2319200.5 | -5.221597165 | -1.493744263 | -0.512812411 |
| 2319360.5 | -4.767112001 | -2.458780599 | -0.938245435 |
| 2319520.5 | -4.089538043 | -3.308660765 | -1.319731728 |
| 2319680.5 | -3.218328586 | -4.001702153 | -1.638654083 |
| 2319840.5 | -2.192021656 | -4.501584421 | -1.878473744 |
| 2320000.5 | -1.057661512 | -4.778999964 | -2.025453126 |
| 2320160.5 | 0.130152877 | -4.813516240 | -2.069478663 |
| 2320320.5 | 1.310809510 | -4.595580898 | -2.004959461 |
| 2320480.5 | 2.420035102 | 4.120487385 | -1.831729013 |
| 2320640.5 | 3.393215370 | -3.429984871 | -1.555817533 |
| 2320800.5 | 4.169686152 | -2.533084949 | -1.189904677 |
| 2320960.5 | 4.697736369 | -1.485572719 | -0.753231843 |
| 2321120.5 | 4.939737420 | -0.347804016 | -0.270783062 |
| 2321280.5 | 4.876532389 | 0.811335845 | 0.228342249 |
| 2321440.5 | 4.510139272 | 1.919994638 | 0.713251041 |
| 2321600.5 | 3.864060410 | 2.909647022 | 1.153950804 |
| 2321760.5 | 2.981016002 | 3.721050853 | 1.523961791 |
| 2321920.5 | 1.918543653 | 4.308866834 | 1.802407871 |
| 2322080.5 | 0.743370741 | 4.644329073 | 1.975301056 |
| 2322240.5 | -0.474400852 | 4.715872450 | 2.035953741 |
| 2322400.5 | -1.666447035 | 4.528061125 | 1.984645446 |

HELIOCENTRIC COORDINATES OF JUPITER

REFERENCE AXES OF 1950.0

| JULIAN DATE | X | Y | Z |
|---|---|---|---|
| 2322560.5 | -2.770222489 | 4.099393361 | 1.827778927 |
| 2322720.5 | -3.731687734 | 3.459573764 | 1.576776347 |
| 2322880.5 | -4.506901117 | 2.646718958 | 1.246918968 |
| 2323040.5 | -5.062744267 | 1.704787048 | 0.856261586 |
| 2323200.5 | -5.377073134 | 0.681359287 | 0.424684032 |
| 2323360.5 | -5.438546792 | -0.374216982 | -0.026910174 |
| 2323520.5 | -5.246319158 | -1.412389093 | -0.477271363 |
| 2323680.5 | -4.809707027 | -2.384766624 | -0.905398901 |
| 2323840.5 | -4.147880667 | -3.245412278 | -1.291102902 |
| 2324000.5 | -3.289562101 | -3.952169659 | -1.615581597 |
| 2324160.5 | -2.272658758 | -4.468125205 | -1.862057865 |
| 2324320.5 | -1.143704747 | -4.763270050 | -2.016506325 |
| 2324480.5 | 0.043069166 | -4.816374966 | -2.068481014 |
| 2324640.5 | 1.227252928 | -4.617010790 | -2.012019958 |
| 2324800.5 | 2.344580948 | -4.167534564 | -1.846554849 |
| 2324960.5 | 3.330220887 | -3.484727477 | -1.577694990 |
| 2325120.5 | 4.123032547 | -2.600646149 | -1.217697420 |
| 2325280.5 | 4.670545625 | -1.562192866 | -0.785404819 |
| 2325440.5 | 4.934085642 | -0.428995977 | -0.305461702 |
| 2325600.5 | 4.893200836 | 0.730532914 | 0.193269185 |
| 2325760.5 | 4.548457918 | 1.844664014 | 0.679983996 |

## TABLE A.III-2
## HELIOCENTRIC COORDINATES OF SATURN

REFERENCE AXES OF 1950.0

| JULIAN DATE | X | Y | Z |
|---|---|---|---|
| 1720960.5 | 0.954892442 | 8.260902289 | 3.349754618 |
| 1721280.5 | -0.934502009 | 8.248793481 | 3.420854353 |
| 1721600.5 | -2.784194740 | 7.892720699 | 3.349295631 |
| 1721920.5 | -4.518613573 | 7.211724662 | 3.139849077 |
| 1722240.5 | -6.070024735 | 6.239547602 | 2.803626488 |
| 1722560.5 | -7.381956578 | 5.021637043 | 2.356989012 |
| 1722880.5 | -8.411285299 | 3.611582817 | 1.820169744 |
| 1723200.5 | -9.128973671 | 2.067515650 | 1.215827832 |
| 1723520.5 | -9.519690539 | 0.448900297 | 0.567721302 |
| 1723840.5 | -9.580690548 | -1.186029110 | -0.100385468 |
| 1724160.5 | -9.320342835 | -2.782163853 | -0.765554976 |
| 1724480.5 | -8.756569310 | -4.288863675 | -1.406290196 |
| 1724800.5 | -7.915308710 | -5.660879688 | -2.002977420 |
| 1725120.5 | -6.829062168 | -6.858943649 | -2.538189815 |
| 1725440.5 | -5.535570431 | -7.850091032 | -2.996876113 |
| 1725760.5 | -4.076659506 | -8.607792298 | -3.366463340 |
| 1726080.5 | -2.497256505 | -9.111969835 | -3.636905603 |
| 1726400.5 | -0.844527119 | -9.348985777 | -3.800714830 |
| 1726720.5 | 0.832937466 | -9.311758922 | -3.853035770 |
| 1727040.5 | 2.485791888 | -9.000115307 | -3.791807050 |
| 1727360.5 | 4.064469652 | -8.420968205 | -3.617864635 |
| 1727680.5 | 5.520023275 | -7.588273087 | -3.334967096 |
| 1728000.5 | 6.805222533 | -6.523110818 | -2.949868483 |
| 1728320.5 | 7.875712147 | -5.253789417 | -2.472399160 |
| 1728640.5 | 8.691260889 | -3.815859117 | -1.915518929 |
| 1728960.5 | 9.217204941 | -2.252005157 | -1.295331803 |
| 1729280.5 | 9.426165716 | -0.611703103 | -0.631016510 |
| 1729600.5 | 9.300020213 | 1.049548932 | 0.055406136 |
| 1729920.5 | 8.831975976 | 2.671602670 | 0.739550281 |
| 1730240.5 | 8.028496795 | 4.191757632 | 1.395446336 |
| 1730560.5 | 6.910747418 | 5.547605652 | 1.996648257 |
| 1730880.5 | 5.515189269 | 6.680504025 | 2.517633059 |
| 1731200.5 | 3.892995953 | 7.539393396 | 2.935389587 |
| 1731520.5 | 2.108093446 | 8.084524786 | 3.231026217 |
| 1731840.5 | 0.233854211 | 8.290583291 | 3.391186925 |
| 1732160.5 | -1.651274508 | 8.148735934 | 3.409071147 |
| 1732480.5 | -3.468723736 | 7.667275198 | 3.284902318 |
| 1732800.5 | -5.144725451 | 6.870737423 | 3.025768481 |
| 1733120.5 | -6.614560913 | 5.797629289 | 2.644861342 |
| 1733440.5 | -7.825619518 | 4.497150250 | 2.160250831 |

HELIOCENTRIC COORDINATES OF SATURN

REFERENCE AXES OF 1950.0

| JULIAN DATE | X | Y | Z |
|---|---|---|---|
| 1733760.5 | -8.739100287 | 3.025442720 | 1.593405451 |
| 1734080.5 | -9.330642357 | 1.441977969 | 0.967727901 |
| 1734400.5 | -9.589992013 | -0.193315393 | 0.307381007 |
| 1734720.5 | -9.519490040 | -1.822334518 | -0.363696920 |
| 1735040.5 | -9.132205894 | -3.390775139 | -1.022690322 |
| 1735360.5 | -8.450256057 | -4.849521487 | -1.648553972 |
| 1735680.5 | -7.503205174 | -6.155492076 | -2.222425599 |
| 1736000.5 | -6.326542532 | -7.272111372 | -2.727879464 |
| 1736320.5 | -4.960283919 | -8.169503551 | -3.151059364 |
| 1736640.5 | -3.447757323 | -8.824467827 | -3.480713130 |
| 1736960.5 | -1.834635837 | -9.220320615 | -3.708162353 |
| 1737280.5 | -0.168226859 | -9.346738273 | -3.827264124 |
| 1737600.5 | 1.503076385 | -9.199719878 | -3.834418595 |
| 1737920.5 | 3.130326431 | -8.781702338 | -3.728640311 |
| 1738240.5 | 4.664647253 | -8.101785505 | -3.511678712 |
| 1738560.5 | 6.058011864 | -7.176007407 | -3.188164605 |
| 1738880.5 | 7.264207036 | -6.027617154 | -2.765762508 |
| 1739200.5 | 8.240019727 | -4.687290757 | -2.255308101 |
| 1739520.5 | 8.946678187 | -3.193216640 | -1.670902314 |
| 1739840.5 | 9.351564475 | -1.590949264 | -1.029920940 |
| 1740160.5 | 9.430175268 | 0.067102782 | -0.352883239 |
| 1740480.5 | 9.168230905 | 1.722719945 | 0.336894465 |
| 1740800.5 | 8.563733325 | 3.313653595 | 1.013955409 |
| 1741120.5 | 7.628835015 | 4.776357882 | 1.651750321 |
| 1741440.5 | 6.391335019 | 6.049351727 | 2.223934363 |
| 1741760.5 | 4.894703532 | 7.076651931 | 2.705773615 |
| 1742080.5 | 3.196455234 | 7.811531821 | 3.075772569 |
| 1742400.5 | 1.365421906 | 8.220127269 | 3.317265093 |
| 1742720.5 | -0.522107728 | 8.284152989 | 3.419679347 |
| 1743040.5 | -2.387122551 | 8.002299990 | 3.379298415 |
| 1743360.5 | -4.152908467 | 7.390098808 | 3.199404546 |
| 1743680.5 | -5.749682064 | 6.478299694 | 2.889800157 |
| 1744000.5 | -7.118274710 | 5.310084044 | 2.465811414 |
| 1744320.5 | -8.212531551 | 3.937589298 | 1.946958726 |
| 1744640.5 | -9.000354358 | 2.418269679 | 1.355506147 |
| 1744960.5 | -9.463550481 | 0.811547232 | 0.715083406 |
| 1745280.5 | -9.596819086 | -0.823938027 | 0.049520776 |
| 1745600.5 | -9.406277254 | -2.432369718 | -0.618043643 |
| 1745920.5 | -8.907862512 | -3.962170722 | -1.265806819 |
| 1746240.5 | -8.125787465 | -5.366959243 | -1.873756162 |

## HELIOCENTRIC COORDINATES OF SATURN

### REFERENCE AXES OF 1950.0

| JULIAN DATE | X | Y | Z |
|---|---|---|---|
| 1746560.5 | -7.091097566 | -6.606177228 | -2.423988421 |
| 1746880.5 | -5.840352544 | -7.645467278 | -2.900920726 |
| 1747200.5 | -4.414457481 | -8.456858944 | -3.291416827 |
| 1747520.5 | -2.857657871 | -9.018830671 | -3.584855175 |
| 1747840.5 | -1.216681972 | -9.316311599 | -3.773166223 |
| 1748160.5 | 0.460033685 | -9.340690688 | -3.850867978 |
| 1748480.5 | 2.123130818 | -9.089974965 | -3.815156626 |
| 1748800.5 | 3.723045884 | -8.569199999 | -3.666095391 |
| 1749120.5 | 5.210612630 | -7.790668081 | -3.406745775 |
| 1749440.5 | 6.537989565 | -6.773915480 | -3.043199260 |
| 1749760.5 | 7.659888167 | -5.545760137 | -2.584639638 |
| 1750080.5 | 8.534892903 | -4.140331868 | -2.043401692 |
| 1750400.5 | 9.126903974 | -2.598966137 | -1.434983630 |
| 1750720.5 | 9.406795286 | -0.969885209 | -0.777984654 |
| 1751040.5 | 9.354317935 | 0.692485144 | -0.093904411 |
| 1751360.5 | 8.960150425 | 2.328753276 | 0.593279199 |
| 1751680.5 | 8.227867434 | 3.876501924 | 1.257824473 |
| 1752000.5 | 7.175504562 | 5.272981970 | 1.873266395 |
| 1752320.5 | 5.836348707 | 6.458473251 | 2.413762399 |
| 1752640.5 | 4.258604843 | 7.380060032 | 2.855653849 |
| 1752960.5 | 2.503707534 | 7.995407438 | 3.179083034 |
| 1753280.5 | 0.643261169 | 8.276026543 | 3.369456760 |
| 1753600.5 | -1.245143961 | 8.209536006 | 3.418545426 |
| 1753920.5 | -3.082771938 | 7.800574161 | 3.325056917 |
| 1754240.5 | -4.794655133 | 7.070230662 | 3.094606681 |
| 1754560.5 | -6.313984161 | 6.054092801 | 2.739096311 |
| 1754880.5 | -7.585454675 | 4.799223289 | 2.275607865 |
| 1755200.5 | -8.567260411 | 3.360554650 | 1.724999970 |
| 1755520.5 | -9.231766687 | 1.797204234 | 1.110413688 |
| 1755840.5 | -9.565294857 | 0.169160247 | 0.455899213 |
| 1756160.5 | -9.567286385 | -1.465184055 | -0.214611675 |
| 1756480.5 | -9.248583046 | -3.050759678 | -0.878073843 |
| 1756800.5 | -8.629442656 | -4.537264770 | -1.513016349 |
| 1757120.5 | -7.737865140 | -5.880178993 | -2.100035753 |
| 1757440.5 | -6.608106177 | -7.041341263 | -2.622093635 |
| 1757760.5 | -5.279304785 | -7.989232928 | -3.064686100 |
| 1758080.5 | -3.794244840 | -8.699047719 | -3.415918930 |
| 1758400.5 | -2.198300885 | -9.152601708 | -3.666508506 |
| 1758720.5 | -0.538617160 | -9.338173631 | -3.809745326 |
| 1759040.5 | 1.136503831 | -9.250404172 | -3.841475907 |

HELIOCENTRIC COORDINATES OF SATURN

REFERENCE AXES OF 1950.0

| JULIAN DATE | X | Y | Z |
|---|---|---|---|
| 1759360.5 | 2.778123439 | -8.890343128 | -3.760144721 |
| 1759680.5 | 4.337242570 | -8.265641161 | -3.566899049 |
| 1760000.5 | 5.765525290 | -7.390824640 | -3.265733752 |
| 1760320.5 | 7.016217406 | -6.287585318 | -2.863649226 |
| 1760640.5 | 8.045290953 | -4.985021546 | -2.370797954 |
| 1760960.5 | 8.812842390 | -3.519758475 | -1.800591331 |
| 1761280.5 | 9.284760782 | -1.935851878 | -1.169728446 |
| 1761600.5 | 9.434651411 | -0.284352367 | -0.498095429 |
| 1761920.5 | 9.245941660 | 1.377617401 | 0.191529050 |
| 1762240.5 | 8.713996837 | 2.988430835 | 0.874036061 |
| 1762560.5 | 7.847950744 | 4.484331996 | 1.522896065 |
| 1762880.5 | 6.672018463 | 5.802763066 | 2.111472430 |
| 1763200.5 | 5.226172761 | 6.886239965 | 2.614559574 |
| 1763520.5 | 3.565218854 | 7.685934657 | 3.009890602 |
| 1763840.5 | 1.756052310 | 8.165132359 | 3.279688701 |
| 1764160.5 | -0.126127054 | 8.302182293 | 3.412032195 |
| 1764480.5 | -2.002412989 | 8.092218424 | 3.401740933 |
| 1764800.5 | -3.795185855 | 7.547353346 | 3.250654001 |
| 1765120.5 | -5.432858957 | 6.695355046 | 2.967276517 |
| 1765440.5 | -6.853785715 | 5.577076657 | 2.565881525 |
| 1765760.5 | -8.008942933 | 4.243093869 | 2.065237640 |
| 1766080.5 | -8.863266524 | 2.750065022 | 1.487168961 |
| 1766400.5 | -9.395749442 | 1.157277288 | 0.855142486 |
| 1766720.5 | -9.598580615 | -0.476279112 | 0.193035304 |
| 1767040.5 | -9.475701576 | -2.094131494 | -0.475830772 |
| 1767360.5 | -9.041158101 | -3.643852555 | -1.129378864 |
| 1767680.5 | -8.317504116 | -5.078053995 | -1.747247886 |
| 1768000.5 | -7.334348665 | -6.355027801 | -2.311116021 |
| 1768320.5 | -6.127046065 | -7.439150122 | -2.804922512 |
| 1768640.5 | -4.735526293 | -8.301108385 | -3.215011202 |
| 1768960.5 | -3.203274167 | -8.918003601 | -3.530216178 |
| 1769280.5 | -1.576454693 | -9.273384199 | -3.741912962 |
| 1769600.5 | 0.096845706 | -9.357261996 | -3.844057484 |
| 1769920.5 | 1.767338117 | -9.166160122 | -3.833234996 |
| 1770240.5 | 3.385327321 | -8.703316930 | -3.708769646 |
| 1770560.5 | 4.901513908 | -7.979148325 | -3.472939634 |
| 1770880.5 | 6.267703385 | -7.011527668 | -3.131130672 |
| 1771200.5 | 7.437849664 | -5.825737401 | -2.691863715 |
| 1771520.5 | 8.369436261 | -4.454442359 | -2.166832051 |
| 1771840.5 | 9.024976892 | -2.937597203 | -1.570917859 |

HELIOCENTRIC COORDINATES OF SATURN

REFERENCE AXES OF 1950.0

| JULIAN DATE | X | Y | Z |
|---|---|---|---|
| 1772160.5 | 9.373657510 | -1.322147618 | -0.922134968 |
| 1772480.5 | 9.393182332 | 0.338591452 | -0.241450525 |
| 1772800.5 | 9.071785516 | 1.986053672 | 0.447589322 |
| 1773120.5 | 8.410224755 | 3.558204767 | 1.119508047 |
| 1773440.5 | 7.423452509 | 4.992077240 | 1.747869451 |
| 1773760.5 | 6.141594209 | 6.227006265 | 2.306564668 |
| 1774080.5 | 4.609867045 | 7.208349923 | 2.771341067 |
| 1774400.5 | 2.887170642 | 7.891307321 | 3.121425317 |
| 1774720.5 | 1.043281384 | 8.244327770 | 3.341035675 |
| 1775040.5 | -0.845149581 | 8.251594552 | 3.420564168 |
| 1775360.5 | -2.699345254 | 7.914198049 | 3.357253088 |
| 1775680.5 | -4.443313514 | 7.249843138 | 3.155278240 |
| 1776000.5 | -6.008336482 | 6.291173226 | 2.825247259 |
| 1776320.5 | -7.336544336 | 5.082986968 | 2.383203786 |
| 1776640.5 | -8.383233699 | 3.678776739 | 1.849298754 |
| 1776960.5 | -9.117827987 | 2.137088626 | 1.246329933 |
| 1777280.5 | -9.523672138 | 0.518121588 | 0.598329347 |
| 1777600.5 | -9.597162470 | -1.119154988 | -0.070657890 |
| 1777920.5 | -9.346573160 | -2.719031420 | -0.737435991 |
| 1778240.5 | -8.790284463 | -4.230288577 | -1.380226313 |
| 1778560.5 | -7.954803440 | -5.607173112 | -1.979164319 |
| 1778880.5 | -6.873156883 | -6.810101493 | -2.516662811 |
| 1779200.5 | -5.583537315 | -7.806035786 | -2.977620607 |
| 1779520.5 | -4.128080423 | -8.568632696 | -3.349530279 |
| 1779840.5 | -2.551774761 | -9.078230911 | -3.622512864 |
| 1780160.5 | -0.901550011 | -9.321729127 | -3.789299857 |
| 1780480.5 | 0.774425352 | -9.292451733 | -3.845203006 |
| 1780800.5 | 2.427293977 | -8.990119360 | -3.788123986 |
| 1781120.5 | 4.008045239 | -8.420978684 | -3.618631478 |
| 1781440.5 | 5.468169910 | -7.598056465 | -3.340094944 |
| 1781760.5 | 6.760511471 | -6.541461171 | -2.958845439 |
| 1782080.5 | 7.840351836 | -5.278654854 | -2.484332922 |
| 1782400.5 | 8.666756012 | -3.844617803 | -1.929249504 |
| 1782720.5 | 9.204190380 | -2.281813148 | -1.309581405 |
| 1783040.5 | 9.424404438 | -0.639834138 | -0.644541170 |
| 1783360.5 | 9.308517781 | 1.025405767 | 0.043680034 |
| 1783680.5 | 8.849176219 | 2.653088786 | 0.730407693 |
| 1784000.5 | 8.052524237 | 4.179642312 | 1.389291441 |
| 1784320.5 | 6.939609493 | 5.541700724 | 1.993473423 |
| 1784640.5 | 5.546913690 | 6.679921410 | 2.517132703 |

## HELIOCENTRIC COORDINATES OF SATURN

### REFERENCE AXES OF 1950.0

| JULIAN DATE | X | Y | Z |
|---|---|---|---|
| 1784960.5 | 3.925999599 | 7.543177606 | 2.937193120 |
| 1785280.5 | 2.141554091 | 8.092032290 | 3.234829773 |
| 1785600.5 | 0.267604043 | 8.301570102 | 3.396808644 |
| 1785920.5 | -1.617114559 | 8.163400068 | 3.416497159 |
| 1786240.5 | -3.434147422 | 7.686210402 | 3.294288132 |
| 1786560.5 | -5.110174199 | 6.894757413 | 3.037377614 |
| 1786880.5 | -6.581131029 | 5.827520491 | 2.658967342 |
| 1787200.5 | -7.795170008 | 4.533457393 | 2.177051510 |
| 1787520.5 | -8.714279876 | 3.068374209 | 1.612991533 |
| 1787840.5 | -9.314634797 | 1.491385707 | 0.990077100 |
| 1788160.5 | -9.585910791 | -0.138171267 | 0.332232757 |
| 1788480.5 | -9.529910215 | -1.763152059 | -0.337025645 |
| 1788800.5 | -9.158871096 | -3.330289207 | -0.995346742 |
| 1789120.5 | -8.493782370 | -4.791256089 | -1.622045221 |
| 1789440.5 | -7.562873537 | -6.103310713 | -2.198426893 |
| 1789760.5 | -6.400296453 | -7.229692223 | -2.708007158 |
| 1790080.5 | -5.044964671 | -8.139860083 | -3.136659446 |
| 1790400.5 | -3.539532609 | -8.809621374 | -3.472711649 |
| 1790720.5 | -1.929509693 | -9.221189618 | -3.707008241 |
| 1791040.5 | -0.262493344 | -9.363221450 | -3.832958089 |
| 1791360.5 | 1.412519787 | -9.230868220 | -3.846585015 |
| 1791680.5 | 3.045829904 | -8.825874396 | -3.746596120 |
| 1792000.5 | 4.587875606 | -8.156826684 | -3.534511640 |
| 1792320.5 | 5.990159937 | -7.239654301 | -3.214902349 |
| 1792640.5 | 7.206100824 | -6.097926195 | -2.795556426 |
| 1792960.5 | 8.192216714 | -4.762752652 | -2.287490064 |
| 1793280.5 | 8.909678806 | -3.272638152 | -1.704941501 |
| 1793600.5 | 9.326001141 | -1.673204205 | -1.065320689 |
| 1793920.5 | 9.416868726 | -0.016615567 | -0.389047979 |
| 1794240.5 | 9.168114419 | 1.639439914 | 0.300781814 |
| 1794560.5 | 8.577722237 | 3.233389593 | 0.978987628 |
| 1794880.5 | 7.657587496 | 4.702151301 | 1.619203396 |
| 1795200.5 | 6.434669357 | 5.984236197 | 2.195103297 |
| 1795520.5 | 4.951154536 | 7.023400696 | 2.681892941 |
| 1795840.5 | 3.263328787 | 7.772491598 | 3.057932532 |
| 1796160.5 | 1.439035447 | 8.196989519 | 3.306293790 |
| 1796480.5 | -0.446134891 | 8.277715623 | 3.416025619 |
| 1796800.5 | -2.313506514 | 8.012269813 | 3.382935076 |
| 1797120.5 | -4.086239709 | 7.414999315 | 3.209777587 |
| 1797440.5 | -5.693906669 | 6.515564075 | 2.905858429 |

HELIOCENTRIC COORDINATES OF SATURN

REFERENCE AXES OF 1950.0

| JULIAN DATE | X | Y | Z |
|---|---|---|---|
| 1797760.5 | -7.076234756 | 5.356368700 | 2.486134170 |
| 1798080.5 | -8.185688825 | 3.989252995 | 1.969959516 |
| 1798400.5 | -8.988748564 | 2.471902437 | 1.379662766 |
| 1798720.5 | -9.465956253 | 0.864423344 | 0.739136207 |
| 1799040.5 | -9.611027725 | -0.773619690 | 0.072576201 |
| 1799360.5 | -9.429532531 | -2.385585489 | -0.596549482 |
| 1799680.5 | -8.937536000 | -3.919357248 | -1.246195753 |
| 1800000.5 | -8.159894642 | -5.328154066 | -1.856135029 |
| 1800320.5 | -7.128379232 | -6.571119533 | -2.408292438 |
| 1800640.5 | -5.880177522 | -7.613783359 | -2.887004883 |
| 1800960.5 | -4.456678950 | -8.428297347 | -3.279165980 |
| 1801280.5 | -2.902391489 | -8.993500337 | -3.574292906 |
| 1801600.5 | -1.263981814 | -9.294861936 | -3.764537451 |
| 1801920.5 | 0.410518381 | -9.324361399 | -3.844663744 |
| 1802240.5 | 2.072331719 | -9.080403602 | -3.812036225 |
| 1802560.5 | 3.672468269 | -8.567870247 | -3.666662781 |
| 1802880.5 | 5.162318986 | -7.798323960 | -3.411305235 |
| 1803200.5 | 6.494427772 | -6.790302879 | -3.051634599 |
| 1803520.5 | 7.623503215 | -5.569616358 | -2.596395865 |
| 1803840.5 | 8.507697607 | -4.169553399 | -2.057547059 |
| 1804160.5 | 9.110165955 | -2.630909422 | -1.450334644 |
| 1804480.5 | 9.400896900 | -1.001718019 | -0.793258638 |
| 1804800.5 | 9.358766453 | 0.663446890 | -0.107870184 |
| 1805120.5 | 8.973699608 | 2.304743343 | 0.581663124 |
| 1805440.5 | 8.248731243 | 3.859037220 | 1.249286833 |
| 1805760.5 | 7.201635615 | 5.262647428 | 1.868126410 |
| 1806080.5 | 5.865678369 | 6.454818658 | 2.411892134 |
| 1806400.5 | 4.289160457 | 7.381826962 | 2.856581212 |
| 1806720.5 | 2.533895878 | 8.001197020 | 3.182241786 |
| 1807040.5 | 0.672249491 | 8.284763099 | 3.374352389 |
| 1807360.5 | -1.217465596 | 8.220574259 | 3.424816735 |
| 1807680.5 | -3.056117344 | 7.813741448 | 3.332520346 |
| 1808000.5 | -4.768728808 | 7.085793907 | 3.103261877 |
| 1808320.5 | -6.288811737 | 6.072610176 | 2.749075839 |
| 1808640.5 | -7.561610360 | 4.821327662 | 2.287093485 |
| 1808960.5 | -8.546024044 | 3.386746751 | 1.738143260 |
| 1809280.5 | -9.215228670 | 1.827723436 | 1.125294184 |
| 1809600.5 | -9.556203337 | 0.203936084 | 0.472509292 |
| 1809920.5 | -9.568472218 | -1.426726646 | -0.196476151 |
| 1810240.5 | -9.262415069 | -3.010049794 | -0.858993137 |

## HELIOCENTRIC COORDINATES OF SATURN

## REFERENCE AXES OF 1950.0

| JULIAN DATE | X | Y | Z |
|---|---|---|---|
| 1810560.5 | -8.657490094 | -4.496688301 | -1.493997961 |
| 1810880.5 | -7.780616266 | -5.842848054 | -2.082415035 |
| 1811200.5 | -6.664797826 | -7.010665328 | -2.607349142 |
| 1811520.5 | -5.347953687 | -7.968430775 | -3.054227720 |
| 1811840.5 | -3.871898575 | -8.690720866 | -3.410899408 |
| 1812160.5 | -2.281451293 | -9.158469090 | -3.667702029 |
| 1812480.5 | -0.623655910 | -9.359012592 | -3.817515213 |
| 1812800.5 | 1.052910394 | -9.286150958 | -3.855814135 |
| 1813120.5 | 2.698791176 | -8.940241555 | -3.780736460 |
| 1813440.5 | 4.264359570 | -8.328345772 | -3.593170450 |
| 1813760.5 | 5.700739967 | -7.464508472 | -3.296900461 |
| 1814080.5 | 6.960889066 | -6.370266677 | -2.898855933 |
| 1814400.5 | 8.000610533 | -5.074923092 | -2.409275532 |
| 1814720.5 | 8.779893876 | -3.615351228 | -1.841681930 |
| 1815040.5 | 9.264634373 | -2.035673239 | -1.212810622 |
| 1815360.5 | 9.428491933 | -0.386730689 | -0.542464363 |
| 1815680.5 | 9.254852377 | 1.274843127 | 0.146786893 |
| 1816000.5 | 8.738832053 | 2.888115732 | 0.830134309 |
| 1816320.5 | 7.889114698 | 4.390055009 | 1.481355664 |
| 1816640.5 | 6.729278506 | 5.718473016 | 2.073967925 |
| 1816960.5 | 5.298221662 | 6.815583966 | 2.582669309 |
| 1817280.5 | 3.649346400 | 7.631859677 | 2.984954323 |
| 1817600.5 | 1.848322680 | 8.129707262 | 3.262713840 |
| 1817920.5 | -0.030492332 | 8.286421599 | 3.403592460 |
| 1818240.5 | -1.908613889 | 8.095930894 | 3.401891112 |
| 1818560.5 | -3.708381978 | 7.569065311 | 3.258880902 |
| 1818880.5 | -5.357662303 | 6.732368829 | 2.982510516 |
| 1819200.5 | -6.793769895 | 5.625721590 | 2.586594239 |
| 1819520.5 | -7.966286198 | 4.299167023 | 2.089626651 |
| 1819840.5 | -8.838623307 | 2.809377295 | 1.513394445 |
| 1820160.5 | -9.388363270 | 1.216185801 | 0.881562896 |
| 1820480.5 | -9.606574296 | -0.420463304 | 0.218389913 |
| 1820800.5 | -9.496440590 | -2.042975200 | -0.452346657 |
| 1821120.5 | -9.071672871 | -3.597986912 | -1.108186748 |
| 1821440.5 | -8.355081261 | -5.037581447 | -1.728526712 |
| 1821760.5 | -7.377002880 | -6.319767963 | -2.294874575 |
| 1822080.5 | -6.173579278 | -7.408768772 | -2.791055965 |
| 1822400.5 | -4.785385405 | -8.275298261 | -3.203392562 |
| 1822720.5 | -3.256342880 | -8.896705134 | -3.520802609 |
| 1823040.5 | -1.632758101 | -9.256987306 | -3.734841662 |

## HELIOCENTRIC COORDINATES OF SATURN

### REFERENCE AXES OF 1950.0

| JULIAN DATE | X | Y | Z |
|---|---|---|---|
| 1823360.5 | 0.037526342 | -9.346718063 | -3.839701632 |
| 1823680.5 | 1.705831300 | -9.162948054 | -3.832194576 |
| 1824000.5 | 3.323211864 | -8.709185935 | -3.711766211 |
| 1824320.5 | 4.841014541 | -7.995529403 | -3.480573666 |
| 1824640.5 | 6.211566652 | -7.038927484 | -3.143624116 |
| 1824960.5 | 7.389097454 | -5.863488307 | -2.708941485 |
| 1825280.5 | 8.330935558 | -4.500730805 | -2.187720614 |
| 1825600.5 | 8.998997288 | -2.989678292 | -1.594427676 |
| 1825920.5 | 9.361559273 | -1.376680976 | -0.946800500 |
| 1826240.5 | 9.395273212 | 0.285164034 | -0.265693284 |
| 1826560.5 | 9.087322163 | 1.937138267 | 0.425297277 |
| 1826880.5 | 8.437536104 | 3.516705676 | 1.100487072 |
| 1827200.5 | 7.460183240 | 4.960077818 | 1.733087793 |
| 1827520.5 | 6.185043774 | 6.205511109 | 2.296523795 |
| 1827840.5 | 4.657303141 | 7.197144657 | 2.766021825 |
| 1828160.5 | 2.935973917 | 7.889134907 | 3.120369433 |
| 1828480.5 | 1.091173996 | 8.249566120 | 3.343601348 |
| 1828800.5 | -0.799706341 | 8.262813988 | 3.426130048 |
| 1829120.5 | -2.657102364 | 7.930315106 | 3.365296221 |
| 1829440.5 | -4.404505414 | 7.270176420 | 3.165421339 |
| 1829760.5 | -5.973038575 | 6.315428105 | 2.837269539 |
| 1830080.5 | -7.304987882 | 5.111135429 | 2.397001422 |
| 1830400.5 | -8.356036268 | 3.710870633 | 1.864814115 |
| 1830720.5 | -9.096172554 | 2.173061717 | 1.263477609 |
| 1831040.5 | -9.509449757 | 0.557633538 | 0.616944908 |
| 1831360.5 | -9.592881674 | -1.076786327 | -0.050839838 |
| 1831680.5 | -9.354808120 | -2.674960154 | -0.716864908 |
| 1832000.5 | -8.813062324 | -4.100417388 | -1.359687224 |
| 1832320.5 | -7.993230842 | -5.566230263 | -1.959817774 |
| 1832640.5 | -6.927172086 | -6.775397092 | -2.499937766 |
| 1832960.5 | -5.651801050 | -7.781034819 | -2.965028635 |
| 1833280.5 | -4.208066762 | -8.556491816 | -3.342463093 |
| 1833600.5 | -2.640062624 | -9.081419068 | -3.622073611 |
| 1833920.5 | -0.994242238 | -9.341821692 | -3.796207751 |
| 1834240.5 | 0.681280778 | -9.330119323 | -3.859783066 |
| 1834560.5 | 2.337383362 | -9.045242504 | -3.810354501 |
| 1834880.5 | 3.924588253 | -8.492776906 | -3.648201303 |
| 1835200.5 | 5.393823475 | -7.685151552 | -3.376434032 |
| 1835520.5 | 6.697461506 | -6.641929741 | -3.001149553 |
| 1835840.5 | 7.790585521 | -5.390290920 | -2.531678647 |

## HELIOCENTRIC COORDINATES OF SATURN

## REFERENCE AXES OF 1950.0

| JULIAN DATE | X | Y | Z |
|---|---|---|---|
| 1836160.5 | 8.632186893 | -3.965230459 | -1.980727391 |
| 1836480.5 | 9.186663390 | -2.409212442 | -1.364293412 |
| 1836800.5 | 9.425693123 | -0.771608246 | -0.701502960 |
| 1837120.5 | 9.330220902 | 0.892161300 | -0.014337091 |
| 1837440.5 | 8.892466949 | 2.521987009 | 0.672843455 |
| 1837760.5 | 8.117816291 | 4.055093111 | 1.334048362 |
| 1838080.5 | 7.026295669 | 5.428798362 | 1.942728743 |
| 1838400.5 | 5.653252545 | 6.583942060 | 2.473154559 |
| 1838720.5 | 4.048867097 | 7.468708485 | 2.902003857 |
| 1839040.5 | 2.276261609 | 8.042406608 | 3.209991157 |
| 1839360.5 | 0.408209266 | 8.278658799 | 3.383310501 |
| 1839680.5 | -1.477263183 | 8.167477358 | 3.414667350 |
| 1840000.5 | -3.301827378 | 7.715884430 | 3.303736019 |
| 1840320.5 | -4.991691083 | 6.947019447 | 3.056993107 |
| 1840640.5 | -6.481731398 | 5.897969188 | 2.686999277 |
| 1840960.5 | -7.718552048 | 4.616728877 | 2.211280212 |
| 1841280.5 | -8.662317620 | 3.158734025 | 1.650978922 |
| 1841600.5 | -9.287380420 | 1.583370384 | 1.029446389 |
| 1841920.5 | -9.581853551 | -0.049185944 | 0.370922394 |
| 1842240.5 | -9.546401316 | -1.680557620 | -0.300582592 |
| 1842560.5 | -9.192580412 | -3.256121176 | -0.962163988 |
| 1842880.5 | -8.541134261 | -4.726437024 | -1.592678871 |
| 1843200.5 | -7.620586008 | -6.048140722 | -2.173151820 |
| 1843520.5 | -6.465830184 | -7.184179659 | -2.686925509 |
| 1843840.5 | -5.116584793 | -8.103884779 | -3.119771580 |
| 1844160.5 | -3.616145261 | -8.783123993 | -3.460011644 |
| 1844480.5 | -2.010404104 | -9.204388858 | -3.698589785 |
| 1844800.5 | -0.346982706 | -9.356785087 | -3.829098892 |
| 1845120.5 | 1.325534282 | -9.235960048 | -3.847776773 |
| 1845440.5 | 2.958238428 | -8.844041832 | -3.753503373 |
| 1845760.5 | 4.502461821 | -8.189686682 | -3.547842521 |
| 1846080.5 | 5.910400755 | -7.288274090 | -3.235149485 |
| 1846400.5 | 7.135935002 | -6.162195927 | -2.822726174 |
| 1846720.5 | 8.135720784 | -4.841134870 | -2.320982733 |
| 1847040.5 | 8.870584375 | -3.362219030 | -1.743559787 |
| 1847360.5 | 9.307213073 | -1.769936501 | -1.107363634 |
| 1847680.5 | 9.420108487 | -0.115680568 | -0.432460178 |
| 1848000.5 | 9.193715630 | 1.543216012 | 0.258234008 |
| 1848320.5 | 8.624567189 | 3.145094135 | 0.939526384 |
| 1848640.5 | 7.723191798 | 4.626327039 | 1.584840301 |

HELIOCENTRIC COORDINATES OF SATURN

REFERENCE AXES OF 1950.0

| JULIAN DATE | X | Y | Z |
|---|---|---|---|
| 1848960.5 | 6.515439682 | 5.924433948 | 2.167443272 |
| 1849280.5 | 5.042803396 | 6.981798684 | 2.661968327 |
| 1849600.5 | 3.361313884 | 7.749644498 | 3.046100356 |
| 1849920.5 | 1.538830255 | 8.191894118 | 3.302266477 |
| 1850240.5 | -0.348752762 | 8.288451276 | 3.419108349 |
| 1850560.5 | -2.221952223 | 8.036656191 | 3.392267468 |
| 1850880.5 | -4.002903916 | 7.450860554 | 3.224450476 |
| 1851200.5 | -5.620307247 | 6.560879693 | 2.924994341 |
| 1851520.5 | -7.013315986 | 5.409361635 | 2.508937147 |
| 1851840.5 | -8.134127569 | 4.048377061 | 1.995719155 |
| 1852160.5 | -8.949220362 | 2.535747145 | 1.407722973 |
| 1852480.5 | -9.439378331 | 0.931565278 | 0.768844678 |
| 1852800.5 | -9.598774893 | -0.704777490 | 0.103234215 |
| 1853120.5 | -9.433435236 | -2.316899661 | -0.565714095 |
| 1853440.5 | -8.959394784 | -3.853021714 | -1.216087903 |
| 1853760.5 | -8.200850420 | -5.266900367 | -1.827911825 |
| 1854080.5 | -7.188533257 | -6.518268974 | -2.383391349 |
| 1854400.5 | -5.958390795 | -7.573006294 | -2.867035384 |
| 1854720.5 | -4.550527469 | -8.403207290 | -3.265730762 |
| 1855040.5 | -3.008311774 | -8.987230720 | -3.568801465 |
| 1855360.5 | -1.377591771 | -9.309742426 | -3.768061237 |
| 1855680.5 | 0.294009317 | -9.361769342 | -3.857866515 |
| 1856000.5 | 1.957745027 | -9.140786563 | -3.835180233 |
| 1856320.5 | 3.564386131 | -8.650855336 | -3.699655853 |
| 1856640.5 | 5.064908823 | -7.902811098 | -3.453743464 |
| 1856960.5 | 6.411335894 | -6.914478382 | -3.102810568 |
| 1857280.5 | 7.557922162 | -5.710945720 | -2.655296499 |
| 1857600.5 | 8.462623028 | -4.324070532 | -2.122942247 |
| 1857920.5 | 9.088476944 | -2.797089647 | -1.520885990 |
| 1858240.5 | 9.405252802 | -1.174982360 | -0.867496861 |
| 1858560.5 | 9.391444322 | 0.487312073 | -0.184093918 |
| 1858880.5 | 9.036316358 | 2.130688324 | 0.505486381 |
| 1859200.5 | 8.341845698 | 3.692818068 | 1.175568553 |
| 1859520.5 | 7.324329625 | 5.110715991 | 1.799623181 |
| 1859840.5 | 6.015307601 | 6.324004214 | 2.351567688 |
| 1860160.5 | 4.461409563 | 7.278669120 | 2.807314090 |
| 1860480.5 | 2.722841140 | 7.930911010 | 3.146411136 |
| 1860800.5 | 0.870430374 | 8.250553002 | 3.353564499 |
| 1861120.5 | -1.018550774 | 8.223459063 | 3.419800347 |
| 1861440.5 | -2.865294333 | 7.852545981 | 3.343081785 |

HELIOCENTRIC COORDINATES OF SATURN

REFERENCE AXES OF 1950.0

| JULIAN DATE | X | Y | Z |
|---|---|---|---|
| 1861760.5 | -4.594424210 | 7.157241956 | 3.128291011 |
| 1862080.5 | -6.138370054 | 6.171560649 | 2.786620321 |
| 1862400.5 | -7.440694788 | 4.941195749 | 2.334519273 |
| 1862720.5 | -8.458194964 | 3.520110364 | 1.792383286 |
| 1863040.5 | -9.161795543 | 1.967038532 | 1.183153844 |
| 1863360.5 | -9.536377489 | 0.342236858 | 0.530974622 |
| 1863680.5 | -9.579768938 | -1.295257331 | -0.139980340 |
| 1864000.5 | -9.301193677 | -2.889725991 | -0.806437860 |
| 1864320.5 | -8.719476399 | -4.390253544 | -1.446713435 |
| 1864640.5 | -7.861332602 | -5.751766968 | -2.041206170 |
| 1864960.5 | -6.760042161 | -6.935667593 | -2.572695473 |
| 1865280.5 | -5.454215380 | -7.909895055 | -3.026415797 |
| 1865600.5 | -3.986421207 | -8.648855028 | -3.390101322 |
| 1865920.5 | -2.402058594 | -9.133493620 | -3.654068507 |
| 1866240.5 | -0.748462048 | -9.351356543 | -3.811272902 |
| 1866560.5 | 0.925895851 | -9.296574497 | -3.857330004 |
| 1866880.5 | 2.572111427 | -8.969808288 | -3.790516350 |
| 1867200.5 | 4.141461720 | -8.378236747 | -3.611786547 |
| 1867520.5 | 5.586003141 | -7.535669236 | -3.324844205 |
| 1867840.5 | 6.859301522 | -6.462785004 | -2.936272661 |
| 1868160.5 | 7.917412441 | -5.187414639 | -2.455693748 |
| 1868480.5 | 8.720166210 | -3.744742906 | -1.895906130 |
| 1868800.5 | 9.232763464 | -2.177308205 | -1.272952049 |
| 1869120.5 | 9.427652382 | -0.534668487 | -0.606058085 |
| 1869440.5 | 9.286613297 | 1.127406611 | 0.082610445 |
| 1869760.5 | 8.802909394 | 2.748356668 | 0.768444475 |
| 1870080.5 | 7.983277818 | 4.264973903 | 1.425190193 |
| 1870400.5 | 6.849447455 | 5.614305952 | 2.026080959 |
| 1870720.5 | 5.438798098 | 6.737225984 | 2.545281239 |
| 1871040.5 | 3.803757912 | 7.582382709 | 2.959541370 |
| 1871360.5 | 2.009630467 | 8.110038010 | 3.249870606 |
| 1871680.5 | 0.130839369 | 8.295343655 | 3.403029123 |
| 1872000.5 | -1.753699856 | 8.130716863 | 3.412664041 |
| 1872320.5 | -3.565319922 | 7.626258073 | 3.279683668 |
| 1872640.5 | -5.230531785 | 6.808186740 | 3.011843420 |
| 1872960.5 | -6.685350370 | 5.716319465 | 2.622884525 |
| 1873280.5 | -7.878309705 | 4.400829925 | 2.131318464 |
| 1873600.5 | -8.772077263 | 2.918614127 | 1.558995804 |
| 1873920.5 | -9.343792023 | 1.329724330 | 0.929653037 |
| 1874240.5 | -9.584379557 | -0.305772201 | 0.267592987 |

HELIOCENTRIC COORDINATES OF SATURN

REFERENCE AXES OF 1950.0

| JULIAN DATE | X | Y | Z |
|---|---|---|---|
| 1874560.5 | -9.497157825 | -1.930310193 | -0.403406388 |
| 1874880.5 | -9.096043692 | -3.490646634 | -1.060911680 |
| 1875200.5 | -8.403645529 | -4.939002564 | -1.684377709 |
| 1875520.5 | -7.449493458 | -6.233632874 | -2.255449784 |
| 1875840.5 | -6.268571464 | -7.339019140 | -2.758100332 |
| 1876160.5 | -4.900168802 | -8.225900361 | -3.178688931 |
| 1876480.5 | -3.386959491 | -8.871277657 | -3.506004641 |
| 1876800.5 | -1.774214127 | -9.258434760 | -3.731309875 |
| 1877120.5 | -0.109091933 | -9.376979609 | -3.848389007 |
| 1877440.5 | 1.560015701 | -9.222917263 | -3.853606927 |
| 1877760.5 | 3.184123219 | -8.798770295 | -3.745985889 |
| 1878080.5 | 4.714316838 | -8.113754907 | -3.527306207 |
| 1878400.5 | 6.102510961 | -7.183998415 | -3.202227035 |
| 1878720.5 | 7.302404068 | -6.032753863 | -2.778411162 |
| 1879040.5 | 8.270836970 | -4.690617635 | -2.266663173 |
| 1879360.5 | 8.969470442 | -3.195817619 | -1.681119146 |
| 1879680.5 | 9.366335988 | -1.594088262 | -1.039267996 |
| 1880000.5 | 9.437597286 | 0.062160518 | -0.361675016 |
| 1880320.5 | 9.169598659 | 1.715068493 | 0.328444925 |
| 1880640.5 | 8.560863448 | 3.303027757 | 1.005844781 |
| 1880960.5 | 7.623812124 | 4.763042985 | 1.644141482 |
| 1881280.5 | 6.385902279 | 6.033795235 | 2.217024893 |
| 1881600.5 | 4.889812426 | 7.059274615 | 2.699740566 |
| 1881920.5 | 3.192336336 | 7.792643381 | 3.070720812 |
| 1882240.5 | 1.361833232 | 8.199826936 | 3.313163641 |
| 1882560.5 | -0.525646514 | 8.262269825 | 3.416323184 |
| 1882880.5 | -2.391143583 | 7.978379889 | 3.376298708 |
| 1883200.5 | -4.157841483 | 7.363415279 | 3.196195722 |
| 1883520.5 | -5.755693605 | 6.447885249 | 2.885660581 |
| 1883840.5 | -7.125076941 | 5.274823931 | 2.459914007 |
| 1884160.5 | -8.219206462 | 3.896438177 | 1.938477377 |
| 1884480.5 | -9.005311580 | 2.370594315 | 1.343780428 |
| 1884800.5 | -9.464721993 | 0.757491880 | 0.699797758 |
| 1885120.5 | -9.592075255 | -0.883224491 | 0.030826584 |
| 1885440.5 | -9.393910121 | -2.494751857 | -0.639514829 |
| 1885760.5 | -8.886928670 | -4.024766592 | -1.289036031 |
| 1886080.5 | -8.096177015 | -5.426500542 | -1.897495506 |
| 1886400.5 | -7.053400570 | -6.659468119 | -2.446963475 |
| 1886720.5 | -5.795827556 | -7.689903556 | -2.922034953 |
| 1887040.5 | -4.365097300 | -8.490709663 | -3.309852982 |

HELIOCENTRIC COORDINATES OF SATURN

REFERENCE AXES OF 1950.0

| JULIAN DATE | X | Y | Z |
|---|---|---|---|
| 1887360.5 | -2.806044538 | -9.041297589 | -3.600110574 |
| 1887680.5 | -1.165667969 | -9.327616264 | -3.785109298 |
| 1888000.5 | 0.507703450 | -9.342199640 | -3.859807820 |
| 1888320.5 | 2.165161597 | -9.084155312 | -3.821841887 |
| 1888640.5 | 3.757893372 | -8.559133872 | -3.671534463 |
| 1888960.5 | 5.237798236 | -7.779366185 | -3.411934311 |
| 1889280.5 | 6.558158525 | -6.763826530 | -3.048911665 |
| 1889600.5 | 7.674506348 | -5.538485502 | -2.591300590 |
| 1889920.5 | 8.545782610 | -4.136541309 | -2.051044490 |
| 1890240.5 | 9.135817097 | -2.598495300 | -1.443290040 |
| 1890560.5 | 9.415110711 | -0.971935692 | -0.786372941 |
| 1890880.5 | 9.362857166 | 0.689111318 | -0.101635355 |
| 1891200.5 | 8.969081367 | 2.325405060 | 0.586988157 |
| 1891520.5 | 8.236692308 | 3.874408929 | 1.253674126 |
| 1891840.5 | 7.183162711 | 5.272963819 | 1.871740045 |
| 1892160.5 | 5.841476766 | 6.460667526 | 2.414996245 |
| 1892480.5 | 4.259958516 | 7.383748553 | 2.859349598 |
| 1892800.5 | 2.500651897 | 7.999011452 | 3.184498492 |
| 1893120.5 | 0.636113365 | 8.277257035 | 3.375475772 |
| 1893440.5 | -1.255194203 | 8.205720814 | 3.423829722 |
| 1893760.5 | -3.094053014 | 7.789371259 | 3.328341586 |
| 1894080.5 | -4.805193696 | 7.050275846 | 3.094969822 |
| 1894400.5 | -6.321637986 | 6.025057943 | 2.736008479 |
| 1894720.5 | -7.588188153 | 4.761646163 | 2.268890616 |
| 1895040.5 | -8.563497446 | 3.315696997 | 1.714795152 |
| 1895360.5 | -9.220770207 | 1.746978318 | 1.097185695 |
| 1895680.5 | -9.547344803 | 0.116080972 | 0.440440423 |
| 1896000.5 | -9.543474669 | -1.518297168 | -0.231314747 |
| 1896320.5 | -9.220622380 | -3.101393260 | -0.895114882 |
| 1896640.5 | -8.599546070 | -4.583743874 | -1.529793008 |
| 1896960.5 | -7.708424774 | -5.921936115 | -2.116361732 |
| 1897280.5 | -6.581226763 | -7.078887664 | -2.638191206 |
| 1897600.5 | -5.256420581 | -8.023865616 | -3.081067975 |
| 1897920.5 | -3.775986586 | -8.732436691 | -3.433216375 |
| 1898240.5 | -2.184614540 | -9.186444130 | -3.685325030 |
| 1898560.5 | -0.528996998 | -9.374026318 | -3.830586568 |
| 1898880.5 | 1.142826626 | -9.289673934 | -3.864750256 |
| 1899200.5 | 2.782105453 | -8.934331966 | -3.786191513 |
| 1899520.5 | 4.340030246 | -8.315555988 | -3.596004068 |
| 1899840.5 | 5.768455070 | -7.447720308 | -3.298116061 |

## HELIOCENTRIC COORDINATES OF SATURN

## REFERENCE AXES OF 1950.0

| JULIAN DATE | X | Y | Z |
|---|---|---|---|
| 1900160.5 | 7.020751976 | -6.352248609 | -2.899420011 |
| 1900480.5 | 8.052909184 | -5.057798940 | -2.409890633 |
| 1900800.5 | 8.825082396 | -3.600379848 | -1.842689667 |
| 1901120.5 | 9.303479252 | -2.023456993 | -1.214292240 |
| 1901440.5 | 9.462051155 | -0.377568431 | -0.544405936 |
| 1901760.5 | 9.284322052 | 1.280842190 | 0.144442847 |
| 1902080.5 | 8.765409683 | 2.891068900 | 0.827507303 |
| 1902400.5 | 7.913849916 | 4.390331209 | 1.478642875 |
| 1902720.5 | 6.752924121 | 5.716631843 | 2.071423631 |
| 1903040.5 | 5.321147062 | 6.812228702 | 2.580546045 |
| 1903360.5 | 3.671562235 | 7.627455625 | 2.983426120 |
| 1903680.5 | 1.869621366 | 8.124437248 | 3.261811873 |
| 1904000.5 | -0.010326372 | 8.280140779 | 3.403180877 |
| 1904320.5 | -1.889609952 | 8.088243815 | 3.401695265 |
| 1904640.5 | -3.690294674 | 7.559485062 | 3.258553585 |
| 1904960.5 | -5.340021290 | 6.720462172 | 2.981696497 |
| 1905280.5 | -6.775998993 | 5.611154939 | 2.584955617 |
| 1905600.5 | -7.947775643 | 4.281665013 | 2.086831810 |
| 1905920.5 | -8.818720008 | 2.788691248 | 1.509111974 |
| 1906240.5 | -9.366367223 | 1.192136307 | 0.875490655 |
| 1906560.5 | -9.581852064 | -0.447902095 | 0.210308923 |
| 1906880.5 | -9.468673421 | -2.073592816 | -0.462511604 |
| 1907200.5 | -9.041052802 | -3.631204428 | -1.120293770 |
| 1907520.5 | -8.322136239 | -5.072328465 | -1.742185831 |
| 1907840.5 | -7.342232756 | -6.354620065 | -2.309541820 |
| 1908160.5 | -6.137280261 | -7.442287281 | -2.806181203 |
| 1908480.5 | -4.747753173 | -8.306394302 | -3.218542702 |
| 1908800.5 | -3.217734000 | -8.924743066 | -3.535676609 |
| 1909120.5 | -1.593831305 | -9.281670300 | -3.749223249 |
| 1909440.5 | 0.075788238 | -9.368059560 | -3.853459667 |
| 1909760.5 | 1.742189889 | -9.181388101 | -3.845345772 |
| 1910080.5 | 3.356405358 | -8.725724955 | -3.724547269 |
| 1910400.5 | 4.870103243 | -8.011728293 | -3.493457860 |
| 1910720.5 | 6.236250617 | -7.056725422 | -3.157258318 |
| 1911040.5 | 7.409893924 | -5.884901242 | -2.724027971 |
| 1911360.5 | 8.349190989 | -4.527521778 | -2.204882078 |
| 1911680.5 | 9.016760679 | -3.023059940 | -1.614081762 |
| 1912000.5 | 9.381349909 | -1.417078878 | -0.969056587 |
| 1912320.5 | 9.419767119 | 0.238271129 | -0.290278672 |
| 1912640.5 | 9.118976601 | 1.885285510 | 0.399074568 |

HELIOCENTRIC COORDINATES OF SATURN

REFERENCE AXES OF 1950.0

| JULIAN DATE | X | Y | Z |
|---|---|---|---|
| 1912960.5 | 8.478175618 | 3.462396199 | 1.073728997 |
| 1913280.5 | 7.510593052 | 4.906599862 | 1.707237289 |
| 1913600.5 | 6.244675451 | 6.156609322 | 2.273226036 |
| 1913920.5 | 4.724289960 | 7.156574607 | 2.746919529 |
| 1914240.5 | 3.007603992 | 7.860039787 | 3.106817496 |
| 1914560.5 | 1.164443431 | 8.233604402 | 3.336315529 |
| 1914880.5 | -0.727800712 | 8.259653362 | 3.424999449 |
| 1915200.5 | -2.589335234 | 7.937770976 | 3.369426853 |
| 1915520.5 | -4.343273067 | 7.284903952 | 3.173388177 |
| 1915840.5 | -5.919981913 | 6.333773869 | 2.847448214 |
| 1916160.5 | -7.260611683 | 5.129600893 | 2.407767264 |
| 1916480.5 | -8.319639986 | 3.726410254 | 1.874681796 |
| 1916800.5 | -9.066030995 | 2.183374839 | 1.271240747 |
| 1917120.5 | -9.483154892 | 0.561406764 | 0.621797869 |
| 1917440.5 | -9.567793529 | -1.079748894 | -0.049225102 |
| 1917760.5 | -9.328562458 | -2.683777537 | -0.718334055 |
| 1918080.5 | -8.784037478 | -4.199389167 | -1.363693826 |
| 1918400.5 | -7.960825849 | -5.581284549 | -1.965601796 |
| 1918720.5 | -6.891791146 | -6.790588491 | -2.506734737 |
| 1919040.5 | -5.614588966 | -7.794916869 | -2.972232994 |
| 1919360.5 | -4.170552175 | -8.568288031 | -3.349709132 |
| 1919680.5 | -2.603837834 | -9.091038423 | -3.629250414 |
| 1920000.5 | -0.960713367 | -9.349798825 | -3.803441630 |
| 1920320.5 | 0.711097216 | -9.337526925 | -3.867408214 |
| 1920640.5 | 2.363046100 | -9.053588216 | -3.818877486 |
| 1920960.5 | 3.946405907 | -8.503887892 | -3.658260684 |
| 1921280.5 | 5.412968292 | -7.701055341 | -3.388758576 |
| 1921600.5 | 6.715850803 | -6.664667162 | -3.016487121 |
| 1921920.5 | 7.810476018 | -5.421462654 | -2.550606009 |
| 1922240.5 | 8.655831110 | -4.005457391 | -2.003413035 |
| 1922560.5 | 9.216217364 | -2.457902212 | -1.390393248 |
| 1922880.5 | 9.463317548 | -0.827145741 | -0.730255346 |
| 1923200.5 | 9.377975866 | 0.832070486 | -0.044720826 |
| 1923520.5 | 8.952017803 | 2.460165040 | 0.642050348 |
| 1923840.5 | 8.190132566 | 3.994811522 | 1.304247964 |
| 1924160.5 | 7.111373781 | 5.373608696 | 1.915444759 |
| 1924480.5 | 5.749924854 | 6.537367921 | 2.449912515 |
| 1924800.5 | 4.154781960 | 7.433841938 | 2.884165026 |
| 1925120.5 | 2.388080072 | 8.021506132 | 3.198579537 |
| 1925440.5 | 0.521996172 | 8.272859850 | 3.378880155 |

## HELIOCENTRIC COORDINATES OF SATURN

### REFERENCE AXES OF 1950.0

| JULIAN DATE | X | Y | Z |
|---|---|---|---|
| 1925760.5 | -1.365544967 | 8.176699733 | 3.417248411 |
| 1926080.5 | -3.195855558 | 7.738955300 | 3.312871576 |
| 1926400.5 | -4.894479233 | 6.981943595 | 3.071842140 |
| 1926720.5 | -6.395517058 | 5.942218475 | 2.706453340 |
| 1927040.5 | -7.644797888 | 4.667454828 | 2.234052240 |
| 1927360.5 | -8.601717434 | 3.212917889 | 1.675668781 |
| 1927680.5 | -9.239859024 | 1.637988688 | 1.054616995 |
| 1928000.5 | -9.546646154 | 0.003040020 | 0.395197488 |
| 1928320.5 | -9.522283125 | -1.633179063 | -0.278423518 |
| 1928640.5 | -9.178227793 | -3.215564790 | -0.943111844 |
| 1928960.5 | -8.535437999 | -4.694070105 | -1.577421737 |
| 1929280.5 | -7.622597628 | -6.024571282 | -2.162029083 |
| 1929600.5 | -6.474460077 | -7.169338950 | -2.679996098 |
| 1929920.5 | -5.130444198 | -8.097351289 | -3.116952374 |
| 1930240.5 | -3.633665501 | -8.784509131 | -3.461208578 |
| 1930560.5 | -2.030131808 | -9.213482834 | -3.703731993 |
| 1930880.5 | -0.367759434 | -9.373495583 | -3.838114015 |
| 1931200.5 | 1.304535801 | -9.260333872 | -3.860612230 |
| 1931520.5 | 2.937622400 | -8.876398896 | -3.770197120 |
| 1931840.5 | 4.482881854 | -8.230717148 | -3.568580261 |
| 1932160.5 | 5.892903514 | -7.338966593 | -3.260250255 |
| 1932480.5 | 7.122224225 | -6.223588122 | -2.852549071 |
| 1932800.5 | 8.128251022 | -4.913970019 | -2.355788421 |
| 1933120.5 | 8.872476730 | -3.446597928 | -1.783364626 |
| 1933440.5 | 9.322021666 | -1.865020513 | -1.151810585 |
| 1933760.5 | 9.451467452 | -0.219478135 | -0.480720513 |
| 1934080.5 | 9.244892267 | 1.433952914 | 0.207516643 |
| 1934400.5 | 8.697950536 | 3.034035294 | 0.888236682 |
| 1934720.5 | 7.819760365 | 4.520560413 | 1.535321674 |
| 1935040.5 | 6.634286173 | 5.829268167 | 2.122342982 |
| 1935360.5 | 5.180859882 | 6.903386649 | 2.623996079 |
| 1935680.5 | 3.513498643 | 7.693523157 | 3.017731692 |
| 1936000.5 | 1.698777598 | 8.162262027 | 3.285410496 |
| 1936320.5 | -0.187762452 | 8.287273499 | 3.414735418 |
| 1936640.5 | -2.066378958 | 8.063131859 | 3.400197335 |
| 1936960.5 | -3.858525688 | 7.501600029 | 3.243401108 |
| 1937280.5 | -5.491975534 | 6.630685999 | 2.952868559 |
| 1937600.5 | -6.904559363 | 5.492259859 | 2.543231274 |
| 1937920.5 | -8.046736701 | 4.138302514 | 2.033813715 |
| 1938240.5 | -8.883169162 | 2.627030422 | 1.447096934 |

HELIOCENTRIC COORDINATES OF SATURN

REFERENCE AXES OF 1950.0

| JULIAN DATE | X | Y | Z |
|---|---|---|---|
| 1938560.5 | -9.393049304 | 1.019367022 | 0.807273495 |
| 1938880.5 | -9.569386504 | -0.624121387 | 0.138955323 |
| 1939200.5 | -9.417599495 | -2.245688739 | -0.533897978 |
| 1939520.5 | -8.953726633 | -3.792219710 | -1.188797159 |
| 1939840.5 | -8.202499955 | -5.216414168 | -1.805288073 |
| 1940160.5 | -7.195482456 | -6.477414020 | -2.365282678 |
| 1940480.5 | -5.969441506 | -7.540987855 | -2.853197625 |
| 1940800.5 | -4.565067582 | -8.379454619 | -3.255972794 |
| 1941120.5 | -3.026020868 | -8.971544235 | -3.563053498 |
| 1941440.5 | -1.398187003 | -9.302315442 | -3.766390050 |
| 1941760.5 | 0.270976532 | -9.363153132 | -3.860466846 |
| 1942080.5 | 1.933069454 | -9.151827952 | -3.842355364 |
| 1942400.5 | 3.539387998 | -8.672607193 | -3.711787813 |
| 1942720.5 | 5.041609733 | -7.936414835 | -3.471252329 |
| 1943040.5 | 6.392570428 | -6.961032948 | -3.126108758 |
| 1943360.5 | 7.547187707 | -5.771316888 | -2.684715847 |
| 1943680.5 | 8.463591353 | -4.399359349 | -2.158544456 |
| 1944000.5 | 9.104559578 | -2.884479827 | -1.562227443 |
| 1944320.5 | 9.439461375 | -1.272957236 | -0.913524875 |
| 1944640.5 | 9.446468106 | 0.382425125 | -0.233238581 |
| 1944960.5 | 9.114348500 | 2.023467206 | 0.455157879 |
| 1945280.5 | 8.444173198 | 3.588538139 | 1.126277407 |
| 1945600.5 | 7.450913308 | 5.015138851 | 1.753809989 |
| 1945920.5 | 6.164441072 | 6.243012153 | 2.311756675 |
| 1946240.5 | 4.629558014 | 7.217734125 | 2.775901275 |
| 1946560.5 | 2.904752782 | 7.894487131 | 3.125410280 |
| 1946880.5 | 1.059547853 | 8.241531658 | 3.344368214 |
| 1947200.5 | -0.829440663 | 8.242826892 | 3.423016702 |
| 1947520.5 | -2.683314665 | 7.899334266 | 3.358488376 |
| 1947840.5 | -4.425945478 | 7.228763884 | 3.154910837 |
| 1948160.5 | -5.988571610 | 6.263828962 | 2.822879257 |
| 1948480.5 | -7.313371317 | 5.049364344 | 2.378421096 |
| 1948800.5 | -8.355710565 | 3.638844119 | 1.841660955 |
| 1949120.5 | -9.085088171 | 2.090832394 | 1.235405013 |
| 1949440.5 | -9.485023135 | 0.465739958 | 0.583806050 |
| 1949760.5 | -9.552177685 | -1.176931012 | -0.088805672 |
| 1950080.5 | -9.294965889 | -2.780779153 | -0.758883977 |
| 1950400.5 | -8.731870942 | -4.294107079 | -1.404399166 |
| 1950720.5 | -7.889679134 | -5.670965714 | -2.005343950 |
| 1951040.5 | -6.801790173 | -6.871732497 | -2.544044462 |

## HELIOCENTRIC COORDINATES OF SATURN

### REFERENCE AXES OF 1950.0

| JULIAN DATE | X | Y | Z |
|---|---|---|---|
| 1951360.5 | -5.506691541 | -7.863380062 | -3.005335852 |
| 1951680.5 | -4.046686139 | -8.619671317 | -3.376688558 |
| 1952000.5 | -2.467033502 | -9.121327754 | -3.648299696 |
| 1952320.5 | -0.815241202 | -9.355867903 | -3.813063627 |
| 1952640.5 | 0.859848479 | -9.317396690 | -3.866537467 |
| 1952960.5 | 2.508944792 | -9.006622385 | -3.806982970 |
| 1953280.5 | 4.083100251 | -8.430911742 | -3.635414634 |
| 1953600.5 | 5.534461689 | -7.604310304 | -3.355633285 |
| 1953920.5 | 6.817019768 | -6.547588280 | -2.974273446 |
| 1954240.5 | 7.887460987 | -5.288357849 | -2.500887785 |
| 1954560.5 | 8.706256900 | -3.861209076 | -1.948050915 |
| 1954880.5 | 9.239069182 | -2.307728398 | -1.331428071 |
| 1955200.5 | 9.458465900 | -0.676238620 | -0.669741392 |
| 1955520.5 | 9.345875607 | 0.978894896 | 0.015433010 |
| 1955840.5 | 8.893641538 | 2.598532949 | 0.700100631 |
| 1956160.5 | 8.106963555 | 4.120715781 | 1.358570429 |
| 1956480.5 | 7.005438317 | 5.483342882 | 1.964492285 |
| 1956800.5 | 5.623854621 | 6.627508806 | 2.492195789 |
| 1957120.5 | 4.011901398 | 7.501277512 | 2.918255962 |
| 1957440.5 | 2.232521617 | 8.063518532 | 3.223142705 |
| 1957760.5 | 0.358814793 | 8.287283260 | 3.392742872 |
| 1958080.5 | -1.530329698 | 8.162123177 | 3.419499015 |
| 1958400.5 | -3.355373705 | 7.694857415 | 3.302937089 |
| 1958720.5 | -5.041422294 | 6.908733899 | 3.049525751 |
| 1959040.5 | -6.522862328 | 5.841511724 | 2.672063734 |
| 1959360.5 | -7.746323780 | 4.542502733 | 2.188606684 |
| 1959680.5 | -8.672257715 | 3.068610473 | 1.620918970 |
| 1960000.5 | -9.275598186 | 1.480498748 | 0.992915544 |
| 1960320.5 | -9.545393474 | -0.160661144 | 0.329301454 |
| 1960640.5 | -9.483617414 | -1.795841181 | -0.345569768 |
| 1960960.5 | -9.103495394 | -3.370170510 | -1.008593991 |
| 1961280.5 | -8.427616877 | -4.834326993 | -1.638566093 |
| 1961600.5 | -7.486035361 | -6.145364754 | -2.216596478 |
| 1961920.5 | -6.314520255 | -7.267061186 | -2.726321979 |
| 1962240.5 | -4.953100655 | -8.169914234 | -3.153960681 |
| 1962560.5 | -3.444962128 | -8.830977928 | -3.488287832 |
| 1962880.5 | -1.835630569 | -9.233710731 | -3.720608338 |
| 1963200.5 | -0.172307901 | -9.367915928 | -3.844762746 |
| 1963520.5 | 1.496747376 | -9.229769284 | -3.857167311 |
| 1963840.5 | 3.122848792 | -8.821909122 | -3.756879280 |

HELIOCENTRIC COORDINATES OF SATURN

REFERENCE AXES OF 1950.0

| JULIAN DATE | X | Y | Z |
|---|---|---|---|
| 1964160.5 | 4.657552618 | -8.153575340 | -3.545683236 |
| 1964480.5 | 6.053409147 | -7.240788813 | -3.228196962 |
| 1964800.5 | 7.264843447 | -6.106552169 | -2.811991249 |
| 1965120.5 | 8.249218773 | -4.781029004 | -2.307707960 |
| 1965440.5 | 8.968133688 | -3.301615713 | -1.729141712 |
| 1965760.5 | 9.389035266 | -1.712751596 | -1.093222567 |
| 1966080.5 | 9.487329010 | -0.065357186 | -0.419868771 |
| 1966400.5 | 9.248676180 | 1.584002554 | 0.268249564 |
| 1966720.5 | 8.670713704 | 3.174632217 | 0.946261749 |
| 1967040.5 | 7.764521198 | 4.644145913 | 1.588051877 |
| 1967360.5 | 6.555782963 | 5.931478428 | 2.167450439 |
| 1967680.5 | 5.085123826 | 6.980348636 | 2.659640698 |
| 1968000.5 | 3.407265055 | 7.742986595 | 3.042722716 |
| 1968320.5 | 1.588805504 | 8.183715117 | 3.299275708 |
| 1968640.5 | -0.295337872 | 8.281855257 | 3.417698747 |
| 1968960.5 | -2.166508677 | 8.033454050 | 3.393109831 |
| 1969280.5 | -3.947340676 | 7.451513341 | 3.227647512 |
| 1969600.5 | -5.566554109 | 6.564681521 | 2.930131206 |
| 1969920.5 | -6.962893163 | 5.414669563 | 2.515162324 |
| 1970240.5 | -8.087797648 | 4.052877008 | 2.001849811 |
| 1970560.5 | -8.906713240 | 2.536789008 | 1.412386940 |
| 1970880.5 | -9.399230978 | 0.926604068 | 0.770674747 |
| 1971200.5 | -9.558380634 | -0.717657798 | 0.101105773 |
| 1971520.5 | -9.389371128 | -2.338488910 | -0.572448941 |
| 1971840.5 | -8.907999156 | -3.882659700 | -1.227435179 |
| 1972160.5 | -8.138916956 | -5.302434929 | -1.843191084 |
| 1972480.5 | -7.113927354 | -6.556302553 | -2.401320769 |
| 1972800.5 | -5.870419774 | -7.609322122 | -2.885905376 |
| 1973120.5 | -4.449990370 | -8.433246953 | -3.283611797 |
| 1973440.5 | -2.897296976 | -9.006648732 | -3.583783953 |
| 1973760.5 | -1.259294299 | -9.315074382 | -3.778527276 |
| 1974080.5 | 0.415403509 | -9.350899431 | -3.862684512 |
| 1974400.5 | 2.077394880 | -9.113138525 | -3.833806254 |
| 1974720.5 | 3.677380843 | -8.607474107 | -3.692195989 |
| 1975040.5 | 5.166988270 | -7.846314794 | -3.440960644 |
| 1975360.5 | 6.499571050 | -6.848821843 | -3.086049517 |
| 1975680.5 | 7.631050357 | -5.640960245 | -2.636308729 |
| 1976000.5 | 8.520914781 | -4.255586926 | -2.103560752 |
| 1976320.5 | 9.133497409 | -2.732483280 | -1.502673803 |
| 1976640.5 | 9.439568467 | -1.118174021 | -0.851556145 |

## HELICCENTRIC COORDINATES OF SATURN

## REFERENCE AXES OF 1950.0

| JULIAN DATE | X | Y | Z |
|---|---|---|---|
| 1976960.5 | 9.418194944 | 0.534633843 | -0.171004414 |
| 1977280.5 | 9.058746673 | 2.168141642 | 0.515660390 |
| 1977600.5 | 8.362857145 | 3.721199204 | 1.183226212 |
| 1977920.5 | 7.346072144 | 5.131756832 | 1.805529073 |
| 1978240.5 | 6.038859243 | 6.339991734 | 2.356696452 |
| 1978560.5 | 4.486638205 | 7.291933871 | 2.812637635 |
| 1978880.5 | 2.748551986 | 7.943265094 | 3.152659431 |
| 1979200.5 | 0.894837595 | 8.262832422 | 3.361024563 |
| 1979520.5 | -0.997120433 | 8.235330746 | 3.428224002 |
| 1979840.5 | -2.847670897 | 7.862629400 | 3.351728411 |
| 1980160.5 | -4.580070234 | 7.163426346 | 3.136057747 |
| 1980480.5 | -6.125328834 | 6.171378469 | 2.792198298 |
| 1980800.5 | -7.426114761 | 4.932409470 | 2.336646143 |
| 1981120.5 | -8.438842247 | 3.501426990 | 1.790166272 |
| 1981440.5 | -9.134291837 | 1.938409999 | 1.176220193 |
| 1981760.5 | -9.497471599 | 0.304815037 | 0.519470298 |
| 1982080.5 | -9.526706893 | -1.339293739 | -0.155445484 |
| 1982400.5 | -9.232137742 | -2.937448855 | -0.824884404 |
| 1982720.5 | -8.633913105 | -4.438363782 | -1.466925323 |
| 1983040.5 | -7.760303279 | -5.796954444 | -2.061865782 |
| 1983360.5 | -6.645888797 | -6.974870403 | -2.592512450 |
| 1983680.5 | -5.329961900 | -7.940626113 | -3.044296879 |
| 1984000.5 | -3.855249179 | -8.669474125 | -3.405272415 |
| 1984320.5 | -2.266969480 | -9.143205392 | -3.666070050 |
| 1984640.5 | -0.612123438 | -9.350018928 | -3.819876213 |
| 1984960.5 | 1.061123816 | -9.284500929 | -3.862453022 |
| 1985280.5 | 2.704058133 | -8.947689004 | -3.792193093 |
| 1985600.5 | 4.268074097 | -8.347192491 | -3.610198252 |
| 1985920.5 | 5.705407811 | -7.497352020 | -3.320376773 |
| 1986240.5 | 6.969985630 | -6.419421181 | -2.929554202 |
| 1986560.5 | 8.018437674 | -5.141737962 | -2.447586598 |
| 1986880.5 | 8.811323499 | -3.699825774 | -1.887452994 |
| 1987200.5 | 9.314604911 | -2.136316230 | -1.265282457 |
| 1987520.5 | 9.501419435 | -0.500509059 | -0.600239827 |
| 1987840.5 | 9.354303684 | 1.152563911 | 0.085768805 |
| 1988160.5 | 8.867486282 | 2.763423686 | 0.768514696 |
| 1988480.5 | 8.048408414 | 4.270112121 | 1.422195864 |
| 1988800.5 | 6.918794212 | 5.611074148 | 2.020591154 |
| 1989120.5 | 5.515201355 | 6.728545178 | 2.538430714 |
| 1989440.5 | 3.888536395 | 7.572212882 | 2.952917852 |

HELIOCENTRIC COORDINATES OF SATURN

REFERENCE AXES OF 1950.0

| JULIAN DATE | X | Y | Z |
|---|---|---|---|
| 1989760.5 | 2.102254392 | 8.102848013 | 3.245293666 |
| 1990080.5 | 0.229200191 | 8.295424997 | 3.402249494 |
| 1990400.5 | -1.652680349 | 8.141219188 | 3.416966144 |
| 1990720.5 | -3.465266913 | 7.648496113 | 3.289600884 |
| 1991040.5 | -5.135175607 | 6.841660150 | 3.027140420 |
| 1991360.5 | -6.598001929 | 5.759026547 | 2.642660483 |
| 1991680.5 | -7.801407152 | 4.449631246 | 2.154142851 |
| 1992000.5 | -8.706835874 | 2.969625951 | 1.583066838 |
| 1992320.5 | -9.289957600 | 1.378779179 | 0.952992918 |
| 1992640.5 | -9.540142316 | -0.262572122 | 0.288291086 |
| 1992960.5 | -9.459318363 | -1.895996547 | -0.386922933 |
| 1993280.5 | -9.060471208 | -3.466873583 | -1.049635665 |
| 1993600.5 | -8.365963415 | -4.925763874 | -1.678572735 |
| 1993920.5 | -7.405830778 | -6.229291489 | -2.254635908 |
| 1994240.5 | -6.216183132 | -7.340618478 | -2.761167568 |
| 1994560.5 | -4.837782743 | -8.229621241 | -3.184084875 |
| 1994880.5 | -3.314806280 | -8.872913820 | -3.511941374 |
| 1995200.5 | -1.693808959 | -9.253937143 | -3.735998686 |
| 1995520.5 | -0.023028567 | -9.363127343 | -3.850314760 |
| 1995840.5 | 1.648207757 | -9.197795791 | -3.851730398 |
| 1996160.5 | 3.270341922 | -8.761962678 | -3.739846411 |
| 1996480.5 | 4.794502187 | -8.066391770 | -3.517070709 |
| 1996800.5 | 6.173377863 | -7.128643528 | -3.188668914 |
| 1997120.5 | 7.362094888 | -5.973095615 | -2.762804904 |
| 1997440.5 | 8.319176060 | -4.630973527 | -2.250593009 |
| 1997760.5 | 9.007711267 | -3.140364190 | -1.666154642 |
| 1998080.5 | 9.396824130 | -1.546084163 | -1.026628137 |
| 1998400.5 | 9.463426668 | 0.100769553 | -0.352059405 |
| 1998720.5 | 9.194164057 | 1.743749519 | 0.334898080 |
| 1999040.5 | 8.587376016 | 3.322601218 | 1.009535413 |
| 1999360.5 | 7.654827790 | 4.775500303 | 1.645935589 |
| 1999680.5 | 6.422899198 | 6.041976026 | 2.218125775 |
| 2000000.5 | 4.932893669 | 7.066384567 | 2.701492836 |
| 2000320.5 | 3.240172093 | 7.801646743 | 3.074357706 |
| 2000640.5 | 1.411940525 | 8.212832560 | 3.319544762 |
| 2000960.5 | -0.476286160 | 8.280093456 | 3.425739819 |
| 2001280.5 | -2.345272165 | 8.000436823 | 3.388415608 |
| 2001600.5 | -4.117149949 | 7.387955879 | 3.210139764 |
| 2001920.5 | -5.720300358 | 6.472412802 | 2.900190261 |
| 2002240.5 | -7.093486054 | 5.296509091 | 2.473589770 |

HELIOCENTRIC COORDINATES OF SATURN

REFERENCE AXES OF 1950.0

| JULIAN DATE | X | Y | Z |
|---|---|---|---|
| 2002560.5 | -8.188867720 | 3.912647938 | 1.949892673 |
| 2002880.5 | -8.973216076 | 2.379556236 | 1.351869653 |
| 2003200.5 | -9.427671193 | 0.758623936 | 0.703998748 |
| 2003520.5 | -9.546936201 | -0.889321491 | 0.031091535 |
| 2003840.5 | -9.337970709 | -2.506449761 | -0.642769432 |
| 2004160.5 | -8.818304205 | -4.039688513 | -1.295026171 |
| 2004480.5 | -8.014197292 | -5.441913325 | -1.905210652 |
| 2004800.5 | -6.958820550 | -6.672648199 | -2.455306668 |
| 2005120.5 | -5.690574352 | -7.698352324 | -2.929933927 |
| 2005440.5 | -4.251662442 | -8.492388278 | -3.316389619 |
| 2005760.5 | -2.686994418 | -9.034821839 | -3.604608424 |
| 2006080.5 | -1.043385168 | -9.312226682 | -3.787115309 |
| 2006400.5 | 0.631078929 | -9.317598262 | -3.859018957 |
| 2006720.5 | 2.287638715 | -9.050382399 | -3.818051620 |
| 2007040.5 | 3.877500804 | -8.516580441 | -3.664641487 |
| 2007360.5 | 5.352553071 | -7.728898568 | -3.402005499 |
| 2007680.5 | 6.666190444 | -6.706918449 | -3.036254939 |
| 2008000.5 | 7.774292694 | -5.477260565 | -2.576504065 |
| 2008320.5 | 8.636399334 | -4.073691874 | -2.034963815 |
| 2008640.5 | 9.217117324 | -2.537098439 | -1.426988770 |
| 2008960.5 | 9.487770748 | -0.915193199 | -0.771022423 |
| 2009280.5 | 9.428301668 | 0.738252247 | -0.088352618 |
| 2009600.5 | 9.029527458 | 2.364800906 | 0.597367349 |
| 2009920.5 | 8.295315680 | 3.903144444 | 1.260749802 |
| 2010240.5 | 7.243757884 | 5.291531921 | 1.875629139 |
| 2010560.5 | 5.907661040 | 6.471047970 | 2.416403073 |
| 2010880.5 | 4.334306618 | 7.389256713 | 2.859540527 |
| 2011200.5 | 2.584016107 | 8.003854976 | 3.185139563 |
| 2011520.5 | 0.727371222 | 8.285937217 | 3.378382853 |
| 2011840.5 | -1.158765200 | 8.222381752 | 3.430682838 |
| 2012160.5 | -2.996011989 | 7.816936703 | 3.340325275 |
| 2012480.5 | -4.709407657 | 7.089797870 | 3.112500010 |
| 2012800.5 | -6.231868838 | 6.075754518 | 2.758722986 |
| 2013120.5 | -7.507629423 | 4.821241469 | 2.295766627 |
| 2013440.5 | -8.494356732 | 3.380803781 | 1.744294983 |
| 2013760.5 | -9.163931039 | 1.813513426 | 1.127425453 |
| 2014080.5 | -9.502134190 | 0.179763851 | 0.469401664 |
| 2014400.5 | -9.507619963 | -1.461349545 | -0.205523732 |
| 2014720.5 | -9.190490211 | -3.054052474 | -0.873980445 |
| 2015040.5 | -8.570679270 | -4.547382367 | -1.514163617 |

HELIOCENTRIC COORDINATES OF SATURN

REFERENCE AXES OF 1950.0

| JULIAN DATE | X | Y | Z |
|---|---|---|---|
| 2015360.5 | -7.676281468 | -5.896204896 | -2.106325134 |
| 2015680.5 | -6.541939618 | -7.061819132 | -2.633092734 |
| 2016000.5 | -5.207385334 | -8.012238144 | -3.079648695 |
| 2016320.5 | -3.716168876 | -8.722248431 | -3.433809311 |
| 2016640.5 | -2.114553877 | -9.173379077 | -3.686058135 |
| 2016960.5 | -0.450567956 | -9.353988164 | -3.829612575 |
| 2017280.5 | 1.226654468 | -9.259464693 | -3.860526414 |
| 2017600.5 | 2.867451257 | -8.892146817 | -3.777692541 |
| 2017920.5 | 4.422507925 | -8.261177892 | -3.582827610 |
| 2018240.5 | 5.843829911 | -7.382538973 | -3.280519214 |
| 2018560.5 | 7.085695085 | -6.279084533 | -2.878273006 |
| 2018880.5 | 8.105635094 | -4.980536023 | -2.386547717 |
| 2019200.5 | 8.865547143 | -3.523451594 | -1.818789698 |
| 2019520.5 | 9.333050406 | -1.951102629 | -1.191441456 |
| 2019840.5 | 9.483129641 | -0.313100785 | -0.523859659 |
| 2020160.5 | 9.300002955 | 1.335400115 | 0.161933198 |
| 2020480.5 | 8.779060120 | 2.935075602 | 0.841730850 |
| 2020800.5 | 7.928641822 | 4.424472461 | 1.489908160 |
| 2021120.5 | 6.771364408 | 5.742737750 | 2.080487827 |
| 2021440.5 | 5.344658118 | 6.832933168 | 2.588480886 |
| 2021760.5 | 3.700212087 | 7.645687153 | 2.991412380 |
| 2022080.5 | 1.902127912 | 8.142800335 | 3.270883096 |
| 2022400.5 | 0.023764077 | 8.300336091 | 3.413976068 |
| 2022720.5 | -1.856526782 | 8.110719951 | 3.414302802 |
| 2023040.5 | -3.660328094 | 7.583450655 | 3.272505135 |
| 2023360.5 | -5.314106422 | 6.744225540 | 2.996101701 |
| 2023680.5 | -6.753488167 | 5.632596434 | 2.598695598 |
| 2024000.5 | -7.926480569 | 4.298624562 | 2.098718590 |
| 2024320.5 | -8.795532011 | 2.799356026 | 1.518064500 |
| 2024640.5 | -9.337940964 | 1.195560839 | 0.880788780 |
| 2024960.5 | -9.544909390 | -0.451513694 | 0.211736336 |
| 2025280.5 | -9.420228339 | -2.082915927 | -0.464668939 |
| 2025600.5 | -8.978720738 | -3.643979502 | -1.125313756 |
| 2025920.5 | -8.244486497 | -5.085675379 | -1.749025234 |
| 2026240.5 | -7.249108677 | -6.365462706 | -2.316994477 |
| 2026560.5 | -6.029944051 | -7.447727199 | -2.813025478 |
| 2026880.5 | -4.628595711 | -8.303879706 | -3.223635030 |
| 2027200.5 | -3.089661384 | -8.912214659 | -3.538042085 |
| 2027520.5 | -1.459797795 | -9.257682677 | -3.748110733 |
| 2027840.5 | 0.212971679 | -9.331729447 | -3.848313563 |

HELIOCENTRIC COORDINATES OF SATURN

REFERENCE AXES OF 1950.0

| JULIAN DATE | X | Y | Z |
|---|---|---|---|
| 2028160.5 | 1.879850154 | -9.132265265 | -3.835746763 |
| 2028480.5 | 3.491879106 | -8.663742691 | -3.710190656 |
| 2028800.5 | 5.000628824 | -7.937295503 | -3.474197899 |
| 2029120.5 | 6.358999952 | -6.970902903 | -3.133195843 |
| 2029440.5 | 7.522169764 | -5.789547848 | -2.695592029 |
| 2029760.5 | 8.448722740 | -4.425326844 | -2.172867171 |
| 2030080.5 | 9.102001957 | -2.917445220 | -1.579629919 |
| 2030400.5 | 9.451697994 | -1.311998987 | -0.933592561 |
| 2030720.5 | 9.475647682 | 0.338604984 | -0.255403391 |
| 2031040.5 | 9.161791903 | 1.976818410 | 0.431761051 |
| 2031360.5 | 8.510344105 | 3.541863902 | 1.102889640 |
| 2031680.5 | 7.535699231 | 4.972062975 | 1.731996961 |
| 2032000.5 | 6.267098404 | 6.207677075 | 2.293273685 |
| 2032320.5 | 4.748367867 | 7.194464863 | 2.762574217 |
| 2032640.5 | 3.036753215 | 7.887392338 | 3.118989649 |
| 2032960.5 | 1.200495077 | 8.254022945 | 3.346338926 |
| 2033280.5 | -0.684837517 | 8.277146646 | 3.434400287 |
| 2033600.5 | -2.540917982 | 7.956216803 | 3.379691167 |
| 2033920.5 | -4.291578127 | 7.307327910 | 3.185662709 |
| 2034240.5 | -5.867422670 | 6.361731554 | 2.862280549 |
| 2034560.5 | -7.209568148 | 5.163156702 | 2.425077506 |
| 2034880.5 | -8.272133554 | 3.764392046 | 1.893851474 |
| 2035200.5 | -9.023376215 | 2.223660813 | 1.291223109 |
| 2035520.5 | -9.445631266 | 0.601262508 | 0.641255339 |
| 2035840.5 | -9.534402362 | -1.043218045 | -0.031728588 |
| 2036160.5 | -9.296984911 | -2.653060092 | -0.704084520 |
| 2036480.5 | -8.750889346 | -4.175990783 | -1.353565738 |
| 2036800.5 | -7.922205124 | -5.565299644 | -1.959852840 |
| 2037120.5 | -6.843998697 | -6.780552467 | -2.504917727 |
| 2037440.5 | -5.554829773 | -7.787974051 | -2.973245727 |
| 2037760.5 | -4.097443577 | -8.560585914 | -3.351948987 |
| 2038080.5 | -2.517648076 | -9.078191501 | -3.630809055 |
| 2038400.5 | -0.863326609 | -9.327323089 | -3.802295587 |
| 2038720.5 | 0.816448634 | -9.301345788 | -3.861637489 |
| 2039040.5 | 2.472063453 | -9.000705781 | -3.806946149 |
| 2039360.5 | 4.053952232 | -8.432891746 | -3.639236869 |
| 2039680.5 | 5.513515324 | -7.612306876 | -3.362419238 |
| 2040000.5 | 6.804170481 | -6.560287066 | -2.983340720 |
| 2040320.5 | 7.882408907 | -5.305103516 | -2.511825176 |
| 2040640.5 | 8.708912968 | -3.881901195 | -1.960692124 |

## HELIOCENTRIC COORDINATES OF SATURN

### REFERENCE AXES OF 1950.0

| JULIAN DATE | X | Y | Z |
|---|---|---|---|
| 2040960.5 | 9.249844868 | -2.332557192 | -1.345753809 |
| 2041280.5 | 9.478392119 | -0.705348962 | -0.685746250 |
| 2041600.5 | 9.376556208 | 0.945741556 | -0.002120333 |
| 2041920.5 | 8.937059339 | 2.562254475 | 0.681381386 |
| 2042240.5 | 8.165155118 | 4.083125482 | 1.339418782 |
| 2042560.5 | 7.080059684 | 5.447244514 | 1.946046383 |
| 2042880.5 | 5.715676825 | 6.596626493 | 2.475987072 |
| 2043200.5 | 4.120301185 | 7.479979382 | 2.906112257 |
| 2043520.5 | 2.355075881 | 8.056312325 | 3.216991389 |
| 2043840.5 | 0.491154302 | 8.298131901 | 3.394327751 |
| 2044160.5 | -1.394269333 | 8.193765368 | 3.430084191 |
| 2044480.5 | -3.222687334 | 7.748425923 | 3.323123543 |
| 2044800.5 | -4.919207516 | 6.983789474 | 3.079243034 |
| 2045120.5 | -6.417017765 | 5.936099735 | 2.710579541 |
| 2045440.5 | -7.660809488 | 4.653108630 | 2.234486379 |
| 2045760.5 | -8.608968406 | 3.190387335 | 1.672094296 |
| 2046080.5 | -9.234681095 | 1.607762027 | 1.046892103 |
| 2046400.5 | -9.525658752 | -0.033660256 | 0.383515324 |
| 2046720.5 | -9.482661772 | -1.674209931 | -0.293434272 |
| 2047040.5 | -9.117833238 | -3.257926611 | -0.960409085 |
| 2047360.5 | -8.453018627 | -4.734117278 | -1.595632519 |
| 2047680.5 | -7.518037378 | -6.058345378 | -2.179577834 |
| 2048000.5 | -6.348994682 | -7.192999015 | -2.695270513 |
| 2048320.5 | -4.986718043 | -8.107522999 | -3.128445631 |
| 2048640.5 | -3.475398612 | -8.778382138 | -3.467584806 |
| 2048960.5 | -1.861512265 | -9.188860557 | -3.703874456 |
| 2049280.5 | -0.193024020 | -9.328849146 | -3.831150034 |
| 2049600.5 | 1.481230615 | -9.194743735 | -3.845881578 |
| 2049920.5 | 3.112133822 | -8.789479226 | -3.747215444 |
| 2050240.5 | 4.650943576 | -8.122655456 | -3.537056461 |
| 2050560.5 | 6.050068646 | -7.210701818 | -3.220170012 |
| 2050880.5 | 7.263989470 | -6.077038927 | -2.804288229 |
| 2051200.5 | 8.250356583 | -4.752194520 | -2.300204386 |
| 2051520.5 | 8.971299211 | -3.273814249 | -1.721832604 |
| 2051840.5 | 9.394964749 | -1.686482397 | -1.086198538 |
| 2052160.5 | 9.497276613 | -0.041235680 | -0.413311411 |
| 2052480.5 | 9.263832532 | 1.605388880 | 0.274153503 |
| 2052800.5 | 8.691815592 | 3.192998932 | 0.951460636 |
| 2053120.5 | 7.791908623 | 4.659825273 | 1.592748882 |
| 2053440.5 | 6.589753892 | 5.945580791 | 2.172127800 |

HELIOCENTRIC COORDINATES OF SATURN

REFERENCE AXES OF 1950.0

| JULIAN DATE | X | Y | Z |
|---|---|---|---|
| 2053760.5 | 5.125929866 | 6.994615225 | 2.664988798 |
| 2054080.5 | 3.454770157 | 7.759566715 | 3.049570440 |
| 2054400.5 | 1.642185199 | 8.204883334 | 3.308495680 |
| 2054720.5 | -0.237705344 | 8.309658111 | 3.430074367 |
| 2055040.5 | -2.106927154 | 8.069362345 | 3.409195453 |
| 2055360.5 | -3.888494137 | 7.496180028 | 3.247665088 |
| 2055680.5 | -5.511113143 | 6.617876811 | 2.953936376 |
| 2056000.5 | -6.913157103 | 5.475402767 | 2.542287763 |
| 2056320.5 | -8.045483397 | 4.119636127 | 2.031600160 |
| 2056640.5 | -8.872918765 | 2.607773982 | 1.443934205 |
| 2056960.5 | -9.374485103 | 0.999858994 | 0.803112755 |
| 2057280.5 | -9.542651115 | -0.644190722 | 0.133471415 |
| 2057600.5 | -9.382002465 | -2.266885372 | -0.541141165 |
| 2057920.5 | -8.907673489 | -3.814841593 | -1.198118051 |
| 2058240.5 | -8.143732559 | -5.239965603 | -1.816650171 |
| 2058560.5 | -7.121602945 | -6.500246644 | -2.378122206 |
| 2058880.5 | -5.878577304 | -7.560228822 | -2.866372055 |
| 2059200.5 | -4.456481198 | -8.391230152 | -3.267839249 |
| 2059520.5 | -2.900515835 | -8.971388355 | -3.571634045 |
| 2059840.5 | -1.258266740 | -9.285611811 | -3.769560172 |
| 2060160.5 | 0.421191056 | -9.325531384 | -3.856131235 |
| 2060480.5 | 2.088150067 | -9.089635484 | -3.828653528 |
| 2060800.5 | 3.692944700 | -8.583569209 | -3.687373422 |
| 2061120.5 | 5.186585254 | -7.820134757 | -3.435516636 |
| 2061440.5 | 6.521744423 | -6.819161821 | -3.079278279 |
| 2061760.5 | 7.653911701 | -5.607487079 | -2.627854222 |
| 2062080.5 | 8.542575979 | -4.218892695 | -2.093459504 |
| 2062400.5 | 9.152500389 | -2.693942406 | -1.491313288 |
| 2062720.5 | 9.455191648 | -1.079659810 | -0.839570148 |
| 2063040.5 | 9.430604339 | 0.571093956 | -0.159138317 |
| 2063360.5 | 9.069000909 | 2.200719701 | 0.526692541 |
| 2063680.5 | 8.372776626 | 3.748596569 | 1.192879781 |
| 2064000.5 | 7.357977754 | 5.153477729 | 1.813549235 |
| 2064320.5 | 6.055191430 | 6.356519997 | 2.363204117 |
| 2064640.5 | 4.509483789 | 7.304764965 | 2.818159725 |
| 2064960.5 | 2.779136786 | 7.954742721 | 3.158079269 |
| 2065280.5 | 0.933095511 | 8.275758985 | 3.367434395 |
| 2065600.5 | -0.952743829 | 8.252402679 | 3.436695148 |
| 2065920.5 | -2.800064829 | 7.885888867 | 3.363076545 |
| 2066240.5 | -4.532983650 | 7.194001185 | 3.150722120 |

## HELIOCENTRIC COORDINATES OF SATURN

## REFERENCE AXES OF 1950.0

| JULIAN DATE | X | Y | Z |
|---|---|---|---|
| 2066560.5 | -6.082627968 | 6.209597303 | 2.810281861 |
| 2066880.5 | -7.390846879 | 4.977891149 | 2.357946103 |
| 2067200.5 | -8.412645630 | 3.552949001 | 1.814097589 |
| 2067520.5 | -9.117349283 | 1.993928672 | 1.201801284 |
| 2067840.5 | -9.488846823 | 0.361683224 | 0.545420571 |
| 2068160.5 | -9.524786836 | -1.283825280 | -0.130457241 |
| 2068480.5 | -9.234761739 | -2.885675097 | -0.801984429 |
| 2068800.5 | -8.638431796 | -4.391757362 | -1.446877855 |
| 2069120.5 | -7.763817707 | -5.755947357 | -2.044977353 |
| 2069440.5 | -6.645658350 | -6.938790844 | -2.578595797 |
| 2069760.5 | -5.323858412 | -7.907847495 | -3.032722720 |
| 2070080.5 | -3.842082092 | -8.637766422 | -3.395110195 |
| 2070400.5 | -2.246561863 | -9.110158205 | -3.656265144 |
| 2070720.5 | -0.585176810 | -9.313374354 | -3.809393528 |
| 2071040.5 | 1.093233279 | -9.242337486 | -3.850358648 |
| 2071360.5 | 2.739448950 | -8.898509798 | -3.777694783 |
| 2071680.5 | 4.304486762 | -8.289989819 | -3.592675975 |
| 2072000.5 | 5.740331699 | -7.431677265 | -3.299417150 |
| 2072320.5 | 7.000823558 | -6.345447384 | -2.904984643 |
| 2072640.5 | 8.042723399 | -5.060284346 | -2.419496746 |
| 2072960.5 | 8.826986824 | -3.612315891 | -1.856191965 |
| 2073280.5 | 9.320263590 | -2.044671110 | -1.231433721 |
| 2073600.5 | 9.496619270 | -0.407057431 | -0.564608172 |
| 2073920.5 | 9.339425383 | 1.245073744 | 0.122142024 |
| 2074240.5 | 8.843276534 | 2.851929605 | 0.804420701 |
| 2074560.5 | 8.015717771 | 4.351547203 | 1.456408462 |
| 2074880.5 | 6.878710200 | 5.682840161 | 2.052047701 |
| 2075200.5 | 5.469445931 | 6.788932093 | 2.566359909 |
| 2075520.5 | 3.839483567 | 7.620466949 | 2.976858916 |
| 2075840.5 | 2.052540975 | 8.139110159 | 3.265097973 |
| 2076160.5 | 0.181310081 | 8.320596425 | 3.418053881 |
| 2076480.5 | -1.696699282 | 8.156737437 | 3.429122324 |
| 2076800.5 | -3.503914565 | 7.656072676 | 3.298581785 |
| 2077120.5 | -5.167459125 | 6.843056120 | 3.033456110 |
| 2077440.5 | -6.623308886 | 5.755924154 | 2.646809503 |
| 2077760.5 | -7.819384975 | 4.443608733 | 2.156604316 |
| 2078080.5 | -8.717349970 | 2.962177803 | 1.584310649 |
| 2078400.5 | -9.293118359 | 1.371289972 | 0.953469379 |
| 2078720.5 | -9.536299526 | -0.268936608 | 0.288382965 |
| 2079040.5 | -9.448933719 | -1.900402432 | -0.386953293 |

HELIOCENTRIC COORDINATES OF SATURN

REFERENCE AXES OF 1950.0

| JULIAN DATE | X | Y | Z |
|---|---|---|---|
| 2079360.5 | -9.043908116 | -3.468819545 | -1.049652216 |
| 2079680.5 | -8.343326003 | -4.924946663 | -1.678511087 |
| 2080000.5 | -7.376941425 | -6.225421818 | -2.254422386 |
| 2080320.5 | -6.180691625 | -7.333290963 | -2.760656265 |
| 2080640.5 | -4.795357369 | -8.218293707 | -3.183037179 |
| 2080960.5 | -3.265382117 | -8.856969869 | -3.510039268 |
| 2081280.5 | -1.637857742 | -9.232656277 | -3.732829126 |
| 2081600.5 | 0.038353511 | -9.335436444 | -3.845283040 |
| 2081920.5 | 1.713446230 | -9.162119145 | -3.844011169 |
| 2082240.5 | 3.337530958 | -8.716413684 | -3.728457257 |
| 2082560.5 | 4.861372266 | -8.009280257 | -3.501071796 |
| 2082880.5 | 6.237097078 | -7.058961770 | -3.167366871 |
| 2083200.5 | 7.419276175 | -5.890832602 | -2.735898823 |
| 2083520.5 | 8.366205658 | -4.537311338 | -2.218277264 |
| 2083840.5 | 9.041239693 | -3.037696408 | -1.629148639 |
| 2084160.5 | 9.414238236 | -1.437840705 | -0.986123153 |
| 2084480.5 | 9.463206032 | 0.210430195 | -0.309606685 |
| 2084800.5 | 9.176101464 | 1.850318807 | 0.377532404 |
| 2085120.5 | 8.552662821 | 3.421635468 | 1.050534858 |
| 2085440.5 | 7.605995771 | 4.863047097 | 1.683608382 |
| 2085760.5 | 6.363604171 | 6.114977070 | 2.251077332 |
| 2086080.5 | 4.867534076 | 7.123002855 | 2.728774707 |
| 2086400.5 | 3.173359094 | 7.841449689 | 3.095563576 |
| 2086720.5 | 1.347881559 | 8.236755173 | 3.334818593 |
| 2087040.5 | -0.534358038 | 8.290134933 | 3.435670989 |
| 2087360.5 | -2.395444550 | 7.999148283 | 3.393839178 |
| 2087680.5 | -4.158847006 | 7.377923053 | 3.211925219 |
| 2088000.5 | -5.754026018 | 6.455990846 | 3.009131955 |
| 2088320.5 | -7.130342348 | 5.275886011 | 2.470438593 |
| 2088640.5 | -8.209816608 | 3.889870093 | 1.945364070 |
| 2088960.5 | -8.988534856 | 2.356272735 | 1.346511828 |
| 2089280.5 | -9.436872400 | 0.735907543 | 0.698094222 |
| 2089600.5 | -9.549019860 | -0.910989057 | 0.024658232 |
| 2089920.5 | -9.331818716 | -2.526903507 | -0.649823745 |
| 2090240.5 | -8.802788895 | -4.058634228 | -1.302695651 |
| 2090560.5 | -7.988174262 | -5.458596156 | -1.913254401 |
| 2090880.5 | -6.921283943 | -6.685664112 | -2.463170866 |
| 2091200.5 | -5.640976296 | -7.705622658 | -2.936736661 |
| 2091520.5 | -4.190257298 | -8.491343690 | -3.320992574 |
| 2091840.5 | -2.615031238 | -9.022754760 | -3.605763296 |

HELIOCENTRIC COORDINATES OF SATURN

REFERENCE AXES OF 1950.0

| JULIAN DATE | X | Y | Z |
|---|---|---|---|
| 2092160.5 | -0.963065289 | -9.286663110 | -3.783623582 |
| 2092480.5 | 0.716801370 | -9.276551479 | -3.849844544 |
| 2092800.5 | 2.375273228 | -8.992472058 | -3.802376287 |
| 2093120.5 | 3.963183721 | -8.441087980 | -3.641892308 |
| 2093440.5 | 5.432175605 | -7.635822178 | -3.371882438 |
| 2093760.5 | 6.735540809 | -6.597041689 | -2.998766248 |
| 2094080.5 | 7.829248414 | -5.352212178 | -2.532001210 |
| 2094400.5 | 8.673181396 | -3.935960615 | -1.984161228 |
| 2094720.5 | 9.232597764 | -2.389971248 | -1.370955706 |
| 2095040.5 | 9.479813713 | -0.762618418 | -0.711149301 |
| 2095360.5 | 9.396068741 | 0.891782355 | -0.026331343 |
| 2095680.5 | 8.973466476 | 2.514238631 | 0.659525867 |
| 2096000.5 | 8.216777932 | 4.042994444 | 1.320785963 |
| 2096320.5 | 7.144801695 | 5.416232877 | 1.931212964 |
| 2096640.5 | 5.791158244 | 6.575615530 | 2.465385146 |
| 2096960.5 | 4.204265959 | 7.469946944 | 2.900182872 |
| 2097280.5 | 2.445542683 | 8.058405365 | 3.216210887 |
| 2097600.5 | 0.586179486 | 8.313590308 | 3.399202334 |
| 2097920.5 | -1.296905941 | 8.223807753 | 3.441124911 |
| 2098240.5 | -3.125726637 | 7.794045795 | 3.340772598 |
| 2098560.5 | -4.826012591 | 7.045480771 | 3.103760003 |
| 2098880.5 | -6.331525031 | 6.013615444 | 2.741934864 |
| 2099200.5 | -7.587400898 | 4.745373553 | 2.272320693 |
| 2099520.5 | -8.552245155 | 3.295611537 | 1.715767367 |
| 2099840.5 | -9.198928723 | 1.723531276 | 1.095507566 |
| 2100160.5 | -9.514254362 | 0.089409018 | 0.435797515 |
| 2100480.5 | -9.497800586 | -1.548072821 | -0.239226594 |
| 2100800.5 | -9.160328290 | -3.133779342 | -0.906415587 |
| 2101120.5 | -8.522092252 | -4.617437798 | -1.544216849 |
| 2101440.5 | -7.611245266 | -5.954492931 | -2.133094074 |
| 2101760.5 | -6.462376363 | -7.106656689 | -2.655817888 |
| 2102080.5 | -5.115179931 | -8.042234052 | -3.097659201 |
| 2102400.5 | -3.613266072 | -8.736276660 | -3.446504960 |
| 2102720.5 | -2.003124021 | -9.170620682 | -3.692919021 |
| 2103040.5 | -0.333228949 | -9.333867026 | -3.830172852 |
| 2103360.5 | 1.346750650 | -9.221350836 | -3.854267132 |
| 2103680.5 | 2.986743305 | -8.835156358 | -3.763968943 |
| 2104000.5 | 4.537169325 | -8.184328756 | -3.560928453 |
| 2104320.5 | 5.949777631 | -7.285262129 | -3.249873695 |
| 2104640.5 | 7.178455461 | -6.161742467 | -2.838673422 |

HELIOCENTRIC COORDINATES OF SATURN

REFERENCE AXES OF 1950.0

| JULIAN DATE | X | Y | Z |
|---|---|---|---|
| 2104960.5 | 8.180422914 | -4.844745128 | -2.338301067 |
| 2105280.5 | 8.917654467 | -3.372243836 | -1.762807087 |
| 2105600.5 | 9.358369932 | -1.788894456 | -1.129247955 |
| 2105920.5 | 9.478646546 | -0.145480237 | -0.457526609 |
| 2106240.5 | 9.264183490 | 1.502010765 | 0.229910331 |
| 2106560.5 | 8.712123178 | 3.093840124 | 0.908592177 |
| 2106880.5 | 7.832701869 | 4.568625547 | 1.552820357 |
| 2107200.5 | 6.650417145 | 5.866114543 | 2.136760363 |
| 2107520.5 | 5.204373177 | 6.930502678 | 2.635792849 |
| 2107840.5 | 3.547508610 | 7.714021762 | 3.028022972 |
| 2108160.5 | 1.744542547 | 8.180388095 | 3.295786182 |
| 2108480.5 | -0.131313870 | 8.307633108 | 3.426952264 |
| 2108800.5 | -2.002590140 | 8.089885458 | 3.415838736 |
| 2109120.5 | -3.792256768 | 7.537836475 | 3.263603530 |
| 2109440.5 | -5.428349329 | 6.677835096 | 2.978070844 |
| 2109760.5 | -6.848003394 | 5.549751084 | 2.573023297 |
| 2110080.5 | -8.000472258 | 4.203910749 | 2.067065832 |
| 2110400.5 | -8.848861060 | 2.697547802 | 1.482232487 |
| 2110720.5 | -9.370564331 | 1.091243129 | 0.842531192 |
| 2111040.5 | -9.556699857 | -0.554299573 | 0.172584299 |
| 2111360.5 | -9.411083085 | -2.180876612 | -0.503487433 |
| 2111680.5 | -8.948849019 | -3.734485803 | -1.162801337 |
| 2112000.5 | -8.194470659 | -5.166405607 | -1.784246701 |
| 2112320.5 | -7.179827501 | -6.434081340 | -2.348948906 |
| 2112640.5 | -5.942643653 | -7.501699404 | -2.840575082 |
| 2112960.5 | -4.525116641 | -8.340458695 | -3.245500673 |
| 2113280.5 | -2.972668762 | -8.928630327 | -3.552882290 |
| 2113600.5 | -1.332848441 | -9.251458220 | -3.754657918 |
| 2113920.5 | 0.345564729 | -9.300970373 | -3.845502301 |
| 2114240.5 | 2.013248847 | -9.075818742 | -3.822788098 |
| 2114560.5 | 3.620923307 | -8.581249283 | -3.686599430 |
| 2114880.5 | 5.119975695 | -7.829212818 | -3.439806944 |
| 2115200.5 | 6.463239136 | -6.838551746 | -3.088180307 |
| 2115520.5 | 7.605968350 | -5.635181030 | -2.640505780 |
| 2115840.5 | 8.507034530 | -4.252189265 | -2.108679267 |
| 2116160.5 | 9.130350058 | -2.729782788 | -1.507743948 |
| 2116480.5 | 9.446519782 | -1.114979745 | -0.855834226 |
| 2116800.5 | 9.434684043 | 0.539058358 | -0.173978084 |
| 2117120.5 | 9.084465570 | 2.174182945 | 0.514298156 |
| 2117440.5 | 8.397850418 | 3.729023383 | 1.183620959 |

HELIOCENTRIC COORDINATES OF SATURN

REFERENCE AXES OF 1950.0

| JULIAN DATE | X | Y | Z |
|---|---|---|---|
| 2117760.5 | 7.390715195 | 5.141444700 | 1.807725021 |
| 2118080.5 | 6.093620737 | 6.351740019 | 2.360743078 |
| 2118400.5 | 4.551722750 | 7.306521426 | 2.818799762 |
| 2118720.5 | 2.823748335 | 7.962509568 | 3.161592429 |
| 2119040.5 | 0.979247192 | 8.289417790 | 3.373713092 |
| 2119360.5 | -0.905538584 | 8.272241051 | 3.445779092 |
| 2119680.5 | -2.752336179 | 7.912536456 | 3.375151276 |
| 2120000.5 | -4.485649674 | 7.228255506 | 3.166061368 |
| 2120320.5 | -6.037204354 | 6.252155548 | 2.829141426 |
| 2120640.5 | -7.349513909 | 5.029084062 | 2.380453442 |
| 2120960.5 | -8.378243181 | 3.612579990 | 1.840187586 |
| 2121280.5 | -9.093278937 | 2.061274673 | 1.231224196 |
| 2121600.5 | -9.478629301 | 0.435516267 | 0.577739680 |
| 2121920.5 | -9.531417197 | -1.205472155 | -0.096004044 |
| 2122240.5 | -9.260319089 | -2.805733341 | -0.766602892 |
| 2122560.5 | -8.683814842 | -4.314016926 | -1.412173658 |
| 2122880.5 | -7.828518257 | -5.684674707 | -2.012791245 |
| 2123200.5 | -6.727688173 | -6.878209438 | -2.550778791 |
| 2123520.5 | -5.419905251 | -7.861610432 | -3.010904765 |
| 2123840.5 | -3.947888821 | -8.608536625 | -3.380511269 |
| 2124160.5 | -2.357450841 | -9.099388498 | -3.649589898 |
| 2124480.5 | -0.696584047 | -9.321316748 | -3.810825161 |
| 2124800.5 | 0.985338343 | -9.268213786 | -3.859625818 |
| 2125120.5 | 2.638302949 | -8.940719578 | -3.794159015 |
| 2125440.5 | 4.212445185 | -8.346277417 | -3.615403891 |
| 2125760.5 | 5.659016700 | -7.499372424 | -3.327282844 |
| 2126080.5 | 6.931357483 | -6.421934316 | -2.936870005 |
| 2126400.5 | 7.985834272 | -5.143359702 | -2.454449864 |
| 2126720.5 | 8.783163952 | -3.700216766 | -1.893446292 |
| 2127040.5 | 9.289981746 | -2.135898767 | -1.270337550 |
| 2127360.5 | 9.480483480 | -0.500089522 | -0.604503145 |
| 2127680.5 | 9.338164520 | 1.152102151 | 0.082057484 |
| 2128000.5 | 8.857623381 | 2.761487043 | 0.765199769 |
| 2128320.5 | 8.046254416 | 4.266759082 | 1.419368977 |
| 2128640.5 | 6.925536293 | 5.607130314 | 2.018629668 |
| 2128960.5 | 5.531572560 | 6.725551728 | 2.537968579 |
| 2129280.5 | 3.914567265 | 7.572277260 | 2.954783955 |
| 2129600.5 | 2.137032107 | 8.108382720 | 3.250409841 |
| 2129920.5 | 0.270721324 | 8.308762042 | 3.411478955 |
| 2130240.5 | -1.607460418 | 8.164140660 | 3.430924875 |

## HELIOCENTRIC COORDINATES OF SATURN

### REFERENCE AXES OF 1950.0

| JULIAN DATE | X | Y | Z |
|---|---|---|---|
| 2130560.5 | -3.420113475 | 7.681791725 | 3.308475271 |
| 2130880.5 | -5.093999628 | 6.884872470 | 3.050578486 |
| 2131200.5 | -6.564180190 | 5.810518229 | 2.669796625 |
| 2131520.5 | -7.777232444 | 4.506980491 | 2.183763640 |
| 2131840.5 | -8.693277275 | 3.030198178 | 1.613856306 |
| 2132160.5 | -9.286734510 | 1.440252438 | 0.983760785 |
| 2132480.5 | -9.545944981 | -0.201893986 | 0.318106663 |
| 2132800.5 | -9.472014462 | -1.837152548 | -0.358723331 |
| 2133120.5 | -9.077419706 | -3.410457274 | -1.023528932 |
| 2133440.5 | -8.384542903 | -4.872061912 | -1.654867400 |
| 2133760.5 | -7.423788093 | -6.178196970 | -2.233410801 |
| 2134080.5 | -6.231744295 | -7.291610402 | -2.742264239 |
| 2134400.5 | -4.849741570 | -8.181935795 | -3.167181144 |
| 2134720.5 | -3.322619651 | -8.825839246 | -3.496671964 |
| 2135040.5 | -1.697616835 | -9.207001888 | -3.722038878 |
| 2135360.5 | -0.023402254 | -9.315984310 | -3.837355455 |
| 2135680.5 | 1.650705817 | -9.150050973 | -3.839423326 |
| 2136000.5 | 3.275364666 | -8.713069092 | -3.727756015 |
| 2136320.5 | 4.801800441 | -8.015552312 | -3.504624008 |
| 2136640.5 | 6.182512742 | -7.074823213 | -3.175154555 |
| 2136960.5 | 7.372177118 | -5.915210247 | -2.747453493 |
| 2137280.5 | 8.328778626 | -4.568188953 | -2.232712895 |
| 2137600.5 | 9.014986163 | -3.072382329 | -1.645270128 |
| 2137920.5 | 9.399762881 | -1.473327367 | -1.002580174 |
| 2138240.5 | 9.460181522 | 0.177101634 | -0.325054959 |
| 2138560.5 | 9.183367319 | 1.821734973 | 0.364284060 |
| 2138880.5 | 8.568423600 | 3.399734149 | 1.040396195 |
| 2139200.5 | 7.628102927 | 4.848890536 | 1.677102058 |
| 2139520.5 | 6.389871653 | 6.108601378 | 2.248272616 |
| 2139840.5 | 4.895947638 | 7.123435330 | 2.729304052 |
| 2140160.5 | 3.202174766 | 7.847152537 | 3.098806935 |
| 2140480.5 | 1.375929771 | 8.246351513 | 3.340173381 |
| 2140800.5 | -0.507474659 | 8.302742953 | 3.442685594 |
| 2141120.5 | -2.369575302 | 8.014436999 | 3.402262453 |
| 2141440.5 | -4.133698918 | 7.396089517 | 3.221722016 |
| 2141760.5 | -5.729554274 | 6.477621419 | 2.910441462 |
| 2142080.5 | -7.097025520 | 5.301713031 | 2.483481522 |
| 2142400.5 | -8.188808674 | 3.920504523 | 1.960336401 |
| 2142720.5 | -8.971756702 | 2.391992358 | 1.363504692 |
| 2143040.5 | -9.427017292 | 0.776562268 | 0.717066421 |

HELIOCENTRIC COORDINATES OF SATURN

REFERENCE AXES OF 1950.0

| JULIAN DATE | X | Y | Z |
|---|---|---|---|
| 2143360.5 | -9.549198074 | -0.866012488 | 0.045412150 |
| 2143680.5 | -9.344877159 | -2.478908554 | -0.627779895 |
| 2144000.5 | -8.830811767 | -4.009865659 | -1.280279627 |
| 2144320.5 | -8.032163982 | -5.412167761 | -1.891780499 |
| 2144640.5 | -6.980929341 | -6.645194082 | -2.444192043 |
| 2144960.5 | -5.714590001 | -7.674732947 | -2.921828151 |
| 2145280.5 | -4.274939382 | -8.473161939 | -3.311534199 |
| 2145600.5 | -2.707041214 | -9.019535124 | -3.602769721 |
| 2145920.5 | -1.058311053 | -9.299609820 | -3.787660087 |
| 2146240.5 | 0.622294420 | -9.305852256 | -3.861034286 |
| 2146560.5 | 2.285012946 | -9.037456426 | -3.820464692 |
| 2146880.5 | 3.879995225 | -8.500392720 | -3.666317695 |
| 2147200.5 | 5.358123634 | -7.707500912 | -3.401823492 |
| 2147520.5 | 6.672082645 | -6.678738866 | -3.033216534 |
| 2147840.5 | 7.777491892 | -5.441570082 | -2.569946468 |
| 2148160.5 | 8.633989827 | -4.030927693 | -2.024717434 |
| 2148480.5 | 9.206690703 | -2.488783929 | -1.413365103 |
| 2148800.5 | 9.467891763 | -0.863596435 | -0.754693083 |
| 2149120.5 | 9.398839922 | 0.790509788 | -0.070209650 |
| 2149440.5 | 8.991533541 | 2.414991520 | 0.616308231 |
| 2149760.5 | 8.250450863 | 3.948713083 | 1.279491105 |
| 2150080.5 | 7.193952317 | 5.330436150 | 1.893341416 |
| 2150400.5 | 5.855016911 | 6.501931742 | 2.432483227 |
| 2150720.5 | 4.280974236 | 7.411509479 | 2.873626747 |
| 2151040.5 | 2.531988375 | 8.017601390 | 3.197108167 |
| 2151360.5 | 0.678235724 | 8.291927660 | 3.388312486 |
| 2151680.5 | -1.204032477 | 8.221758831 | 3.438771572 |
| 2152000.5 | -3.037108261 | 7.810906809 | 3.346768325 |
| 2152320.5 | -4.746546526 | 7.079307754 | 3.117366716 |
| 2152640.5 | -6.265425408 | 6.061317170 | 2.761893213 |
| 2152960.5 | -7.537732366 | 4.803015166 | 2.296972780 |
| 2153280.5 | -8.520627457 | 3.358896862 | 1.743261195 |
| 2153600.5 | -9.185485834 | 1.788356822 | 1.124037590 |
| 2153920.5 | -9.517792561 | 0.152368947 | 0.463826945 |
| 2154240.5 | -9.516130241 | -1.489341016 | -0.212814015 |
| 2154560.5 | -9.190626663 | -3.080533281 | -0.882289563 |
| 2154880.5 | -8.561354691 | -4.570132681 | -1.522715523 |
| 2155200.5 | -7.656875944 | -5.913174254 | -2.114350649 |
| 2155520.5 | -6.512524340 | -7.071130442 | -2.639812150 |
| 2155840.5 | -5.168712555 | -8.012182103 | -3.084276693 |

## HELIOCENTRIC COORDINATES OF SATURN

## REFERENCE AXES OF 1950.0

| JULIAN DATE | X | Y | Z |
|---|---|---|---|
| 2156160.5 | -3.669615076 | -8.711438039 | -3.435628370 |
| 2156480.5 | -2.062064198 | -9.151003128 | -3.684528849 |
| 2156800.5 | -0.394557217 | -9.319918225 | -3.824429054 |
| 2157120.5 | 1.283603403 | -9.214016528 | -3.851541061 |
| 2157440.5 | 2.923010622 | -8.835783490 | -3.764806814 |
| 2157760.5 | 4.474755839 | -8.194323527 | -3.565910058 |
| 2158080.5 | 5.891075319 | -7.305467587 | -3.259350602 |
| 2158400.5 | 7.126168664 | -6.191962701 | -2.852560123 |
| 2158720.5 | 8.137253976 | -4.883643775 | -2.356019898 |
| 2159040.5 | 8.885877216 | -3.417488458 | -1.783340160 |
| 2159360.5 | 9.339470281 | -1.837457321 | -1.151260900 |
| 2159680.5 | 9.473128610 | -0.194011235 | -0.479528631 |
| 2160000.5 | 9.271538587 | 1.456813728 | 0.209403110 |
| 2160320.5 | 8.730924831 | 3.054892799 | 0.890904392 |
| 2160640.5 | 7.860810187 | 4.538153850 | 1.538978801 |
| 2160960.5 | 6.685289501 | 5.845400180 | 2.127377503 |
| 2161280.5 | 5.243427418 | 6.919712141 | 2.630992267 |
| 2161600.5 | 3.588374305 | 7.712191008 | 3.027441915 |
| 2161920.5 | 1.785131124 | 8.185815855 | 3.298739877 |
| 2162240.5 | -0.092574003 | 8.318637000 | 3.432719038 |
| 2162560.5 | -1.966513283 | 8.105204701 | 3.423820732 |
| 2162880.5 | -3.759042319 | 7.556703984 | 3.273379491 |
| 2163200.5 | -5.397932683 | 6.699962218 | 2.989411460 |
| 2163520.5 | -6.820409869 | 5.575214226 | 2.585857778 |
| 2163840.5 | -7.976077475 | 4.232950790 | 2.081406273 |
| 2164160.5 | -8.828559110 | 2.730342012 | 1.498085195 |
| 2164480.5 | -9.355901076 | 1.127699042 | 0.859821157 |
| 2164800.5 | -9.549943200 | -0.514673382 | 0.191115943 |
| 2165120.5 | -9.414951225 | -2.139004054 | -0.484053357 |
| 2165440.5 | -8.965837376 | -3.691886333 | -1.143056015 |
| 2165760.5 | -8.226294408 | -5.125401334 | -1.765149456 |
| 2166080.5 | -7.227098577 | -6.397727985 | -2.331797202 |
| 2166400.5 | -6.004683137 | -7.473436890 | -2.826851010 |
| 2166720.5 | -4.599943004 | -8.323642623 | -3.236666985 |
| 2167040.5 | -3.057197206 | -8.926084179 | -3.550185920 |
| 2167360.5 | -1.423269265 | -9.265157999 | -3.758988352 |
| 2167680.5 | 0.253332093 | -9.331929071 | -3.857335395 |
| 2168000.5 | 1.923155510 | -9.124151369 | -3.842209639 |
| 2168320.5 | 3.536512936 | -8.646320699 | -3.713367584 |
| 2168640.5 | 5.044211643 | -7.909762062 | -3.473406558 |

HELIOCENTRIC COORDINATES OF SATURN

REFERENCE AXES OF 1950.0

| JULIAN DATE | X | Y | Z |
|---|---|---|---|
| 2168960.5 | 6.398436478 | -6.932744338 | -3.127845701 |
| 2169280.5 | 7.553952370 | -5.740709940 | -2.685264909 |
| 2169600.5 | 8.469415980 | -4.366603092 | -2.157500897 |
| 2169920.5 | 9.108608528 | -2.850723705 | -1.559645666 |
| 2170240.5 | 9.442014872 | -1.240108448 | -0.909849432 |
| 2170560.5 | 9.448637951 | 0.412289224 | -0.229053004 |
| 2170880.5 | 9.117828751 | 2.048765521 | 0.459415083 |
| 2171200.5 | 8.451045622 | 3.608554634 | 1.130456866 |
| 2171520.5 | 7.463355176 | 5.030154788 | 1.758097135 |
| 2171840.5 | 6.184361517 | 6.254299680 | 2.316666709 |
| 2172160.5 | 4.658211897 | 7.227420139 | 2.782227599 |
| 2172480.5 | 2.942393819 | 7.905269661 | 3.134117216 |
| 2172800.5 | 1.105205370 | 8.256255198 | 3.356427339 |
| 2173120.5 | -0.777977016 | 8.263969978 | 3.439205278 |
| 2173440.5 | -2.629248653 | 7.928507253 | 3.379187839 |
| 2173760.5 | -4.372969845 | 7.266344538 | 3.179957798 |
| 2174080.5 | -5.940192980 | 6.308865717 | 2.851525330 |
| 2174400.5 | -7.272250721 | 5.099819096 | 2.409436300 |
| 2174720.5 | -8.323245076 | 3.692108413 | 1.873557682 |
| 2175040.5 | -9.061350729 | 2.144313638 | 1.266698152 |
| 2175360.5 | -9.468979333 | 0.517316064 | 0.613220538 |
| 2175680.5 | -9.541989000 | -1.128653436 | -0.062213171 |
| 2176000.5 | -9.288229013 | -2.736355564 | -0.735673958 |
| 2176320.5 | -8.725761706 | -4.253278772 | -1.384760574 |
| 2176640.5 | -7.881183182 | -5.632896706 | -1.989179703 |
| 2176960.5 | -6.788235909 | -6.835327128 | -2.531058028 |
| 2177280.5 | -5.486283671 | -7.827416546 | -2.995050923 |
| 2177600.5 | -4.018777492 | -8.582811457 | -3.368453913 |
| 2177920.5 | -2.432057390 | -9.082073036 | -3.641301483 |
| 2178240.5 | -0.774354170 | -9.312695554 | -3.806411568 |
| 2178560.5 | 0.905107503 | -9.269023630 | -3.859383512 |
| 2178880.5 | 2.556884827 | -8.952116783 | -3.798570135 |
| 2179200.5 | 4.131967934 | -8.369655286 | -3.625064308 |
| 2179520.5 | 5.582398965 | -7.535970912 | -3.342738958 |
| 2179840.5 | 6.862005485 | -6.472198136 | -2.958343434 |
| 2180160.5 | 7.927356694 | -5.206458957 | -2.481623146 |
| 2180480.5 | 8.738984820 | -3.773969404 | -1.925417233 |
| 2180800.5 | 9.262871975 | -2.216959084 | -1.305689580 |
| 2181120.5 | 9.472175953 | -0.584293191 | -0.641446814 |
| 2181440.5 | 9.349132646 | 1.069322457 | 0.045508327 |

HELIOCENTRIC COORDINATES OF SATURN

REFERENCE AXES OF 1950.0

| JULIAN DATE | X | Y | Z |
|---|---|---|---|
| 2181760.5 | 8.887017900 | 2.684674208 | 0.731043090 |
| 2182080.5 | 8.091980742 | 4.200011488 | 1.389430091 |
| 2182400.5 | 6.984480605 | 5.553701771 | 1.994383197 |
| 2182720.5 | 5.599984287 | 6.687510565 | 2.520385223 |
| 2183040.5 | 3.988533785 | 7.550276524 | 2.944229045 |
| 2183360.5 | 2.212854587 | 8.101598619 | 3.246622198 |
| 2183680.5 | 0.345052431 | 8.315238882 | 3.413711592 |
| 2184000.5 | -1.537387160 | 8.181592552 | 3.438249515 |
| 2184320.5 | -3.356177988 | 7.708132350 | 3.319993741 |
| 2184640.5 | -5.037244789 | 6.918374049 | 3.065507040 |
| 2184960.5 | -6.515114212 | 5.849857698 | 2.687503342 |
| 2185280.5 | -7.736169014 | 4.551186942 | 2.203758779 |
| 2185600.5 | -8.660610859 | 3.078511150 | 1.635741733 |
| 2185920.5 | -9.263148352 | 1.491932404 | 1.007159394 |
| 2186240.5 | -9.532592439 | -0.147773633 | 0.342588261 |
| 2186560.5 | -9.470639133 | -1.781873548 | -0.333696448 |
| 2186880.5 | -9.090149268 | -3.355706275 | -0.998628262 |
| 2187200.5 | -8.413234120 | -4.820010653 | -1.630973765 |
| 2187520.5 | -7.469428803 | -6.131644479 | -2.211702131 |
| 2187840.5 | -6.294136751 | -7.253892792 | -2.724176082 |
| 2188160.5 | -4.927369335 | -8.156580976 | -3.154253131 |
| 2188480.5 | -3.412698439 | -8.816123162 | -3.490353054 |
| 2188800.5 | -1.796342012 | -9.215546381 | -3.723509741 |
| 2189120.5 | -0.126344140 | -9.344502412 | -3.847413030 |
| 2189440.5 | 1.548172491 | -9.199286574 | -3.858449499 |
| 2189760.5 | 3.177712685 | -8.782886284 | -3.755753329 |
| 2190080.5 | 4.713119028 | -8.105070998 | -3.541274139 |
| 2190400.5 | 6.106352028 | -7.182510619 | -3.219858450 |
| 2190720.5 | 7.311469283 | -6.038891801 | -2.799334961 |
| 2191040.5 | 8.285987653 | -4.705093946 | -2.290639094 |
| 2191360.5 | 8.992384748 | -3.219408675 | -1.707974232 |
| 2191680.5 | 9.399475978 | -1.627225950 | -1.068745080 |
| 2192000.5 | 9.484093609 | 0.019831388 | -0.393261943 |
| 2192320.5 | 9.232959370 | 1.665056316 | 0.295658525 |
| 2192640.5 | 8.644489810 | 3.248234060 | 0.973265031 |
| 2192960.5 | 7.730376415 | 4.707807307 | 1.613693881 |
| 2193280.5 | 6.516681586 | 5.983685192 | 2.191077670 |
| 2193600.5 | 5.044106872 | 7.020598162 | 2.680915106 |
| 2193920.5 | 3.367116485 | 7.771720901 | 3.061600529 |
| 2194240.5 | 1.551739362 | 8.202131819 | 3.315942373 |

HELIOCENTRIC COORDINATES OF SATURN

REFERENCE AXES OF 1950.0

| JULIAN DATE | X | Y | Z |
|---|---|---|---|
| 2194560.5 | -0.327897992 | 8.291601869 | 3.432459062 |
| 2194880.5 | -2.193989167 | 8.036249762 | 3.406248181 |
| 2195200.5 | -3.969873238 | 7.448778562 | 3.239288966 |
| 2195520.5 | -5.584632288 | 6.557282339 | 2.940147267 |
| 2195840.5 | -6.976915171 | 5.402884808 | 2.523167923 |
| 2196160.5 | -8.097669488 | 4.036628860 | 2.007312297 |
| 2196480.5 | -8.911693564 | 2.516040342 | 1.414809308 |
| 2196800.5 | -9.398061750 | 0.901726595 | 0.769769608 |
| 2197120.5 | -9.549573663 | -0.745686416 | 0.096895610 |
| 2197440.5 | -9.371473244 | -2.368055589 | -0.579608826 |
| 2197760.5 | -8.879732765 | -3.911573074 | -1.236879793 |
| 2198080.5 | -8.099193722 | -5.328051909 | -1.854005106 |
| 2198400.5 | -7.061908184 | -6.575832276 | -2.412458621 |
| 2198720.5 | -5.805859973 | -7.620222801 | -2.896319815 |
| 2199040.5 | -4.373630785 | -8.433422825 | -3.292308748 |
| 2199360.5 | -2.811021418 | -8.994461893 | -3.589835710 |
| 2199680.5 | -1.165953968 | -9.289249739 | -3.781067989 |
| 2200000.5 | 0.512447917 | -9.310568434 | -3.860959600 |
| 2200320.5 | 2.174687341 | -9.057982322 | -3.827243241 |
| 2200640.5 | 3.771615909 | -8.537722246 | -3.680408825 |
| 2200960.5 | 5.255073552 | -7.762639634 | -3.423710586 |
| 2201280.5 | 6.578564427 | -6.752286209 | -3.063230562 |
| 2201600.5 | 7.698109712 | -5.533076941 | -2.607985135 |
| 2201920.5 | 8.573360719 | -4.138426965 | -2.070031453 |
| 2202240.5 | 9.168991292 | -2.608740830 | -1.464523865 |
| 2202560.5 | 9.456348559 | -0.991136460 | -0.809671200 |
| 2202880.5 | 9.415306836 | 0.661216452 | -0.126543320 |
| 2203200.5 | 9.036220774 | 2.290265900 | 0.561327717 |
| 2203520.5 | 8.321807792 | 3.834861906 | 1.228603419 |
| 2203840.5 | 7.288715749 | 5.233216979 | 1.849087771 |
| 2204160.5 | 5.968462667 | 6.426018115 | 2.396973437 |
| 2204480.5 | 4.407390119 | 7.360026616 | 2.848333619 |
| 2204800.5 | 2.665293005 | 7.991806411 | 3.182724839 |
| 2205120.5 | 0.812530953 | 8.291082425 | 3.384696769 |
| 2205440.5 | -1.074179612 | 8.243417534 | 3.445055493 |
| 2205760.5 | -2.916377524 | 7.851755778 | 3.361676634 |
| 2206080.5 | -4.638463630 | 7.135849031 | 3.139488871 |
| 2206400.5 | -6.172291891 | 6.130168804 | 2.789819922 |
| 2206720.5 | -7.460860902 | 4.881075803 | 2.329377615 |
| 2207040.5 | -8.460669166 | 3.443408492 | 1.778936955 |

HELIOCENTRIC COORDINATES OF SATURN

REFERENCE AXES OF 1950.0

| JULIAN DATE | X | Y | Z |
|---|---|---|---|
| 2207360.5 | -9.142740470 | 1.876867546 | 1.161890176 |
| 2207680.5 | -9.492485947 | 0.242613933 | 0.502836986 |
| 2208000.5 | -9.508672319 | -1.399629097 | -0.173654433 |
| 2208320.5 | -9.201801446 | -2.993820614 | -0.844040694 |
| 2208640.5 | -8.592196831 | -4.489150557 | -1.486520038 |
| 2208960.5 | -7.708073979 | -5.840936447 | -2.081479636 |
| 2209280.5 | -6.583825943 | -7.011010666 | -2.611724239 |
| 2209600.5 | -5.258635016 | -7.967816086 | -3.062573558 |
| 2209920.5 | -3.775370819 | -8.686402995 | -3.421909835 |
| 2210240.5 | -2.179668811 | -9.148417862 | -3.680215675 |
| 2210560.5 | -0.519112783 | -9.342100081 | -3.830610317 |
| 2210880.5 | 1.157513789 | -9.262290856 | -3.868886992 |
| 2211200.5 | 2.800933734 | -8.910468482 | -3.793558604 |
| 2211520.5 | 4.362076481 | -8.294824525 | -3.605919689 |
| 2211840.5 | 5.792822451 | -7.430381102 | -3.310126881 |
| 2212160.5 | 7.046858608 | -6.339120726 | -2.913287956 |
| 2212480.5 | 8.080777321 | -5.050073182 | -2.425539533 |
| 2212800.5 | 8.855610247 | -3.599394894 | -1.860140278 |
| 2213120.5 | 9.338508172 | -2.030426106 | -1.233574729 |
| 2213440.5 | 9.504232853 | -0.393153362 | -0.565396888 |
| 2213760.5 | 9.336889595 | 1.256951352 | 0.122185856 |
| 2214080.5 | 8.831782090 | 2.860452045 | 0.804864827 |
| 2214400.5 | 7.997079844 | 4.355963759 | 1.456987511 |
| 2214720.5 | 6.855069382 | 5.682782010 | 2.052585109 |
| 2215040.5 | 5.442680908 | 6.784094753 | 2.566671954 |
| 2215360.5 | 3.810956394 | 7.610561634 | 2.976742331 |
| 2215680.5 | 2.023225454 | 8.123870462 | 3.264313391 |
| 2216000.5 | 0.151963183 | 8.299771507 | 3.410309008 |
| 2216320.5 | -1.725429962 | 8.130093707 | 3.426069785 |
| 2216640.5 | -3.531436491 | 7.623385688 | 3.293821450 |
| 2216960.5 | -5.193182207 | 6.804077688 | 3.026532388 |
| 2217280.5 | -6.646537133 | 5.710354307 | 2.637213347 |
| 2217600.5 | -7.839145144 | 4.391149572 | 2.143811169 |
| 2217920.5 | -8.732244638 | 2.902734176 | 1.567882502 |
| 2218240.5 | -9.301344098 | 1.305285183 | 0.933208821 |
| 2218560.5 | -9.535906431 | -0.340270127 | 0.264479120 |
| 2218880.5 | -9.438255142 | -1.974914866 | -0.413857829 |
| 2219200.5 | -9.021971173 | -3.543576169 | -1.078504601 |
| 2219520.5 | -8.310052678 | -4.996524954 | -1.707974872 |
| 2219840.5 | -7.333070571 | -6.290276510 | -2.283045154 |

HELIOCENTRIC COORDINATES OF SATURN

REFERENCE AXES OF 1950.0

| JULIAN DATE | X | Y | Z |
|---|---|---|---|
| 2220160.5 | -6.127589757 | -7.388214830 | -2.787069323 |
| 2220480.5 | -4.735016421 | -8.260872224 | -3.206128269 |
| 2220800.5 | -3.200441369 | -8.885746017 | -3.529015245 |
| 2221120.5 | -1.571403335 | -9.247155727 | -3.747244140 |
| 2221440.5 | 0.103125726 | -9.336255150 | -3.855095727 |
| 2221760.5 | 1.773603601 | -9.151012217 | -3.849639309 |
| 2222080.5 | 3.390699541 | -8.696122598 | -3.730724283 |
| 2222400.5 | 4.905990063 | -7.982922724 | -3.500970291 |
| 2222720.5 | 6.272624489 | -7.029390531 | -3.165795782 |
| 2223040.5 | 7.446094817 | -5.860254587 | -2.733498326 |
| 2223360.5 | 8.385234957 | -4.507133697 | -2.215357865 |
| 2223680.5 | 9.053504112 | -3.008580178 | -1.625711595 |
| 2224000.5 | 9.420547582 | -1.409897166 | -0.981946524 |
| 2224320.5 | 9.463986909 | 0.237395196 | -0.304356374 |
| 2224640.5 | 9.171346769 | 1.876568838 | 0.384192010 |
| 2224960.5 | 8.541965479 | 3.447285054 | 1.058835735 |
| 2225280.5 | 7.588664540 | 4.887854677 | 1.693577103 |
| 2225600.5 | 6.338886080 | 6.138160934 | 2.262442035 |
| 2225920.5 | 4.834963190 | 7.143128795 | 2.740902917 |
| 2226240.5 | 3.133192560 | 7.856461417 | 3.107464903 |
| 2226560.5 | 1.301479810 | 8.244169141 | 3.345227969 |
| 2226880.5 | -0.584451869 | 8.287328744 | 3.443186352 |
| 2227200.5 | -2.445930159 | 7.983811801 | 3.397128565 |
| 2227520.5 | -4.206236542 | 7.348778326 | 3.210034449 |
| 2227840.5 | -5.794921374 | 6.413138913 | 2.891649942 |
| 2228160.5 | -7.151685414 | 5.220608632 | 2.457446846 |
| 2228480.5 | -8.229199493 | 3.824322607 | 1.927337412 |
| 2228800.5 | -8.994534409 | 2.283244126 | 1.324247667 |
| 2229120.5 | -9.429340626 | 0.658679623 | 0.672685413 |
| 2229440.5 | -9.529046369 | -0.988780964 | -0.002556638 |
| 2229760.5 | -9.301381383 | -2.601712385 | -0.677473208 |
| 2230080.5 | -8.764520442 | -4.127625877 | -1.329654537 |
| 2230400.5 | -7.945110440 | -5.520080072 | -1.938824964 |
| 2230720.5 | -6.876419323 | -6.739218626 | -2.487138911 |
| 2231040.5 | -5.596783427 | -7.751909217 | -2.959301127 |
| 2231360.5 | -4.148395061 | -8.531702684 | -3.342601454 |
| 2231680.5 | -2.576342621 | -9.058769828 | -3.626932275 |
| 2232000.5 | -0.927789629 | -9.319867869 | -3.804812943 |
| 2232320.5 | 0.748774307 | -9.308334856 | -3.871421921 |
| 2232640.5 | 2.404242148 | -9.024111317 | -3.824637498 |

## HELIOCENTRIC COORDINATES OF SATURN

## REFERENCE AXES OF 1950.0

| JULIAN DATE | X | Y | Z |
|---|---|---|---|
| 2232960.5 | 3.989587246 | -8.473798281 | -3.665092427 |
| 2233280.5 | 5.456591737 | -7.670757300 | -3.396246908 |
| 2233600.5 | 6.758657545 | -6.635240586 | -3.024477118 |
| 2233920.5 | 7.851766212 | -5.394505945 | -2.559162118 |
| 2234240.5 | 8.695719211 | -3.982833983 | -2.012738311 |
| 2234560.5 | 9.255850296 | -2.441457486 | -1.400739932 |
| 2234880.5 | 9.504860912 | -0.818395433 | -0.741819189 |
| 2235200.5 | 9.424381224 | 0.832368211 | -0.057473072 |
| 2235520.5 | 9.006688508 | 2.452391281 | 0.628516342 |
| 2235840.5 | 8.256441016 | 3.980704497 | 1.290806445 |
| 2236160.5 | 7.192065159 | 5.356270107 | 1.903445731 |
| 2236480.5 | 5.846510307 | 6.521020194 | 2.441091290 |
| 2236800.5 | 4.267049905 | 7.423336299 | 2.880450034 |
| 2237120.5 | 2.513854188 | 8.021640455 | 3.201819096 |
| 2237440.5 | 0.657231599 | 8.287626104 | 3.390534850 |
| 2237760.5 | -1.226310173 | 8.208618088 | 3.438113162 |
| 2238080.5 | -3.058767879 | 7.788644710 | 3.342891217 |
| 2238400.5 | -4.765505012 | 7.048024206 | 3.110063996 |
| 2238720.5 | -6.279625084 | 6.021558558 | 2.751126902 |
| 2239040.5 | -7.545347775 | 4.755693981 | 2.282851591 |
| 2239360.5 | -8.520152970 | 3.305142231 | 1.725987988 |
| 2239680.5 | -9.175727189 | 1.729414092 | 1.103877807 |
| 2240000.5 | -9.497887066 | 0.089580301 | 0.441115094 |
| 2240320.5 | -9.485689943 | -1.554530285 | -0.237650364 |
| 2240640.5 | -9.149969837 | -3.146553597 | -0.908696376 |
| 2240960.5 | -8.511560586 | -4.635179942 | -1.549963654 |
| 2241280.5 | -7.599438612 | -5.975161904 | -2.141554368 |
| 2241600.5 | -6.448957768 | -7.127905860 | -2.666052364 |
| 2241920.5 | -5.100382077 | -8.061886337 | -3.108746110 |
| 2242240.5 | -3.597858181 | -8.752816542 | -3.457730261 |
| 2242560.5 | -1.988406988 | -9.183404859 | -3.703861218 |
| 2242880.5 | -0.320800581 | -9.343168282 | -3.840740718 |
| 2243200.5 | 1.355395319 | -9.228424179 | -3.864749827 |
| 2243520.5 | 2.990652259 | -8.842262700 | -3.775065771 |
| 2243840.5 | 4.536238082 | -8.194465881 | -3.573655186 |
| 2244160.5 | 5.944916310 | -7.301446969 | -3.265276329 |
| 2244480.5 | 7.171671590 | -6.186281504 | -2.857523782 |
| 2244800.5 | 8.174601053 | -4.878812008 | -2.360912614 |
| 2245120.5 | 8.916072090 | -3.415718156 | -1.788959724 |
| 2245440.5 | 9.364170124 | -1.840413373 | -1.158205053 |

HELIOCENTRIC COORDINATES OF SATURN

REFERENCE AXES OF 1950.0

| JULIAN DATE | X | Y | Z |
|---|---|---|---|
| 2245760.5 | 9.494401117 | -0.202633006 | -0.488115559 |
| 2246080.5 | 9.291567591 | 1.442413991 | 0.199183950 |
| 2246400.5 | 8.751682151 | 3.035448044 | 0.879407417 |
| 2246720.5 | 7.883714619 | 4.515164512 | 1.526876546 |
| 2247040.5 | 6.710901275 | 5.820944998 | 2.115583794 |
| 2247360.5 | 5.271298299 | 6.896154994 | 2.620530955 |
| 2247680.5 | 3.617256297 | 7.691810644 | 3.019268198 |
| 2248000.5 | 1.813557809 | 8.170227352 | 3.293484426 |
| 2248320.5 | -0.065859484 | 8.308097259 | 3.430421673 |
| 2248640.5 | -1.942398949 | 8.098436502 | 3.423867787 |
| 2248960.5 | -3.738026895 | 7.551252399 | 3.274634166 |
| 2249280.5 | -5.380139262 | 6.692991793 | 2.990525323 |
| 2249600.5 | -6.805332951 | 5.564188206 | 2.585555304 |
| 2249920.5 | -7.962409554 | 4.215910579 | 2.078617284 |
| 2250240.5 | -8.814273247 | 2.706083525 | 1.492035995 |
| 2250560.5 | -9.338504388 | 1.095933789 | 0.850120498 |
| 2250880.5 | -9.526836939 | -0.553214235 | 0.177818079 |
| 2251200.5 | -9.383861033 | -2.182583395 | -0.500432609 |
| 2251520.5 | -8.925258191 | -3.737934627 | -1.161597411 |
| 2251840.5 | -8.175833044 | -5.170901354 | -1.784678760 |
| 2252160.5 | -7.167569581 | -6.439722065 | -2.351096708 |
| 2252480.5 | -5.937911718 | -7.509490209 | -2.844860940 |
| 2252800.5 | -4.528389052 | -8.352116175 | -3.252611430 |
| 2253120.5 | -2.983567826 | -8.946206224 | -3.563614283 |
| 2253440.5 | -1.350208379 | -9.276974804 | -3.769764635 |
| 2253760.5 | 0.323480156 | -9.336209628 | -3.865607240 |
| 2254080.5 | 1.988543939 | -9.122276558 | -3.848370770 |
| 2254400.5 | 3.595977257 | -8.640160400 | -3.718015533 |
| 2254720.5 | 5.097437884 | -7.901545039 | -3.477297877 |
| 2255040.5 | 6.446030151 | -6.924928703 | -3.131851727 |
| 2255360.5 | 7.597208454 | -5.735749203 | -2.690278856 |
| 2255680.5 | 8.509863527 | -4.366454922 | -2.164222550 |
| 2256000.5 | 9.147717924 | -2.856409540 | -1.568381749 |
| 2256320.5 | 9.481212817 | -1.251616836 | -0.920476524 |
| 2256640.5 | 9.489463648 | 0.395725466 | -0.241159184 |
| 2256960.5 | 9.161832142 | 2.028382159 | 0.446399721 |
| 2257280.5 | 8.499549060 | 3.585972857 | 1.117234040 |
| 2257600.5 | 7.517210316 | 5.007330754 | 1.745491588 |
| 2257920.5 | 6.243734879 | 6.233378763 | 2.305575902 |
| 2258240.5 | 4.722466559 | 7.210481437 | 2.773521290 |

HELIOCENTRIC COORDINATES OF SATURN

REFERENCE AXES OF 1950.0

| JULIAN DATE | X | Y | Z |
|---|---|---|---|
| 2258560.5 | 3.010129212 | 7.894019328 | 3.128509803 |
| 2258880.5 | 1.174468766 | 8.251759290 | 3.354360858 |
| 2259200.5 | -0.709354248 | 8.266512746 | 3.440780821 |
| 2259520.5 | -2.563289004 | 7.937621920 | 3.384168348 |
| 2259840.5 | -4.311280392 | 7.280994637 | 3.187836711 |
| 2260160.5 | -5.883864157 | 6.327680686 | 2.861622885 |
| 2260480.5 | -7.221891890 | 5.121268033 | 2.420974255 |
| 2260800.5 | -8.279039333 | 3.714575997 | 1.885695184 |
| 2261120.5 | -9.023053105 | 2.166152812 | 1.278559479 |
| 2261440.5 | -9.435911550 | 0.536956475 | 0.623950091 |
| 2261760.5 | -9.513139056 | -1.112561741 | -0.053374564 |
| 2262080.5 | -9.262503007 | -2.724842274 | -0.729323672 |
| 2262400.5 | -8.702329138 | -4.246999431 | -1.381285124 |
| 2262720.5 | -7.859669848 | -5.632007031 | -1.988696375 |
| 2263040.5 | -6.768509525 | -6.839377604 | -2.533404409 |
| 2263360.5 | -5.468122035 | -7.835514391 | -2.999881244 |
| 2263680.5 | -4.001732308 | -8.593978717 | -3.375380545 |
| 2264000.5 | -2.415608469 | -9.095595973 | -3.650009872 |
| 2264320.5 | -0.758174746 | -9.328187391 | -3.816672018 |
| 2264640.5 | 0.921025912 | -9.286369915 | -3.871037108 |
| 2264960.5 | 2.572256191 | -8.971541400 | -3.811571051 |
| 2265280.5 | 4.146405872 | -8.391850067 | -3.639550578 |
| 2265600.5 | 5.595753626 | -7.562121521 | -3.359060556 |
| 2265920.5 | 6.874703750 | -6.503819418 | -2.977007781 |
| 2266240.5 | 7.940607843 | -5.245084617 | -2.503174085 |
| 2266560.5 | 8.754803446 | -3.820795060 | -1.950288620 |
| 2266880.5 | 9.283938089 | -2.272514262 | -1.334065992 |
| 2267200.5 | 9.501568175 | -0.648181978 | -0.673148777 |
| 2267520.5 | 9.389964056 | 0.998589077 | 0.011104452 |
| 2267840.5 | 8.942000134 | 2.609733519 | 0.695041158 |
| 2268160.5 | 8.162946066 | 4.124568595 | 1.353395037 |
| 2268480.5 | 7.071907812 | 5.482287914 | 1.960253172 |
| 2268800.5 | 5.702616169 | 6.625065055 | 2.490303575 |
| 2269120.5 | 4.103248492 | 7.501594602 | 2.920306242 |
| 2269440.5 | 2.335014661 | 8.070753256 | 3.230668677 |
| 2269760.5 | 0.469368417 | 8.304910414 | 3.406936600 |
| 2270080.5 | -1.416048205 | 8.192311937 | 3.440954431 |
| 2270400.5 | -3.242337494 | 7.738054972 | 3.331473587 |
| 2270720.5 | -4.934584100 | 6.963663500 | 3.084181455 |
| 2271040.5 | -6.426375265 | 5.905582344 | 2.711268003 |

HELIOCENTRIC COORDINATES OF SATURN

REFERENCE AXES OF 1950.0

| JULIAN DATE | X | Y | Z |
|---|---|---|---|
| 2271360.5 | -7.662834991 | 4.612215858 | 2.230359839 |
| 2271680.5 | -8.602667442 | 3.140042562 | 1.663007118 |
| 2272000.5 | -9.219160552 | 1.549882388 | 1.033174845 |
| 2272320.5 | -9.500023124 | -0.096442049 | 0.365853050 |
| 2272640.5 | -9.446323477 | -1.738930164 | -0.314155461 |
| 2272960.5 | -9.070859908 | -3.321605269 | -0.983213247 |
| 2273280.5 | -8.396243868 | -4.794071048 | -1.619571121 |
| 2273600.5 | -7.452920584 | -6.112460927 | -2.203841476 |
| 2273920.5 | -6.277320797 | -7.239859479 | -2.719246672 |
| 2274240.5 | -4.910305509 | -8.146337675 | -3.151696017 |
| 2274560.5 | -3.395969050 | -8.808804726 | -3.489774521 |
| 2274880.5 | -1.780729092 | -9.210854640 | -3.724720391 |
| 2275200.5 | -0.112570222 | -9.342681546 | -3.850425775 |
| 2275520.5 | 1.559655956 | -9.201057947 | -3.863460657 |
| 2275840.5 | 3.186919678 | -8.789356584 | -3.763113239 |
| 2276160.5 | 4.720720322 | -8.117609265 | -3.551445552 |
| 2276480.5 | 6.113849255 | -7.202599236 | -3.233364881 |
| 2276800.5 | 7.321255223 | -6.067971974 | -2.816706944 |
| 2277120.5 | 8.301062979 | -4.744324900 | -2.312316223 |
| 2277440.5 | 9.015799249 | -3.269191649 | -1.734089199 |
| 2277760.5 | 9.433940452 | -1.686778372 | -1.098925080 |
| 2278080.5 | 9.531944888 | -0.047420340 | -0.426589170 |
| 2278400.5 | 9.296264659 | 1.593200194 | 0.260511351 |
| 2278720.5 | 8.724840074 | 3.175583711 | 0.937901801 |
| 2279040.5 | 7.828513442 | 4.638706010 | 1.579936131 |
| 2279360.5 | 6.632146666 | 5.922780941 | 2.160887994 |
| 2279680.5 | 5.174998319 | 6.972469264 | 2.656254453 |
| 2280000.5 | 3.510048373 | 7.740385315 | 3.044228594 |
| 2280320.5 | 1.702059099 | 8.190545164 | 3.307203250 |
| 2280640.5 | -0.175647945 | 8.301273014 | 3.433106768 |
| 2280960.5 | -2.045458580 | 8.067080267 | 3.416359796 |
| 2281280.5 | -3.830291071 | 7.499167057 | 3.258287692 |
| 2281600.5 | -5.458348599 | 6.624440249 | 2.966918146 |
| 2281920.5 | -6.867177178 | 5.483228144 | 2.556212442 |
| 2282240.5 | -8.006580075 | 4.126112480 | 2.044885397 |
| 2282560.5 | -8.840227782 | 2.610406260 | 1.455024970 |
| 2282880.5 | -9.346086942 | 0.996736378 | 0.810704321 |
| 2283200.5 | -9.515935437 | -0.653992699 | 0.136709314 |
| 2283520.5 | -9.354222946 | -2.283128484 | -0.542556112 |
| 2283840.5 | -8.876498702 | -3.836231364 | -1.203950090 |

HELIOCENTRIC COORDINATES OF SATURN

REFERENCE AXES OF 1950.0

| JULIAN DATE | X | Y | Z |
|---|---|---|---|
| 2284160.5 | -8.107623153 | -5.264450568 | -1.826240776 |
| 2284480.5 | -7.079961628 | -6.525364546 | -2.390533477 |
| 2284800.5 | -5.831695699 | -7.583397336 | -2.880517309 |
| 2285120.5 | -4.405317335 | -8.409992579 | -3.282600744 |
| 2285440.5 | -2.846414612 | -8.983783905 | -3.586022802 |
| 2285760.5 | -1.202865742 | -9.290667613 | -3.782909814 |
| 2286080.5 | 0.475957008 | -9.323529858 | -3.868211745 |
| 2286400.5 | 2.140167807 | -9.082047185 | -3.839669638 |
| 2286720.5 | 3.740279804 | -8.572667789 | -3.697840648 |
| 2287040.5 | 5.228034459 | -7.808575970 | -3.446111488 |
| 2287360.5 | 6.557186766 | -6.809624245 | -3.090699760 |
| 2287680.5 | 7.684307273 | -5.602303288 | -2.640675553 |
| 2288000.5 | 8.569726447 | -4.219759555 | -2.108011644 |
| 2288320.5 | 9.178733708 | -2.701765525 | -1.507625770 |
| 2288640.5 | 9.483061957 | -1.094493157 | -0.857353199 |
| 2288960.5 | 9.462609539 | 0.550058693 | -0.177785699 |
| 2289280.5 | 9.107292008 | 2.175114773 | 0.508081795 |
| 2289600.5 | 8.418856856 | 3.720759261 | 1.175447940 |
| 2289920.5 | 7.412428996 | 5.126218052 | 1.798580232 |
| 2290240.5 | 6.117498659 | 6.332766878 | 2.351972836 |
| 2290560.5 | 4.578042480 | 7.287128826 | 2.811744587 |
| 2290880.5 | 2.851503205 | 7.945091344 | 3.157178324 |
| 2291200.5 | 1.006455063 | 8.274939425 | 3.372242507 |
| 2291520.5 | -0.881028453 | 8.260189680 | 3.446880039 |
| 2291840.5 | -2.732001813 | 7.901088450 | 3.377826489 |
| 2292160.5 | -4.469721627 | 7.214532508 | 3.168787783 |
| 2292480.5 | -6.024587925 | 6.232603576 | 2.830028427 |
| 2292800.5 | -7.338071504 | 5.000258313 | 2.377586550 |
| 2293120.5 | -8.364895720 | 3.571982386 | 1.832012144 |
| 2293440.5 | -9.074110661 | 2.007818066 | 1.216772423 |
| 2293760.5 | -9.449277669 | 0.369774505 | 0.556766129 |
| 2294080.5 | -9.487704386 | -1.281172314 | -0.122951969 |
| 2294400.5 | -9.198996226 | -2.887469770 | -0.798220322 |
| 2294720.5 | -8.603231370 | -4.396747532 | -1.446574128 |
| 2295040.5 | -7.728998494 | -5.762974319 | -2.047804862 |
| 2295360.5 | -6.611481878 | -6.947074444 | -2.584293085 |
| 2295680.5 | -5.290756427 | -7.917099221 | -3.041147531 |
| 2296000.5 | -3.810415454 | -8.648111432 | -3.406212768 |
| 2296320.5 | -2.216541918 | -9.121981637 | -3.670028988 |
| 2296640.5 | -0.556912698 | -9.327238219 | -3.825806785 |

HELIOCENTRIC COORDINATES OF SATURN

REFERENCE AXES OF 1950.0

| JULIAN DATE | X | Y | Z |
|---|---|---|---|
| 2296960.5 | 1.119698574 | -9.259005583 | -3.869434887 |
| 2297280.5 | 2.764201293 | -8.919006720 | -3.799512964 |
| 2297600.5 | 4.327887489 | -8.315607225 | -3.617401342 |
| 2297920.5 | 5.763181391 | -7.463890182 | -3.327284927 |
| 2298240.5 | 7.024478404 | -6.385749714 | -2.936248348 |
| 2298560.5 | 8.069120117 | -5.109975465 | -2.454352922 |
| 2298880.5 | 8.858550333 | -3.672272599 | -1.894693842 |
| 2299200.5 | 9.359691583 | -2.115111616 | -1.273393666 |
| 2299520.5 | 9.546634861 | -0.487234940 | -0.609464033 |
| 2299840.5 | 9.402775928 | 1.157222267 | 0.075462395 |
| 2300160.5 | 8.922808219 | 2.759903241 | 0.757517186 |
| 2300480.5 | 8.114036503 | 4.260161079 | 1.411347038 |
| 2300800.5 | 6.997450489 | 5.597751626 | 2.011188805 |
| 2301120.5 | 5.608311425 | 6.715950071 | 2.532128234 |
| 2301440.5 | 3.995801680 | 7.564970761 | 2.951522012 |
| 2301760.5 | 2.221477895 | 8.105433462 | 3.250491607 |
| 2302080.5 | 0.356442351 | 8.311444214 | 3.415314133 |
| 2302400.5 | -1.522614973 | 8.172803884 | 3.438502168 |
| 2302720.5 | -3.338097421 | 7.695944379 | 3.319389675 |
| 2303040.5 | -5.016317771 | 6.903401621 | 3.064120014 |
| 2303360.5 | -6.491816876 | 5.831907783 | 2.685043411 |
| 2303680.5 | -7.710665570 | 4.529443432 | 2.199642953 |
| 2304000.5 | -8.632475095 | 3.051754483 | 1.629186272 |
| 2304320.5 | -9.231140905 | 1.458847263 | 0.997315137 |
| 2304640.5 | -9.494571590 | -0.188175162 | 0.328731342 |
| 2304960.5 | -9.423709174 | -1.829773247 | -0.351943148 |
| 2305280.5 | -9.031084278 | -3.410082783 | -1.021102524 |
| 2305600.5 | -8.339102584 | -4.878467188 | -1.656875027 |
| 2305920.5 | -7.378249471 | -6.190534088 | -2.239622766 |
| 2306240.5 | -6.185370065 | -7.308692463 | -2.752248038 |
| 2306560.5 | -4.802115229 | -8.202369492 | -3.180351531 |
| 2306880.5 | -3.273575875 | -8.848062625 | -3.512310880 |
| 2307200.5 | -1.647179953 | -9.229457322 | -3.739365428 |
| 2307520.5 | 0.028034132 | -9.337494759 | -3.855670218 |
| 2307840.5 | 1.702153466 | -9.170110939 | -3.858234934 |
| 2308160.5 | 3.325345078 | -8.732046179 | -3.746891206 |
| 2308480.5 | 4.848781273 | -8.034820812 | -3.524314855 |
| 2308800.5 | 6.225499080 | -7.096688865 | -3.196036591 |
| 2309120.5 | 7.411236622 | -5.942564669 | -2.770444392 |
| 2309440.5 | 8.365332380 | -4.603976613 | -2.258803719 |

HELIOCENTRIC COORDINATES OF SATURN

REFERENCE AXES OF 1950.0

| JULIAN DATE | X | Y | Z |
|---|---|---|---|
| 2309760.5 | 9.051814975 | -3.119018109 | -1.675286973 |
| 2310080.5 | 9.440764042 | -1.532169465 | -1.036961434 |
| 2310400.5 | 9.509930359 | 0.106168031 | -0.363668617 |
| 2310720.5 | 9.246525818 | 1.740570765 | 0.322268613 |
| 2311040.5 | 8.649031997 | 3.312041675 | 0.996565744 |
| 2311360.5 | 7.728813894 | 4.760085058 | 1.633763263 |
| 2311680.5 | 6.511267056 | 6.025411768 | 2.208293711 |
| 2312000.5 | 5.036195478 | 7.053172489 | 2.695800310 |
| 2312320.5 | 3.357140080 | 7.796488383 | 3.074624650 |
| 2312640.5 | 1.539464793 | 8.219924179 | 3.327325875 |
| 2312960.5 | -0.342840643 | 8.302449082 | 3.442044921 |
| 2313280.5 | -2.211452849 | 8.039381113 | 3.413495270 |
| 2313600.5 | -3.988627903 | 7.442882725 | 3.243378216 |
| 2313920.5 | -5.602096805 | 6.540849868 | 2.940125717 |
| 2314240.5 | -6.989433121 | 5.374555267 | 2.518098196 |
| 2314560.5 | -8.101219607 | 3.995749411 | 1.996529087 |
| 2314880.5 | -8.902357822 | 2.463166877 | 1.398163338 |
| 2315200.5 | -9.372251309 | 0.838685055 | 0.747674484 |
| 2315520.5 | -9.504301560 | -0.815978483 | 0.070267126 |
| 2315840.5 | -9.304694715 | -2.441862085 | -0.609450654 |
| 2316160.5 | -8.790708459 | -3.984711575 | -1.268345667 |
| 2316480.5 | -7.988802585 | -5.396331305 | -1.885398120 |
| 2316800.5 | -6.932687001 | -6.635394667 | -2.442116517 |
| 2317120.5 | -5.661517692 | -7.667786936 | -2.922752379 |
| 2317440.5 | -4.218356843 | -8.466587051 | -3.314354173 |
| 2317760.5 | -2.648979903 | -9.011857556 | -3.606729212 |
| 2318080.5 | -1.000989890 | -9.290426436 | -3.792392407 |
| 2318400.5 | 0.676896517 | -9.295761721 | -3.866548210 |
| 2318720.5 | 2.335516190 | -9.027939827 | -3.827109416 |
| 2319040.5 | 3.925945050 | -8.493672684 | -3.674740066 |
| 2319360.5 | 5.400233390 | -7.706368583 | -3.412913425 |
| 2319680.5 | 6.712222392 | -6.686210931 | -3.047980274 |
| 2320000.5 | 7.818482144 | -5.460232430 | -2.589240277 |
| 2320320.5 | 8.679414593 | -4.062343045 | -2.049001017 |
| 2320640.5 | 9.260556451 | -2.533239299 | -1.442595410 |
| 2320960.5 | 9.534099602 | -0.920067569 | -0.788303562 |
| 2321280.5 | 9.480689268 | 0.724360269 | -0.107098963 |
| 2321600.5 | 9.091596953 | 2.342842571 | 0.577778752 |
| 2321920.5 | 8.370599427 | 3.875384035 | 1.241389598 |
| 2322240.5 | 7.334993124 | 5.261529404 | 1.858025825 |

HELIOCENTRIC COORDINATES OF SATURN

REFERENCE AXES OF 1950.0

| JULIAN DATE | X | Y | Z |
|---|---|---|---|
| 2322560.5 | 6.016187346 | 6.443419797 | 2.402456546 |
| 2322880.5 | 4.459617376 | 7.369167459 | 2.851321133 |
| 2323200.5 | 2.723558398 | 7.996320266 | 3.184605886 |
| 2323520.5 | 0.876679634 | 8.295066695 | 3.387069217 |
| 2323840.5 | -1.005586827 | 8.250717588 | 3.449420259 |
| 2324160.5 | -2.845535701 | 7.865033341 | 3.369058181 |
| 2324480.5 | -4.568001515 | 7.156134973 | 3.150241008 |
| 2324800.5 | -6.104869991 | 6.156992760 | 2.803654593 |
| 2325120.5 | -7.398724410 | 4.912750055 | 2.345464467 |
| 2325440.5 | -8.405260454 | 3.477337628 | 1.796023526 |
| 2325760.5 | -9.094382846 | 1.909906859 | 1.178450271 |

## APPENDIX IV

## THE TWO CHOICES OF EPICYCLE AND DEFERENT FOR A PLANET

Let a point C move around the earth on a circle of radius $r_1$. Let $A_1$ denote the angle from the x-axis around to the point C at any time. Then the x and y coordinates of C, $x_C$ and $y_C$, say, are

$$x_C = r_1 \cos A_1,$$

$$y_C = r_1 \sin A_1.$$

Let a planet P move around the point C on an epicyclic circle of radius $r_2$. Let $A_2$ denote the angle between the x-axis and the direction of the line CP. Then the coordinates of P relative to C are

$$x_P - x_C = r_2 \cos A_2,$$

$$y_P - y_C = r_2 \sin A_2.$$

Therefore the coordinates of P relative to the earth are

$$x_P = r_1 \cos A_1 + r_2 \cos A_2,$$

$$y_P = r_1 \sin A_1 + r_2 \sin A_2.$$

The expressions for $x_P$ and $y_P$ are unaltered if we interchange the subscripts 1 and 2. Thus there are two choices of epicycle and deferent that give exactly the same description of the geocentric motion.

There is still another instructive way of seeing the interchangeability of deferent and epicycle. Suppose we interchange the positions of planet and earth in Figure III.2. Then, if we put the sun at the center of the epicycle, we have a correct deferent-epicycle combination for the motion of the earth as seen by an inhabitant of an outer planet. The combination is also the 'pictorial' one, that is, it shows the pictorial relations of the earth, sun, and planet.

# APPENDIX V

## SOME CLASSES OF LATE BABYLONIAN OBSERVATIONS THAT ARE NOT USED IN THE ANALYSIS

As I explained in Section IV.5, I have some reservations about whether the Babylonian records of double conjunctions have been read correctly. Because of these reservations, I decided not to use the double conjunctions in the quantitative studies. In order to preserve them, I have reduced the records of double conjunctions to tabular form and I have listed them in Table A.V-1.

The reader should remember that I use the term double conjunction to mean a conjunction of the moon with a planet, or a conjunction of two planets with each other.

As I explained in Section I.3, Jupiter and Saturn move so slowly that we cannot use observations of them in studying accelerations within the solar system. Therefore I have put the late Babylonian observations of Jupiter and Saturn into this appendix also. Tables A.V-2 and A.V-3 give the synodic phenomena and the normal star conjunctions, respectively, for Jupiter. Tables A.V-4 and A.V-5 do the same for Saturn.

The reader should consult Chapter IV, particularly Sections IV.5, IV.6, and IV.7, for a discussion of the meanings of the columns and the comments in the tables. He can find the references where the data are published with the aid of Table IV.2. Only a few specific comments on the tables in this appendix are needed.

In Table A.V-2, in the year SE 153, one time interval seems wrong. The interval between opposition and the second turning point is consistent with Table III.6, but the interval from the first turning point to opposition looks long. The date of the first turning point should perhaps be SE 153 VII 6 rather than SE 153 VI 6.

In Table A.V-4, for the observations of the year SE 32, the Julian dates listed disagree considerably with those in the published edition. I believe that the editor (Schaumberger) accidentally converted the dates using the calendar for SE 33 rather than for SE 32. The editor also gives several events in the translation that are not in Table A.V-4 for this year. The reader who consults the published source should note that the editor put these events in brackets. They are not in the surviving text, and the editor determined them by calculation.

In Table A.V-1, it is interesting to note that there was nearly a quadruple conjunction of Venus, Mars, Jupiter, and Saturn near the middle of the year SE -211.

TABLE A.V-1

DOUBLE CONJUNCTIONS FROM LATE BABYLONIAN SOURCES

| Babylonian Date (Seleucid Era) | Tentative Julian Date | Time of Day | Body Conjoined | Position Relative to Body Conjoined | Comments |
|---|---|---|---|---|---|
| A. MOON AS INNER BODY | | | | | |
| -211 IV 1 | -522 Jul 3 | E | Mercury | West, 3 am. | See Table IV.5. |
| -211 VII 23 | -522 Oct 23 | M | Jupiter | West, 3 am. | |
| - 67 VIII 22[a] | -378 Nov 18 | M | Saturn | Over, $2\frac{1}{2}$ am. | *num du* |
| - 67 VIII 26 | -378 Nov 22 | M? | Venus | West, 3 am. | *su du* |
| 38 VII 3[a] | -273 Oct 8 | E | Venus | Over, 5 am. | |
| 38 VII 5 | -273 Oct 10 | BN | Mars | West $1\frac{1}{2}$ am., over $2\frac{1}{2}$ am. | Cloudy |
| 38 VII 10 | -273 Oct 15 | BN | Saturn | East, $\frac{1}{2}$ am. | *num du*, cloudy |
| 38 VII 26 | -273 Nov 1 | M | Jupiter | East, 2 am. | *num du* |
| 38 XII 26 | -272 Mar 28 | M | Venus | East, $1\frac{1}{2}$ am. | *num du* |
| 79 VII 1 | -232 Oct 3 | E | Venus | East, 4 am. | *num du* |
| 79 VII 21 | -232 Oct 24 | M | Saturn | East, $1\frac{1}{2}$ am. | *num du* |
| 79 VIII 8 | -232 Nov 8 | BN | Jupiter | Under $1\frac{2}{3}$ am., back $\frac{1}{2}$ am. | Cloudy |
| 79 VIII 19 | -232 Nov 20 | M | Saturn | Over, 1 am., back $\frac{1}{2}$ am. | |
| 79 IX 2 | -232 Dec 2 | E | Venus | West 4 u., under 3 u. | |
| 79 IX 6 | -232 Dec 6 | BN | Jupiter | East, $2\frac{1}{2}$ am. | *num du* |
| 79 IX 16 | -232 Dec 17 | M | Saturn | West, 3 am. | *su du* |
| 79 X 3 | -231 Jan 2 | E | Jupiter | West, 1 am. | *su du* |
| 79 X 15 | -231 Jan 15 | M | Saturn | East, $1\frac{1}{2}$ am. | |
| 79 XI 5 | -231 Feb 2 | BN | Jupiter | West, 4 am. | Cloudy |
| 79 XI 13 | -231 Feb 10 | BN | Saturn | West, 1 am. | *su du* |
| 79 XII 11 | -231 Mar 10 | E? | Saturn | West, $2\frac{1}{2}$ am. | *su du* |
| B. MERCURY AS INNER BODY | | | | | |
| -211 X 5 | -522 Dec 31/32 | — | Venus | East, $\frac{1}{2}$ am. | |
| 35 I 27 | -276 May 11 | E | Jupiter | Over, | Rest of text missing |
| 179 IV 16 | -132 Jul 25/26 | — | Venus | West, 1 am. | Text broken |
| C. VENUS AS INNER BODY | | | | | |
| -211 VII 29 | -522 Oct 29 | M | Jupiter | North, 2 u. | |
| -211 VIII 2 | -522 Oct 31 | M | Saturn | Under, 8 u. | |
| 79 X ? | -231 Jan | E | Jupiter | West, 2 u. | *su du*; day missing |
| D. MARS AS INNER BODY | | | | | |
| -211 VII 11 | -522 Oct 11 | M | Jupiter | Approached within 2 u. | |
| E. JUPITER AS INNER BODY | | | | | |
| -211 VII 12 | -522 Oct 12 | M | Saturn | East, 1 am.[b] | |

[a]The translator supplied the day of the month by calculation.

[b]Note that Venus, Mars, Jupiter, and Saturn are all close together at this time.

## TABLE A.V-2
## SYNODIC PHENOMENA INVOLVING JUPITER, FROM LATE BABYLONIAN SOURCES

| Babylonian Date (Seleucid Era) | Tentative Julian Date | Event | Comments |
|---|---|---|---|
| -256 I 12 | -567 May 3/4 | Opposition | |
| -211 V 22 | -522 Aug 22 | Last, evening, west of Virgo | |
| -211 VI 22 | -522 Sep 22 | First, morning, east of Virgo | |
| -211 X 27 | -521 Jan 22/23 | Turning point, in Libra | |
| -210 II 25 | -521 May 18/19 | Turning point, at the beginning of Virgo | |
| -210 VI 4 | -521 Aug 23 | Last, evening, east of Libra | |
| - 67 VIII 16 | -378 Nov 12 | First, morning, visible for 11° 30′ | |
| 10 I 19 | -301 May 11 | First, morning, in Taurus, visible for 12° | Text is broken; this may belong to year 81. |
| 10 V 22 | -301 Sep 8/9 | Turning point | |
| 19 II 16 | -292 May 26/27 | Event missing, probably a turning point | |
| 19 IV 16 | -292 Jul 24 | Rose in opposition | |
| 19 X 26 | -291 Jan 28 | Last, evening | |
| 19 XII 2 | -291 Mar 4/5 | Event missing, probably first visibility | |
| 69 II 3 | -242 May 2 | First, morning, in Aries | See 140 II 3. |
| 69 VIII 3 | -242 Oct 26/27 | Opposition | |
| 79 VII 14 | -232 Oct 16/17 | Turning point | |
| 79 XI 27 | -231 Feb 24 | Last, evening, in Pisces | |
| 79 XII 29 | -231 Mar 29 | First, morning, in Pisces | |
| 140 II 3? | -171 Apr 27? | First, morning, visible for 12° 30′ | See 69 II 3. Day of month was not repeated and may not be the same in in both years. |
| 153 VI 6 | -158 Sep 1/2 | Turning point, 1 am. east of γ Geminorum | Month seems wrong. |
| 153 IX 4 | -158 Nov 26/27 | Opposition | |
| 153 XI 6 | -157 Jan 25/26 | Turning point, 1½ am. | Star and direction missing |
| 154 IX 19 | -157 Dec 30/31 | Opposition | |
| 154 XI 16 | -156 Feb 24/25 | Turning point, 3 am. before γ Cancri | |

## TABLE A.V-3
## CONJUNCTIONS OF JUPITER FROM LATE BABYLONIAN SOURCES

| Babylonian Date (Seleucid Era) | | | Tentative Julian Date | Time of Day | Body Conjoined | Position Relative to Body Conjoined | Comments |
|---|---|---|---|---|---|---|---|
| -256 | III | 1 | -567 Jun 20/21 | | α Sco | Over | Distance missing |
| 38 | VII | 26 | -273 Nov 1 | M | α Vir | Over; distance missing | See Table IV.4. |
| 57 | III | 15 | -254 Jun 26 | M | | Probably over; distance missing | Object can probably be calculated. |
| 142 | I | 18 | -169 Apr 21 | E | μ Gem | Over, 10 u. | |
| 142 | II | 1 | -169 May 3 | E | γ Gem | Over, 3½ am. | |
| 142 | IV | 12 | -169 Jul 12 | M | β Gem | Under, 3 am. | There are records of other conjunctions, but the text is damaged. |
| 153 | V | 7 | -158 Aug 5 | M | μ Gem | Over, 4 u. | |
| 153 | IX | 10 | -158 Dec 2 | E | μ Gem | | Data missing |
| 153 | XII | 20 | -157 Mar 11 | E | η Gem | Over, 7 u. | |
| 153 | XII$_2$ | 27[a] | -157 Apr 17 | E | γ Gem | Over, 3 am. | |
| 165[b] | IV | 18 | -146 Aug 3 | M | η Gem | Over, 6 u. | |
| 165 | IV | 30 | -146 Aug 15 | M | μ Gem | Over, 6 u. | |
| 165 | V | 18 | -146 Sep 2 | M | γ Gem | Over, 3 am. | |
| 165 | IX | 1 | -146 Dec 11 | E | γ Gem | Over, 3 am. | |
| 165 | X | 1 | -145 Jan 9 | E | μ Gem | Over, 8 u. | |
| 165 | XI | 3 | -145 Feb 9 | E | μ Gem | Over, 8 u. | |
| 165 | XII | 13 | -145 Mar 21 | E | γ Gem | Over, 2-2/3 am. | |

[a]The translator gives this date as 154 I 27.

[b]The translator apparently converted dates in the year 165 on the assumption that 164 was not intercalary, but that 165 was intercalary with a month VI$_2$.

## TABLE A.V-4
## SYNODIC PHENOMENA INVOLVING SATURN, FROM LATE BABYLONIAN SOURCES

| Babylonian Date (Seleucid Era) | | | Tentative Julian Date | Event | Comments |
|---|---|---|---|---|---|
| -211 | VI | 3 | -522 Sep 2 | Last, evening, at beginning of Virgo | |
| -211 | VII | 13 | -522 Oct 13 | First, morning, east of Virgo | |
| -210 | V | 28 | -521 Aug 18 | Last, evening | Month seems wrong. |
| 32 | II | 28/29 | -279 Jun 14/16 | Opposition | |
| 32 | IX | 29 | -278 Jan 10 | First, morning in Sagittarius | There seems to be the beginning of an observational remark for year 91. |
| 38 | XII | 19 | -272 Mar 21 | First, morning | |
| 59 | I | 16 | -252 May 4/5 | Opposition | |
| 59 | III | 18 | -252 Jul 5/6 | Turning point, $1\frac{1}{2}$ am. over $\alpha$ Scorpii | |
| 59 | VII | 14 | -252 Oct 27/28 | Last, evening | Not watched for in 118 |
| 59 | VIII | 18 | -252 Nov 29/30 | Event missing, probably first, morning | |
| 79 | IX | 2 | -232 Dec 2/3 | Turning point, 8 u. over $\rho$ Leonis | |
| 80 | $XII_2$ | 25 | -230 Apr 12/13 | Turning point, $1\frac{1}{2}$ am. west of $\beta$ Leonis | |
| 81 | IV | 14 | -230 Jul 27/28 | Visible for 16°, probably last, evening | May apply to 140. |
| 81 | V | 25 | -230 Sep 6 | First, morning, at the end of Leo | See 140 VI 23. |
| 81 | IX | 19 | -230 Dec 27/28 | Turning point, 2 am. 8 u. east of $\beta$ Virginis | |
| 81 | XI | 16 | -229 Feb 22/23 | Opposition | |
| 140 | VI | 23 | -171 Sep 12 | Visible for 16° 30′ in the morning | See 81 V 25. |
| 177 | II | 4 | -134 May 8/9 | Opposition | |
| 177 | IV | 8 | -134 Jul 10/11 | Turning point, 1 am. east of $\alpha$ Scorpii | |
| 177 | VIII | 4 | -134 Nov 2 | Last, evening, in Scorpio | Not seen in 236? |
| 177 | IX | 6 | -134 Dec 5 | First, morning, at the end of Scorpio | See 236 IX 4. |
| 178 | I | 6 | -133 Mar 31/32 | Turning point, 3 am. west | Reference star uncertain |
| 236 | IX | 4 | - 75 Dec 9 | Visible for 15° in the morning | See 177 IX 6. |

TABLE A.V-5

CONJUNCTIONS OF SATURN FROM LATE BABYLONIAN SOURCES

| Babylonian Dates (Seleucid Era) | Tentative Julian Date | Time of Day | Body Conjoined | Position Relative to Body Conjoined | Comments |
|---|---|---|---|---|---|
| 79 IX 16 | -232 Dec 17 | M | α Leo | East, $2\frac{1}{2}$ or $3\frac{1}{2}$ am. | See Table IV.4. |
| 79 XI 24 | -231 Feb 21/22 | | α Leo | Over, 6 u. | Cloudy |
| 81 III 22 | -230 Jul 6 | E | β Leo | Under, 4 am. | Retrograde |
| 81 VI 20 | -230 Oct 1 | M | β Vir | Over | Distance missing |
| 81 XI 19 | -229 Mar 7 | E | β Vir | Over, 14 u. | Retrograde |

# APPENDIX VI

## OBSERVATIONS OF JUPITER AND SATURN FOUND IN PTOLEMY'S SYNTAXIS

### 1. Observations of Jupiter

The general remarks that were made in Section V.7 about Ptolemy's observations of Mars also apply to his observations of Jupiter and Saturn. He does not name the person who made the observation at Alexandria approximately four centuries before his own time. It is probably the person who made the observations of Mercury on -236 October 30 and -244 November 19 that were described in Section V.5.

1. Chapter XI.1. Ptolemy observed Jupiter at mean opposition at 1 hour before the midnight between EN 880 XI 1 and 2. Hence the observation was made on 133 May 17. Jupiter was at longitude 233°11′.

2. Chapter XI.1. Ptolemy observed Jupiter at mean opposition at 2 hours before the midnight between EN 884 II 13 and 14, and hence on 136 August 31. Jupiter was at 337°54′.

3. Chapter XI.1. Ptolemy observed Jupiter at mean opposition at 5 hours after the midnight between EN 885 III 20 and 21, so that the observation was made on 137 October 8. Jupiter was at longitude 14°23′.

Ptolemy gives the longitude in sexagesimal notation in all these records, as I have written it. The hours are equal.

4. Chapter XI.2. Ptolemy observed Jupiter at 5 equal hours after the midnight between EN 886 XII 26 and 27, so that the observation was made on 139 July 11. When it was compared with $\alpha$ Tauri, Jupiter was found to be at longitude 75 3/4 degrees. The moon was in conjunction with Jupiter but to the south. Since the observation was made almost at the epoch of Ptolemy's star table he probably used the tabular longitude of $\alpha$ Tauri,

which is 42 2/3 degrees. Thus, according to the record, Jupiter and the moon were both 33 1/12 degrees east of $\alpha$ Tauri. The reader should note that Ptolemy gives the position of Jupiter and the moon by means of the usual fractions in this record.

5. Chapter XI.3. An old observation was made in Alexandria at the morning between EN 507 XI 17 and 18, and hence on -240 September 4. Jupiter occulted $\delta$ Cancri.

The observations are summarized in Table A.VI-1, which has the same format as Tables V.5, V.6, and V.7 in the main text, except that I have omitted the reliability. In particular, the values of the Julian day number are in terms of Greenwich mean solar time.

## 2. Observations of Saturn

Ptolemy continues the pattern for Saturn that he had used for Mars and Jupiter. Again he does not name the observer from four centuries earlier. The style suggests that it is the person who made the early observation of Jupiter on -240 September 4 and the observations of Mercury on -236 October 30 and -244 November 19.

1. Chapter XI.5. Ptolemy observed Saturn at mean opposition in the evening between EN 874 IX 7 and 8, so that the observation was made on 127 March 26. Saturn was at longitude 181°13′.

2. Chapter XI.5. Ptolemy made several observations of Saturn near opposition. From them, he concluded that mean opposition occurred at 4 hours after noon on EN 880 XI 18 = 133 June 3, when the longitude of Saturn was 249°40′.

3. Chapter XI.5. In the same way, Ptolemy concluded that Saturn was at mean opposition at noon on EN 883 XII 24 = 136 July 8, when its longitude was 284°14′.

4. Chapter XI.6. Ptolemy observed Saturn at 4 equal hours before the midnight between EN 886 VI 6 and 7, and hence on 138 December 22. When it was

compared with $\alpha$ Tauri, it was seen to be at longitude 309 1/15 degrees, and hence it was 93°36′ east of $\alpha$ Tauri. It was also $\frac{1}{2}$ degree east of the center of the moon, since it was that distance east of the northern horn of the crescent.

5. Chapter XI.7. An old observation was made in Alexandria in the evening of EN 519 V 14 = -228 March 1. Saturn was 2 digits† below $\gamma$ Virginis.

The observations are summarized in Table A.VI-2, which has the same format as the preceding table.

---

†In this context, 1 digit is probably 1/12 degree.

‡Throughout this appendix, where I have written "Ptolemy observed", or the equivalent, please read "Ptolemy claims to have observed". I do not believe that Ptolemy actually made these observations, for reasons that are explained in Chapter XIII.

## TABLE A.VI-1

## OBSERVATIONS OF JUPITER PRESERVED BY PTOLEMY

| Date | Hour | JSD – 1 000 000 | Position[a] |
|---|---|---|---|
| 133 May 17 | 23 | 769 773.372 | Mean opposition, longitude 233°11′ |
| 136 Aug 31 | 22 | 770 975.334 | Mean opposition, longitude 337°54′ |
| 137 Oct 8 | 05 | 771 377.617 | Mean opposition, longitude 14°23′ |
| 139 Jul 11 | 05 | 772 018.628 | 33 1/12 degrees east of α Tau and in conjunction with the moon |
| -240 Sep 4 | M[b] | 633 644.818 | Occulting δ Cnc |

[a]East is measured parallel to the ecliptic, not the equator.

[b]Morning, which probably means about 45 minutes before sunrise.

## TABLE A.VI-2

## OBSERVATIONS OF SATURN PRESERVED BY PTOLEMY

| Date | Hour | JSD – 1 000 000 | Position[a] |
|---|---|---|---|
| 127 Mar 26 | E[b] | 767 529.208 | Mean opposition, longitude 181°13′ |
| 133 Jun 3 | 16 | 769 790.082 | Mean opposition, longitude 249°40′ |
| 136 Jul 8 | 12 | 770 920.920 | Mean opposition, longitude 284°14′ |
| 138 Dec 22 | 20 | 771 818.248 | 93°36′ east of α Tauri, ½ degree east of the moon |
| -228 Mar 1 | E[b] | 637 841.195 | 2 digits below γ Virginis |

[a]East is measured parallel to the ecliptic, not the equator.

[b]Evening, which probably means about 45 minutes after sunset.

## APPENDIX VII

## OBSERVATIONS OF JUPITER AND SATURN PRESERVED BY IBN YUNIS

Ibn Yunis has given us three observations that involve either Jupiter or Saturn without involving any of the closer planets.

1. Page 94. This observation was made by al-Mahani at Baghdad. It is part of the same record as item 1 in Section VII.5. Venus was close to Saturn, and Saturn in turn was 2/3 degrees from $\alpha$ Leonis, presumably in longitude. Calculation shows that Saturn was about this amount west of the star. The time is dawn on EY 227 V 10, the day numbered 2 034 682. The estimated hour is given in Table VII.5.

2. Page 156. This observation is by Habash at Baghdad. He saw Jupiter in conjunction with $\alpha$ Leonis on the 4th feria, EH 250 VII 30, EY 233 V 21, September† 6 of the year 1175 of Alexander. The reader should remember from Section VII.5 that what ibn Yunis calls the year 1 of Alexander is really the year 1 of Seleucus, which began in -311; the year is hence 864 of the common era. The record says that Jupiter was slightly north of $\alpha$ Leonis, and calculation puts it about 20′ north.

The record does not state an hour. At the time of the observation, Jupiter was about 30° west of the sun, so that the observation could only have been made near the end of the night. It is legitimate to assign the conventional time of 45 minutes before sunrise.

The Muslim form of the date agrees with the other forms if 622 July 16 is taken as the era.

3. Page 210. This observation is by ibn Yunis at Cairo. He observed Jupiter and Saturn at dawn on

---

†The original text has the Syrian equivalent of September.

the 6th feria, EH 398 II 23, EY 376 VIII 28, November[†] 7 of the year 1319 of Alexander. Hence the date is 1007 November 7 in the Julian calendar, which was the 6th feria. On the basis of the observation, of which he does not give the details, ibn Yunis concludes that conjunction occurred at noon on the day stated. He puts Jupiter about 40′ south of Saturn, while calculation puts Jupiter about 47′ to the south.

These three observations are summarized in Table A.VII-1, which has the same format as the corresponding tables in Sections VII.4, VII.5, and VII.6.

[†] The original text has the Syrian equivalent of November.

TABLE A.VII-1

OBSERVATIONS OF JUPITER AND SATURN PRESERVED BY IBN YUNIS

| Date | Place | Julian day number -2 000 000 | Hour, local mean time | Configuration |
|---|---|---|---|---|
| EY 227 V 10[a] | Baghdad | 34 682 | 4.85 | Saturn was 2/3 degree west of α Leo. |
| 864 Sep 6 | Baghdad | 36 883 | 4.92 | Jupiter was in conjunction with α Leo. |
| 1007 Nov 7 | Cairo | 89 175 | 11.76 | Jupiter and Saturn were in conjunction. |

[a]Equals 858 August 28.

# REFERENCES

## REFERENCES


Aaboe, A., Scientific astronomy in antiquity, Philosophical Transactions of the Royal Society of London, A. 276, pp. 21-42, 1974.

al-Battani, Abu Abd Allah Muhammad al-Harrani, Az-Zij as-Sabi, ca. 920. There is an edition with commentary and a translation into Latin by C.A. Nallino, under the title Al-Battani sive Albatenii Opus Astronomicum, Pubblicazioni del Reale Osservatorio de Brera, Milano, in 3 volumes, 1899-1907. Volume 1 has the Latin translation, with commentary. Volume 2 has al-Battani's tables, with commentary. Volume 3 has the Arabic text. Only volume 1 is used in this work, and page references used here are to this volume.

al-Biruni, Abu al-Raihan Muhammad bin Ahmad, Al-athar al-Baqiya an al-Qurun al-Khaliya, 1000. There is a translation into English, under the title The Chronology of Ancient Nations, by C. Edward Sachau, published for the Oriental Translation Fund of Great Britain and Ireland by William H. Allen and Co., London, 1879. Page numbers used in citation are taken from this translation.

al-Biruni, Abu al-Raihan Muhammad bin Ahmad, Kitab Tahdid Nihayat al-Amakin Litashih Masafat al-Masakin, 1025. There is a translation into English by Jamil Ali, with the title translated as The Determination of the Coordinates of Positions for the Correction of Distances between Cities, published by the American University of Beirut, Beirut, Lebanon, 1967. The pagination used here is that of this translation.

Allen, C.W., editor, Astrophysical Quantities, Second Edition, p. 120, Athlone Press, London, 1962.

American Ephemeris and Nautical Almanac, U. S. Government Printing Office, Washington, D.C., published annually.

Becvar, A., Atlas of the Heavens, Catalogus 1950.0, Czechoslovak Academy of Sciences, Prague, or Sky Publishing Corp., Cambridge, Mass., 1964.

Bernoldus, Chronicon, 1100. There is an edition by G.H. Pertz in Monumenta Germaniae Historica, Scriptores, v. V, G.H. Pertz, editor, Hahn's, Hannover, 1844.

Brown, E.W., with the assistance of H. B. Hedrick, Chief Computer, Tables of the Motion of the Moon, in 3 volumes, Yale University Press, New Haven, Connecticut, 1919. The reader should also note Complement to the Tables of the Motion of the Moon, Containing the Remainder Terms for the Century 1800-1900, and Errata in the Tables, with the same authors and publisher, 1926.

Bury, J.B., The History of the Decline and Fall of the Roman Empire by Edward Gibbon, Edited in 7 volumes, Methuen and Co., London, Third Edition, 1908.

Carmody, F.J., The Astronomical Works of Thabit b. Qurra, University of California Press, Berkeley and Los Angeles, 1960.

Censorinus, Liber de Die Natali, 238. There is an edition, with a parallel translation into French, by Desiré Nisard, in Celse, Vitruve, Censorin, et Frontin, Dubochet et Cie., Paris, 1846.

Clemence, G.M., First-order theory of Mars, Astronomical Papers Prepared for the Use of the American Ephemeris and Nautical Almanac, XI, Part 2, U.S. Government Printing Office, Washington, D.C., 1949.

Clemence, G.M., Theory of Mars — completion, Astronomical Papers Prepared for the Use of the American Ephemeris and Nautical Almanac, XVI, Part 2, pp. 259-333, 1961.

Clemence, G.M., The concept of ephemeris time, Journal for the History of Astronomy, 2, pp. 73-79, 1971.

Copernicus, Nicolaus, De Revolutionibus Orbium Caelestium Libri Sex, 1543. There is an edition, including a facsimile of Copernicus's holograph, by Fritz Kubach, Verlag von R. Oldenbourg, Munich, in 2 volumes: volume 1, facsimile of the holograph, 1944; volume 2, a critical edition of the text, 1949.

Delambre, J.B.J., Histoire de l'Astronomie Ancienne, Chez Mme. Veuve Courcier, Paris, in 2 volumes, 1817.

Delambre, J.B.J., Histoire de l'Astronomie du Moyen Age, Chez Mme. Veuve Courcier, Paris, 1819.

de Sitter, W., On the secular accelerations and the fluctuations of the longitude of the moon, the sun, Mercury, and Venus, Bulletin of the Astronomical Institute of the Netherlands, IV, pp. 21-38, 1927.

Dinsmoor, W.B., The Archons of Athens in the Hellenistic Age, Harvard University Press, Cambridge, Massachusetts, 1931

Dreyer, J.L.E., History of the Planetary Systems from Thales to Kepler, 1905. Page references used here are to the Dover edition, issued under the title History of Astronomy from Thales to Kepler, Dover Publications, Inc., New York, 1953.

Eckert, W.J., Brouwer, Dirk, and Clemence, G.M., Coordinates of the five outer planets 1653-2060, Astronomical Papers Prepared for the Use of the American Ephemeris and Nautical Almanac, XII, U.S. Government Printing Office, Washington, D.C., 1951.

Eckert, W.J., Jones, Rebecca, and Clark, H.K., Construction of the lunar ephemeris, in _An Improved Lunar Ephemeris, 1952-1959_, issued as a _Joint Supplement to the American Ephemeris and Nautical Almanac and to the Nautical Almanac and Astronomical Ephemeris_, U.S. Government Printing Office, Washington, D.C., 1954.

Epping, J. and Strassmaier, J.N., Neue babylonische Planeten-Tafeln, _Zeitschrift für Assyriologie und Verwandte Gebiete_, _5_, pp. 341-366, 1890.

Epping, J. and Strassmaier, J.N., Neue babylonische Planeten-Tafeln, Part II, _Zeitschrift für Assyriologie und Verwandte Gebiete_, _6_, pp. 89-102, 1891a.

Epping, J. and Strassmaier, J.N., Neue babylonische Planeten-Tafeln, Part III, _Zeitschrift für Assyriologie und Verwandte Gebiete_, _6_, pp. 217-228, 1891b.

Epping, J. and Strassmaier, J.N., Babylonische Mondbeobachtungen aus den Jahren 38 und 79 der Seleuciden-Aera, _Zeitschrift für Assyriologie und Verwandte Gebiete_, _7_, pp. 220-254, 1892.

_Explanatory Supplement to The Astronomical Ephemeris and The American Ephemeris and Nautical Almanac_, H.M. Stationery Office, London, 1961.

Fotheringham, J.K. (assisted by Gertrude Longbottom), The secular acceleration of the moon's mean motion as determined from the occultations in the _Almagest_, _Monthly Notices of the Royal Astronomical Society_, _75_, pp. 377-394, 1915.

Fotheringham, J.K., The secular acceleration of the sun as determined from Hipparchus' equinox observations; with a note on Ptolemy's false equinox, _Monthly Notices of the Royal Astronomical Society_, _78_, pp. 406-423, 1918.

Fotheringham, J.K., A solution of ancient eclipses of the sun, _Monthly Notices of the Royal Astronomical Society_, _81_, pp. 104-126, 1920.

Fotheringham, J.K., Trepidation (in two parts), Monthly Notices of the Royal Astronomical Society, 87, part 1: pp. 142-167, part 2: pp. 182-196, 1927.

Gaubil, le Père, Une Histoire de l'Astronomie Chinoise, which is Part 2 of Observations Mathématiques, Astronomiques, Géographiques, Chronométriques, et Physiques Tirées des Anciens Livres Chinois, E. Souciet, editor, Rollin Père, Paris, 1732.

Geminus, Eisagoge Eis ta Phainomena, ca. -100. There is an edition under the title Elementa Astronomiae, with a parallel translation into German, by K. Manitius, B.G. Teubner, Leipzig, 1898. The date -100 is a guess; the real date may be as late as +200.

Goldstein, B.R., Ibn al-Muthanna's Commentary on the Astronomical Tables of al-Khwarizmi, Yale University Press, New Haven, Connecticut, 1967. This is a translation into English, with notes, of a commentary on al-Khwarizmi's tables that was written sometime in the 10th century.

Goldstein, B.R., Some medieval reports of Venus and Mercury transits, Centaurus, 14, pp. 49-59, 1969.

Graves, Robert, The Greek Myths, in two volumes, Penguin Books, Baltimore, Maryland, 1955.

Halley, Edmond, Some account of the ancient state of the city of Palmyra; with short remarks on the inscriptions found there, Philosophical Transactions of the Royal Society, 19, pp. 160-175, 1695.

Haskins, C.H., Studies in the History of Medieval Science, Harvard University Press, Cambridge, Massachusetts, 1924.

Heiberg, J.L., Preface to Claudii Ptolemaei Opera Astronomica Minora, which is v. 2 of Claudii Ptolemaei Opera Quae Exstant Omnia, B. G. Teubner, Leipzig, 1907.

Hemenway, P.D., The Washington 6-inch transit circle, Sky and Telescope, 31, 72-77, 1966.

Hers, J., R.T.A. Innes and the variable rotation of the earth, Monthly Notes of the Astronomical Society of Southern Africa, 30, pp. 129-134, 1971.

Hewsen, R.H., Science in seventh-century Armenia: Ananias of Sirak, Isis, 59, pp. 32-45, 1968.

Hill, G.W., Tables of Jupiter, constructed in accordance with the methods of Hansen, Astronomical Papers Prepared for the Use of the American Ephemeris and Nautical Almanac, VII, Part 1, U.S. Government Printing Office, Washington, D.C., 1895.

Hill, G.W., Tables of Saturn, constructed in accordance with the methods of Hansen, Astronomical Papers Prepared for the Use of the American Ephemeris and Nautical Almanac, VII, Part 2, U.S. Government Printing Office, Washington, D.C., 1895a.

Hoang, P., Catalogue des Éclipses de Soleil et de Lune, Imprimerie de la Mission Catholique, Shanghai, 1925.

ibn Yunis, Az-Zij al-Kabir al-Hakimi, 1008. There is an edition of part of this work, with a parallel translation into French by J.J.A. Caussin de Perceval, under the title Le Livre de la Grande Table Hakémite, with the author's name spelled as Ebn Iounis, Imprimerie de la République, Paris, 1804. Page citations used here are to this edition.

Jeffreys, Sir Harold, The Earth, Fifth Edition, Cambridge University Press, Cambridge, 1970.

Jones, Charles W., Bedae Opera de Temporibus, published by The Mediaeval Academy of America, Cambridge, Mass., printed by George Banta Publishing Co., Menasha, Wisconsin, 1943.

Kennedy, E.S., A survey of Islamic astronomical tables, Transactions of the American Philosophical Society, 46, pp. 123-175, 1956.

Kugler, F.X., Sternkunde und Sterndienst in Babel, v. 1, Aschendorffsche Verlagsbuchhandlung, Münster, Westphalia, 1907.

Kugler, F.X., Sternkunde und Sterndienst in Babel, v. 2, Part 1, Aschendorffsche Verlagsbuchhandlung, Münster, Westphalia, 1909. (I have seen every year from 1909 to 1913 used in citing this volume. The cover of the copy that I have used gives 1909 as the date, but the title page of the same copy has "1909/10".)

Kugler, F.X., Sternkunde und Sterndienst in Babel, Ergänzungen zum Ersten und Zweiten Buch, II Teil, Aschendorffsche Verlagsbuchhandlung, Münster, Westphalia, 1914.

Kugler, F.X., Sternkunde und Sterndienst in Babel, v. 2, Part 2, Aschendorffsche Verlagsbuchhandlung, Münster, Westphalia, 1924.

Langdon, S. and Fotheringham, J.K., The Venus Tablets of Ammizaduga, Oxford University Press, Oxford, 1928.

Loiselianos, Annales, ca. 814. There is an edition by M. Bouquet in Recueil des Historiens des Gaules et de la France, v. V, Chez Martin, Coignard, Mariette, Guerin, et Guerin, Paris, 1744.

MacPike, E.F., Correspondence and Papers of Edmond Halley, Clarendon Press, Oxford, 1932.

Markowitz, W., Sudden changes in rotational acceleration of the earth and secular motion of the pole, in Earthquake Displacement Fields and the Rotation of the Earth, L. Mansinha, D.E. Smylie, and A.E. Beck, editors, D. Reidel Publishing Co., Dordrecht, Holland, pp. 69-81, 1970.

Martin, C.F., A study of the rate of rotation of the earth from occultations of stars by the moon, 1627-1860, a dissertation presented to Yale University, 1969. (Expected to be published in the Astronomical Papers Prepared for the Use of the American Ephemeris and Nautical Almanac.)

Meritt, B.D., The Athenian Year, University of California Press, Berkeley and Los Angeles, California, 1961.

Miller, G.R., The flux of tidal energy out of the deep oceans, Journal of Geophysical Research, 71, pp. 2485-2489, 1966.

Morrison, L.V., The rotation of the earth from AD 1663-1972 and the constancy of G, Nature, 241, 519-520, 1973.

Muller, P.M. and Stephenson, F.R., The accelerations of the earth and moon from early astronomical observations, paper presented during a symposium on "Biological Rhythms and the Rotation of the Earth" at the University of Newcastle upon Tyne, January, 1974.

Munk, W.H. and MacDonald, G.J.F., The Rotation of the Earth, Cambridge University Press, Cambridge, 1960.

Nasr, Seyyed Hossein, Science and Civilization in Islam, Harvard University Press, Cambridge, Massachusetts, 1968.

Neugebauer, O., The 'Metonic cycle' in Babylonian astronomy, Chapter XXI in Studies and Essays in the History of Science and Learning Offered in Homage to George Sarton, M.F. Ashley Montagu, ed., Henry Schuman, New York, 1946.

Neugebauer, O., Solstices and equinoxes in Babylonian astronomy during the Seleucid period, Journal of Cuneiform Studies, 2, pp. 209-222, 1948.

Neugebauer, O., Astronomical Cuneiform Texts, in 3 volumes, Lund Humphries, London, 1955.

Neugebauer, O., The Exact Sciences in Antiquity, Brown University Press, Providence, Rhode Island, 1957.

Neugebauer, O., The Astronomical Tables of Al-Khwarizmi, Ejnar Munksgaard, Kobenhavn, 1962. This is a translation into English, with notes and commentary, of the Latin translation made by Adelard of Bath.

Neugebauer, O. and Sachs, A., Some atypical astronomical cuneiform texts, Part I, Journal of Cuneiform Studies, 21, pp. 183-218, 1967.

Neugebauer, O. and Sachs, A., Some atypical astronomical cuneiform texts, Part II, Journal of Cuneiform Studies, 22, pp. 92-113, 1969.

Neugebauer, P.V. and Weidner, E.F., Ein astronomischer Beobachtungstext aus dem 37. Jahre Nebukadnezars II. (-567/66), Berichet uber die Verhandlungen der Königlichen Sachsischen Akademie der Wissenschaften zu Leipzig, Philologie-Historie Klasse, Bd. 67, Heft 2, pp. 29-89, 1915.

Newcomb, S., Researches on the motion of the moon, Washington Observations, U.S. Naval Observatory, Washington, D.C., 1875.

Newcomb, S., Tables of the motion of the earth on its axis and around the sun, Astronomical Papers Prepared for the Use of the American Ephemeris and Nautical Almanac, VI, Part 1, U.S. Government Printing Office, Washington, D.C., 1895.

Newcomb, S., Tables of Mercury, Astronomical Papers Prepared for the Use of the American Ephemeris and Nautical Almanac, VI, Part 2, U.S. Government Printing Office, Washington, D.C., 1895a.

Newcomb, S., Tables of Venus, Astronomical Papers Prepared for the Use of the American Ephemeris and Nautical Almanac, VI, Part 3, U.S. Government Printing Office, Washington, D.C., 1895b.

Newcomb, S., Tables of Mars, Astronomical Papers Prepared for the Use of the American Ephemeris and Nautical Almanac, VI, Part 4, U.S. Government Printing Office, Washington, D.C., 1898.

Newcomb, S., *Astronomy for Everybody*, Garden City Publishing Co., Garden City, New York, 1902.

Newcomb, S., Researches on the motion of the moon, Part II, *Astronomical Papers Prepared for the Use of the American Ephemeris and Nautical Almanac*, *v. IX*, U.S. Government Printing Office, Washington, D.C., 1912.

Newton, R.R., *Ancient Astronomical Observations and the Accelerations of the Earth and Moon*, The Johns Hopkins Press, Baltimore and London, 1970.

Newton, R.R., *Medieval Chronicles and the Rotation of the Earth*, The Johns Hopkins Press, Baltimore and London, 1972.

Newton, R.R., Astronomical evidence concerning non-gravitational forces in the earth-moon system, *Astrophysics and Space Science*, *16*, pp. 179-200, 1972a.

Newton, R.R., The earth's acceleration as deduced from al-Biruni's solar data, *Memoirs of the Royal Astronomical Society*, *76*, pp. 99-128, 1972b.

Newton, R.R., The authenticity of Ptolemy's parallax data - Part I, *Quarterly Journal of the Royal Astronomical Society*, *14*, pp. 367-388, 1973.

Newton, R.R., The historical acceleration of the earth, *Geophysical Surveys*, *1*, pp. 123-145, 1973a.

Newton, R.R., The authenticity of Ptolemy's parallax data - Part II, *Quarterly Journal of the Royal Astronomical Society*, *15*, pp. 7-27, 1974.

Newton, R.R., Two uses of ancient astronomy, *Philosophical Transactions of the Royal Society of London*, *276A*, pp. 99-116, 1974a.

Newton, R.R., The authenticity of Ptolemy's eclipse and star data, Quarterly Journal of the Royal Astronomical Society, 15, pp. 107-121, 1974b.

Oesterwinter, C. and Cohen, C.J., New orbital elements for moon and planets, Celestial Mechanics, 5, pp. 317-395, 1972.

Oppolzer, T.R. von, Canon der Finsternisse, Kaiserlich-Königlichen Hof- und Staatsdruckerei, Wien, 1887. There is a reprint, with the explanation of the tables translated into English by O. Gingerich, by Dover Publishing Co., New York, 1962.

Parker, R.A., The Calendars of Ancient Egypt, Univeristy of Chicago Press, Chicago, Illinois, 1950.

Parker, R.A. and Dubberstein, W.H., Babylonian Chronology, 626 B.C. – A.D. 75, Brown University Press, Providence, Rhode Island, 1956.

Pauly-Wissowa [1894]: This is a conventional citation for Wissowa, Georg, Paulys Real-Encyclopädie der Classischen Altertumswissenschaft, J.B. Metzler, Stuttgart, 1894 and later years. This encyclopedia was continued by still others after Wissowa, and there is what is called the Second Series. For simplicity, I cite all volumes as Pauly-Wissowa [1894]. This citation includes the volumes originally prepared by Pauly alone.

Pritchett, W.K. and Neugebauer, O., The Calendars of Athens, Harvard University Press, Cambridge, Massachusetts, 1947.

Ptolemy, C., 'E Mathematike Syntaxis, ca. 142. There is an edition by J.L. Heiberg in C. Ptolemaei Opera Quae Exstant Omnia, B.G. Teubner, Leipzig, 1898. There is an edition with parallel French translation by N.B. Halma, Henri Grand Libraire, Paris, 1813. There is a translation into German by K. Manitius, B.G. Teubner, Leipzig, 1913. The chapter numbering used here is that of Heiberg and Manitius, which differs in places from that used by Halma. I often designate this as the Syntaxis.

Ross, F.E., New elements of Mars and tables for correcting the heliocentric positions derived from Astronomical Papers, Vol. VI, Part IV, Astronomical Papers Prepared for the Use of the American Ephemeris and Nautical Almanac, IX, Part 2, U.S. Government Printing Office, Washington, D.C., 1917.

Sachs, A., A classification of the Babylonian astronomical tablets of the Seleucid period, Journal of Cuneiform Studies, 2, pp. 271-290, 1948.

Sachs, A., Sirius dates in Babylonian astronomical texts of the Seleucid Period, Journal of Cuneiform Studies, 6, pp. 105-114, 1952.

Sachs, A., Late Babylonian Astronomical and Related Texts, with the cooperation of J. Schaumberger, Brown University Press, Providence, Rhode Island, 1955.

Salam, H. and Kennedy, E.S., Solar and lunar tables in early Islamic astronomy, Journal of the American Oriental Society, 87, pp. 492-497, 1967.

Schaumberger, J., Drei babylonische Planetentafeln der Seleukidenzeit, Orientalia, 2, pp. 97-116, 1933a.

Schaumberger, J., Drei planetarische Hilfstafeln, Analecta Orientale, 6, pp. 3-12, 1933b.

Schmeidler, F., Methods in meridian astronomy, Vistas in Astronomy, volume 6, Arthur Beer, editor, pp. 69-91, 1965.

Schoch, Carl, Astronomical and calendarial tables, 1928. This work was printed as Chapter XV in the reference Langdon and Fotheringham [1928].

Smith, S., Chronology: Babylonian and Assyrian, Encyclopaedia Britannica, v. 5, Encyclopaedia Britannica, Inc., Chicago, as published in 1958.

Smithsonian Astrophysical Observatory, Star Catalogue, Positions and Proper Motions of 258,997 Stars for the Epoch and Equinox of 1950.0, in 4 volumes, Smithsonian Institution, Washington, D.C., 1966.

Spencer Jones, H., The rotation of the earth, and the secular accelerations of the sun, moon, and planets, Monthly Notices of the Royal Astronomical Society, 99, pp. 541-558, 1939.

Syntaxis = Ptolemy [ca. 142].

Tannery, Paul, Recherches sur l'Histoire de l'Astronomie Ancienne, Gauthier-Villars et Fils, Paris, 1893.

Thabit (Thabit bin Qurra), De Anno Solis, ca. 880. I have seen only the translation of this into Latin made by Gerard of Cremona about 1170. There is an edition of this translation, with notes and commentary, by Francis J. Carmody, in The Astronomical Works of Thabit B. Qurra, University of California Press, Berkeley and Los Angeles, 1960. Page references are to this edition.

Thorndike, Lynn, A History of Magic and Experimental Science, in 4 volumes, MacMillan Co., New York, 1923.

Times Atlas of the World, Mid-Century Edition, in 5 volumes, The Times Office, London, 1955.

van Flandern, T.C., The secular acceleration of the moon, Astronomical Journal, 75, pp. 657-658, 1970.

van Flandern, T.C., A determination of the rate of change of G, paper presented to the American Geophysical Union, Washington, D.C., April 8, 1974.

van der Waerden, B.L., Greek astronomical calendars and their relation to the Athenian civil calendar, Journal of Hellenic Studies, 80, pp. 168-180, 1960.

Weir, John D., The Venus Tablets of Ammizaduga, Nederlands Instituut voor het Nabije Oosten, Leiden, 1972.

Wylie, A., Chinese Researches, Shanghai, 1897. The name of the publisher is not stated.

# INDEX

## A

B



C

## D



## E

F



G



H



I

## J



## K



## L



## M

## N



## O



## P

## T



## U



## V

**Library of Congress Cataloging in Publication Data**

Newton, Robert R.
Ancient planetary observations and the validity of ephemeris time.

Bibliography: p.
Includes Index.
1. Planets—Observations—History. 2. Sun—Observations—History. 3. Ephemerides—History. I. Title.
QB7.N48 523.4'093 75-44392
ISBN 0-8018-1842-7